Thermodynamics and Heat Power

SIXTH EDITION

Irving Granet, P.E.

late, Queensborough Community College of City University of New York

Maurice Bluestein, Ph.D.

Indiana University–Purdue University, Indianapolis

PRENTICE HALL

Upper Saddle River, New Jersey   Columbus, Ohio

Library of Congress Cataloging-in-Publication Data

```
Granet, Irving.
    Thermodynamics and heat power/Irving Granet,
    Maurice Bluestein. — 6th ed.
    p. cm.
    ISBN 0-13-021539-2
    1. Thermodynamics. 2. Heat engineering.
    I. Bluestein, Maurice. II. Title.
  TJ265 .G697 2000
  621.042-dc21
```

99-32661
CIP

Editor: Stephen Helba
Production Editor: Louise N. Sette
Production Supervision: York Production Services
Design Coordinator: Karrie Converse-Jones
Cover Designer: Allen Bumpus
Cover art: Allen Bumpus
Production Manager: Deidra M. Schwartz
Marketing Manager: Chris Bracken

This book was set in Times Roman by York Graphic Services, Inc., and was printed and bound by R. R. Donnelley & Sons Company. The cover was printed by Phoenix Color Corp.

©2000, 1996, 1990, 1985, 1980, 1974 by Prentice-Hall, Inc.
Pearson Education
Upper Saddle River, New Jersey 07458

Printed in the United States of America

10 9 8 7 6 5 4 3 2 1

ISBN:0-13-021539-2

Prentice-Hall International (UK) Limited, *London*
Prentice-Hall of Australia Pty. Limited, *Sydney*
Prentice-Hall of Canada, Inc., *Toronto*
Prentice-Hall Hispanoamericana, S. A., *Mexico*
Prentice-Hall of India Private Limited, *New Delhi*
Prentice-Hall of Japan, Inc., *Tokyo*
Prentice-Hall (Singapore) Pte. Ltd., *Singapore*
Editora Prentice-Hall do Brasil, Ltda., *Rio de Janeiro*

This book is dedicated to the memory of Irving Granet

Contents

Preface

The sixth edition marks a generational shift in the history of this widely used and highly regarded text. It is with deep regret that I announce the passing of Irving Granet, the original author of this and other noteworthy textbooks in engineering technology, in 1998. His educational achievements have been outstanding, and he will long be remembered for his contributions. I was asked to assume authorship of this textbook, which I have used in my thermodynamics classes for 10 years and have reviewed in the past. I hope to continue Irving's vision of what a good textbook should be.

In the short time available to me for the preparation of this latest edition, I have concentrated on small improvements using the suggestions of several reviewers from other institutions. I have stressed ease of usage and clarity of presentation. One major change has been in the heat-transfer chapter 11, where material and problems dealing with combined modes of heat transfer have been added. In addition to corrections of typographical errors in previous editions, the following important changes have been made:

- A computer disk with steam tables has been included for use with DOS or IBM-type computers.
- A table of contents for the supplemental tables has been added to Appendix 3 to facilitate finding the desired chart.
- Units have been clarified, with particular attention to the difference between pound mass and pound force and the use of the correction factor g_c.
- Updated photos of equipment have been provided where available from manufacturers.
- The BASIC and EXCEL exercises have been removed; they are outdated and, according to reviewers, have been rarely used.
- A better explanation for the operation of the heat pump has been added, particularly regarding how the unit is switched between heating and cooling.
- A diagram of how a refrigerator works that appeared in an older edition has been corrected and restored.
- Additional heat-transfer problems have been provided, with a separate section of problems on combined modes of heat transfer.

I want to encourage those using this new edition to contact me with suggestions for future editions. I thank the editors at Prentice Hall for their support and encouragement. I particularly want to thank my wife, Maris, for love and support over many wonderful years.

Maurice Bluestein
Indianapolis, Indiana
bluestei@engr.iupui.edu

Acknowledgments

The following organizations kindly supplied illustrations and other materials. Their contributions and the contributions of many others are gratefully acknowledged. In alphabetical order, they are:

Alco Power, Inc.

Allison Engine Company, d.b.a. Rolls Royce Allison

American Ref-Fuel Company of Hempstead

American Chain and Cable Corp.

Babcock and Wilcox Corp.

Carrier Corp.

Chevrolet Motor Division, General Motors Corp.

Chrysler Corp.

Colt Industries, Fairbanks Morse Power Systems Division

Combustion Engineering Inc.

Consolidated Edison Corp.

Cummins Engine Co. Inc.

Curtiss–Wright Corp.

Detroit Stoker Corp.

Dunham–Bush Inc.

DuPont Chemicals

Ford Motor Company

Foster Wheeler Corp.

Garrett Corp.

General Electric Corp.

Grumman Aerospace Corp.

Grumman Energy Systems Corp.

Indianapolis Power and Light Company

Lockheed Missile and Space Co., Inc.

Long Island Lighting Co.

McGraw-Hill, Inc.

New York State Electric and Gas Corp.

Patterson-Kelley Co.; Division of HARSCO Corp.

Power

Power Authority of the State of New York

Pratt and Whitney Aircraft Group; United Technologies Corp.

Renwal Products Corp.

Riley Stoker Corp.

Sandia National Labs

Solar Energy Research Institute

Subaru of America, Inc.

Tubular Exchanger Manufacturers Association

Volkswagen of America, Inc.

Western Precipitation Division of Joy Industrial Equipment Co.

Westinghouse Electric Corp.

Worthington Corp.

Symbols

Symbol	Definition	British Engineering	SI
		Units	
a	Acceleration	ft/s^2	m/s^2
a, A	Area	ft^2	m^2
A, B, C, D, E	Constants		
C_D	Discharge coefficient	dimensionless	
C_v	Velocity coefficient	dimensionless	
c	Specific heat	$\text{Btu/lb}_\text{m}\cdot°\text{R}$	$\text{kJ/kg}\cdot°\text{K}$
c_p	Specific heat at constant pressure	$\text{Btu/lb}_\text{m}\cdot°\text{R}$	$\text{kJ/kg}\cdot°\text{K}$
c_v	Specific heat at constant volume	$\text{Btu/lb}_\text{m}\cdot°\text{R}$	$\text{kJ/kg}\cdot°\text{K}$
c_n	Specific heat of any process	$\text{Btu/lb}_\text{m}\cdot°\text{R}$	$\text{kJ/kg}\cdot°\text{K}$
C_p	Total specific heat at constant pressure	$\text{Btu/}°\text{R}$	$\text{kJ/}°\text{K}$
C_v	Total specific heat at constant volume	$\text{Btu/}°\text{R}$	$\text{kJ/}°\text{K}$
COP	Coefficient of performance	dimensionless	
d, D	Diameter	ft	m
e	Base of natural logarithms	dimensionless	
F	Force	lb_f	N
F_A	Geometric factor	dimensionless	
F_e	Emissivity factor	dimensionless	
g	Acceleration of gravity	ft/s^2	m/s^2
g_c	Gravitational constant	$32.174\ \text{ft}\cdot\text{lb}_\text{m}/\text{lb}_\text{f}\cdot\text{s}^2$	
Gr	Grashof number	dimensionless	
h	Height	ft	m
H	Enthalpy	Btu	kJ
h	Specific enthalpy	Btu/lb_m	kJ/kg
h_f	Specific enthalpy — saturated liquid	Btu/lb_m	kJ/kg
h_g	Specific enthalpy — saturated vapor	Btu/lb_m	kJ/kg

Symbol	Description	US Units	SI Units
h_{fg}	Specific enthalpy of vaporization $(h_g - h_f)$	Btu/lb$_m$	kJ/kg
h^0	Stagnation enthalpy	Btu/lb$_m$	kJ/kg
h	Heat-transfer coefficient	Btu/hr·ft^2·°F	kW/m^2·°K
h_r	Heat-transfer coefficient — radiation	Btu/hr·ft^2·°F	kW/m^2·°K
i	Current	Amperes	Amperes
J	Mechanical equivalent of heat	778 ft·lb$_f$/Btu	
K	Proportionality constant		
k	Spring constant	lb$_f$/in.	N/m
k	Thermal conductivity	Btu/hr·ft·°F	kW/m·°K
k	c_p/c_v	dimensionless	
K.E.	Kinetic energy	ft·lb$_f$/lb$_m$	kJ/kg
l, L	Length	ft	m
m	Mass	lb$_m$	kg
$\dot{m}$	Mass flow rate	lb$_m$/s	kg/s
M	Mach number	dimensionless	
mep	Mean effective pressure	lb$_f$/in^2	kPa
MW	Molecular weight	lb$_m$/lb$_m$·mole	kg/kg·mole
n	Polytropic exponent	dimensionless	
n	Number of particles	dimensionless	
n	Number of moles	dimensionless	
Nu	Nusselt number	dimensionless	
n	Number of moles	mass/MW	mass/MW
p	Pressure	lb$_f$/in^2	kPa
p_m	Mixture pressure	lb$_f$/in^2	kPa
p_m	Mean effective pressure	lb$_f$/in^2	kPa
p_r	Reduced pressure	dimensionless	
p_r	Relative pressure	dimensionless	
P.E.	Potential energy	ft lb$_f$/lb$_m$	kJ/g
Pr	Prandtl number	dimensionless	
Q	Heat interchange	Btu	kJ
q	Specific heat interchange	Btu/lb$_m$	kJ/kg
$\dot{Q}$	Heat transfer	Btu/hr	kW
Q_r	Radiant heat transfer	Btu/hr	kW
R	Universal gas constant	ft lb$_f$/lb$_m$·°R	kJ/kg·°K
R	Electrical resistance	ohms	ohms
Re	Reynolds number	dimensionless	
R_t	Thermal resistance	°F·hr/Btu	°C/W
r	Radius	ft	m
r_e	Expansion ratio	dimensionless	
r_c	Compression ratio	dimensionless	
$r_{c.o.}$	Cutoff ratio	dimensionless	
r_p	Pressure ratio	dimensionless	
s	Specific entropy	Btu/lb$_m$·°R	kJ/kg·°K
S	Total entropy	Btu/°R	kJ/K

s_f	Specific entropy of saturated liquid	Btu/lb$_m$·°R	kJ/kg·°K
s_g	Specific entropy of saturated vapor	Btu/lb$_m$·°R	kJ/kg·°K
s_{fg}	Specific entropy of vaporization $(s_g - s_f)$	Btu/lb$_m$·°R	kJ/kg·°K
sg	Specific gravity	dimensionless	
T	Temperature, absolute	°R	°K
T_c	Critical temperature	°R	°K
t	Temperature	°F	°C
t	Time	s (seconds)	s
T_r	Reduced temperature	dimensionless	
$(\Delta t)_m$	Logarithmic temperature difference	°F	°C
U	Internal energy	Btu	kJ
U	Overall heat transfer coefficient	Btu/hr·ft^2·°F	kW/m^2·°K
u	Specific internal energy	Btu/lb$_m$	kJ/kg
u_f	Specific internal energy — saturated liquid	Btu/lb$_m$	kJ/kg
u_g	Specific internal energy — saturated liquid	Btu/lb$_m$	kJ/kg
u_{fg}	Specific internal energy of vaporization $(u_g - u_f)$	Btu/lb$_m$	kJ/kg
V	Velocity	ft/s	m/s
V_a	Acoustic velocity	ft/s	m/s
V	Volume	ft^3	m^3
v	Specific volume	ft^3/lb$_m$	m^3/kg
v_r	Reduced specific volume	dimensionless	
v_r	Relative specific volume	dimensionless	
v_c	Critical specific volume	ft^3/lb$_m$	m^3/kg
v_f	Specific volume of saturated liquid	ft^3/lb$_m$	m^3/kg
v_g	Specific volume of saturated vapor	ft^3/lb$_m$	m^3/kg
v_{fg}	Specific volume of vaporization $(v_g - v_f)$	ft^3/lb$_m$	m^3/kg
w	Weight	lb$_f$	N
W	Weight	lb$_f$	kN
W	Humidity ratio	dimensionless	
W	Work	ft lb$_f$	kJ
$\mathbf{w}$	Work per unit mass	ft lb$_f$/lb$_m$	kJ/kg
x	Mole fraction	dimensionless	
x	Quality	dimensionless	
x	Length	ft	m
z	Elevation above reference plane	ft	m
Z	Compressibility factor	dimensionless	
α	Absortivity	dimensionless	
γ	Specific weight	lb$_f$/ft^3	kn/m^3
Δ	Small change of variable	dimensionless	
ε	Emissivity	dimensionless	
η	Efficiency	dimensionless	
μ	Viscosity	lb$_m$/ft^2·hr	N·s/m^2
ρ	Density	lb$_m$/ft^3	kg/m^3
ρ	Reflectivity	dimensionless	

σ	Stefan – Boltzmann constant	Btu/hr$\cdot$ft$^2\cdot{}^\circ$R^4	W/m$^2\cdot{}^\circ$K^4
τ	Transmissivity	dimensionless	
θ	A function of	dimensionless	
θ	Relative humidity	dimensionless	

In addition to the symbols listed above, the following notation is used:

Superscript ^{0} refers to the stagnation property.

Superscript $*$ refers to the state where M = 1.

Subscript $_m$ refers to the mixture property.

1

Fundamental Concepts

Learning Goals

After reading and studying the material in this chapter, you should be able to:

1. Define thermodynamics as the study of energy and the conversion of energy from one form to another.

2. Use the observable external characteristics that are known as properties to describe a system.

3. Establish and convert from one system of temperature measurement to another, and understand the four methods of measuring temperature.

4. Use both the English and SI systems of units.

5. Use elementary kinetic theory of gases to establish the concepts of pressure, temperature, density, specific weight, specific volume, and Avogadro's law.

6. Use the concept of pressure in both English and SI units. Gage and absolute pressure definitions are important ideas that are necessary in engineering applications.

7. Use the concept that fluids exert pressures that can be expressed in terms of the height and specific weight of the column of fluid.

8. Describe the various methods of measuring pressure and the methods used to calibrate pressure-measuring devices.

1.1 INTRODUCTION

Thermodynamics is the study of energy, heat, work, the properties of the media employed, and the processes involved. Thermodynamics is also the study of the conversion of one form of energy to another. Because energy can be derived from electrical, chemical, nuclear, or other means, thermodynamics plays an important role in all branches of engineering, physics, chemistry, and the biological sciences.

In defining the word *thermodynamics,* we have used the terms *energy, heat,* and *work.* It is necessary to examine these terms in detail, and this will be done in subsequent chapters. In this chapter, certain fundamental concepts are defined and basic ideas are developed for future use.

The role of thermodynamics in modern life is of great importance. For example, Figure 1.1 shows a modern jet engine for use on commercial aircraft. This engine is capable of producing a

FIGURE 1.1 Modern aircraft gas turbine.
(Courtesy of Pratt and Whitney Aircraft Group, United Technologies Corp.)

thrust of 17,400 lb. The thermodynamic principles that apply to this jet engine are essentially the same as those that apply to the automotive turbine engine shown in Figure 1.2. This engine is rated at 123 hp. Figure 1.3 shows it installed in a car. Compare the size of this unit with the aircraft gas turbine shown in Figure 1.1 and you see the large size differential that exists when considering a given application of thermodynamics.

Other thermodynamic systems that will be considered include the direct conversion of solar energy to electricity and the use of solar energy for the heating of water. Figure 1.4 shows a fast-food restaurant with solar collectors to heat water for kitchen and rest room use.

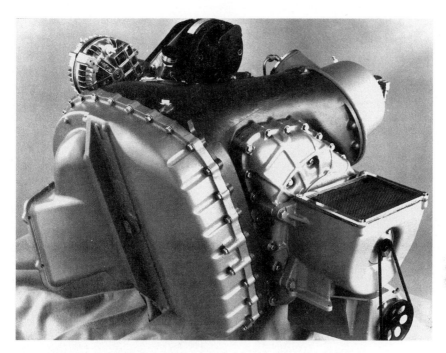

FIGURE 1.2 Automotive gas turbine engine.
(Courtesy of Chrysler Corp.)

FIGURE 1.3 Automotive gas turbine installation.
(Courtesy of Chrysler Corp.)

FIGURE 1.4 Solar collectors for hot water heating.

1.2 PROPERTIES OF A SYSTEM

In physics, when studying the motion of a rigid body (i.e., a body that is not deformed or only slightly deformed by the forces acting on it), extensive use is made of *free-body diagrams.* Briefly, a free-body diagram is an outline of a body (or a portion of a body) showing *all* the external forces acting on it. A free-body diagram is one example of the concept of a system. As a general concept applicable to all situations, we can define a system as a grouping of matter taken in any convenient or arbitrary manner. In addition, a thermodynamic system will invariably have energy transferred from it or to it and can also have energy stored in it. From this definition, it will be noted that we are at liberty to choose the grouping, but once having made a choice, we must take into account *all* energies involved. In addition to the systems mentioned earlier, Figure 1.5 shows a modern diesel engine designed for trucks. This engine is capable of developing 600 hp at 2000 rpm. Note the absence of wires and hoses hanging from it. It is possible to consider the entire engine as a system or a portion of the engine as a system.

Let us consider a given system and then ask ourselves how we can distinguish changes that occur in the system. It is necessary to have external characteristics that permit us to measure and evaluate system changes. If these external characteristics do not change, we should be able to state that the system has not changed. Some of these measurements that can be made

FIGURE 1.5 A 600-hp diesel engine.
(Courtesy of Cummins Engine Company.)

on a system are temperature, pressure, volume, and position. These observable external characteristics are called *properties.* When all properties of a system are the same at two different times, we can say that we cannot distinguish any difference in the system at these times. The properties of a system enable us to uncover differences in the system after it has undergone a change. Therefore, the complete description of a system is given by its properties. The condition of the system, that is, its position, energy content, and so on, is called the *state* of the system. Thus, its properties determine its state. Those properties that depend on the size and total mass of a system are termed *extensive properties;* that is, they depend on the extent of the system. An *intensive property* is independent of the size of the system. Pressure and temperature are examples of intensive properties. In addition, there are properties that are known as *specific properties* because they are given per unit mass or per defined mass in the system. Specific properties are intensive properties.

It has already been noted that a given state of a system is reproduced when all its properties are the same. Because a given set of properties determines the state of a system, the state is reproduced regardless of the history or path the system may undergo to achieve the state. For example, consider a weight that is lifted vertically from one position to another. This weight can be brought to the same position by first lifting it vertically part of the way, then moving it horizontally to the right, then lifting it another part of the way, then moving it horizontally to the left, and finally, lifting it vertically to the desired point. In this example, the state of the system at the end of the two processes is the same, and the path the system took did not affect its state after the change occurred.

As we shall see in Chapter 2, a consequence of the foregoing is that the change in energy of a system between two given states is the same, regardless of the method of attaining the state. In mathematical terminology, energy is a state function, not a path function.

The properties temperature and pressure are used throughout this book, and it is necessary to have a good understanding of them. The following sections of this chapter deal in detail with these properties.

CALCULUS ENRICHMENT

As we have stated, a property has a unique and singular value when the system in question is in a given state. This value does not depend on the intermediate states that the system has experienced. Thus, a property is not a function of the system's path. The change of a property is a function only of the initial and final states of the system. Mathematically, this can be written as

$$\int_{\phi_1}^{\phi_2} d\phi = \phi_2 - \phi_1 \qquad \textbf{(a)}$$

where ϕ denotes the property. Notice that the value of the property change is a function only of the limits of the integral. The differential, $d\phi$, is known as an *exact differential*. There are quantities such as heat and work that are not exact differentials, and it becomes necessary to specify the path in order to evaluate them. These quantities are known as *path functions*.

1.3 TEMPERATURE

The *temperature* of a system is a measure of the random motion of the molecules in the system. If there are different temperatures within the body (or bodies composing the system), the question arises as to how the temperature at a given location is measured and how this measurement is interpreted. Let us examine this question in detail, because similar questions will also have to be considered when other properties of a system are studied. In air at room pressure and temperature, there are approximately 2.7×10^{19} molecules per cubic centimeter. If we divide the cube whose dimensions are 1 centimeter (cm) on a side into smaller cubes, each of whose sides is one thousandth of a centimeter, there will be approximately 2.7×10^{10} molecules in each of the smaller cubes, still an extraordinarily large number. Although we speak of temperature at a point, we really mean the average temperature of the molecules in the neighborhood of the point.

Let us now consider two volumes of inert gases separated from each other by a third volume of inert gas. By inert, we mean that the gases will not react chemically with each other. If the first volume is brought into contact with the second volume and left there until no observable change occurs in any physical property, the two volumes are said to be in *thermal equilibrium.* Should the third volume then be brought in contact with the second and no noticeable change in physical properties observed, the second and third volumes can also be said to be in thermal equilibrium. For the assumed conditions of this experiment, it can be concluded that the three volumes are in thermal equilibrium. Based on this discussion, the three volumes can also be stated to be at the same temperature. This simple experiment can be repeated under the same conditions for solids, liquids, and gases, with the same result every time. The results of all these experiments are summarized and embodied in the *zeroth law of thermodynamics,* which states that two systems having equal temperatures with a third system also have equal temperature with each other. As an alternative definition of the zeroth law, we can say that if two bodies are each in thermal equilibrium with a third body, they are in thermal equilibrium with each other. The importance of this apparently obvious statement was recognized after the first law was given its name, and consequently, it was called the zeroth law to denote that it precedes the first law. It should be noted that a thermometer measures only its own temperature, and for it to be an accurate indication of the temperature of a second system, the thermometer and the second system must be in thermal equilibrium. As a consequence of the zeroth law, we can measure the temperatures of two bodies by a third body (a thermometer) without bringing the bodies in contact with each other.

The common scales of temperature are called the Fahrenheit and Celsius (centigrade) temperatures and are defined by using the ice point and boiling point of water at atmospheric pressure. In the Celsius temperature scale, the interval between the ice point and the boiling point is divided into 100 equal parts. In addition, as shown in Table 1.1, the Celsius ice point is zero and the Fahrenheit ice point is 32. The conversion from one scale to the other is directly derived from Table 1.1 and results in the following relations:

$$°C = \tfrac{5}{9}(°F - 32) \tag{1.1}$$

and

$$°F = \tfrac{9}{5}(°C) + 32 \tag{1.2}$$

TABLE 1.1
Defining Points for Temperature Scales

	°F	°C	°K	°R
Atmospheric boiling point	212	100	373	672
Ice point	32	0	273	492
Absolute zero	−460	−273	0	0

The ability to extrapolate to temperatures below the ice point and above the boiling point of water and to interpolate in these regions is provided by the *International Scale of Temperature.* This agreed-on standard utilizes the boiling and melting points of different elements and establishes suitable interpolation formulas in the various temperature ranges between these elements. The data for these elements are given in Table 1.2.

TABLE 1.2

Temperature Data

Element	Melting or Boiling Point at 1 atm	Temperature	
		°C	°F
Oxygen	Boiling	−182.97	−297.35
Sulfur	Boiling	444.60	832.28
Antimony	Melting	630.50	1166.90
Silver	Melting	960.8	1761.4
Gold	Melting	1063.0	1945.4
Water	Melting	0	32
	Boiling	100	212

ILLUSTRATIVE PROBLEM 1.1

Determine the temperature at which the same value is indicated on both Fahrenheit and Celsius thermometers.

SOLUTION

Using Equation (1.1) and letting °C = °F yields

$$°F = \tfrac{5}{9}(°F − 32)$$

$$\tfrac{4}{9}°F = \frac{−160}{9}$$

$$°F = −40$$

Therefore, both Fahrenheit and Celsius temperature scales indicate the same temperature at −40°.

By using the results of Illustrative Problem 1.1, it is possible to derive an alternative set of equations to convert from the Fahrenheit to the Celsius temperature scale. When this is done, we obtain

$$°F = \tfrac{9}{5}(40 + °C) − 40 \tag{1.3}$$

and

$$°C = \tfrac{5}{9}(40 + °F) − 40 \tag{1.4}$$

The symmetry of Equations (1.3) and (1.4) makes them relatively easy to remember and use.

Let us consider the case of a gas that is confined in a cylinder (with a constant cross-sectional area) by a piston that is free to move. If heat is now removed from the system, the piston will move down, but due to its weight, it will maintain a constant pressure on the gas. This procedure can be carried out for several gases, and if volume is plotted as a function of temperature, we obtain a family of straight lines that intersect at zero volume (Figure 1.6). This unique temperature is known as the absolute zero temperature, and the accepted values on the Fahrenheit and Celsius temperature scales are $-459.69°$ and $-273.16°$, respectively, with the values $-460°$ and $-273°$, respectively, used for most engineering calculations. It is also possible to define an absolute temperature scale that is independent of the properties of any substance, and we consider this point later in this book.

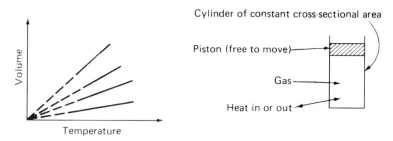

FIGURE 1.6 Gas thermometer.

The absolute temperature scale begins at absolute zero temperature, is always positive, and more accurately represents the concept that temperature is a measure of the molecular motion of matter. Thus, molecular motion ceases at absolute zero temperature. In the English system of units, absolute temperature is given in degrees Rankine. In the SI system, the absolute temperature is given in degrees Kelvin. Thus, we define

$$\boxed{\text{degrees Rankine} = °R = °F + 460} \tag{1.5}$$

and
$$\boxed{\text{degrees Kelvin} = °K = °C + 273} \tag{1.6}$$

The relation among degrees Rankine, degrees Fahrenheit, degrees Kelvin, and degrees Celsius is also shown in Table 1.1.

As noted earlier, the state of a system is uniquely determined by its properties. Thus, the accurate measurement of these properties is of great importance from both a theoretical and practical standpoint. Temperatures are measured in many ways, but in general, all the methods of measuring temperature can be categorized into four classes depending on the basic physical phenomena used to make the measurement:

1. Methods utilizing the expansion of gases, liquids, or solids.

2. Methods utilizing the change in electrical resistance of an element.

3. Methods utilizing the change in electrical potential of an element.

4. Methods utilizing the optical changes of a sensor.

The most common device used to measure temperature is the familiar *liquid-in-glass thermometer,* which consists of a reservoir of liquid and a long glass stem with a fine-line capillary. The operation of this type of thermometer is based on the coefficient of expansion of the liquid (usually mercury) being greater than the coefficient of expansion of the glass. For accurate measurements,

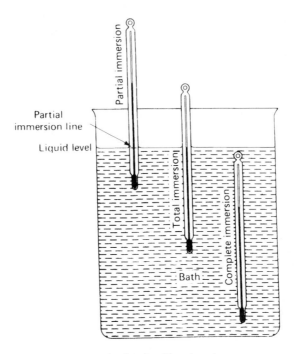

FIGURE 1.7 Methods of calibrating thermometers.

these thermometers are calibrated by either partial, total, or complete immersion in a suitable bath, as shown in Figure 1.7. If the thermometer is calibrated by one method but is used in a different way, it is necessary to make corrections to the readings for the difference in usage. Advantages of the liquid-in-glass thermometers are low cost, simplicity, good reliability, and long life.

Another device that is used to measure temperature or temperature differences depends on the expansion of materials and is called the *bimetallic element*. This element usually consists of two thin, flat strips placed side by side and welded together. The composite strip can be used flat or coiled into a helix or spiral. Changes in temperature cause the strip to change its curvature, and the motion produced can be used to move a pointer. The flat bimetallic strip is commonly used in room thermostats, where the motion of one end is used to close or open an electrical contact. The action of a bimetallic strip is shown in Figure 1.8.

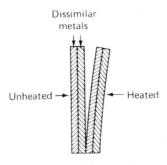

FIGURE 1.8 Bimetallic strip.

FIGURE 1.9 Industrial resistance thermometer.
(Courtesy of American Chain and Cable Corp.)

Resistance thermometers (Figure 1.9) are commonly used in industry to measure process temperatures. The basic principle of this type of instrument is that the change in electrical resistance of a sensor due to a change in its temperature is easily measurable. The electrical resistivity of some metals increases very nearly in direct proportion to an increase of temperature. Thus, the measured change in resistance of a sensor can be converted to a temperature change. Metals used for the sensors include nickel, copper, and platinum. Because of their calibration stability, high temperature coefficient, and moderate cost, nickel resistance units are normally recommended for temperature ranges between -100 and $+500°$F. The resistance of the sensing element is usually measured by a *Wheatstone bridge* (shown schematically in Figure 1.10). When the indicator is nulled to read zero by using the variable resistance r_b,

$$\frac{r_e}{r_b} = \frac{r_1}{r_2} \tag{1.7}$$

where the various resistances are as shown in Figure 1.10. Errors in using the Wheatstone bridge make the circuit shown in Figure 1.10 unsatisfactory for highly accurate work. These errors are due to the contact resistance of the variable resistor, resistance changes in lead wires due to temperature gradients along them, and self-heating of the sensor due to the supply current. Modifications of the basic Wheatstone bridge have been made to compensate for and correct these faults. These circuits will be found by the interested student in the references in Appendix 1.

Thermistors are also included as resistance elements. The name *thermistor* is derived from *therm*ally sensitive re*sistors,* because their resistance varies rapidly with temperature. Thermistors are included in the class of solids known as *semiconductors,* which have electrical conductivities

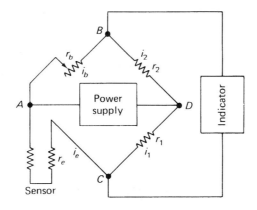

FIGURE 1.10 Wheatstone bridge.

between those of conductors and insulators. The advantages of the thermistor over the resistance thermometer and the thermocouple (to be discussed later) are as follows:

1. A temperature coefficient of resistance approximately 10 times that of metals, with a correspondingly greater sensitivity to temperature change.

2. A much higher resistivity than the metals so that small units may have high resistance, virtually eliminating the lead and contact resistance problem.

3. No need for cold-end or lead material compensation, because the thermistor resistance is a function of its absolute temperature.

For a limited temperature range, the thermistor combines all the best features of resistance thermometers and thermocouples, and has greater sensitivity than either.

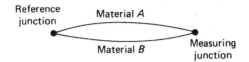

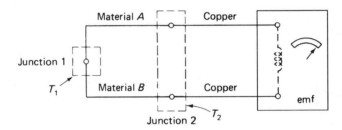

FIGURE 1.11 Elementary thermocouple circuit.
[(b) From R. P. Benedict, Fundamentals of Temperature, Pressure, and Flow Measurements *(New York: John Wiley & Sons, Inc., 1969), with permission.]*

When two wires of different materials are joined at their ends and their junctions are at different temperatures, a net thermal electromotive force (EMF) is generated that induces a net electric current. This is shown schematically in Figure 1.11a. The thermocouple is used as a thermometer by placing one junction in contact with the body whose temperature is to be measured and measuring the voltage produced at the other junction with a millivoltmeter, as shown schematically in Figure 1.11b. The practical reduction of the thermocouple to use as a temperature-measuring device in industry depends on three so-called laws:

1. If each section of wire in the circuit is homogeneous, that is, if there is no change in composition or physical properties along its length, the EMF in the circuit depends only on the nature of the metals and the temperatures of the junctions.

2. If both of the junctions involving a particular homogeneous metal are at the same temperature, this metal makes no net contribution to the EMF. Thus, if the complete circuit consists of iron, constantan (60% copper, 40% nickel alloy), and copper, but both of the junctions involving copper are at the same temperature, we can consider the circuit as if it consisted entirely of iron and constantan, with only two junctions.

3. If all junctions of the circuit except one are held at constant temperature, the EMF in the circuit will be a function of the temperature of the remaining junction and can be used to measure that temperature. It is customary to prepare tables giving this EMF as a function of temperature for the case where the reference junction (or junctions) is held at 0°C (32°F).

Figure 1.11b shows a thermocouple with two *continuous* dissimilar wires from the measuring function to the reference junction. From the reference junction, copper wires and the potentiometer (or millivoltmeter) complete the circuit. For the case of more than one thermocouple to be monitored, a circuit of the type shown in Figure 1.12* can be used. It is important to note that

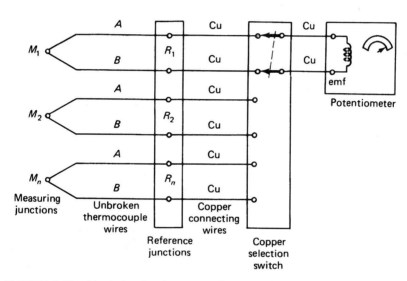

FIGURE 1.12   Ideal circuit when more than one thermocouple is involved.

each thermocouple consists of two continuous wires between the measuring junction and the reference junction. Rather than use a circuit with multiple junctions, it is possible to use the circuit shown in Figure 1.13, which has a single reference junction. A typical industrial circuit using a potentiometer that is constructed to compensate automatically for the reference junction temperature is shown in Figure 1.14.

---

* Figures 1.12 through 1.15 are from *Fundamentals of Temperature, Pressure, and Flow Measurements* by R. P. Benedict, John Wiley & Sons, Inc., New York, 1969, with permission.

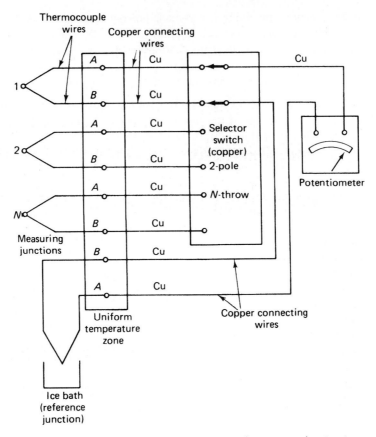

FIGURE 1.13 Single reference junction-thermocouple circuit.

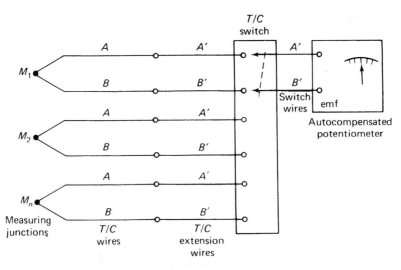

FIGURE 1.14 Typical industrial thermocouple circuit.

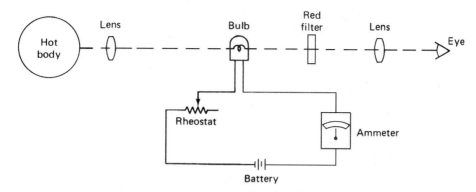

FIGURE 1.15 Schematic of the optical pyrometer.

All bodies radiate energy at a rate proportional to the fourth power of their absolute temperature. This is the well-known Stefan–Boltzmann law (discussed in some detail in Chapter 11). For the moment we concern ourselves with the optical pyrometer, which provides us with a method of converting this radiation to a temperature measurement. The optical pyrometer, shown schematically in Figure 1.15, consists of a telescope within which there is mounted a red glass filter and a small light bulb. In practice, the telescope is aimed at the body whose temperature is to be measured. The filament of the bulb appears black against the bright background being measured. The current through the bulb is varied by adjusting the rheostat until the brightness of the filament matches the brightness of the body being measured. By prior calibration, the reading of the ammeter is directly convertible to temperature. The advantage of this device is that no part of it is in contact with the body, and the optical pyrometer can be used to measure temperature above the melting points of either resistance thermometers or thermocouples.

A summary of commercially available measuring devices is given in Table 1.3, showing their temperature ranges and characteristics.

1.4 FORCE AND MASS

1.4a The English System

Force is very often defined in elementary physics texts as the push or pull exerted on a body. Although this definition serves to satisfy our daily experience, it is not satisfactory when dealing with the motion of bodies that are subjected to resultant forces that are not zero. The following paragraphs of this section deal with the concept of force in a consistent manner, and we attempt to clarify the confusion concerning the units of force and mass.

The basis of much of the physical sciences rests on the work of Newton. For the present, two physical laws that are attributed to him will be used:

1. Law of universal gravitation.

2. Second law of motion.

The first of these, the *law of universal gravitation,* states simply that the force of attraction experienced by two bodies is proportional to the product of their masses and inversely proportional to the square of the distance (*d*) separating them. At this point, we define the *mass* of a body as

TABLE 1.3
Summary of Information Regarding Commercially Available Temperature-Measuring Instruments or Systems

Kind of instrument	Glass stem (liquid-in-glass)		Filled systems				Thermocouples			
	Mercury	Other liquids	Bimetal	Vapor pressure	Liquid-filled	Gas-filled	Iron/constantan	Chromel/alumel	Copper/constantan	Platinum/platinum-rhodium
Low temp limit	−38°F (−39°C)	−320°F (−196°C) (pentane)	−100°F (−73°C)	−127°F (−20°C)	−125°F (−87°C)	−452°F (4°K)	About −320°F (−196°C)	About −300°F (−184°C)	−424°F (20°K)	Commonly 32°F (0°C)
High temp limit	1100°F (593°C)		1000°F (538°C)	600°F (316°C)	1200°F (649°C)	1000°F (538°C)	1400°F (760°C) in oxidizing atm. 1800°F (982°C) in reducing atm. 2000°F (1093°C) spot readings.	2100° to 2300°F (1149° to 1260°C). 2400°F (1316°C) spot readings. (Not to be used in reducing atm.)	660°F (349°C) 930°F (499°C) spot readings.	2600° to 2800°F (1427° to 1538°C) 3216°F (1769°C) spot readings. (Not to be used in reducing atm.)
Thermoelectric power, microvolts per degree C							64 at 760°C	38 at 1100°C	41 at 25°C	12 at 1200°C (alloy 10% Rh) 14 at 1200°C (alloy 13% Rh)
Remarks	Low price. Reasonable accuracy. Easily broken. Short scale length.		Easier to read than liquid-in-glass. Sometimes damaged by overheating.	Normally has nonlinear scale. Has "cross-ambient" effect. Cheapest of the filled systems.	Can be compensated accurately for ambient temperature variations.	Suitable for wide ranges. Linear scale. Requires large bulb.	The couple most widely used in industry. Poor reproducibility above 1600°F (871°C)	Chromel is 90 Ni, 10 Cr (chromel P); Alumel is 94 Ni, 3 Mn, 2Al, 1 Si.	Constantan is 57 Cu, 43 Ni	Expensive
	No auxiliary equipment required						Can give very rapid response, matched only by the radiation and optical pyrometers.			

Kind of instrument	Resistance thermometers				Radiation pyrometer	Optical pyrometer	Pyrometric cones	Melting pellets
	Platinum	Copper	Nickel	Thermistors				
Low temp limit	−300°F (−184°C)	−220°F (−140°C)	−300°F (−184°C)	−76°F (−60°C) (most types)	About room temperature	1400°F (760°C) (normally)	1085°F (585°C)	113°F (45°C)
High temp limit	1400° to 1800°F (760° to 982°C)	250°F (121°C)	600°F (316°C)	750°F (399°C) (glass coated)	As high as desired	As high as desired	3659°F (2015°C)	2500°F (1371°C)
$(R_{100°C} - R_{0°C})/100\,R_{0°C}$	0.0039	0.0043	0.0066	−0.0098 (one type)	Can operate a recorder and automatic controls.	Cannot operate a recorder or automatic controls.	Indication somewhat affected by heating rate.	
Remarks	May drift at the higher temperatures.			High sensitivity. Only fair stability.	Absorbers of radiation, such as windows or dirt, can affect accuracy. Emissivity of source important. Very rapid response.			A single cone or pellet can indicate only its rated temperature, and is normally used only once.
	Give greater accuracy than thermocouples, for the same investment. High sensitivity makes them suitable for narrow range instruments.							

Source: Temperature, Its Measurement and Control in Science and Industry, Wolfe (ed.), © 1955 by Litton Educational Publishing, Inc.: reprinted by permission of Van Nostrand Reinhold Co., New York. Published by Krieger Publishing.

the quantity of matter contained in the body. Thus, if the earth is assumed to be spherical and its mass center is taken to be at its geometrical center, a body on the surface will experience a constant force. This force is given the name *weight*. Because the earth is an oblate spheroid, a body at different locations at the surface will have different weights. Also, the surface of the earth is not smooth, so weight is a function of elevation.

So far, it would appear that the foregoing concepts are both clear and relatively simple. In its simplest form, it can be stated that the weight of a body, in a given location, is proportional to its mass. By choosing the constant of proportionality to be unity, **the mass of an object in pounds at the earth's surface may, for most practical purposes, be assumed to be numerically equal to the weight of the body.** The difficulty arises from the fact that force can also be defined by *Newton's second law of motion* in terms of the fundamental units of length, time, and mass. This relation can be stated as follows:

$$F \propto ma \tag{1.8}$$

where a is the acceleration of the body. This proportionality can be written more explicitly by defining a force of 1 lb as that force necessary to give a mass of 1 lb an acceleration of 32.174 ft/s². Equation (1.8) then becomes

$$F = \frac{ma}{g_c} \tag{1.9}$$

where g_c is numerically equal to 32.174. It is in this constant of proportionality, g_c, that the confusion occurs, as well as in the ambiguous use of the word *pound*. To differentiate between the pound force and pound mass, we will use the notation lb_f and lb_m. Also, note that $g_c = 32.174 \ lb_m \cdot ft/lb_f \cdot s^2$. Use of g_c provides a benefit in that at sea level, the number of pound mass is the same as the number of pound force. Thus, thermodynamic parameters can be stated as per pound regardless of whether it is a pound mass or a pound force. Table 1.4 lists some of the most common combinations of units that can be used. To avoid the obvious confusion that this multitude of units can cause, the definition of pound mass and pound weight, as given in this section, is applied in conjunction with Equation (1.9) throughout this book. As a consequence of these considerations, weight and mass, at a location in which the local gravitational attraction is expressed as g, can be interrelated in the following manner:

TABLE 1.4
Common Combinations of Units

Length	Time	Mass	Force	g_c
foot	second	pound	pound	32.174 ft/s² × pound mass/pound force
foot	hour	pound	pound	4.17 × 10⁸ ft/hr² × pound mass/pound force
foot	second	pound	poundal	1.0 ft/s² × pound mass/poundal
foot	second	slug	pound	1.0 ft/s² × slug/pound force
centimeter	second	gram	dyne	1.0 cm/s² × gram/dyne
meter	second	kilogram	newton	1.0 m/s² × kilogram/newton
meter	second	kilogram	kilogram	9.81 m/s² × kilogram mass/kilogram force
meter	second	gram	gram	981 cm/s² × gram mass/gram force

$$w = \frac{mg}{g_c}$$

We also note that mass can be operationally defined as the inertia a body has in resisting acceleration.

ILLUSTRATIVE PROBLEM 1.2

A mass of 1 lb weighs 1 lb on the earth. How much will it weigh on the moon? Assume that the diameter of the moon is 0.273 when the earth's diameter is taken to be unity and that the mass of the moon is 0.0123 in relation to the earth's mass.

SOLUTION

The force exerted on the mass by the moon will determine its "weight" on the moon. In general, the law of universal gravitation can be written as

$$F = K\frac{m_1 m_2}{d^2}$$

where K is a proportionality constant. Using the subscripts e for earth and m for moon, we have

$$F_e = \frac{K m_e m}{(r_e)^2}$$

because r_e, the radius of earth, is the distance separating the mass center of the body m on the earth and the mass center of the earth. Similarly,

$$F_m = \frac{K m_m m}{(r_m)^2}$$

$$\frac{F_e}{F_m} = \frac{m_e}{m_m}\left(\frac{r_m}{r_e}\right)^2 = \frac{1}{0.0123}(0.273)^2 = 6.06$$

Thus, a body on earth feels a force (weight) approximately six times the force it would feel on the moon. The solution to the problem is that a mass of 1 lb will weigh approximately $\frac{1}{6}$ lb on the moon.

1.4b The SI System

For the engineer, the greater confusion has been the units for mass and weight. As we have noted, the literature abounds with units such as slugs, pounds, mass, pound force, poundal, kilogram force, kilogram mass, dyne, and so on. In the SI system, the base unit for *mass* (not weight or force) is the kilogram (kg), which is equal to the mass of the international standard kilogram located at the International Bureau of Weights and Measures. It is used to specify the quantity of matter in a body. The mass of a body never varies, and it is independent of gravitational force.

The SI *derived* unit for force is the newton (N). The unit of force is defined from Newton's law of motion: force is equal to mass times acceleration ($F = ma$). By this definition, 1 newton applied to a mass of 1 kilogram gives the mass an acceleration of 1 meter per second squared ($N = kg \cdot m/s^2$). The newton is used in all combinations of units that include force, for example, pressure or stress (N/m^2), energy ($N \cdot m$), and power ($N \cdot m/s = W$). By this procedure, the unit of force is not related to gravity, as was the older kilogram force.

Table 1.5 gives the seven base units of the SI system. Several observations concerning this table should be noted. *The unit of length is the meter and the kilogram is a unit of mass, not weight.* Also, symbols are never pluralized, never written with a period, and uppercase and lowercase symbols *must* be used as shown *without exception*.

Table 1.6 gives the derived units with and without symbols often used in engineering. These derived units are formed by the algebraic combination of base and supplementary units. Note that

TABLE 1.5

Base SI Units

Quantity	Base SI Unit	Symbol
Length	meter	m
Mass	kilogram	kg
Time	second	s
Electric current	ampere	A
Thermodynamic temperature	kelvin	°K
Amount of substance	mole	mol
Luminous intensity	candela	cd

TABLE 1.6

Derived SI Units

Quantity	Name	Symbol	Formula	Expressed in Terms of Base Units
Acceleration	acceleration	m/s^2	m/s^2	m/s^2
Area	square meter	m^2	m^2	m^2
Density	kilogram per cubic meter	kg/m^3	kg/m^3	$kg \cdot m^{-3}$
Energy or work	joule	J	$N \cdot m$	$m^2 \cdot kg \cdot s^{-2}$
Force	newton	N	$m \cdot kg \cdot s^{-2}$	$m \cdot kg \cdot s^{-2}$
Moment	newton-meter	$N \cdot m$	$N \cdot m$	$m^2 \cdot kg \cdot s^{-2}$
Moment of inertia of area	—	m^4	m^4	m^4
Plane angle	radian	rad	rad	rad
Power	watt	W	J/s	$m^2 \cdot kg \cdot s^{-3}$
Pressure or stress	pascal	Pa	N/m^2	$N \cdot m^{-2}$
Rotational frequency	revolutions per second	rev/s	s^{-1}	s^{-1}
Temperature	degree Celsius	°C	°C	°C
Torque (*see* moment)	newton-meter	$N \cdot m$	$N \cdot m$	$m^2 \cdot kg \cdot s^{-2}$
Velocity (speed)	meter per second	m/s	m/s	$m \cdot s^{-1}$
Volume	cubic meter	m^3	m^3	m^3

where the name is derived from a person, the first letter of the symbol appears as a capital; for example, newton is N. Otherwise, the convention is to make the symbol lowercase.

Weight has been defined as a measure of gravitational force acting on a material object at a specified location. Thus, weight is a force that has both a mass component and an acceleration component (gravity). Gravitational forces vary by approximately 0.5% over the earth's surface. For nonprecision measurements, these variations normally can be ignored. Thus, a constant mass has an approximate constant weight on the surface of the earth. The agreed standard value (standard acceleration) of gravity is 9.806 650 m/s². Figure 1.16 illustrates the difference between mass (kilogram) and force (newton).

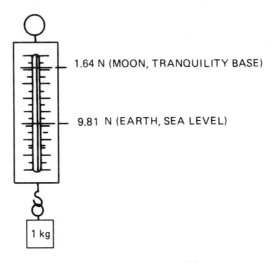

FIGURE 1.16   Mass and force.

The term *mass* or *unit mass* should be used to indicate only the quantity of matter in an object, and the old practice of using weight in such cases should be avoided in engineering and scientific practice. However, because the determination of an object's mass will be accomplished by the use of a weighing process, the common usage of the term *weight* instead of *mass* is expected to continue but should be avoided.

Based on the foregoing, Equation (1.10) can be written in SI units as

$$\boxed{w = mg}$$ **(1.10a)**

ILLUSTRATIVE PROBLEM 1.3

A body has a mass of 5 kg. How much will it weigh on earth?

SOLUTION

The weight of the body will be its mass multiplied by the local acceleration of gravity, that is,

$$w = mg$$

Thus,

$$w = 5 \text{ kg} \times 9.81 \, \frac{\text{m}}{\text{s}^2} = 49.05 \, \frac{\text{kg} \cdot \text{m}}{\text{s}^2}$$

and because $1 \text{ N} = 1 \text{ kg} \cdot \text{m/s}^2$,

$$w = 49.05 \text{ N}$$

ILLUSTRATIVE PROBLEM 1.4

A body has a mass of 10 kg. If the local gravitational acceleration is 9.5 m/s², determine (a) its weight and (b) its horizontal acceleration if it is acted on by a 10 N horizontal force.

SOLUTION

(a) From Equation (1.10a),

$$w = (10 \text{ kg}) \left(9.5 \, \frac{\text{m}}{\text{s}^2} \right) = 95.0 \, \frac{\text{kg} \cdot \text{m}}{\text{s}^2} = 95.0 \text{ N}$$

(b) Because $F = ma$, $a = F/m$. Therefore,

$$a = \frac{10 \, \dfrac{\text{kg} \cdot \text{m}}{\text{s}^2}}{10 \text{ kg}} = 1 \, \frac{\text{m}}{\text{s}^2}$$

For the SI system to be universally understood, it is most important that the symbols for the SI units and the conventions governing their use be adhered to strictly. Care should be taken to use the correct case for symbols, units, and their multiples (e.g., K for Kelvin, k for kilo, m for milli, M for mega). As noted earlier, unit names are never capitalized except at the beginning of a sentence. SI unit symbols derived from proper names are written with the first letter in uppercase; all other symbols are written in lowercase, for example, m (meter), s (second), °K (Kelvin), Wb (weber). Also, unit names form their plurals in the usual manner. Units symbols are always written in singular form, for example, 350 megapascals or 350 MPa, 50 milligrams or 50 mg. Because the unit symbols are standardized, the symbols should always be used in preference to the unit names. Unit symbols are not followed by a period unless they occur at the end of a sentence, and the numerical value associated with a symbol should be separated from that symbol by a space, for example, 1.81 mm, not 1.81mm. The period is used only as a decimal marker. Because the comma is used by some countries as a decimal marker, the SI system does not use the comma. A space is used to separate large numbers in groups of three. Thus, 3 807 747 and 0.030 704 254 indicate this type of grouping. Note that for numerical values less than 1, the decimal point is preceded by a zero. For a number of four digits, the space can be omitted.

In addition, certain style rules should be adhered to:

1. When a product is to be indicated, use a space between unit names (e.g., newton meter).

2. When a quotient is indicated, use the word *per* (e.g., meter per second).

3. When a product is indicated, use the square, cubic, and so on (e.g., square meter).

4. In designating the product of units, use a centered dot (e.g., N·s, kg·m).

5. For quotients, use a solidus (/) or a negative exponent (e.g., m/s or m·s^{-1}). The solidus (/) should not be repeated in the same expression unless ambiguity is avoided by the use of parentheses. Thus, one should use m/s^2 or m·s^{-2} but not m/s/s; also, use m·kg/(s^3·A) or m·kg·s^{-3}·A^{-1} but not m·kg/s^3/A.

One of the most useful features of the older metric system and the current SI system is that multiples and submultiples of units are in terms of factors of 10. Thus, the prefixes given in Table 1.7 are used in conjunction with SI units to form names and symbols of multiples of SI units. Certain general rules apply to the use of these prefixes:

TABLE 1.7

Factors of 10 for SI Units

Prefix	Symbol	Factor	
tera	T	10^{12}	1 000 000 000 000
giga	G	10^{9}	1 000 000 000
mega	M	10^{6}	1 000 000
kilo	k	10^{3}	1 000
hecto	h	10^{2}	100
deka	da	10^{1}	10
deci	d	10^{-1}	0.1
centi	c	10^{-2}	0.01
milli	m	10^{-3}	0.001
micro	μ	10^{-6}	0.000 001
nano	n	10^{-9}	0.000 000 001
pico	p	10^{-12}	0.000 000 000 001
femto	f	10^{-15}	0.000 000 000 000 001
atto	a	10^{-18}	0.000 000 000 000 000 001

1. The prefix becomes part of the name or symbol without separation (e.g., kilometer, megagram).

2. Compound prefixes should not be used; use GPa, not kMPa.

3. In calculations, use powers of 10 in place of prefixes.

4. Try to select a prefix so that the numerical value will fall between 0.1 and 1000. This rule may be disregarded when it is better to use the same multiple for all items. It is also

recommended that prefixes representing 10 raised to a power that is a multiple of 3 be used (e.g., 100 mg, not 10 cg).

5. The prefix is combined with the unit to form a new unit, which can be provided with a positive or negative exponent. Therefore, mm^3 is $(10^{-3} \text{ m})^3$ or 10^{-9} m^3.

6. Where possible, avoid the use of prefixes in the denominator of compound units. The exception to this rule is the prefix k in the base unit kg (kilogram).

There are certain units outside the SI system that may be used together with the SI units and their multiples. These are recognized by the International Committee for Weights and Measures as having to be retained because of their practical importance.

It is almost universally agreed that when a new language is to be learned, the student should be completely immersed and made to "think" in the new language. This technique has been proved most effective by the Berlitz language schools and the Ulpan method of language teaching. A classic joke about this is the American traveling in Europe who was amazed that 2-year-old children were able to speak "foreign" languages. In dealing with the SI system, the student should not think in terms of customary units and then perform a mental conversion, it is better to learn to think in terms of the SI system, which will then become a second language. However, there will be times when it may be necessary to convert from customary U.S. units to SI units. To facilitate such conversions, Table 1.8 gives some commonly used converison factors.

TABLE 1.8
Conversions from Conventional to SI Units

Multiply	By	To Obtain
atmospheres (760 torr)	2.992×10^1	inches mercury (32°F)
atmospheres	1.033×10^4	kilograms/square meter[a]
atmospheres	1.013×10^2	kilopascals
bars	9.869×10^{-1}	atmospheres
bars	1.000×10^2	kilopascals
British thermal units (Btu)	3.927×10^{-4}	horsepower-hours
British thermal units (Btu)	1.056	kilojoules
British thermal units (Btu)	2.928×10^{-4}	kilowatt-hours
British thermal units (Btu)	1.221×10^{-8}	megawatt-days
Btu/hour-square foot	3.153×10^{-4}	watts/square centimeter
Btu/hour-square foot-°F	5.676×10^{-4}	watts/square centimeter-°C
Btu/minute	2.356×10^{-2}	horsepower
Btu/minute	1.757×10^1	watts
calories	4.1868	joules
cubic feet	2.832×10^{-2}	cubic meters
cubic feet	2.832×10^1	liters
cubic feet/minute	4.720×10^{-4}	cubic meters/second
cubic meters	8.107×10^{-4}	acre-feet
cubic meters	3.531×10^1	cubic feet
cubic meters	2.642×10^2	gallons (U.S.)

TABLE 1.8 (Cont'd)

cubic meters/second	2.119×10^3	cubic feet/minute
cubic meters/second	1.585×10^4	gallons/minute
degrees Celsius	$(9/5)C + 32$	degrees Fahrenheit
degrees Fahrenheit	$5/9(F - 32)$	degrees Celsius
feet	3.048×10^{-1}	meters
feet of H_2O (39.2°F)	3.048×10^2	kilograms/square meter
feet of H_2O (39.2°F)	4.335×10^{-1}	pounds force/square inch
feet/second	3.048×10^{-1}	meters/second
foot-pound (force)	1.356	joules
foot-pounds (force)/minute	2.260×10^{-2}	watts
gallons	3.785×10^{-3}	cubic meters
gallons/minute	6.309×10^{-5}	cubic meters/second
horsepower	4.244×10^1	British thermal units/minute
horsepower	7.457×10^{-1}	kilowatts
horsepower-hours	2.547×10^3	British thermal units
horsepower-hours	7.457×10^{-1}	kilowatt-hours
inches of H_2O (39.2°F)	2.491×10^{-1}	kilopascals
inches mercury (32°F)	3.342×10^{-2}	atmospheres
inches mercury (32°F)	3.453×10^2	kilograms/square meter
inches mercury (32°F)	3.386	kilopascals
inches mercury (32°F)	4.912×10^{-1}	pounds force/square inch
joules	7.376×10^{-1}	foot-pounds (force)
joules	1.000	watt-seconds
joules	2.387×10^{-1}	calories
kilograms	2.205	pounds force
kilograms	1.102×10^{-3}	tons (short)
kilograms/cubic meter	6.243×10^{-2}	pounds force/cubic foot
kilograms/square meter	9.678×10^{-5}	atmospheres[a]
kilograms/square meter	3.281×10^{-3}	feet of H_2O (at 39.2°F)[a]
kilograms/square meter	2.896×10^{-3}	inches mercury (32°F)[a]
kilograms/square meter	1.422×10^{-3}	pounds force/square inch[a]
kilojoules	9.471×10^{-1}	British thermal units
kilopascals	4.015	inches H_2O (at 39.2°F)
kilopascals	1.450×10^{-1}	pounds force/square inch
kilopascals	2.953×10^{-1}	inches mercury (32°F)
kilopascals	1.000×10^{-2}	bars
kilopascals	9.869×10^{-3}	atmospheres (760 torr)
kilowatts	1.341	horsepower
kilowatt-hours	3.413×10^3	British thermal units
kilowatt-hours	1.341	horsepower-hours
kilowatt-hours	4.167×10^{-5}	megawatt-days
liters	3.531×10^{-2}	cubic feet

TABLE 1.8 (Cont'd)

megawatt-days	8.189×10^7	British thermal units
megawatt-days	2.400×10^4	kilowatt-hours
meters	3.281	feet
newtons	2.248×10^{-1}	pounds force
pounds	4.536×10^{-1}	kilograms
pounds (force)	4.448	newtons
pounds mass/cubic feet	1.602×10^1	kilograms/cubic meter
pounds force/square inch	2.307	feet of H_2O (at 39.2°F)
pounds force/square inch	2.036	inches mercury (32°F)
pounds force/square inch	7.031×10^2	kilograms/square meter[a]
pounds force/square inch	6.895	kilopascals
square feet	9.290×10^{-2}	square meters
square meters	2.471×10^{-4}	acres
square meters	1.076×10^1	square feet
tonnes (metric tons)	2.205×10^3	pounds force
tons (short)	9.072×10^2	kilograms
watts	5.688×10^{-2}	Btu/minute
watts	4.427×10^1	foot-pounds force/minute
watt-seconds	1.000	joules
watts-square centimeter	3.171×10^3	Btu/hour-square foot
watts/square centimeter-°C	1.762×10^3	Btu/hour-square foot-°F

[a] Force units with $g_c = 9.81$.

ILLUSTRATIVE PROBLEM 1.5

Table 1.8 lists the conversion factor for obtaining square meters from square feet as 9.290×10^{-2}. Starting with the definition that 1 in. equals 2.54 cm (0.0254 m), derive this converison factor.

SOLUTION

In solving this type of conversion problem, it must be kept in mind that units need to be consistent and that we can use the familiar rules of algebra to manipulate units.

Proceeding as noted, we have

$$1 \text{ in.} = 0.0254 \text{ m}$$

or

$$1 = 0.0254 \frac{\text{m}}{\text{in.}} \tag{a}$$

Because 1 ft = 12 in.,

$$1 = 12 \frac{\text{in.}}{\text{ft}} \tag{b}$$

If we now multiply Equation (a) by Equation (b), we obtain

$$1 = 0.0254 \, \frac{m}{in.} \times 12 \, \frac{in.}{ft}$$

Because inches cancel, we have

$$1 = 0.0254 \times 12 \, \frac{m}{ft}$$

We can now square both sides to obtain

$$(1)^2 = (0.0254 \times 12)^2 \, \frac{m^2}{ft^2} = 9.290 \times 10^{-2} \, \frac{m^2}{ft^2}$$

or the desired result,

$$1 \, ft^2 = 9.290 \times 10^{-2} \, m^2$$

which states that $ft^2 \times 9.290 \times 10^{-2} = m^2$.

The foregoing can also be obtained in a single chain-type calculation as

$$\left(0.0254 \, \frac{m}{in.} \times \frac{12 \, in.}{ft} \right)^2 = 9.290 \times 10^{-2} \, \frac{m^2}{ft^2}$$

1.5 ELEMENTARY KINETIC THEORY OF GASES

The following derivation, based on elementary kinetic gas theory, is very useful and will serve later to introduce the theory of ideal or perfect gases. Imagine a cube, as shown in Figure 1.17, with sides l. In this cube, it will be assumed that there are n identical molecules (particles), each having a mass m. These particles are further assumed to be perfectly elastic spheres each traveling with the same velocity. By *perfectly elastic*, it is implied that the collision of a molecule with another molecule or the walls of the container causes the molecule to rebound without loss of energy or momentum. The size of the molecule will be taken to be negligibly small compared to dimensions of the system. If the velocity in the x direction is essentially constant over the length l and is denoted by V_x, the molecule shown in Figure 1.17 will travel a distance equal numerically to V_x in a unit time. In this same time, the molecule will collide elastically with the walls (parallel to the y–z plane) a number of times equal to V_x divided by l. This last statement has considered only the x-directed velocity of the molecule. It is also assumed that the molecules do not collide with each other.

At this point, it will be necessary to involve the concepts of impulse and momentum for a rigid particle that has only translatory motion. On the basis of Newton's laws of motion for such a particle, it is possible to write Equation (1.9) as follows:

$$F = \frac{ma}{g_c} = \frac{m \, \Delta V}{g_c \, \Delta t}; \qquad F \, \Delta t = \frac{m \, \Delta V}{g_c} \qquad \textbf{(1.11)}$$

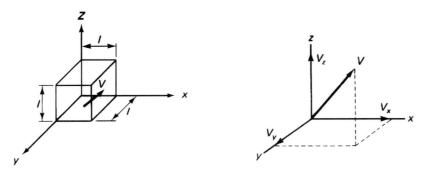

FIGURE 1.17 Kinetic theory derivation.

[Equation (1.11) is written in English units.] The symbol Δt is used to indicate a small interval of time, and ΔV is the change in velocity during this time interval. The term $F\Delta t$ is known as the *impulse* imparted to the particle by the force F acting for the time Δt, and the term $(m\Delta V/g_c)$ is called the *momentum change* that the particle undergoes during this same time interval.

Let us consider the collision of a particle with a wall. As shown in Figure 1.18, the particle travels in the x direction with a velocity V_x prior to striking the wall. Because the particle has been assumed to be perfectly elastic, it will collide with the wall and rebound with a speed equal to V_x, but it will be traveling in a direction opposite to its initial direction. Thus, its velocity will be $-V_x$. The wall is stationary, and as a result of the impact, a reactive force F is set up. Because the molecule has a momentum equal to mV_x/g_c just before colliding with the wall and a momentum of $-mV_x/g_c$ after the collision, the change in momentum per molecule will be $2mV_x/g_c$. As stated earlier, the number of collisions with the walls in a unit time will be V_x/l, and we must consider that the n identical particles each undergo the same collision with the walls. The total number of collisions will therefore be nV_x/l.

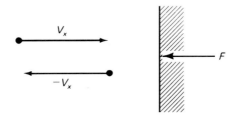

FIGURE 1.18 Particle collision with wall.

Multiplying the total number of impacts per unit of time by the momentum change per impact yields the total change of momentum per unit of time for the system:

$$\frac{2mV_x}{g_c}\left(\frac{nV_x}{l}\right) \tag{1.12}$$

Equation (1.12) also expresses the *force* exerted on the wall. *Pressure* is the normal force exerted per unit surface area. For present purposes, it will be reasonable to assume that the force expressed by Equation (1.12) is distributed uniformly over the y–z planes. Because these consist of two faces, each l^2 in area, the pressure exerted on the y–z plane is

$$\frac{\dfrac{2mV_x}{g_c}\left(\dfrac{nV_x}{l}\right)}{2l^2} = \frac{mnV_x^2}{g_c l^3} \tag{1.13}$$

However, it was previously assumed that the molecule had the same velocity in each of the three component directions. Therefore, in the unit time interval under consideration, there is an equal probability that the particle will hit one of the three sets of parallel faces composing the cube. If this were not so, there would be unequal pressures on the various faces of the tube. Because the volume of the cube (V) equals l^3, Equation (1.14) follows from Equations (1.11) and (1.13) combined:

$$pV = \frac{1}{3}\left(\frac{mnV_x^2}{g_c}\right) \tag{1.14}$$

or

$$pV = \tfrac{1}{3}(mnV_x^2) \qquad \text{(SI)} \tag{1.14a}$$

Equation (1.14) is used repeatedly later (in slightly different form) and has special significance. However, for the present, only certain limited conclusions will be deduced from it. It should be remembered that this equation was derived on the basis of certain simplifying and therefore restrictive assumptions. In addition to the assumptions stated, it is important to note that the rotations and vibrations of the molecule have been neglected. Also, molecule collisions with other molecules have been ignored. This corresponds to assuming that the dimensions of the unit cube are small compared with the mean free path of the gas particles, that is, the distance molecules travel between molecule–molecule collisions. Although these assumptions and others inherent to this derivation lead to varying degrees of deviation of real gases from Equation (1.14), it is amazing that this equation can be used with a reasonable degree of accuracy for engineering purposes to describe the behavior of many gases over a wide range of conditions.

On the basis of experimental and theoretical investigations by Joule and others, it can be shown that for ideal gases, the pressure of these gases increases equally for equal temperature increments. This is equivalent to stating that the product mV_x^2 is a constant for all gases at a given temperature. As a consequence of this conclusion, Equation (1.14) yields *Avogadro's law,* which states that equal volumes of gases at the same temperature and pressure contain the same number of molecules (particles). It is also possible to deduce another concept from Equation (1.14), specifically, density. The product mn found in the numerator of Equation (1.14) is the total mass of the molecules in the volume under consideration. The total mass divided by the total volume yields the property of *density* (ρ). The reciprocal of density is called *specific volume*. It is the total volume that a unit mass occupies under a given set of conditions. It is denoted by v.

We summarize some of the conclusions of the foregoing analysis (the first three conclusions are limited to gases and the last three are general definitions):

1. *Pressure* (which is defined as the normal force per unit area) results from the collisions of the molecules with the walls of the container and is a function of the number of impacts per unit time, the mass of the particle, and the velocity of the particle.

2. *Avogadro's law* states that equal volumes of gases at the same pressure and temperature contain the same number of particles.

3. *Temperature* can be taken to be a measure of the translational energy of the particles. In a more general sense, it can be said to be a measure of the molecular activity (and internal energy) of the gases.

4. *Density* is defined as the mass per unit volume (ρ).

5. The reciprocal of density is the volume per unit mass and is called *specific volume* (v).

6. *Specific weight* (γ) is defined as the weight per unit volume. Note that this term, which equals $\rho g/g_c$, does not exist in SI as a basic unit.

7. *Specific gravity* (sg) is defined as the ratio of the density of a substance to the density of water at 4°C. It can also be defined as the ratio of the specific weight of a substance to the specific weight of water at 4°C. The density of water at 4°C is 1000 kg/m³. In the older cgs (centimeter-gram-second) system, the density of water at 4°C is 1 g/cm³, making specific gravity numerically equal to density in this system.

1.6 PRESSURE

We have seen that when a gas is confined in a container, molecules of the gas strike the sides of the container, and these collisions with the walls of the container cause the molecules of the gas to exert a force on the walls. When the component of the force that is perpendicular (normal) to the wall is divided by the area of the wall, the resulting normal force per unit area is called the *pressure*. To define the pressure at a point, it is necessary to consider the area in question to be shrinking steadily. Pressure at a point is defined to be the normal force per unit area in the limit as the area goes to zero. Mathematically,

$$p = \left(\frac{\Delta F}{\Delta A}\right)_{\lim(\Delta A \Rightarrow 0)} \tag{1.15}$$

In common engineering units, pressure is expressed as pounds force per square inch or pounds force per square foot, which are usually abbreviated as psi and psf. In SI units, pressure is expressed in N/m², which is known as the *pascal* (Pa). The *bar* is 10^5 Pa. These units are the ones used most frequently in the literature, but certain others are also used because of the manner in which the pressure measurements are made.

Most mechanical pressure gages, such as the Bourdon gage (see Figure 1.33), measure the pressure above local atmospheric pressure. This pressure is called gage pressure and is measured in units of pounds force per square foot or pounds force per square inch, that is, psfg or psig. The relation of gage pressure to absolute pressure is shown in Figure 1.19 and is

$$\boxed{\text{absolute pressure} = \text{atmospheric pressure} + \text{gage pressure}} \tag{1.16}$$

Absolute pressure is indicated as psia or psfa. It is usual to take the pressure of the atmosphere as being equal to 14.7 psia. However, in Europe, 1 "ata" has been defined and is used as 1 kg/cm² for convenience, but this unit is not 14.7 psia. It is more nearly 14.2 psia. Care should be exercised

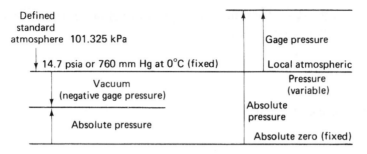

FIGURE 1.19 The relation of gage to absolute pressure.

when using the unit of pressure that is expressed as ata. In SI, the equivalent of 14.696 (14.7) psia is 101.325 kPa. The terms *gage* and *vacuum* are not used in SI, because all pressures are absolute in this system.

Referring again to Figure 1.19, it will be noted that pressure below atmospheric pressure is called vacuum. By reference to this figure, we obtain the relation between absolute pressure and vacuum as

$$\boxed{\text{absolute pressure} = \text{atmospheric pressure} - \text{vacuum}} \tag{1.17}$$

ILLUSTRATIVE PROBLEM 1.6

Assume that a fluid whose specific weight is constant and equal to γ lb$_f$/ft^3 is placed in a uniform tube until its height in the tube is h feet above the base of the column. Determine the pressure at the base of the column.

SOLUTION

From our definition of pressure as force divided by area, we must find the force being exerted on the base by the column of fluid. This force will equal the weight of the fluid,

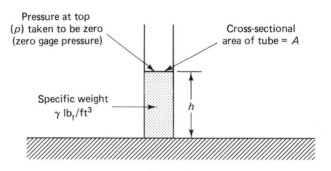

FIGURE 1.20 Illustrative Problem 1.6.

which in turn will equal the volume of fluid multiplied by the specific weight of the fluid. Denoting the weight to be w, the cross-sectional area to be A, and the liquid height to be h, as shown in Figure 1.20, we have $w = \gamma V$, where V is the volume of fluid. The volume of fluid is equal to $A \times h$, giving us

$$w = \gamma A h \tag{1.18}$$

which is the force on the base. Dividing w by A yields the pressure

$$p = \frac{\gamma A h}{A} = \gamma h \tag{1.19}$$

Because it is usual to measure distance up from the base, our result is

$$\boxed{p = -\gamma h} \tag{1.19a}$$

In English units,

$$\boxed{p = \frac{-\rho g}{g_c} h} \tag{1.20}$$

In SI units,

$$\boxed{p = -\rho g h} \tag{1.20a}$$

CALCULUS ENRICHMENT

The relation between the pressure and height of a fluid can be derived considering Figure A below:

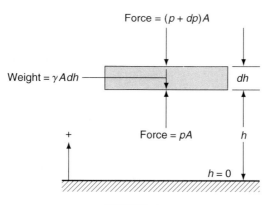

FIGURE A

Summing the forces in the vertical direction and noting that *up* is positive yields

$$dp = -\gamma dh \tag{a}$$

where γ is also ρg. Thus,
$$dp = -\rho g dh \tag{b}$$

In order to solve Equation (a) or (b), it is necessary to know p as a function of h. In general, we can integrate these equations:

$$p(h) - p(0) = -\int_0^h \rho g dh \tag{c}$$

For the special case of a liquid having a constant density, Equation (c) can be integrated, starting at the surface, where $p = 0$, and noting that h is measured positive upward. This gives us the result that

$$p = -\rho g h \tag{d}$$

or
$$p = -\gamma h \tag{e}$$

Equations (d) and (e) are the same as Equations (1.19a) and (1.20a).

Let us consider the case of a fluid whose density is a linear function of pressure. We can write this in terms of the height as

$$\rho = \rho_0(1 - \alpha h) \tag{f}$$

where ρ_0 corresponds to h_0 and α is the rate at which density decreases with elevation. Placing (f) into (c) gives us

$$p(h) - p(0) = -\int_0^h \rho_0(1 - \alpha h)\, dh \tag{g}$$

Integrating
$$p(h) - p(0) = -\left[\rho_0 h - \frac{\alpha \rho_0 h^2}{2} \right]_0^h \tag{h}$$

Simplifying
$$p(h) - p(0) = -\rho_0 h \left[1 - \frac{\alpha h}{2} \right] \tag{i}$$

In meteorology, the decrease of temperature with height is called the lapse rate. The lapse rate is obviously related to Equations (c) and (i).

The most common fluid used to measure pressure differences, atmospheric pressure, and vacuum pressure is mercury. At approximately room temperature, the specific gravity of mercury is very nearly 13.6. It will usually be assumed that the specific gravity of mercury is 13.6 unless otherwise stated. The effect of temperature on the specific gravity of mercury is given in Figure 1.21.

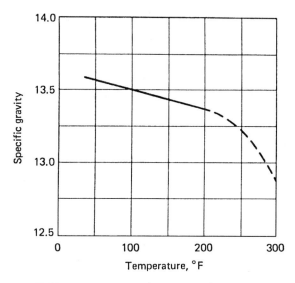

FIGURE 1.21 Specific gravity of mercury.

A tube of glass open at the top has a 1-in. level of mercury in it. Take the specific gravity of mercury (Hg) to be 13.6, and determine the pressure on the base of the column.

SOLUTION

To solve this problem, let us first convert the unit of weight of grams per cubic centimeter to pounds per cubic foot. In 1 lb_f, there are approximately 454 g. Also, there are 2.54 cm in 1 in. Thus, 1 g is $1/454$ lb_f, and 1 ft has 12×2.54 cm. To convert 1 g per cubic centimeter to pounds per cubic foot, we apply the following dimensional reasoning:

$$\frac{g}{cm^3} \times \frac{lb}{g} \times \frac{cm^3}{ft^3} = \frac{lb}{ft^3} = \gamma$$

Continuing yields

$$\frac{1 \, g}{cm^3} \times \frac{1}{454} \frac{lb_f}{g} \times (2.54 \times 12)^3 \frac{cm^3}{ft^3} = \frac{62.4 \, lb_f}{ft^3} = \gamma$$

Thus, the conversion factor from grams per cubic centimeter to pounds per cubic foot is 62.4. The student is strongly urged to check all computations for dimensional consistency when doing problems.

Applying Equation (1.19) gives us

$$p = (\tfrac{1}{12} \, ft)(13.6 \times 62.4) \frac{lb_f}{ft^3} = 70.72 \frac{lb_f}{ft^2}$$

In pounds force per square inch,

$$p \frac{lb_f}{ft^2} \times \frac{ft^2}{in^2} \times \frac{1}{144} = psi$$

Therefore,

$$p = \frac{1}{12} \times 13.6 \times 62.4 \times \frac{1}{144} \frac{ft^2}{in^2} = 0.491 \text{ psi}$$

This is a gage pressure. If the local atmospheric pressure is 14.7 psia,

$$p = 0.491 + 14.7 = 15.19 \text{ psia}$$

The value of 0.491 psi/in. Hg is a useful conversion factor.

ILLUSTRATIVE PROBLEM 1.8

The density of mercury is 13.595 kg/m^3. Determine the pressure at the base of a column of mercury that is 25.4 mm high.

SOLUTION

Using Equation (1.20a), we obtain

$$p = -\rho g h = 13.595 \frac{kg}{m^3} \times 9.806 \frac{m}{s^2} \times 0.0254 \text{ m} = 3386.1 \text{ Pa}$$

Note: 1 psi $\simeq$ 6895 Pa. Therefore, 3386.1/6895 = 0.491 psi, which checks with Illustrative Problem 1.7.

In vacuum work, it is common to express the absolute pressure in a vacuum chamber in terms of millimeters of mercury. Thus, a vacuum may be expressed as 10^{-5} mm Hg. If this is expressed in pounds force per square inch, it would be equivalent to 0.000000193 psi, with the assumption that the density of mercury (for vacuum work) is 13.6 g/cm^3. Another term, *torr,* is also found in the technical literature. A torr is defined as 1 mm Hg. Thus, 10^{-5} torr is the same as 10^{-5} mm Hg. Another unit of pressure used in vacuum work is the micron, μ. A micron is defined as one thousandth of 1 mm Hg, so 10^{-3} mm Hg is equal to 1 μ.

ILLUSTRATIVE PROBLEM 1.9

A mercury manometer (vacuum gage) reads 26.5 in. of vacuum when the local barometer reads 30.0 in. Hg at standard temperature. Determine the absolute pressure in psia.

SOLUTION

$$p = (30.0 - 26.5) \text{ in. Hg absolute}$$

and from Illustrative Problem 1.7, 1 in. Hg exerts a pressure of 0.491 psi. Thus,

$$p = (30 - 26.5)(0.491) = 1.72 \text{ psia}$$

If the solution is desired in psfa, it is necessary to convert from square inches to square feet. Because there are 12 in. in 1 ft, there will be 144 in² in 1 ft². Thus, the conversion to psfa requires that psia be multiplied by 144. This conversion is often necessary to keep the dimensions of equations consistent.

ILLUSTRATIVE PROBLEM 1.10

A column of fluid is 10 m high. If its density is 2000 kg/m³, determine the pressure at the base of the column if it is located in a local gravity of 9.6 m/s².

SOLUTION

Applying Equation (1.20a) gives us

$$p = -\rho g h = -2000 \frac{\text{kg}}{\text{m}^3} \times 9.6 \frac{\text{m}}{\text{s}^2} \times -10 \text{ m} = 192 \text{ kPa}$$

The height is negative, because it is measured up from the base.

ILLUSTRATIVE PROBLEM 1.11

Using the data of Illustrative Problem 1.9, determine the absolute pressure in kPa.

SOLUTION

As before, $p = (30 - 26.5) = 3.5$ in. Hg absolute. Using "standard" temperature yields

$$\frac{3.5 \text{ in.} \times \dfrac{\text{ft}}{12 \text{ in.}} \times (13.6 \times 62.4) \dfrac{\text{lb}_f}{\text{ft}^3}}{\left(12 \dfrac{\text{in.}}{\text{ft}}\right)^2 \times \left(0.0254 \dfrac{\text{m}}{\text{in.}}\right)^2} \times \frac{\text{kg}}{2.2 \text{ lb}_f} \times \left(9.806 \frac{\text{N}}{\text{kg}}\right) = 11.875 \text{ kPa}$$

Note: $11.875 \times 1000/6895 = 1.72$ psia, which checks with Illustrative Problem 1.9.

Modern power stations use hundreds of pressure and temperature gages to monitor and control their operations. Figure 1.22 shows a portion of the control room of a modern power station. It is evident from this figure that a large number of monitors, recorders, and controllers are used to ensure trouble-free operation of these plants.

FIGURE 1.22 View of the control room of the Petersburg, Indiana, plant.
(Courtesy of the Indianapolis Power and Light Company.)

In performing any measurement, it is necessary to have a standard of comparison in order to calibrate the measuring instrument. In the following paragraphs, five pressure standards currently used as the basis for all pressure measurement work will be discussed. Table 1.9 summarizes these standards, the pressure range in which they are used, and their accuracy.* Figure 1.23 shows the basic pressure measurement concept on which all pressure standards are based.

TABLE 1.9
Characteristics of Pressure Standards

Type	Range	Accuracy
Dead-weight piston gage	0.01 to 10,000 psig	0.01 to 0.05% of reading
Manometer	0.01 to 100 psig	0.02 to 0.2% of reading
Micromanometer	0.0002 to 20 in. H_2O	1% of reading to 0.001 in. H_2O
Barometer	27 to 31 in. Hg	0.001 to 0.03% of reading
McLeod gage	0.01 μm to 1 mm Hg	0.5 to 3% of reading

---

* The remainder of this section is based on the material in Irving Granet, *Fluid Mechanics for Engineering Technology,* 3rd. ed. (Englewood Cliffs, N.J.: Prentice-Hall, Inc., 1989); reprinted with permission of the publisher.

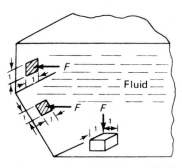

Pressure is the normal force F (lb/in^2, lb/ft^2, dynes/cm^2) exerted on a unit area of a surface bounding a fluid.

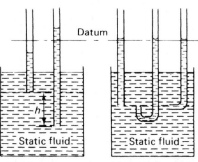

Fluid pressure varies with depth, but it is the same in all directions at a given depth.

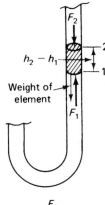

Variation of fluid pressure with elevation is found by balancing the forces on a static-fluid element (F_1 is equal to F_2 plus the weight of the element). For a constant-density fluid, the pressure difference $p_2 - p_1$ is equal to the specific weight γ times $(h_2 - h_1)$.

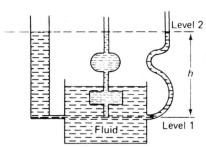

Pressure is independent of the shape and size of the vessel. The pressure difference between level 1 and level 2 is always $p_1 - p_2 = \gamma h$, where γ is the specific weight of the constant-density fluid in the vessel.

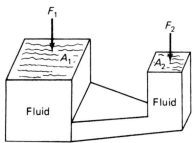

Constant pressure transmission in a confined fluid can be used to multiply force by the relation $p = F_1/A_1 = F_2/A_2$.

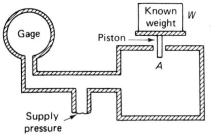

When the known weight is balanced, the gage pressure is $p = W/A$; this is the basic principle of dead-weight testing.

FIGURE 1.23   Basic pressure-measurement concepts.

1.6a   Dead-Weight Piston Gage

The dead-weight free-piston gage consists of an accurately machined piston inserted into a close-fitting cylinder. Masses of known weight are loaded on one end of the free piston, and pressure is applied to the other end until enough force is developed to lift the piston–weight combination. When the piston is floating freely between the cylinder limit stops, the gage is in equilibrium with the unknown system pressure. The dead-weight pressure can then be defined as

$$p_{\text{dw}} = \frac{F_e}{A_e} \tag{1.21}$$

where F_e is the equivalent force of the piston–weight combination, dependent on such factors as local gravity and air buoyancy, and A_e is the equivalent area of the piston–cylinder combination, dependent on such factors as piston–cylinder clearance, pressure level, and temperature.

A fluid film provides the necessary lubrication between the piston and cylinder. In addition, the piston, or less frequently the cylinder, may be rotated or oscillated to reduce friction even further. Because of fluid leakage, system pressure must be continuously trimmed upward to keep the piston–weight combination floating. This is often achieved by decreasing the system volume using a pressure–volume apparatus (as shown in Figure 1.24). As long as the piston is freely balanced, system pressure is defined by Equation (1.21).

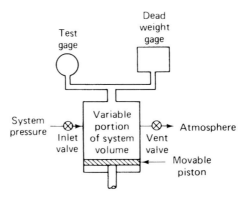

FIGURE 1.24 Pressure volume regulator to compensate for fluid leakage in a dead-weight gage.

Corrections must be applied to the indication of the dead-weight piston gage p_i to obtain the accurate system pressure p_{dw}. The two most important corrections concern air buoyancy and local gravity. The effective area of the dead-weight piston gage is usually taken as the mean of the cylinder and piston areas, but temperature affects this dimension. The effective area increases between 13 and 18 ppm (parts per million)/°F for commonly used materials, and a suitable correction for this effect may also be applied.

1.6b Manometer

We have already shown that the pressure at the base of a column of liquid is simply a function of the height of the column and the specific weight of the liquid. Therefore, the height of a column of liquid of known specific weight can be and is used to measure pressure and pressure differences. Instruments that utilize this principle are known as manometers, and the study of these pressure-measuring devices is known as manometry. By properly arranging a manometer and selecting the fluid judiciously, it is possible to measure extremely small pressures, very large pressures, and pressure differences. A simple manometer is shown in Figure 1.25, where the right arm is exposed to the atmosphere while the left arm is connected to the unknown pressure. As shown, the fluid is depressed in the left arm and raised in the right arm until no unbalanced pressure forces remain. It has already been demonstrated that the pressure at a given level in either arm must be the same so that we can select any level as reference and write a relation for the pressure. Actually, it is much easier and more convenient to select the interface between the manometer fluid and the unknown fluid as a common reference level. In Figure 1.25, the pressure at elevation AA is the same in both arms of the manometer. Starting with the open manometer arm (right), we have atmospheric pressure p_a acting on the fluid. As one proceeds down the arm, the pressure increases until we arrive at level $AA,$ where the pressure is $p_a + \gamma h$. This pressure must equal the unknown pressure on the connected arm p_u. Therefore,

$$p_u = p_a + \gamma h \qquad (1.22)$$

or

$$p_u - p_a = \gamma h \qquad (1.23)$$

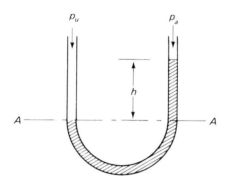

FIGURE 1.25 U-tube manometer.

Temperature and capillary effects must be considered for accurate pressure measurements. To minimize the effect of a variable meniscus, which can be caused by dirt, the method of approaching equilibrium, tube bore, and so on, the tubes are always tapped before reading, and the measured liquid height is always based on readings taken at the center of the meniscus in each leg of the manometer. To reduce the capillary effect itself, the use of large-bore tubes (more than $\frac{3}{8}$ in. diameter) is most effective.

To achieve greater accuracy and sensitivity in manometers, several different arrangements have been used. Perhaps the simplest of these is the inclined manometer. Consider a relatively large reservoir of liquid connected to a small-bore tube that makes an angle θ with the horizontal. The pressure or pressure differential to be measured is connected to the large reservoir, while the inclined tube is open ended. Schematically, this is shown in Figure 1.26. The unknown pressure p_u is given by

$$p_u = p_a + \gamma h \qquad (1.24)$$

or

$$p_u - p_a = \gamma h \qquad (1.25)$$

$$p_u - p_a = \gamma h' \sin \theta \qquad (1.26)$$

Because θ is fixed, a scale placed along the tube can be calibrated to read directly in units of h of a fluid. Usually, this is done by directly reading inches of water for $p_u - p_a$.

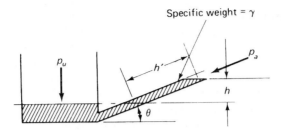

FIGURE 1.26 Inclined manometer.

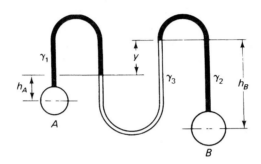

FIGURE 1.27 Three-fluid manometer.

Another method of achieving greater sensitivity and accuracy is to use a manometer with more than one fluid. The manometer shown in Figure 1.27 can be used for this purpose by properly selecting the manometer fluid. Starting at level A, we have

$$p_A - h_A\gamma_1 - y\gamma_3 + h_B\gamma_2 = p_B \qquad (1.27)$$

or

$$p_A - p_B = (-h_B\gamma_2 + h_A\gamma_1) + y\gamma_3 \qquad (1.28)$$

In the usual case, the manometer is connected to different positions on the same pipe. Thus, A and B would be at the same level, and γ_1 can be taken to be equal to γ_2. Also, $h_B - h_A = y$. Thus,

$$p_A - p_B \simeq y(\gamma_3 - \gamma_1) \quad \text{or} \quad y(\gamma_3 - \gamma_2) \qquad (1.29)$$

For small differences in $p_A - p_B$, it is apparent that the manometer fluid (γ_3) should have a specific weight very nearly equal to the specific weight of the fluid in the pipes. For large pressure differences, one can use a heavy fluid such as mercury to increase $\gamma_3 - \gamma_2$ and reduce the manometer reading.

A word on the reference pressures employed in manometry is pertinent at this point in the discussion. If atmospheric pressure is used as a reference, the manometer yields gage pressures. Because of the variability of air pressure, gage pressures vary with time, altitude, latitude, and temperature. If, however, a vacuum is used as reference, the manometer yields absolute pressures directly, and it may serve as a barometer. In any case, the absolute pressure is always equal to the sum of the gage and ambient pressures; by ambient pressure, we mean the pressure surrounding the gage, which is usually atmospheric pressure.

1.6c Micromanometer

While the manometer is a useful and convenient tool for pressure measurements, it is unfortunately limited when making low-pressure measurements. To extend its usefulness in the low-pressure range, micromanometers have been developed that have extended the useful range of low-pressure manometer measurements to pressures as low as 0.0002 in. H_2O.

One type is the *Prandtl-type micromanometer,* in which capillary and meniscus errors are minimized by returning the meniscus of the manometer liquid to a null position before measuring the applied pressure difference. As shown in Figure 1.28, a reservoir, which forms one side of the manometer, is moved vertically to locate the null position. This position is reached when the meniscus falls within two closely scribed marks on the near-horizontal portion of the micromanometer tube. Either the reservoir or the inclined tube is then moved by a precision lead-screw arrangement to determine the micromanometer liquid displacement (Δh), which corresponds to the applied pressure difference. The Prandtl-type micromanometer is generally accepted as a pressure standard within a calibration uncertainty of 0.001 in. H_2O.

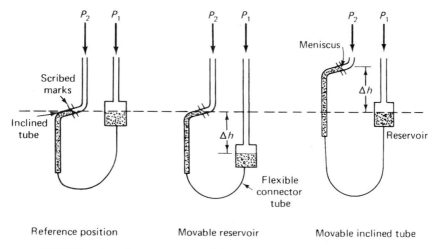

FIGURE 1.28 Two variations of the Prandtl-type micromanometer.

Another method for minimizing capillary and meniscus effects in manometry is to measure liquid displacements with micrometer heads fitted with adjustable, sharp index points. Figure 1.29 shows a manometer of this type; the micrometers are located in two connected, transparent containers. In some commercial micromanometers, contact with the surface of the manometric liquid may be sensed visually by dimpling the surface with the index point, or even by electrical contact. Micrometer-type micromanometers also serve as pressure standards within a calibration uncertainty of 0.001 in. H_2O.

An extremely sensitive, high-response micromanometer uses air as the working fluid and thus avoids all the capillary and meniscus effects usually encountered in liquid manometry. In this device, as shown in Figure 1.30, the reference pressure is mechanically amplified by centrifugal action in a rotating disk. The disk speed is adjusted until the amplified reference pressure just balances the unknown pressure. This null position is recognized by observing the lack of movement

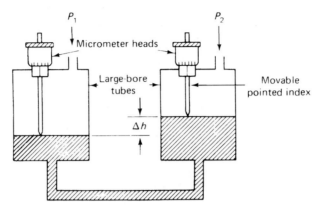

FIGURE 1.29 Micrometer-type manometer.

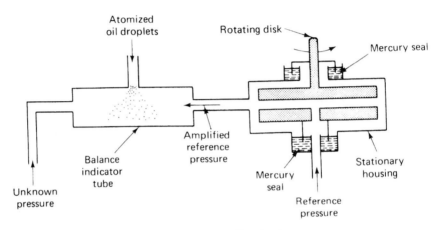

FIGURE 1.30 Air-type centrifugal micromanometer.

of minute oil droplets sprayed into a glass indicator tube. At balance, the air micromanometer yields the applied pressure difference as

$$\Delta p_{\text{micro}} = Kpn^2 \qquad (1.30)$$

where p is the reference air density, n is the rotational speed of the disk, and K is a constant that depends on disk radius and annular clearance between the disk and the housing. Measurements of pressure differences as small as 0.0002 in. H_2O can be made with this type of micromanometer within an uncertainty of 1%.

1.6d Barometers

The reservoir or cistern barometer consists of a vacuum-reference mercury column immersed in a large-diameter, ambient-vented mercury column that serves as a reservoir. The most common cistern barometer in general use is the Fortin type, in which the height of the mercury surface in the

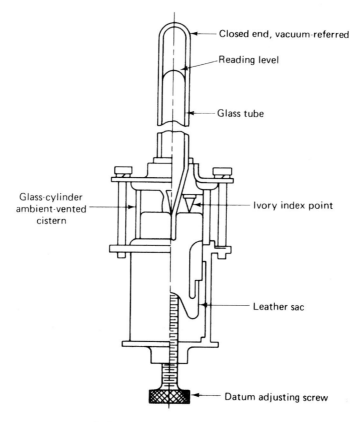

Closed end, vacuum-referred

Reading level

Glass tube

Glass-cylinder ambient-vented cistern

Ivory index point

Leather sac

Datum adjusting screw

FIGURE 1.31 Fortin barometer.

cistern can be adjusted. The operation of this instrument can best be explained with reference to Figure 1.31.

The datum-adjusting screw is turned until the mercury in the cistern makes contact with the ivory index, at which point the mercury surface is aligned with zero on the instrument scale. Next, the indicated height of the mercury column in the glass tube is determined. The lower edge of a sighting ring is lined up with the top of the meniscus in the tube. A scale reading and a vernier reading are taken and combined to yield the indicated height at the barometer temperature.

Because atmospheric pressure on the mercury in the cistern is exactly balanced by the weight per unit area of the mercury column in the glass tube,

$$p_{\text{baro}} = \gamma_{\text{Hg}} h_{\text{to}} \tag{1.31}$$

The referenced specific weight of mercury, γ_{Hg}, depends on such factors as temperature and local gravity; the referenced height of mercury, h_{to}, depends on such factors as thermal expansion of the scale and the mercury.

Other factors may also contribute to the uncertainty of h_{to}. Proper illumination is essential to define the location of the crown of the meniscus. Precision meniscus sighting under optimum viewing conditions can approach ±0.001 in. With proper lighting, contact between the ivory index and the mercury surface in the cistern can be detected to much better than ±0.001 in.

To keep the uncertainty in h_{to} within 0.01% ($\approx$0.003 in. Hg), the mercury temperature must be taken within $\pm 1°F$. Scale temperature need not be known to better than $\pm 10°F$ for comparable accuracy. Uncertainties caused by nonequilibrium temperature conditions can be avoided by installing the barometer in a uniform temperature room.

The barometer tube must be vertically aligned for accurate pressure determination. This is accomplished by a separately supported ring encircling the cistern; adjustment screws control the horizontal position.

Depression of the mercury column in commercial barometers is accounted for in the initial calibration setting at the factory. The quality of the barometer is largely determined by the bore of the glass tube. Barometers with a bore of $\frac{1}{4}$ in. are suitable for readings of about 0.01 in. Hg, whereas barometers with a bore of $\frac{1}{2}$ in. are suitable for readings down to 0.002 in. Hg.

1.6e   McLeod Gage

The McLeod gage is used in making low-pressure measurements. This instrument (shown in Figure 1.32) consists of glass tubing arranged so that a sample of gas at unknown pressure can be trapped and then isothermally compressed by a rising mercury column. This amplifies the unknown pressure and allows measurement by conventional manometric means. All the mercury is initially contained in the area below the cutoff level. The gage is first exposed to the unknown gas pressure, p_1; the mercury is then raised in tube A beyond the cutoff, trapping a gas sample of initial volume $V_1 = \bar{V} + ah_c$. The mercury is continuously forced upward until it reaches the zero level in the reference capillary B. At this time, the mercury in the measuring capillary C reaches

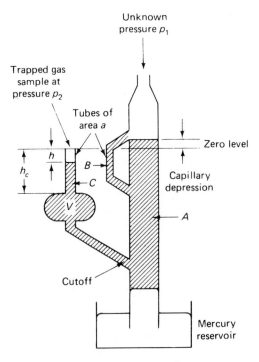

FIGURE 1.32 McLeod gage.

a level h, where the gas sample is at its final volume, $V_2 = ah$, and at the amplified pressure, $p_2 = p_1 + h$. Then,

$$p_1\overline{V}_1 = p_2\overline{V}_2 \tag{1.32}$$

$$p_1 = \frac{ah^2}{\overline{V}_1 - ah} \tag{1.33}$$

If $ah \ll \overline{V}_1$, as is usually the case,

$$p_1 = \frac{ah^2}{\overline{V}_1} \tag{1.34}$$

The larger the volume ratio $(\overline{V}_1/\overline{V}_2)$, the greater the amplified pressure p_2 and manometer reading h. Therefore, it is desirable that measuring tube C have a small bore. Unfortunately, for tube bores less than 1 mm, the compression gain is offset by reading uncertainty caused by capillary effects.

Reference tube B is introduced to provide a meaningful zero for the measuring tube. If the zero is fixed, Equation (1.34) indicates that manometer indication h varies nonlinearly with initial pressure p_1. A McLeod gage with an expanded scale at the lower pressures exhibits a higher sensitivity in this region. The McLeod pressure scale, once established, serves equally well for all the permanent gases (those whose critical pressure is appreciably less than room temperature).

There are no corrections to be applied to the McLeod gage reading, but certain precautions should be taken. Moisture traps must be provided to avoid taking any condensable vapors into the gage. Such vapors occupy a larger volume at the initial low pressures than they occupy in the liquid phase at the high reading pressures. Thus, the presence of condensable vapors always causes pressure readings to be too low. Capillary effects, while partially counterbalanced by using a reference capillary, can still introduce significant uncertainties, because the angle of contact between mercury and glass can vary $\pm 30°$. Finally, because the McLeod gage does not give continuous readings, steady-state conditions must prevail for the measurements to be useful.

In the earlier portions of this section, we discussed five pressure standards that can be used for either calibration or measurement of pressure in static systems. In this section, we discuss some common devices that are used for making measurements using an elastic element to convert fluid energy to mechanical energy. Such a device is known as a *pressure transducer*. Examples of mechanical pressure transducers having elastic elements only are dead-weight free-piston gages, manometers, Bourdon gages, bellows, and diaphragm gages.

Electrical transducers have an element that converts their displacement to an electrical signal. Active electrical transducers generate their own voltage or current output as a function of displacement. Passive transducers require an external signal. The piezoelectric pickup is an example of an active electrical transducer. Electric elements employed in passive electrical pressure transducers include strain gages, slide-wire potentiometers, capacitance pickups, linear differential transformers, and variable-reluctance units.

In the Bourdon gage, the elastic element is a small-volume tube that is fixed at one end but free at the other end to allow displacement under the deforming action of the pressure difference across the tube walls. In the most common model, shown in Figure 1.33, a tube with an oval cross section is bent in a circular arc. Under pressure, the tube tends to become circular, with a subsequent increase in the radius of the arc. By an almost frictionless linkage, the free end of the tube rotates a pointer over a calibrated scale to give a mechanical indication of pressure.

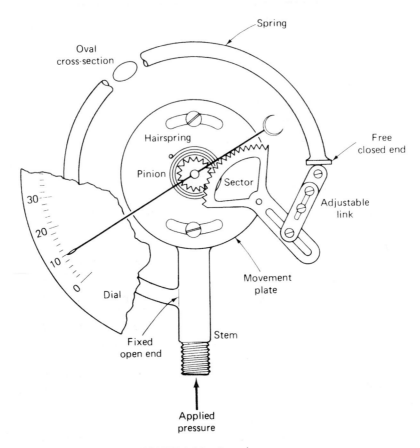

FIGURE 1.33 Bourdon gage.

The reference pressure in the case containing the Bourdon tube is usually atmospheric, so that the pointer indicates gage pressures. Absolute pressures can be measured directly without evacuating the complete gage casing by biasing a sensing Bourdon tube against a reference Bourdon tube, which is evacuated and sealed as shown in Figure 1.34. Bourdon gases are available for a wide range of absolute gage and differential pressure measurements within a calibration uncertainty of 0.1% of the reading.

Another common elastic element used in pressure transducers is the bellows, shown as a gage element in Figure 1.35. In one arrangement, pressure is applied to one side of a bellows, and the resulting deflection is partially counterbalanced by a spring. In a differential arrangement, one pressure is applied to the inside of one sealed bellows, and the pressure difference is indicated by a pointer.

A final elastic element to be mentioned because of its widespread use in pressure transducers is the diaphragm. One such arrangement is shown in Figure 1.36. Such elements may be flat, corrugated, or dished plates; the choice depends on the strength and amount of deflection desired. In high-precision instruments, a pair of diaphragms is used back to back to form an elastic capsule. One pressure is applied to the inside of the capsule; the other pressure is external. The calibration of this differential transducer is relatively independent of pressure magnitude.

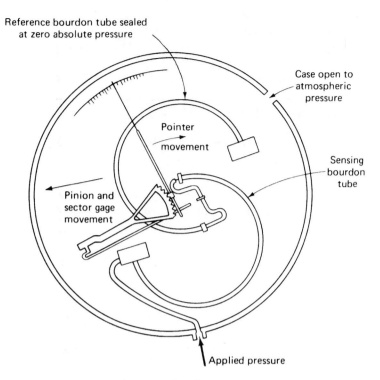

FIGURE 1.34 Bourdon gage for absolute pressure measurement.

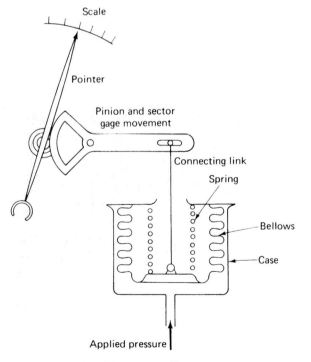

FIGURE 1.35 Bellows gage.   1.6 PRESSURE   **47**

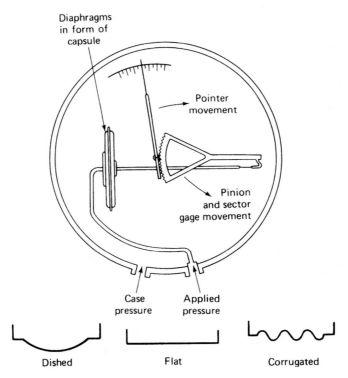

FIGURE 1.36 Diaphragm-based pressure transducer.

Thus far, we have discussed a few mechanical pressure transducers. In many applications, it is more convenient to use transducer elements that depend on the change in the electrical parameters of the element as a function of the applied pressure. The only active electrical pressure transducer in common use is the piezoelectric transducer. Sound-pressure instrumentation makes extensive use of piezoelectric pickups in such forms as hollow cylinders and disks. Piezoelectric pressure transducers are also used in measuring rapidly fluctuating or transient pressures. In a recently introduced technique called electrocalibration, the transducer is calibrated by electric field excitation rather than by physical pressure. The most common passive electrical pressure transducers are the variable-resistance types.

The strain gage is probably the most widely used pressure transducer element. Strain gages operate on the principle that the electrical resistance of a wire varies with its length under load. In unbounded strain gages, four wires run between electrically insulated pins located on a fixed frame and other pins located on a movable armature, as shown in Figure 1.37. The wires are installed under tension and form the legs of a bridge circuit. Under pressure, the elastic element (usually a diaphragm) displaces the armature, causing two of the wires to elongate while reducing the tension in the remaining two wires. The resistance change causes a bridge imbalance proportional to the applied pressure.

The bonded strain gage takes the form of a fine wire filament set in cloth, paper, or plastic and fastened by a suitable cement to a flexible plate, which takes the load of the elastic element. This is shown in Figure 1.38. Two similar strain gage elements are often connected in a bridge circuit to

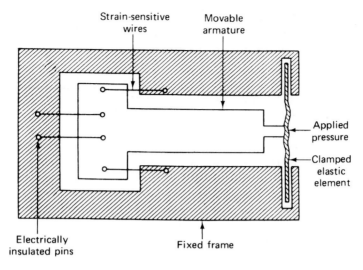

FIGURE 1.37 Typical unbonded strain gage.

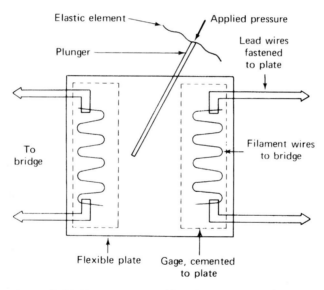

FIGURE 1.38 Typical bonded strain gage.

balance unavoidable temperature effects. The nominal bridge output impedance of most strain gage pressure transducers is 350 ohms (Ω), nominal excitation voltage is 10 V (ac or dc), and natural frequency can be as high as 50 Hz. Transducer resolution is infinite, and the usual calibration uncertainty of such gages is within 1% of full scale.

Many other forms of electrical pressure transducers are in use in industry, and a few of these will be discussed briefly. These elements fall under the categories of potentiometer, variable capacitance, linear variable differential transformer (LVDT), and variable-reluctance transducers. The potentiometer operates as a variable-resistance pressure transducer. In one arrangement, the elastic element is a helical Bourdon tube, while a precision, wire-wound potentiometer serves as the electric element. As pressure is applied to the open end of the Bourdon tube, it unwinds, causing the wiper (connected directly to the closed end of the tube) to move over the potentiometer.

In the variable-capacitance pressure transducer, the elastic element is usually a metal diaphragm that serves as one plate of a capacitor. Under an applied pressure, the diaphragm moves with respect to a fixed plate. By means of a suitable bridge circuit, the variation in capacitance can be measured and related to pressure by calibration.

The electric element in an LVDT is made up of three coils mounted in a common frame to form the device shown in Figure 1.39. A magnetic core centered in the coils is free to be displaced by a bellows, Bourdon, or diaphragm elastic element. The center coil, the primary winding of the transformer, has an ac excitation voltage impressed across it. The two outside coils form the secondaries of the transformer. When the core is centered, the induced voltages in the two outer coils are equal and 180° out of phase; this represents the zero-pressure position. However, when the core is displaced by the action of an applied pressure, the voltage induced in one secondary increases while that in the other decreases. The output voltage difference varies essentially linearly with pressure for the small core displacements allowed in these transducers. The voltage difference is measured and related to the applied pressure by calibration.

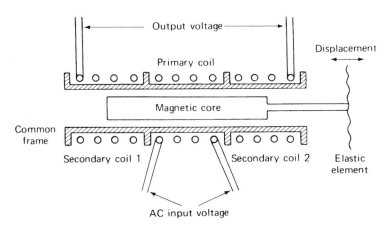

FIGURE 1.39   Linear variable differential transformer.

The last elastic element to be discussed is the variable-reluctance pressure transducer. The basic element of the variable-reluctance pressure transducer is a movable magnetic vane in a magnetic field. In one type, the elastic element is a flat magnetic diaphragm located between two magnetic output coils. Displacement of the diaphragm changes the inductance ratio between the output coils and results in an output voltage proportional to pressure.

1.7   REVIEW

After defining thermodynamics as the study of the conversion of one form of energy to another, we started our study by setting forth certain concepts concerning systems and properties. A *system* is any grouping of matter taken in a convenient or arbitrary manner, whereas a property is an observable characteristic that is determined by the state of a system and, in turn, aids in determining the state of a system. The condition of a system, described by its position, energy, and so on, is called the *state* of the system. We studied in detail the properties of temperature and pressure and the methods used to measure them. Due to the use of both the English and SI systems of units, we are required to have facility in each. This caused us to spend some time in establishing and defining the basic rules and styles of these systems. Using elementary kinetic theory, we were able to quantify the concepts of pressure, temperature, density, specific weight, specific volume, and Avogadro's number.

Some of the material covered in this chapter may have been covered in other courses and may appear to be elementary. It cannot be emphasized strongly enough that this material must be fully mastered before proceeding further. You are urged to be sure of all the units used in any equation and to be equally sure of the meaning of all terms. Problems at the end of the chapters are included to develop a better level of understanding through the application of the text material. Just reading and studying is not sufficient. The ability to apply the material studied indicates that you have really mastered it and are not just repeating material from memory.

KEY TERMS

Terms used for the first time in this chapter are:

Avogadro's law equal volumes of gases at the same temperature and pressure contain the same number of particles.

density mass per unit of volume.

micron (μ) 10^{-3} mm of mercury. A unit of pressure used in vacuum systems.

pressure normal force per unit of area.

property an observable characteristic of state that is determined by the state and, in turn, aids in determining the state of the system. It is not dependent on the path or the means of attaining the state.

specific gravity the ratio of the density of a substance to the density of water at 4°C. It can also be defined as the ratio of the specific weight of a substance to the specific weight of water at 4°C. In the older cgs system, the density of water at 4°C is 1 g/cm^3, making specific gravity numerically equal to density in this system.

specific volume the reciprocal of density, that is, the volume per unit mass.

specific weight weight per unit of volume.

state the condition of a system that fixes the position and the energy stored in the system and is identified by the properties of the system.

system a grouping of matter taken in any convenient or arbitrary manner.

temperature the temperature of a system is a measure of the random motion of the molecules of the system. The Celsius, Fahrenheit, Kelvin, and Rankine temperature scales represent the conventional and absolute temperature scales, respectively. Celsius and Kelvin are SI units, and Fahrenheit and Rankine are English units.

thermodynamics the study of energy, heat, work, the properties of the media employed, and the processes used; also the study of the conversion of one form of energy to another.

torr a unit of pressure equal to 1 mm of Hg.

weight the force of attraction between a body and the earth or another planet or moon.

zeroth law of thermodynamics when two bodies are in thermal equilibrium with a third body, they are in thermal equilibrium with one another.

EQUATIONS DEVELOPED IN THIS CHAPTER

temperature conversion $\qquad$ $°C = \frac{5}{9}(°F - 32)$ $\qquad$ **(1.1)**

temperature conversion $\qquad$ $°F = \frac{9}{5}(°C) + 32$ $\qquad$ **(1.2)**

temperature conversion	$°F = \frac{9}{5}(40 + °C) - 40$	**(1.3)**
temperature conversion	$°C = \frac{5}{9}(40 + °F) - 40$	**(1.4)**
absolute temperature	degrees Rankine $= °R = °F + 460$	**(1.5)**
absolute temperature	degrees Kelvin $= °K = °C + 273$	**(1.6)**
force–mass relation	$F = \dfrac{ma}{g_c}$	**(1.9)**
weight–mass relation	$w = \dfrac{mg}{g_c}$	**(1.10)**
weight–mass relation (SI)	$w = mg$	**(1.10a)**
ideal gas relation	$pV = \dfrac{1}{3}\left(\dfrac{mnV_x^2}{g_c}\right)$	**(1.14)**
ideal gas relation (SI)	$pV = \frac{1}{3}mnV_x^2$	**(1.14a)**
absolute pressure	absolute pressure = atmospheric pressure + gage pressure	**(1.16)**
absolute pressure	absolute pressure = atmospheric pressure − vacuum	**(1.17)**
pressure–height relation	$p = -\gamma h$	**(1.19a)**
pressure–height relation	$p = \dfrac{-\rho g}{g_c}h$	**(1.20)**
pressure–height relation	$p = -\rho g h$	**(1.20a)**

QUESTIONS

1.1 Indicate which of the following are or are not properties: (a) location; (b) velocity; (c) time; (d) weight; (e) size.

1.2 Would you expect that your weight would be (a) greater, (b) less, or (c) equal to your earth weight on the planet Jupiter? Why?

1.3 State on which factor or factors the pressure at the bottom of a column of liquid depends: (a) its height; (b) the specific weight of the fluid; (c) the local acceleration of gravity; (d) the cross-sectional area of the column.

1.4 What temperature does a thermometer measure?

1.5 What condition or conditions must exist if a thermometer is to measure a person's temperature accurately?

1.6 If a thermometer measures a temperature difference of 10°C, what is the temperature difference in °K?

1.7 If a thermocouple measures a temperature difference of 5°F, what is the temperature difference in °R?

1.8 Comment on the fact that commodities such as bread and meat are sold by the kilo (kilogram) in many countries of the world.

1.9 The unit of 100 is not a recommended unit in the SI system. Comment on whether it would or would not have been desirable to define the pascal as N/cm^2.

1.10 If an equation is not dimensionally consistent, is it necessarily incorrect? Why?

1.11 If the average velocity of the molecules in a box is doubled, would you expect that its pressure would (a) double, (b) stay the same, (c) be four times greater, or (d) be indeterminate?

1.12 What is temperature a measure of?

1.13 How are density and specific volume related?

1.14 A student states that the use of a digital readout will increase the accuracy of an instrument. Comment on this statement.

1.15 What must be done to establish the accuracy of any instrument?

PROBLEMS*

Unless indicated otherwise, use $g = 32.17$ ft/s^2 or 9.806 m/s^2, and use $p_{atm} = 14.696$ psia or 101.325 Pa. Also, use $g_c = 32.174$ lb$_m$·ft/lb$_f$·s^2.

Problems Involving Conversion of Units

1.1 Convert a length of 15 ft to meters.

1.2 Convert 3.85 inches to centimeters.

1.3 Convert 90.68 centimeters to inches.

1.4 Convert 5787 pounds to tons.

1.5 Convert 6.3 tons to pounds.

1.6 Convert 35.4 inches to feet.

1.7 Convert 3.87 feet to inches.

1.8 Convert 8.6 pounds to grams.

1.9 Convert 462 grams to pounds.

1.10 Convert 2 yards 2 feet 5 inches to centimeters.

1.11 Convert 550 mi/hr to m/s.

1.12 Convert a flow rate of 2000 ft^3/min to L/s.

1.13 Convert 200 gal/hr (gph) to m^3/s.

1.14 Convert 200 gph to L/s.

1.15 Convert 200 gph to L/min.

1.16 Convert 200 L/min to m^3/s.

1.17 Comvert 200 gpm to L/min.

1.18 Convert 50.7 cm^3/s to m^3/s.

1.19 Convert 861 gpm to L/min.

1.20 Convert 1.71 m^3/s to L/min.

1.21 Convert 100 m^3/s to gpm.

1.22 Convert 3.52 ft^3/s to gpm.

1.23 Convert 3.52 ft^3/s to m^3/s.

1.24 Convert 150 gpm to m^3/s.

1.25 Convert 6.52 m^3/s to gpm.

---

* More difficult problems are preceded by an asterisk in this chapter and in all following chapters.

1.26 Convert 3.65 m^3/s to L/s.

1.27 Convert 1.2 ft^3/s to gpm.

1.28 Convert 4.8 ft^3/s to L/s.

1.29 Convert 101 kPa (kN/m^2) to lb$_f$/in^2.

1.30 A rectangle has the dimensions of 12 in. × 10 in. Calculate its area in square meters.

1.31 A rectangle has the dimensions of 2 m × 3.5 m. Determine the area of the rectangle in square inches.

1.32 A box has the dimensions 12 in. × 12 in. × 6 in. How many cubic meters are there in the box?

1.33 A cylinder has a diameter of 10 cm and a height of 25 cm. Determine its volume in cubic meters.

1.34 A cylinder has a diameter of 5 in. and a height of 10 in. Determine its volume in cubic meters.

1.35 A cube is made 8 in. on a side. Calculate its volume in cubic meters.

1.36 An automobile engine has a displacement of 240 in^3. Determine its displacement in liters.

1.37 A car is claimed to obtain a fuel usage in Europe of 12 km/L. Determine its equivalent fuel usage in mi/gal.

1.38 A driveway requires 38 yd^3 of concrete. How many m^3 of concrete does this correspond to?

1.39 A man weighs 175 lb and stands 6 ft 0 in. tall. What is his weight in newtons and his height in meters?

1.40 A barrel of oil contains 55 gallons. If a gallon is a volume of 231 in^3, determine the volume of 15 barrels in cubic meters.

*1.41 A gallon is a measure of volume that equals 231 in^3, so determine the number of liters in a gallon. Note that the liter is sometimes called the metric quart. Does your answer support this name?

1.42 A 1500-m race is often called the "metric mile." What part of a mile (5280 ft) is the metric mile?

1.43 A car is operated at 50 mi/hr. How many feet per second does this correspond to?

1.44 A car travels at 64 km/hr. What is the speed of the car in mi/hr?

1.45 A car weighs 2650 lb and is 18 ft long. What is its weight in kilonewtons and its length in meters?

1.46 A block of metal weighing 1200 lb is hung from the end of a vertical rod, and it causes the rod to lengthen 0.001 in. Express these data in terms of newtons and meters.

1.47 One gigapascal (GPa) is: (a) 10^6 Pa; (b) 10^8 Pa; (c) 10^3 Pa; (d) 10^9 Pa; (e) none of these.

Problems Involving Temperature

1.48 Convert 20, 40, and 60°C to equivalent degrees Fahrenheit.

1.49 Change 0, 10, and 50°F to equivalent degrees Celsius.

1.50 Convert 500°R, 500°K, and 650°R to degrees Celsius.

1.51 Derive a relation between degrees Rankine and degrees Kelvin, and based on the results, show that (°C + 273)1.8 = °F + 460.

*1.52 If a Fahrenheit temperature is twice that of a Celsius temperature, determine both temperatures.

*1.53 If a Celsius temperature is $\frac{2}{3}$ a Fahrenheit temperature, determine both temperatures.

*1.54 A Fahrenheit and a Celsius thermometer are used to measure the temperature of a fluid. If the Fahrenheit reading is 1.5 times that of the Celsius reading, what are both readings?

*1.55 An arbitrary temperature scale is proposed in which 20° is assigned to the ice point and 75° is assigned to the boiling point. Derive an equation relating this scale to the Celsius scale.

*1.56 For the temperature scale proposed in Problem 1.55, what temperature corresponds to absolute zero?

*1.57 A new thermometer scale on which the freezing point of water at atmospheric pressure would correspond to a marking of 200 and the boiling point of water at atmospheric pressure would correspond to a marking of minus 400 is proposed. What would the reading of this new thermometer be if a Fahrenheit thermometer placed in the same environment read 80°?

Problems Involving Mass and Weight

1.58 A mass of 5 kg is placed on a planet whose gravitational force is 10 times that of earth. What does it weigh on this planet?

1.59 A mass of 10 kg weighs 90 N. The acceleration of gravity at this location is: (a) $\frac{1}{9}$ m/s^2, (b) 9.0 m/s^2; (c) 90 m/s^2; (d) 10 m/s^2; (e) none of these.

1.60 At a location on earth where $g = 32.2$ ft/s^2, a body weighs 161 lb. At another location where $g = 32.0$, the same body will weight: (a) 161 lb; (b) 162 lb; (c) 160 lb; (d) 163 lb; (e) none of these.

1.61 A body at mean sea level weighs 100 lb ($g = g_c$). Estimate the weight of this body at an elevation of 7500 ft. Assume that the mean diameter of the earth is 12,742 km.

*1.62 A mass of 100 kg weighs 980.6 N at sea level. Estimate the weight of this body at the top of a mountain 5 km high. Assume that the mean diameter of the earth is 12,742 km.

*1.63 A balance-type scale is used to weigh a sample on the moon. If the value of g is $\frac{1}{6}$ of earth's gravity and the "standard" weights (earth weights) add up to 20 lb, what is the mass of the body?

*1.64 A balance-type of scale is used to weigh a sample on the moon. If the value of g is $\frac{1}{6}$ of earth's gravity and the "standard" weights (earth weights) add up to 100 N, what is the mass of the body?

*1.65 Solve Problem 1.64 if a spring balance is used that was calibrated on earth and reads 20 lb.

*1.66 Solve Problem 1.64 if a spring balance is used that was calibrated on earth and reads 100 N.

*1.67 A mass of 100 kg is hung from a spring in a local gravitational field where $g = 9.806$ m/s^2, and the spring is found to deflect 25 mm. If the same mass is taken to a planet where $g = 5.412$ m/s^2, how much will the spring deflect if its deflection is directly proportional to the applied force?

*1.68 The mass of the planet Mars is 0.1069 relative to earth, and its diameter is 0.523 relative to earth. What is the weight of a pound of mass on Mars?

*1.69 A body having an unknown mass "weighs" 1 lb on the planet Jupiter. Jupiter has a mass 318.35 times that of earth, and its diameter is 10.97 times larger: (a) What will the body weigh on earth? (b) What is the mass of the body?

*1.70 Solve Problem 1.69 for a mass of 10 kg.

*1.71 Solve Problem 1.69 for a weight of 10 N on Jupiter.

Problems Involving Newton's Second Law, $F = ma$

1.72 A force of 100 N acts horizontally on a 10-kg body. What is its horizontal acceleration?

1.73 An unbalanced force in pounds equal numerically to the weight of a body in pounds causes the body to accelerate. What is its acceleration?

1.74 If a body is accelerating at the rate of 5 m/s^2 when acted on by a net force of 10 N, determine its mass.

1.75 A body has a mass of 321.7 lb and is accelerated at the rate of 5 ft/s^2. Determine the magnitude of the unbalanced force acting on it.

1.76 A mass of 2 kg has an acceleration of 1.2 m/s^2. Calculate the magnitude of the unbalanced force acting on it.

1.77 What is the acceleration of a body if a net force of 25 lb acts on it if its mass is 160.8 lb?

Problems Involving Specific Weight, Specific Volume, and Specific Gravity

1.78 Determine the density and specific volume of the contents of a 10-ft^3 tank if the contents weigh 250 lb.

1.79 A tank contains 500 kg of a fluid. If the volume of the tank is 0.5 m^3, what is the density of the fluid, and what is the specific volume?

1.80 Calculate the specific weight of a fluid if a gallon of it weighs 20 lb. Also, calculate its density.

1.81 The specific gravity of a fluid is 1.2. Calculate its density and its specific weight in SI units.

1.82 A fluid has a density of 0.90 g/cm^3. What is its specific weight in both the English and SI systems?

1.83 It is proposed by gasoline dealers to sell gasoline by the liter. If gasoline has a specific gravity of 0.85, what is the weight of 60 L of gasoline?

1.84 A household oil tank can hold 275 gal of oil. If oil has a specific weight of 8800 N/m^3, how many pounds of oil will there be in a full tank?

1.85 A car has a fuel tank that holds 14 gal of gasoline. If the gasoline has a specific gravity of 0.70, calculate the weight of gasoline in units of pounds.

1.86 An oil has a specific gravity of 0.8. Determine its specific weight, density, and specific volume in SI units.

1.87 An oil has a specific weight of 0.025 lb$_f$/in^3. Calculate its specific gravity.

1.88 A can weighs 12 N when empty, 212 N when filled with water at 4°C, and 264 N when filled with an oil. Calculate the specific gravity of the oil.

1.89 The density of a liquid is 48.6 lb$_m$/ft^3. Calculate the specific weight and the specific gravity of the liquid.

1.90 A shipping company requires that packages have a base perimeter plus the height not to exceed 108 in. By calculation, a student determines that the optimum package size should be an 18-in. square base and a 36-in. height. If the company further specifies that the weight must not exceed 70 lb$_f$, determine the maximum specific weight of the contents of the package.

1.91 A body has a specific gravity of 1.48 and a volume of 7.24 × 10^{-4} m^3. What is the weight of the body?

1.92 Oil has a specific gravity of 0.83. What is the weight of a liter of this oil in SI units?

1.93 What is the weight in pounds of a gallon of oil that has a specific gravity of 0.86?

1.94 A cylinder 6 in. in diameter and 10 in. in height contains oil that has a density of 850 kg/m^3. Determine the weight of the oil in English units.

1.95 A liquid has a density of 1100 kg/m^3. Determine its specific gravity and its specific weight.

1.96 The fuel tank of a car holds 60 L of gasoline. Assuming that the gasoline has a specific gravity of 0.74, determine the weight of the gasoline in the tank.

1.97 If the density of a liquid is 780 kg/m^3, calculate its specific gravity and its specific weight.

1.98 A liquid has a specific weight of 200 lb$_f$/ft^3. Calculate the volume needed to have a weight of 390 lb.

*1.99 In the older cgs system, the density in g/cm^3 is numerically equal to the specific gravity of the fluid. Show that this is indeed true.

Problems Involving Pressure

1.100 A skin diver descends to a depth of 60 ft in fresh water. What is the pressure on the diver's body? The specific weight of fresh water can be taken as 62.4 lb$_f$/ft^3.

1.101 A skin diver descends to a depth of 25 m in a salt lake where the density is 1026 kg/m^3. What is the pressure on the diver's body at this depth?

1.102 If a Bourdon gage reads 25 psi when the atmosphere is 14.7 psia, what is the absolute pressure in pascals?

1.103 A column of fluid is 1 m high. The fluid has a density of 2500 kg/m^3. What is the pressure at the base of the column?

1.104 A column of fluid is 25 in. high. If the specific weight of the fluid is 60.0 lb$_f$/ft^3, what is the pressure in psi at the base of the column?

1.105 Determine the density and specific volume of the contents of a 20-ft^3 tank if the contents weigh 250 lb.

1.106 A tank contains 500 kg of a fluid. If the volume of the tank is 2.0 m^3, what is the density of the fluid, and what is the specific gravity?

*1.107 Convert 14.696 psia to 101.325 kPa.

1.108 A pressure gage indicates 25 psi when the barometer is at a pressure equivalent to 14.5 psia. Compute the absolute pressure in psi and feet of mercury when the specific gravity of mercury is 13.0.

1.109 Same as Problem 1.108, but the barometer stands at 750 mm Hg and its specific gravity is 13.6.

1.110 A vacuum gage reads 8 in. Hg when the atmospheric pressure is 29.0 in. Hg. If the specific gravity of mercury is 13.6, compute the absolute pressure in psi.

1.111 A vacuum gage reads 10 in. Hg when the atmospheric pressure is 30 in. Hg. Assuming the density of mercury to be 13.595 kg/m^3, determine the pressure in pascals.

1.112 A U-tube mercury manometer, open on one end, is connected to a pressure source. If the difference in the mercury levels in the tube is 6.5 in., determine the unknown pressure in psfa.

*1.113 Two sources of pressure M and N are connected by a water–mercury differential gage as shown in Figure P1.113. What is the difference in pressure between M and N in psi?

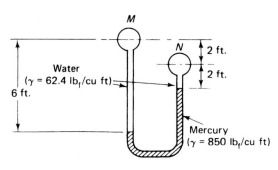

FIGURE P1.113

*1.114 Determine the difference in pressure between A and B in Figure P1.114 if the specific weight of water is 62.4 lb$_f$/ft^3.

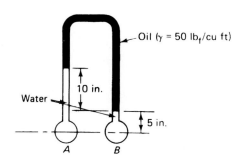

FIGURE P1.114

*1.115 If the liquid in pipe B in Problem 1.114 is carbon tetrachloride, whose specific weight is 99 lb$_f$/ft^3, determine the pressure difference between A and B.

*1.116 In a U-tube manometer, one end is closed, trapping atmospheric air in the column. The other end is connected to a pressure supply of 5 psig. If the level of mercury in the closed end is 2 in. higher than that in the open end, what is the pressure of the trapped air?

*1.117 For the arrangement shown in Figure P1.117, determine $p_A - p_B$.

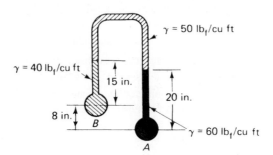

FIGURE P1.117

2

Work, Energy, and Heat

Learning Goals

After reading and studying the material in this chapter, you should be able to:

1. State the definitions of work, energy, and heat.

2. Use the fact that both work and heat are forms of energy in transition.

3. Apply the convention that heat into a system is to be taken as a positive quantity and that work out of a system is also to be taken as a positive quantity. This convention is taken from the customary power-producing cycle in which heat into a system is used to generate useful work.

4. Show that internal energy is a form of energy in which a body is said to possess internal energy by virtue of the motion of the molecules of the body.

5. Differentiate between a nonflow or closed system and a flow or open system.

6. Understand the origin of the term *flow work,* and apply it to a flow system.

7. Use the fact that the work of a quasi-static, nonflow system is the area under the pressure–volume curve.

2.1   INTRODUCTION

In Chapter 1, certain concepts were arrived at by considering the motion of gas particles in an enclosure. Briefly, pressure was found to involve the principle of momentum interchange with the container walls, temperature was associated with the motion of the particles, and density was taken to be a measure of the number of particles per unit volume. This simple analysis followed the history of a single particle, and it was subsequently generalized to all the particles in the enclosure. This type of analysis is representative of a microscopic description of the processes occurring within the boundaries of the defined system, because the history of a single particle was followed in detail. Rather than pursue further the microscopic concept of matter, we shall be concerned with the macroscopic, or average, behavior of the particles composing a system. The macroscopic viewpoint essentially assumes that it is possible to describe the average behavior of these particles at a given time and at some subsequent time after changes have occurred to the system. The system changes of concern to us in this study are temperature, pressure, density, work, energy, velocity, and position. The power of the macroscopic approach lies in its ability to describe the changes that have occurred to the system without having to detail all the events of the processes involved.

2.2 WORK

The *work* done by a force is the product of the displacement of the body multiplied by the component of the force in the direction of the displacement. Thus, in Figure 2.1, the displacement of the body on the horizontal plane is x, and the component of the force in the direction of the displacement is $(F \cos \theta)$. The work done is $(F \cos \theta)x$. The constant force $(F \cos \theta)$ is plotted as a function of x in Figure 2.1b, and it should be noted that the resulting figure is a rectangle. The area of this rectangle (shaded) is equal to the work done, because it is $(F \cos \theta)(x)$. If the force varies so that it is a function of the displacement, it is necessary to consider the variation of force with displacement in order to find the work done. Figure 2.1c shows a general plot of force as a function of displacement. If the displacement is subdivided into many small parts, Δx, and for each of these small parts F is assumed to be very nearly constant, it is apparent that the sum of the small areas $(F)(\Delta x)$ will represent the total work done when the body is displaced from x_1 to x_2. Thus, the area under the curve of F as a function of x represents the total work done if F is the force component in the direction of the displacement x.

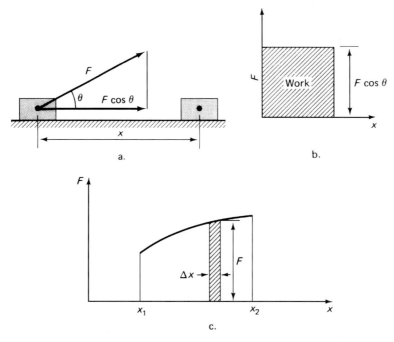

FIGURE 2.1   Work.

ILLUSTRATIVE PROBLEM 2.1

A spring is slowly compressed by a varying force F until it reaches an equilibrium position. Assuming that the force on the spring is proportional to the spring displacement, determine the work done on the spring. Assume that the constant of proportionality k is constant and expressed in pounds force per foot of spring deflection or newtons per meter of spring deflection.

SOLUTION

As shown in Figure 2.2a, the system consists of a spring and a force F directed along the axis of the spring. A plot of F as a function of x is shown in Figure 2.2b. It will be noted that the force–displacement relation is a linear one, in which the force equals kx at all times. Thus, the work done is represented by the shaded triangular area of Figure 2.2b. Because the base of the triangle is l and the height is kl, the work done is $\frac{1}{2}(l)(kl)$ or $\frac{1}{2}(k)(l^2)$.

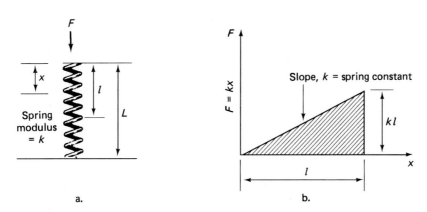

FIGURE 2.2   Illustrative Problems 2.1 and 2.2.

CALCULUS ENRICHMENT

The work done on a spring by a force can be found by integration of the expression for work, namely,

$$W = \int F \cdot dx \qquad \textbf{(a)}$$

where the force and displacement are related by the spring constant k as,

$$F = kx \qquad \textbf{(b)}$$

Using the spring equation in conjunction with Equation (a) between the limits l_1 and l_2,

$$W = \int_{l_1}^{l_2} kx \, dx \qquad \textbf{(c)}$$

Integrating, $\qquad\qquad W = k(l_2^2 - l_1^2)/2 \qquad \textbf{(d)}$

If the initial length is zero, we obtain the same result as before, namely, $(1/2)kl^2$.

ILLUSTRATIVE PROBLEM 2.2

A spring has a spring constant k of 100 lb_f/in. deflection. How much work is done when the spring is compressed 2 in.?

SOLUTION

Referring to Figure 2.2 and the results of Illustrative Problem 2.1, we have

$$\text{work} = \frac{1}{2} kl^2$$

For this problem,

$$\text{work} = \frac{1}{2} ((100) \, lb_f/in.) \, (2)^2 \, in.^2 = 200 \, in.\cdot lb_f$$

ILLUSTRATIVE PROBLEM 2.3

If the spring constant in Illustrative Problem 2.2 is 20 kN/m, how much work is done when the spring is compressed 75 mm?

SOLUTION

As before,

$$\text{work} = \frac{1}{2} kl^2 = \frac{1}{2} (20 \times 10^3) \, N/m \times (0.075)^2 (m)^2$$

$$= 56.25 \, N\cdot m = 56.25 \, J$$

2.3 ENERGY

At this point, let us define *energy* in terms of work. Energy can be described as the capacity to do work. At first, it may appear that this definition is too restrictive when applied to electrical and magnetic systems. Yet in all instances, the observed effects on a system can (in principle and ideally) be converted to mechanical work. Because work has been defined as the product of a force multiplied by a displacement, it is not stored in a system. It represents a form of energy that must be crossing the boundaries (real or imaginary) of the system and can properly be placed in the category of energy in transition.

To distinguish between the transfer of energy as work to or from a system, we shall adopt the convention that *the work done by a system on its surroundings is positive, and the work done by the surroundings on the system is negative.* For the student, it is best to think of this convention regarding the *useful work out of a system as a conventional, desirable quantity and, therefore, positive.* Thus, in Illustrative Problem 2.1, the spring has work done on it by a force. If we consider the spring as our system, the work is negative; if we consider the system as the external variable force that is compressing the spring, the work is positive. The student will note that the decision whether the work term is positive or negative requires that the system be carefully defined.

The work that a system can perform on its surroundings is not an intrinsic property of the system. The manner in which the process is carried out will determine the effect on the surroundings. As stated earlier, work is a transitory effect and is neither a property of a system nor stored in a system. There is one process, however, that does permit the evaluation of the work done, because the path is uniquely defined. This process is frictionless and quasi-static, and we shall find it useful in subsequent discussions. To describe this process, we first define *equilibrium state* in the manner given by Hatsopoulos and Keenan: *A state is an equilibrium state if no finite rate of change can occur without a finite change, temporary or permanent, in the state of the environment.** The term *permanent change of state* refers to one that is not canceled out before completion of the process. Therefore, the frictionless *quasi-static process* can be identified as a succession of equilibrium states. Involved in this definition is the concept of a process carried out infinitely slowly, so that it is in equilibrium at all times. The utility of the frictionless, quasi-static process lies in our ability to evaluate the work terms involved in it, because its path is uniquely defined.

Energy or work done per unit time is called power or the rate of energy change. Energy, work, and power units as well as their conversions are detailed in Table 1.8.

2.4   INTERNAL ENERGY

To this point, we have considered the energy in a system that arises from the work done on the system. However, it was noted in Chapter 1 and earlier in this chapter that a body possesses energy by virtue of the motion of the molecules of the body. In addition, it possesses energy due to the internal attractive and repulsive forces between particles. These forces become the mechanism for energy storage whenever particles become separated, such as when a liquid evaporates or the body is subjected to a deformation by an external energy source. Also, energy may be stored in the rotation and vibration of the molecules. Additional amounts of energy are involved with the electron configuration within the atoms and with the nuclear particles. The energy from all such sources is called the *internal energy* of the body and is designated by the symbol U. Per unit mass (m), the specific internal energy is denoted by the symbol u, where $mu = U$. Thus,

$$mu = U$$

(2.1)

or

$$u = \frac{U}{m}$$

(2.2)

From a practical standpoint, the measurement of the absolute internal energy of a system in a given state presents an insurmountable problem and is not essential to our study of thermodynamics. We are concerned with changes in internal energy, and the arbitrary datum for the zero of internal energy will not enter into these problems.

---

* G. N. Hatsopoulos and J. H. Keenan, *A Single Axiom for Classical Thermodynamics,* ASME paper 61-WA-100, 1961.

Just as it is possible to distinguish the various forms of energy, such as work and heat, in a mechanical system, it is equally possible to distinguish the various forms of energy associated with electrical, chemical, and other systems. For the purpose of this book, these forms of energy, work, and heat are not considered. The student is cautioned that if a system includes any forms of energy other than mechanical, these items must be included. For example, the energy that is dissipated in a resistor as heat when a current flows through it must be taken into account when all the energies of an electrical system are being considered.

2.5 POTENTIAL ENERGY

Let us consider the following problem, illustrated in Figure 2.3, where a body of mass m is in a locality in which the local gravitational field is constant and equal to g. A force is applied to the body, and the body is raised a distance Z from its initial position. The force is assumed to be only infinitesimally greater than the mass, so the process is carried out on a frictionless, quasi-static basis. In the absence of electrical, magnetic, and other extraneous effects, determine the work done on the body. The solution to this problem is obtained by noting that the equilibrium of the body requires that a force be applied to it equal to its weight. The weight of the body is given from Chapter 1 as mg/g_c in English units. In moving through a distance Z, the work done by this force will then equal

$$\text{work} = \frac{mg}{g_c} Z \tag{2.3}$$

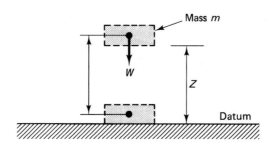

FIGURE 2.3 Potential energy.

The work done on the body can be returned to the external environment by simply reversing this frictionless, quasi-static process, a feature that is discussed in detail in Chapter 3. Returning to Equation (2.3), we conclude that this system has had work done on it equal to $(mg/g_c)(Z)$ and that, in turn, the system has stored in it an amount of energy in excess of the amount it had in its initial position. The energy added to the system in this case is called *potential energy*. Thus,

$$\text{potential energy (P.E.)} = \frac{mg}{g_c} Z \qquad \text{ft·lb}_f \tag{2.4}$$

In terms of SI units, Equations (2.3) and (2.4) become

$$\text{work} = mgZ \qquad \textbf{(2.5)}$$

and

$$\boxed{\text{potential energy (P.E.)} = mgZ \quad \text{joules}} \qquad \textbf{(2.6)}$$

A feature of importance of potential energy is that a system can be said to possess it only with respect to an arbitrary initial or datum plane.

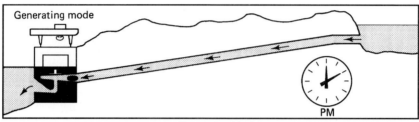

a. Generating Mode

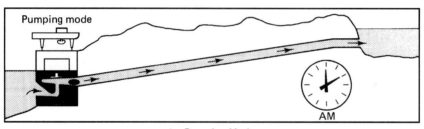

b. Pumping Mode

FIGURE 2.4 Pumped storage concept.

An interesting application of the concept of potential energy storage is the pumped storage hydroelectric power plant.* The principle of operation can be illustrated using Figure 2.4. In this system, reversible turbine–generator units are used as follows: At those times when excess generating capacity is available from other generating stations, the water can be pumped from the lower to the upper reservoir. When additional generating capacity is required, the water is allowed to run downhill, passing through the generating station where it operates the reversible turbine–generator as a turbine to rotate the generator and produce electricity. Thus, the pumped storage station helps to smooth out fluctuations in the load demand, resulting in steady, more efficient operation of other stations. In this application, the potential energy of the water stored in the upper reservoir provides the energy to the system as needed. Dams also use the potential energy of water to generate electrical power.

---

* See, for example. J. Tillinghast, "Power Systems: The Place of Hydro and Pumped Storage," *Mechanical Engineering,* July 1969, pp. 24–28.

ILLUSTRATIVE PROBLEM 2.4

A pumped storage plant uses water pumped to an elevation of 600 ft above the turbo-generators. How much work is done lifting the water to this elevation? Assume that local gravity is g, which is numerically equal to g_c.

SOLUTION

Because local gravity is g, which is equal to g_c, for each pound mass of water,

$$\text{P.E.} = \frac{mg}{g_c}(Z) = 1\ \text{lb}_m \times \frac{32.174\ \text{ft/s}^2}{32.174\ \text{lb}_m \cdot \text{ft}/\text{lb}_f \cdot \text{s}^2} \times 600\ \text{ft} = 600\ \text{ft} \cdot \text{lb}_f$$

Note that the use of the term g/g_c is equivalent to a conversion factor lb_f/lb_m.

ILLUSTRATIVE PROBLEM 2.5

A pump delivers water from a well that is 50 m deep. Determine the change in potential energy per kg of water. Use $g = 9.81\ \text{m/s}^2$.

SOLUTION

The change in potential energy is mgZ, where Z is 50 m, using the bottom of the well as the datum. Therefore,

$$\text{P.E.} = mgZ$$

$$= 1\ \text{kg} \times 9.81\ \frac{\text{m}}{\text{s}^2} \times 50\ \text{m}$$

$$= 490.5\ \frac{\text{kg} \cdot \text{m}^2}{\text{s}^2}$$

$$= 490.5\ \text{N} \cdot \text{m} \qquad \text{because N} = \frac{\text{kg} \cdot \text{m}}{\text{s}^2}$$

$$= 490.5\ \text{J}$$

ILLUSTRATIVE PROBLEM 2.6

If the pumped-storage plant of Illustrative Problem 2.4 has a flow of 10,000 gal/min, determine the power generated. Assume that the local gravity is $g = g_c$ and that the density of water is $62.4\ \text{lb}_m/\text{ft}^3$.

SOLUTION

In Illustrative Problem 2.4, we determined the work done to be 600 ft·lb$_f$/lb$_m$. The power generated is the energy stored per pound mass (also equal to the potential energy) multiplied by the pounds mass per minute that flows.

One gallon is a volumetric measure equal to 231 in^3. Thus,

$$10,000 \text{ gal/min} \times \frac{231 \text{ in}^3}{1728 \text{ in}^3/\text{ft}^3} = 1337 \text{ ft}^3/\text{min}$$

(because the flow is given as 10,000 gal/min).

The mass flow is $\dot{m} = \rho A V = 62.4 \text{ lb}_m/\text{ft}^3 \times 1337 \text{ ft}^3/\text{min} = 83{,}429 \text{ lb}_m/\text{min}$. The power generated $= 83{,}429 \text{ lb}_m/\text{min} \times 600 \text{ ft·lb}_f/\text{lb}_m = 50{,}057{,}000 \text{ ft·lb}_f/\text{min}$. Finally, $50{,}057{,}000/33{,}000 = 1517$ hp, since 1 hp $= 33{,}000$ ft·lb$_f$/min.

ILLUSTRATIVE PROBLEM 2.7

Determine the horsepower required by the pump in Illustrative Problem 2.5 if the water has a density of 1000 kg/m^3 and the pump delivers 1000 kg/min from the well.

SOLUTION

From Illustrative Problem 2.5, we found the potential energy to be 490.5 N·m above the bottom of the well. Therefore,

$$\text{power} = (\text{P.E.})\dot{m}$$

$$= 490.5 \frac{\text{N·m}}{\text{kg}} \times 1000 \text{ kg/min} \times \frac{1}{60 \text{ s/min}}$$

$$= 8175 \frac{\text{N·m}}{\text{s}} = 8175 \frac{\text{J}}{\text{s}} = 8175 \text{ W}$$

Because 1 hp = 746 W,

$$\frac{8175}{745} \frac{\text{W}}{\text{W/hp}} = 10.96 \text{ hp}$$

2.6 KINETIC ENERGY

Let us consider another situation in which a body of mass m is at rest on a frictionless plane (Figure 2.5). If the force F is applied to the mass, it will be accelerated in the direction of the force. After moving through a distance x, the velocity of the body will have increased from 0 to V_2. The only effect of the work done on the body will be to increase its velocity.

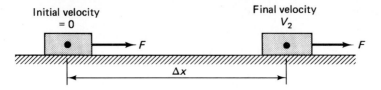

FIGURE 2.5 Kinetic energy.

Because velocity is the time rate of change of displacement,

$$V = \frac{x}{\Delta t} \qquad (2.7)$$

Also, acceleration is the time rate of change of velocity,

$$a = \frac{\Delta V}{\Delta t} \qquad (2.8)$$

Combining Equation (2.7) with Equation (2.8) by eliminating the Δt terms yields

$$V \, \Delta V = ax \qquad (2.9)$$

If we now multiply both sides of Equation (2.9) by the mass, m, we obtain

$$mV \, \Delta V = max \qquad (2.10)$$

The left side of Equation (2.10) can be evaluated by referring to Figure 2.6, where V is plotted against V. It will be noted that $V \, \Delta V$ is the area of the shaded rectangle. Summing these small rectangles up to V_2 represents the area of the triangle whose base is V_2 and whose altitude is V_2. This area is $\frac{1}{2}V_2^2$ making the left side of Equation (2.10), $mV_2^2/2$.

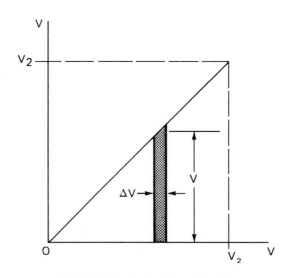

FIGURE 2.6 Evaluation of $V \, \Delta V$.

The right side of Equation (2.10) is interpreted by noting that $ma = F$ (in SI units). Therefore, max is the work done by the force F to move the body a distance x. Thus, we can write

$$\frac{mV_2^2}{2} = Fx \tag{2.11}$$

The term $mV^2/2$ is called the *kinetic energy* of the body. At the beginning of a process, it is $mV_1^2/2$, and at the end, it is $mV_2^2/2$. Equation (2.11) therefore permits us to evaluate the energy possessed by a body of mass m having a velocity V relative to a stationary reference. It is usual to use the earth as the reference.

In English units, Equation (2.11) becomes

$$\text{kinetic energy (K.E.)} = \frac{mV_2^2}{2g_c} \text{ or, in general, } \frac{mV^2}{2g_c} \tag{2.12}$$

In terms of SI units,

$$\text{kinetic energy (K.E.)} = \frac{mV^2}{2} \tag{2.12a}$$

ILLUSTRATIVE PROBLEM 2.8

A mass of 10 lb is slowed from a velocity of 88 ft/s to 10 ft/s. What is the change in the kinetic energy of the system if the body is considered to be the system?

SOLUTION

The kinetic energy of the body before it is slowed down is

$$\text{K.E.} = \frac{mV^2}{2g_c} = \frac{10 \text{ lb}_m \times \left(88\dfrac{\text{ft}}{\text{s}}\right)^2}{2 \times 32.174 \left(\dfrac{\text{ft} \cdot \text{lb}_m}{\text{s}^2 \cdot \text{lb}_f}\right)} = 1203.5 \text{ ft} \cdot \text{lb}_f$$

After slowing down,

$$\text{K.E.} = \frac{mV^2}{2g_c} = \frac{10 \text{ lb}_m \times \left(10\dfrac{\text{ft}}{\text{s}}\right)^2}{2 \times 32.174 \left(\dfrac{\text{ft} \cdot \text{lb}_m}{\text{s}^2 \cdot \text{lb}_f}\right)} = 15.5 \text{ ft} \cdot \text{lb}_f$$

The change in kinetic energy $\Delta(\text{K.E.})$ therefore is

$$\Delta \text{K.E.} = 1203.5 - 15.5 = 1188 \text{ ft} \cdot \text{lb}_f$$

ILLUSTRATIVE PROBLEM 2.9

A car having a mass of 1500 kg is slowed from 50 km/h to 30 km/h (Figure 2.7). What is the change in its kinetic energy if $g = 9.81$ m/s^2?

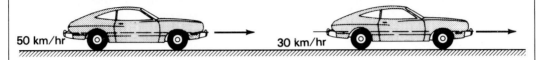

FIGURE 2.7 Illustrative Problem 2.9.

SOLUTION

The car's initial kinetic energy is $mV^2/2$. Therefore,

$$\text{K.E.} = \frac{1500 \text{ kg}}{2} \times \left(\frac{50 \times 1000 \text{ m/h}}{3600 \text{ s/h}} \right)^2$$

$$= 144\ 676 \text{ J} = 144.68 \text{ kJ}$$

After slowing down, we have

$$\text{K.E.} = \frac{mV^2}{2} = \frac{1500 \text{ kg}}{2} \times \left(\frac{30 \times 1000 \text{ m/h}}{3600 \text{ s/h}} \right)^2$$

$$= 52\ 083 \text{ J} = 52.08 \text{ kJ}$$

The change in kinetic energy therefore is

$$\Delta \text{K.E.} = 144.68 - 52.08 = 92.6 \text{ kJ}$$

Note that g did not enter the problem in this system of units.

ILLUSTRATIVE PROBLEM 2.10

A body has a mass of 10 kg and falls freely from rest (Figure 2.8). After falling 10 m, what will its kinetic energy be? Also, what will its velocity be just before impact? Neglect air friction.

SOLUTION

Because there are no losses in the system, we conclude that the initial potential energy plus the initial kinetic energy must equal the sum of the final potential energy plus the final kinetic energy. Therefore,

$$\text{P.E.}_1 + \text{K.E.}_1 = \text{P.E.}_2 + \text{K.E.}_2 \qquad \textbf{(a)}$$

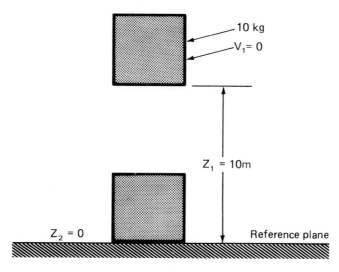

FIGURE 2.8 Illustrative Problem 2.10.

By selecting our reference plane as shown in Figure 2.8, K.E.$_1$ = 0 and P.E.$_2$ = 0. Equation (a) reduces to

$$\text{P.E.}_1 = \text{K.E.}_2 \qquad \qquad \textbf{(b)}$$

Substituting the data of the problem into Equation (b) gives

$$10 \text{ kg} \times 10 \text{ m} \times 9.81 \text{ m/s}^2 = \frac{10 \text{ kg} \times V^2 (\text{m/s})^2}{2}$$

Solving yields

$$V^2 = 2 \times 9.81 \times 10$$

and

$$V = 14.0 \text{ m/s}$$

Note that this result is independent of the mass of the body. The kinetic energy is readily found from Equation (b) to be equal to 10 kg × 10 m × 981 m/s² = 981 N·m.

2.7 HEAT

When a heat interaction occurs in a system, two distinct events are observed. The first is an interchange of energy, and the second is that this interchange would not have taken place if there were no temperature difference between the system and its surroundings. Therefore, we may define *heat* as the energy in transition across the boundaries of a system due to a temperature difference between the system and its surroundings. In this definition of heat, the transfer of mass across the boundaries of the system is excluded. It should be noted that this indicates a similarity between heat and work. Both are energies in transition, and neither is a property of the system in question. Just as in work, heat can transfer quasi-statically to or from a system. The difference in temperature between the system and its surroundings for quasi-static heat transfer can be only an infinitesimal amount at any time. Once again, it is necessary to adopt a convention for the energy interchanged by a system with its surroundings. *We shall use the convention that heat to a system from its surroundings is positive and that heat out of a system is negative.* To learn these conventions, it is convenient to consider the typical situation in which heat is transferred to a system to obtain useful work from the system. This sets the convention that heat into a system is positive and work out of the system is also positive. Positive in this sense means either desirable or conventional from the viewpoint of conventional power cycles. For refrigeration cycles, the opposite of this convention will be more useful.

It is important to recognize the difference between heat and temperature. Temperature is a measure of the energy contained in the molecules of a system due to their motion. When the temperature of a system is greater than that of its surroundings, some of that molecular energy is transferred to the surroundings in what we call heat. Thus, temperature is a property of a system in a given state, whereas heat is associated with a change in the state of a system.

Because work and heat are both forms of energy in transition, it follows that the units of work should be capable of being expressed as heat units, and vice versa. In the English system of units the conversion factor between work and heat, sometimes called *mechanical equivalent of heat,* is 778.169 ft·lb$_f$/Btu and is conventionally given the symbol J. We shall use this symbol to designate 778 ft·lb$_f$/Btu, because this is sufficiently accurate for engineering applications

FIGURE 2.9 Modern coal-burning central station: Milliken Station.
(Courtesy of New York State Electric and Gas Corp.)

of thermodynamics. In the SI system, this conversion factor is not necessary, because the joule (N·m) is the basic energy unit.

Figure 2.9 shows a large, modern central station that burns coal to generate electricity. This station consists of two 150-MW units, having a heat rate of 9451 Btu per kilowatt-hour while operating at 1800 psi. This plant is one of the most efficient fossil-fired generating units in the country.

2.8 FLOW WORK

At this time, let us look at two systems, namely, the *nonflow* or *closed system* and the *steady-flow* or *open system.* The nonflow system has boundaries across which both heat and work can penetrate, but no mass can cross these boundaries. In the steady-flow system, both mass and energy can cross the system boundaries. The term *steady* denotes a process or system that is not time dependent. When a fluid is caused to flow in a system, it is necessary that somewhere in the system work must have been supplied. At this time, let us evaluate the net amount of work required to push the fluid into and out of the system. Consider the system shown in Figure 2.10, where a fluid is flowing steadily across the system boundaries as shown. At the inlet section ①, the pressure is p_1, the area is A_1, the mass flow rate is $\dot{m}$, and the fluid density is ρ_1 (or its reciprocal specific volume, $1/v_1$); at the outlet section ②, the pressure is p_2, the area is A_2, the mass flow rate is still $\dot{m}$, and the fluid density is ρ_2 (or its reciprocal specific volume, $1/v_2$). Let us now consider a plug of fluid of length l_1 entering the system such that the amount of fluid contained in the plug is numerically $\dot{m}$. The force acting on the inlet cross-sectional area A_1 is p_1A_1. To push the plug into the system, it is necessary for this force to move the plug a distance equal to l_1. In so doing, the work done will be $p_1A_1l_1$. However, A_1l_1 is the volume of the plug containing a mass m. Using this we find the work, W, to be

$$W = p_1A_1l_1 = m(p_1v_1) \qquad \text{ft·lb}_f \text{ or N·m} \tag{2.13}$$

The work per unit mass is $W/m = \mathbf{w}$

$$\mathbf{w} = p_1v_1 \qquad \text{ft·lb}_f/\text{lb}_m \text{ or N·m/kg} \tag{2.14}$$

If we now consider the outlet section using the same reasoning, we have

$$W = m(p_2v_2) \qquad \text{ft·lb}_f \text{ or N·m} \tag{2.15}$$

or

$$\mathbf{w} = p_2v_2 \qquad \text{ft·lb}_f/\text{lb}_m \text{ or N·m/kg} \tag{2.16}$$

Each of the pv terms is known as the flow work. The net flow work becomes

$$\text{net flow work} = p_2v_2 - p_1v_1 \qquad \text{ft·lb}_f/\text{lb}_m \text{ or N·m/kg} \tag{2.17}$$

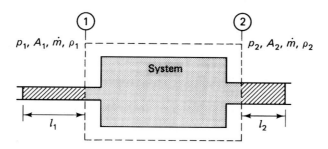

FIGURE 2.10 Steady-flow system: flow work.

or in thermal units,

$$\text{net flow work} = \frac{p_2 v_2}{J} - \frac{p_1 v_1}{J} \qquad \text{Btu/lb}_m \qquad\qquad (2.18)$$

or

$$\text{net flow work} = p_2 v_2 - p_1 v_1 \qquad \text{N·m/kg} \qquad\qquad (2.19)$$

Equations (2.18) and (2.19) are interpreted to mean that the difference in the pv terms represents the amount of work that is done on a system to introduce a unit mass into it minus the work done on its environment as it leaves the system. However, a word of caution is necessary at this time. Any fluid in any system has both properties, pressure and specific volume, and therefore the product pv can always be evaluated. *The product pv represents flow work only in the steady-flow system.* Thus, even though the term pv/J appears in the nonflow process, it cannot and does not represent flow work, because the system is by definition stationary. Flow work exists only to cause fluid to cross the boundaries of a flow system.

In the foregoing derivations, certain assumptions were made, and these are repeated here for emphasis. When applied to a flow situation, the term *steady* means that the condition at any section of the system is independent of time. Even though the velocity, specific volume, and temperature of the fluid can vary in any arbitrary manner across the stream, they are not permitted to vary with time. The mass entering the system per unit time must equal the mass leaving the system in the same period of time; otherwise, the system would either store or be depleted of fluid.

ILLUSTRATIVE PROBLEM 2.11

At the entrance to a steady-flow device, it is found that the pressure is 100 psia and that the density of the fluid is 62.4 lb_m/ft^3. At the exit, the pressure is 50 psia and the corresponding density 30 lb/ft^3. Determine the flow work term at the entrance and exit of the device.

SOLUTION

Refer to Figure 2.10. At the entrance,

$$\frac{pv}{J} = \frac{100\,\dfrac{\text{lb}_f}{\text{in}^2} \times 144\,\dfrac{\text{in}^2}{\text{ft}^2} \left(\dfrac{1}{62.4\,\text{lb}_m/\text{ft}^3}\right)}{778\,\dfrac{\text{ft·lb}_f}{\text{Btu}}} = 0.297\ \text{Btu/lb}_m$$

At the exit,

$$\frac{pv}{J} = \frac{50\,\dfrac{\text{lb}_f}{\text{in}^2} \times 144\,\dfrac{\text{in}^2}{\text{ft}^2} \times \left(\dfrac{1}{30\,\text{lb}_m/\text{ft}^3}\right)}{778\,\dfrac{\text{ft·lb}_f}{\text{Btu}}} = 0.308\ \text{Btu/lb}_m$$

ILLUSTRATIVE PROBLEM 2.12

Determine the flow work at the entrance and exit of a steady-flow device in which the entrance pressure is 200 kPa and the density of the fluid is 1000 kg/m³. At the exit, the pressure is 100 kPa and the density 250 kg/m³.

SOLUTION

At the entrance,

$$pv = \left(200 \text{ kPa} \times 1000 \frac{\text{Pa}}{\text{kPa}}\right)\frac{\text{N}}{\text{m}^2} \times \frac{1}{100 \text{ kg/m}^3} = 200 \frac{\text{N·m}}{\text{kg}}$$

At the exit,

$$pv = \left(100 \text{ kPa} \times 1000 \frac{\text{Pa}}{\text{kPa}}\right)\frac{\text{N}}{\text{m}^2} \times \frac{1}{250 \text{ kg/m}^3} = 400 \frac{\text{N·m}}{\text{kg}}$$

Because 1 N·m/kg is 1 J/kg, the answers are in J/kg.

2.9 NONFLOW WORK

Let us now consider the case of the piston and cylinder arrangement shown in Figure 2.11. We assume that the piston is in equilibrium with the contents and, initially, a distance l above the end of the cylinder. The piston will now be permitted to compress the contents of the cylinder frictionlessly and quasi-statically. After the piston has moved a distance Δl, a very small change in distance, the pressure in the cylinder will have increased from its initial value of p to a value of $p + \Delta p$, where Δp indicates a very small change in pressure. Because this process was specified to be quasi-static, it is possible to evaluate the work as the product of the average force multiplied by the displacement. The average forces is $\frac{1}{2}[pA + (p + \Delta p)A]$, and the displacement is Δl. Thus,

$$\mathbf{w} \text{ (per unit mass)} = \frac{p + (p + \Delta p)}{2} (A\Delta l) \tag{2.20}$$

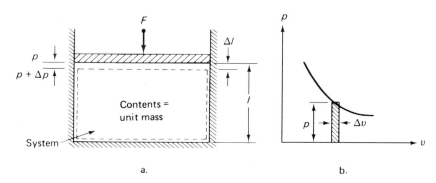

FIGURE 2.11   Quasi-static nonflow compression process.

However, the product of $A\Delta l$ is the small change in volume that the piston has swept out. Replacing $A\Delta l$ by Δv and saying that the product of two very small numbers ($\Delta v\,\Delta p$) is indeed so small as to be negligible, we have

$$\boxed{\mathbf{w} = p\Delta v}$$ per unit mass of working fluid   **(2.21)**

Equation (2.21) can be interpreted by referring to Figure 2.11b. It will be noted that the term $p\Delta v$ represents a small element of area, and that the total work can be evaluated by summing all the $p\Delta v$ terms. Therefore, we can conclude that the work done in a quasi-static, frictionless, nonflow process is the area under the pv curve.

ILLUSTRATIVE PROBLEM 2.13

A process is carried out in a nonflow, quasi-static manner so that the pressure volume relationship of the fluid is given by

$$pv = \text{constant}$$

where p is the pressure and v is the specific volume of the fluid. Determine the work done on the fluid if it undergoes a process in which its specific volume goes from v_1 to v_2.

SOLUTION

To solve this problem, it is necessary to sum all values of $p\,\Delta v$ over the entire range of the problem. As shown in Figure 2.12b, this corresponds to obtaining the area under the p–v curve between the limits of v_2 and v_1. Therefore, per unit mass,

$$\mathbf{w} = \sum p\,\Delta v$$

where the symbol Σ means the sum of all such values. To carry out this summation, we must first express p as a function of v. From the given p,v relation, $p = \text{constant}/v$, where the constant is $pv = p_1v_1 = p_2v_2$. When this is substituted in the $p\Delta v$ expression, we obtain

$$\mathbf{w} = \text{constant}\sum \frac{\Delta v}{v}$$

where the total work is the sum of these terms. The summation referred to can best be illustrated by plotting $1/v$ as a function of v. The shaded area of Figure 2.13 is simply $\Delta v/v$. Thus, the total work per pound of fluid is the area under this curve between v_1 and v_2 multiplied by a constant. Using the methods of calculus, we find the summation to be numerically equal to $\ln(v_2/v_1)$. Note that $\ln x = \log_e x = 2.3026 \log_{10} x$. Because the constant was either p_1v_1 or p_2v_2,

$$\mathbf{w} = p_1v_1 \ln \frac{v_2}{v_1} \frac{\text{ft}\cdot\text{lb}_\text{f}}{\text{lb}_\text{m}} \quad \text{or} \quad \frac{\text{N}\cdot\text{m}}{\text{kg}}$$

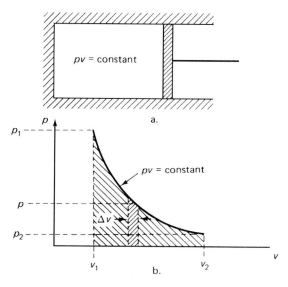

FIGURE 2.12   Illustrative Problem 2.13.

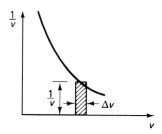

FIGURE 2.13   Evaluation of $\Sigma\,(\Delta v/v)$.

or $\qquad\qquad \mathbf{w} = p_2 v_2 \ln \dfrac{v_2}{v_1} \dfrac{\text{ft·lb}_f}{\text{lb}_m} \quad \text{or} \quad \dfrac{\text{N·m}}{\text{kg}}$

We shall have occasion to perform this type of summation in connection with other situations, and the procedure should be understood at this time for ease of application in the future.

CALCULUS ENRICHMENT

The work of the nonflow, quasi-static process, where the equation of path is known, is:

$$W = \int_{v_1}^{v_2} p \, dv \qquad\qquad \textbf{(a)}$$

(a) If the given relationship between p and v is $pv = $ constant, then $p = $ constant$/v$. Using this with Equation (a) gives us

$$W = \text{constant} \int_{v_1}^{v_2} dv/v \qquad \text{(b)}$$

The integral of dv/v between the limits of v_2 and v_1 is $\ln(v_2/v_1)$, or

$$W = (\text{constant}) \ln(v_2/v_1) \qquad \text{(c)}$$

but the constant equals pv. Therefore,

$$W = (pv) \ln(v_2/v_1) \qquad \text{(d)}$$

(b) If the given relationship between p and v is $pv^n = $ constant, then $p = $ constant$/v^n$. Again, using this with Equation (a),

$$W = \int_{v_1}^{v_2} p \, dv = \int_{v_1}^{v_2} (\text{constant}/v^n) \, dv$$

Integrating,

$$W = \text{constant} \left[\frac{v_2^{1-n} - v_1^{1-n}}{1 - n} \right] \qquad \text{(e)}$$

The constant can be determined from either $p_1 v_1^n$ or $p_2 v_2^n$. Using both of these, the work is found to be

$$W = \frac{p_2 v_2 - p_1 v_1}{1 - n} \qquad \text{(f)}$$

Note that Equation (f) is valid for every value of n **except** 1.0. For $n = 1.0$, the results of part (a) must be used.

ILLUSTRATIVE PROBLEM 2.14

For the process described in Illustrative Problem 2.13, the initial pressure is found to be 100 psia and the initial specific volume 2 ft^3/lb. If the final specific volume is 1 ft^3/lb, how much work was done on the fluid per pound of fluid?

SOLUTION

From Illustrative Problem 2.13, $\mathbf{w} = p_1 v_1 \ln(v_2/v_1)$. For the proper units, it is necessary that pressure be expressed as psfa when the volume is in cubic feet. Thus, 100 psia $= (100)(144)$ psfa, and $v_1 = 2$ ft^3/lb.

$$\mathbf{w} = (100 \times 144 \times 2)\left(\ln \frac{1}{2} \right)$$

This is evaluated by noting that $\ln \frac{1}{2} = \ln 1 - \ln 2$. The $\ln 2$ is 0.693, and the $\ln 1$ is zero. Therefore,

$$\mathbf{w} = -100 \times 144 \times 2 \times 0.693 = 19{,}958 \text{ ft·lb}_f/\text{lb}_m$$

This computation is also readily done using the ln function of any scientific calculator. In thermal units,

$$\mathbf{w} = \frac{-19{,}958 \text{ ft·lb}_f/\text{lb mass}}{778 \text{ ft·lb}_f/\text{Btu}} = -25.7 \text{ Btu/lb}_m$$

The minus sign in the answer indicates work into the system. As an exercise, it is left to the student to solve this problem by graphically evaluating the area under the pressure–volume curve.

ILLUSTRATIVE PROBLEM 2.15

For the process $pv = $ constant, a gas compression is carried out from an initial pressure of 200 kPa to a final pressure of 800 kPa. If the initial specific volume is 0.1 m³/kg, determine the work done per kilogram of gas.

SOLUTION

Because $p_1 v_1 = p_2 v_2$,

$$v_2 = \frac{p_1 v_1}{p_2} = \frac{200}{800} \times 0.1 \text{ m}^3/\text{kg} = 0.025 \text{ m}^3/\text{kg}$$

and

$$\mathbf{w} = p_1 v_1 \ln \frac{v_2}{v_1}$$

$$= 200 \text{ kPa} \times 0.1 \text{ m}^3/\text{kg} \times \ln \frac{0.025}{0.100}$$

$$= -27.7 \text{ kJ/kg} \qquad \text{(into the system)}$$

2.10   REVIEW

This chapter has been devoted to developing qualitatively and quantitatively the concepts of work, energy, and heat. A careful study of the contents of this chapter will be invaluable later on in our study. In addition to defining work, energy, and heat, we also showed that the work of a nonflow, quasi-static process can be taken to be the area under the $p–v$ curve. The term *steady* implies a system that does not vary with time. In the steady-flow system, both mass and energy are permitted to cross the boundaries of the system, but such a system is not time dependent. We also noted that when a fluid is caused to flow in a system, somewhere in the system work must have been supplied to sustain this flow. Using this concept, we demonstrated that the difference in the pv terms equals the amount of work that is done on a system to introduce a unit mass into it minus the work done on its environment as it leaves the system. Even though all systems have pressure and specific volume, the *pv* term represents flow work only in a flow system. For a nonflow, quasi-static, frictionless process, the work done is in the area under the *pv* curve.

KEY TERMS

Terms used for the first time in this chapter are:

closed system see *nonflow system.*

energy the capacity to do work.

equilibrium state a state in which no finite rate of change can occur without a finite change, temporary or permanent, in the state of the environment.

flow work the product of the pressure and specific volume of a fluid in a given state in a flow process.

heat a form of energy in transition to or from a system due to the fact that temperature differences exist.

internal energy the energy possessed by a body by virtue of the motion of the molecules of the body and the internal attractive and repulsive forces between molecules.

kinetic energy the energy possessed by a body due to its motion.

mechanical equivalent of heat in the English system of units, 1 Btu = 778 ft·lb$_f$. This is termed the *mechanical equivalent of heat.*

nonflow system a system in which there is no mass crossing the boundaries.

potential energy the energy possessed by a body due to its location with respect to an arbitrary reference plane.

quasi-static process a frictionless process carried out infinitely slowly so that it is in equilibrium at all times.

work the product of force and distance in which the distance is measured in the direction of the force; a form of energy in transition that is not stored in a system.

EQUATIONS DEVELOPED IN THIS CHAPTER

internal energy	$mu = U$	(2.1)
internal energy	$u = \dfrac{U}{m}$	(2.2)
potential energy	$\text{P.E.} = \dfrac{mg}{g_c} Z \quad \text{ft·lb}_f$	(2.4)
potential energy (SI)	$\text{P.E.} = mgZ \quad \text{joules}$	(2.6)
kinetic energy	$\text{K.E.} = \dfrac{mV^2}{2g_c} \quad \text{ft·lb}_f/\text{lb}_m$	(2.12)
kinetic energy (SI)	$\text{K.E.} = \dfrac{mV^2}{2} \quad \text{joules}$	(2.12a)
net flow work	$\text{net flow work} = \dfrac{p_2 v_2}{J} - \dfrac{p_1 v_1}{J} \quad \text{Btu/lb}_m$	(2.18)
net flow work	$\text{net flow work} = p_2 v_2 - p_1 v_1 \; \text{N·m/kg}$	(2.19)
work of a nonflow, quasi-static process	$\mathbf{w} = p\Delta v$	(2.21)

QUESTIONS

2.1 A person picks up a package, walks with it for a distance, and then places it on a table. Discuss our concept of work for each of these actions.

2.2 A weight is held stationary at arm's length by a student. Is energy being expended, and is work being done?

2.3 A spring is compressed and placed into an acid solution, where it dissolves. What has happened to the work done on the spring?

2.4 Both heat and work represent forms of energy. Discuss the differences between them.

2.5 What is the value of the mechanical equivalent of heat in the SI system?

2.6 What distinguishes a nonflow system from a flow system?

2.7 Can you describe any quasi-static processes?

2.8 Of what utility is the quasi-static process concept?

2.9 The term $p\Delta v$ is the area under the p–v curve. How do you interpret this area?

2.10 In discussing potential energy, what is a necessary requirement?

2.11 Why do we say that a body that is moving possesses energy?

2.12 All substances are subjected to pressure and have a specific volume. Because they all have the product pv, do they all have flow work?

2.13 State the convention that we have adopted for net heat and net work.

PROBLEMS

Unless indicated otherwise, use $g_c = 32.174$ ft·lb$_m$/lb$_f$·s^2, and 32.174 ft/s^2 or 9.81 m/s^2 for g.

Problems Involving Potential Energy, Kinetic Energy, and Work

2.1 A 2-kg mass is elevated to a distance of 6 m above a reference plane. Determine its potential energy with respect to the plane.

2.2 A mass of 10 lb is placed at an elevation of 30 ft above a reference plane. Another mass of 15 lb is placed at an elevation of 40 ft above a second reference plane that is 8 ft below the first plane.
(a) Compare the potential energy of each mass with respect to its own reference plane.
(b) Compare the potential energy of each mass with respect to the second reference plane.

2.3 A body of mass 10 lb is placed 10 ft above an arbitrary plane. If the local gravitational field is equivalent to an acceleration of 16.1 ft/s^2, how much work (ft·lb) was done lifting the body above the plane?

2.4 A body of mass of 10 kg is placed 3 m above a plane. How much work was done to lift the body above the plane?

2.5 A body having a mass of 3 lb is moving with a velocity of 30 ft/s relative to the earth.
(a) Calculate it kinetic energy.
(b) If the body is resisted by a constant force of 8 lb, how far will it move before coming to rest?

2.6 A body has a mass of 5 kg. If its velocity is 10 m/s, what is its kinetic energy?

2.7 From what height would the body in Problem 2.6 have to fall to attain its velocity?

2.8 A body weighs 25 lb. If it is moving at 10 ft/s, evaluate its kinetic energy.

2.9 From what height would the body in Problem 2.8 have to fall to achieve its velocity?

2.10 A 500-kg mass is lifted 2 m above a datum plane.
   (a) Determine its potential energy.
   (b) If it is dropped, what will its velocity be at the instant just before it strikes the earth?

2.11 A jet of air is discharged from a nozzle. If the jet has a velocity of 100 m/s, determine its kinetic energy in joules if 1 kg leaves the nozzle.

2.12 A body weighing 10 lb is lifted 100 ft. What is its change in potential energy? What velocity will it possess after falling 100 ft?

2.13 A body weighing 10 lb ($g = g_c$) is moving with a velocity of 50 ft/s. From what height would it have to fall to achieve this velocity? What is its kinetic energy?

2.14 A body weighing 100 N is lifted 3 m. What will its velocity be after a free fall of 3 m?

2.15 Water flows over the top of a dam and falls freely until it reaches the bottom some 600 ft below. What is the velocity of the water just before it hits the bottom? What is its kinetic energy per pound mass at this point?

2.16 An object is dragged from rest a distance of 40 ft up an inclined plane that is smooth. During this time, it is elevated 10 ft and acquires a velocity of 10 ft/s. If its mass is 25 lb:
   (a) Determine the work done in the process.
   (b) Determine the change in potential energy.
   (c) Determine the increase in kinetic energy.

2.17 A truck having a weight of 50 kN is moving with a velocity of 25 m/s. Determine its kinetic energy.

2.18 If an object has a kinetic energy of 1 kJ when moving at a velocity of 2 m/s, determine its kinetic energy when its velocity is: (a) 4 m/s; (b) 6 m/s.

2.19 A force of 100 lb deflects a spring 5 in. Calculate the spring constant and work done.

2.20 A spring is deflected 250 mm from its free length by a force of 500 N. Calculate the spring constant and the work done.

2.21 A spring is used in a shock absorber. It is found to absorb 2000 J of energy after deflecting 0.35 m. Determine the spring constant.

2.22 A spring is compressed 5 in. from its equilibrium position. If the spring modulus k is 10 lb_f/in. of deflection, how much work was done in foot pounds to deflect the spring?

2.23 A spring is deflected by a weight a distance of 100 mm. If the modulus of the spring is 100 N/m, what is the mass that was placed on the spring?

*2.24 A 100-N weight causes a spring to stretch an unknown distance, h. The addition of a weight of 250 N causes the spring to deflect an additional distance 100 mm. Determine the spring constant.

2.25 A spring is initially compressed 75 mm by a 400-N force. If it is compressed an additional distance of 50 mm, determine the work required for this additional compression.

2.26 A 5-lb mass ($g = g_c$) falls 25 ft until it strikes a spring. If the spring modulus k is 25 lb_f/ft of deflection, how far will the spring be deflected? Assume that all the energy of the falling body just goes to compress the spring.

2.27 A 10-kg mass falls 3 m until it strikes a spring. If the spring modulus k is 1000 N/m, how far will the spring be deflected if all the energy of the falling body goes into compressing the spring?

2.28 In Figure P2.28, a force of 1000 N moves the block 5 m along the plane. How much work was done by the force on the block?

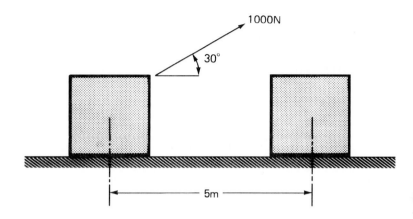

2.29 What is the least amount of work required to move the block 3 m up in the plane shown in Figure P2.29? Assume the plane to be frictionless.

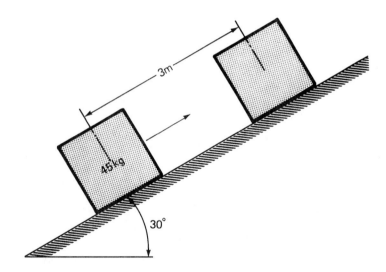

2.30 How much work must be done on a car to decrease its velocity from 90 km/hr to 45 km/hr? The car has a mass of 1000 kg.

2.31 An automobile has a weight of 3000 lb and a velocity of 30 mi/hr. What constant braking force is required to bring it to a stop in 190 ft?

*2.32 Show that the kinetic energy of a fluid flowing in a pipe varies inversely as the fourth power of the pipe diameter.

*2.33 If $50 \times 10^6 \ lb_m/hr$ of water flows over the dam of Problem 2.15, what ideal power can be generated by a power plant located at the base of the fall? Express your answer in kilowatts and horsepower.

*2.34 If 25×10^6 kg/hr flows over a waterfall that is 200 m high, what ideal power in kilowatts can be generated by a power plant located at the base of the fall?

Problems Involving the p–v Concept

2.35 In a constant-pressure process, steam at 100 psia is presented to the piston of a pump and causes it to travel 3 in. If the cross-sectional area of the piston is 5 in^2, how much work was done by the steam on the piston?

2.36 A pressure of 500 kPa is presented to the piston of a pump and causes it to travel 100 mm. The cross-sectional area of the piston is 1000 mm^2. How much work is done by the steam on the piston if the steam pressure is constant?

2.37 A gas expands according to the equation $pv = 100$, where p is the pressure in pounds per square foot absolute and v is the specific volume in cubic feet per pound. If the pressure of the gas drops from 100 to 50 psfa, how much work was done by the gas?

2.38 A gas expands according to the equation $pv = 1000$, where p is the pressure in kPa and v is the specific volume in m^3/kg. If the gas pressure drops from 1000 to 500 kPa, how much work was done by the gas?

*2.39 A gas undergoes a process that starts at a pressure of 25 psia and an initial volume of 4 ft^3. Work is done on the gas in the amount of 12 Btu. If the p–v curve is given by $pV = $ constant, determine the final pressure of the gas.

*2.40 A gas that is contained in a cylinder undergoes two processes that are in series. The first process, from 1 to 2, consists of a constant-pressure compression at 50 psia, and the volume goes from 0.5 to 0.25 ft^3. The second process consists of a constant-volume heating in which the pressure doubles to 100 psia. Sketch the processes on a p–v diagram, and determine the total work of the combined process.

*2.41 A gas undergoes two processes that are in series. The first process is an expansion that is carried out according to the relation $pv = $ constant, and the second process is a constant-pressure process that returns the gas to the initial volume of the first process. The start of the first process is at 400 kPa and 0.025 m^3 with the expansion to 200 kPa. Sketch the processes on a p–v diagram, and determine the work of the combined process.

*2.42 A gas undergoes two processes in series. The first is a constant-pressure expansion that takes the gas from 100 psia and 1 to 2 ft^3. The second process is a constant-volume process during which the pressure decreases to 30 psia. Sketch the processes on a p–v diagram, and determine the work of the combined process.

*2.43 A gas expands according to the equation, $pv = 100$, where p is the pressure in pounds force per square foot and v is the specific volume in cubic feet per pound mass. The initial pressure of the gas is 100 psfa, and the final pressure is 50 psfa. The gas is then heated at constant volume back to its original pressure of 100 psfa. Determine the work of the combined process.

*2.44 A gas expands according to the equation $p = -250V + 600$, where V is in m^3 and p is in kPa. If the initial pressure is 575 kPa, the initial volume is 0.1 m^3, and the final volume is 0.4 m^3, determine the work done. (Hint: Plot a p–v diagram.)

*2.45 If the gas in Problem 2.44 expands at constant pressure from its initial pressure, determine the work done.

Cyclic Processes

A cycle consists of a series of thermodynamic processes during which the working fluid can be made to undergo changes involving energy transitions and is subsequently returned to its original state. On a p–v diagram, it is a closed figure.

*2.46 A cycle consists of three processes. The first is a constant-pressure compression at 200 kPa from an initial volume of 0.70 m^3 to a final volume of 0.2 m^3. The second process takes place at constant volume with the pressure increasing to 600 kPa. The third process completes the cycle and consists of a straight line from the end of the second process to the beginning of the first process. Sketch the cycle in p–v coordinates, and calculate the net work of the cycle.

*2.47 A cycle consists of three processes. The first process consists of a constant-pressure compression from 14.7 psia and 26 to 10 ft^3. The second process is carried out at constant volume to a final pressure of 35 psia. The cycle is closed with a straight line from the end of the second process to the start of the first process. Sketch the cycle on p–v coordinates, and calculate the net work of the cycle.

*2.48 A cycle consists of three processes. The first is an expansion that is carried out according to the relation pv = constant from an initial condition of 500 kPa and 0.2 m^3 to a final pressure of 250 kPa. The second process is a constant-pressure compression that returns the gas to the volume that it had at the beginning of the first process. The cycle is closed with a constant-volume process that returns the gas to its state at the beginning of the first process. Sketch the cycle on p–v coordinates, and calculate the net work of the cycle.

3

The First Law of Thermodynamics

Learning Goals

After reading and studying the material in this chapter, you should be able to:

1. State the first law of thermodynamics, or energy conservation, for both nonflow and flow systems.

2. Define enthalpy, and show that it is a property and therefore does not depend on whether a system is a nonflow or flow system.

3. Apply the first law of thermodynamics to several nonflow systems, such as the adiabatic system, the constant-volume system, and the constant-pressure system.

4. Derive and apply the continuity equation in its several forms to express the conservation of mass in a steady-flow system.

5. Derive the energy equation from the first law of thermodynamics for a steady-flow system.

6. Think of the area under the pressure–volume diagram as the work of compression or expansion and the area behind the pressure–volume diagram as the work of the compressor or expander.

7. Understand the shortcomings of the Bernoulli equation.

8. Define the term *specific heat,* and show that the specific heats at constant volume and constant pressure are properties.

9. Apply the first law of thermodynamics to the analysis of the steam or gas turbine, pipe flow, boilers, nozzles, throttling, heat exchangers, and the filling of a tank.

3.1 INTRODUCTION

We have discussed properties, state functions, systems, work, energy, and heat in the previous chapters. In this chapter, we will consider the conservation of energy as the basis for the first law of thermodynamics. While Newton first proposed the equivalence of heat and work, it is only in relatively recent times, as a result of both observation and analysis, that the principle of conservation has become one of the cornerstones of thermodynamics. Most of this chapter will be concerned with the application of the first law of thermodynamics to both nonflow and flow systems and to develop energy equations applicable to these systems. The first law of thermodynamics will enable us to determine the energy quantities passing to or from a system as work and/or heat and also to determine the changes in the energy stored in the system.

3.2 THE FIRST LAW OF THERMODYNAMICS

The first law of thermodynamics can be expressed in the following equivalent statements:

1. The first law of thermodynamics is essentially the statement of the principle of the conservation of energy for thermodynamical systems. As such, it may be expressed by stating that the variation of energy of a system during any transformation is equal to the amount of energy that the system receives from its environment.*

2. Energy can be neither created nor destroyed but only converted from one form to another.**

3. If a system is caused to change from an initial state to a final state by adiabatic means only, the work done is the same for all adiabatic paths connecting the two states.†

In statement 3, by Zemansky, the term adiabatic is used. In general, we define an *adiabatic* transformation of a system as a process the system is caused to undergo, with no energy interchange as heat occurring during the process.

For the purpose of this book, the concept of the conservation of energy, explicitly stated in the Fermi and Obert definitions, is essentially the first law of thermodynamics. By combining them, we have for the statement of the **first law of thermodynamics** that *energy can neither be created nor destroyed but only converted from one form to another.* Because we shall not concern ourselves with nuclear reactions at this time, it will not be necessary to invoke the interconvertibility of energy and mass in our present study.

A system has already been defined as a grouping of matter taken in any convenient or arbitrary manner. However, when dealing with fluids in motion, it is more convenient to utilize the concept of an arbitrary volume in space, known as a *control volume,* that can be bounded by either a real or imaginary surface, known as a *control surface.* By correctly noting all the forces acting on the fluid within the control volume, the energies crossing the control surface, and the mass crossing the control surface, it is possible to derive mathematical expressions that will evaluate the flow of the fluid relative to the control volume. For a system in which fluids are flowing steadily, a monitoring station placed anywhere within the control volume will indicate no change in the fluid properties or energy quantities crossing the control surface with time, even though these quantities can and will vary from position to position within the control volume. As noted by Keenan, *"the first step in the solution of a problem in thermodynamics is the description of a system and its boundaries."*§

---

* E. Fermi, *Thermodynamics* (New York: Dover Publications, 1956), p. 11.

** E. F. Obert, *Concepts of Thermodynamics* (New York: McGraw-Hill Book Company, 1960), p. 59.

---

† M. W. Zemansky, *Heat and Thermodynamics,* 4th ed. (New York: McGraw-Hill Book Company, 1957), p. 59.

---

§ J. H. Keenan, *Thermodynamics* (New York: John Wiley & Sons, Inc., 1941), p. 14.

3.3 THE NONFLOW SYSTEM

The statements made in Section 3.2 about the first law of thermodynamics are, in essence, equivalent to each other in that they express the concept of energy conservation. To explore some of the implications and applications of the first law, let us examine the *nonflow* or *closed system*. This system will have boundaries across which both heat and work can penetrate, but no mass will be permitted to cross them. Note that the boundaries can move with the system as the process proceeds along if no mass flow crosses the boundaries. An example is a piston-and-cylinder arrangement in which the piston compresses the working fluid in the cylinder and heat may cross the boundary (e.g., by cooling of the cylinder) at the same time. This system is shown schematically in Figure 3.1. For convenience, it is assumed that there is a unit mass in the cylinder. Assuming also that there are no (or negligible) changes in elevation during the process and no directed (net) velocity of the working fluid, we may properly neglect energy terms relating to potential and kinetic energy. Writing the first law in words which state that the energy in state *a* plus or minus any additions or depletions from the system must equal the energy in state *b* yields the following:

$$u_1 + q - \mathbf{w} = u_2 \tag{3.1}$$

Rearranging yields

$$\boxed{u_2 - u_1 = q - \mathbf{w}} \tag{3.2a}$$

where both q and $\mathbf{w}$ are used to denote the net heat and net work, respectively, entering or leaving the system per unit mass of fluid and u denotes the internal energy per unit mass.

Note that heat added is always positive and will increase the energy in the system. If work is added to the system from the surroundings, this will also increase the energy in the system. For consistency, because from Section 2.3 work added is considered negative, the work term in Equation (3.1) must have a negative sign.

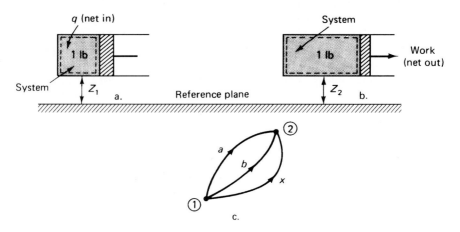

FIGURE 3.1 The nonflow system.

It is interesting and instructive to note that the piston–cylinder arrangement is treated as a nonflow process by the selection of the boundaries of the system, as shown by the dashed lines in Figure 3.1a and b. However, when the piston and cylinder are part of an internal combustion engine, as shown in Figure 3.2, the selection of the system boundary at the exterior of the engine as shown leads us to a steady-flow system where fuel and air enter while heat and work cross the boundary. This concept will be treated in further detail when we concern ourselves with power cycles in Chapters 8 and 9.

Equation (3.2a) is useful in establishing that internal energy is a property of a system and is not dependent on the path taken to place the system in a given configuration. This is true even though both q and $\mathbf{w}$ are path functions; that is, their values depend on the method of placing a system in a given configuration. For proof of the fact that internal energy is a property, the following reasoning is used.* Consider a system initially at state 1 that is caused to undergo a change to a second state 2 via path a. From Equation (3.2a) and Figure 3.1c, we have

$$u_2 - u_1 = q_a - \mathbf{w}_a \tag{3.2b}$$

where the subscript a is used to denote path a. If path b is now followed between the same points 1 and 2, Equation (3.2a) becomes

$$u_2 - u_1 = q_b - \mathbf{w}_b \tag{3.2c}$$

This procedure can be carried out for any path (x), and it must follow that when Equation (3.2a) is applied,

$$u_2 - u_1 = q_x - \mathbf{w}_x \tag{3.2d}$$

Because the left sides of Equations (3.2b), (3.2c), and (3.2d) are equal, it follows that the right sides of these equations

$$q_x - \mathbf{w}_x = q_a - \mathbf{w}_a = q_b - \mathbf{w}_b \tag{3.3}$$

are also equal, and that $u_2 - u_1$ is fixed only by the end states of the system and is independent of the process. It can therefore be concluded that internal energy is a property. As such, it is a state function and independent of the path of any process. There is one point about which the student should make careful note. The proof of the fact that internal energy is a property is extremely important. In the calculus, an area can be obtained by integration. This process can be performed for those functions (paths) that can be evaluated as being continuous and dependent only on the end states. Such a function is said to be mathematically exact, and it is possible to perform all operations of the calculus on it. Opposed to this concept is the function whose value between two end states is determined not by the end states but by the path taken to achieve the end state. For example, the work done in moving a given block from one position on a plane to another position on the plane will depend on the amount of work done against friction. If the table is rough, more work is required to go from the initial to the final position than if the table were smooth. Mathematically, such a function is said to be inexact, and in general, it is not possible to evaluate this function directly by the methods of the calculus unless the path is defined. *Work and heat are inexact (path functions), whereas internal energy is exact (a function of the end states only)*.

---

* This material is a modification of the proof in Keenan, *Thermodynamics,* p. 12 and following.

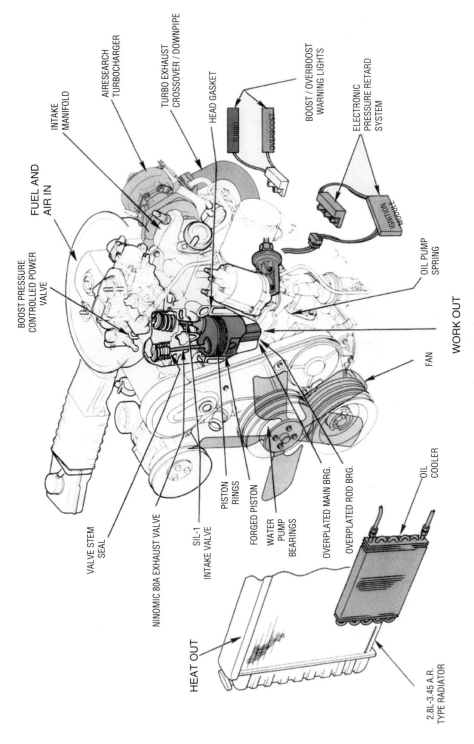

FIGURE 3.2 A 2.3-liter turbocharged engine. (*Courtesy of Ford Motor Co.*)

FUEL AND AIR IN

BOOST PRESSURE CONTROLLED POWER VALVE

INTAKE MANIFOLD

AIRESEARCH TURBOCHARGER

TURBO EXHAUST CROSSOVER / DOWNPIPE

HEAD GASKET

BOOST / OVERBOOST WARNING LIGHTS

TURBO

OVERBOOST

ELECTRONIC PRESSURE RETARD SYSTEM

IGNITION MODULE

OIL PUMP SPRING

WORK OUT

FAN

VALVE STEM SEAL

NINOMIC 80A EXHAUST VALVE

PISTON RINGS

SIL-1 INTAKE VALVE

FORGED PISTON

WATER PUMP BEARINGS

OVERPLATED MAIN BRG.

OVERPLATED ROD BRG.

OIL COOLER

HEAT OUT

2.8L-3.45 A.R. TYPE RADIATOR

ILLUSTRATIVE PROBLEM 3.1

If a nonflow, constant-volume process has 10 Btu/lb$_m$ as heat added to the system, what is the change in internal energy per pound of working fluid?

SOLUTION

For a constant-volume process, we can consider that the piston in Figure 3.1 has not moved. Alternatively, we can consider that a tank having a fixed volume has heat added to it. Under these conditions, the mechanical work done on or by the system must be zero. The application of Equation (3.2a) to this system yields

$$u_2 - u_1 = q$$

and it must be concluded that all the energy crossing the boundary as heat has been converted to internal energy of the working fluid. Therefore,

$$u_2 - u_1 = 10 \text{ Btu/lb}_m$$

ILLUSTRATIVE PROBLEM 3.2

The working fluid in a nonflow system undergoes an adiabatic change. Determine the work done in this process.

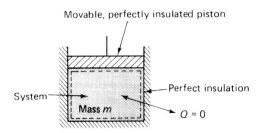

Movable, perfectly insulated piston

System

Mass m

Perfect insulation

$Q = 0$

FIGURE 3.3 Illustrative Problem 3.2.

SOLUTION

An adiabatic process is defined to be one during which no energy interchange *as heat* crosses the boundary of a system. As shown in Figure 3.3, this is the same as considering the system boundaries to be perfectly insulated. Note, however, that the piston can and does move. Application of Equation (3.2a) to this situation yields

$$u_2 - u_1 = -\mathbf{w}$$

Thus, the energy interchange as work to or from the system per unit mass of working substance equals the change in internal energy of the working fluid per unit mass of fluid. The negative sign is taken to mean that work into the system (negative work by convention) will cause an increase in the internal energy of the working fluid and that work out of the system (positive work) will cause a decrease in the internal energy of the working fluid.

In Chapter 2, we concluded that the work done in a quasi-static frictionless, nonflow process is the area under the p–v curve. Using this concept and Equation (2.21) for the work in Equation (3.2a) gives us the following useful relation:

$$u_2 - u_1 = q - p\Delta v$$

$$(3.4)$$

Equation (3.4) is written per unit mass of working fluid, and the student is cautioned to understand the reasoning behind this equation fully so that it will not be misapplied.

Either Equation (3.2a) or Equation (3.4) is called the *nonflow energy equation* and expresses the first law as applied to a nonflow process. Equation (3.4) is more restrictive and, in the strictest sense, is applicable only to frictionless, quasi-static, nonflow processes.

ILLUSTRATIVE PROBLEM 3.3

Solve Illustrative Problem 3.1 by the direct application of Equation (3.4).

SOLUTION

$$u_2 - u_1 = q - p\Delta v$$

But $v_2 - v_1$ (*or* Δv) is zero. Therefore,

$$u_2 - u_1 = q \quad \text{(as before)}$$

There is one nonflow process that will be found to be quite important in our subsequent discussion: the quasi-static, *nonflow, constant-pressure* process. Using Equation (3.4), we will now derive an expression relating heat, work, and internal energy for this process. Refer to the situation illustrated in Figures 3.1 and 3.4 where heat and work can cross the system boundary. Transposing terms in Equation (3.4) yields

$$u_2 - u_1 + p\Delta v = q$$

$$(3.4a)$$

However, p, the pressure, has been defined to be constant. Therefore,

$$u_2 - u_1 + p(v_2 - v_1) = q$$

$$(3.4b)$$

The condition of constant pressure permits us to write $p = p_1 = p_2$. Thus,

$$u_2 - u_1 + p_2v_2 - p_1v_1 = q$$

$$(3.4c)$$

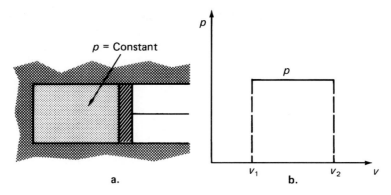

FIGURE 3.4   Quasi-static, nonflow, constant-pressure process.

Equation (3.4c) is consistent in terms of SI units. If the internal energy is in Btu per pound, pressure is in pounds per square foot, and specific volume is in cubic feet per pound, we can regroup terms and place them into consistent thermal units of Btu per pound as follows:

$$\left(u_2 + \frac{p_2 v_2}{J} \right) - \left(u_1 + \frac{p_1 v_1}{J} \right) = q \tag{3.4d}$$

The composite term $u + pv/J$ is a property that we will find to have great utility when flow processes are considered. At present, we will simply define h to be enthalpy expressed in Btu/lb$_\mathrm{m}$ and to be given by

$$h = u + \frac{pv}{J} \tag{3.5}$$

The conversion factor J is carried along as a reminder to the student of the necessity to use consistent units in the English system at all times. Returning to Equation (3.4d), we now have the following energy equation for the nonflow, quasi-static, constant-pressure process:

$$q = h_2 - h_1 = \Delta h \tag{3.4e}$$

So far, three nonflow, quasi-static processes have been considered. To summarize the energy equations for these processes,

1. *Constant-volume process:*

$$q = u_2 - u_1 = \Delta u \tag{3.6}$$

2. *Adiabatic process:*

$$-\mathbf{w} = u_2 - u_1 = \Delta u \tag{3.7}$$

3. *Constant-pressure process:*

$$q = h_2 - h_1 = \Delta h \tag{3.4e}$$

Further, the work of a quasi-static, nonflow process is:

$$\mathbf{w} = p\Delta v \qquad (2.21)$$

Enthalpy is defined as

$$h = u + \frac{pv}{J} \qquad (3.5)$$

or

$$h_2 - h_1 = \left(u_2 + \frac{p_2 v_2}{J}\right) - \left(u_1 + \frac{p_1 v_1}{J}\right) \qquad (3.8)$$

In all the foregoing, it has been assumed that the mass of fluid was unity. However, the equations derived can be used for any mass simply by multiplying each term in the energy equation by m, the mass involved. The enthalpy definition used in this book will include the conversion factor J for convenience and consistency, especially when using English units.

ILLUSTRATIVE PROBLEM 3.4

A rigid container contains 10 lb of water. (a) If 100 Btu as heat is added to the water, what is its change in internal energy per pound of water? (b) If the 100 Btu is added by the mechanical fraction of a paddle wheel stirring the water, what is the change in internal energy of the water? Discuss both processes. Refer to Figure 3.5.

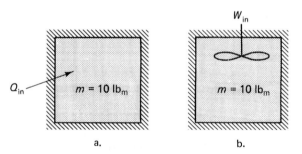

FIGURE 3.5 Illustrative Problem 3.4.

SOLUTION

(a) The nonflow energy equation applied to this process yields $q = u_2 - u_1$. Per pound mass of working substance, $q = 100/10$, or 10 Btu/lb$_m$. Therefore, $u_2 - u_1 = 10$ Btu/lb$_m$.

(b) In this process, energy crosses the boundary of the system by means of frictional work. As far as the system is concerned, we note only that energy has crossed its boundaries. The similarity of the terms *work* and *heat* lies in the fact that both are energy in transition. Thus, for the present problem, the contents of the tank will not distinguish between the energy if it is added as heat or the energy added as frictional work. As for part (a), $u_2 - u_1 = 10$ Btu/lb$_m$.

3.4   THE STEADY-FLOW SYSTEM

3.4a   Conservation of Mass—The Continuity Equation

In the steady-flow system, both mass and energy are permitted to cross the boundaries of the system, but by denoting the process to be steady, we limit ourselves to those systems that are not time dependent. Because we are considering steady-flow systems, we can express the principle of conservation of mass for these systems as requiring the mass of fluid in the control volume at any time be constant. In turn, this requires that the net mass flowing into the control volume must equal the net mass flowing out of the control volume at any instant of time. To express these concepts in terms of a given system, let us consider the system schematically in Figure 3.6.

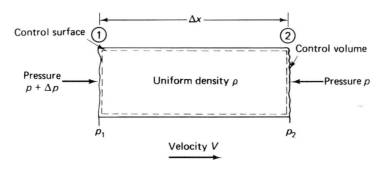

FIGURE 3.6   Elementary flow system.

Let us assume that at a certain time, fluid starts to enter the control volume by crossing the control surface at ① and that, after a small interval of time, the flowing fluid fills the pipe for a short distance Δx. If it is further assumed that in this short section of uniform pipe no heat is added or work is exchanged and that the fluid density stays constant, we can evaluate the amount of fluid that flowed between ① and ②. The mass contained between these sections is equal to the volume contained between the sections multiplied by the density of the fluid. The volume is $A\Delta x$, and the density is ρ. Therefore, the contained mass is $\rho A\Delta x$. The distance between stations, Δx, is simply $V\Delta t$, where V is the velocity of the fluid and Δt is the flow time required to fill the pipe between ① and ②. Substituting this for Δx yields

$$m = \rho AV\Delta t \tag{3.9a}$$

or

$$\dot{m} = \rho AV = \rho_1 A_1 V_1 = \rho_2 A_2 V_2 \tag{3.9b}$$

where $\dot{m}$ is the mass rate of flow per unit time, $m/\Delta t$. Also,

$$\dot{m} = \frac{AV}{v}$$

(3.9c)

where the specific volume replaces the density, $v = 1/\rho$.

   Equation (3.9b) is known as the *continuity equation,* and as written for a pipe or duct, we have made the assumption that the flow is normal to the pipe cross-section and that the velocity V is either constant across the section or is the average value over the cross-section of the pipe. We shall use the continuity equation quite frequently in our study, and it is important to note that it expresses the fact that the mass flow into the control volume must equal the mass flow out of the control volume in steady flow.

ILLUSTRATIVE PROBLEM 3.5

At the entrance to a steady-flow device, it is found that the pressure is 100 psia and that the density of the fluid is constant at 62.4 lb_m/ft^3. If 10,000 ft^3/min of this fluid enters the system and the exit area is 2 ft^2, determine the mass flow rate and the exit velocity (see Figure 3.7).

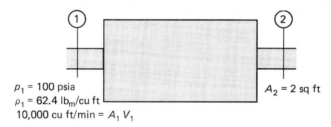

p_1 = 100 psia
p_1 = 62.4 lb_m/cu ft
10,000 cu ft/min = $A_1 V_1$

A_2 = 2 sq ft

FIGURE 3.7   Illustrative Problem 3.5.

SOLUTION

Let us first calculate the mass flow into the system. From Equation (3.9b), $\dot{m} = \rho_1 A_1 V_1$. We are told that 10,000 ft^3/min of this fluid enters the system, which is the same as saying $A_1V_1 = 10,000$ ft^3/min. Therefore, $\dot{m} = \rho_1 A_1 V_1 = 62.4 \times 10,000 = 624,000$ lb_m/min. Because this mass flow must also leave the system,

$$\dot{m} = \rho_2 A_2 V_2$$

Thus,

$$624,000 \, \frac{lb_m}{min} = 62.4 \, \frac{lb_m}{ft^3} \times 2 \, ft^2 \times V_2 \, \frac{ft}{min}$$

$$V_2 = 5000 \, ft/min$$

ILLUSTRATIVE PROBLEM 3.6

If the fluid entering the system shown in Figure 3.7 has a density of 1000 kg/m³ and 2000 m³/min enters the system, determine the mass flow rate and the exit velocity if the exit area is 0.5 m². Assume that the density is constant.

SOLUTION

Proceeding as in Illustrative Problem 3.5, $m = \rho A V$; $\dot{m} = \rho_1 A_1 V_1$. Therefore,

$$\dot{m} = \left(1000 \, \frac{\text{kg}}{\text{m}^3} \right) \times 2000 \, \frac{\text{m}^3}{\text{min}}$$

$$= 2 \times 10^6 \, \frac{\text{kg}}{\text{min}}$$

Because $\dot{m} = \rho_2 A_2 V_2$,

$$2 \times 10^6 \, \frac{\text{kg}}{\text{min}} = 1000 \, \frac{\text{kg}}{\text{m}^3} \times 0.5 \, \text{m}^2 \times V_2$$

and
$$V_2 = 4000 \, \text{m}/\text{min}$$

ILLUSTRATIVE PROBLEM 3.7

A hose is 1 in. in diameter and has water whose density is 62.4 lb$_\text{m}$/ft³ flowing steadily in it at a velocity of 100 ft/s. Determine the mass flow of water in the hose (see Figure 3.8).

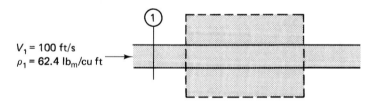

$V_1 = 100$ ft/s
$\rho_1 = 62.4$ lb$_\text{m}$/cu ft

FIGURE 3.8   Illustrative Problem 3.7.

SOLUTION

From the continuity equation,

$$\dot{m} = \rho A V = 62.4 \, \frac{\text{lb}_\text{m}}{\text{ft}^3} \times \frac{\pi (1 \, \text{in.})^2}{4 \times 144 \, \dfrac{\text{in}^2}{\text{ft}^2}} \times 100 \, \frac{\text{ft}}{\text{s}} = 34.0 \, \text{lb}_\text{m}/\text{s}$$

Note that the dimension of the area has been used in square feet for dimensional consistency.

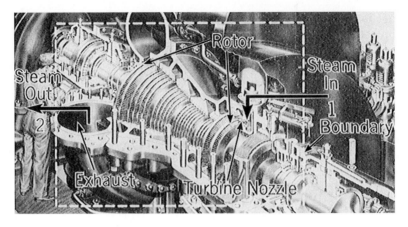

FIGURE 3.9 Steam turbine flow system.
[From V.M. Faires, Elementary Thermodynamics *(New York: Macmillan, Inc., 1957), p. 27, with permission.]*

Figure 3.9 shows a steam turbine as a flow system. Our present interest is in the flow of steam into and out of the noted boundaries of this system. The actions and interactions within the boundaries of this system will be studied later, but for us at this time, the boundaries can be thought of as defining a "black box." Only the fluid (steam) interfaces are considered. The continuity equation is applied to the inlet and outlet steam flow, as in the following example.

ILLUSTRATIVE PROBLEM 3.8

A steam turbine has an inlet steam flow of 50,000 lb_m/hr of steam whose specific volume is 0.831 ft^3/lb_m. The inlet diameter is 6 in. At the outlet, the pipe diameter is 8 in., and the specific volume of the steam is 1.825 ft^3/lb_m. Determine the velocity at inlet and outlet of the turbine in ft/s.

SOLUTION

Referring to Figure 3.9, we have at the inlet $\dot{m}_1 = 50{,}000$ lb_m/hr and $v_1 = 0.831$ ft^3/lb_m. Therefore,

$$\dot{m}_1 = \frac{A_1 V_1}{v_1}$$

$$V_1 = \frac{\dot{m}_1 v_1}{A_1} = \frac{50{,}000\,\dfrac{lb_m}{hr}}{60\,\dfrac{min}{hr} \times 60\,\dfrac{s}{min}} \times \frac{0.831\,\dfrac{ft^3}{lb_m}}{\dfrac{\pi(6\ in.)^2}{4(144\ in.^2/ft^2)}} = 58.8\ ft/s$$

At the outlet, the value of $\dot{m}_2$ is the same, that is, 50,000 lb_m/hr. Therefore,

$$V_2 = \frac{\dot{m}_2 v_2}{A_2} = \frac{50{,}000 \, \dfrac{\text{lb}_m}{\text{hr}}}{60 \, \dfrac{\text{min}}{\text{hr}} \times 60 \, \dfrac{\text{s}}{\text{min}}} \times \frac{1.825 \, \text{ft}^3/\text{lb}_m}{\dfrac{\pi}{4}\left(\dfrac{8 \, \text{in.}}{12 \, \text{in.}/\text{ft}}\right)^2}$$

$$V_2 = 72.6 \, \text{ft/s}$$

Again note the conversion factors of $(12 \text{ in.})^2/\text{ft}^2$ and 60 min/hr $\times$ 60 s/min to obtain the desired units of velocity.

ILLUSTRATIVE PROBLEM 3.9

A steam turbine has an inlet steam flow of 10^4 kg/hr whose specific volume is 0.05 m³/kg. The inlet diameter is 100 mm, and the outlet diameter is 200 mm. If the outlet specific volume is 0.10 m³/kg, determine the inlet and outlet velocities.

SOLUTION

$$\dot{m}_1 = \dot{m}_2 = 10^4 \, \text{kg/hr}$$

Because $\dot{m} = AV/v$, $V = \dot{m}v/A$. Therefore,

$$V_1 = \frac{\dot{m}_1 v_1}{A_1} = \frac{10^4 \, \text{kg/hr}}{(60 \times 60) \, \text{s/hr}} \times \frac{0.05 \, \dfrac{\text{m}^3}{\text{kg}}}{\dfrac{\pi}{4}(0.1)^2 \, \text{m}^2} = 17.68 \, \text{m/s}$$

At the outlet,

$$V_2 = \frac{\dot{m}_2 v_2}{A_2} = \frac{10^4 \, \text{kg/hr}}{(60 \times 60) \, \text{s/hr}} \times \frac{0.10 \, \dfrac{\text{m}^3}{\text{kg}}}{\dfrac{\pi}{4}(0.2)^2 \, \text{m}^2} = 8.84 \, \text{m/s}$$

3.4b The Steady-Flow Energy Equation

Figure 3.10 shows a steady-flow system in which it is assumed that each form of energy can enter and leave the system. At the entrance, $\dot{m}$ lb of fluid per second enters and the same amount leaves at the exit. At the entrance, the fluid has a pressure of p_1, a specific volume of v_1, an internal energy of u_1, and a velocity of V_1. At the exit, we have similar quantities expressed as p_2, v_2, and V_2. The fluid enters and leaves at different elevations, and work and heat cross the boundary in both directions.

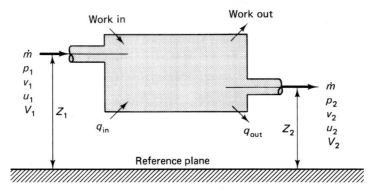

FIGURE 3.10 Steady-flow system.

In applying the first law to a system in which both mass and energy can cross the boundaries, it is necessary to adhere to the mathematical conventions chosen for positive and negative quantities of energy. It is also imperative that all pertinent energy terms be included in any analysis. To summarize, Table 3.1 identifies six energy terms that apply to various situations for steady-flow systems.

We express the first law for the system shown in Figure 3.10 by stating that all the energy entering the system must equal all the energy leaving the system. This energy balance is set up in Table 3.2. By equating all the terms in Table 3.2 in English units, we obtain

$$
\frac{Z_1}{J}\left(\frac{g}{g_c}\right) + \frac{V_1^2}{2g_cJ} + u_1 + \frac{p_1v_1}{J} + \frac{\mathbf{w}_{\text{in}}}{J} + q_{\text{in}}
$$
$$
= \frac{Z_2}{J}\left(\frac{g}{g_c}\right) + \frac{V_2^2}{2g_cJ} + u_2 + \frac{p_2v_2}{J} + \frac{\mathbf{w}_{\text{out}}}{J} + q_{\text{out}} \qquad \textbf{(3.10)}
$$

TABLE 3.1
Energy Terms

Item	Value	
	Btu/unit mass	kJ/unit mass
Potential energy	$\dfrac{Zg}{Jg_c}$	Zg
Kinetic energy	$\dfrac{V^2}{2g_cJ}$	$\dfrac{V^2}{2}$
Internal energy	u	u
Flow work	$\dfrac{pv}{J}$	pv
Work	$\dfrac{\mathbf{w}}{J}$	$\mathbf{w}$
Heat	q	q

TABLE 3.2

Energy Balance

	Energy in		Energy out	
	Btu/lb$_m$	kJ/kg	Btu/lb$_m$	kJ/kg
Potential energy	$\dfrac{Z_1 g}{J g_c}$	$Z_1 g$	$\dfrac{Z_2 g}{J g_c}$	$Z_2 g$
Kinetic energy	$\dfrac{V_1^2}{2 g_c J}$	$\dfrac{V_1^2}{2}$	$\dfrac{V_2^2}{2 g_c J}$	$\dfrac{V_2^2}{2}$
Internal energy	u_1	u_1	u_2	u_2
Flow work	$\dfrac{p_1 v_1}{J}$	$p_1 v_1$	$\dfrac{p_2 v_2}{J}$	$p_2 v_2$
Work	$\dfrac{\mathbf{w}_{in}}{J}$	$\mathbf{w}_{in}$	$\dfrac{\mathbf{w}_{out}}{J}$	$\mathbf{w}_{out}$
Heat	q_{in}	q_{in}	q_{out}	q_{out}

In SI units,

$$Z_1 g + \frac{V_1^2}{2} + u_1 + p_1 v_1 + \mathbf{w}_{in} + q_{in}$$
$$= Z_2 g + \frac{V_2^2}{2} + u_2 + p_2 v_2 + \mathbf{w}_{out} + q_{out} \tag{3.10a}$$

Equation (3.10) is quite general and expresses the first law for this system. It is sometimes called the *steady-flow energy* or the *general energy equation*. We note that it is also quite proper to call it the *energy equation* applied to a steady-flow system.

If at this point we note that both the heat and the work terms can be combined to form individual terms of net heat and net work, and being careful of the mathematical signs of these net terms, we can write for English units,

$$\frac{Z_1}{J}\left(\frac{g}{g_c}\right) + \frac{V_1^2}{2 g_c J} + u_1 + \frac{p_1 v_1}{J} + q = \frac{Z_2}{J}\left(\frac{g}{g_c}\right) + \frac{V_2^2}{2 g_c J} + u_2 + \frac{p_2 v_2}{J} + \frac{\mathbf{w}}{J} \tag{3.11}$$

In SI units,

$$Z_1 g + \frac{V_2^2}{2} + u_1 + p_1 v_1 + q = Z_2 g + \frac{V_2^2}{2} + u_2 + p_2 v_2 + \mathbf{w} \tag{3.11a}$$

In Equation (3.11) all terms are written as Btu/lb$_m$, and q and $\mathbf{w}$ represent *net values* per unit mass of working fluid. If we now note further that the grouping of terms $u + pv/J$ appears on both sides of Equation (3.11), and that this combined term is a property that we have already called enthalpy, then from the definition of the term *enthalpy* we have

$$h = u + \frac{pv}{J} \tag{3.5}$$

(or $h = u + pv$ in SI units). Using Equations (3.5) and (3.11) yields, for English units,

$$\frac{Z_1}{J}\left(\frac{g}{g_c}\right) + \frac{V_1^2}{2g_cJ} + h_1 + q = \frac{Z_2}{J}\left(\frac{g}{g_c}\right) + \frac{V_2^2}{2g_cJ} + h_2 + \frac{w}{J} \tag{3.12}$$

By regrouping the terms in Equation (3.12), we obtain

$$q - \frac{w}{J} = (h_2 - h_1) + \frac{Z_2 - Z_1}{J}\left(\frac{g}{g_c}\right) + \frac{V_2^2 - V_1^2}{2g_cJ} \tag{3.13}$$

where each quantity is in Btu/lb$_m$.
  In terms of SI units,

$$q - w = h_2 - h_1 + g(Z_2 - Z_1) + \frac{V_2^2 - V_1^2}{2} \tag{3.13a}$$

where each quantity is in terms of J/kg or kJ/kg.
  Note that when Z is in meters and V is in meters per second, the potential energy and kinetic energy terms will be in J/kg, the equivalent of m^2/s^2. Heat, work, and enthalpy per unit mass are usually given in kJ/kg. Thus, a factor of 1000 is often needed to make all the terms in Equation (3.13a) compatible.
  The student should note that Equations (3.10) through (3.13) are essentially the same. To apply these equations intelligently, it is important that each of the terms be completely understood. Although they are not difficult, the student may encounter trouble at this point because of a lack of understanding of these energy equations and the basis for each of the terms in them.
  Before illustrating the use of these energy equations, let us consider the following situation. A fluid is flowing steadily in a device in which it undergoes a compression. Let us further assume that this process is frictionless and quasi-static. Its work term can be written as the sum of the work done on the fluid plus the flow work. Kinetic and potential energy terms are assumed to be negligible. Thus, the work for this process becomes

$$w = p_1v_1 - p_2v_2 - p\Delta v \tag{3.14}$$

Let us now plot a pv diagram for the fluid, as shown in Figure 3.11. As shown in Figure 3.11a, the term $p\Delta v$ is the area under the curve between the limits 1 and 2. Subtracting p_1v_1 and adding p_2v_2 (graphically) results in $v\Delta p$ and is shown on Figure 3.11b. Mathematically,

$$-v\Delta p = p_2v_2 - p_1v_1 + p\Delta v \tag{3.15}$$

It will be recalled that for the quasi-static, frictionless, nonflow system, $p\Delta v$ evaluated the work done on the working fluid. For the flow system, flow work of an amount of p_1v_1 enters the system,

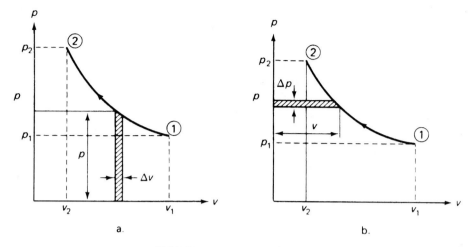

FIGURE 3.11 A p–v diagram.

and flow work of an amount of p_2v_2 leaves the system. We can therefore interpret the meaning of the areas on Figure 3.11 or the terms of Equation (3.15) as follows: The work of the quasi-static, frictionless flow system (neglecting the kinetic and potential energy terms) is the algebraic sum of the work to induct the fluid into the system plus the work of compression (or expansion) less the flow work to deliver the fluid to the downstream exit. It is sometimes convenient to think of $p\Delta v$ as the work of *compression* (or expansion) and $v\Delta p$ as the work of the *compressor* (or expander). It is extremely important to note that both $p\Delta v$ and $v\Delta p$ evaluate work only for a quasi-static process or system. For systems that are real, they do not evaluate work. However, for a large class of systems, they approximate the work terms, and their use is justified both from this approximation and the relative ease with which they can be evaluated.

CALCULUS ENRICHMENT

From the general energy equation [Equations (3.13) and (3.13a)], we can write,

$$\delta q - \delta w = dh + d(\text{K.E.}) + d(\text{P.E.}) \tag{a}$$

where δ (delta) denotes an inexact quantity, that is, one that is a function of the path. Because h, enthalpy, is defined to equal $u + pv$,

$$h = u + pv; \, dh = du + p\,dv + v\,dp \tag{b}$$

Substitute (b) into (a), and

$$\delta q - \delta w = du + p\,dv + v\,dp + d(\text{K.E.}) + d(\text{P.E.}) \tag{c}$$

From the energy equation for the nonflow system,

$$\delta q - \delta u = p\,dv \tag{d}$$

Using (d) and (c),

$$\delta q - \delta w = du + \delta q - \delta u + v\,dp + d(\text{K.E.}) + d(\text{P.E.}) \tag{e}$$

and

$$\delta w = -v \, dp \qquad \textbf{(f)}$$

if the K.E. and P.E. terms are negligible.

From (f), we conclude that the work of a quasi-static flow process is the area *behind* the p–v curve.

3.4c The Bernoulli Equation

In fluid mechanics, use is sometimes made of the *Bernoulli equation.* Because of the frequent mis-application of this equation, a brief discussion of it will now be undertaken. Let us consider a system in which the flow is steady, there is no change in internal energy, no work is done on or by the system, no energy as heat crosses the boundaries of the system, and the fluid is incompressible. We further assume that all processes are ideal in the sense that they are frictionless. For this system, the energy equation reduces to the following in mechanical units of foot pounds per pound mass:

$$Z_1 \frac{g}{g_c} + \frac{V_1^2}{2g_c} + p_1 v_1 = Z_2 \frac{g}{g_c} + \frac{V_2^2}{2g_c} + p_2 v_2 \qquad \textbf{(3.16)}$$

or in SI units,

$$Z_1 g + \frac{V_1^2}{2} + p_1 v_1 = Z_2 g + \frac{V_2^2}{2} + p_2 v_2 \qquad \textbf{(3.16a)}$$

Equation (3.16) is the usual form of the Bernoulli equation found in elementary physics and fluid mechanics texts. Each term can represent a height, because the dimension of foot pounds per pound corresponds numerically (and dimensionally) to a height. Hence, the term *head* is frequently used to denote each of the terms in Equation (3.16). From the derivation of this equation, we note that its use is restricted to situations in which the flow is steady, there is no friction, no shaft work is done on or by the fluid, the flow is incompressible, and there is no change in internal energy during the process. These restrictions are severe, and only under the simplest situations can we hope to apply the Bernoulli equation successfully. It becomes even more difficult to justify the procedure of adding heat and work terms to this equation while maintaining all the other restrictions. The student is cautioned against use of this equation without a thorough understanding of its restrictions. In every case, it is preferable to write the complete and correct form of the energy equation first and then to make those assumptions that can be justified for each problem.

3.4d Specific Heat

The term *specific heat* is defined as the ratio of energy as *heat* transferred during a particular process per unit mass of fluid involved divided by the corresponding change of temperature of the fluid that occurs during this process. Because heat can be transferred to or from a fluid and algebraic signs have been adopted for the direction of heat transfer, it is entirely possible for a process to have a negative specific heat. The student should not confuse the specific heat of a process with the specific heat property.

This definition of specific heat is important for two processes, because they serve to define a new property of a fluid. These processes are the constant-pressure process (flow or nonflow) and the constant-volume process. Recall from earlier work that the constant-pressure, nonflow process is characterized by the energy equation that $q = h_2 - h_1 = \Delta h$. For the steady-flow process without change in elevation, in the absence of external work, and with negligible changes in kinetic energy, the steady-flow energy equation leads to the same result. Thus, for *any constant-pressure process* (flow or nonflow with the conditions noted), the specific heat is defined as

$$c_p = \left(\frac{q}{\Delta T}\right)_p = \left(\frac{\Delta h}{\Delta T}\right)_p \qquad (3.17)$$

or

$$(c_p \Delta T = \Delta h)_p \qquad (3.17a)$$

where the subscript p indicates a constant-pressure process.

For the constant-volume process (which can only be a nonflow process), $q = u_2 - u_1 = \Delta u$. Therefore,

$$c_v = \left(\frac{q}{\Delta T}\right)_v = \left(\frac{\Delta u}{\Delta T}\right)_v \qquad (3.18)$$

or

$$(c_v \Delta T = \Delta u)_v \qquad (3.18a)$$

where the subscript v indicates a constant-volume process.*

These specific heats are the properties of the fluid and depend only on the state of the fluid. Typical units of specific heat are Btu/lb$_m$·°R in English units and kJ/kg·°K in SI units. A further discussion of specific heats is given in Chapter 6, where the internal energy and enthalpy of the ideal gas are functions of temperature only, making the definition of Equations (3.17) and (3.18) general for the ideal gas and not restricted to only constant-pressure or constant-volume processes.

ILLUSTRATIVE PROBLEM 3.10

A gas initially at 100°F and having $c_p = 0.22$ Btu/lb$_m$·°R and $c_v = 0.17$ Btu/lb$_m$·°R is placed within a cylinder. If 800 Btu as heat is added to 10 lb$_m$ of the gas in a nonflow, constant-pressure process, determine the final gas temperature. Also, determine the work done on or by the gas. Refer to Figure 3.12.

---

* Equations (3.17) and (3.18) are more correctly written mathematically as

$$c_p \equiv \left(\frac{\partial h}{\partial t}\right)_p \quad \text{and} \quad c_v \equiv \left(\frac{\partial u}{\partial t}\right)_v$$

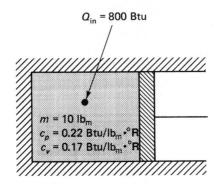

FIGURE 3.12 Illustrative Problem 3.10.

SOLUTION

For the nonflow, constant-pressure process, $q = \Delta h = h_2 - h_1$. However, Equation (3.17) allows us to evaluate Δh in terms of temperature and the specific heat at constant pressure. Thus,

$$q = h_2 - h_1 = c_p(T_2 - T_1)$$

Using the data given and q on a unit mass basis,

$$q = \frac{800 \text{ Btu}}{10 \text{ lb}_\text{m}} = 0.22 \, (T_2 - T_1)$$

so $T_2 - T_1 = 363.6°\text{F}$. Because the initial temperature is $100°\text{F}$, the final temperature is $363.6°\text{F} + 100 = 463.6°\text{F}$.

To obtain the work, we apply Equation (3.2) to the nonflow process:

$$q - \text{w} = u_2 - u_1$$

The term $u_2 - u_1$ can be evaluated in terms of temperature and the specific heat at constant volume from Equation (3.18). Using this, we obtain

$$-\text{w} = (u_2 - u_1) - q = c_v(T_2 - T_1) - q$$

Using q on a unit mass basis and noting that $T_2 - T_1 = 363.6°\text{F}$,

$$-\text{w} = 0.17 \text{ Btu/lb}_\text{m}\cdot°\text{R}(363.6°\text{R}) - \frac{800 \text{ Btu}}{10 \text{ lb}_\text{m}}$$

$$-\text{w} = 61.8 - 80 = -18.2 \text{ Btu/lb}_\text{m}$$

or

$$\text{w} = 18.2 \text{ Btu/lb}_\text{m}$$

Thus, $18.2 \text{ Btu/lb}_\text{m}$ or 182 Btu as work is taken out of the system due to work done by the gas, because there is 10 lb_m in the system.

3.5 APPLICATIONS OF THE FIRST LAW OF THERMODYNAMICS

At this point in our study, we will apply the first law to several steady-flow situations that we will have applications for in later sections of this book. For the present, we restrict ourselves to seven steady-flow processes:

1. The steam or gas turbine.

2. Pipe flow.

3. The boiler.

4. The flow in nozzles.

5. The throttling process.

6. The heat exchanger.

7. Filling a tank.

3.5a The Turbine

As our first illustration of the applicability of the first law to steady-flow processes, let us consider the steam turbine shown in Figure 3.13. In this device, steam enters and expands in fixed nozzles to a high velocity. The high-velocity steam is then directed over the turbine blades, where it does work on the turbine wheel. The steam then exhausts from the turbine. The purpose of this machine is to obtain shaft work, and certain features about it should be noted. First, the shaft of the turbine is horizontal. Second, as the steam expands, its specific volume increases (see Chapter 5), and to keep the exit velocity nearly equal to the entering velocity, the exit pipe area is proportionally greater than the inlet pipe area. The turbine is suitably insulated to minimize heat losses to the surroundings and also to eliminate the possibility of injuring operating personnel working in the vicinity of the hot turbine casing.

Figure 3.14 shows the turbine generator at the E.W. Stout plant of Indianapolis, Indiana. Whether we are dealing with a single-stage turbine as in Figure 3.13 or a large central station as in Figure 3.14, our overall analysis will be essentially the same.

With the foregoing in mind, let us now apply the first law to the turbine (either steam or gas) shown schematically in Figure 3.15. The first law is given by

$$q - \frac{w}{J} = h_2 - h_1 + \frac{Z_2 - Z_1}{J}\left(\frac{g}{g_c}\right) + \frac{V_2^2 - V_1^2}{2g_c J} \tag{3.19}$$

or

$$q - w = h_2 - h_1 + g(Z_2 - Z_1) + \frac{V_2^2 - V_1^2}{2} \tag{3.19a}$$

Because the shaft of the machine is horizontal, $[(Z_2 - Z_1)/J]\,(g/g_c)$ can be taken to be zero; that is,

$$\left(\frac{Z_2 - Z_1}{J}\right)^{\!\!\!\diagup 0}\left(\frac{g}{g_c}\right)$$

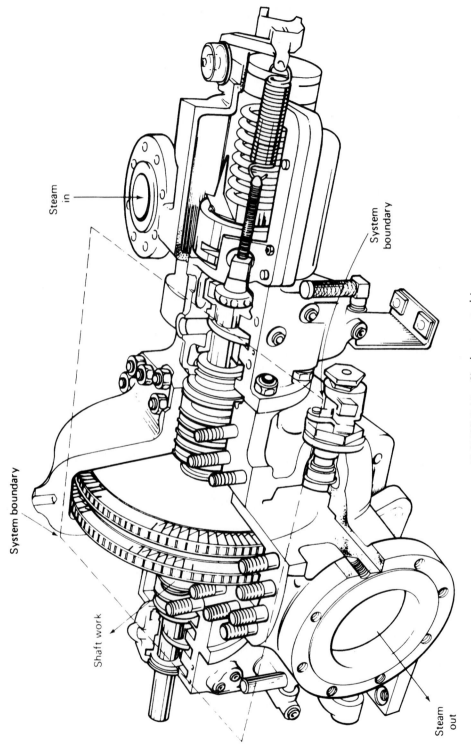

Steam in

System boundary

System boundary

Shaft work

Steam out

FIGURE 3.13 Single-stage turbine. *(Courtesy of Worthington Corp.)*

FIGURE 3.14 Turbine generator, E.W. Stout plant, Indianapolis, Indiana. *(Courtesy of the Indianapolis Power and Light Company.)*

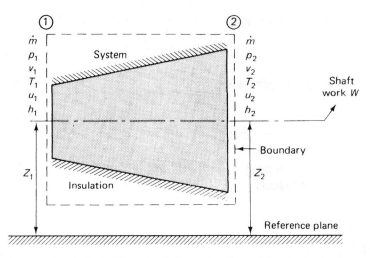

FIGURE 3.15 Schematic of a turbine.

where $\nearrow^0$ is taken to mean goes to zero. Also as noted, the inlet and outlet velocities are kept nearly equal, leading us to conclude that the kinetic energy difference term goes to zero:

$$\left(\frac{V_2^2 - V_1^2}{2g_c J}\right)^{\!\!\nearrow^0}$$

Finally, the insulation of the turbine would effectively prevent heat losses to the surroundings:

$$q^{\nearrow^0}$$

FIGURE 3.16 Industrial gas turbine.
(Courtesy of Pratt and Whitney Aircraft Group, United Technologies Corp.)

Equation (3.19), as applied to this device, becomes

$$\frac{\mathbf{w}}{J} = h_1 - h_2$$

or from Equation (3.19a),

$$\mathbf{w} = h_1 - h_2$$

All the assumptions made for the steam turbine are equally applicable to the gas turbine shown in Figure 3.16. In this industrial unit, the turbine also drives the compressor. Exhaust gases leave at the right of the figure via an exhaust elbow. This unit will be discussed in detail later on as a prime mover in a power-producing cycle.

ILLUSTRATIVE PROBLEM 3.11

A large steam turbine operates with a steam flow rate of 150,000 lb_m/hr. The conditions at inlet and outlet are tabulated below. Determine the power produced (a) if heat losses are negligible and (b) if heat losses equal 50,000 Btu/hr.

	Inlet	Outlet
Pressure (psia)	1000	1
Temperature (°F)	1000	101.74
Velocity (ft/s)	125	430
Inlet position (ft)	+ 10	0
Enthalpy (Btu/lb$_m$)	1505.4	940.0

SOLUTION

Let us start this problem by referring to Figure 3.15 and using the form of energy equation given by Equation (3.12),

$$\frac{Z_1}{J}\left(\frac{g}{g_c}\right) + \frac{V_1^2}{2g_cJ} + h_1 + q = \frac{Z_2}{J}\left(\frac{g}{g_c}\right) + \frac{V_2^2}{2g_cJ} + h_2 + \frac{\mathbf{w}}{J}$$

where all terms are per unit mass.
(a) $q = 0$, and assuming that $g = g_c$,

$$\frac{10}{778} + \frac{(125)^2}{2 \times 32.17 \times 778} + 1505.4 + 0$$

$$= 0 + \frac{(430)^2}{2 \times 32.17 \times 778} + 940.0 + \frac{\mathbf{w}}{J} \quad \textbf{(a)}$$

$$0.013 + 0.312 + 1505.4 + 0 = 0 + 3.694 + 940.0 + \frac{\mathbf{w}}{J} \quad \textbf{(b)}$$

or

$$\frac{\mathbf{w}}{J} = 562.03 \text{ Btu}/lb_m$$

(Had we neglected potential and kinetic terms, $w/J = 565.4$ Btu/lb$_m$, a difference of 0.6%, which is quite small.)

The total work of the turbine is 150,000 lb$_m$/hr $\times$ 562.03 Btu/lb$_m$ = 84,304,500 Btu/hr. In terms of horsepower,

$$\frac{84{,}304{,}500 \text{ Btu/hr} \times 778 \text{ ft·lb}_f/\text{Btu}}{60 \text{ min/hr} \times 33{,}000 \text{ ft·lb}_f/\text{hp}} = 33{,}125.7 \text{ hp}$$

In terms of kilowatts,

$$33{,}125.7 \text{ hp} \times 0.746 \text{ kW/hp} = 24{,}711.8 \text{ kW}$$

(b)

$$\frac{50{,}000 \text{ Btu/hr}}{150{,}000 \text{ lb}_m/\text{hr}} = 0.333 \text{ Btu/lb}_m \text{ (heat loss)}$$

Using this with Equation (b), and noting that it is negative by convention, yields

$$0.013 + 0.312 + 1505.4 - 0.333 = 0 + 3.694 + 940.0 + \frac{\mathbf{w}}{J}$$

or

$$\frac{\mathbf{w}}{J} = 561.70 \text{ Btu/lb}_m$$

which differs from part (a) by 0.06%. For all practical purposes, we could have neglected the potential, kinetic, and heat loss terms.

ILLUSTRATIVE PROBLEM 3.12

A steam turbine operates with an inlet velocity of 40 m/s and an inlet enthalpy of 3433.8 kJ/kg. At the outlet, which is 2 m lower than the inlet, the enthalpy is 2675.5 kJ/kg and the velocity is 162 m/s. If the heat loss from the turbine is 1 kJ/kg, determine the work output per kilogram.

SOLUTION

If we refer to Illustrative Problem 3.11, it will be noted that these problems are similar except for units. Using Equation (3.11a) yields

$$Z_1 g + \frac{V_1^2}{2} + h_1 + q = Z_2 g + \frac{V_2^2}{2} + h_2 + \mathbf{w}$$

For the units of this problem,

$$\frac{2 \text{ m} \times 9.81 \text{ m/s}^2}{1000 \text{ J/kJ}} + \frac{(40 \text{ m/s})^2}{2 \times 1000 \text{ J/kJ}} + 3433.8 \frac{\text{kJ}}{\text{kg}} - 1 \frac{\text{kJ}}{\text{kg}}$$

$$= 0 + \frac{(162 \text{ m/s})^2}{2 \times 1000 \text{ J/kJ}} + 2675.5 \frac{\text{kJ}}{\text{kg}} + \mathbf{w}$$

$$0.02 \, \frac{\text{kJ}}{\text{kg}} + 0.8 \, \frac{\text{kJ}}{\text{kg}} + 3433.8 \, \frac{\text{kJ}}{\text{kg}} - 1 \, \frac{\text{kJ}}{\text{kg}} = 0 + 13.12 \, \frac{\text{kJ}}{\text{kg}} + 2675.5 \, \frac{\text{kJ}}{\text{kg}} + \mathbf{w}$$

or

$$\mathbf{w} = 745.0 \, \frac{\text{kJ}}{\text{kg}}$$

ILLUSTRATIVE PROBLEM 3.13

A turbine (gas) receives air at 150 psia and 1000°R and discharges to a pressure of 15 psia. The actual temperature at discharge is 600°R. If c_p of the gas can be taken to be constant over this temperature range and equal to 0.24 Btu/lb$_m$·°R, determine the work output of the turbine per pound of working fluid. At inlet conditions, the specific volume is 2.47 ft³/lb$_m$, and at outlet conditions, it is 14.8 ft³/lb$_m$. The data are also shown in Figure 3.17.

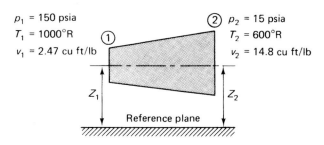

FIGURE 3.17   Illustrative Problem 3.13.

SOLUTION

We have already discussed the turbine in detail and concluded that the first law yields

$$\frac{\mathbf{w}}{J} = h_1 - h_2$$

Equation (3.17) enables us to obtain $h_1 - h_2$, because for the gas,

$$h_1 - h_2 = c_p(T_1 - T_2)$$

Therefore,

$$\frac{\mathbf{w}}{J} = h_1 - h_2 = c_p(T_1 - T_2) = (0.24)(1000 - 600)$$

and

$$\frac{\mathbf{w}}{J} = 96 \, \text{Btu/lb}_m$$

Note that the specific volume and pressure given do not enter the solution of the problem. However, a pressure differential is required to cause the gas to flow.

ILLUSTRATIVE PROBLEM 3.14

Even though the turbine of Illustrative Problem 3.13 may be well insulated, there will be some heat loss. If the heat loss is found by experiment to be equal to 1.1 Btu/lb$_m$ of gas, determine the work output of the turbine per pound mass of gas.

SOLUTION

We must again return to the point in our discussion where we took q^0 and not make this assumption. Doing this gives us

$$q - \frac{w}{J} = h_2 - h_1$$

or

$$\frac{w}{J} = q + (h_1 - h_2)$$

Because q is out of the system, it is a negative quantity, and $h_1 - h_2$ is the same as for Illustrative Problem 3.13. Thus,

$$\frac{w}{J} = -1.1 + 96.0 = 94.9 \text{ Btu/lb}_m$$

In other words, a heat loss decreases the work output of the turbine, as we have seen earlier.

3.5b Pipe Flow

As our next illustration of the first law, we will apply the law to the flow of fluids in pipes. Figure 3.18 shows a large chemical processing plant, and the amount and complexity of the piping in installations of this type is self-evident.

FIGURE 3.18 Large chemical processing plant.
(Courtesy of Foster Wheeler Corp.)

A gas flows in a pipe whose pressure and temperature at one section are 100 psia and 950°F. At a second section of the pipe, the pressure is 76 psia and the temperature is 580°F. The specific volume of the inlet gas is 4.0 ft³/lb$_m$, and at the second section, it is 3.86 ft³/lb$_m$. Assume that the specific heat at constant volume is 0.32 Btu/lb·°R. If no shaft work is done and the velocities are small, determine the magnitude and direction of the heat transfer. Assume the pipe to be horizontal, and neglect velocity terms. The data are also shown in Figure 3.19.

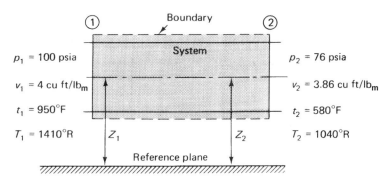

FIGURE 3.19 Illustrative Problem 3.15.

SOLUTION

We first write out the equation of the first law and then apply it to this process:

$$\frac{Z_1}{J}\frac{g}{g_c} + \frac{V_1^2}{2g_cJ} + u_1 + \frac{p_1v_1}{J} + q = \frac{Z_2}{J}\frac{g}{g_c} + \frac{V_2^2}{2g_cJ} + u_2 + \frac{p_2v_2}{J} + \frac{\mathbf{w}}{J}$$

Because the pipe is horizontal and velocity terms are to be neglected,

$$\left(\frac{Z_1 - Z_2}{J}\right)\left(\frac{g}{g_c}\right)^{\!0}, \quad \left(\frac{V_2^2 - V_1^2}{2g_cJ}\right)^{\!0}$$

Also, $\mathbf{w}/J^0$, no work crosses the boundaries of the system. The energy equation is reduced to

$$u_1 + \frac{p_1v_1}{J} + q = u_2 + \frac{p_2v_2}{J}$$

However, Equation (3.18) permits us to express the internal energy change for the gas in terms of the temperature change as

$$u_2 - u_1 = c_v(T_2 - T_1) \qquad \text{for constant } c_v$$

Therefore,

$$q = c_v(T_2 - T_1) + \frac{p_2v_2}{J} - \frac{p_1v_1}{J}$$

By inserting numerical quantities, we obtain

$$q = 0.32(1040 - 1410) + \frac{76 \times 144 \times 3.86}{778} - \frac{100(144)(4)}{778}$$

$$= -118.4 + 54.3 - 74.0 = -138.1 \text{ Btu/lb}_m$$

Thus, 138.1 Btu/lb$_m$ is transferred *from* the gas.

ILLUSTRATIVE PROBLEM 3.16

If the pipe referred to in Illustrative Problem 3.15 was a vertical run of pipe such that section ② was 100 ft above section ① determine the direction and magnitude of the heat transfer.

SOLUTION

Using the reference plane of Figure 3.19 to coincide with the elevation of the pipe at section 1 makes $Z_1 = 0$. The energy equation becomes

$$u_1 + \frac{p_1 v_1}{J} + q = u_2 + \frac{p_2 v_2}{J} + \frac{Z_2}{J}\left(\frac{g}{g_c}\right)$$

and

$$q = u_2 - u_1 + \frac{p_2 v_2}{J} - \frac{p_1 v_1}{J} + \frac{Z_2}{J}\left(\frac{g}{g_c}\right)$$

Using $u_2 - u_1 = c_v(T_2 - T_1)$ and $Z_2 = 100$ ft gives us

$$q = 0.32(1040 - 1410) + \frac{76 \times 144 \times 3.86}{778} - \frac{100\,(144)\,(4)}{778} + \frac{110}{778}\left(\frac{g}{g_c}\right)$$

Letting $g/g_c = 1$ yields

$$q = -118.4 + 54.3 - 74.0 + 0.13 = -138.0 \text{ Btu/lb}_m$$

For this problem, neglecting the elevation term leads to an insignificant error.

3.5c The Boiler

The next application of the first law will be to the boiler or steam generator. The basic purpose of the steam generator is the turning of water into steam by the application of heat. Figure 3.20 shows a large steam-generating unit, which as can be seen, consists of a combination of many elements. In this unit, pulverized coal is burned in the furnace with air that has been preheated in the air heater. In addition to generating superheated steam, this unit reheats steam from the high-pressure turbine exhaust and returns it to the low-pressure turbine. The energy input to this system comes from the fuel, air, and feed water, and the useful output is steam. Because the purpose of the unit is to generate steam, the energy in the stack gas, unburned fuel, and heat transfer to the surroundings all represent losses that decrease the useful steam output. Again, we shall return to the steam

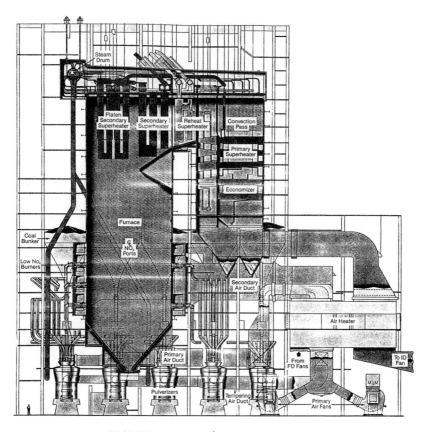

FIGURE 3.20 Modern steam generator.
(Reprinted with permission from STEAM, copyright by The Babcock and Wilcox Company, 1992.)

generator in Chapter 8, where it is studied in some detail. For the present, the following problem will serve to illustrate the application of the first law to this unit.

ILLUSTRATIVE PROBLEM 3.17

A steam boiler is required to produce 10,000 lb_m/hr of superheated steam at 1000°F and 1000 psia ($h = 1505.9$ Btu/lb_m) from feed water supplied at 1000 psia and 100°F ($h = 70.68$ Btu/lb_m). How much energy has been added to the water to convert it to steam at these conditions? The data are shown in Figure 3.21.

SOLUTION

As indicated in Figure 3.21, we can consider this system as a single unit with feed water entering and steam leaving. If well designed, this unit will be thoroughly insulated, and heat losses will be reduced to a negligible amount. Also, no work will be added to the fluid during the time it is passing through the unit, and kinetic energy differences will be assumed to be negligibly small. In large units, the inlet and outlet may be as much as

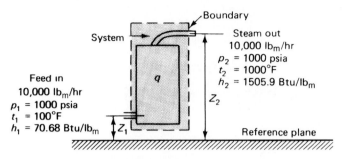

FIGURE 3.21   Illustrative Problem 3.17.

100 ft apart. However, 100/778 is 0.129 Btu/lb$_m$, which is quite small compared to the 1000 Btu/lb$_m$ or more involved in this problem. On this basis, differences in elevation will also be considered negligible. Once again, the energy equation is

$$\frac{Z_1}{J}\left(\frac{g}{g_c}\right) + \frac{V_1^2}{2g_cJ} + u_1 + \frac{p_1v_1}{J} + q = \frac{Z_2}{J}\left(\frac{g}{g_c}\right) + \frac{V_2^2}{2g_cJ} + u_2 + \frac{p_2v_2}{J} + \frac{\mathbf{w}}{J}$$

and for this problem

$$\left(\frac{Z_1 - Z_2}{J}\right)\left(\frac{g}{g_c}\right)^{\!\!0}, \qquad \left(\frac{V_2^2 - V_1^2}{2g_cJ}\right)^{\!\!0}$$

Therefore,

$$u_1 + \frac{p_1v_1}{J} + q = u_2 + \frac{p_2v_2}{J}$$

or because

$$h = u + \frac{pv}{J}$$

$q = h_2 - h_1$ (this is the net value, because we assumed no heat losses)

Using the data given, we have

$$q = (1505.9 - 70.68) = 1435.2 \text{ Btu/lb}_m$$

For 10,000 lb$_m$/hr,

$$10{,}000 \times 1435.2 = 14.35 \times 10^6 \text{ Btu/hr}$$

are required.

   Note that the working fluid (water) is our system and that the solution gives us the energy added to the water. This is not the energy released by combustion.

3.5d Nozzle

A nozzle is a static device that is used to convert the energy of a fluid into kinetic energy. Basically, the fluid enters the nozzle at a high pressure and leaves at a lower pressure. In the process of expanding, velocity is gained as the fluid progresses through the nozzle. No work is done on or by the fluid in its passage through the nozzle.

ILLUSTRATIVE PROBLEM 3.18

Steam is expanded in a nozzle from an initial enthalpy of 1220 Btu/lb$_m$ to a final enthalpy of 1100 Btu/lb$_m$. (a) If the initial velocity of the steam is negligible, what is the final velocity? (b) If the initial velocity is 1000 ft/s, what is the final velocity? See Figure 3.22.

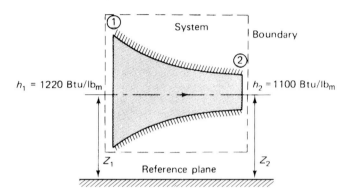

FIGURE 3.22 Illustrative Problem 3.18.

SOLUTION

The energy equation for a steady-flow device is

$$\frac{Z_1}{J}\left(\frac{g}{g_c}\right) + \frac{V_1^2}{2g_cJ} + u_1 + \frac{p_1v_1}{J} + q = \frac{Z_2}{J}\left(\frac{g}{g_c}\right) + \frac{V_2^2}{2g_cJ} + u_2 + \frac{p_2v_2}{J} + \frac{\mathbf{w}}{J}$$

For this device, differences in elevation are negligible. No work is done on or by the fluid, friction is negligible, and due to the speed of the fluid flowing and the short length of the nozzle, heat transfer to or from the surroundings is also negligible. Under these circumstances,

$$\cancel{\left(\frac{Z_2 - Z_1}{J}\right)\left(\frac{g}{g_c}\right)}^0, \quad \cancel{q}^0, \quad \cancel{\left(\frac{\mathbf{w}}{J}\right)}^0$$

Therefore,

$$u_1 + \frac{p_1v_1}{J} + \frac{V_1^2}{2g_cJ} = u_2 + \frac{p_2v_2}{J} + \frac{V_2^2}{2g_cJ}$$

or

$$\boxed{h_1 - h_2 = \frac{V_2^2 - V_1^2}{2g_cJ}}$$

(a) For negligible entering velocity

$$h_1 - h_2 = \frac{V_2^2}{2g_c J}$$

and

$$V_2 = \sqrt{2g_c J(h_1 - h_2)}$$

Substituting the data of the problem gives us

$$V_2 = \sqrt{2 \times 32.17 \ \text{lb}_m \cdot \text{ft}/\text{lb}_f \cdot \text{s}^2 \times 778 \ \text{ft} \cdot \text{lb}_f/\text{Btu} \times (1220 - 1100)\frac{\text{Btu}}{\text{lb}_m}}$$

$$= 2451 \ \text{ft/s}$$

(b) If the initial velocity is appreciable,

$$h_1 - h_2 + \frac{V_1^2}{2g_c J} = \frac{V_2^2}{2g_c J}$$

Again, inserting numerical values yields

$$(1220 - 1100)\frac{\text{Btu}}{\text{lb}_m} + \frac{(1000)^2\left(\dfrac{\text{ft}}{\text{s}}\right)^2}{2 \times 32.17 \ \text{lb}_m \cdot \text{ft}/\text{lb}_f \cdot \text{s}^2 \times 778 \ \dfrac{\text{ft} \ \text{lb}_f}{\text{Btu}}}$$

$$= \frac{V_2^2\left(\dfrac{\text{ft}}{\text{s}}\right)^2}{2 \times 32.17 \ \text{lb}_m \cdot \text{ft}/\text{lb}_f \cdot \text{s}^2 \times 778 \ \dfrac{\text{ft} \cdot \text{lb}_f}{\text{Btu}}}$$

$$120 + 19.98 = \frac{V_2^2}{2 \times 32.17 \times 778}$$

$$V_2 = 2647 \ \text{ft/s}$$

Note that in this part of the problem, the entering velocity was nearly 40% of the final velocity, yet neglecting the entering velocity makes approximately a $7\frac{1}{2}\%$ error in the answer. It is quite common to neglect the entering velocity in many of these problems.

ILLUSTRATIVE PROBLEM 3.19

Assume steam enters a nozzle with an enthalpy of 3450 kJ/kg and leaves with an enthalpy of 2800 kJ/kg. If the initial velocity of the steam is negligible, what is the final velocity?

SOLUTION

Refer to Illustrative Problem 3.18, and note that in SI units,

$$\frac{V_2^2}{2} = h_1 - h_2$$

and

$$V_2 = \sqrt{2(h_1 - h_2)}$$

Substitution yields

$$V_2 = \sqrt{2 \times 1000 \text{ J/kJ} (3450 - 2800)} = 1140.2 \text{ m/s}$$

(Note the need for 1000 J/kJ to obtain m/s.)

3.5e The Throttling Process

The next device that we consider is the case of an obstruction placed in a pipe (deliberately or due to the presence of a valve). One deliberate local obstruction that is often placed in a pipe is an orifice that is used to meter the quantity of fluid flowing. When used in this manner, the orifice usually consists of a thin plate inserted into a pipe and clamped between flanges. The hole in the orifice plate is concentric with the pipe, and static pressure taps are provided upstream and downstream of the orifice. Due to the presence of the orifice, the flow is locally constricted, and a measurable drop in static pressure occurs across the orifice. Static taps placed as shown in Figure 3.23 are usually used to measure the flow. The advantages of this device are its relatively small size, the ease of installation in a pipe, and the fact that standard installations can be used without the need for calibration. However, the orifice meter behaves in the same manner as a partly open valve and causes a relatively high pressure drop. This effect leads to the descriptive term of *throttling* for partly open valves, orifices, or other obstructions in pipes. In effect, the full flow is throttled back to some lesser flow by the obstruction.

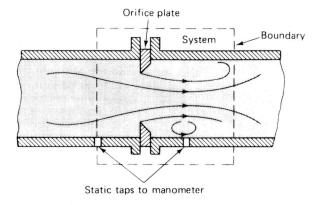

FIGURE 3.23 Orifice as a meter.

A fluid is flowing in a pipe. At some section of the pipe, there is an obstruction that causes an appreciable local pressure loss. Derive the energy equation for this process after flow has become uniform in the downstream section of the pipe. As noted earlier, this process is known as a throttling process and is characteristic of valves and orifices placed in pipelines. Refer to Figure 3.24.

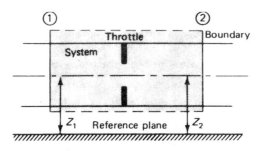

FIGURE 3.24 Illustrative Problem 3.20.

SOLUTION

For the conditions of this problem, we can take differences in elevation to be negligible, differences in kinetic energy terms to be relatively small, and no work or heat to cross the system boundary. The complete energy equation is

$$\frac{Z_1}{J}\left(\frac{g}{g_c}\right) + \frac{V_1^2}{2g_cJ} + u_1 + \frac{p_1v_1}{J} + q = \frac{Z_2}{J}\left(\frac{g}{g_c}\right) + \frac{V_2^2}{2g_cJ} + u_2 + \frac{p_2v_2}{J} + \frac{\mathbf{W}}{J}$$

and, with the assumption made for this process,

$$u_1 + \frac{p_1v_1}{J} = u_2 + \frac{p_2v_2}{J}$$

or

$$\boxed{h_1 = h_2}$$

In words, a throttling process is carried out at constant enthalpy. One assumption should be verified; that is, the kinetic energy differences at inlet and outlet are indeed negligible. This can easily be done by using the constant-enthalpy condition and using the final properties of the fluid and the cross-sectional area of the duct to determine an approximate final velocity. The throttling process is discussed again in Chapter 5, where we will find a very useful application of throttling.

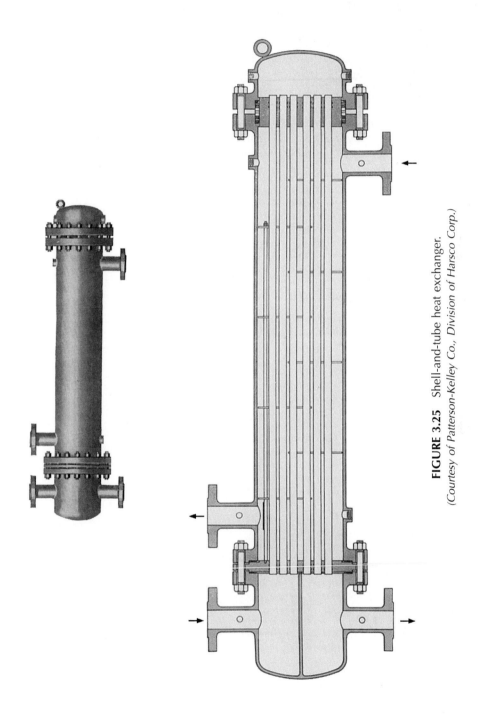

FIGURE 3.25 Shell-and-tube heat exchanger.
(Courtesy of Patterson-Kelley Co., Division of Harsco Corp.)

3.5f The Heat Exchanger

The heat exchanger, as its name implies, is used to transfer heat from one fluid to another when the fluids are not mixed. In Chapter 11, the heat exchanger is considered in detail, when the problem of calculating the heat transfer is studied. Figure 3.25 shows a typical shell-and-tube heat exchanger. The "tube-side" fluid flows inside the tubes, and the "shell-side" fluid flows outside the tubes but is contained by the shell of the heat exchanger.

Let us now analyze the heat exchanger by referring to the sketch of Figure 3.26. The two fluids flow at rates $\dot{m}_1$ and $\dot{m}_2$, respectively. The temperatures at inlet and outlet are as shown. For the purposes of our present study, we will assume that potential and kinetic terms in the energy equation are negligible. Also, we will assume that there are no heat losses from the unit. This requires all of the heat transferred from one fluid to be received by the other. That is,

$$\dot{Q}_1 = \dot{Q}_2 \tag{3.20}$$

But

$$\dot{Q}_1 = \dot{H}_{12} - \dot{H}_{11} = \dot{m}_1 c_{p1}(t_{12} - t_{11}) \tag{3.20a}$$

and

$$\dot{Q}_2 = \dot{H}_{22} - \dot{H}_{21} = \dot{m}_2 c_{p2}(t_{22} - t_{21}) \tag{3.20b}$$

Equating (3.20a) and (3.20b) yields a "heat balance" for the exchanger:

$$\dot{m}_1 c_{p1}(t_{12} - t_{11}) = \dot{m}_2 c_{p2}(t_{22} - t_{21}) \tag{3.21}$$

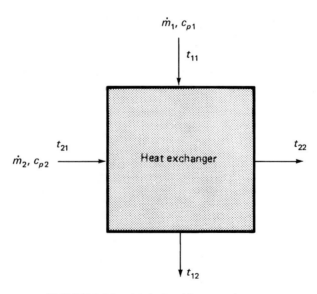

FIGURE 3.26   Analysis of heat exchanger.

ILLUSTRATIVE PROBLEM 3.21

A heat exchanger is used to cool 400 lb_m/min of lubricating oil. Hot oil enters at 215°F and leaves at 125°F. The specific heat of the oil is 0.85 Btu/lb_m·°R. Cooling water enters the unit at 60°F and leaves at 90°F. If the specific heat of the water is 1.0 Btu/lb_m·°R, determine the required water flow rate if heat losses are negligible.

SOLUTION

Refer to Figure 3.26:

$$\dot{Q}_{oil} = \dot{m}c_p\Delta t$$
$$= 400 \ lb_m/min \times 0.85 \ Btu/lb_m\cdot°R \times (125 - 215) \ °R$$
$$= -30,600 \ Btu/min \ (out \ of \ oil)$$

Because the heat out of the oil is the heat into the water,

$$\dot{m}_w c_{pw}(\Delta t)_w = 30,600 \ Btu/min$$
$$\dot{m}_w \times 1 \times (90 - 60) = 30,600$$
$$\dot{m}_w = 1020 \ lb_m/min$$

3.5g Filling a Tank

Consider the following situation, namely, that an evacuated tank is connected to a large steam source, as shown in Figure 3.27. The valve is opened, and steam is allowed to flow into the tank until the pressure in the tank equals the pressure in the line. Assume that the tank is perfectly insulated and that no shaft work occurs during the process. Also, assume that the kinetic and potential energy terms are negligible. In order to write the energy equation for this process, we will first define our system as consisting of the tank and the valve up to the supply side of the valve. The

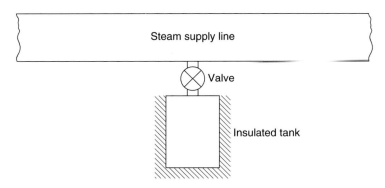

FIGURE 3.27 Filling a tank.

process of charging the closed tank is really an example of an open or flow system in transient flow that can be solved using the information that we have already developed. The supply steam is throttled through the valve, and its enthalpy remains constant during the process. This assumes that the mass of steam withdrawn from the supply line is small enough to leave the conditions in the supply line constant. The change in energy of the mass in the tank will be $m_2 u_2 - m_1 u_1$, where m_1 and m_2 are the initial and final masses in the tank and u_1 and u_2 are the initial and final internal energy, respectively. The initial mass in the tank, m_1, is taken as zero, because the tank was specified as being evacuated at the start of the process. The total energy into the tank will equal $m_2 h_i$, where h_i is the enthalpy of the source. Equating these energy terms gives us

$$m_2 u_2 = m_2 h_i \tag{3.22}$$

or

$$\boxed{u_2 = h_i} \tag{3.23}$$

From Equation (3.23), we see that the final internal energy of the mass in the tank equals the enthalpy of the steam in the supply line. As we will see in Chapters 5 and 6, the final temperature in the tank will be greater than the temperature in the supply line.

3.6 REVIEW

Using the material developed in Chapter 2, namely, the concepts of energy, work, heat, and the statement of the first law of thermodynamics as being the conservation of energy, we analyzed both nonflow and steady-flow mechanical systems. For the nonflow or closed system, one in which no mass crosses the boundary, we were able to write an energy equation that related the internal energy and heat entering a system to the internal energy and work leaving a system. This equation then permitted us to analyze several nonflow situations, including the adiabatic, constant-volume, and constant-pressure systems. As a consequence of our analysis of the constant-pressure system, we defined a new property, which we called enthalpy, to be the sum of the internal energy plus the product of the pressure and the specific volume of a substance. We also showed that the work of a nonflow, quasi-static process can be taken to be the area under the p–v curve.

The term *steady* implies a system that does not vary with time. For the steady-flow system, this requires that the rate at which mass enters must equal the rate at which mass leaves the system. We then developed the continuity equation, which expresses this fact mathematically. Then we considered the energy terms required to write the first law of thermodynamics for the steady-flow system. These terms were potential energy, kinetic energy, and flow work, which we considered in Chapter 2. Using these terms with the already developed concepts of internal energy, heat, and work enabled us to write an energy equation which we called the steady-flow energy equation or, more simply, the general energy equation. Before we applied this equation to several industrial situations, we developed two new properties from the definition of the term *specific heat*. These properties were the specific heat at constant pressure and the specific heat at constant volume, and we were able to show their relation to the properties enthalpy and internal energy, respectively.

Our final part of this chapter was the application of the general energy equation to several common industrial situations: the boiler, the turbine, the nozzle, throttling, pipe flow, the heat exchanger, and the filling of a tank. By making certain simplifications, it was possible to analyze each system quantitatively.

In addition to the foregoing, it should be apparent that the successful application of the material of this chapter requires a systematic approach to problem solving. It is strongly suggested that the following procedure be followed to analyze all problems:

1. Read, visualize, and understand the problem.

2. Draw a schematic sketch of the problem, and indicate all known quantities.

3. Write out the full energy equation.

4. On the basis of the statement of the problem or knowledge of the device in question, determine those terms that can be omitted or neglected.

5. Solve the problem by performing the necessary algebraic and arithmetic steps. Check carefully for dimensional consistency.

6. Using the solution as a first approximation, check the validity of the assumptions made in step 4.

There are no shortcuts to an understanding of thermodynamics, and it will be found that the procedure outlined here will prove to be invaluable.

KEY TERMS

Terms used for the first time in this chapter are:

adiabatic process a process in which no energy as heat enters or leaves during the specified state change.

Bernoulli equation a very restrictive form of the energy equation that is frequently used in fluid mechanics.

boiler a device used to convert the chemical energy in a fuel to an increase in the energy of a working fluid by the combustion of the fuel.

continuity equation an equation expressing the conservation of mass in a steady-flow system.

energy equation see *general energy equation.*

enthalpy the sun of the internal energy and the *pv* term.

first law of thermodynamics an expression of the conservation of energy in a system.

general energy equation the quantitative statement of the first law of thermodynamics applied to a steady-flow system.

nonflow energy equation the quantitative expression of the first law of thermodynamics applied to a nonflow system.

nozzle a static device used to increase the velocity of a fluid.

specific heat the ratio of the energy transferred as heat to the corresponding temperature change of the substance. The specific heat at constant volume and the specific heat at constant pressure are both properties.

steady-flow energy equation see *general energy equation.*

throttling an irreversible, adiabatic, steady-flow process in which the fluid is caused to flow through an obstruction in a pipe with a resulting drop in pressure.

turbine a machine consisting of a rotor and stator whose purpose is to convert the energy in a working fluid to useful work.

EQUATIONS DEVELOPED IN THIS CHAPTER

nonflow energy equation $\qquad u_2 - u_1 = q - \mathbf{w}$ (3.2a)

nonflow energy equation $\qquad u_2 - u_1 = q - p\Delta v$ (3.4)

enthalpy $\qquad h = u + \dfrac{pv}{J}$ (3.5)

continuity equation $\qquad \dot{m} = \rho AV = \rho_1 A_1 V_1 = \rho_2 A_2 V_2$ (3.9b)

continuity equation $\qquad \dot{m} = \dfrac{AV}{v}$ (3.9c)

general energy equation

$$\frac{Z_1}{J}\left(\frac{g}{g_c}\right) + \frac{V_1^2}{2g_c J} + u_1 + \frac{p_1 v_1}{J} + \frac{\mathbf{w}_{in}}{J} + q_{in}$$
$$= \frac{Z_2}{J}\left(\frac{g}{g_c}\right) + \frac{V_2^2}{2g_c J} + u_2 + \frac{p_2 v_2}{J} + \frac{\mathbf{w}_{out}}{J} + q_{out} \quad \text{(3.10)}$$

general energy equation

$$Z_1 g + \frac{V_1^2}{2} + u_1 + p_1 v_1 + \mathbf{w}_{in} + q_{in}$$
$$= Z_2 g + \frac{V_2^2}{2} + u_2 + p_2 v_2 + \mathbf{w}_{out} + q_{out} \quad \text{(3.10a)}$$

general energy equation

$$\frac{Z_1}{J}\left(\frac{g}{g_c}\right) + \frac{V_1^2}{2g_c J} + u_1 + \frac{p_1 v_1}{J} + q$$
$$= \frac{Z_2}{J}\left(\frac{g}{g_c}\right) + \frac{V_2^2}{2g_c J} + u_2 + \frac{p_2 v_2}{J} + \frac{\mathbf{w}}{J} \quad \text{(3.11)}$$

general energy equation

$$Z_1 g + \frac{V_1^2}{2} + u_1 + p_1 v_1 + q$$
$$= Z_2 g + \frac{V_2^2}{2} + u_2 + p_2 v_2 + \mathbf{w} \quad \text{(3.11a)}$$

general energy equation $\qquad q - \dfrac{\mathbf{w}}{J} = (h_2 - h_1) + \dfrac{Z_2 - Z_1}{J}\left(\dfrac{g}{g_c}\right) + \dfrac{V_2^2 - V_1^2}{2g_c J}$ (3.13)

general energy equation $\qquad q - \mathbf{w} = h_2 - h_1 + g(Z_2 - Z_1) + \dfrac{V_2^2 - V_1^2}{2}$ (3.13a)

specific heat at constant pressure	$c_p = \left(\dfrac{q}{\Delta T}\right)_p = \left(\dfrac{\Delta h}{\Delta T}\right)_p$	(3.17)

specific heat at constant volume	$c_v = \left(\dfrac{q}{\Delta T}\right)_v = \left(\dfrac{\Delta u}{\Delta T}\right)_v$	(3.18)

nozzle	$h_1 - h_2 = \dfrac{V_2^2 - V_1^2}{2g_c J}$	Illustrative Problem 3.23

throttling	$h_1 = h_2$	Illustrative Problem 3.26

filling a tank	$u_2 = h_i$	(3.23)

QUESTIONS

3.1 Does the first law of thermodynamics apply to each of the following situations? (a) An atom bomb; (b) a system that is time dependent; (c) a chemical system that is exothermic, that is, giving off heat as the reaction proceeds; (d) the tides on the earth; (e) the sum and the reactions that occur on the sun; (f) a permanent magnet.

3.2 An automobile engine consists of a number of pistons and cylinders. If a complete cycle of the events that occur in each cylinder can be considered to consist of a number of nonflow events, can the engine be considered to a nonflow device?

3.3 Can you name or describe some adiabatic processes?

3.4 What is unique about the nonflow, constant-pressure process?

3.5 Is enthalpy the sum of internal energy plus flow work?

3.6 In words, what does the continuity equation express?

3.7 What are the limitations of the continuity equation?

3.8 All substances are subjected to pressure and have a specific volume. Because this means that they all have the product pv, do they all have flow work?

3.9 Do all substances have enthalpy?

3.10 The term $p\Delta v$ is the area under a p–v curve, and the term $-v\Delta p$ is the area behind the p–v curve. How do you interpret each of these areas?

3.11 Discuss the problems associated with the Bernoulli equation.

3.12 With all of the problems associated with the Bernoulli equation, why is it still used?

3.13 When a fluid is vaporized, the temperature does not change during the process as heat is added. What is the specific heat for this process?

3.14 Can you name another common process that is carried out at constant temperature and therefore has a specific heat that is similar to the one in Question 3.13?

3.15 What is the importance of the fact that c_p and c_v are properties?

3.16 A number of assumptions were made when the seven applications of the energy equation were discussed. Is it always necessary to make these assumptions when any of these situations are analyzed?

3.17 In all seven applications, we took the difference in elevations to be negligible. Would this also be true if you were analyzing the hydroelectric power plants at Niagara Falls?

PROBLEMS

Unless indicated otherwise, use $g_c = 32.174 \ \text{lb}_m \cdot \text{ft/lb}_f \cdot s^2$ and $g = 32.174 \ \text{ft/s}^2$ or $9.81 \ \text{m/s}^2$.

Problems Involving Nonflow Systems

3.1 The internal energy of a nonflow system increases by 70 Btu when 82 Btu of work is done by the system on its surroundings. How much heat has been transferred to or from the system?

3.2 If the internal energy of a nonflow system increases by 90 kJ while the system does 125 kJ of work on the surroundings, determine the heat transfer to or from the system.

3.3 A nonflow process is carried out when 1000 Btu as heat are added to 10 lb_m of hydrogen at constant volume. What is the change in specific internal energy? How much work is done by the gas?

3.4 If a nonflow process is carried out so that 30 Btu/lb_m of work is removed from the process while 100 Btu/lb_m is added as heat, determine the change (increase or decrease) in the specific internal energy of the fluid.

3.5 If 10 kg of a gas is heated by the addition of 5 kJ in a nonflow process, what is the change in its specific internal energy if the process is carried out at constant volume?

3.6 If a nonflow process is carried out so that 30 kJ/kg of work is removed while 75 kJ/kg is added as heat, determine the change in internal energy of the fluid.

3.7 A constant-pressure, nonflow process is carried out at 125 kPa. During the process, the volume changes from 0.15 to 0.05 m^3 and 25 kJ of heat is rejected. Evaluate (a) the work done and (b) the change in internal energy of the system.

3.8 A closed gaseous system undergoes a reversible process during which 20 Btu/lb_m is rejected as heat and the specific volume changes from 0.5 to 0.2 ft^3/lb_m. If the pressure is constant at a value of 40 psia, determine the change in specific internal energy.

3.9 In a constant-pressure, nonflow process, 10 lb_m of gas has 500 Btu added to it. During the process, the internal energy decreases by 25 Btu/lb_m. How much work was done by the gas per pound of gas?

3.10 A constant-pressure, nonflow system receives heat at a constant pressure of 350 kPa. The internal energy of the system increases by 180 kJ while the temperature increases by 170°C, and the work done is 75 kJ. Determine c_p and the change in volume if there is 1.5 kg in the cylinder.

3.11 A constant-pressure, nonflow process is carried out at a pressure of 200 kPa. If 50 kJ of heat is removed while the volume changes from 0.2 to 0.1 m^3, what is the change in specific internal energy of the working fluid? Assume 0.5 kg of fluid.

3.12 A gas that is contained in a cylinder is made to undergo a constant-pressure expansion from a volume of 0.1 m^3 to a volume of 0.4 m^3. If the piston moves to keep the pressure constant at 400 kPa, determine the work in kilojoules.

3.13 A nonflow system has a mass of working fluid of 5 kg. The system undergoes a process in which 140 kJ as heat is rejected to the surroundings and the system does 80 kJ of work. Assuming that the initial specific internal energy of the system is 500 kJ/kg, determine the final specific internal energy in kJ/kg.

3.14 A nonflow system has a mass of 5 lb in it. The system undergoes a process during which 200 ft·lb_f of heat is transferred to the surroundings. If the system does 1000 ft·lb_f of work on the surroundings, determine the change in specific internal energy of the fluid.

3.15 A nonflow process is carried out adiabatically. If 55 kJ of work is removed from 4 kg of fluid, what is the change in the total internal energy and the specific internal energy of the fluid?

3.16 A nonflow process is carried out adiabatically. What is the change in specific internal energy of the fluid if 55,000 ft·lb$_f$ of work is removed from 8 lb$_m$ of fluid in this process?

3.17 During a certain nonflow process, 100 Btu/lb$_m$ is added as heat to the working fluid while 25,000 ft·lb$_f$/lb$_m$ is extracted as work. Determine the change in internal specific energy of the fluid.

3.18 If the enthalpy of a gas is 240 Btu/lb$_m$ when the pressure is 100 psia and the specific volume is 3.70 ft^3/lb$_m$, determine the specific internal energy of the gas.

3.19 What is the specific internal energy of a gas whose enthalpy is 500 kJ/kg? At this condition, the pressure is 100 kPa and the specific volume 0.1 m^3/kg.

3.20 A closed system receives 1 kJ as heat, and its temperature rise is 10°K. If 20 kg is in the system, what is the specific heat of the process?

3.21 Determine the specific heat of an adiabatic process. Is this the specific heat at a constant volume or the specific heat at constant pressure?

3.22 Two kilograms of a gas receive 200 kJ as heat at constant volume. If the temperature of the gas increases by 100°C, determine the c_v of the process.

3.23 Heat is supplied to a gas in a rigid container. If the container has 0.6 lb$_m$ of gas in it and 100 Btu is added as heat, determine the change in temperature of the gas and the change in its specific internal energy. For this gas, c_v is 0.35 Btu/lb$_m$·°R.

3.24 If 100 kJ/kg of heat is added to 10 kg of a fluid while 25 kJ/kg is extracted as work, determine the change in the specific internal energy of the fluid for a nonflow process.

3.25 Heat is supplied to a gas that is contained in a rigid container. If 0.2 kg of gas has 100 kJ as heat added to it, determine the change in temperature of the gas and the change in its specific internal energy. Use $c_v = 0.7186$ kJ/kg·°K and $c_p = 1.0062$ kJ/kg·°K.

3.26 If c_p and c_v of air are, respectively, 1.0062 and 0.7186 kJ/kg·°K and 1 MJ as heat is added to 10 kg in a nonflow, constant-pressure process, what is the final temperature of the gas, and how much work is done by the gas? Assume that the initial temperature of the gas is 50°C.

3.27 The c_p and c_v of air are 0.24 and 0.17 Btu/lb$_m$·°R, respectively. If 1000 Btu is added as heat to 20 lb$_m$ of air in a nonflow, constant-pressure process, what is the final temperature? How much work is done by the gas? The initial temperature is 100°F.

3.28 In a certain process, the working fluid is cooled until 100 Btu/lb$_m$ has been extracted as heat. During this process, 100 ft·lb$_f$/lb$_m$ of fluid is added as work. If the working substance undergoes a change in temperature from 100°F to 50°F, what is the specific heat of the process?

3.29 Air is adiabatically compressed in a nonflow process from a pressure and temperature of 14.7 psia and 70°F to 200 psia and 350°F. If c_v of the air is 0.171 Btu/lb$_m$·°R, determine the change in specific internal energy of the air and the work done.

3.30 If a nonflow system has 6 kg in it and undergoes a process in which 175 kJ as heat is transferred to its surroundings, determine the final specific internal energy in kJ/kg if the work done by the system is 100 kJ. Assume that the initial specific internal energy is 400 kJ/kg.

3.31 A nonflow, cylinder–piston apparatus contains 0.5 lb$_m$ of a gas. If 10 Btu is supplied as work to compress the gas as its temperature increases from 70°F to 150°F, determine the energy interchange as heat. For the gas, c_v can be taken as 0.22 Btu/lb$_m$·°R.

3.32 A nonflow system undergoes a process in which 42 kJ of heat is rejected. If the pressure is kept constant at 125 kPa while the volume changes from 0.20 to 0.06 m^3, determine the work done and the change in internal energy.

*3.33 Four individual nonflow processes numbered as 1, 2, 3, and 4 are carried out. For each of the processes, fill in the following table. Assume all values are in kJ.

Process	q	w	u_2	u_1	Δu
1	15	8		6	
2	22		16	4	
3	−20		10		15
4		18	14	4	

*3.34 A closed cycle consists of the three processes listed in the table below. Fill in the missing values if all the quantities are in Btu/lb$_m$. (*Hint:* A closed cycle returns to its starting point.)

Process	q	w	Δu
$1 \rightarrow 2$	200	x	100
$2 \rightarrow 3$	z	100	t
$3 \rightarrow 1$	y	300	300

Problems Involving the Continuity Equation

3.35 Ten cubic feet of fluid per second flows in a pipe 1 in. in diameter. If the fluid is water whose density is 62.4 lb$_m$/ft^3, determine the weight rate of flow in pounds per hour.

3.36 A gallon is a volume measure of 231 in^3. Determine the velocity in a pipe 2 in. in diameter when water flows at the rate of 20 gal/min in the pipe.

3.37 When a fluid of constant density flows in a pipe, show that for a given mass flow the velocity is inversely proportional to the square of the pipe diameter.

3.38 Water having a density of 1000 kg/m^3 flows in a pipe that has an internal diameter of 50 mm. If 0.5 m^3/s flows in the pipe, determine the mass flow rate in kilograms per hour.

3.39 What is the velocity of the water flowing in Problem 3.38?

3.40 Air flows through a 12 in. × 12 in. duct at a rate of 800 cfm. Determine the mean flow velocity in the duct.

3.41 Air flows through a rectangular duct 0.5 m × 0.75 m at the rate of 200 m^3/min. Determine the average velocity of the air.

3.42 Water flows through the pipe shown at the rate of 30 ft^3/s. Determine the velocity at sections 1 and 2. Also calculate the mass flow rate per second in the pipe at these sections. The diameters shown in Figure P3.42 are internal diameters.

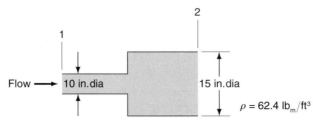

FIGURE P3.42

3.43 Air, having a specific weight of 11.8 N/m³, flows in the pipe shown in Figure P3.43 with a velocity of 10 m/s at section 1. Calculate the weight flow rate per second of the air and its specific weight at section 2 if the air has a velocity of 3 m/s at section 2.

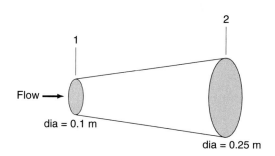

Flow →
dia = 0.1 m
dia = 0.25 m

FIGURE P3.43

3.44 If the water level in the tank remains constant, determine the velocity at section 2 as shown in Figure P3.44. Assume the water to be incompressible.

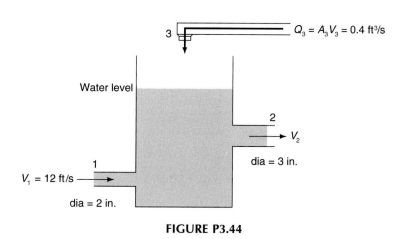

$Q_3 = A_3 V_3 = 0.4$ ft³/s

Water level

V_2
dia = 3 in.

$V_1 = 12$ ft/s
dia = 2 in.

FIGURE P3.44

3.45 Water flows through a pipe at the rate of 1500 L/min. At one section, the pipe is 250 mm in diameter, and later on, it reduces to a diameter of 100 mm. What is the average velocity in each section of the pipe?

3.46 A pump can fill a tank in 1 min. If the tank has a capacity of 18 gal, determine the average velocity in the 1.25-in. diameter pipe that is attached to the pump.

3.47 For the liquid rocket shown in Figure P3.47, calculate V_2 for steady operation.

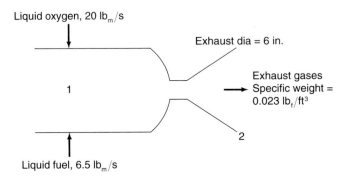

Liquid oxygen, 20 lb_m/s

Exhaust dia = 6 in.

1

Exhaust gases
Specific weight =
0.023 lb_f/ft^3

2

Liquid fuel, 6.5 lb_m/s

FIGURE P3.47

3.48 Water flows through the nozzle shown in Figure P3.48 at the rate of 100 lb_m/s. Calculate the velocity at sections 1 and 2.

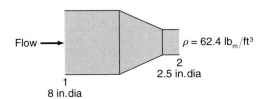

Flow

ρ = 62.4 lb_m/ft^3

2
2.5 in.dia

1
8 in.dia

FIGURE P3.48

Problems Involving the Steady-Flow Energy Equation

3.49 A water turbine operates from a water supply that is 100 ft above the turbine inlet. It discharges to the atmosphere through a 6-in. diameter pipe with a velocity of 25 ft/s. If the reservoir is infinite in size, determine the power out of the turbine if the density of water is 62.4 lb_m/ft^3.

3.50 A water pump is placed at the bottom of a well. If it is to pump 1.0 ft^3/s to the surface, 100 ft away, through a 4-in.-inside-diameter pipe, determine the power required in foot pounds per second and horse-power if the density of water is 62.4 lb_m/ft^3.

3.51 Air is compressed until its final volume is half its initial volume. The initial pressure is 100 psia, and the final pressure is 35 psia. If the initial specific volume is 1 ft^3/lb_m, determine the difference in the pv/J term in Btu/lb_m. Is this difference "flow work" for this process?

3.52 Determine the power of a pump if 2000 kg/min of water is compressed from 100 kPa to 1 MPa. The water density can be taken to be 1000 kg/m^3, and its temperature does not change. The inlet to the pump is 100 mm in diameter, and the outlet is 150 mm in diameter. The inlet is 50 m below the outlet.

3.53 A steady-flow device is operated with an entering pressure of 50 psia and a specific volume of 0.8 ft^3/lb_m. At the exit of the device, the pressure is 15 psia and the specific volume 3.2 ft^3/lb_m. Determine the change in flow work in this device.

3.54 If a fluid flows past a section of pipe with a pressure of 100 kPa and having a specific volume of 10^{-3} m^3/kg, determine its flow work at this section.

3.55 At the entrance to a steady-flow device, the pressure is 350 kPa and the specific volume 0.04 m³/kg. At the outlet, the pressure is 1 MPa and the specific volume 0.02 m³/kg. Determine the flow work change in this device.

3.56 The enthalpy of a substance is found to be 1000 Btu/lb$_m$ at 1000°F above an arbitrary datum of 0°F. If c_v is a constant and equal to 0.5 Btu/lb$_m$·°R, determine the difference in the pv/J terms at zero and 1000°F. Does this difference represent flow work?

*3.57 A steam turbine is supplied with steam at 700°F and 500 psia. The steam is expanded adiabatically until the final condition is saturated vapor at 300°F. If the properties of steam at the initial and final states are as given here, determine the work output of the turbine per pound of steam.

500 psia and 700°F	300°F (saturated vapor)
h = 1356.7 Btu/lb$_m$	h = 1180.2 Btu/lb$_m$
v = 1.3040 ft³/lb$_m$	v = 6.472 ft³/lb$_m$
u = 1236.0 Btu/lb$_m$	u = 1100.0 Btu/lb$_m$

*3.58 A fluid at 1 MPa has a specific volume of 0.2 m³/kg and an entering velocity of 200 m/s. The heat loss is 10 kJ/kg, and the fluid does 180 kJ/kg of work. The fluid leaves the device with a pressure of 200 kPa, a specific volume of 1 m³/kg, and a velocity of 600 m/s. What is the change in the specific internal energy of the fluid?

*3.59 A steam turbine operates adiabatically with inlet and other conditions as given here. Determine the work out of the turbine. State all assumptions.

At 0.5 MPa and 500°C	At 0.1 MPa (saturated vapor)
h = 3483.9 kJ/kg	h = 2675.5 kJ/kg
v = 0.7109 m³/kg	v = 1.6940 m³/kg
u = 3128.4 kJ/kg	u = 2506.1 kJ/kg

3.60 A steam turbine receives steam at 110 ft/s velocity and 1525 Btu/lb$_m$ enthalpy. The steam leaves at 810 ft/s and 1300 Btu/lb$_m$ enthalpy. What is the work out in Btu per pound mass flowing?

3.61 A turbine is operated with an enthalpy at entrance of 1340 Btu/lb$_m$ and an exit enthalpy of 1285 Btu/lb$_m$. If the entrance velocity to the turbine is 150 ft/s and the exit velocity is 500 ft/s, determine the work out of the turbine.

3.62 Solve Problem 3.61 if there is a heat loss of 5.5 Btu/lb$_m$ of fluid by heat transfer from the turbine casing.

3.63 A gas turbine receives an air–fuel mixture having an enthalpy of 550 Btu/lb$_m$ of gas. If 10 Btu/lb$_m$ is lost by heat transfer from the turbine and the enthalpy of the leaving gas is 50 Btu/lb$_m$, how much work can be obtained from the turbine?

3.64 An air compressor takes in air having an enthalpy of 100 kJ/kg. The air is compressed to a final condition in which the enthalpy is 180 kJ/kg. As the air travels through the compressor, it loses 20 kJ/kg as heat. Determine the work per kilogram of the compressor.

*3.65 Air having the following properties passes through a steady flow device. Determine the work in kJ/kg if the heat loss is 20 kJ/kg.

	Inlet	Outlet
pressure	700 kPa	170 kPa
density	3.0 kg/m^3	0.60 kg/m^3
velocity	70 m/s	200 m/s
internal energy	2100 kJ/kg	2025 kJ/kg

3.66 A gas turbine has gas entering it at 1100°F and leaving at 800°F. If the specific heat c_p = 0.265 Btu/lb$_m$·°R, determine the work out of the turbine. Neglect losses and velocity effects.

3.67 Solve Problem 3.61 if the entering velocity is 100 ft/s, the leaving velocity is 300 ft/s, and there is a heat loss of 4.8 Btu/lb$_m$ from the turbine.

*3.68 Assume 4.8 kg/s of steam enters a turbine. The inlet of the turbine is 2.5 m higher than the outlet. The inlet velocity is 132 m/s, and the outlet velocity is 327 m/s. Heat is lost from the casing of the turbine at the rate of 9.2 kJ/s. At the inlet of the unit, the enthalpy is 3127.4 kJ/kg, and at the outlet, the enthalpy is 2512 kJ/kg. Determine the power out of the turbine in kilowatts.

3.69 Gas enters a turbine at 600°C and leaves at 350°C. If the specific heats of the gas are c_p = 0.8452 kJ/kg·°K and c_v = 0.6561 kJ/kg·°K, determine the work out of the turbine. State all assumptions.

3.70 Assume the entering velocity in Problem 3.69 is 86 m/s and the leaving velocity is 232 m/s. If the outlet of the turbine is 3.2 m higher than the inlet, determine the work out of the turbine if the process is adiabatic.

3.71 Air flows in a compressor at the rate of 1.1 kg/s. The air enters the compressor at a pressure of 100 kPa, a temperature of 25°C, and a velocity of 72 m/s. It leaves with a pressure of 225 kPa, a temperature of 84°C, and a velocity of 125 m/s. A cooling jacket surrounds the compressor to cool the air, and it removes 15 kJ/kg as heat from the air. If c_p = 1.0041 kJ/kg·°K and c_v = 0.7172 kJ/kg·°K, calculate the power to operate the compressor.

3.72 Air expands through a nozzle from 1000 psia and 500°F to 600 psia and 0°F. If c_p of air can be taken to be constant and equal to 0.24 Btu/lb$_m$·°R, what is the final velocity? Assume that the initial velocity is zero.

3.73 If the initial velocity in Problem 3.72 is 9.25 ft/s, what is the final velocity?

3.74 Steam expands in a nozzle from an initial enthalpy of 1300 Btu/lb$_m$ to a final enthalpy of 980 Btu/lb$_m$. Determine the final velocity if the entering velocity and heat losses are negligible.

3.75 Solve Problem 3.74 if the initial velocity is 1100 ft/s.

3.76 Air expands in a nozzle from 7 MPa and 250°C to 3.5 MPa and 50°C. If c_p = 1.0062 kJ/kg·K and c_v = 0.7186 kJ/kg·°K, determine the velocity out of the nozzle. State all assumptions.

3.77 Solve Problem 3.76 if the initial velocity is 325 m/s.

3.78 A nozzle has steam flowing through it. The steam enters the nozzle at 250 psia with a velocity of 100 ft/s. The initial specific volume is 2.31 ft^3/lb$_m$ and the initial internal energy is 1114.1 Btu/lb$_m$. The steam expands to 15.0 psia with a specific volume of 27.34 ft^3/lb$_m$ and a final internal energy of 1080.1 Btu/lb$_m$. Calculate the velocity at the exit of the nozzle.

3.79 Steam flows through a nozzle. If the properties are as given here, determine the exit velocity if the nozzle is well insulated. Neglect the inlet velocity.

	Inlet	Outlet
pressure	300 psia	100 psia
temperature	500°F	350°F
internal energy	1159.5 Btu/lb$_m$	1115.4 Btu/lb$_m$
enthalpy	1257.5 Btu/lb$_m$	1200.4 Btu/lb$_m$

3.80 Solve Problem 3.79 for an initial velocity of 600 ft/s.

3.81 Steam is generated in a boiler. If the initial water enters the boiler as saturated water at 1000 psia ($h = 542.4$ Btu/lb$_m$) and leaves as superheated steam at 950 psia and 1000°F ($h = 1507.4$ Btu/lb$_m$), how much heat is added to the steam to generate 1 lb of steam at these conditions?

3.82 Steam is generated in a boiler. Water enters the boiler as saturated liquid at 7 MPa ($h = 1267.00$ kJ/kg) and leaves as superheated steam at 6 MPa and 500°C ($h = 3422.2$ kJ/kg). Neglecting potential and kinetic terms, determine the amount of heat that is added to generate 1 kg of steam under these conditions. Assume that there are no heat losses.

*3.83 Water flows in a heat exchanger, and heat is transferred from the water to heat air. At the inlet, the pressure of the water is 100 psia and at the outlet it is 80 psia. The initial specific volume is 0.017736 ft^3/lb$_m$, and the final specific volume is 0.01757 ft^3/lb$_m$. The initial enthalpy is 298.61 Btu/lb$_m$, the initial internal energy is 298.28 Btu/lb$_m$, the final enthalpy is 282.21 Btu/lb$_m$, and the final internal energy is 281.95 Btu/lb$_m$. Determine the heat transferred from the water. State all assumptions.

*3.84 Air flows in steady flow in a pipeline whose internal diameter is constant. At a particular section of the pipe, the pressure, temperature, and velocity are 150 psia, 500°R, and 30 ft/s. The specific volume corresponding to these conditions is 1.24 ft^3/lb$_m$. At a section farther downstream, the pressure is 300 psia, the velocity is 45 ft/s, and the temperature is 1500°R. Determine the specific volume of the air at the downstream section and also the magnitude and direction of the heat transfer in Btu per pound mass. Assume that c_p is 0.24 Btu/lb$_m$·°R and that it is constant.

*3.85 Dry saturated steam at 200 psia flows adiabatically in a pipe. At the outlet of the pipe, the pressure is 100 psia and the temperature 400°F. The initial specific volume of the steam is 2.289 ft^3/lb$_m$, and its internal energy is 1114.6 Btu/lb$_m$. The final enthalpy is 1227.5 Btu/lb$_m$, and the final specific volume is 4.934 ft^3/lb$_m$. How much energy per pound mass was transferred? Was this into or out of the system? Was this heat or work?

*3.86 A heat exchanger is used to heat petroleum distillates for use in a chemical process. Hot water enters the unit at 180°F and leaves at 105°F. The distillate enters at 85°F and leaves at 148°F. If the specific heat of water is taken to be 1.0 Btu/lb$_m$·°R and the distillate has a specific heat of 0.82 Btu/lb$_m$·°R, determine the pounds of water required per pound of distillate flowing.

4

The Second Law of Thermodynamics

Learning Goals

After reading and studying the material in this chapter, you should be able to:

1. Understand that work can be converted into heat but that the conversion of heat into useful work may not always be possible.

2. Define a heat engine as a continuously operating system across whose boundaries flow only heat and work.

3. Define thermal efficiency as the ratio of the useful work delivered by a heat engine or cycle to the heat input to the engine or cycle.

4. Understand what is meant by the statement that a reversible process is any process performed so that the system and all its surroundings can be restored to their initial states by performing the process in reverse.

5. State the second law of thermodynamics as: "Heat cannot of itself pass from a lower temperature to a higher temperature."

6. Understand that all natural processes are irreversible, and cite some of the effects that cause irreversibility.

7. Explain the four processes that constitute the Carnot cycle.

8. Deduce from the Carnot cycle three important general conclusions concerning the limits of the efficiency of a heat engine.

9. Define the new property that is introduced in this chapter that we have called entropy.

10. Understand that entropy is also a measure of the unavailability of energy that occurs in an irreversible process.

11. Calculate the change in entropy for a process in which there is a temperature change, such as the mixing of two fluids.

12. Understand that the entropy of an isolated system increases or, in the limit, remains the same, which for a given internal energy we interpret to mean that the state having the greatest entropy will be the most probable state that the system will assume. This state is called stable equilibrium.

4.1 INTRODUCTION

Thus far, we have considered various forms of energy (including energy in transition as both work and heat) without regard to any limitations on these quantities. It has been assumed that work and heat are mutually interchangeable forms of energy, and it may have appeared to the student that the distinction made between these quantities was arbitrary and possibly not necessary. In this chapter, the interconvertibility of these quantities is explored with the object of determining any possible limitations and to express these limitations quantitatively, if they exist. As an example of the point in question, consider the motion of a block sliding along a rough horizontal plane. For motion to proceed along the plane, it is necessary that work be done on the body. All this work subsequently appears as heat at the interface between the block and the plane. There is no question that work has been converted into heat, but can the heat generated in this process be converted into an equivalent amount of work? Let us assume (incorrectly) that this heat can be converted into work without any losses in the process. We know that the energy as heat increases the motion of the individual molecules of the body. By increasing the molecular motion within the body, we have, in a general sense, done work, but it is also possible to distinguish that this form of work is not the same as the external work put into the process. The original transitional energy as work has been converted into heat, and this heat can be expressed as molecular work. However, this form of energy will not be available to return the body to its original state. From this simple example, we note that work can be converted into heat but that the conversion of heat into useful work may not always be possible. Even though the first law states that energy is conserved, it does not furnish the necessary information to enable us to determine whether energy has become unavailable.

It is now necessary to define certain terms. The first of these is the concept of a *heat engine*. As defined by Keenan:

> *A heat engine may be defined as a continuously operating system across whose boundaries flow only heat and work. It may be used to deliver work to external devices, or it may receive work from external devices and cause heat to flow from a low level of temperature to a high level of temperature. This latter type of heat engine is known as a refrigerator.*[*]

In essence, this definition of a heat engine can be taken to be the definition of a thermodynamic *cycle,* which we shall understand to be a series of thermodynamic processes during which the working fluid can be made to undergo changes involving only heat and work interchanges and is then returned to its original state.

The purpose of the conventional engineering thermodynamic cycle is, of course, to convert heat into work. In an air-conditioning or refrigeration cycle, work is used to remove heat from an area in which it is undesirable. Other special cycles exist but are not treated in this text. Associated with the concept of a cycle is the term *efficiency.* Because the usual purpose of a cycle is to produce useful work, the *thermal efficiency* of a cycle is defined as the ratio of the *net work* of the cycle to the *heat added* to the cycle; that is,

$$\eta = \frac{\text{net work output}}{\text{heat added}} \times 100 \qquad\qquad \textbf{(4.1)}$$

---

[*] J. H. Keenan, *Thermodynamics* (New York: John Wiley & Sons, Inc., 1941), p. 58.

Note that the heat term is the heat added and not the net heat of the cycle. For power-producing cycles, the heat is usually added from some high-temperature source. Using the notation that Q_{in} is the heat added to the cycle and that Q_r is the heat rejected by the cycle, the first law applied to the cycle will yield $W/J = Q_{in} - Q_r$. Therefore,

$$\eta = \frac{Q_{in} - Q_r}{Q_{in}} \times 100 = \left(1 - \frac{Q_r}{Q_{in}}\right) \times 100 \qquad \textbf{(4.2)}$$

For cycles whose purpose is not the production of useful work, other standards of comparison have been devised and are in use.

If we consider the large central station plant, such as that shown in Figure 4.1, the importance of obtaining high thermal efficiencies is at once obvious. An examination of Equation (4.2) leads us to the conclusion that minimizing the heat rejection of a cycle leads to the maximum conversion of heat to work. This leads us to two questions: (1) Must there be a rejection of heat from a cycle, and if so, (2) what is the best mode of cycle operation to minimize the heat rejected in order to obtain maximum thermal efficiency? These questions will be partially answered in this chapter, and we shall return to them when we study practical engine cycles.

4.2 REVERSIBILITY—THE SECOND LAW OF THERMODYNAMICS

In Section 4.1, the illustration of a block sliding along a horizontal plane was used to introduce the concept that heat and work are not always mutually convertible without losses. This same body moving along the plane will also serve to answer the following question: By reversing each step of the process that caused the body to move along the plane, is it possible to restore the body to its original state, and at the same time, will the surroundings also be restored to the condition that existed before the start of the original process? To answer this question, let us once again consider the forward motion of the body along the plane. We have stated that as the block moves, heat is generated at the interface between the block and the plane. This energy is transferred to the body and the plane and will tend to raise the temperature of the body and its surroundings. When the block reaches the end of the plane, let us reverse the force system acting on the body and attempt to restore it to its original position. As the body moves back along the path, heat will again be generated at the interface between the body and the plane. Obviously, the heat generated on the return path is *in addition* to the heat generated at the interface during forward motion of the body. To the casual observer who viewed the block before the beginning of motion and then viewed it some time after motion had ceased, it would appear that the body had not moved and that it, as well as the surroundings, had been restored to its original state. This is not true. A net transfer of energy has taken place to the body and its surroundings, and they are not in their original state. Even though the net effect has been an infinitesimal change in the temperature of the body and its surroundings, it is a real effect that precludes us from saying that the system *and its surroundings* have been restored to their original state. We also notice that each step of the forward motion of the body was not identically reversed because of this effect on the surroundings. The heat generated during the forward motion of the block was not returned to the system as work during the return motion. On the contrary, more heat was generated during the return motion, and even if both the plane and the body were perfectly insulated from their surroundings, none of the heat generated would have been returned to the system as work.

FIGURE 4.1 Aerial view of the Charles Poletti Power Project in Astoria, New York. *(Courtesy of the Power Authority of the State of New York.)*

The process we have considered is illustrative of an irreversible process. To formalize the concepts of reversibility and irreversibility, the following definition of *reversible process* is used:

> *A reversible process is any process performed so that the system and all its surroundings can be restored to their initial states by performing the process in reverse.*

All processes of a reversible cycle must, therefore, also be reversible. The student should note that the concept of the frictionless, quasi-static process introduced earlier basically implies that such a process is reversible.

The *second law of thermodynamics* is an expression of empirical fact that all forms of energy are not necessarily equivalent in their ability to perform useful work. There are many statements and corollaries of the second law that can be found in the literature on thermodynamics. For the present, the statements of Clausius and Kelvin–Planck will serve to express the second law fully.*

The Clausius Statement

It is impossible to construct a device that operates in a cycle and whose sole effect is to transfer heat from a cooler body to a hotter body. (Heat cannot, of itself, pass from a lower temperature to a higher temperature.)

The Kelvin–Planck Statement

It is impossible to construct a device that operates in a cycle and produces no other effect than the production of work and exchange of heat with a single reservoir.

One of the many consequences of the second law of thermodynamics is the conclusion that all natural processes are irreversible. It has already been shown that the presence of friction will cause a process to be irreversible. Some processes that are irreversible are the following:

1. *Any process* in which work is transformed into internal energy via the agency of friction or inelastic action.

2. *Any process* in which inelastic molecular action occurs.

3. *Any process* that transfers heat from one portion of a system to another by virtue of a finite temperature difference.

4. *Any process* that causes temperature differences between parts of the same system.

5. *Any process* involving combustion or chemical reactions.

6. *Any process* that is not performed quasi-statically; thus, to be reversible, a process must proceed at an infinitesimally slow rate.

It is important for the student to fully understand where the irreversibilities occur in the listed processes. Also, by observing the effects on the environment, other irreversible processes will become apparent.

---

* J. R. Howell and R. D. Buckius, *Fundamentals of Engineering Thermodynamics,* 2nd ed. (New York: McGraw-Hill Book Company, 1992), p. 284.

The next question we ask is: Under what conditions will a process be reversible? The answer is that in reality, no process is reversible. However, as an abstract ideal, the reversible process is extremely useful, and this ideal can be achieved only if the process is frictionless and quasi-static—and then only for an isothermal or adiabatic process. The quasi-static process is always in thermodynamic equilibrium and is carried out with infinite slowness so that at any step in the process, it can be reversed and all steps retraced. Also, when such a process is specified to be either isothermal or adiabatic, temperature differences within the system or in parts of the system are precluded. To be general, we must exclude other irreversible effects such as magnetic hysteresis and electrical currents.

4.3 THE CARNOT CYCLE

The material discussed so far in this chapter has served to define a cycle, its efficiency, and the concept of a reversible process. It would appear quite natural at this point to combine all these concepts and to discuss reversible cycles and their efficiency. Historically, these concepts were first enunciated by Nicolas Leonard Sadi Carnot in 1824, and the reversible thermodynamic cycle that he proposed now bears his name. It is interesting to note that Carnot did his work approximately 175 years ago. In this short span of human history, scientific thermodynamics has become a reality. Figure 4.2 shows some important names and dates in the development of the science of thermodynamics.

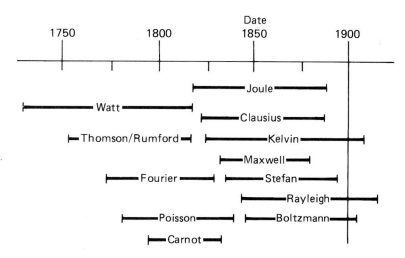

FIGURE 4.2 Some important dates and persons in the development of thermodynamics.

As noted in the preceding section, the two frictionless, quasi-static processes that are reversible are the isothermal (constant temperature) and adiabatic (no energy as heat crosses the boundary). Carnot proposed a reversible cycle composed of two reversible isothermal processes and two reversible adiabatic processes, and on the basis of this cycle, he was able to reach certain general conclusions. Let us consider the cycle that has been named for him by describing each step of the

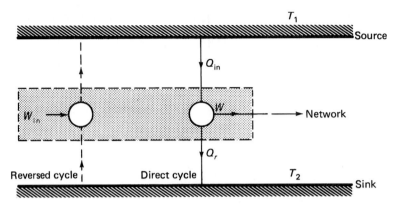

FIGURE 4.3 Elements of Carnot cycle.

cycle. Figure 4.3 (solid line) is a schematic of a direct engine cycle. For the Carnot cycle, we define the following sequence of events:

1. Heat is taken from an infinite reservoir (source) at T_1 isothermally and reversibly. Basically, this is equivalent to a quasi-static reception of heat into the cycle without temperature differences.

2. The energy received from step 1 is permitted to produce work by expanding reversibility and adiabatically in an ideal frictionless engine. During this step, net work is produced, but no energy as heat is permitted to cross the boundaries of the system even though the pressure and temperature of the working fluid may have changed.

3. At this point in the cycle, the working fluid is at temperature T_2, and we shall want to return it to its starting point. To do this, we first reject heat at constant temperature (T_2) reversibly and isothermally to an infinite sink.

4. The final step in the cycle is to cause the working fluid to be adiabatically and reversibly compressed to its initial state.

Note that in every analysis of the Carnot cycle, T_1 always represents the higher temperature and T_2 always represents the lower temperature.

For a noncondensing gas, the steps of the cycle are portrayed on pressure–volume coordinates in Figure 4.4.

The Carnot cycle just described is a reversible cycle, and it is therefore possible to reverse each step in turn and thus reverse the cycle. Such a reversed cycle would effectively take work as an input and pump heat from T_2 to T_1. The reversed cycle is known as a *heat pump* and is discussed further in Chapter 10. It should be noted that the Carnot cycle is not unique, and it is not the only reversible cycle that can be devised. Actually, many reversible cycles have been proposed as prototypes of real cycles. The power of the Carnot cycle is that the following general conclusions can be deduced from it:

1. *No engine operating between fixed sour (T_1) and sink (T_2) temperatures and continuously delivering work can be more efficient than a reversible engine operating between these same temperature limits.*

144 THE SECOND LAW OF THERMODYNAMICS

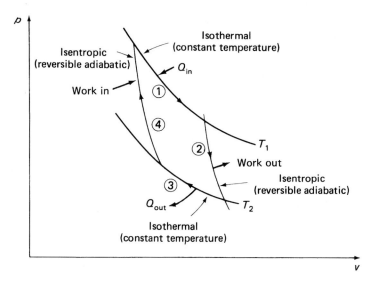

FIGURE 4.4 Carnot cycle on p–v coordinates; noncondensible gas.

To prove this proposition, let us take two engines, engine 1 with an efficiency *greater* than the efficiency of a reversible engine, and engine 2, which is a reversible engine, and operate them between a hot reservoir at T_1 and a cold reservoir at T_2. If these engines are arranged as shown in Figure 4.5, we have engine 1 operating directly to produce 100 units of work. The output of engine 1 is now used to drive engine 2, the reversible engine, in the reversed direction. Because the work of engine 1 is used as the input to engine 2, there is no net work out of the combined system. Assuming that engine 1 has an efficiency of 50% and engine 2 has an efficiency of 40%, we find that the input to engine 1 is 200 units and that it rejects 100 units of heat. Engine 2, the reversible engine with an assumed efficiency of 40%, has 150 units of heat entering from the cold reservoir at T_2 and rejects 250 units to the hot reservoir at T_1. Therefore, we have 50 units of heat being delivered from T_2, the lower temperature, to T_1, the higher temperature, *with no net work being put into*

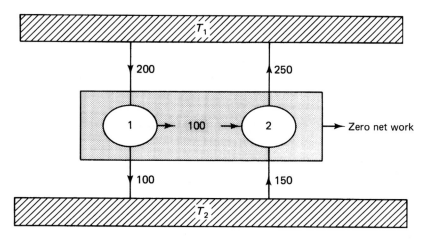

FIGURE 4.5 Proof of Carnot principle 1.

the combined cycle. This directly violates the Clausius statement of the second law of thermodynamics, which states that heat cannot *of itself* pass from a lower to a higher temperature. Therefore, we may reject the assumption that an engine can be more efficient than a reversible engine operating between the same temperature limits. Thus, Carnot principle 1 is proved.

2. *The efficiency of all reversible cycles operating between the same temperature limits is the same.*

The proof of Carnot principle 2 is essentially the same as that used for Carnot principle 1 and is not given in detail. The student should note that this principle, combined with the first, proves that the reversible cycle and its associated processes indeed serve to establish the index of performance for heat engine cycles.

3. *The thermal efficiency of a reversible engine is a function solely of the upper and lower temperatures of the cycle and is not a function of the working substances used in the cycle.*

This third principle is somewhat different in its viewpoint, and part of the mathematical reasoning is quite abstract. We can argue this point qualitatively in the following manner. Let us assume that the efficiency of a reversible engine is a function of the working substance used in the cycle. By using two reversible cycles, as for principle 1, we can place a different working fluid in each of the cycles. One reversible cycle would be more efficient than the other, and by the identical reasoning used in principle 1, we would arrive at a violation of the Clausius statement. Thus, the efficiency of a reversible engine cycle cannot be a function of the working substance used in the cycle. By continuing this line of reasoning, we are also directed to the conclusion that the efficiency of a reversible engine is a function only of the upper and lower temperatures used in the cycle.

To establish the temperature function, we can resort to reasoning similar to that used by Fermi and Dodge.* Consider the three heat reservoirs shown in Figure 4.6 and maintained at temperatures

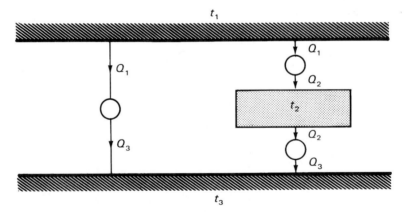

FIGURE 4.6 Derivation of the absolute temperature.

* E. Fermi, *Thermodynamics* (New York: Dover Publications, 1956); B. F. Dodge, *Chemical Engineering Thermodynamics* (New York: McGraw-Hill Book Company, 1944).

t_1, t_2, and t_3, respectively, on some arbitrary absolute temperature scale. Assume that three Carnot heat engines operate between these temperatures. Because the efficiency of the cycle is assumed to be some function of the temperature limits of the cycle, it follows from Equation (4.2) that for each engine, Q_r/Q_{in} is also a function of t_1 and t_2 for the respective temperatures and heat quantities associated with each engine. Therefore,

$$\frac{Q_1}{Q_2} = \phi_1(t_1, t_2) \tag{4.3a}$$

$$\frac{Q_2}{Q_3} = \phi_2(t_2, t_3) \tag{4.3b}$$

$$\frac{Q_1}{Q_3} = \phi_3(t_1, t_3) \tag{4.3c}$$

The symbol ϕ is interpreted to mean "a function of"; thus, from Equation (4.3a), Q_1/Q_2 is a function of t_1 and t_2. Dividing Equation (4.3c) by Equation (4.3b) and comparing with Equation (4.3a) yields

$$\frac{Q_1}{Q_2} = \phi_1(t_1, t_2) = \frac{\phi_3(t_1, t_3)}{\phi_2(t_2, t_3)} \tag{4.4}$$

The left side of Equation (4.4) indicates that ϕ_3/ϕ_2 is a function only of t_1 and t_2. Thus, the function of t_3 must cancel out of Equation (4.4), and we obtain

$$\frac{Q_1}{Q_2} = \phi_1(t_1, t_2) = \frac{\phi_3(t_1)}{\phi_2(t_2)} \tag{4.5}$$

At this point, it becomes impossible to determine the function in Equation (4.5) analytically, because it is entirely arbitrary and many temperature functions can satisfy it. Kelvin proposed that the temperature function in Equation (4.5) be taken as

$$\boxed{\frac{Q_1}{Q_2} = \frac{T_1}{T_2}} \tag{4.6}$$

Equation (4.6) is taken to be the definition of the *absolute thermodynamic temperature* scale. This temperature scale is the same as the absolute temperature scale that was defined by the ideal gas (see Chapters 1 and 6). Thus, the temperature functions given by Equations (4.5a) and (4.6) are simply the absolute temperatures.

If Equation (4.6) is inverted,

$$\frac{Q_2}{Q_1} = \frac{T_2}{T_1} \tag{4.6a}$$

Subtracting unity from each side of Equation (4.6a) gives us

$$\frac{Q_2}{Q_1} - 1 = \frac{T_2}{T_1} - 1 \tag{4.7}$$

Simplifying Equation (4.7) yields

$$\frac{Q_1 - Q_2}{Q_1} = \frac{T_1 - T_2}{T_1} = 1 - \frac{T_2}{T_1} \tag{4.7a}$$

In terms of the previous notation,

$$\eta = \frac{Q_{in} - Q_r}{Q_{in}} \times 100 = \frac{T_1 - T_2}{T_1} \times 100 = \left(1 - \frac{T_2}{T_1}\right) \times 100 \tag{4.7b}$$

From Equation (4.7b), we conclude the following:

1. The efficiency of a reversible-engine cycle is a function only of the upper and lower temperatures of the cycle.

2. Increasing the upper temperature while the lower temperature is kept constant increases the efficiency of the cycle.

3. Decreasing the temperature at which heat is rejected while keeping the upper temperature of the cycle constant increases the efficiency of the cycle.

As noted before, the temperature scale that is defined by Equation (4.6) is called the *absolute thermodynamic scale,* because it does not depend on the working substance.

ILLUSTRATIVE PROBLEM 4.1

A reversible engine operations between 1000°F and 80°F. (a) What is the efficiency of the engine? (b) If the upper temperature is increased to 2000°F while the lower temperature is kept constant, what is the efficiency of the engine? (c) If the lower temperature of the cycle is increased to 160°F while the upper temperature is kept at 1000°F, what is the efficiency of the cycle?

SOLUTION

Referring to Figure 4.7 and converting all temperatures to absolute temperatures, we obtain the following:

(a) $\dfrac{T_1 - T_2}{T_1} = \dfrac{1460 - 540}{1460} = 0.63 = 63\%$

(b) $\dfrac{T_1 - T_2}{T_1} = \dfrac{2460 - 540}{2460} = 0.78 = 78\%$

(c) $\dfrac{T_1 - T_2}{T_1} = \dfrac{1460 - 620}{1460} = 0.575 = 57.5\%$

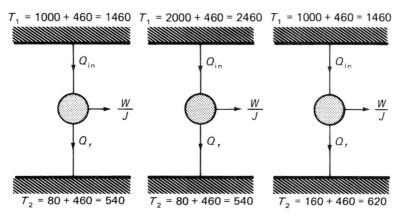

$T_1 = 1000 + 460 = 1460$ $T_1 = 2000 + 460 = 2460$ $T_1 = 1000 + 460 = 1460$

Q_{in} $\dfrac{W}{J}$ Q_r

$T_2 = 80 + 460 = 540$ $T_2 = 80 + 460 = 540$ $T_2 = 160 + 460 = 620$

FIGURE 4.7   Illustrative Problem 4.1.

ILLUSTRATIVE PROBLEM 4.2

Assume that 100 units of heat enter the reversible engine in Illustrative Problem 4.1. If the cycle is reversed, determine the amount of work into the cycle and the heat removed from the reservoir at T_2.

SOLUTION

The quantities of energy in the reversed cycle must equal those of the direct cycle. Several approaches are possible in solving this problem. We can use the efficiency obtained in Illustrative Problem 4.1. Therefore,

$$\frac{W/J}{Q_{in}} = 0.63 \quad \text{and} \quad \frac{W}{J} = 100\,(0.63) = 63 \text{ units out}$$

$$Q_{in} - Q_r = \frac{W}{J}$$

Therefore,

$$100 - 63 = 37 = Q_r$$

For the reversed cycle, we have only to note Q_r (or 37 units) is taken into the system from the low-temperature reservoir, 63 units of work enter the system, and 100 units are returned to the high-temperature reservoir.

ILLUSTRATIVE PROBLEM 4.3

It is desired to heat a house in the winter with a heat pump when the outside air is at 15°F. If the inside of the house is maintained at 70°F while the house loses 125,000 Btu/hr, what is the minimum horsepower input required?

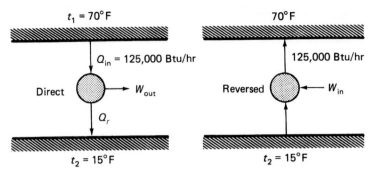

FIGURE 4.8 Illustrative Problem 4.3.

SOLUTION

Refer to Figure 4.8, and note that the reversed cycle is evaluated by considering the direct cycle first. From Equation (4.6),

$$\frac{Q_{in}}{T_1} = \frac{Q_r}{T_2}$$

where $Q_{in} = 125,000$ Btu/hr, $T_1 = 70 + 460 = 530°R$, and $T_2 = 15 + 460 = 475°R$:

$$Q_r = Q_{in} \times \frac{T_2}{T_1} = 125,000 \times \frac{475}{530}$$
$$= 112,028 \text{ Btu/hr}$$
$$\text{work} = Q_{in} - Q_r = 125,000 - 112,028$$
$$= 12,972 \text{ Btu/hr}$$

Therefore, the reversed cycle (heat pump) requires at least 12,972 Btu/hr of work input. This is,

$$\frac{12,972 \dfrac{\text{Btu}}{\text{hr}} \times 778 \dfrac{\text{ft·lb}_f}{\text{Btu}}}{60 \dfrac{\text{min}}{\text{hr}} \times 33,000 \dfrac{\text{ft·lb}_f}{\text{min·hp}}} = 5.1 \text{ hp}$$

ILLUSTRATIVE PROBLEM 4.4

A Carnot engine operates between a source temperature of 1000°F and a sink temperature of 100°F. If the engine is to have an output of 50 hp, determine the heat supplied, the efficiency of the engine, and the heat rejected. Refer to Figure 4.9.

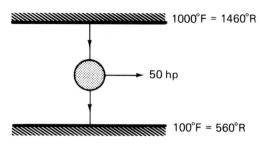

1000°F = 1460°R

50 hp

100°F = 560°R

FIGURE 4.9 Illustrative Problem 4.4.

SOLUTION

$$50 \text{ hp} = \frac{50 \times 33{,}000}{778} = 2120.8 \text{ Btu/min as output}$$

$$\begin{pmatrix} \text{because 1 hp is equivalent} \\ \text{to } 33{,}000 \text{ ft} \cdot \text{lb}_f/\text{min} \end{pmatrix}$$

$$\eta = \left(1 - \frac{T_2}{T_1}\right) \times 100 = \left(1 - \frac{560}{1460}\right) \times 100 = 61.6\%$$

$$= \frac{W/J}{Q_{in}} = \frac{2120.8}{Q_{in}}$$

$$Q_{in} = \frac{2120.8}{0.616} = 3443 \text{ Btu/min}$$

$$Q_r = 3443(1 - 0.616) = 1322 \text{ Btu/min}$$

ILLUSTRATIVE PROBLEM 4.5

A Carnot engine operates between a source temperature of 700°C and a sink temperature of 20°C. Assuming that the engine will have an output of 65 hp, determine the heat supplied, the efficiency of the engine, and the heat rejected.

SOLUTION

We can also refer to Figure 4.9, noting that $T_1 = 700 + 273 = 973°\text{K}$ and $T_2 = 20 + 273 = 293°\text{K}$. The efficiency of the Carnot engine is

$$\eta = \frac{T_1 - T_2}{T_1} \times 100 = \frac{973 - 293}{973} \times 100 = 69.9\%$$

The work output of the engine is $65 \times 0.746 = 48.49$ kJ/s. Because efficiency is work out/heat in,

$$\text{heat in} = Q_{\text{in}} = \frac{48.49}{0.699} = 69.37 \text{ kJ/s}$$

$$\text{heat rejected} = Q_{\text{out}} - \text{work out}$$
$$= 69.37 - 48.49 = 20.88 \text{ kJ/s}$$

As a check,

$$Q_r = Q_{\text{in}}(1 - \eta) = 69.37(1 - 0.699) = 20.88 \text{ kJ/s}$$

ILLUSTRATIVE PROBLEM 4.6 USING CALCULUS ENRICHMENT

If you have the choice of lowering the lowest temperature in a Carnot cycle by 1° or of raising the upper temperature by 1°, which would you do? Base your answer on the one that will yield the higher efficiency.

SOLUTION

We start the solution to this problem by writing the efficiency of the Carnot cycle as

$$\eta = (1 - T_2/T_1)$$

Differentiating the efficiency, first with respect to T_1 while holding T_2 constant and then with respect to T_2 while holding T_1 constant, yields

$$d\eta = -(T_2)(-dT_1/T_1^2) = (dT_1/T_1)(T_2/T_1) \quad \textbf{(a)}$$

and
$$d\eta = -dT_2/T_1 \quad \textbf{(b)}$$

If we now refer to Equation (b), we see that a 1° decrease in T_2 gives us an efficiency increase of $1/T_1$. From Equation (a), we see that a 1° increase in T_1 gives us an increase in efficiency of $(1/T_1)(T_2/T_1)$. This is always smaller than the result that we obtained for the 1° decrease in T_2, because the ratio of (T_2/T_1) is always less than unity. Thus, we conclude that if it is feasible, we should strive to decrease the lower temperature of the cycle as much as possible to increase the cycle efficiency. In most cases, the lower temperature is determined by ambient conditions, and it may not be possible to lower it to any extent.

ILLUSTRATIVE PROBLEM 4.7

Two Carnot engines are operated in series with the exhaust of the first engine being the input to the second engine. The upper temperature of this combination is 700°F, and the lower temperature is 200°F. If each engine has the same thermal efficiency, determine the exhaust temperature of the first engine (the inlet temperature to the second engine). Refer to Figure 4.10.

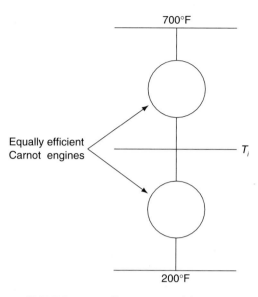

700°F

Equally efficient
Carnot engines

T_i

200°F

FIGURE 4.10 Illustrative Problem 4.7.

SOLUTION

The statement of the problem has the efficiency of both engines being equal. Denoting the efficacy of each engine as η_1 and η_2, respectively,

$$\eta_1 = (T_1 - T_i)/T_1$$

and

$$\eta_2 = (T_i - T_2)/T_2$$

Equating these efficiencies and using the given data,

$$(1160 - T_i)/1160 = (T_i - 660)/T_i$$

Solving,

$$1160T_i - T_i^2 = 1160T_i - (660)(1160)$$

or

$$T_i^2 = (660)(1160)$$

We can generalize this result as $T_i = \sqrt{T_1 T_2}$

Finally, for this problem, $T_i = \sqrt{(660)(1160)}$

and $T_i = 875°R = 415°F$

4.4 ENTROPY

If we refer to the working fluid and the changes that occur to it (for any reversible cycle operating between the same temperature limits), we have established that

$$\frac{Q_{in}}{T_1} = \frac{Q_r}{T_2}$$

In other words, the heat reception or rejection for the fluid in *any* reversible cycle divided by the temperature at which the heat is interchanged is a constant. The specific reversible paths that constitute the cycle do not change the value of these quantities. If the value of these quantities was a function of the specific reversible paths chosen, we could readily show that a violation of the second law would result.

The uniqueness of these ratios leads us to the conclusion that they may represent state functions of the fluid, and as such, we may define them for a reversible process as properties. In fact, it is correct to make this assumption. To generalize the foregoing conclusion, we define the quantity S as referring to this new property and call it *entropy*. On a unit mass basis, the specific entropy is s. The defining equation for entropy is given by

$$\Delta S = \frac{Q}{T} \qquad \text{reversible process} \qquad \Delta S \text{ in Btu/}^\circ\text{R or kJ/}^\circ\text{K}$$

$$\Delta s = \frac{q}{T} \qquad \text{reversible process} \qquad \Delta s \text{ in Btu/lb}_\text{m}\cdot^\circ\text{R or kJ/kg}\cdot^\circ\text{K}$$

(4.8)

Entropy is defined as a differential, because it is associated with a transfer of heat, which changes the state of the substance. Thus, the absolute value of the entropy of a substance may be assigned a zero value at any arbitrary state, as has been done in the development of the tables in Appendix 3.

To show that entropy is a property of the state and not a function of the path chosen, let us consider the following situation: A reversible cycle operates between states a and b as indicated by the path a, 1, b, 2, a in Figure 4.11. Let us also indicate a second possible return path, b, 3, a. For the first path (a, 1, b, 2, a),

$$\Delta S_{1,2} = \left[\frac{Q}{T}\right]_{a,b}^{\text{path 1}} + \left[\frac{Q}{T}\right]_{b,a}^{\text{path 2}}$$

(4.9)

where the symbol $\left[\ \right]_{a,b}^{\text{path 1}}$ indicates the sum of the Q/T items from a to b via path 1 and $\left[\ \right]_{b,a}^{\text{path 2}}$ denotes the same from b to a via path 2. For path b, 3, a, as the return path,

$$\Delta S_{1,3} = \left[\frac{Q}{T}\right]_{a,b}^{\text{path 1}} + \left[\frac{Q}{T}\right]_{b,a}^{\text{path 3}}$$

(4.10)

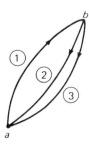

FIGURE 4.11 Proof that entropy is a property.

All the processes in these two cycles are reversible, which permits us to write the foregoing relations for known paths. Let us once again repeat that the two frictionless, quasi-static processes that are reversible are the adiabatic and isothermal. For the reversible isothermal process, Q is not a function of T, and the reversible adiabatic process requires Q to be zero. Thus, if a cycle is composed of these reversible processes, ΔS of the cycle is zero. We may generalize this statement for all reversible cycles and say that the summation of ΔS around the reversible cycle must equal zero. Using this fact and equating Equations (4.9) and (4.10) yields

$$\left[\frac{Q}{T}\right]_{b,a}^{\text{path 2}} = \left[\frac{Q}{T}\right]_{b,a}^{\text{path 3}} \qquad\qquad \textbf{(4.11)}$$

Thus, we can conclude that the function $(Q/T)_{\text{reversible}}$ represents a property that is a function only of the state of the fluid and is independent of the reversible path taken to reach the particular state. The importance of this statement is not just in the proof of the fact that entropy is a state function; it is important also in that it provides a means of calculating the change in entropy for any process. All that is necessary is a knowledge of the initial and final states of the process, because we can always (at least in principle) consider a reversible process between the same initial and final states. It must be emphasized that Q/T can be used to evaluate only the entropy change for a reversible process.

CALCULUS ENRICHMENT

In order to establish entropy as a property, let us first consider a mathematical relation known as *the inequality of Clausius.* Consider Figure A, which shows a reversible engine operating in a cycle with a constant upper temperature reservoir of T_0 and rejecting heat as a temperature of T to a second engine operating in a cycle. The second engine converts all of the rejected heat of the first engine into useful work. If we take both engines as our system, the total work output will equal the sum of the work outputs of the individual engines. Thus,

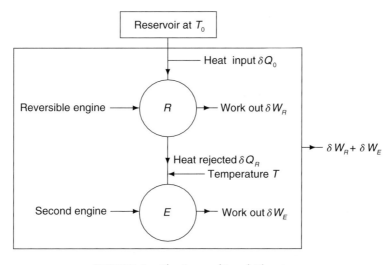

FIGURE A   The inequality of Clausius.

$$\delta W = \delta W_R + \delta W_E \qquad \text{(a)}$$

For the work of the first engine, we have

$$\delta W_R = \delta Q_0 (1 - T/T_0) \qquad \text{(b)}$$

and for the second engine,

$$\delta W_E = \delta Q_R \qquad \text{(c)}$$

Adding Equations (b) and (c) and simplifying yields the total work,

$$\delta W = (T_0/T)\, \delta Q_R \qquad \text{(d)}$$

We can write Equation (d) for the complete cycle using the notation that $\oint$ indicates an integration around the complete cycle as

$$\oint \delta W = T_0 \oint (\delta Q_R/T) \qquad \text{(e)}$$

If Figure A is studied, it will be seen that this arrangement cannot produce net work output, because it would violate the Kelvin–Planck statement of the second law. We therefore conclude that the only way this arrangement can operate is with a *net work input* and a heat flow *into* the reservoir. Equation (e), with T_0 constant, leads us to conclude

$$\oint \delta W \le 0 \qquad \text{(f)}$$

Combining this conclusion with Equation (e) yields

$$\oint (\delta Q/T) \le 0 \qquad \text{(g)}$$

This relation, Equation (g), is known as *the inequality of Clausius.*

For a reversible cycle, it is readily shown the equality in Equation (g) holds. Thus,

$$\oint (\delta Q/T)_{\text{rev}} = 0 \qquad \text{(h)}$$

In Chapter 1, we showed that a quantity that is a property is a function only of the state of a system. Such a quantity is also an exact differential and can be represented mathematically as

$$\oint dx = 0 \qquad \text{(i)}$$

We therefore conclude that Equation (h) defines a property, S, that we have called entropy, where the property entropy is defined as

$$dS = (\delta Q/T)_{\text{rev}} \qquad \text{(j)}$$

Therefore, for a reversible process, we have

$$\Delta S = S_2 - S_1 = \int_1^2 (dQ/T)_{\text{rev}} \qquad \text{(k)}$$

Because entropy is a state function, we may use it to portray graphically any equilibrium state of a fluid. Let us plot on temperature–entropy coordinates the reversible processes constituting the Carnot cycle. For convenience, we use a unit mass of working fluid, and the entropy coordinate becomes the specific entropy s. Let us consider each step of the cycle and interpret each step with the aid of Figure 4.12.

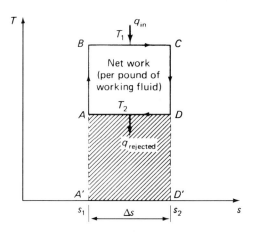

FIGURE 4.12 Direct Carnot cycle.

Energy as heat enters the cycle at the constant upper temperature T_1 with an increase in entropy from B to C. This is represented by the horizontal line BC. Because $\Delta s = q_{in}/T_1$, $q_{in} = T_1 \Delta s$, which is equivalent to the area under the line BC or area $A'BCD'$. The fluid is then expanded via a reversible adiabatic process, which cannot give rise to an increase in entropy because no heat enters or leaves the system during its course. The reversible adiabatic process is therefore carried out at constant entropy. Because of this fact, it is called an *isentropic* process. A reversible adiabatic process is isentropic, but not all isentropic processes are reversible adiabatic processes. Therefore, the expansion of this step is the vertical line CD. The next step in the cycle is the isothermal rejection of heat represented by the horizontal line DA. Because $\Delta s = q_r/T_2$, $T_2 \Delta s = q_r$, and the heat rejected by the cycle is represented by area $A'ADD'$, which is the area under the line DA. The final step in the cycle is the isentropic (reversible adiabatic) compression along path AB to return the cycle to its starting point. Because the area $A'BCD'$ is the heat into the cycle and area $A'ADD'$ is the heat rejected, the net work of the cycle is the area $ABCD$, or the area enclosed by the paths of the cycle on T–s coordinates. It should be noted that we have considered 1 lb of working fluid, because s is the specific entropy.

ILLUSTRATIVE PROBLEM 4.8

To vaporize 1 lb_m of saturated water into saturated steam at 200 psia, 843.7 Btu is required. If the temperature of this process is constant and equal to 381.86°F, what is the change in entropy for the process?

SOLUTION

For the reversible isothermal process, we can write $\Delta s = q/T$. Thus,

$$\Delta s = \frac{843.7}{381.86 + 460} \quad \text{or} \quad \Delta s = 1.002 \, \frac{\text{Btu}}{\text{lb}_m \cdot {}^\circ \text{R}}$$

ILLUSTRATIVE PROBLEM 4.9

If the process described in Illustrative Problem 4.8 is the heat reception portion of a Carnot cycle, what is the efficiency of the cycle if the lowest temperature of the cycle is 50°F? How much work is done per pound of fluid? How much energy is rejected?

SOLUTION

$$\eta = \frac{T_1 - T_2}{T_1} \times 100 = \left(1 - \frac{T_2}{T_1}\right) \times 100$$

$$= \left(1 - \frac{460 + 50}{460 + 381.86}\right) \times 100$$

$$= (1 - 0.606) \times 100 = 39.4\%$$

$$\frac{w}{J} = q_{\text{in}} \, \eta = (0.394)(843.7) = 332.4 \text{ Btu/lb}_m$$

$$q_r = q_{\text{in}} - \frac{w}{J} = 843.7 - 332.4 = 511.3 \text{ Btu/lb}_m$$

As an alternative solution and referring to Figure 4.12, we can write the following:

$$q_{\text{in}} = T_1 \Delta S \tag{4.12a}$$

$$q_r = T_2 \Delta s \tag{4.12b}$$

$$\frac{w}{J} = q_{\text{in}} - q_r = (T_1 - T_2)\Delta s \tag{4.12c}$$

$$q_r = T_2 \Delta s = (460 + 50)(1.002) = 511 \text{ Btu/lb}_m$$

$$\frac{w}{J} = q_{\text{in}} - q_r = 843.7 - 511 = 332.7 \text{ Btu/lb}_m$$

$$\eta = \frac{w/J}{q_{\text{in}}} \times 100 = \frac{332.7}{843.7} \times 100 = 39.4\%$$

ILLUSTRATIVE PROBLEM 4.10

Determine the change in entropy at 1.4 MPa for the vaporization of 1 kg of saturated water to saturated steam. Compare your answers to the *Steam Tables* in Appendix 3. ($h_{fg} = 1959.7$ kJ/kg, $s_{fg} = 4.1850$ kJ/kg·K, $t = 195.07°C$)

SOLUTION

As in Illustrative Problem 4.9, we can consider the vaporization process to be isothermal. Therefore,

$$\Delta s = \frac{h_{fg}}{T} = \frac{1959.7}{195.07 + 273}$$
$$= 4.1867 \text{ kJ/kg·°K}$$

If we use 273.16 to obtain temperature in degrees Kelvin,

$$\Delta s = 4.1853 \text{ kJ/kg·°K}$$

This compares very closely to the *Steam Tables* value.

It has already been noted that the cycle efficiency can be improved by raising the upper temperature of the cycle or lowering its lowest temperature. From Equation (4.12b), it is noted that the heat rejected during this cycle is equal to the change in entropy during the heat rejection process multiplied by the temperature at which the heat is rejected. In a general sense, the entropy change of the process becomes a measure of the amount of heat that becomes unavailable (rejected) during the cycle. For an irreversible process, we have already indicated that some energy becomes unavailable because of friction during the process. If it is assumed that the irreversible process can be restored to its original state by the input of additional work to the system via a reversible path, it can be shown that the increase in entropy during the irreversible process can be used to evaluate the least amount of work necessary to restore the system to its original state. Thus, a greater entropy change requires more work to restore the system. In this sense, entropy is also a measure of the unavailability of energy that occurs in an irreversible process.

ILLUSTRATIVE PROBLEM 4.11

One hundred Btu enters a system as heat at 1000°F. How much of this energy is unavailable with respect to a receiver at 50°F? Also, how much of this energy is unavailable with respect to a receiver at 0°F?

SOLUTION

Let us assume that a Carnot engine cycle operates between the two temperatures in each case. Figure 4.13 shows this problem on *T–S* coordinates. The shaded area under the 1460°R line represents the input of 100 Btu. Thus,

$$T_1 \Delta S = Q_{in}$$

or
$$\Delta S = \frac{Q_{in}}{T_1} = \frac{100}{1460} = 0.0685 \frac{Btu}{°R}$$

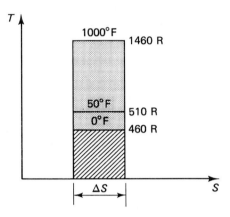

FIGURE 4.13   Illustrative Problem 4.11.

The crosshatched area represents heat rejected, or

$$Q_r = T_2 \Delta S$$
$$= (510)(0.0685) = 34.9 \text{ Btu}$$

The crosshatched area for a receiver at 0°F is

$$Q_r = (460)(0.0685) = 31.5 \text{ Btu}$$

Note from this problem that the maximum work output is

$$\text{work (max.)} = Q_{in} - T_2 \Delta S$$

and the *unavailable energy* is the difference between the heat into the cycle and the maximum work output, or Q_r. Also, by lowering T_2, we decrease the energy rejected and increase the energy available.

ILLUSTRATIVE PROBLEM 4.12

If 1 kJ enters a system as heat at 500°C, how much of this energy is unavailable with respect to (a) a receiver at 20°C and (b) a receiver at 0°C?

SOLUTION

Refer to Illustrative Problem 4.11.

$$\Delta S = \frac{Q_{in}}{T_1} = \frac{1000 \text{ J}}{500 + 273} = 1.2937 \frac{J}{°K}$$
$$Q_r = T_2 \Delta S$$

Therefore,

$$\text{(a)} \quad Q_r = (273 + 20)(1.2937)$$
$$= 379.05 \text{ J for the receiver at } 20°C$$
$$\text{(b)} \quad Q_r = (273 + 0)(1.2937)$$
$$= 353.18 \text{ J for the receiver at } 0°C$$

Note that ΔS for the process is independent of the receiver temperature.

The available energy for a process that receives heat is the amount of net work that would be obtained from the most efficient engine cycle of which the process could be made a part. Similarly, the unavailable energy for a process is the amount of energy that would be rejected from the most efficient engine cycle of which the process could be made a part. Because the atmosphere is essentially the lowest temperature reservoir to which heat is rejected, 77°F (25°C) and 1 atm is often taken as the heat rejection condition. This has been called the *dead state*. If a state is in equilibrium with the dead state, it can no longer be used to obtain useful work.

Any process can be considered to have a characteristic specific heat associated with it. Therefore, the heat transfer during such a process can be written as

$$q = c\Delta T \tag{4.13}$$

where it is reversible or irreversible.

Using Equation (4.13) and the definition of entropy given in Equation (4.8), $s_2 - s_1 =$ summation of $c\Delta T/T$ for a reversible process between its temperature limits. This relation can be used to evaluate the change in entropy for a process taking place between temperatures T_1 and T_2, and it will correctly evaluate the change in entropy for any process between the prescribed limits as long as we restrict it to apply to the system and not to the surroundings. Thus, for a constant c, it can be shown that the summation of $c\Delta T/T$ between the limits of T_1 and T_2 yields

$$\boxed{s_2 - s_1 = c \ln \frac{T_2}{T_1}} \tag{4.14}$$

where the symbol ln is the natural logarithm to the base e.

ILLUSTRATIVE PROBLEM 4.13

If 6 lb of a gas undergoes a constant-pressure change from 1440°F to some second temperature, determine the final temperature if the entropy change is −0.7062 Btu/°R. Assume the specific heat, $c_p = 0.361$ Btu/lb$_m$·°R for this gas.

SOLUTION

If we multiply both sides of Equation (4.14) by the mass m,

$$m\Delta s = \Delta S = mc_p \ln \frac{T_2}{T_1}$$

For this problem,

$$-0.7062 \frac{\text{Btu}}{°\text{R}} = 6 \text{ lb}_m \times 0.361 \frac{\text{Btu}}{\text{lb}_m°\text{R}} \times \ln \frac{T_2}{1440 + 460}$$

or

$$\ln \frac{T_2}{1440 + 460} = \frac{-0.7062}{6 \times 0.361} = -0.3260$$

From the definition of ln,

$$\frac{T_2}{1440 + 460} = e^{-0.3260}$$

$$T_2 = (1440 + 460)(0.7218)$$

and

$$T_2 = 1371.4°\text{R or } 911.4°\text{F}$$

ILLUSTRATIVE PROBLEM 4.14

If 1 lb_m of water at 500°F is adiabatically mixed with 1 lb_m of water at 100°F, determine the change in entropy. Assume that the specific heat of the hot and cold streams can be considered constant and equal to unity. Also, the specific heat of the mixture can be taken to be unity. Refer to Figure 4.14.

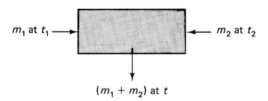

m_1 at t_1 m_2 at t_2

$(m_1 + m_2)$ at t

FIGURE 4.14 Illustrative Problem 4.14.

SOLUTION

To solve this problem, it is necessary first to establish the equilibrium temperature of the final mixture. We apply the first law of energy balance to the diagram, $m_1 c_1 t_1 + m_2 c_2 t_2 = (m_1 + m_2)c_{mix}t$, where t is the resulting mixture temperature. Therefore,

$$(1)(1)(500) + (1)(1)(100) = (1 + 1)(1)t \quad \text{and} \quad t = 300°F$$

For this problem, the hot stream is cooled from 500°F to 300°F. Because we may state that heat Q was removed during this process, Δs would be expected to be negative. By applying Equation (4.14),

$$\Delta s = 1 \ln\left(\frac{300 + 460}{500 + 460}\right) = \ln\left(\frac{760}{960}\right) = -\ln\left(\frac{960}{760}\right) = -\ln 1.263$$
$$= -0.233 \text{ Btu/lb}_m\cdot°R$$

The cold stream is heated from 100°F to 300°F; therefore,

$$\Delta s = 1 \ln\left(\frac{760}{560}\right) = \ln 1.357 = 0.305 \text{ Btu/lb}_m\cdot°R$$

The net change is $0.305 - 0.233 = +0.072 \text{ Btu/lb}_m\cdot°R$. An alternative solution to this problem is found by assuming an arbitrary temperature lower than any other temperature in the system. The change in entropy of each fluid is determined with respect to this arbitrary temperature, and the net change in entropy is the algebraic sum of the values found in this manner. An illustration of the procedure is found in Chapter 7, as well as in Illustrative Problem 4.15. If the specific heat of the substances being mixed varies with temperature, the appropriate average specific heats for the temperature intervals should be used.

ILLUSTRATIVE PROBLEM 4.15

Solve Illustrative Problem 4.14 using 0°F as a reference temperature for entropy.

SOLUTION

As a first step, we calculate the final mixture temperature as was done in Illustrative Problem 4.14 and find it to be 300°F. Next, let us calculate the initial entropy of each fluid above the 0°F base. For the "hot" fluid,

$$\Delta s = c \ln\frac{T_2}{T_1} = 1 \ln\left(\frac{500 + 460}{0 + 460}\right)$$
$$= 1 \ln 2.087 = 0.736 \text{ Btu/lb}_m\cdot°R$$

For the "cold" fluid,

$$s = c \ln\frac{T_2}{T_1} = 1 \ln\left(\frac{100 + 460}{0 + 460}\right)$$
$$= 1 \ln 1.217 = 0.196 \text{ Btu/lb}_m\cdot°R$$

At the final mixture temperature of 300°F, the entropy of each stream above 0°F is, for the "hot" fluid,

$$s = c \ln\frac{T_2}{T_1} = c \ln\left(\frac{300 + 460}{0 + 460}\right)$$
$$= 1 \ln 1.652 = 0.502 \text{ Btu/lb}_m\cdot°R$$

and for the "cold" fluid,

$$s = c \ln \frac{T_2}{T_1} = c \ln \left(\frac{300 + 460}{0 + 460} \right)$$

$$= 1 \ln 1.652 = 0.502 \ \text{Btu/lb}_m \cdot {}^\circ\text{R}$$

The change in entropy of the "hot" fluid is $0.502 - 0.736 = -0.234 \ \text{Btu/lb}_m \cdot {}^\circ\text{R}$. The change in entropy of the "cold" fluid is $0.502 - 0.196 = 0.306 \ \text{Btu/lb}_m \cdot {}^\circ\text{R}$.

$$\text{total change} = 0.306 - 0.234 = 0.072 \ \text{Btu/lb}_m \cdot {}^\circ\text{R}$$

The advantage of this alternative procedure is that by using a convenient, arbitrary datum below the lowest temperature in the system, we avoid negative logarithms. Either method is correct, and the choice of one over the other is purely personal preference.

The fact that entropy is a property leads us to inquire whether there are possible relations between entropy and other properties of a fluid. If they exist, they should prove to be very valuable, because it would then be possible to compute one from the other without resorting to experiment. Let us recall the energy equation applied to the reversible nonflow process:

$$q = \Delta u + \frac{p \Delta v}{J} \tag{4.15}$$

Because the process is assumed to be reversible, we may replace q by $T\Delta s$. Thus,

$$T\Delta s = \Delta u + \frac{p \Delta v}{J} \tag{4.16}$$

In Equation (4.16), $T\Delta s$ evaluates the energy as heat only if the process is reversible, and similarly, $p\Delta v/J$ evaluates only the work of a reversible process. However, each term of this equation consists of properties of the fluid that are not functions of the path. Thus, by applying the energy equation and the second law to a reversible, nonflow process, we have been able to arrive at a general equation involving only property terms. The only restriction on this equation occurs when each of the various terms is interpreted as either heat or work.

By referring to the reversible steady-flow process, we can arrive at another relationship between the properties. Alternatively, the same result can be obtained as follows: By definition,

$$h_2 - h_1 = (u_2 - u_1) + \frac{p_2 v_2 - p_1 v_1}{J} \tag{4.17}$$

or

$$\Delta h = \Delta u + \frac{\Delta(pv)}{J} \tag{4.18}$$

The change in the product pv/J equals

$$\frac{(p + \Delta p)(v + \Delta v) - pv}{J} \tag{4.19}$$

Carrying out the multiplication of the terms of Equation (4.19) yields

$$\frac{\Delta(pv)}{J} = \Delta p\,v + \Delta v\,p + \Delta p\Delta v \qquad \textbf{(4.20)}$$

The product $(\Delta p\Delta v)$ is the product of two small terms and can be considered negligible. Thus,

$$\frac{\Delta(pv)}{J} = \frac{\Delta p\,v + p\Delta v}{J} \qquad \textbf{(4.21)}$$

Replacing $\Delta(pv)/J$ in Equation (4.18) with its equivalent from Equation (4.21) yields

$$\Delta h = \Delta u + \frac{\Delta p\,v + \Delta v\,p}{J} \qquad \textbf{(4.22)}$$

By the substitution of Equation (4.22) into Equation (4.16) and rearranging, we have the desired result:

$$\boxed{T\Delta s = \Delta h - \frac{v\Delta p}{J}} \qquad \textbf{(4.23)}$$

We can thus relate entropy to enthalpy by means of Equation (4.23). Once again, each term in this equation is a property, but the interpretation of such terms as $T\Delta s$ as a heat quantity or $v\Delta p$ as a work quantity can be valid only for a reversible process.

 The student should note that Equations (4.15) through (4.23) can be written in terms of SI units simply by omitting the conversion factor J wherever it appears. The final result would be

$$\boxed{T\Delta s = \Delta h - v\Delta p} \qquad \textbf{(4.23a)}$$

CALCULUS ENRICHMENT

Applying the first and second laws to a nonflow system undergoing an internally reversible process, we obtain the following relation:

$$\delta q = \delta w + du \qquad \textbf{(a)}$$

For this process,

$$\delta w = p\,dv \qquad \textbf{(b)}$$

For the heat transfer, dq, we can write,

$$dq = T\,ds \qquad \textbf{(c)}$$

Substitution of (b) and (c) into Equation (a) yields,

$$T\,ds = du + p\,dv \qquad \textbf{(d)}$$

Equation (d) is known as the first of the Gibbs or $T\,ds$ equations.

Starting with the definition of enthalpy, we can obtain the second $T\,ds$ equation:

$$du = d(h - pv) = dh - p\,dv - v\,dp \tag{e}$$

Equating du in Equation (d) with du in Equation (e) gives us the desired second $T\,ds$ relation, namely,

$$T\,ds = dh - v\,dp \tag{f}$$

The $T\,ds$ equations involve only properties, and even though they were obtained by considering a nonflow, internally reversible system, they are valid for all systems and processes. Solving these equations for ds gives us

$$ds = du/T + p\,dv/T \tag{g}$$

and

$$ds = dh/T - v\,dp/T \tag{h}$$

By integration of either ds equation, we can obtain the change in entropy for a process knowing the p, v, T relation and the relation of temperature to either h or u.

It has been established that entropy is a property and can be evaluated by considering reversible paths connecting the given end states of a process. Let us consider the situation in which two processes start out from a given state. The first one is carried out reversibly until a second state is reached. The second process is carried out irreversibly until the same pressure as the first process has stopped at is reached. These processes operate between a fixed source at T_1 and a fixed sink at T_2 as part of a Carnot cycle. For the reversible process,

$$\frac{Q_{in}}{T_1} = \frac{Q_r}{T_2}$$

For the irreversible process, less net work is produced, with the consequence that more energy must be rejected as heat. Thus, Q_r/T_2 for the irreversible process can at best equal the equivalent ratio of the reversible process, or as is the case in all instances, it must be greater. Using this qualitative argument we can arrive at another important consequences of the second law:

*The entropy of an isolated system increases or in the limit remains the same.**

This *principle of the increase of entropy* also serves as a criterion of irreversibility. Thus, if we find during a process that the entropy of an isolated system increases, we must conclude that the process is irreversible. Another consequence of the principle of the increase of entropy is that at a given internal energy, that state having the greatest entropy will be the most probable state that the system will assume. At this most probable state, the system is said to be in stable equilibrium.

The foregoing can be summarized by the following simple equation:

$$\boxed{\Delta S \geq 0 \qquad \text{for isolated systems}} \tag{4.24}$$

which is interpreted to state that for any reversible change in an isolated system, the total entropy remains unchanged, and for any irreversible change, the total entropy increases.

---

* Keenan, *Thermodynamics.*

CALCULUS ENRICHMENT

From the inequality of Clausius, we can now develop the *principle of the increase of entropy*. Consider points A and B in Figure A that are connected by two processes,

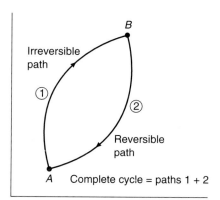

FIGURE A   The principle of the increase of entropy.

one reversible and the other irreversible. From the inequality of Clausius, we can write that

$$\oint (\delta Q/T) = \int_A^B (\delta Q/T) + \int_B^A (\delta Q/T) < 0 \qquad \textbf{(a)}$$

because the cycle is irreversible. For the reversible path, we can use the definition of entropy to give us,

$$\int_A^B (\delta Q/T) + S_A - S_B < 0 \qquad \textbf{(b)}$$

Transposing terms,

$$S_B - S_A \geq \int_A^B (\delta Q/T) \qquad \textbf{(c)}$$

Note that the equality holds for a reversible process and that the inequality holds for an irreversible process. Equation (c) is the general statement of the *principle of the increase of entropy*. For a reversible adiabatic process, the change in entropy will be zero, while for an irreversible adiabatic process, the entropy will increase. For a system that is isolated,

$$S_B - S_A \geq 0 \qquad \textbf{(e)}$$

and we conclude that *the entropy of an isolated system increases or in the limit remains the same*. Thus, if the changes in an isolated system are entirely reversible, the entropy will remain constant.

4.5 REVIEW

The reasoning in this chapter is more abstract than in previous chapters and therefore may present more difficulty for the student. Starting with the definition of thermal efficiency, we then continued with the statement of the second law of thermodynamics. For our purposes, the second law of thermodynamics can be stated as: "Heat cannot, of itself, flow from a lower temperature to a higher temperature." Based on the second law and the concept of reversibility, we were able to reason out the three general conclusions that Carnot arrived at:

1. No engine operating between fixed source (T_1) and sink (T_2) temperatures and continuously delivering work can be more efficient than a reversible engine operating between these same temperature limits.

2. The efficiency of all reversible cycles operating between the same temperature limits is the same.

3. The thermal efficiency of a reversible engine is a function solely of the upper and lower temperatures of the cycle and is not a function of the working substance used in the cycle.

In 1824, Sadi Carnot proposed a reversible cycle consisting of two constant-temperature and two adiabatic processes. The utility of this cycle lies in the ease with which we were able to calculate its thermal efficiency using the temperature function proposed by Kelvin. This is the absolute thermodynamic temperature, which coincides with our earlier definition of absolute temperature. We thus became able to determine the limiting efficiency of any cycle operating between upper and lower temperature limits by calculating the thermal efficiency of the Carnot cycle between the same temperature limits. This definition of the absolute thermodynamic temperature combined with the Carnot cycle also enabled us to define a new property called entropy. The property entropy was also shown to be a measure of the unavailability of energy that occurs in an irreversible process. Using the energy equation for a reversible, nonflow process enabled us to derive some general relations among entropy, enthalpy, and internal energy.

KEY TERMS

Terms used for the first time in this chapter are:

absolute thermodynamic temperature an arbitrary temperature function used with the Carnot cycle. For practical purposes, it is identical to the absolute temperature scales of Kelvin or Rankine.

Carnot cycle a reversible cycle proposed by Sadi Carnot in 1824 that consists of two constant-temperature and two adiabatic processes.

cycle a series of thermodynamic processes during which the working fluid can be made to undergo changes involving energy transitions and is subsequently returned to its original state.

entropy a property of a substance; also a measure of the unavailability that occurs in an irreversible process.

heat engine a continuously operating system across whose boundaries flow only heat and work.

isentropic a process carried out at constant entropy. A reversible adiabatic process is isentropic.

isothermal a process carried out at constant temperature.

principle of the increase of entropy the entropy of an isolated system increases or in the limit remains the same.

reversible process any process performed so that the system and all its surroundings can be restored to their initial states by performing the process in reverse.

second law of thermodynamics heat cannot, of itself, pass from a lower temperature to a higher temperature.

thermal efficiency the ratio of the net work of a cycle to the heat added to the cycle.

EQUATIONS DEVELOPED IN THIS CHAPTER

thermal efficiency

$$\eta = \frac{\text{net work output}}{\text{heat added}} \times 100 \qquad (4.1)$$

thermal efficiency

$$\eta = \frac{Q_{in} - Q_r}{Q_{in}} \times 100 = \left(1 - \frac{Q_r}{Q_{in}}\right) \times 100 \qquad (4.2)$$

Kelvin temperature function

$$\frac{Q_1}{Q_2} = \frac{T_1}{T_2} \qquad (4.6)$$

efficiency of a reversible cycle

$$\eta = \frac{T_1 - T_2}{T_1} \times 100 = \left(1 - \frac{T_2}{T_1}\right) \times 100 \qquad (4.7a)$$

entropy

$$\Delta S = \frac{Q}{T} \quad \text{or } \Delta s = \frac{q}{T} \qquad \text{reversible process} \qquad (4.8)$$

entropy change

$$s_2 - s_1 = c \ln \frac{T_2}{T_1} \qquad (4.14)$$

general property relation

$$T\Delta s = \Delta u + \frac{p\Delta v}{J} \qquad (4.16)$$

general property relation

$$T\Delta s = \Delta h - \frac{v\Delta p}{J} \qquad (4.23)$$

general property relation (SI)

$$T\Delta s = \Delta h - v\Delta p \qquad (4.23a)$$

entropy increase principle

$$\Delta S \geq 0 \qquad \text{for all isolated systems} \qquad (4.24)$$

QUESTIONS

4.1 Can a heat engine do anything other than deliver work?

4.2 Define a cycle.

4.3 Thermal efficiency is defined to be the ratio of net work to heat added in a cycle. Would you think that this is an appropriate definition to be used when a refrigerator is being discussed?

4.4 Do you know of any process in nature that is reversible?

4.5 There are four distinct events that occur in the Carnot cycle. Starting with the heat reception event, name and describe each one.

4.6 There are three conclusions reached from the Carnot cycle regarding reversible cycles. What are they?

4.7 What was the contribution by Kelvin?

4.8 What two factors determine the limiting efficiency of any cycle?

4.9 How does the combination of the work of Kelvin and Carnot help in the design of power cycles?

4.10 The statement has been made that entropy is a property. What does this mean to you?

4.11 Other than being a property, what else does entropy represent?

4.12 It is possible to derive a relation among entropy, enthalpy, and internal energy that is perfectly general. Under what conditions can you associate heat and work to these terms?

4.13 What is the principle of the increase of entropy?

4.14 Why is the principle of the increase of entropy important?

PROBLEMS

Problems Involving Thermal Efficiency

4.1 An engine cycle is operated to produce 12.5 Btu/min as work. If 100 Btu/min as heat enters the cycle, determine the heat rejected and the efficiency of the cycle.

4.2 A reversible engine produces 10 hp as work while 1270 Btu/min enter the engine as heat. Determine the energy rejected and the thermal efficiency of the cycle.

4.3 A heat engine produces 75 hp for a heat addition of 9000 Btu/min. Determine the thermal efficiency and heat rejected by this engine.

4.4 A Carnot power plant operates between a high temperature reservoir at 1500°F and a low temperature reservoir at 70°F. Determine the heat addition if the plant produces 1000 MW of power.

4.5 An engine produces 10 kJ of work while 80 kJ enters the engine cycle as heat. Determine the energy rejected and the thermal efficiency of the cycle if it is reversible.

4.6 An inventor claims to have an engine that has a thermal efficiency of 90%. Comment on the claim.

4.7 A Diesel engine uses 450 lb_m of fuel per hour. If the burning of the fuel releases 18,000 Btu/lb_m and the engine produces 840 hp, determine its thermal efficiency.

4.8 An internal combustion engine uses 0.38 lb_m/hr of gasoline for each horsepower that it produces. If the gasoline has an energy content of 19,500 Btu/lb_m, determine the thermal efficiency of the engine.

*4.9 An automobile engine has an efficiency of 23%. If the engine produces 100 hp as its output, determine the heat input to the engine. If gasoline has a heat content of 20,750 Btu/lb_m and a specific gravity of 0.74, determine the gallons of gasoline used per hour at this rating.

*4.10 An automobile has an efficiency of 22%. It is rated to give an output of 100 kW. Determine the heat input to the engine and the fuel consumption in liters per hour if the gasoline has a heat content of 4.9×10^4 kJ/kg and a specific gravity of 0.82.

4.11 A gas turbine has a thermal efficiency of 25% and develops 10,000 hp. If the fuel releases 18,500 Btu/lb_m, determine the rate of fuel usage.

4.12 A power plant burns 900 kg of coal per hour and produces 480 kW of power. If each kg of coal releases 5.9 MJ, determine the efficiency of the plant.

4.13 A gas expands isothermally in a cylinder, and it is found to deliver 1 kJ of work. During the process, it is found that 1 kJ of heat is added to the gas to keep it isothermal. It would appear that all the heat has been converted to useful work in this process. Is this possible? Base your answer on both the first and second laws of thermodynamics.

Problems Involving Reversible Heat Engines and the Carnot Cycle

4.14 A Carnot cycle is operated between 1000°F and 500°F. If the upper temperature is increased to 1100°F, what is the efficiency of the cycle? If the lower temperature is decreased to 400°F, what is the efficiency of the cycle?

4.15 A Carnot cycle operates between 900°C and 100°C. Determine the efficiency of the cycle, the heat rejected, and the useful work output if 100 kJ enters the cycle as heat.

4.16 A Carnot engine operates between 940°F and 60°F and produces 80 hp. How much heat is supplied to the engine, how much heat is rejected by the engine, and what is the thermal efficiency of the engine?

4.17 A reversible power cycle has a lower reservoir temperature of 300°K. The cycle produces 42 kW as power and rejects 180 kW to the cold reservoir. Determine the temperature of the high temperature reservoir.

4.18 A Carnot engine operates between 1350°F and 125°F. If it rejects 55 Btu as heat, determine the work output.

4.19 A Carnot engine produces 40 hp when operated between temperatures of 1500°R and 500°R. Calculate the heat supplied per hour to the engine.

4.20 A Carnot cycle operates between 1800°F and 200°F. Determine the efficiency of the cycle, the heat rejected, and the useful work output if 500 Btu/min enters the cycle as heat.

4.21 Is it possible for a reversible engine cycle to produce 100 hp if it receives 300 Btu/s at 1000°F and rejects heat at 500°F?

4.22 An inventor tests an engine cycle that she has built and finds that its thermal efficiency is 58%. If the engine operates between 540°F and 60°F, should she release her findings to the press or retest the engine?

4.23 An inventor claims that he can obtain 155 hp as work from an engine with a heat input equivalent to 500 hp. Comment on the validity of his claim if the engine is to operate between 250°F and 10°F. Also comment on the probability of achieving these conditions.

*4.24 Two engines are operated between the same temperature limits, and each engine rejects 1000 kJ as heat to the sink. If engine A is irreversible and has a claimed efficiency of 50% and engine B is reversible and has a claimed efficiency of 40%, prove numerically that a violation of the second law will occur under these conditions.

*4.25 Two reversible engines operate in series between temperatures of 1000°F and 100°F. If the first engine receives 1000 Btu, determine the total work output of this arrangement and the temperature at the inlet to the second engine. Assume that both engines have the same thermal efficiency.

*4.26 A Carnot cycle has an efficiency of 32%. Assuming that the lower temperature is kept constant, determine the percent increase of the upper temperature of the cycle if the cycle efficiency is raised to 48%.

*4.27 A reversible cycle is shown on a T–S diagram in Figure P4.27. Calculate the efficiency of this cycle.

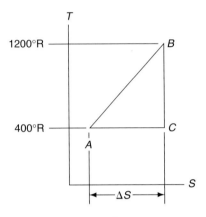

FIGURE P4.27

*4.28 A house is cooled by the removal of 68,000 Btu/hr using a reversed Carnot cycle. Assume that the house is kept at 72°F and that the outside temperature is 90°F. If the motor used is rated at 6.5 hp and electricity costs 14 cents per kW·hr, determine the cost per day to operate this cycle. Repeat this problem for a 4-hp motor, and compare your results. Note that this is an ideal cycle and that all processes are ideal.

*4.29 A reversible engine has an efficiency of 40%. If it is operated in such a manner that 15% of the work is lost as friction in its moving parts, what would the efficiency be?

4.30 A reversible engine operates between 900°F and 200°F. What is the maximum efficiency of this cycle? If the rating of this engine is 1 hp, determine the heat rejected per minute.

4.31 If the cycle in Problem 4.30 is reversed, how much heat is rejected to the upper temperature reservoir?

4.32 A reversible engine operates between 750°C and 80°C. Determine the maximum efficiency of the cycle and the heat rejected if the engine is rated at 1 hp.

4.33 If the cycle in Problem 4.32 is reversed, how much heat is rejected to the upper temperature reservoir?

4.34 In order to keep a freezer at 20°F while the room temperature is 72°F, 13,500 Btu/hr must be removed. If the freezer operates as a reversible cycle, determine the minimum work input required.

4.35 A Carnot engine operates between a source temperature of 900°F and a sink temperature of 100°F. If the engine produces 100,000 ft·lb of work, determine its thermal efficiency, the heat supplied, and the heat rejected.

4.36 The ocean is an almost infinite energy source. Off the coast of Hawaii, the surface temperature is 86°F, while at a depth of 2200 ft, the temperature is 43°F. Determine the maximum efficiency of a reversible power plant that is operated between these temperature limits. This cycle is described in Chapter 9 as the OTEC (ocean thermal energy conversion) plant.

4.37 A Carnot engine receives 1000 Btu as heat at 1000°F and operates with a sink temperature of 150°F. Determine the efficiency, work out, and heat rejected by this engine.

4.38 A Carnot engine operates between an upper temperature of 700°C and a lower temperature of 30°C. If the engine can produce 150 kJ of work, determine its thermal efficiency, the heat supplied, and the heat rejected.

4.39 A Carnot engine develops 30 hp while rejecting 100 MJ/hr as heat to a receiver at 20°C. Determine the upper temperature of the cycle and its efficiency.

4.40 A Carnot engine develops 20 hp while rejecting 70,000 Btu/hr as heat to a receiver at 80°F. Determine the efficiency of the engine and the source temperature.

4.41 A reversed Carnot engine operates between 150°F and 60°F; 100 Btu/min is extracted from the cold body. Determine the horsepower required to operate the engine.

4.42 A Carnot engine operates with an efficiency of 30% and, when reversed, removes 1000 Btu/hr as heat from the cold sink. Compute the heat rejected to the hotter source.

4.43 A reversed Carnot engine operates between 80°C and 20°C; 100 kJ/min as heat is extracted from the cold body. Determine the horsepower required to operate the engine.

4.44 A reversed Carnot engine operates with an efficiency of 30%, removing 1 MJ as heat from the cold sink. Calculate the heat rejected to the hotter source.

4.45 A reversed Carnot cycle operates between temperatures of 100°F and 0°F. If the motor input is 5 hp, determine the heat removed from the cold sink.

4.46 A reversed Carnot cycle is used to heat a room. If 9000 Btu/hr is required, how much energy must be put in electrically if the room is to be at 70°F and a deep well at 40°F is used as the source of working fluid for the cycle?

4.47 A reversed Carnot engine operates between 40°C and 5°C. If the motor input is 4 hp, determine the heat removed from the cold sink.

4.48 A reversed Carnot engine is used to heat a room. If 10 kW is required to replace heat losses from the room, how much energy must be put into the motor if the room is to be kept at 22°C and a deep well at 4°C is used as the source of the working fluid of the cycle?

4.49 A steam power plant can operate at 1400°F when the water used in the condenser is at 40°F. Calculate the maximum efficiency that this plant can have.

4.50 A Carnot engine operates with an efficiency of 40%. If the upper temperature is 500°C, determine the lower temperature of the cycle.

4.51 A Carnot engine develops 25 hp while operating with an efficiency of 32%. If the engine rejects heat to the atmosphere at 70°F, determine the upper temperature of operation and the heat rejected.

4.52 A reversible engine cycle operates with an efficiency of 35%. If the engine develops 19 kW while rejecting heat to a reservoir at 5°C, determine the heat input, the heat rejected, and the upper temperature of the cycle.

4.53 A reversed Carnot engine is used to cool a large room in the summer. The room is to be kept at 24°C when the outside temperature is 35°C. It is estimated that 150 kJ/min is the heat transfer from the outside to the room. Calculate the power required to operate the engine.

Problems Involving Entropy

4.54 A Carnot engine receives 5800 Btu/min as heat at a temperature of 400°F. If the engine develops 50 hp, what is the receiver temperature? What is the change in entropy during the heat input portion of the cycle?

4.55 What is the change in entropy during the heat rejection portion of the engine cycle described in Problem 4.54?

*4.56 The *Steam Tables* show that at 100°F, the heat of vaporization, h_{fg}, is 1037.0 Btu/lb$_m$. Determine the change in entropy, s_{fg}, and compare the calculated value to the value given for s_{fg} in the *Steam Tables*. (*Note:* The *Steam Tables* use $-459.67°F$ as absolute zero.)

4.57 To change a pound of saturated water to saturated steam at 32°F, 1075.4 Btu/lb$_m$ is required. Determine the change in entropy for this process.

4.58 Water is heated from 32°F to 212°F. If the specific heat of this process is taken to be constant and equal to unity, what is the change in entropy for this process per pound of water?

4.59 If the process described in Problem 4.58 is reversed, what is the change in entropy per pound of water?

4.60 For the engine of Problem 4.35, determine the change in entropy for the heat reception and heat rejection portions of the cycle.

*4.61 A Carnot engine cycle is operated as described in Problem 4.35. If there are irreversibilities in the cycle that cause an entropy change in the heat rejection portion of the cycle to be 10% greater than in the heat reception portion, determine the thermal efficiency of the cycle.

4.62 One kilogram of air is heated at constant volume from 100°C to 400°C. If c_v is 0.7186 kJ/kg·K, determine the change in entropy of the air.

4.63 If 1 lb$_m$ of air is heated at constant volume from 70°F by the addition of 200 Btu as heat, calculate the change in entropy for the process. Assume $c_v = 0.171$ Btu/lb$_m$·°R and remains constant for the entire process.

4.64 If 1 lb$_m$ of air is heated at constant pressure from 70°F by the addition of 200 Btu as heat, calculate the change in entropy for the process. Assume $c_p = 0.24$ Btu/lb$_m$·°R and remains constant for the entire process.

4.65 If the process described in Problem 4.62 is carried out at constant pressure with $c_p = 1.0062$ kJ/kg·°K, determine the change in entropy of the process.

*4.66 A Carnot engine has an entropy change during the isothermal heat addition portion of the cycle of 0.15 Btu/lb$_m$·°R when it operates between temperatures of 1500°F and 500°F. How much work does this engine deliver per pound of working fluid?

4.67 One pound of air is heated at constant volume. If the value of $c_v = 0.171$ Btu/lb$_m$·°R and is constant, determine the entropy change when the air is heated from 200°F to 800°F.

4.68 If the process in Problem 4.67 is carried out at constant pressure and $c_p = 0.24$ Btu/lb$_m$·°R, determine the entropy change.

4.69 A Carnot engine operates between 200°F and 1800°F. If the entropy change during the heat addition portion of the cycle equals 0.25 Btu/lb$_m$·°R, calculate the work per lb$_m$ of working fluid.

4.70 A Carnot power cycle operates between 1200°K and 300°K. If 4.8 MJ of heat is supplied to the cycle, determine the heat rejected and the change in entropy during the heat reception portion of the cycle.

4.71 One lb$_m$ of air is heated at a constant temperature of 100°F by the addition of 250 Btu. Calculate the change in entropy for this process.

4.72 One kilogram of air is heated at a constant temperature of 40°C by the addition of 200 kJ. Determine the change in entropy for this process.

*4.73 Five pounds of water at 200°F is mixed with 2 lb at 100°F. What is the total change in entropy for this process? Assume that all specific heats are constant and equal to unity. Work this problem by considering each stream separately.

4.74 A process is carried out at constant volume. It is found that the change in entropy for the process is 0.05 Btu/lb$_m$·°R. If c_v is constant and equal to 0.171 Btu/lb$_m$·°R and the upper temperature of the process is 250°F, determine the initial temperature of the process.

4.75 A gas is cooled from 250°C, and it is found that the entropy change is −0.1427 kJ/kg·°K. Assuming that the appropriate specific heat for the process is 1.4876 kJ/kg·°K, determine the temperature after cooling.

4.76 A gas is heated at constant pressure from 100°F. If $c_p = 0.31$ Btu/lb$_m$·°R and the entropy change is 0.2107 Btu/lb$_m$·°R, determine the upper temperature.

4.77 A process is carried out at constant volume. It is found that the entropy change is 1 kJ/kg·°K. If c_v is constant and equal to 0.7186 kJ/kg·°K and the lower temperature of the process is 20°C, determine the upper temperature.

*4.78 An *isentropic* process is carried out with the working fluid expanding from 700°F to 100°F. The specific heat at constant pressure is 0.24 Btu/lb$_m$·°R. During the process the internal energy changes by 102.6 Btu/lb$_m$. Determine the change in entropy for this process.

4.79 A constant-temperature process is carried out with a concurrent change in entropy of 0.1 Btu/lb$_m$·°R at 600°F. Determine the unavailable portion of the energy received with respect to a receiver at 50°F and at 100°F.

4.80 A constant-temperature process is carried out with a concurrent change in entropy of 0.2 kJ/kg·°K at 400°C. What is the unavailable portion of the energy received with respect to a receiver at 20°C and at 45°C?

*4.81 During a constant-pressure process, it is found that the change in entropy is equal to $\frac{1}{2}c_p$. If c_p is constant and the initial temperature is 450°F, determine the final temperature for the process.

*4.82 One pound of air undergoes two processes that are in series. The first process consists of an isothermal expansion during which the air is supplied with 22 Btu at a temperature of 386°F. This process is then followed by a reversible adiabatic process that takes the gas to a temperature of 134°F. Determine the change in entropy for these combined processes.

5

Properties of Liquids and Gases

Learning Goals

After reading and studying the material in this chapter, you should be able to:

1. Define the word *phase,* and distinguish among the three phases of matter.

2. Understand and use the fact that the vaporization process is carried out at constant temperature and that less heat is required to vaporize a unit mass of water as the pressure is raised.

3. Define the state that is known as the critical state.

4. Understand and use the nomenclature used in the *Steam Tables.*

5. Understand the use of the triple point as the reference state for zero internal energy and zero entropy.

6. Define the term *quality,* and us it to determine the properties in the wet region.

7. Obtain the properties of steam in the subcooled, saturated, and superheated states.

8. Describe the various graphical representations of the properties of steam, especially the temperature—entropy and the enthalpy—entropy (Mollier) charts.

9. Represent state paths for various processes on the Mollier chart.

10. Show how the throttling process can be used to determine the quality of wet steam.

11. Be aware that industry uses computer systems to determine the properties of steam either by use of equations or by use of readily available commercial programs, and be able to use the enclosed computer disk to obtain the properties of steam in both SI and USCS units.

5.1 INTRODUCTION

In Chapter 1, a brief, introductory study of the properties of a gas was presented. In this chapter, the properties of liquids and gases are investigated in some detail, because the state of a system can be described in terms of its properties. Based on the observable properties of pressure, temperature, and volume, it is possible to derive other properties that can also suffice to describe the state of a system. The relationship among the pressure, volume, and temperature of a system, when

176

expressed mathematically, is called the *equation of state* for the substance in question. A state is an equilibrium state if no finite rate of change of state can occur without a finite change, temporary or permanent, in the state of the environment. The ideal gas relation derived in Chapter 1 and used extensively throughout Chapter 6 is an example of such an equation of state.

A *phase* of a substance can be defined as that part of a pure substance which consists of a single, homogeneous aggregate of matter. The three common phases that are usually spoken of are *solid, liquid,* and *gaseous.* When dealing with the gaseous phase, a distinction is made between a gas and a vapor that is somewhat artificial but in common usage. The term *vapor* is applied to the gaseous phase that is in contact with saturated liquid or is not far removed from the saturated state, and the term *gas* is used for the vapor that is either at very low pressure or far removed from the saturated state.

5.2   LIQUIDS AND VAPORS

The distinction between vapor and liquid is usually made (in an elementary manner) by stating that both will take up the shape of their containers, but that the liquid will present a free surface if it does not completely fill its container. The vapor (or gas) will always fill its container.

With the foregoing in mind, let us consider the following system: A container is filled with water, and a movable, frictionless piston is placed on the container, as shown in Figure 5.1. As heat is added to the system, the temperature of the system will increase. Note that the pressure on the system is being kept constant by the weight of the piston. The continued addition of heat will cause the temperature of the system to increase until the pressure of the vapor generated exactly balances the pressure of the atmosphere plus the pressure due to the weight of the piston.

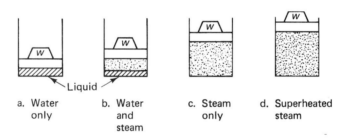

FIGURE 5.1   Heating water and steam at constant pressure.

At this point, the vapor and liquid are said to be *saturated.* As more heat is added, the liquid that was at saturation will start to vaporize. The two-phase mixture of vapor and liquid has only one degree of freedom, and as long as liquid is present, vaporization will continue at a constant temperature. As long as liquid is present, the mixture is said to be *wet,* and both the liquid and vapor are saturated. After all the liquid is vaporized, only the vapor is present, and the further addition of heat will cause the temperature of the vapor to increase at constant system pressure. This state is called the *superheat state,* and the vapor is said to be *superheated.* If this process is carried out at various pressures, a singular curve of temperature of saturation as a function of pressure will be generated. Such a curve is called a *vapor pressure* or *saturation curve* for the substance.

For a constant rate of heat input, it is possible to plot a curve of temperature as a function of time or heat added at a given system pressure, say 100 psia. This curve is shown as a solid line in Figure 5.2. If the rate of heating is kept constant and the identical system is made to undergo this process again but with a system pressure of 1000 psia, the dashed curve of Figure 5.2 will represent

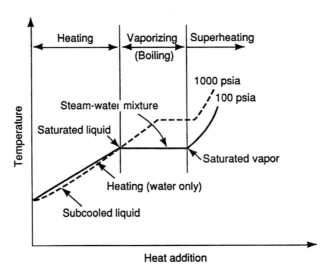

FIGURE 5.2   Heating of water.

the heat-added temperature history of the system. The decreased slope at 1000 psia during the heating portion indicates an increased specific heat of the liquid (water) due to the increased pressure. At the end of the heating portion, the liquid is saturated at a higher temperature. The water is subsequently vaporized at constant temperature, and the length of the horizontal portions of the curves is proportional to the heat necessary to vaporize the fluid at constant pressure (the latent heat of vaporization). At 1000 psia, the figure shows that less heat is required to vaporize a unit mass of fluid than at 100 psia. After all the water is vaporized, the specific heat of the vapor at the higher pressure is higher.

   Repeating the experiment a number of times and measuring pressure, temperature, and specific volume enables us to plot curves of $T–v$, $p–T$, and $p–v$. Figure 5.3(a) shows a typical $T–v$

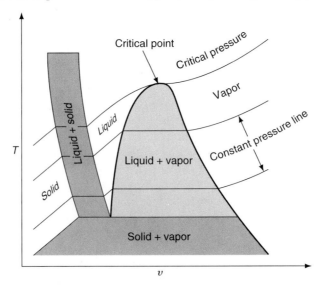

FIGURE 5.3(a)   A typical temperature–specific volume ($T–v$) diagram. (Not to scale.)

diagram (not to scale) for a substance that expands on freezing, that is, when it goes from liquid to solid. From this diagram, we can see that the water can exist in several states:

1. A pure solid state, usually called ice.

2. A pure liquid state.

3. A pure gaseous state (vapor), usually called steam.

4. Equilibrium mixtures of liquid and vapor states, liquid and solid states, and solid and vapor states.

Figures 5.3(b) and 5.3(c) show typical $p–T$ and $p–v$ diagrams (not to scale) for the vapor–liquid states of a typical substance such as water. The *triple point* is the state in which all three phases are in equilibrium. The *critical point* is the state in which it is impossible to distinguish

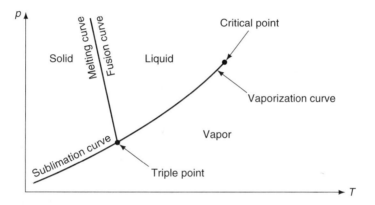

FIGURE 5.3(b) A typical pressure–temperature ($p–T$) diagram. (Not to scale.)

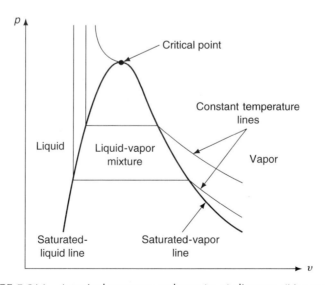

FIGURE 5.3(c) A typical pressure–volume ($p–v$) diagram. (Not to scale.)

between the liquid and vapor phases. At the critical point, the properties of the liquid and vapor phases are identical. The temperature and pressure at this point are known, respectively, as the critical temperature and the critical pressure.

All the data can be presented as a single, three-dimensional figure, as shown in Figure 5.4 (not to scale). Figures 5.3(a), 5.3(b), and 5.3(c) can be obtained from Figure 5.4 by projection onto the T–v, p–T, and p–v planes. Note that the triple line shown in Figure 5.4 projects as a point when it is viewed parallel to the v axis. Due to this, the projection of the triple line is called the triple point.

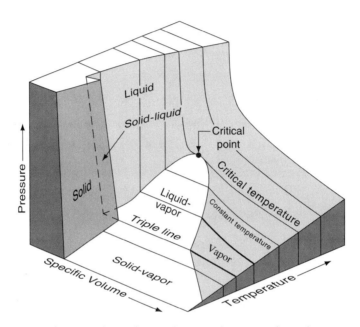

FIGURE 5.4 A typical p–v–T figure for a substance that expands on freezing. (Not to scale.)

5.3 THERMODYNAMIC PROPERTIES OF STEAM

Water has been used as a thermodynamic working fluid for many centuries, and it has been the subject of extensive research to establish its thermodynamic properties in the last two centuries. Figure 5.5 shows a modern *once-through steam generator* in which water that is subcooled (i.e., the pressure is greater than the saturation pressure corresponding to its temperature) enters at one end of a continuous tube and superheated steam is discharged at the other end of the tube. In the design shown in Figure 5.5, the water–steam flow is through the furnace walls, primary horizontal superheater, and finally, the first and second sections of secondary (pendant) superheater. The water–steam flow is through a multitude of tubes that discharge into a common header at the outlet. A unit such as this can handle pressures above or below the critical pressure.

The need to know the thermodynamic properties of water from the subcooled liquid state to states above the critical point is obvious if units such as that shown in Figure 5.5 are to be properly designed. Research on thermodynamic properties of water has been carried out throughout the world, and as a result of this research, extensive tables of the properties of water have been

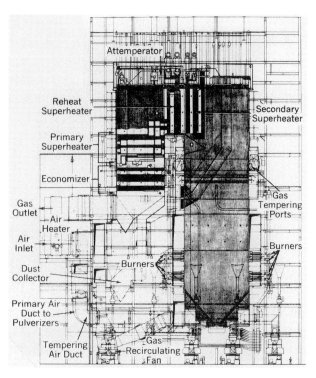

FIGURE 5.5 Universal-pressure boiler for pulverized-coal firing. Superheater outlet pressure, 3650 psi; primary and reheat steam temperatures, 1003°F; capacity, 9,300,000 lb of steam per hour. *(Courtesy of Babcock and Wilcox Co.)*

published.* There are nine tables of thermodynamic properties in the *Steam Tables* (as this work is commonly called) that cover the properties of ice, water, and steam. These tables include the transport properties, viscosity, and thermal conductivity. The triple point, that is, the state in which the solid, liquid, and vapor are in equilibrium, is used as the reference state for zero internal energy and zero entropy. The use of this reference state gives rise to negative values of internal energy, enthalpy, and entropy of the liquid state at 32°F. The fact that these properties are negative should not cause any difficulty, because we are invariably interested in differences in properties.

Tables 1 and 2 of the *Steam Tables* are for the saturated state. Table 1 presents the data with temperature as the independent variable; Table 2 presents essentially the same data with pressure as the independent variable. All the tabulated data are arranged to facilitate linear interpolation between values. The nomenclature used in the *Steam Tables* is shown in Table 5.1. It should be noted that all the tabulated properties are specific properties and are tabulated per unit mass. Table 1, the temperature table, goes to the critical temperature of 705.44°F, and Table 2 tabulates data up to the corresponding critical pressure of 3203.6 psia. Portions of these tables are shown in Figures 5.6 (5.6a for SI units) and 5.7 (5.7a for SI units). The SI table goes to 374.136°C and 22.09 MPa.

* J. H. Keenan, F. G. Keyes, P. G. Hill, and J. G. Moore, *Steam Tables—Thermodynamic Properties of Water Including Vapor, Liquid, and Solid Phases* (New York: John Wiley & Sons, Inc., 1969). These tables are in English units. A separate volume bearing the same title with the additional parenthetical phrase (*SI Units*) has also been published by the same authors, 1969, 1978.

TABLE 5.1
Nomenclature Used in the Steam Tables

	English units	SI units
t = temperature	°F	°C
p = absolute pressure	psia or in. Hg	MPa
v = specific volume	ft³/lb$_m$	m³/kg
h = enthalpy	Btu/lb$_m$	kJ/kg
s = entropy	Btu/lb$_m$·°R	kJ/kg·°K
u = internal energy	Btu/lb$_m$	kJ/kg
Subscripts		
f = property of the saturated liquid		
g = property of the saturated vapor		
fg = property change due to evaporation		

Note that the change in property going from saturated liquid to saturated vapor (the fg subscript) is given for internal energy, enthalpy, and entropy, but not for specific volume. The relation for the saturation state properties, that is, f, g, and fg, is given by

$$
\begin{aligned}
h_g &= h_f + h_{fg} \\
v_g &= v_f + v_{fg} \\
s_g &= s_f + s_{fg} \\
u_g &= u_f + u_{fg}
\end{aligned}
$$

(5.1)

and, in addition, the definition of enthalpy in terms of internal energy, pressure, and specific volume must also be satisfied. Thus,

$$ h = u + \frac{pv}{J} $$

(5.2)

or in SI units,

$$ h = u + pv $$

(5.2a)

The following examples will serve to illustrate the use of Tables 1 and 2.

ILLUSTRATIVE PROBLEM 5.1

Determine the enthalpy of saturated steam at 90°F using the tabulated properties of pressure, specific volume, and internal energy in Table 1. Compare the result with the tabulated value of h.

TABLE 1

Saturation Temperatures

Temp. °F	Press. Lb_f Sq. In.	Specific Volume		Internal Energy			Enthalpy			Entropy		
		Sat. Liquid	Sat. Vapor	Sat. Liquid	Evap.	Sat. Vapor	Sat. Liquid	Evap.	Sat. Vapor	Sat. Liquid	Evap.	Sat. Vapor
t	p	v_f	v_g	u_f	u_{fg}	u_g	h_f	h_{fg}	h_g	s_f	s_{fg}	s_g
90	0.6988	0.016099	467.7	58.07	982.2	1040.2	58.07	1042.7	1100.7	0.11165	1.8966	2.0083
91	0.7211	0.016102	454.0	59.06	981.5	1040.6	59.07	1042.1	1101.2	0.11346	1.8922	2.0056
92	0.7439	0.016105	440.9	60.06	980.8	1040.9	60.06	1041.5	1101.6	0.11527	1.8877	2.0030
93	0.7674	0.016108	428.2	61.06	980.2	1041.2	61.06	1041.0	1102.0	0.11708	1.8833	2.0003
94	0.8914	0.016111	415.9	62.06	979.5	1041.5	62.06	1040.4	1102.4	0.11888	1.8788	1.9977
95	0.8162	0.016114	404.0	63.06	978.8	1041.9	63.06	1039.8	1102.9	0.12068	1.8744	1.9951
96	0.8416	0.016117	392.4	64.05	978.1	1042.2	64.06	1039.2	1103.3	0.12248	1.8700	1.9925
97	0.8677	0.016121	381.3	65.05	977.5	1042.5	65.05	1038.7	1103.7	0.12427	1.8657	1.9899
98	0.8945	0.016124	370.5	66.05	976.8	1042.8	66.05	1038.1	1104.2	0.12606	1.8613	1.9874
99	0.9220	0.016127	360.1	67.05	976.1	1043.2	67.05	1037.5	1104.6	0.12785	1.8569	1.9848

FIGURE 5.6 Extract from saturation table.*

* Figures 5.6, 5.7, 5.10., and 5.11 are extracted from *Steam Tables* by J. H. Keenan, F. G. Keyes, P. G. Hill, and J. G. Moore, John Wiley & Sons, Inc., 1969, with permission. More detailed values are found in the Appendix 3. The "a" tables are from the *SI Units Edition*, 1978.

TABLE 1

Saturation Temperatures

Temp. °C	Press. MPa	Specific Volume		Internal Energy			Enthalpy			Entropy		
		Sat. Liquid	Sat. Vapor	Sat. Liquid	Evap.	Sat. Vapor	Sat. Liquid	Evap.	Sat. Vapor	Sat. Liquid	Evap.	Sat. Vapor
t	p	$10^3 v_f$	$10^3 v_g$	u_f	u_{fg}	u_g	h_f	h_{fg}	h_g	s_f	s_{fg}	s_g
30	0.004246	1.0043	32 894	125.78	2290.8	2416.6	125.79	2430.5	2556.3	0.4369	8.0164	8.4533
31	0.004496	1.0046	31 165	129.96	2288.0	2418.0	129.97	2428.1	2558.1	0.4507	7.9822	8.4329
32	0.004759	1.0050	29 540	134.14	2285.2	2419.3	134.15	2425.7	2559.9	0.4644	7.9483	8.4127
33	0.005034	1.0053	28 011	138.32	2282.4	2420.7	138.33	2423.4	2561.7	0.4781	7.9146	8.3927
34	0.005324	1.0056	26 571	142.50	2279.5	2422.0	142.50	2421.0	2563.5	0.4917	7.8811	8.3728
35	0.005628	1.0060	25 216	146.67	2276.7	2423.4	146.68	2418.6	2565.3	0.5053	7.8478	8.3531
36	0.005947	1.0063	23 940	150.85	2273.9	2424.7	150.86	2416.2	2567.1	0.5188	7.8147	8.3336
37	0.006281	1.0067	22 737	155.03	2271.1	2426.1	155.03	2413.9	2568.9	0.5323	7.7819	8.3142
38	0.006632	1.0071	21 602	159.20	2268.2	2427.4	159.21	2411.5	2570.7	0.5458	7.7492	8.2950
39	0.006999	1.0074	20 533	163.38	2265.4	2428.8	163.39	2409.1	2572.5	0.5592	7.7167	8.2759

FIGURE 5.6a Extract from saturation table (SI units).

TABLE 2

Saturation Pressures

Press. Lb$_f$ Sq. In. p	Temp. °F t	Specific Volume		Internal Energy			Enthalpy			Entropy		
		Sat. Liquid v_f	Sat. Vapor v_g	Sat. Liquid u_f	Evap. u_{fg}	Sat. Vapor u_g	Sat. Liquid h_f	Evap. h_{fg}	Sat. Vapor h_g	Sat. Liquid s_f	Evap. s_{fg}	Sat. Liquid s_g
115	338.12	0.017850	3.884	308.95	798.8	1107.7	309.33	881.0	1190.4	0.48786	1.1042	1.5921
116	338.77	0.017858	3.852	309.62	798.2	1107.8	310.01	880.5	1190.5	0.48870	1.1027	1.5914
117	339.41	0.017865	3.821	310.29	797.6	1107.9	310.68	880.0	1190.7	0.48954	1.1012	1.5907
118	340.04	0.017872	3.790	310.96	797.1	1108.1	311.34	879.5	1190.8	0.49037	1.0996	1.5900
119	340.68	0.017879	3.760	311.62	796.6	1108.2	312.01	879.0	1191.0	0.49119	1.0981	1.5893
120	341.30	0.017886	3.730	312.27	796.0	1108.3	312.67	878.5	1191.1	0.49201	1.0966	1.5886
121	341.93	0.017894	3.701	312.92	795.5	1108.4	313.32	877.9	1191.3	0.49282	1.0951	1.5880
122	342.55	0.017901	3.672	313.57	794.9	1108.5	313.97	877.4	1191.4	0.49363	1.0937	1.5873
123	343.17	0.017908	3.644	314.21	794.4	1108.6	314.62	876.9	1191.6	0.49443	1.0922	1.5866
124	343.78	0.017915	3.616	314.85	793.9	1108.7	315.26	876.4	1191.7	0.49523	1.0907	1.5860

FIGURE 5.7 Extract from saturation table.

TABLE 2

Saturation Pressures

Press. MPa p	Temp. °C t	Specific Volume		Internal Energy			Enthalpy			Entropy		
		Sat. Liquid $10^3 v_f$	Sat. Vapor $10^3 v_g$	Sat. Liquid u_f	Evap. u_{fg}	Sat. Vapor u_g	Sat. Liquid h_f	Evap. h_{fg}	Sat. Vapor h_g	Sat. Liquid s_f	Evap. s_{fg}	Sat. Vapor s_g
1.00	179.91	1.1273	194.44	761.68	1822.0	2583.6	762.81	2015.3	2778.1	2.1387	4.4478	6.5865
1.02	180.77	1.1284	190.80	765.47	1818.8	2584.2	766.63	2012.2	2778.9	2.1471	4.4326	6.5796
1.04	181.62	1.1296	187.30	769.21	1815.6	2584.8	770.38	2009.2	2779.6	2.1553	4.4177	6.5729
1.06	182.46	1.1308	183.92	772.89	1812.5	2585.4	774.08	2006.2	2780.3	2.1634	4.4030	6.5664
1.08	183.28	1.1319	180.67	776.52	1809.4	2585.9	777.74	2003.3	2781.0	2.1713	4.3886	6.5599
1.10	184.09	1.1330	177.53	780.09	1806.3	2586.4	781.34	2000.4	2781.7	2.1792	4.3744	6.5536
1.12	184.89	1.1342	174.49	783.62	1803.3	2586.9	784.89	1997.5	2782.4	2.1869	4.3605	6.5473
1.14	185.68	1.1353	171.56	787.11	1800.3	2587.4	788.40	1994.6	2783.0	2.1945	4.3467	6.5412
1.16	186.46	1.1364	168.73	790.56	1797.4	2587.9	791.86	1991.8	2783.6	2.2020	4.3332	6.5351
1.18	187.23	1.1375	165.99	793.94	1794.4	2588.4	795.28	1989.0	2784.2	2.2093	4.3199	6.5292

FIGURE 5.7a Extract from saturation table (SI units).

SOLUTION

From Table 1 (Figure 5.6),

$$p = 0.6988 \text{ psia}$$
$$v_g = 467.7 \text{ ft}^3/\text{lb}_m$$
$$u_g = 1040.2 \text{ Btu}/\text{lb}_m$$

Because

$$h = u + \frac{pv}{J} \quad \text{and} \quad h_g = u_g + \frac{pv_g}{J}$$

therefore,

$$h_g = 1040.2 \text{ Btu}/\text{lb}_m + \frac{0.6988 \text{ lb}_f/\text{in}^2 \times 144 \text{ in}^2/\text{ft}^2 \times 467.7 \text{ ft}^3/\text{lb}_m}{778 \text{ ft·lb}_f/\text{Btu}}$$

$$= 1100.7 \text{ Btu}/\text{lb}_m$$

This value is in agreement with 1100.7 Btu/lb$_m$ for h_g from Table 1.

ILLUSTRATIVE PROBLEM 5.2

Determine the enthalpy of saturated steam at 30°C using the tabulated properties of pressure, specific volume, and internal energy in Table 1. Compare the result with the tabulated value of h_g.

SOLUTION

From Table 1 (Figure 5.6a),

$$p = 0.004246 \text{ MPa} = 4.246 \text{ kPa}$$
$$v_g = 32894/10^3 \text{ m}^3/\text{kg} = 32.894 \text{ m}^3/\text{kg}$$
$$u_g = 2416.6 \text{ kJ}/\text{kg}$$

Because

$$h = u + pv \quad \text{and} \quad h_g = u_g + pv_g,$$

therefore,

$$h_g = 2416.6 \text{ kJ}/\text{kg} + 4.246 \text{ kN}/\text{m}^2 \times 32.894 \text{ m}^3/\text{kg}$$
$$= 2556.27 \text{ kJ}/\text{kg}$$

The tabulated value of h_g from Table 1 is 2556.3 kJ/kg. (The student should carefully note the units used.)

Determine the enthalpy, entropy, specific volume, and internal energy of saturated steam at 118 psia. Assume that the values for 115 and 120 psia are available, and perform the necessary interpolations. Compare the results with the tabulated values for 118 psia.

SOLUTION

The necessary interpolations are best done in tabular form as shown:

p	h_g	
115	1190.4	Table 2
118	1190.8	$(h_g)_{118} = 1190.8$
120	1191.1	

$\frac{3}{5}(1191.1 - 1190.4) = 0.42$
$(h_g)_{118} = 1190.4 + 0.42$
$\qquad = 1190.8 \text{ Btu/lb}_m$

p	v_g	
115	3.884	Table 2
118	3.792	$(v_g)_{118} = 3.790$
120	3.730	

$\frac{3}{5}(3.884 - 3.730) = 0.09$
$(v_g)_{118} = 3.884 - 0.09 = 3.792 \text{ ft}^3/\text{lb}_m$

p	s_g	
115	1.5921	Table 2
118	1.5900	$(s_g)_{118} = 1.5900$
120	1.5886	

$\frac{3}{5}(1.5921 - 1.5886) = 0.0021$
$(s_g)_{118} = 1.5921 - 0.0021 = 1.5900$

p	u_g	
115	1107.7	Table 2
118	1108.06	$(u_g)_{118} = 1108.1$
120	1108.3	

$\frac{3}{5}(1108.3 - 1107.7) = 0.36$
$(u_g)_{118} = 1107.7 - 0.36 = 1108.06$

The interpolation process that was done in tabular form for this illustrative problem can also be demonstrated by referring to Figure 5.8 for the specific volume. It will be seen that the results of this problem and the tabulated values are essentially in exact agreement and that linear interpolation is satisfactory in these tables.

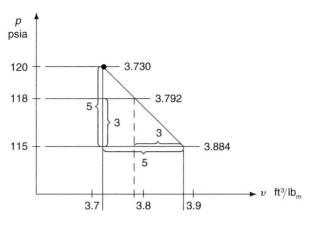

FIGURE 5.8 Illustrative Problem 5.3.

ILLUSTRATIVE PROBLEM 5.4

Determine h_{fg} for saturated steam at 115 psia using the pressure, volumes, and internal data given in Table 2 (Figure 5.7).

SOLUTION

By definition,

$$h_g = u_g + \frac{pv_g}{J}$$

$$h_f = u_f + \frac{pv_f}{J}$$

and $\quad h_{fg} = h_g - h_f = (u_g - u_f) + \dfrac{p(v_g - v_f)}{J} = u_{fg} + \dfrac{p(v_g - v_f)}{J}$

From Table 2 at 115 psia,

$$u_{fg} = 798.8 \text{ Btu/lb}_m$$
$$u_g = 3.884 \text{ ft}^3/\text{lb}_m$$
$$v_f = 0.017850 \text{ ft}^3/\text{lb}_m$$

Note that if the table had v_{fg}, we could have read it directly because $v_g - v_f = v_{fg}$. Proceeding gives

$$h_{fg} = 798.8 \text{ Btu/lb}_m + \frac{115 \text{ lb}_f/\text{in}^2 \times 144 \text{ in}^2/\text{ft}^2(3.884 - 0.017850) \text{ ft}^3/\text{lb}_m}{778 \text{ ft·lb}_f/\text{Btu}}$$

$$= 798.8 + 82.3 = 881.1 \text{ Btu/lb}_m$$

The tabulated value is 881.0 Btu/lb$_m$, and the agreement is satisfactory.

Determine h_{fg} for saturated steam at 1.0 MPa using the pressure, volumes, and internal energy data given in Table 2 (Figure 5.7a).

SOLUTION

Refer to Illustrative Problem 5.4. From Table 2 at 1.0 MPa,

$$u_{fg} = 1822.0 \text{ kJ/kg}$$
$$v_g = 194.44/10^3 = 0.19444 \text{ m}^3/\text{kg}$$
$$v_f = 1.1273/10^3 = 0.0011273 \text{ m}^3/\text{kg}$$
$$p = 1.0 \text{ MPa}$$

Because $v_{fg} = v_g - v_f$,

$$v_{fg} = 0.19444 - 0.0011273 = 0.1933127 \text{ m}^3/\text{kg}$$
$$h_{fg} = v_{fg} + p(v_{fg})$$
$$= 1822.0 \text{ kJ/kg} + (1000 \text{ kN/m}^2)(0.1933127 \text{ m}^3/\text{kg})$$
$$= 2015.3 \text{ kJ/kg}$$

The tabulated value is 2015.3 kJ/kg, which is in exact agreement with the value calculated.

ILLUSTRATIVE PROBLEM 5.6

Determine h_{fg} at 115 psia by considering this process to be a reversible, constant-temperature process where $T\Delta s = \Delta h$.

SOLUTION

For constant-temperature, reversible vaporization, $h_{fg} = \Delta h = T\Delta s = Ts_{fg}$. Therefore, $(388.12 + 460)1.1042 = 881.3$ Btu/lb$_m$, which is also in good agreement with the tabular values. Use of $-459.67°F$ for absolute zero, which is the value used in the table, gives almost exact agreement.

Between the saturated liquid and the saturated vapor, there exists a mixture of vapor plus liquid (the *wet* region). To denote the state of a liquid–vapor mixture, it is necessary to introduce a term describing the relative quantities of liquid and vapor in the mixture. The *quality* of a mixture (*x*) is defined as the ratio of the mass of vapor to the mass of the mixture. Thus, in 1 lb of mixture, there must be $(1 - x)$ lb of liquid. Another term is used to describe the wet region; *percent moisture*. A mixture whose quality is 80% has 20% moisture by weight and is simply said to have 20% moisture.

Consider a mixture weighing 1 lb and having a quality *x*. In this mixture, there is *x* lb of vapor and $(1 - x)$ lb of liquid. The enthalpy of this mixture per pound of mixture is the sum of the enthalpies of the components. Therefore, the enthalpy of the liquid portion of the mixture is

$$h_l = (1 - x)h_f$$

and the enthalpy of the vapor portion is

$$h_v = xh_g$$

The sum of these terms is the enthalpy of the mixture h_x:

$$h_x = (1 - x)h_f + xh_g \tag{5.3a}$$

By the same reasoning, the entropy, internal energy, and specific volume of the wet mixture are given as

$$s_x = (1 - x)s_f + xs_g \tag{5.3b}$$

$$u_x = (1 - x)u_f + xu_g \tag{5.3c}$$

$$v_x = (1 - x)v_f + xv_g \tag{5.3d}$$

It is sometimes more convenient to express Equations (5.3) in terms of the property change during vaporization. By noting that the *fg* property is equal to the *g* value minus the *f* value, it is possible to rewrite Equations (5.3) as follows:

$$h_x = h_f + xh_{fg} \tag{5.4a}$$

$$s_x = s_f + xs_{fg} \tag{5.4b}$$

$$u_x = u_f + xu_{fg} \tag{5.4c}$$

$$v_x = v_f + xv_{fg} \tag{5.4d}$$

The relationship amount these quantities can be demonstrated by referring to Figure 5.9, which is a *T–s* diagram for water. The saturation curve is indicated, and a point below the saturation curve is in the wet region. The *T–s* diagram is discussed further later in this chapter, but for the present, it should be noted that the expressions for entropy in Equations (5.3) and (5.4) can be directly determined from Figure 5.9.

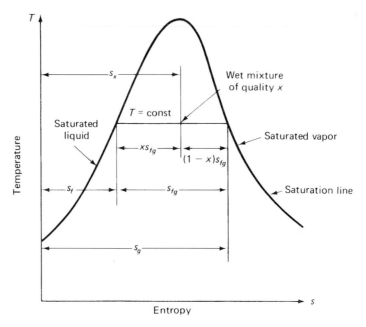

FIGURE 5.9 *T–s* diagram showing relation of properties and quality.
[From F. P. Durham, Thermodynamics, 2nd ed. (Englewood Cliffs, N.J.:
Prentice-Hall, Inc., 1959), p. 43, with permission.]

ILLUSTRATIVE PROBLEM 5.7

A wet steam mixture at 120 psia is found to have a quality of 80%. Determine its entropy, enthalpy, internal energy, and specific volume.

SOLUTION

Using Table 2 and a quality of 80% ($x = 0.8$), we have

$$s_x = s_f + xs_{fg} = 0.49201 + 0.8(1.0966) = 1.3693 \text{ Btu/lb}_m\cdot°\text{R}$$
$$h_x = h_f + xh_{fg} = 312.67 + 0.8(878.5) = 1015.47 \text{ Btu/lb}_m$$
$$u_x = u_f + xu_{fg} = 312.27 + 0.8(796.0) = 949.07 \text{ Btu/lb}_m$$
$$v_x = v_f + xv_{fg} = 0.017886 + 0.8(3.730 - 0.017886)$$
$$= 2.988 \text{ ft}^3/\text{lb}_m$$

As a check on the solution obtained, we calculate

$$u_x = h_x - \frac{p_x v_x}{J}$$

$$= 1015.47 \text{ Btu/lb}_m - \frac{120 \frac{\text{lb}_f}{\text{in}^2} \times 144 \frac{\text{in}^2}{\text{ft}^2} \times 2.988 \frac{\text{ft}^3}{\text{lb}_m}}{778 \text{ ft·lb}_f/\text{Btu}}$$

$$= 949.10 \text{ Btu/lb}_m$$

which agrees with the 949.07 Btu/lb$_m$.

ILLUSTRATIVE PROBLEM 5.8

A wet mixture at 1.0 MPa is found to have a quality of 85%. Determine its entropy, enthalpy, internal energy, and specific volume.

SOLUTION

Using $x = 0.85$ yields

$s_x = s_f + xs_{fg} = 2.1387 \text{ kJ/kg·°K} + (0.85)(4.4487 \text{ kJ/kg·°K})$
   $= 5.9201 \text{ kJ/kg·°K}$

$h_x = h_f + xh_{fg} = 762.81 \text{ kJ/kg} + (0.85)(2015.3 \text{ kJ/kg})$
   $= 2475.82 \text{ kJ/kg}$

$u_x = u_f + xu_{fg} = 761.68 \text{ kJ/kg} + (0.85)(1822.0 \text{ kJ/kg})$
   $= 2310.38 \text{ kJ/kg}$

$v_x = v_f + xv_{fg} = [1.1273 \text{ m}^3/\text{kg}$
$\qquad\qquad\qquad + (0.85)(194.44 \text{ m}^3/\text{kg} - 1.1273 \text{ m}^3/\text{kg})] \times 10^{-3}$
   $= 165.44/10^3 = 0.16544 \text{ m}^3/\text{kg}$

As a check,

$$u_x = h_x - p_x v_x = 2475.82 \text{ kJ/kg} - \frac{1.0 \times 10^6 (0.16544 \text{ m}^3/\text{kg})}{10^3 \text{ kPa/MPa}}$$

$$= 2310.38 \text{ kJ/kg}$$

Again, the agreement is exact. Note the units used for consistency.

ILLUSTRATIVE PROBLEM 5.9

A mixture of wet steam at 90°F is found to have an enthalpy of 900 Btu/lb$_m$. What is its quality?

SOLUTION

For the wet mixture, $h_x = h_f + xh_{fg}$. Solving for x gives us

$$x = \frac{h_x - h_f}{h_{fg}}$$

Using the data from Table 1 (Figure 5.6), we have

$$x = \frac{900 \text{ Btu/lb}_m - 58.07 \text{ Btu/lb}_m}{1042.7 \text{ Btu/lb}_m} = 0.807 = 80.7\%$$

ILLUSTRATIVE PROBLEM 5.10

A mixture of wet steam at 30°C is found to have an enthalpy of 2000.0 kJ/kg. Determine its quality.

SOLUTION

Because $h_x = h_f + xh_{fg}$,

$$x = \frac{h_x - h_f}{h_{fg}} = \frac{2000.0 \text{ kJ/kg} - 125.79 \text{ kJ/kg}}{2430.5 \text{ kJ/kg}} = 0.771 \quad \text{or} \quad 77.1\%$$

Table 3 of the *Steam Tables* gives the properties of the superheated vapor, and it occupies 89 pages in the main table. The beginning of this table is for water vapor at low pressure, which can be treated as an ideal gas. For the present, we will not be concerned with this section of Table 3. The main portion of the table extends to pressures of 15,000 psia and 2400°F. In the SI tables, the values go to 100 MPa and 1300°C. For the superheated region, it is necessary to specify two variables, such as pressure and temperature or enthalpy and pressure, in order to specify the state of the vapor. Table 3 is organized using pressure and temperature as independent variables, and it lists values of specific volume, internal energy, enthalpy, and entropy corresponding to specified values of temperature and pressure. Referring to Figure 5.10 (Figure 5.10a gives SI units), which is an extract from Table 3, it will be noted that the topmost line lists the different pressures in psia and in parentheses displays the saturation temperature corresponding to the pressure. The second horizontal line down lists column headings. All values are specific values, and the temperature is listed in °F. The third horizontal line shown lists saturation (Sat.) values corresponding to the pressure listed on the first line of the table. These values are for convenience so that it is not necessary to turn back to Tables 1 and 2 for saturation properties. The vertical column at the leftmost and rightmost parts of the table gives temperature in °F in bold letters. The desired thermodynamic properties at a given temperature are read horizontally underneath the appropriate pressure heading. It will be noted that at the upper part of the table, values are given in italics for certain temperatures with a horizontal line

TABLE 3

Vapor

p (t Sat.)	320 (423.39)				325 (424.84)				330 (426.27)			
t	v	u	h	s	v	u	h	s	v	u	h	s
Sat.	1.4493	1118.6	1204.4	1.5058	1.4274	1118.6	1204.5	1.5044	1.4061	1118.7	2204.6	1.5031
410	1.4110	1110.8	1194.4	1.4944	1.3852	1110.0	1193.3	1.4917	1.3602	1109.2	1192.2	1.4890
420	1.4398	1116.6	1201.9	1.5030	1.4139	1115.9	1200.9	1.5004	1.3887	1118.1	1199.9	1.4978
430	1.4677	1122.2	1209.2	1.5112	1.4416	1121.5	1208.2	1.5087	1.4163	1120.8	1207.3	1.5061
440	1.4949	1127.7	1216.2	1.5191	1.4686	1127.0	1215.4	1.5166	1.4430	1126.4	1214.5	1.5142
450	1.5213	1133.0	1223.1	1.5267	1.4948	1132.4	1222.3	1.5243	1.4691	1131.8	1221.5	1.5219
460	1.5472	1138.1	1229.8	1.5340	1.5205	1137.6	1229.0	1.5316	1.4945	1137.0	1228.2	1.5293
470	1.5726	1143.2	1236.3	1.5411	1.5456	1142.6	1235.6	1.5387	1.5194	1142.1	1234.9	1.5364
480	1.5975	1148.1	1242.7	1.5479	1.5702	1147.6	1242.0	1.5456	1.5438	1147.1	1241.4	1.5434
490	1.6219	1152.9	1249.0	1.5545	1.5944	1152.4	1248.3	1.5523	1.5678	1152.0	1247.7	1.5501

p (t Sat.)	335 (427.68)				340 (429.07)				345 (430.45)			
t	v	u	h	s	v	u	h	s	v	u	h	s
Sat.	1.3854	1118.8	1204.7	1.5017	1.3653	1118.9	1204.8	1.5004	1.3457	1118.9	1204.9	1.4991
410	1.3360	1108.3	1191.2	1.4863	1.3124	1107.5	1190.1	1.4837	1.2894	1106.6	1189.0	1.4810
420	1.3643	1114.3	1198.9	1.4952	1.3406	1113.6	1197.9	1.4926	1.3175	1112.8	1196.9	1.4901
430	1.3917	1120.1	1206.4	1.5036	1.3678	1119.4	1205.5	1.5012	1.3445	1118.7	1204.5	1.4987
440	1.4182	1125.7	1213.6	1.5117	1.3941	1125.0	1212.8	1.5093	1.3707	1124.4	1211.9	1.5069
450	1.4441	1131.1	1220.7	1.5195	1.4198	1130.5	1219.8	1.5171	1.3962	1129.9	1219.0	1.5148
460	1.4693	1136.4	1227.5	1.5270	1.4448	1135.8	1226.7	1.5247	1.4210	1135.2	1226.0	1.5224
470	1.4940	1141.5	1234.2	1.5342	1.4693	1141.0	1233.4	1.5319	1.4453	1140.4	1232.7	1.5297
480	1.5182	1146.6	1240.7	1.5412	1.4933	1146.0	1240.0	1.5389	1.4691	1145.5	1239.3	1.5368
490	1.5419	1151.5	1247.1	1.5479	1.5168	1151.0	1246.4	1.5458	1.4924	1150.5	1245.8	1.5436

FIGURE 5.10 Extract from superheat table.

TABLE 3

Vapor

p (t Sat.)

t	2.00 (212.42) 10³v	u	h	s	2.05 (213.67) 10³v	u	h	s	2.10 (214.90) 10³v	u	h	s
Sat.	99.63	2600.3	2799.5	6.3409	97.25	2600.7	28000.0	6.3318	94.98	2601.0	2800.5	6.3229
200	95.27	2570.6	2761.1	6.2608	92.52	2567.6	2757.3	6.2427				
205	97.06	2582.8	2777.0	6.2941	94.29	2580.1	2773.4	6.2765	91.65	2577.2	2769.7	6.2591
210	98.80	2594.7	2792.3	6.3259	96.01	2592.1	2788.9	6.3089	93.35	2589.5	2785.5	6.2920
215	100.50	2606.2	2807.2	6.3566	97.69	2603.7	2804.0	6.3399	95.01	2601.3	2800.8	6.3235
220	102.15	2617.4	2821.7	6.3861	99.32	2615.1	2818.7	6.3699	96.62	2612.8	2815.7	6.3538
225	103.77	2628.3	2835.8	6.4147	100.92	2626.1	2833.0	6.3987	98.20	2623.9	2830.2	6.3831
230	105.36	2638.9	2849.6	6.4423	102.48	2636.9	2847.0	6.4267	99.74	2634.9	2844.3	6.4113
235	106.91	2649.4	2863.2	6.4691	104.02	2647.5	2860.7	6.4538	101.25	2645.5	2858.2	6.4387
240	108.45	2659.6	2876.5	6.4952	105.52	2657.8	2874.1	6.4801	102.74	2656.0	2871.7	6.4653
245	109.95	2669.7	2889.6	6.5205	107.01	2668.0	2887.3	6.5057	104.20	2666.2	2885.0	6.4911

p (t Sat.)

t	2.15 (216.10) 10³v	u	h	s	2.20 (217.29) 10³v	u	h	s	2.25 (218.45) 10³v	u	h	s
Sat.	92.81	2601.4	2800.9	6.3141	90.73	2601.7	2801.3	6.3056	88.75	2602.0	2801.7	6.2972
200												
205	89.12	2574.4	2766.0	6.2419	86.71	2571.5	2762.2	6.2248				
210	90.81	2586.8	2782.0	6.2753	88.38	2584.1	2778.5	6.2587	86.06	2581.3	2774.9	6.2423
215	92.45	2598.8	2797.6	6.3073	90.01	2596.3	2794.3	6.2912	87.66	2593.7	2790.9	6.2753
220	94.05	2610.4	2812.6	6.3380	91.58	2608.1	2809.5	6.3223	89.23	2605.7	2806.4	6.3068
225	95.60	2621.7	2827.3	6.3676	93.12	2619.5	2824.4	6.3523	90.75	2617.3	2821.5	6.3372
230	97.13	2632.8	2841.6	6.3962	94.63	2630.7	2838.9	6.3812	92.23	2628.6	2836.1	6.3664
235	98.62	2643.6	2855.6	6.4238	96.10	2641.6	2853.0	6.4092	93.69	2639.6	2850.4	6.3947
240	100.08	2654.1	2869.3	6.4506	97.54	2652.2	2866.8	6.4362	95.11	2650.4	2864.4	2.4220
245	101.52	2664.5	2882.7	6.4767	98.96	2662.7	2880.4	6.4626	96.51	2660.9	2878.1	6.4486

FIGURE 5.10a Extract from superheat table (SI units).

separating the main portion of the table from the italicized values. The values in italics (above the horizontal dividing line) are for vapor temperature *below* the saturation temperature corresponding to the pressure listed. These states are metastable states (they are not equilibrium states), and we will not deal with them at all in our study. They should be ignored at this time by the student.

The SI tables are arranged in the same manner as the English tables. Pressures are in megapascals (MPa), temperatures are in degrees Celsius (°C), and the unit mass is the kilogram (kg). The energy unit is the kilojoule (kJ) and the volume unit is $10^3 \times$ cubic meters/kilogram ($10^3 \times m^3/kg$). Figure 5.10a is an extract from the superheat table.

ILLUSTRATIVE PROBLEM 5.11

Determine the specific volume, internal energy, enthalpy, and entropy of superheated steam at 330 psia and 450°F.

SOLUTION

The values of temperature and pressure are listed in Table 3 (Figure 5.10) and can be read directly.

$$v = 1.4691 \text{ ft}^3/\text{lb}_m$$
$$u = 1131.8 \text{ Btu}/\text{lb}_m$$
$$h = 1221.5 \text{ Btu}/\text{lb}_m$$
$$s = 1.5219 \text{ Btu}/\text{lb}_m \cdot °R$$

ILLUSTRATIVE PROBLEM 5.12

Determine the specific volume, internal energy, enthalpy, and entropy of superheated steam at 2.0 MPa and 240°C.

SOLUTION

Reading directly from Figure 5.10a, we have

$$v = 108.45/10^3 = 0.10845 \text{ m}^3/\text{kg}$$
$$u = 2659.6 \text{ kJ/kg}$$
$$h = 2876.5 \text{ kJ/kg}$$
$$s = 6.4952 \text{ kJ/kg} \cdot °\text{K}$$

ILLUSTRATIVE PROBLEM 5.13

Determine the specific volume, internal energy, enthalpy, and entropy of superheated steam at 330 psia and 455°F.

SOLUTION

Because the data in Figure 5.10 do not give the properties at this temperature, it is necessary to interpolate between 450°F and 460°F. Thus, at 330 psia,

t	v
460	1.4945
455	1.4818
450	1.4691

t	u
460	1137.0
455	1134.4
450	1131.8

t	h
460	1228.2
455	1224.9
450	1221.5

$$v_{455} = 1.4691 + \tfrac{1}{2}$$
$$(1.4945 - 1.4691)$$

$$u_{455} = 1131.8 + \tfrac{1}{2}$$
$$(1137.0 - 1131.8)$$

$$h_{455} = 1221.5 + \tfrac{1}{2}$$
$$(1228.2 - 1221.5)$$

t	s
460	1.5293
455	1.5256
450	1.5219

$$s_{455} = 1.5219 + \tfrac{1}{2}$$
$$(1.5293 - 1.5219)$$

Thus,

$$v = 1.4818 \ \text{ft}^3/\text{lb}_\text{m}$$
$$u = 1134.4 \ \text{Btu}/\text{lb}_\text{m}$$
$$h = 1224.9 \ \text{Btu}/\text{lb}_\text{m}$$
$$s = 1.5256 \ \text{Btu}/\text{lb·°R}$$

ILLUSTRATIVE PROBLEM 5.14

Determine the specific volume and enthalpy of superheated steam at 465°F and 337 psia.

SOLUTION

It will be noted from Figure 5.10 that neither the temperature nor the pressure values of the problem are tabulated in Table 3. Thus, it becomes necessary to interpolate around both the listed temperature and pressure for the desired properties. We will first obtain the properties at 337 psia and 460°F and then at 337 psia and 470°F. Proceeding with the calculation, at 460°F,

p	v
340	1.4448
337	1.4595
335	1.4693

p	h
340	1226.7
337	1227.2
335	1227.5

$$v_{337} = 1.4693 - \tfrac{2}{5}(1.4693 - 1.4448)$$

$$h_{337} = 1227.5 - \tfrac{2}{5}(1227.5 - 1226.7)$$

and at 470°F,

p	v
340	1.4693
337	1.4841
335	1.4940

p	h
340	1233.4
337	1233.9
335	1234.2

$$v_{337} = 1.4940 - \tfrac{2}{5}(1.4940 - 1.4693)$$

$$h_{337} = 1234.2 - \tfrac{2}{5}(1234.2 - 1233.4)$$

Therefore, at 337 psia and 465°F,

t	v
470	1.4841
465	1.4718
460	1.4595

t	h
470	1233.9
465	1230.7
460	1227.5

$$v_{465} = 1.4595 + \tfrac{1}{2}(1.4841 - 1.4595)$$

$$h_{465} = 1227.5 + \tfrac{1}{2}(1233.9 - 1227.5)$$

The desired values of specific volume and enthalpy at 465°F and 337 psia are 1.4718 ft³/lb$_m$ and 1230.7 Btu/lb$_m$.

An extract from Table 4 of the *Steam Tables* entitled *Liquid* is shown in Figure 5.11 (Figure 5.11a gives SI units). The state of the liquid is the *subcooled* state, in which the pressure on the liquid exceeds the saturation pressure corresponding to the temperature of the liquid. In this region, the tabulated properties of specific volume, internal energy, enthalpy, and entropy are exhibited as functions of pressure and temperature similar to the format used for the superheated vapor. It will be noted from Figure 5.11 that italicized values are shown for certain values of pressure and temperature. Once again, these are metastable states, because they correspond to temperatures of the liquid above the saturation temperature. They are not to be used within the scope of this text.

TABLE 4

Liquid

p (t Sat.)		0				500 (467.13)				1000 (544.75)		
t	v	u	h	s	v	u	h	s	v	u	h	s
Sat.					0.019748	447.70	449.53	0.64904	0.021591	538.39	542.38	0.74320
32	0.016022	−0.01	−0.01	0.00003	0.015994	0.00	1.49	0.00000	0.015967	0.03	2.99	0.00005
50	0.016024	18.06	18.06	0.0360	0.015998	18.02	19.50	0.03599	0.015972	17.99	20.94	0.03592
100	0.016130	68.05	68.05	0.12963	0.016106	67.87	69.36	0.12932	0.016082	67.70	70.68	0.12901
150	0.016343	117.95	117.95	0.21504	0.016318	117.66	119.17	0.21457	0.016293	117.38	120.40	0.21410
200	0.016635	168.05	168.05	0.29402	0.016608	167.65	169.19	0.29341	0.016580	167.26	170.32	0.29281
250	0.017003	218.52	218.52	0.36777	0.016972	217.99	219.56	0.36702	0.016941	217.47	220.61	0.36628
300	0.017453	269.61	269.61	0.43732	0.017416	268.92	270.53	0.43641	0.017379	268.24	271.46	0.43552
350	0.018000	321.59	321.59	0.50359	0.017954	320.71	322.37	0.50249	0.017909	319.83	323.15	0.50140
400	0.018668	374.85	374.85	0.56740	0.018608	373.68	375.40	0.56604	0.018550	372.55	375.98	0.56472
450	0.019503	429.96	429.96	0.62970	0.019420	428.40	430.19	0.62798	0.019340	426.89	430.47	0.62632
500	0.02060	488.1	488.1	0.6919	0.02048	485.9	487.8	0.6896	0.02036	483.8	487.5	0.6874
510	0.02087	500.3	500.3	0.7046	0.02073	497.9	499.8	0.7021	0.02060	495.6	499.4	0.6997
520	0.02116	512.7	512.7	0.7173	0.02100	510.1	512.0	0.7146	0.02086	507.6	511.5	0.7121
530	0.02148	525.5	525.5	0.7303	0.02130	522.6	524.5	0.7273	0.02114	519.9	523.8	0.7245
540	0.02182	538.6	538.6	0.7434	0.02162	535.3	357.3	0.7402	0.02144	532.4	536.3	0.7372
550	0.02221	552.1	552.1	0.7569	0.02198	548.4	550.5	0.7532	0.02177	545.1	549.2	0.7499
560	0.02265	566.1	566.1	0.7707	0.02237	562.0	564.0	0.7666	0.02213	558.3	562.4	0.7630
570	0.02315	580.8	580.8	0.7851	0.02281	576.0	578.1	0.7804	0.02253	571.8	576.0	0.7763
580					0.02332	590.8	592.9	0.7946	0.02298	585.9	590.1	0.7899
590					0.02392	606.4	608.6	0.8096	0.02349	600.6	604.9	0.8041
600									0.02409	616.2	620.6	0.8189
610									0.02482	632.9	637.5	0.8345

FIGURE 5.11   Extract from subcooled table.

TABLE 4

Liquid

p (t Sat.) MPa	0				2.5 (223.99)				5.0 (263.99)			
t	10^3v	u	h	s	10^3v	u	h	s	10^3v	u	h	s
Sat.					1.1973	959.1	962.1	2.5546	1.2859	1147.8	1154.2	2.9202
0	1.0002	−0.03	−0.03	−0.0001	0.9990	−0.00	2.50	−0.0000	0.9977	0.04	5.04	0.0001
20	1.0018	83.95	83.95	0.2966	1.0006	83.80	86.30	0.2961	0.9995	83.65	88.65	0.2956
40	1.0078	167.56	167.56	0.5725	1.0067	167.25	169.77	0.5715	1.0056	166.95	171.97	0.5705
60	1.0172	251.12	251.12	0.8312	1.0160	250.67	253.21	0.8298	1.0149	250.23	255.30	0.8285
80	1.0291	334.87	334.87	1.0753	1.0280	334.29	336.86	1.0737	1.0268	333.72	338.85	1.0720
100	1.0436	418.96	418.96	1.3069	1.0423	418.24	420.85	1.3050	1.0410	417.52	422.72	1.3030
120	1.0604	503.57	503.57	1.5278	1.0590	502.68	505.33	1.5255	1.0576	501.80	507.09	1.5233
140	1.0800	588.89	588.89	1.7395	1.0784	587.82	590.52	1.7369	1.0768	586.76	592.15	1.7343
160	1.1024	675.19	675.19	1.9434	1.1006	673.90	676.65	1.9404	1.0988	672.62	678.12	1.9375
180	1.1283	762.72	762.72	2.1410	1.1261	761.16	763.97	2.1375	1.1240	759.63	765.25	2.1341
200	1.1581	851.8	851.8	2.3334	1.1555	849.9	852.8	2.3294	1.1530	848.1	853.9	2.3255
210	1.1749	897.1	897.1	2.4281	1.1720	895.0	898.0	2.4238	1.1691	893.0	898.8	2.4195
220	1.1930	943.0	943.0	2.5221	1.1898	940.7	943.7	2.5174	1.1866	938.4	944.4	2.5128
230	1.2129	989.6	989.6	2.6157	1.2092	987.0	990.1	2.6105	1.2056	984.5	990.6	2.6055
240	1.2347	1037.1	1037.1	2.7091	1.2305	1034.2	1037.2	2.7034	1.2264	1031.4	1037.5	2.6979
250	1.2590	1085.6	1085.6	2.8027	1.2540	1082.3	1085.4	2.7964	1.2493	1079.1	1085.3	2.7902
260	1.2862	1135.4	1135.4	2.8970	1.2804	1131.6	1134.8	2.8898	1.2749	1127.9	1134.3	2.8830
270	1.3173	1186.8	1186.8	2.9926	1.3102	1182.4	1185.7	2.9844	1.3036	1178.2	1184.3	2.9766
280	1.3535	1240.4	1240.4	3.0904	1.3447	1235.1	1238.5	3.0808	1.3365	1230.2	1236.8	3.0717
290	1.3971	1297.0	1297.0	3.1918	1.3855	1290.5	1294.0	3.1801	1.3750	1284.4	1291.3	3.1693
300	1.4520	1358.1	1358.1	3.2992	1.4357	1349.6	1353.2	3.2843	1.4214	1341.9	1349.0	3.2708
310									1.4803	1404.1	1411.5	3.3789

FIGURE 5.11a Extract from subcooled table (SI units).

A useful first approximation is often made for the enthalpy of the subcooled liquid by assuming it to be essentially incompressible. Then

$$\boxed{h - h_f \simeq \frac{(p - p_f)v_f}{J}} \tag{5.5}$$

where p_f and v_f are the values of pressure and specific volume corresponding to the saturation condition at the temperature of the fluid. In other words, the change in enthalpy is approximately equal to $v\Delta p$, the work done on the fluid during a process carried out at nearly constant volume. In SI units,

$$h - h_f \simeq (p - p_f)v_f \tag{5.5a}$$

ILLUSTRATIVE PROBLEM 5.15

Determine the enthalpy, specific volume, internal energy, and entropy of subcooled water at 300°F and 1000 psia.

SOLUTION

From Table 4 (Figure 5.11), we find the values are directly tabulated. Therefore,

$$v = 0.017379 \text{ ft}^3/\text{lb}_m$$
$$u = 268.24 \text{ Btu}/\text{lb}_m$$
$$h = 271.46 \text{ Btu}/\text{lb}_m$$
$$s = 0.43552 \text{ Btu}/\text{lb}_m \cdot °R$$

ILLUSTRATIVE PROBLEM 5.16

Determine the enthalpy of the subcooled water in Illustrative Problem 5.15 using the approximation of Equation (5.5), and compare the result obtained with the tabulated value.

SOLUTION

It is necessary to obtain the saturation values corresponding to 300°F. This is done by reading Table A.1 in Appendix 3, which gives us $p_f = 66.98$ psia, $v_f = 0.017448 \text{ ft}^3/\text{lb}_m$, and $h_f = 269.73 \text{ Btu/lb}_m$. From Equation (5.5),

$$h = h_f + \frac{(p - p_f)v_f}{J}$$

$$= 269.73 \text{ Btu/lb}_m + \frac{(1000 - 66.98) \text{ lb}_f/\text{in}^2 (0.017448) \text{ ft}^3/\text{lb}_m \times 144 \text{ in}^2/\text{ft}^2}{778 \text{ ft·lb}_f/\text{Btu}}$$

$$= 272.74 \text{ Btu/lb}_m$$

The difference between this value and the value found in Illustrative Problem 5.15 expressed as a percentage is

$$\text{percent of error} = \frac{272.74 - 271.46}{271.46} \times 100 = 0.47\%$$

For this pressure and temperature, the approximation expressed by Equation (5.5) is obviously within the accuracy required in most engineering calculations.

5.4 COMPUTERIZED PROPERTIES

It is obvious from the previous section that a great deal of effort is involved in the interpolation in the *Steam Tables* and that the potential for errors is ever present. Values for the thermodynamic properties of steam are available in the form of computer programs that can provide the desired properties. The National Institute of Standards and Testing (NIST), formerly the National Bureau of Standards (NBS), has such a program that was used to generate the NBS/NRC *Steam Tables*. A modified version of this program is included in the floppy disk in the back inside cover of this text.* Any combination of two independent properties that define a state may be entered, and the program will return all of the properties of that state. The user has the choice of SI units or USCS (U.S. conventional system) units. In DOS, the program is accessed by A: {Enter} and STEAM.EXE {Enter}. Note that entropy units appearing in the tables may be given using either common or absolute temperature units, because they refer to a differential temperature that has the same numerical value in either unit designation.

Because the properties are computed from equations, they do not contain any interpolation errors.

ILLUSTRATIVE PROBLEM 5.17

Use the computer program to solve Illustrative Problem 5.1.

SOLUTION

The problem is to determine the enthalpy of saturated steam at 90°F. We select USCS units and enter 90°F and the quality, *x*, of 1.000. The program gives us the properties of the saturated liquid, the saturated vapor, and *T, p, v, h, s, u,* and *x* for the desired input. The agreement with Table 1 is very good.

```
            Saturation Properties:
            ─────────────────────────────
            T =     90.000    deg F
            P =     0.6987    psia
```

---

* The authors are indebted to Dr. John R. Howell of the University of Texas at Austin for making this program available. It is from *Fundamentals of Engineering Thermodynamics* by J. R. Howell and R. O. Buckius, 2nd ed., 1992, McGraw-Hill Book Co., with permission.

```
               z             zl            zg
v(ft3/lbm)         0.01610        467.66
h(BTU/lbm)         58.026         1100.3
s(BTU/lbmF)        0.1116         2.0078
u(BTU/lbm)         58.023         1039.9

    Thermo Properties:
    _____

       T =      90.000      deg F
       P =       0.6987     psia
       v =     467.66       ft3/lbm
       h =    1100.3        BTU/lbm
       s =       2.0078     BTU/lbmF
       u =    1039.9        BTU/lbm
       x =       1.0000

    Region:  saturated
```

ILLUSTRATIVE PROBLEM 5.18

Use the computer program to solve Illustrative Problem 5.3.

SOLUTION

This problem is similar to Illustrative Problem 5.17. The entries to the program are pressure (118 psia) and quality ($x = 1.000$). The results from the computer program are in excellent agreement with the interpolated values.

```
          Saturation Properties:
          _____

          T =    340.06     deg F
          P =    118.00     psia
               z             zl            zg
v(ft3/lbm)         0.01787        3.7891
h(BTU/lbm)         311.39         1190.7
s(BTU/lbmF)        0.4904         1.5899
u(BTU/lbm)         311.00         1108.0

    Thermo Properties:
    _____

       T =     340.06      deg F
       P =     118.00      psia
       v =       3.7891    ft3/lbm
       h =    1190.7       BTU/lbm
       s =       1.5899    BTU/lbmF
       u =    1108.0       BTU/lbm
       x =       1.0000

    Region:  saturated
```

SOLUTION

The input to the program is USCS units; $p = 120$ psia, and $x = 0.8$. The results are in essentially exact agreement with the results of Illustrative Problem 5.7.

```
Saturation Properties:
_____

 T =     341.32    deg F
 P =     120.00    psia
      z            zl          zg
v(ft3/lbm)       0.01789      3.7291
h(BTU/lbm)       312.71       1191.0
s(BTU/lbmF)      0.4921       1.5886
u(BTU/lbm)       312.32       1108.2

     Thermo Properties:
  _____

     T =    341.32     deg F
     P =    120.00     psia
     v =    2.9869     ft3/lbm
     h =    1015.3     BTU/lbm
     s =    1.3693     BTU/lbmF
     u =    949.01     BTU/lbm
     x =    0.8000

  Region: saturated
```

SOLUTION

The input to this problem is 1000 kPa (1 MPa) and $x = 0.85$. The printout agrees with the calculated values in Illustrative Problem 5.8.

```
Saturation Properties:
_____

 T =     179.92    deg C
 P =     1000.0    kPa
      z            zl          zg
v(m3/kg)        0.001127      0.1944
h(kJ/kg)        762.88        2777.7
s(kJ/kgK)       2.1388        6.5859
u(kJ/kg)        761.75        2583.3
```

```
         Thermo Properties:
        _____
        T =    179.92    deg C
        P =    1000.0    kPa
        v =    0.1654    m3/kg
        h =    2475.5    kJ/kg
        s =    5.9189    kJ/kgK
        u =    2310.1    kJ/kg
        x =    0.8500

        Region: saturated
```

ILLUSTRATIVE PROBLEM 5.21

Solve Illustrative Problem 5.9 using the computerized properties of steam.

SOLUTION

This problem has USCS inputs of 90°F and $h = 900$ Btu/lb$_m$. The solution yields the quality and all of the other properties at this state.

```
        Saturation Properties:
       _____
       T =    90.000    deg F
       P =    0.6987    psia
           z            zl          zg
       v(ft3/1bm)     0.01610     467.66
       h(BTU/1bm)     58.026      1100.3
       s(BTU/1bmF)    0.1116      2.0078
       u(BTU/1bm)     58.023      1039.9

         Thermo Properties:
        _____
        T =    90.000    deg F
        P =    0.6987    psia
        v =    377.78    ft3/1bm
        h =    900.00    BTU/1bm
        s =    1.6433    BTU/1bmF
        u =    851.15    BTU/1bm
        x =    0.8078

        Region: saturated
```

ILLUSTRATIVE PROBLEM 5.22

Use the computer program to solve Illustrative Problem 5.12.

SOLUTION

The input for this problem is SI units of 240°C and 2000 kPa (2.0 MPa). The output of the program yields the desired properties and also the saturated properties at 240°C.

```
            Saturation Properties:
            _____

         T =    240.00     deg C
         P =    3344.7      kPa
            z               zl          zg
         v(m3/kg)           0.001229    0.05974
         h(kJ/kg)           1037.2      2803.0
         s(kJ/kgK)          2.7013      6.1423
         u(kJ/kg)           1033.1      2603.1

            Thermo Properties:
            _____

             T =    240.00     deg C
             P =    2000.0      kPa
             v =    0.1084      m3/kg
             h =    2875.6      kJ/kg
             s =    6.4937      kJ/kgK
             u =    2658.8      kJ/kg

        Region:  superheated
```

ILLUSTRATIVE PROBLEM 5.23

Solve Illustrative Problem 5.14 using the computerized properties.

SOLUTION

The inputs in USCS units are 465°F and 337 psia. The computer printout yields v, h, s, and u in an instant, while the calculated value in Illustrative Problem 5.14 requires six interpolations. Not only is the work less, but the potential for error in the interpolation is eliminated.

```
            Saturation Properties:
            _____

         T =    465.00     deg F
         P =    489.83     psia
            z               zl          zg
         v(ft3/lbm)         0.01971     0.9476
         h(BTU/lbm)         447.07      1205.0
         s(BTU/lbmF)        0.6464      1.4661
         u(BTU/lbm)         445.28      1119.1
```

```
            Thermo Properties:
            _____

                T  =    465.00      deg F
                P  =    337.00      psia
                v  =    1.4713      ft3/1bm
                h  =    1230.2      BTU/1bm
                s  =    1.5293      BTU/1bmF
                u  =    1138.4      BTU/1bm

            Region: superheated
```

ILLUSTRATIVE PROBLEM 5.24

Use the computer program to solve Illustrative Problem 5.15.

SOLUTION

With USCS inputs of 300°F and 1000 psia, the program yields the following results. Note that in all of the computer programs, the region (superheated, saturated, or subcooled) is also given.

```
        Saturation Properties:
        _____

        T  =    300.00      deg F
        P  =     66.963     psia
              z              zl          zg
        v(ft3/1bm)        0.01745      6.4725
        h(BTU/1bm)        269.78       1180.2
        s(BTU/1bmF)       0.4373       1.6356
        u(BTU/1bm)        269.57       1099.9

            Thermo Properties:
            _____

                T  =    300.00      deg F

                P  =    1000.0      psia

                v  =    0.01738     ft3/1bm

                h  =    271.51      BTU/1bm

                s  =    0.4356      BTU/1bmF

                u  =    268.30      BTU/1bm

            Region: subcooled
```

The computer program for steam gives us the accuracy that is equivalent to the tables in Appendix 3 and should be used to avoid the tedious calculations entailed when using the tables. However, it is important to obtain an understanding and mastery of basics first and then to use the computerized tables as an invaluable tool.

5.5 THERMODYNAMIC DIAGRAMS

Tables of thermodynamic properties provide accurate data for various substances. However, diagrams and charts based on the data of these tables are both useful and desirable. One most important fact must be borne in mind. Thermodynamic data such as those given by the *Steam Tables* are equilibrium data, and charts plotted from these data can represent only the equilibrium states. The path of a process that is not an equilibrium path cannot be drawn on these charts.

Figure 5.12 is a *T–s* diagram for steam showing the liquid and vapor phases. In the wet region (below the saturation curve), lines of constant temperature and lines of constant pressure are horizontal lines. For convenience, lines of constant moisture (constant *x*) and lines of constant volume are also shown in the wet region. In the superheat region above the saturation curve, lines of constant pressure start at the saturation curve, rise steeply, and are almost vertical. Lines of constant enthalpy are nearly horizontal away from the saturation curve. Near the saturation curve, and

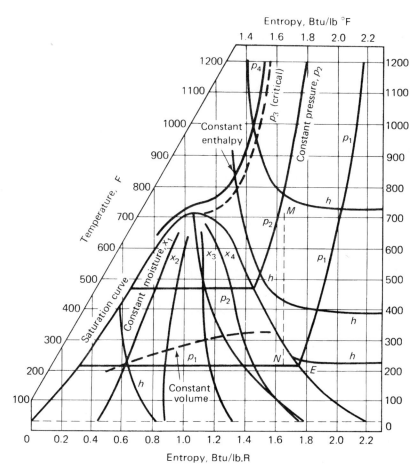

FIGURE 5.12 Outline of a temperature entropy diagram for steam.
[From B. F. Dodge, Chemical Engineering Thermodynamics (New York: McGraw-Hill Book Co., 1944).]

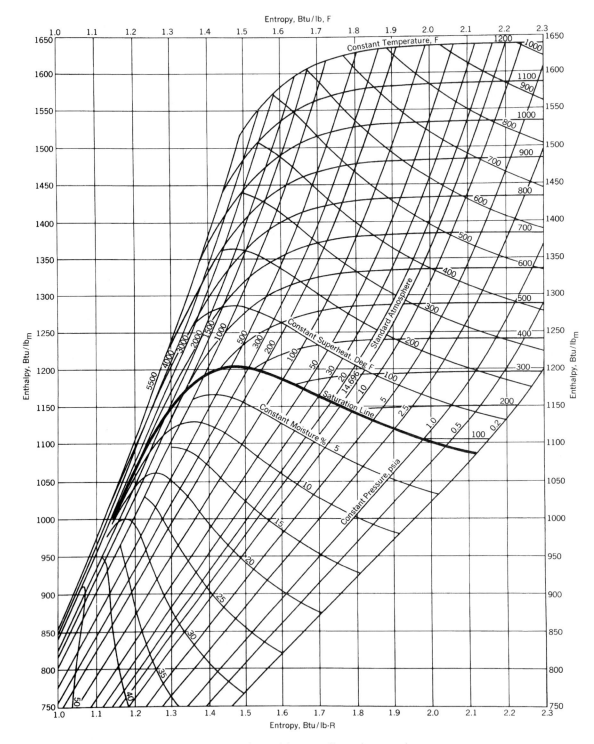

FIGURE 5.13 Outline of *h–s* (Mollier) diagram for steam.
(Courtesy of Babcock and Wilcox Corp.)

especially near the critical point, there is a marked change in curvature of the constant-enthalpy lines, which approach the vertical as the critical pressure is approached. On the large *T–s* diagram that is found in the *Steam Tables,* curves of constant superheat are also shown in the superheat region. A curve of constant superheat corresponds to a curve plotted parallel to the saturation curve and a fixed number of degrees above the saturation curve. Thus, the term *200 degrees of superheat* means superheated steam whose temperature corresponds to saturation temperature plus 200°F. While the *T–s* chart is useful in portraying processes, it is not as useful nor as widely used as the *h–s* diagram (or *Mollier chart* as the *h–s* diagram is usually called).

The Mollier chart is a plot on *h–s* coordinates of the thermodynamic properties of a substance. Figure 5.13 shows the outline of a portion of the Mollier chart for steam. Below the saturation line, lines of constant pressure (which are also lines of constant temperature) and lines of constant moisture are shown. Above the saturation line, curves of constant temperature, constant pressure, and constant superheat are shown. As noted for the *T–s* diagram, the curves of constant superheat correspond to curves plotted parallel to the saturation curve and a fixed number of degrees above the saturation curve. As we shall see, the Mollier chart is particularly suited to obtaining properties, in describing flow, or describing constant-pressure processes. If a process is not reversible, then only its end states can be shown on a thermodynamic diagram such as the *T–s* diagram or *Mollier chart.* A Mollier chart similar to the one in the *Steam Tables* is found in Appendix 3.

In addition to the *T–s* and *h–s* charts, other diagrams of the thermodynamic properties also have utility. Figure 5.14 shows the specific heat of steam at constant pressure as a function of temperature and pressure. It can be proved by theoretical reasoning that the specific heats at the critical point are infinite, and it will be seen from Figure 5.14 that in the regions of the critical temperature, there is a sharp rise in specific-heat values.

The wide fluctuations in the region of the critical point are more evident from Figure 5.15, which also shows the specific heat at constant pressure for a saturated liquid. At the critical point, the saturated vapor and saturated liquid are indistinguishable, and both curves merge to infinity.

The effect of pressure on the specific heat at constant pressure for a liquid (water) is shown in Figure 5.16. It should be noted that the specific heat at constant pressure and at given temperature decreases with increasing pressure. At a given pressure, it increases with temperature. It should be noted that Figures 5.12 through 5.16 exists in the (*SI Units*) *Edition* of the *Steam Tables.*

ILLUSTRATIVE PROBLEM 5.25

Determine the enthalpy of saturated steam at 90°F using the Mollier chart. Compare the result with Illustrative Problem 5.1.

SOLUTION

On the chart in Appendix 3, it is necessary to estimate the 90°F point on the saturation line. From the chart or the table in the upper left of the chart, we note that 90°F is between 1.4 and 1.5 in. of mercury. Estimating the intersection of this value with the saturation curve yields $h_g = 1100$ Btu/lb$_m$. This is in good agreement with Illustrative Problem 5.1.

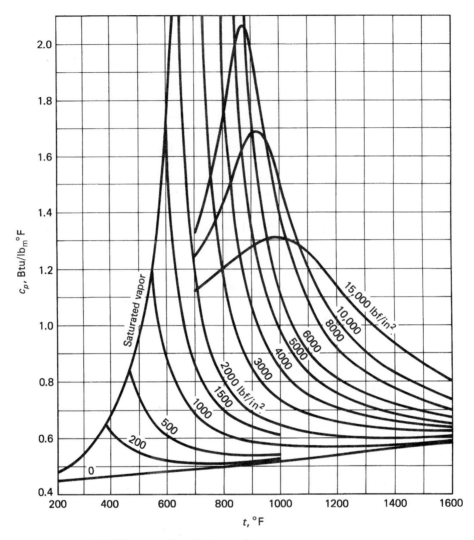

FIGURE 5.14   Specific heat of steam at constant pressure.
[From J. H. Keenan, F. G. Keyes, P. G. Hill, and J. G. Moore,
Steam Tables *(New York: John Wiley & Sons, Inc., 1969), with permission.]*

ILLUSTRATIVE PROBLEM 5.26

Determine the enthalpy of a wet steam mixture at 120 psia having a quality of 80% by using the Mollier chart. Compare the results with Illustrative Problem 5.7.

SOLUTION

The Mollier chart has lines of constant moisture in the wet region which correspond to $(1 - x)$. Therefore, we read at 20% moisture (80% quality) and 120 psia an enthalpy of 1015 Btu/lb$_m$, which also agrees well with the calculated value in Illustrative Problem 5.7.

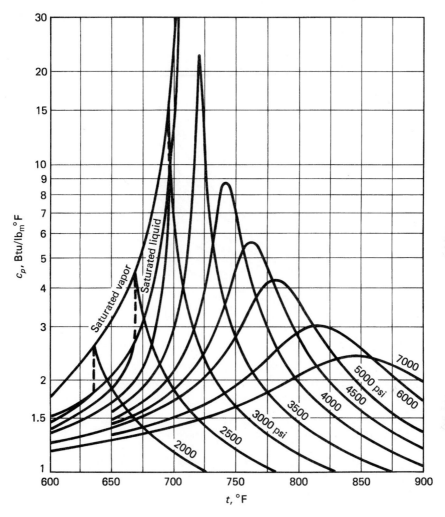

FIGURE 5.15 Specific heat at constant pressure near the critical point.
*[From J. H. Keenan, F. G. Keyes, P. G. Hill, and J. G. Moore, Steam Tables
(New York: John Wiley & Sons, Inc., 1969), with permission.]*

ILLUSTRATIVE PROBLEM 5.27

Using the Mollier chart, determine the quality of a wet steam mixture having an enthalpy of 900 Btu/lb$_m$ and a temperature of 90°F. Compare the result with Illustrative Problem 5.9.

SOLUTION

Entering the Mollier chart at 900 Btu/lb$_m$ and estimating 90°F (near the 1.5-in. Hg dashed line) yields a constant moisture percentage of 19.2%. The quality is therefore $(1 - 0.192)(100) = 80.8\%$. We again show good agreement with the calculated value.

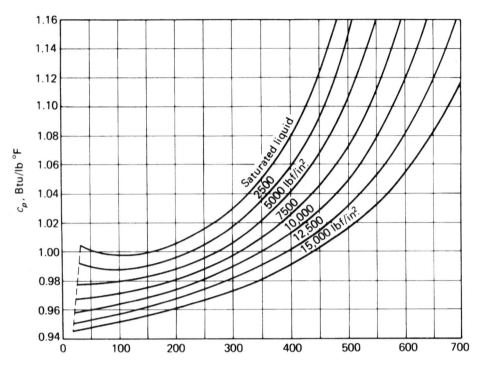

FIGURE 5.16 Specific heat of water at constant pressure.
[From J. H. Keenan, F. G. Keyes, P. G. Hill, and J. G. Moore, Steam Tables *(New York: John Wiley & Sons, Inc., 1969), with permission.]*

ILLUSTRATIVE PROBLEM 5.28

Determine the enthalpy of steam at 330 psia and 450°F using the Mollier chart. Compare the results with Illustrative Problem 5.11.

SOLUTION

From the chart, $h = 1220$ Btu/lb$_m$, compared to 1221.5 Btu/lb$_m$ found in Illustrative Problem 5.11.

ILLUSTRATIVE PROBLEM 5.29

Determine the enthalpy and entropy of steam at 2.0 MPa and 240°C using the Mollier chart.

SOLUTION

We note that the steam is superheated. From the Mollier chart in SI units, $h = 2880$ kJ/kg and $s = 6.5$ kJ/kg·°K. These values compare with $h = 2876.5$ kJ/kg and s = 6.4952 kJ/kg·°K found in Illustrative Problem 5.12. For most purposes, this accuracy would be acceptable.

ILLUSTRATIVE PROBLEM 5.30

Determine the enthalpy of steam at 465°F and 337 psia using the Mollier chart. Compare the results with Illustrative Problem 5.14.

SOLUTION

Because neither pressure nor temperature is shown directly, it is necessary to estimate to obtain the desired value. Reading the chart in this manner, we obtain $h = 1231$ Btu/lb$_m$. In Illustrative Problem 5.14, we obtained $h = 1230.7$ Btu/lb$_m$. The agreement is good, and the savings in laborious interpolations is considerable.

ILLUSTRATIVE PROBLEM 5.31

Use the Mollier chart to solve Illustrative Problem 5.2.

SOLUTION

Reading the chart at 30°C and saturation gives us $h_g = 2556$ kJ/kg, which is in excellent agreement with the tables.

ILLUSTRATIVE PROBLEM 5.32

Use the Mollier chart to solve Illustrative Problem 5.8.

SOLUTION

Reading the chart in the wet region at 1.0 MPa and $x = 0.85$ (moisture content of 15%) gives us

$$h_x = 2476 \text{ kJ/kg}$$
$$s_x = 5.92 \text{ kJ/kg·°K}$$

The chart does not give us u_x or v_x directly.

ILLUSTRATIVE PROBLEM 5.33

Solve Illustrative Problem 5.10 using the Mollier chart.

SOLUTION

Locate 30°C on the saturation line. Now follow a line of constant pressure, which is also a line of constant temperature in the wet region, until an enthalpy of 2000 kJ/kg is reached. At this point, we find the moisture content to be 23%, or $x = 77\%$.

ILLUSTRATIVE PROBLEM 5.34

Solve Illustrative Problem 5.12 using the Mollier chart.

SOLUTION

We enter the chart in the superheat region at 2.0 MPa and 240°C to read the enthalpy and entropy. This procedure gives

$$h = 2877 \text{ kJ/kg}$$
$$s = 6.495 \text{ kJ/kg·°K}$$

The other properties cannot be obtained directly from the chart.

Use of the Mollier chart does not permit us to obtain the specific volume or internal energy directly. Special charts are available that plot enthalpy as ordinate and specific volume as abscissa. These charts are particularly useful in steam turbine computations.

5.6 PROCESSES

Thus far, we have concerned ourselves with the properties of a substance in a given state. These properties are useful in actual processes to describe the path of the fluid or to establish the end states once the process has been specified. Some of the processes of interest are the throttling process (pipe flow and flow measurement), the constant-volume process (accumulator), reversible and irreversible compressions and expansions (pumps and turbines), and the constant-pressure process (heaters and boilers). For each of these processes, it is necessary to know the energy equation for the path and one of the end states.

5.6a Throttling

The throttling process, as already noted in Chapter 2, is found to occur when an obstruction occurs in a pipe that locally disturbs the flow. This process is an irreversible adiabatic process whose path equation we have seen to be a constant-enthalpy path. Figure 5.17 shows this process on both *Ts* and *hs* coordinates. The end states, 1 and 2, are known. The path is shown as a dashed line on both diagrams, because it is an irreversible path. As will be noted from the diagram, steam that is initially wet will become drier, and depending on the initial and final states, it will become superheated after throttling. The final pressure is always less than the initial pressure.

The superheating of initially wet steam in a throttling process provides the basis for a device known as a *throttling calorimeter* that is used to determine the quality of wet steam flowing in a pipe. Figure 5.18 is a diagram of the throttling calorimeter showing its installation in a vertical run of pipe. The steam enters the sampling tube and is expanded in the orifice to the main body of the calorimeter. The initial pressure in the pipe is monitored by a pressure gage, and the final temperature and pressure after expansion are monitored by a thermometer and manometer. To eliminate the velocity terms in the energy equation, the unit is sized so that the entry flow area and the flow area at the point of temperature measurement are made approximately equal. For reliable operation of the calorimeter, the final state should have a superheat of at least 10°F. The calculations for the throttling process are greatly simplified by use of the Mollier chart, as will be seen from Illustrative Problem 5.35.

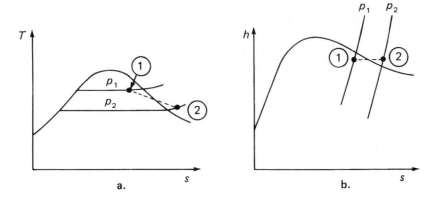

FIGURE 5.17 Throttling process.

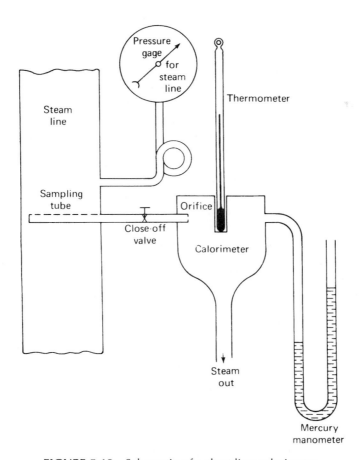

FIGURE 5.18 Schematic of a throttling calorimeter.

Steam flows in a pipe at 150 psia. If a throttling calorimeter installed in the pipe has a thermometer reading of 250°F and the calorimeter is essentially 14.7 psia, determine the moisture in the steam flowing in the pipe.

SOLUTION

As already noted, $h_1 = h_2$ for this process. On the Mollier chart, h_2 is found to be 1170 Btu/lb$_m$ at 14.7 psia and 250°F. Proceeding to the left on the chart, the constant-enthalpy value of 1170 Btu/lb$_m$ to 150 psia yields a moisture of 3% or a quality of 97%.

If we use the tables to obtain the solution to this problem, we would first obtain h_2 from the superheated vapor tables as 1168.8 Btu/lb$_m$. Because $h_x = h_f + xh_{fg}$, we obtain x as

$$x = \frac{h_x - h_f}{h_{fg}} \quad \text{where } h_f \text{ and } h_{fg} \text{ are values of 150 psia}$$

$$= \frac{1168.8 - 330.75}{864.2} = 0.97$$

Very often, it is necessary to perform multiple interpolations if the tables are used, and the Mollier chart yields results within the required accuracy for most engineering problems and saves considerable time.

We can also use the computerized properties to solve this problem. We first enter the 250°F and 14.7 psia to obtain h of 1168.7 Btu/lb$_m$. We then continue by entering h of 1168.7 Btu/lb$_m$ and p of 150 psia. The printout gives us x of 0.9699 or 97%. While the computer solution is quick and easy to use, you should still sketch out the problem on an h–s or T–s diagram to show the path of the process.

```
           Saturation Properties:
         ────────────────────────────────
         T =     250.00     deg F
         P =      29.814    psia
            z             zl          zg
    v(ft3/lbm)         0.01700      13.830
    h(BTU/lbm)         218.62       1164.1
    s(BTU/lbmF)        0.3678       1.7001
    u(BTU/lbm)         218.52       1087.8

           Thermo Properties:
         ────────────────────────────────
         T =     250.00     deg F
         P =      14.700    psia
         v =      28.417    ft3/lbm
         h =      1168.7    BTU/lbm
         s =      1.7831    BTU/lbmF
         u =      1091.4    BTU/lbm

      Region: superheated
```

```
              Saturation Properties:
      ──────────────────────────────────────
        T =     358.49    deg F
        P =     150.00    psia
           z               zl         zg
        v(ft3/lbm)        0.01809    3.0152
        h(BTU/lbm)        330.77     1194.7
        s(BTU/lbmF)       0.5142     1.5702
        u(BTU/lbm)        330.27     1111.0

           Thermo Properties:
      ──────────────────────────────────────
           T =     358.49    deg F
           P =     150.00    psia
           v =     2.9248    ft3/lbm
           h =     1168.7    BTU/lbm
           s =     1.5384    BTU/lbmF
           u =     1087.5    BTU/lbm
           x =     0.9699

        Region:  saturated
```

5.6b Constant-Volume Process (Isometric Process)

The constant-volume process is a nonflow process that we can consider to occur when a fluid is heated in a closed tank. Figure 5.19 shows a constant-volume process in which wet steam ① is heated at constant volume in a closed tank and goes to the superheated state at ②. The energy equation for this process has already been derived as $q = u_2 - u_1$. Because the Mollier chart does not have lines of constant internal energy, it is not suited for calculations of this process. The T–s chart does not have lines of constant internal energy, and it is not too useful in calculations involving the constant-volume process. A heat exchanger utilizing U-bend removable elements is shown in Figure 5.20. This unit can be used to heat stored materials, to provide pressurization of a system, and to heat process fluids in chemical plants.

ILLUSTRATIVE PROBLEM 5.36

A closed tank contains 1 lb_m of saturated liquid at 150°F. If the tank is heated until it is filled with saturated steam, determine the final pressure of the steam and the heat added. The tank has a volume of 10 ft³.

SOLUTION

Because the tank volume is 10 ft³, the final specific volume of the steam is 10 ft³/lb_m. Interpolations in Table A.2 yield a final pressure of 42 psia. The heat added is simply the difference in internal energy between the two states.

$$q = u_2 - u_1 = (1093.0 - 117.95) = 975.05 \text{ Btu/lb}_m \text{ added}$$

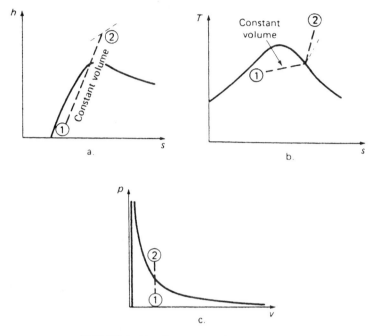

FIGURE 5.19   Constant-volume process.

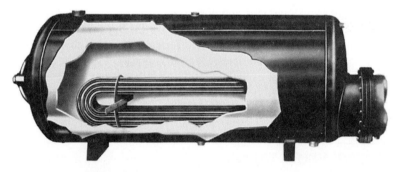

FIGURE 5.20   Tank-type storage water heater.
(Courtesy of Patterson-Kelley Co., Division of Harsco Corp.)

ILLUSTRATIVE PROBLEM 5.37

A closed tank has a volume of 60 ft^3. It contains 15 ft^3 of steam and 45 ft^3 of water at 14.7 psia. Heat is added until the pressure increases to 800 psia. Determine the heat added.

SOLUTION

The mass in the tank is constant, and the heat added will be the change in internal energy of the contents of the tank between the two states. The initial mass in the tank is found as follows:

$$m_f = \frac{V_f}{v_f} = \frac{45}{0.016715} = 2692.19 \text{ lb}_m$$

$$m_g = \frac{V_g}{v_g} = \frac{15}{26.80} = 0.56 \text{ lb}_m$$

$$\text{Total mass} = 2692.75 \text{ lb}_m$$

The initial internal energy is the sum of the internal energy of the liquid plus the vapor:

$$U_g = m_g u_g = 0.56 \times 1077.6 = \qquad 603.5$$
$$U_f = m_f u_f = 2692.19 \times 180.1 = \underline{484,863.4}$$
$$\text{Total internal energy} = 485,466.9 \text{ Btu}$$

Because the mass in the tank is constant, the final specific volume must equal the initial specific volume, or

$$v_x = \frac{60}{2692.75} = 0.022282 \text{ ft}^3/\text{lb}_m$$

But $v_x = v_f + x v_{fg}$. Therefore, using Table A.2 at 800 psia,

$$x = \frac{v_x - v_f}{v_{fg}} = \frac{0.022282 - 0.02087}{0.5691 - 0.02087} = 0.0025756$$

The final amount of vapor is

$$0.0025756 \times 2692.75 = 6.935 \text{ lb}_m$$

The final amount of liquid is

$$2692.75 - 6.935 = 2685.815 \text{ lb}_m$$

The final internal energy is found as before:

$$U_g = m_g u_g = 6.935 \times 1115.0 = \qquad 7732.5$$
$$U_f = m_f u_f = 2685.15 \times 506.6 = \underline{1,360,633.9}$$
$$1,368,366.4 \text{ Btu}$$

The difference is

$$1,368,366.4 - 485,466.9 = 882,899.5 \text{ Btu}$$

Per unit mass, the heat added is

$$\frac{882,899.5}{2692.75} = 327.88 \text{ Btu}/\text{lb}_m$$

The small quantity of vapor mass necessitates the unusual accuracy needed to solve this problem.

5.6c Adiabatic Processes

The adiabatic process is one of the most important processes that we shall consider, because most compressions and expansions can be idealized as adiabatic processes. Ideally, these processes would be carried out isentropically or approach isentropic conditions. Figure 5.21 shows the h–s diagram for two differing types of expansions and two differing types of compressions. The solid lines on this figure represent isentropic (reversible adiabatic) paths, while the dashed lines are used to represent irreversible processes. All the processes shown in Figure 5.21 are steady-flow processes. Figures 5.22 and 5.23 show a vertical multiple plunger compressor and a single-stage turbine, respectively, as typical of some of the equipment used industrially.

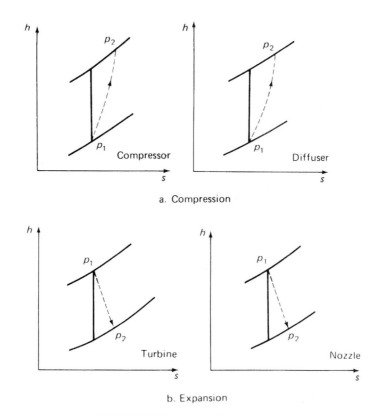

FIGURE 5.21   Adiabatic processes.

ILLUSTRATIVE PROBLEM 5.38

Steam is expanded isentropically without change in elevation and with negligible kinetic energy differences between the inlet and outlet section of a turbine. If the initial pressure is 800 psia and the initial temperature is 600°F, determine the change in enthalpy if the end-state pressure is 200 psia.

SOLUTION

As shown in Figure 5.21b, the process described in this problem is a vertical line on the Mollier chart. For 800 psia and 600°F, the Mollier chart yields $h_1 = 1270$ Btu/lb$_m$ and $s_1 = 1.485$. Proceeding vertically down the chart at constant s to 200 psia yields a final enthalpy $h_2 = 1148$ Btu/lb$_m$. The change in enthalpy for the process is $1270 - 1148 = 122$ Btu/lb$_m$.

We may also solve this problem using the *Steam Tables* in Appendix 3. Thus, the enthalpy at 800 psia and 600°F is 1270.4 Btu/lb$_m$, and its entropy is 1.4861 Btu/lb$_m$·°R. Because the process is isentropic, the final entropy at 200 psia must be 1.4861. From the saturation table, the entropy of saturated steam at 200 psia is 1.5464, which indicates the final steam condition must be wet because the entropy of the final steam is less than the entropy of saturation. Using the wet steam relation yields

$$s_x = s_f + x s_{fg} \qquad 1.4861 = 0.5440 + x(1.0025)$$

$$x = \frac{1.4861 - 0.544}{1.0025} = 0.94$$

Therefore, the final enthalpy is

$$h_x = h_f + x h_{fg} = 355.6 + 0.94(843.7) = 1148.7 \text{ Btu/lb}_m$$

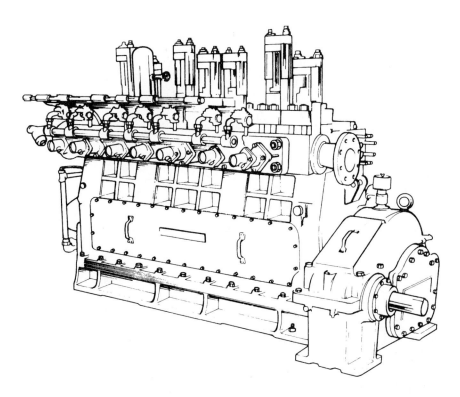

FIGURE 5.22 A 300-hp vertical multiple-plunger pump capable of a maximum discharge of 600 gal/min and a maximum discharge pressure of 2400 psig.
(Courtesy of Worthington Corp.)

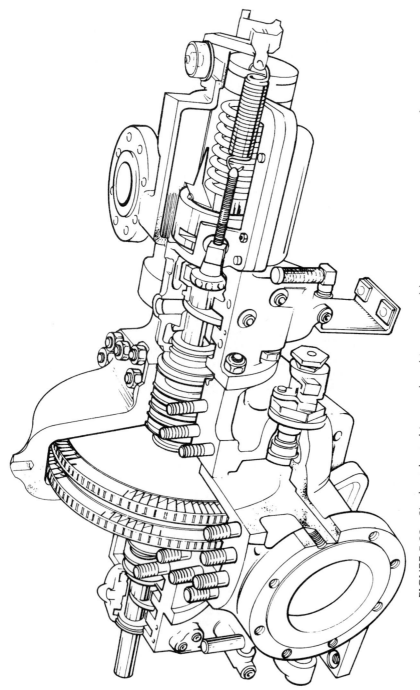

FIGURE 5.23 Single-stabe turbine used to drive pumps, blowers, compressors, generators, fans, and so on provides a compact, economical, and reliable plant accessory. *(Courtesy of Worthington Corp.)*

The change in enthalpy is 1270.4 − 1148.7 = 121.7 Btu/lb$_m$. Note the agreement with the Mollier chart solution.

We can also use the computer program to solve this problem. For 600°F and 800 psia, h = 1270.0 Btu/lb$_m$ and s = 1.4857 Btu/lb$_m$·°R. Now using p = 200 psia and s = 1.4857, we obtain h = 1148.1 Btu/lb$_m$. The change in enthalpy is 1270.0 − 1148.1 = 121.9 Btu/lb$_m$. Note the effort saved using either the Mollier chart or the computer program.

```
Saturation Properties:
─────────────────────────────
  T =    600.00    deg F
  P =    1541.7    psia
     z            zl         zg
v(ft3/1bm)      0.02362    0.2675
h(BTU/1bm)      616.59     1166.2
s(BTU/1bmF)     0.8129     1.3316
u(BTU/1bm)      609.85     1089.9

   Thermo Properties:
   ──────────────────────────
    T =    600.00   deg F
    P =    800.00   psia
    v =    0.6773   ft3/1bm
    h =    1270.0   BTU/1bm
    s =    1.4857   BTU/1bmF
    u =    1169.7   BTU/1bm

   Region: superheated


Saturation Properties:
─────────────────────────────
 T =       381.87    deg F
 P =       200.00    psia
     z            zl         zg
v(ft3/1bm)      0.01839    2.2883
h(BTU/1bm)      355.60     1199.0
s(BTU/1bmF)     0.5440     1.5462
u(BTU/1bm)      354.92     1114.3

   Thermo Properties:
   ──────────────────────────
    T =    381.87   deg F
    P =    200.00   psia
    v =    2.1512   ft3/1bm
    h =    1148.1   BTU/1bm
    s  =   1.4857   BTU/1bmF
    u =    1068.5   BTU/1bm
    x =    0.9396

   Region: saturated
```

ILLUSTRATIVE PROBLEM 5.39

If the process described in Illustrative Problem 5.38 is carried out adiabatically but not reversibly between the same initial conditions and the same final pressure, determine the final state of the steam if only 80% of the isentropic enthalpy difference is realized.

SOLUTION

Again referring to Figure 5.21, it will be seen that the final temperature and enthalpy will both be higher than for the isentropic case. The change in enthalpy is $0.8 \times 122 = 97.6$ Btu/lb$_m$. Therefore, the final enthalpy is $1270 - 97.6 = 1172.4$ Btu/lb$_m$, and the final pressure is 200 psia. The Mollier chart indicates the final state to be in the wet region, with a 3.1% moisture content and an entropy of 1.514 Btu/lb$_m$·°R.

ILLUSTRATIVE PROBLEM 5.40

A turbine expands steam isentropically from an initial pressure of 1 MPa and an initial temperature of 250°C to a final pressure of 0.1 MPa. Neglecting kinetic and potential energy terms, determine the change in enthalpy of the steam.

SOLUTION

Using the Mollier chart, $h_1 = 2942$ kJ/kg. Proceeding as shown in Figure 5.21b, that is, vertically at constant entropy to a pressure of 0.1 MPa, gives us a final enthalpy of 2512 kJ/kg. The change in enthalpy is $2942 - 2512 = 430$ kJ/kg.

ILLUSTRATIVE PROBLEM 5.41

Steam is expanded isentropically in a nozzle from an initial condition of 800 psia and 600°F to a final pressure of 200 psia. Determine the final velocity of the steam as it leaves the nozzle.

SOLUTION

The conditions given correspond to those in Illustrative Problem 5.38. Therefore, the isentropic change in enthalpy for this process is 122 Btu/lb$_m$. We now recall from Chapter 3 that the change in velocity in a nozzle with negligible entering velocity as given from Illustrative Problem 3.22 is

$$V_2 = \sqrt{2g_c J(h_1 - h_2)}$$

Substituting the value of 122 Btu/lb$_m$ for $h_1 - h_2$ yields

$$V_2 = \sqrt{2 \times 32.17 \text{ lb}_m\cdot\text{ft/lb}_f\cdot\text{s}^2 \times 778 \text{ ft}\cdot\text{lb}_f/\text{Btu} \times (122 \text{ Btu/lb}_m)}$$
$$= 2471.2 \text{ ft/s}$$

ILLUSTRATIVE PROBLEM 5.42

If the entering steam in Illustrative Problem 5.41 is expanded irreversibly in the nozzle to the same final pressure and to saturated vapor, determine the final velocity as the steam leaves the nozzle.

SOLUTION

Because the process is irreversible, we cannot show it on the Mollier diagram. However, the analysis of Illustrative Problem 3.22 for the nozzle is still valid, and all that is needed is the enthalpy at the beginning and the end of the expansion. From Illustrative Problem 5.38, h_1 is 1270 Btu/lb$_m$. For h_2 we locate the state point on the Mollier diagram as being saturated vapor at 200 psia. This gives us 1199 Btu/lb$_m$. Using the results of Illustrative Problem 3.22 yields

$$V_2 = \sqrt{2g_c J(h_1 - h_2)}$$
$$= \sqrt{2 \times 32.17 \text{ lb}_m \cdot \text{ft/lb}_f \cdot \text{s}^2 \times 778 \text{ ft} \cdot \text{lb}_f/\text{Btu} \times (1270 - 1199)\text{Btu/lb}_m}$$
$$= 1885.2 \text{ ft/s}$$

As would be expected, the final velocity in the irreversible process is less than the final velocity that we obtained for the isentropic expansion. It is left as an exercise for the student to use the computer program to solve this problem.

5.6d   Constant-Pressure Process (Isobaric Process)

The constant-pressure process is an idealization that can be used to describe the addition of heat to the working fluid in a boiler or the combustion process in a gas turbine. Figure 5.24 shows the p–v, T–s, and h–s diagrams for a vapor undergoing an irreversible, constant-pressure process. The paths are shown as dashed lines to denote that the process is irreversible. A large industrial gas turbine is shown in Figure 5.25 with both compressor and turbine blading clearly visible and a portion of the combustion section also visible. A large exhaust duct is provided for exhausting the hot gases from the unit. Both the heat-addition portion and heat-rejection portion of the gas turbine cycle are considered as constant-pressure flow processes.

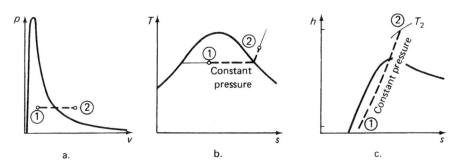

FIGURE 5.24   Constant-pressure process.

FIGURE 5.25 Large industrial gas turbine unit.
(Courtesy of Pratt and Whitney Aircraft Group, United Technologies Corp.)

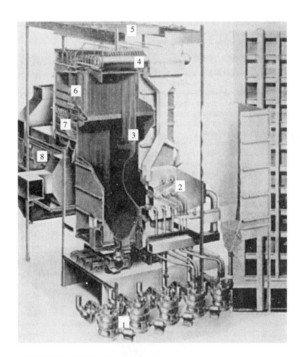

FIGURE 5.26 Combined cycle steam generator.
(Courtesy of Pratt and Whitney Aircraft Group, United Technologies Corp.)

Because the exhaust temperature of a gas turbine is relatively high, the exhaust gases can also be used as a heat source. Figure 5.26 shows a combined cycle unit in which the exhaust of the gas turbine is used to generate steam for a steam turbine generator. Supplemental oil or gas firing keeps steam pressure and temperature at the required level. Because a substantial number of Btu are recovered from the gas-turbine exhaust, this unit operates at relatively high efficiency. This combination of gas turbine and heat-recovery steam generator is responsible for much of the current interest in on-site generating systems for commercial buildings and small industrial plants.

ILLUSTRATIVE PROBLEM 5.43

Steam that is initially saturated is superheated in a boiler at constant pressure. Determine the final state if the initial pressure is 500 psia and the final temperature is 800°F. How much heat per pound of steam was added?

SOLUTION

From the saturation table, 500 psia corresponds to a temperature of 467.13°F, and the saturated vapor has an enthalpy of 1205.3 Btu/lb$_m$. At 500 psia and 800°F, the superheated vapor has an enthalpy of 1412.1 Btu/lb$_m$. Because this process is a steady-flow process at constant pressure, the energy equation becomes $q = h_2 - h_1$, assuming that differences in the kinetic energy and potential energy terms are negligible. Therefore, $q = 1412.1 - 1205.3 = 206.8$ Btu/lb$_m$ added.

5.6e   Constant-Temperature Process (Isothermal Process)

At the exhaust of a steam turbine, the steam is usually wet. This steam is subsequently condensed in a unit appropriately known as a condenser. Because the steam is initially wet, this process is carried out essentially at constant temperature (isothermally). This process is also one of constant pressure in the wet region. Figure 5.27 shows the isothermal process on various diagrams, and Figure 5.28 shows a schematic of a shell-and-tube condenser, indicating the steam flow and cooling water paths.

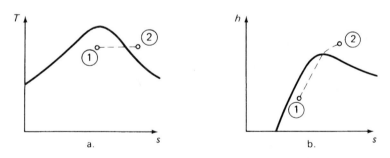

FIGURE 5.27   Isothermal process.

ILLUSTRATIVE PROBLEM 5.44

Steam is initially wet, having a moisture content of 3% at 1 psia. If it is condensed to saturated liquid, how much heat is removed?

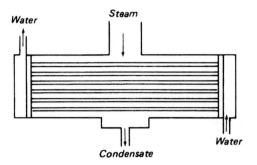

Water Steam Water Condensate

FIGURE 5.28 Shell-and-tube condenser.

SOLUTION

From the saturation table at 1 psia, $h_f = 69.74$ Btu/lb$_m$, $h_{fg} = 1036.0$ Btu/lb$_m$, and $h_g = 1105.8$ Btu/lb$_m$. Because the condensation process is carried out at constant pressure, the energy equation is $q = \Delta h$. The initial enthalpy is

$$h_x = h_f + xh_{fg} = 69.74 + (0.97)(1036.0) = 1074.66 \text{ Btu/lb}_m$$

The final enthalpy is $h_f = 69.74$. The enthalpy difference (Δh) is $1074.66 - 69.74 = 1004.92$ Btu/lb$_m$ removed during the condensation process.

The computer solution is given below where $\Delta h = 1074.3 - 69.725 = 1004.6$ Btu/lb$_m$.

```
Saturation Properties:
_____

     T =    101.71    deg F
     P =    1.0000    psia

        z              zl        zg
  v(ft3/lbm)        0.01614    333.55
  h(BTU/lbm)        69.725     1105.4
  s(BTU/lbmF)       0.1326     1.9774
  u(BUT/lbm)        69.722     1043.6

     Thermo Properties:
_____

        T =    101.71    deg F
        P =    1.0000    psia
        v =    323.55    ft3/lbm
        h =    1074.3    BTU/lbm
        s =    1.9221    BTU/lbmF
        u =    1014.4    BTU/lbm
        x =    0.9700

     Region: saturated
```

A summary of all the foregoing processes on the Mollier diagram is shown in Figure 5.29. The dashed paths are used to indicate that these are not equilibrium paths, and as such, they cannot really be drawn on an equilibrium diagram.

For the irreversible expansion between the same pressure limits, the final enthalpy (and temperature) is higher than for the same process performed reversibly. The constant-pressure process is also one of constant temperature in the wet region. Finally, throttling is the irreversible, constant-enthalpy process shown in Figure 5.29.

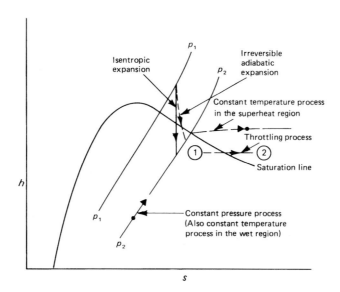

FIGURE 5.29 Processes on a Mollier diagram.

5.7 REVIEW

Although this chapter is entitled "Properties of Liquids and Gases," it has been devoted primarily to the properties of water, because the most detailed and complete information is known about water. This information exists in the *Steam Tables* in extensive tabulations and charts. A great deal of time was spent in exercises to become familiar with the nomenclature and the format of the tables and charts. Once familiar with this resource, we were able to apply the tables and charts to solve various problems. Unfortunately, there are great amounts of arithmetic computation associated with the application of the data in the *Steam Tables*. The equations for the properties of steam have been programmed and are available in the disk included with the text. In addition, it was shown that these computer programs will yield the desired properties with relative ease.

Whether the *Steam Tables* or a computer program is used, it is important to write the energy equation for the process and to sketch out the process on an appropriate *T–s* or *h–s* diagram. This procedure will lead to a better understanding of the problem and eliminate a great deal of unnecessary numerical calculations.

KEY TERMS

Terms used for the first time in this chapter are:

critical point a limiting state of a fluid in which the density of saturated liquid equals the density of saturated vapor. Also, the latent heat of vaporization is zero at this state. At the critical point, the saturated liquid and saturated vapor states are identical.

critical pressure the pressure at the critical point.

critical temperature the temperature at the critical point.

gas a vapor that is either at very low pressure or far removed from the saturated state.

Mollier chart a chart having enthalpy–entropy coordinates.

percent moisture the percent by weight of moisture in a wet mixture; equal to 100 − quality, where quality is expressed as a percentage.

phase that part of a pure substance which consists of a single-homogeneous aggregate of matter; for our purposes, these may be characterized as solid, liquid, and gas.

quality (*x*) fraction by weight of vapor in a mixture.

saturation the state of a liquid or vapor in which the vapor and liquid coexist in equilibrium in any proportion. Saturation temperature and saturation pressure refer, respectively, to the properties in the saturation state.

Steam Tables J. H. Keenan, F. G. Keyes, P. G. Hill, and J. G. Moore, *Steam Tables—Thermodynamic Properties of Water Including Vapor, Liquid, and Solid Phases* (New York: John Wiley & Sons, Inc., 1969, 1978).

subcooled that liquid state in which the pressure exceeds the saturation pressure corresponding to the temperature of the liquid.

subscripts *f* for saturated liquid, *g* for saturated vapor, and *fg* for a property change due to vaporization.

superheated vapor a vapor whose temperature is greater than the saturation temperature corresponding to the pressure.

throttling calorimeter a device that uses the throttling process to determine the quality of steam.

USCS United States conventional system of units.

vapor the gaseous phase that is in contact with saturated liquid or is not far from the saturated state.

wet a mixture of vapor plus liquid.

EQUATIONS DEVELOPED IN THIS CHAPTER

saturation properties	$h_g = h_f + h_{fg}$	
	$v_g = v_f + v_{fg}$	
	$s_g = s_f + s_{fg}$	
	$u_g = u_f + u_{fg}$	**(5.1)**
definition of enthalpy	$h = u + \dfrac{pv}{J}$	**(5.2)**
definition of enthalpy (SI)	$h = u + pv$	**(5.2a)**
wet mixture property	$h_x = h_f + x h_{fg}$	**(5.4a)**
wet mixture property	$s_x = s_f + x s_{fg}$	**(5.4b)**
wet mixture property	$u_x = u_f + x u_{fg}$	**(5.4c)**
wet mixture property	$v_x = v_f + x v_{fg}$	**(5.4d)**
approximation for the enthalpy of subcooled liquid	$h - h_f \simeq \dfrac{(p - p_f)v_f}{J}$	**(5.5)**

QUESTIONS

5.1 Define the word *phase,* and list the three common phases of matter.

5.2 State the difference between a gas and a vapor.

5.3 As the saturation pressure is increased, the saturation temperature: (a) increases; (b) decreases; (c) remains the same?

5.4 What is meant by the *critical point?*

5.5 Define the subscripts *f, g,* and *fg.* To what states do these subscripts refer?

5.6 Define the term *wet.* How is this term related to percent moisture?

5.7 How many properties are required to determine: (a) the saturation state; (b) the superheat state; (c) the wet state?

5.8 What are the principal coordinates of the Mollier diagram?

5.9 How is an isentropic process shown on a Mollier diagram?

5.10 How do you show a constant-temperature line in the wet region on the Mollier diagram?

5.11 Can the path of an irreversible process be shown on a *T–s* or a Mollier diagram?

5.12 What is known about the path of a throttling process on the Mollier diagram?

5.13 What is the ideal way that a compressor or an expander should be operated?

5.14 Using a sketch of the Mollier diagram show constant-pressure, constant-temperature, isentropic expansion, irreversible expansion, and throttling processes.

PROBLEMS

Wherever appropriate, sketch processes on a *T–s* or *h–s* diagram. Also, use the tables and Mollier charts in Appendix 3 for all problems.

Problems Involving the Use of the Tables and Charts

5.1 Determine the enthalpy, entropy, specific volume, and internal energy of saturated steam at 100 and 1000 psia.

5.2 Determine the enthalpy, entropy, specific volume, and internal energy of saturated steam at 1.0 and 1.1 MPa.

5.3 Determine the enthalpy, entropy, specific volume, and internal energy of saturated water at 1.0 and 1.1 MPa.

5.4 Determine the pressure, specific volume, and enthalpy of saturated water at 350°F and 500°F.

5.5 Determine the internal energy of steam at 10 MPa and 500°C.

5.6 Determine h_{fg} at 400°F.

5.7 Determine s_{fg} at 150°F.

5.8 Determine u_{fg} at 212°F knowing h_{fg} and v_{fg}. Check your answer against the tabulated value.

5.9 Determine the pressure, specific volume, and entropy of saturated water at 35°C.

5.10 Determine the enthalpy, entropy, specific volume, and internal energy of wet steam at 100 psia if it has 5% moisture.

5.11 Determine the enthalpy, specific volume, entropy, and internal energy of wet steam at 1.12 MPa if the quality is 90%.

5.12 Steam at 200 psia has an enthalpy of 1050 Btu/lb$_m$. Determine its quality.

5.13 If the enthalpy of steam at 1.0 MPa is 1400 kJ/kg, what is its quality?

5.14 Steam at 150 psia has a quality of 97%. What is the specific volume of this wet mixture?

5.15 The internal energy of wet steam is 2000 kJ/kg. If the pressure is 1.18 MPa, what is the quality of the mixture?

5.16 Determine the internal energy for a steam–water mixture for $t = 400°F$ and $x = 0.6$ in Btu/lb_m.

5.17 Determine the specific volume of wet steam at 200°C if the quality is 60%.

5.18 Calculate the specific volume of wet steam at 250°F if the mixture has a quality of 80%.

5.19 A steam–water mixture has a specific volume of 0.19 m³/kg. If the pressure of the mixture is 44 kPa, determine the quality and temperature of the mixture.

5.20 The temperature of a steam–water mixture is 270°F. If the specific volume of the mixture is 4.0 ft³/lb$_m$, determine the quality and pressure of the mixture.

5.21 A tank contains 2 kg of water and 0.2 kg of vapor at 220°C. Determine the quality, the volume of the tank, the volume of the water, and the pressure in the tank.

5.22 Saturated steam has an enthalpy of 2782.0 kJ/kg. What is its pressure?

5.23 If the enthalpy of saturated water is 100 Btu/lb_m, determine its pressure, temperature, and specific volume.

5.24 Steam at 500 psia has a specific volume of 0.800 ft³/lb$_m$. Determine its enthalpy and temperature.

5.25 Steam at 2.0 MPa has a specific volume of 0.08 m³/kg. Determine its temperature and enthalpy.

5.26 Steam at 300 psia is at 600°F. Determine its specific volume.

5.27 Steam at 600 psia has a specific volume of 1.500 ft³/lb$_m$. Determine its enthalpy.

5.28 Steam at 900°F has an enthalpy of 1465 Btu/lb_m. Determine its pressure.

5.29 Steam at 1000 psia has an enthalpy of 1250 Btu/lb_m. Determine its specific volume.

5.30 Steam at 1.00 MPa is at 400°C. Determine its specific volume.

5.31 Steam is at 2.0 MPa and 225°C. Determine its enthalpy.

5.32 Steam at 200 psia is at 610°F. Determine its enthalpy.

5.33 Steam is at 2.5 MPa and 227°C. Determine its enthalpy.

5.34 Determine the internal energy of steam at 190 psia and 910°F.

5.35 Calculate the internal energy of a mixture of steam and water for $t = 200°C$ and $x = 0.8$, in kJ/kg.

5.36 Determine the specific volume, enthalpy, and internal energy of steam having a pressure of 5 psia and a quality of 90%.

5.37 What is the internal energy of steam at 2.17 MPa and 225°C?

5.38 Determine the enthalpy, specific volume, internal energy, and entropy of subcooled water at 500°F and 4000 psia.

5.39 Determine the enthalpy of subcooled water at 500°F and 4000 psia using Equation (5.5). Compare your result with Problem 5.38.

5.40 Steam at 245°C has a specific volume of 0.10 m³/kg. Determine its pressure.

*5.41 Steam at 500 psia has a specific volume of 1.7500 ft³/lb$_m$. Determine its temperature and enthalpy.

Problems Involving the Use of the Computerized Tables

5.42–5.82 Solve assigned problems 5.1–5.41 using the computerized tables on the computer disk.

Problems Involving Processes

Use the *Steam Tables* and Mollier chart as applicable to solve the following problems.

5.83   A tank contains 200 lb_m of water and 20 lb_m of steam at 400 psia. Determine the volume of the tank and the quality of the mixture.

5.84   A tank contains a steam–water mixture. The tank volume is 1.5 m^3 and the pressure is 700 kPa. Calculate the mass of the mixture in the tank if the quality is 0.85.

5.85   A tank having a volume of 30 ft^3 is half filled with water, and the remainder is filled with water vapor at 100 psia. Calculate the quality of the mixture and the enthalpy of the mixture in Btu/lb_m.

5.86   A 20-ft^3 drum contains saturated steam at 400°F. What is the pressure in the drum, and what is the mass of vapor in the drum?

5.87   Solve Problem 5.86 if the drum contains saturated water.

5.88   A 5-m^3 drum contains saturated steam at 300°C. Determine the pressure in the drum and the mass of steam in the drum.

5.89   A 3-m^3 drum contains saturated water at 150°C. Determine the pressure in the drum and the mass of water in the drum.

*5.90   A steam drum has a volume of 70 ft^3. If 70% of the volume is occupied by the vapor and the contents of the drum are at 500 psia, determine the weight of liquid and vapor in the drum.

5.91   Ten pounds of steam–water mixture occupies a steam drum. If the quality of the mixture is 65%, what is the volume of the drum? The pressure is 100 psia.

5.92   A steam drum has a volume of 7 m^3. If the quality is 80% and the drum is at 1.1 MPa, determine the weight of liquid and vapor in the drum.

*5.93   A rigid, closed vessel contains a mixture of liquid and vapor at 400°F. If there is 45% liquid and 55% vapor by volume in the tank, determine the quality and pressure in the vessel.

5.94   A mixture of steam and water at 230°F has a volume of 150 ft^3. If there is 12 lb_m in the mixture, determine its quality.

5.95   Dry saturated steam at 350°C is cooled to 300°C in a tank. Determine the final quality.

5.96   Dry saturated steam at 350°F is cooled in a tank until the temperature is 300°F. Determine the final quality.

*5.97   Dry saturated steam at 200°C expands to a specific volume of 1.1 m^3/kg at constant temperature. Determine the final pressure.

*5.98   How much heat must be added to 1 lb_m of saturated steam at 500 psia in a closed tank to convert it to superheated steam at 900 psia and 1000°F?

*5.99   Saturated steam fills a 100-ft^3 container at 100 psia. If the container is cooled until the contents are at 50 psia, what is the quality of the final mixture?

*5.100 A closed, rigid vessel contains vapor and liquid at 400°F. If there is initially 45% liquid and 55% vapor by volume in the tank, determine the final percent by volume if the tank is cooled to 300°F.

5.101   One pound of superheated steam is initially at 100 psia and 400°F. If it is cooled at constant volume, determine the pressure at which it becomes saturated.

*5.102 A rigid (constant volume) tank has a volume of 2.0 ft^3 and contains saturated water vapor at 50 psia. Heat is transferred from the tank until the pressure is 20 psia. Calculate the heat transfer from the tank.

*5.103 A closed container has 20 lb_m of liquid and 20 lb_m of vapor in it at 400°F. Determine the heat that must be added to raise the temperature to 425°F.

*5.104 A boiler drum contains a steam–water mixture at 100 psia. At this time, there is 10,000 lb$_m$ of water in the drum and 5 lb$_m$ of vapor. If the contents of the drum are now heated until the pressure is 200 psia, how much heat was added?

5.105 Steam at 50 psia undergoes a process in which its temperature goes from 500°F to 700°F. What is the change in enthalpy for this process?

5.106 One pound of saturated water at 200°F is converted to saturated steam at 100 psia. Determine the enthalpy difference between these two states.

5.107 If saturated steam at 500 psia is superheated to a temperature of 1000°F, determine the change in enthalpy between these two states. Assume the final pressure of the steam to be 500 psia.

5.108 If the final pressure of the steam in Problem 5.107 is 450 psia, determine the difference in enthalpy between the two states given.

5.109 Saturated steam at 3.0 MPa is superheated to 600°C. Determine the enthalpy change for this process if it is carried out at constant pressure.

5.110 Steam flowing in a pipe is throttled by a partly open valve. The initial steam conditions are 600 psia and 800°F. Determine the final temperature if the final pressure of the steam is 100 psia.

5.111 Wet steam at a pressure of 200 psia and a quality of 90% is throttled to a final pressure of 80 psia. What is the temperature and quality of the final condition? Use the *Steam Tables* to obtain your solution.

5.112 Solve Problem 5.111 using the Mollier chart.

5.113 After throttling, it is found that the pressure and temperature of steam are 30 psia and 300°F, respectively. If the pressure before throttling is 400 psia, what is the quality of the initial mixture? Use the *Steam Tables.*

5.114 Solve Problem 5.113 using the Mollier chart.

*5.115 Superheated steam is expanded isentropically from 900 psia and 700°F to saturated vapor. Determine the difference in specific enthalpy for this expansion. Use the *Steam Tables.*

5.116 Solve Problem 5.115 using the Mollier chart.

5.117 One pound mass of steam is expanded isentropically from 500 psia and 800°F to 10 psia. Use the Mollier chart to determine the final enthalpy.

5.118 Solve Problem 5.117 using the *Steam Tables.*

5.119 Steam expands isentropically from 300 psia and 620°F to saturation. Determine the final enthalpy using the Mollier chart.

5.120 Solve Problem 5.119 using the *Steam Tables.*

5.121 Steam is used for heating a room. Assuming that the required heating load is 10,000 Btu/hr and the steam enters at 20 psia and 250°F and is condensed, how many pounds of steam per hour are required? Assume that pressure losses are negligible and that the liquid is just saturated.

5.122 Calculate the "average" specific heat at constant pressure for superheated steam between 400°F and 1000°F if the pressure is 200 psia. The definition of average for this case is $\Delta h / \Delta T$.

5.123 Steam expands isentropically in a turbine from 500 psia and 1000°F to 14.7 psia. Determine the difference in enthalpy between the initial and final conditions.

5.124 If the expansion in Problem 5.123 is not carried out isentropically but is expanded irreversibly to the same final pressure and has 40°F of superheat, determine the difference in enthalpy between the initial and final conditions. Use the Mollier chart.

5.125 One pound of superheated steam at 200 psia and 800°F expands irreversibly and adiabatically to 14.7 psia and 250°F. Determine the change in enthalpy between the initial and final states.

5.126 Wet steam leaves the exhaust of a turbine and is subsequently condensed. Assuming the wet steam to be at 0.5 psia with 15% moisture, determine the heat extracted in the condenser if the condensation process takes the mixture to saturated liquid.

5.127 Heat is added to steam in a closed cylinder. If a movable piston is placed on one end of the cylinder, it is possible (in principle) to carry out the process at constant pressure. If the initial steam is at 250°F and is saturated, how much heat is added if the final temperature is 500°F?

5.128 Two kilograms of saturated water at 1.0 MPa are heated to saturated vapor at constant pressure. Determine the heat transfer for this process.

5.129 Two lb$_m$ of saturated water at 200 psia is vaporized to saturated steam at constant pressure. Determine the change in volume for this change.

5.130 Steam having a quality of 0.80 is contained in a rigid vessel at a pressure of 300 psia. Heat is added until the temperature reaches 500°F. Determine the final pressure.

5.131 Steam at 340 kPa and 600°C undergoes a constant pressure process to 900°C. Calculate the work done per kg of steam.

5.132 Determine the change in internal energy of water vapor per lb$_m$ as it undergoes a constant pressure process from 30 psia and 500°F to 700°F.

5.133 A kilogram of saturated water at 200°C expands isothermally until it changes into a saturated vapor. Determine the work done.

5.134 Dry saturated steam at 350°F is heated at constant pressure until it reaches a temperature of 600°F. Determine the work for this process in Btu/lb$_m$.

5.135 In a constant pressure piston–cylinder process, steam is compressed from 1 MPa and 500°C to a saturated vapor. Calculate the work done to compress the steam and the heat transfer per kg.

5.136 Saturated steam at 500°F expands isothermally to 100 psia. Determine the change in enthalpy between the initial and final states.

*5.137 Saturated water enters a boiler at 500 psia. It is vaporized and superheated to a final condition of 500 psia and 1000°F. From this state, it enters a turbine where it is expanded isentropically to 1 in. Hg. What fraction of the energy required to produce the steam is obtained from the turbine if the turbine is 100% efficient?

5.138 Steam is expanded isentropically in an ideal nozzle from a state of 1000 psia and 1000°F to a final state of 20 psia. Determine the final velocity of the steam from the nozzle if the initial velocity is negligible. Use the Mollier chart.

5.139 Solve Problem 5.138 if the initial velocity in the nozzle is 950 ft/s.

5.140 Steam is expanded isentropically in a nozzle from 1 MPa and 500°C to saturated vapor. If the initial velocity is negligible, determine the final velocity from the nozzle. Use the Mollier chart.

5.141 Steam expands irreversibly in a nozzle. If the initial state is 900 psia and 600°F and the steam is expanded to 50 psia and a moisture content of 12%, determine the final velocity. Assume that the initial velocity is negligible. Use the Mollier chart.

5.142 Steam expands irreversibly in a nozzle. This initial state is 1 MPa and 500°C. The final state is 20 KPa and a moisture content of 2%. If the initial velocity can be neglected, determine the final velocity from the nozzle. Use the Mollier chart.

Use the computerized properties to solve any of the following problems.

5.143–5.202 Solve assigned problems 5.83–5.142 using the computerized properties.

6

The Ideal Gas

Learning Goals

After reading and studying the material in this chapter, you should be able to:

1. Understand the terms *ideal* and *perfect* as applied to gases.

2. Express the units of the gas constant in both SI and English systems, and use the molecular weight to determine the gas constant for a given gas.

3. Calculate the average specific heat of a gas for a gas over a temperature range.

4. Use the *Gas Tables* to solve gas processes.

5. Write the general expression for the entropy change of an ideal gas from which the equation of path for an isentropic process is obtained.

6. Derive relations for constant-volume, constant-pressure, constant-temperature, isentropic, and polytropic processes.

7. Define Mach number.

8. Calculate the states in adiabatic flow processes and in isentropic flow.

9. Calculate the mass flow rate in an ideal nozzle in terms of the initial and throat properties.

10. Calculate actual nozzle performance.

11. Calculate processes for real gases.

6.1 INTRODUCTION

In Chapter 1, we derived a simple equation for the pressure–volume relation of a gas based on elementary considerations. Because many oversimplifications were involved in this derivation, it could not be reasonably expected that the result obtained in this manner would even closely approach the behavior of a real gas. Surprisingly, it has been found that the equation can be used to represent the behavior of a large class of actual gases with an accuracy usually sufficient for engineering applications. In any case, this equation can be used to predict qualitatively the behavior of most gases, and the results so obtained can be employed as a guide for design or performance purposes.

Figures 6.1 and 6.2, respectively, show a modern, high-performance automotive engine and an aircraft gas turbine. Based on certain idealizations, which will be investigated in some detail in

FIGURE 6.1 A 2.8-L, V-6 engine mounted transversely in a front-wheel-drive automobile.
(Courtesy of Chevrolet Motor Division, General Motors Corp.)

FIGURE 6.2 Aircraft gas turbine—thrust range is 46,950 to 54,500 lb. It powers the Boeing 747, McDonnell Douglas DC-10, Airbus A300, and Airbus A310.
(Courtesy of Pratt and Whitney Aircraft Group, United Technologies Corp.)

a later chapter, it is possible to analyze the performance of these engines to obtain meaningful design criteria using the simplified equation of state.

The expressions *ideal gas* and *perfect gas* appear in many textbooks on thermodynamics, and unfortunately, some confusion has developed regarding the exact definition of these terms. For clarity and consistency, these terms will be given identical meanings and defined as gases having equations of state that correspond to Equation (6.7) or (6.8).

6.2 BASIC CONSIDERATIONS

The first observations concerning the equation of state of a gas were made by Robert Boyle (1629–1691), an English chemist, in the year 1662. He observed experimentally that *the volume of a given quantity of gas varies inversely with absolute pressure if the temperature of the gas is held constant.*

We can write *Boyle's law* as

$$\frac{p_1}{p_2} = \frac{V_2}{V_1} \qquad \text{or} \qquad p_1 V_1 = p_2 V_2 = \text{constant} \tag{6.1}$$

Because $T = $ constant, we can also write Equation (6.1) as

$$pV = C \tag{6.1a}$$

where C represents a constant for a given temperature. A family of curves for different temperatures is shown in Figure 6.3. Each of these curves is an equilateral hyperbola, because they are of the form $xy = C$.

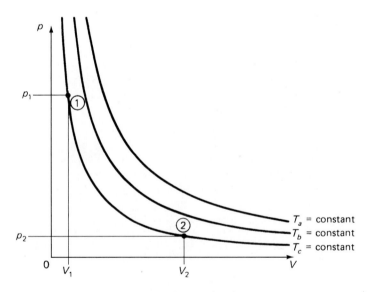

FIGURE 6.3 Boyle's law.

240 THE IDEAL GAS

ILLUSTRATIVE PROBLEM 6.1

A gas occupies a volume of 100 ft³ and is at a pressure of 100 psia. If the pressure is reduced to 30 psia, what volume will the gas occupy? Assume that the gas temperature is kept constant.

SOLUTION

For a constant temperature process, Boyle's law is

$$p_1 V_1 = p_2 V_2$$

Therefore,

$$V_2 = \frac{p_1}{p_2} V_1$$

and

$$V_2 = \frac{100}{30} (100 \text{ ft}^3) = 333 \text{ ft}^3$$

ILLUSTRATIVE PROBLEM 6.2

A gas occupies a volume of 2 m³ at a pressure of 1 MPa. If the pressure is increased to 8 MPa, what volume will the gas occupy if the gas temperature is kept constant?

SOLUTION

For the constant-temperature process, we again have from Boyle's law

$$p_1 V_1 = p_2 V_2$$

and

$$V_2 = \frac{p_1}{p_2} V_1 = \frac{10^6}{8 \times 10^6} (2 \text{ m}^3) = 0.25 \text{ m}^3$$

In 1787, approximately 100 years after Boyle's law was formulated, the French physicist Jacques Charles (1747–1832) investigated and published his experiments that showed *the volume of a gas was directly proportional to its absolute temperature if the pressure was held constant.* He also demonstrated that *the pressure of a gas was directly proportional to the absolute temperature if the volume was held constant.* Mathematically, Charles' laws can be written as

$$\frac{V}{T} = \text{constant (pressure constant)} \tag{6.2}$$

$$\frac{P}{T} = \text{constant (volume constant)} \tag{6.3}$$

ILLUSTRATIVE PROBLEM 6.3

A given mass of gas occupies 150 ft^3 at 32°F. If heat is added while the gas pressure is kept constant, determine the volume the gas occupies when its temperature is 100°F.

SOLUTION

For the conditions given, $T_1 = 32 + 460 = 492°R$ and $T_2 = 100 + 460 = 560°R$. Thus, for a constant-pressure process,

$$\frac{V_1}{V_2} = \frac{T_1}{T_2} \quad \text{or} \quad V_2 = V_1 \frac{T_2}{T_1} = \frac{(150 \text{ ft}^3)(560)}{492} = 170.7 \text{ ft}^3$$

ILLUSTRATIVE PROBLEM 6.4

If after the process performed in Illustrative Problem 6.3 the gas is contained at constant volume and its absolute temperature is increased by 25%, what percent increase in its absolute pressure will occur?

SOLUTION

If for this process $T_2 = 1.25T_1$,

$$\frac{T_2}{T_1} = 1.25$$

Therefore,

$$\frac{p_1}{p_2} = \frac{T_1}{T_2} \quad \text{or} \quad \frac{p_2}{p_1} = \frac{T_2}{T_1}$$

Thus, $p_2/p_1 = T_2/T_1 = 1.25$, and the absolute gas pressure increases by 25%.

ILLUSTRATIVE PROBLEM 6.5

A gas is cooled at constant pressure from 100°C to 0°C. If the initial volume is 4 m^3, what will its final volume be?

SOLUTION

$$V_2 = V_1 \frac{T_2}{T_1} = (4 \text{ m}^3) \left(\frac{0 + 273}{100 + 273} \right) = 2.93 \text{ m}^3$$

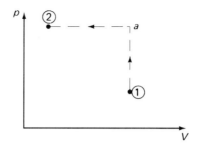

FIGURE 6.4 Ideal gas law derivation.

By combining Boyle's law and Charles' laws into a single relation, we can obtain a general gas law. We can do this by considering a gas that goes from one state, state ①, to a second state, state ②. To go from state ① to state ②, we shall first assume that the pressure is increased from state ① to the pressure of state ② by heating at constant volume. The gas is then cooled at constant pressure (the pressure at ②) until its volume reaches the volume of state ②. These events are portrayed in Figure 6.4. From this figure, we can write for the constant-volume path 1–a,

$$\frac{p_a}{p_1} = \frac{T_a}{T_1} \tag{6.4}$$

and for path $a-2$, the constant-pressure path,

$$\frac{T_a}{T_2} = \frac{V_a}{V_2} \tag{6.5}$$

Solving for T_a from Equation (6.4) and substituting into Equation (6.5) yields

$$\frac{p_a T_1}{p_1} = \frac{T_2 V_a}{V_2} \tag{6.6}$$

but $p_a = p_2$ and $V_a = V_1$. Using this and rearranging gives us

$$\frac{p_1 V_2}{T_1} = \frac{p_2 V_2}{T_2} \tag{6.7}$$

Because we selected points ① and ② arbitrarily, it follows that the term pV/T must be a constant for a given gas. For convenience, it is easier to express this constant on the basis of a unit mass of gas. Denoting the *gas constant* per unit mass by the symbol R, we have

$$\boxed{pV = pmv = mRT} \tag{6.8}$$

It will be recalled from Chapter 1 that we identified the velocity term in the equation derived on the basis of elementary kinetic theory as being proportional to the temperature of the gas (the random motion of the gas molecules). Using this, we can also obtain the equation of state of a gas as Equation (6.8).

In English engineering units, p is the pressure in pounds force per square foot absolute, V is the total volume of the gas in cubic feet, T is the absolute temperature in degrees Rankine, m is the mass of gas in pounds mass, and R is a constant of proportionality in ft $lb_f/lb_m \cdot °R$. If m is eliminated in Equation (6.8), it becomes

$$\boxed{pv = RT} \qquad (6.9)$$

in which p, R, and T correspond to the terms in Equation (6.8) and v represents specific volume. In SI units, p is in kilopascals, T is in Kelvin, m is in kilograms when R is in kJ/kg·°K, and v is in cubic meters per kilogram.

R for actual gases is not a constant and varies from gas to gas. However, it has been found experimentally that most gases at very low pressure or with high degrees of superheat exhibit nearly constant values of R. As further indicated in Table 6.1, the product of molecular weight (MW) and R for most gases is nearly constant and, in usual English engineering units, equals 1545.3 ft lb_f/lb_m mole·°R. For any other combination of pressure, volume, and temperature units, it is a relatively straightforward calculation to derive R. Some values of this constant in other systems are listed in Table 6.2. For the purposes of this book, a value of MW $\times R$ of 1545 ft lb_f/lb_m mole·°R is used. In SI units, the value of this product is 8.314 kJ/kg mole·°K.

TABLE 6.1

Gas Constant Data

Gas	Gas MW (approx.)	Product MW $\times R$
Air	29	1545 ft $lb_f/lb_m \cdot$ mole·°R
Ammonia (NH_3)	17	1520
Carbon dioxide (CO_2)	44	1532
Carbon monoxide (CO)	28	1545
Hydrogen (H_2)	2	1535
Nitrogen (N_2)	28	1537
Oxygen (O_2)	32	1543

Source: G. A. Hawkins, *Thermodynamics,* 2nd ed. (New York: John Wiley & Sons, Inc., 1951).

TABLE 6.2

Molecular Weight $\times R$

1545.3 ft lb_f/lb_m mole·°R
0.082 liter·atm/g mole·°K
8.3143 $\times 10^7$ erg/g mole·°K
8.314 J/g mole·°K
1.986 Btu/lb_m mole·°R
1.986 cal/g mole·°K

The pound mole, or more briefly, the *mole,* is the amount of a substance whose mass is equal to its molecular weight in pounds mass. Similarly, the kilogram mole is the amount of a substance whose mass is its molecular weight in kilograms. The number of moles of a gas (n) equals the mass of the gas divided by its molecular weight. Thus, 28 lb_m of nitrogen is 1 lb·mole of nitrogen. Two moles of nitrogen would have a mass of 2×28, or 56 lb_m. Molecular weights and gas constant values for many gases can be found in Appendix 3, Table A13.

ILLUSTRATIVE PROBLEM 6.6

Nitrogen at 200 psig is used to fill a container of 120 in^3. The filling process is very slow, and the contents of the tank attain the room temperature of 73°F. How much gas is there in the container?

SOLUTION

Let us first put each of the given variables into a consistent set of units:

$$p = (200 + 14.7)(144) \text{ psfa } (lb_f/ft^2)$$
$$T = (460 + 73)°R$$
$$V = \frac{120}{1728} ft^3$$
$$R = \frac{1545}{28} ft \, lb_f/lb_m \cdot °R \quad \text{(because the molecular weight of nitrogen is 28)}$$

Applying Equation (6.9), $pv = RT$, $v = RT/p$, and

$$v = \frac{(1545/28)(460 + 73)}{(200 + 14.7)(144)}$$
$$= 0.951 \text{ ft}^3/lb_m$$

The mass of gas is the total volume divided by the specific volume. Thus,

$$\frac{(120/1728) \text{ ft}^3}{0.951 \text{ ft}^3/lb_m} = 0.073 \text{ lb}_m$$

The same result is obtained by direct use of Equation (6.8); $pV = mRT$, $m = pV/RT$, and

$$m = \frac{(200 + 14.7)(144)(120/1728)}{(1545/28)(460 + 73)} = 0.073 \text{ lb}_m$$

ILLUSTRATIVE PROBLEM 6.7

If the gas in Illustrative Problem 6.6 is now heated until the temperature is 200°F, what is the pressure?

SOLUTION

Apply Equation (6.8), and note that the volume is constant.

$$\frac{p_1 V_1}{T_1} = \frac{p_2 V_2}{T_2} \quad \text{and} \quad p_2 = p_1 \frac{T_2}{T_1} \quad \text{because} \quad V_1 = V_2$$

Therefore,

$$p_2 = (200 + 14.7)\left(\frac{460 + 200}{460 + 73}\right) = 266 \text{ psia}$$

ILLUSTRATIVE PROBLEM 6.8

Carbon dioxide (MW = 44) occupies a tank at 100°C. If the volume of the tank is 0.5 m^3 and the pressure is 500 kPa, determine the mass of gas in the tank.

SOLUTION

For CO_2,

$$R = \frac{8.314 \text{ kJ/kg·mole·°K}}{44 \text{ kg/kg·mole}} = 0.1890 \text{ kJ/kg·°K}$$

Applying Equation (6.8), $pV = mRT$, and

$$m = \frac{pV}{RT} = \frac{(500)(0.5)}{0.1890(273 + 100)} = 3.546 \text{ kg}$$

It is sometimes convenient to express the volume of a gas in terms of the volume that 1 mole of gas will occupy at a given temperature and pressure. We can readily obtain this value by considering Equation (6.9). Let us multiply both sides of this equation by MW, the molecular weight of a given gas. Thus,

$$(MW)pv = (MW)RT \tag{6.10}$$

Because $R = 1545/MW$,

$$pv(MW) = 1545T \tag{6.11}$$

Because $v(MW)$ equals $V(MW)/m$ and $m/(MW)$ is the number of moles, n, 1 mole of a gas would have a volume, called V_{molar}, of

$$V_{molar} = \frac{1545T}{p} \quad \text{or} \quad \frac{8.314T}{p} \quad \text{in SI units} \tag{6.12}$$

It is apparent from Equation (6.12) that for a given pressure and temperature, a mole of *any* gas will occupy the *same volume*. Quite often, the term *standard state* appears in the engineering and scientific literature, and some confusion exists as to its specific meaning. This is due to the lack of general agreement on the definition of temperature and pressure at this state. The most common standard state in use is 32°F and 14.7 psia. For this state, it will be found that 1 lb_m mole occupies 358 ft^3 or 22.4 L/g mole. The use of the term *standard state* should be avoided unless the conditions of this state are specified.

6.3 THE SPECIFIC HEAT

In the general case of a fluid involved in a thermodynamic process, both heat and work energy interchanges are involved, and concurrently, there is a temperature change of the working fluid. The term *specific heat* has been defined in Chapter 3 as the ratio of the heat transferred per unit mass by the working fluid in a process to the corresponding change in temperature of the fluid. Mathematically,

$$c = \frac{q}{\Delta T} \tag{6.13}$$

Several features of Equation (6.13) have already been noted in Chapter 3 and are repeated here for emphasis. Heat is transferred energy and, by convention, is positive if added to a system and negative if extracted from a system. Also, the addition or removal of energy as work does not enter into the definition of specific heat. These concepts lead us to the conclusion that the specific heat of a process can be zero, positive, negative, or even infinite. Two processes proved themselves to be of particular interest; the constant-pressure process, and the constant-volume process. For these processes, the respective specific heats are

$$c_p = \left(\frac{q}{\Delta T}\right)_{p \text{ constant}} \quad \text{(flow or nonflow)}$$

$$c_v = \left(\frac{q}{\Delta T}\right)_{v \text{ constant}} \tag{6.14}$$

Throughout this chapter, it will be necessary to invoke both the first and second laws to specify the path and state conditions as a gas undergoes a change of state. In Chapter 3, the first law was applied to the steady-flow and nonflow, constant-pressure processes, and the energy relation $q = \Delta h$ resulted in both processes. Although a complete discussion of the temperature and pressure dependence of the specific heat of a real gas is beyond the scope of this book, it may be noted that a gas whose equation of state is given by Equation (6.9) has both its internal energy and enthalpy independent of pressure and that these properties depend solely on temperature. Therefore, for the ideal gas for any process, flow or nonflow,

$$c_p = \frac{\Delta h}{\Delta T} \tag{6.15a}$$

and

$$c_v = \frac{\Delta u}{\Delta T} \tag{6.15b}$$

Note that these equations are valid for any ideal gas in any process. Thus, Δh always equals $c_p \Delta T$ even for a constant-volume process and Δu always equals $c_v \Delta T$ even for a constant-pressure process.

CALCULUS ENRICHMENT

Because temperature and specific volume are independent, measurable properties, they can be used to establish the thermodynamic state of a substance. Therefore, we can state that the internal energy, u, is given as a function of T and v. Mathematically,

$$u = u(T,v) \tag{a}$$

The total differential of u is given in terms of the partial derivatives as

$$du = \left(\frac{\delta u}{\delta T}\right)_v dT + \left(\frac{\delta u}{\delta v}\right)_T dv \tag{b}$$

The first partial derivative in Equation (a) is the change in internal energy per degree of temperature change for a constant volume process, and it is called the *specific heat at constant volume*. Thus,

$$c_v \equiv \left(\frac{\delta u}{\delta T}\right)_v \tag{c}$$

In a series of famous experiments, Sir James Prescott Joule (1818–1889), an English physicist, determined that the internal energy of a gas depended only on its temperature. Joule performed his experiments on an apparatus such as the one shown schematically in Figure A. Basically, it consisted of two tanks connected by a valve that were placed in a tank of water. The temperature of the water was monitored with a thermometer. One tank was

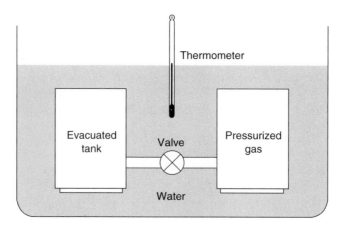

FIGURE A Schematic of Joule's experiment.

evacuated while a gas was sealed in the other thank. The valve was slowly opened, and after equilibrium was attained, the temperature of the water was found to be unchanged. There was no heat transfer to or from the system, and no work was performed on or by the system. From the First Law for a closed system, we conclude that there was no change in the internal energy of the system even though both the pressure and specific volume of the gas changed. Joule concluded that the internal energy of the gas was a function only of temperature, making Equation (a),

$$u = u(T) \tag{d}$$

Equation (d) can also be derived mathematically for a gas whose equation of state corresponds to Equation (6.9).

Equation (c) can be changed as it is applied to an ideal gas as

$$c_v = \frac{du}{dT} \quad \text{(ideal gas)} \tag{e}$$

We can consider the change in enthalpy of an ideal gas in the same manner as we considered the internal energy. Thus,

$$h = h(T, p) \tag{f}$$

where the independent variables chosen are pressure and temperature. The change in enthalpy is

$$dh = \left(\frac{\delta h}{\delta T}\right)_p dT + \left(\frac{\delta h}{\delta p}\right)_T dp \tag{g}$$

The first partial derivative in Equation (g) is the change in enthalpy per degree of temperature change for a constant pressure process, and it is called the specific heat at constant pressure. Thus,

$$c_p \equiv \left(\frac{\delta h}{\delta T}\right)_p \tag{h}$$

Because the definition of enthalpy is

$$h \equiv u + pv \tag{i}$$

we can write for the ideal gas that,

$$h = u + RT \tag{j}$$

Therefore, because R is a constant and u is only a function of temperature, it follows that for the ideal gas the enthalpy is a function only of temperature. The specific heat at constant pressure for the ideal gas is

$$c_p = \frac{dh}{dT} \tag{k}$$

When internal energy or enthalpy data are available (e.g., in the Keenan and Kaye *Gas Tables*),* an "average" value of c_p and c_v can readily be obtained.

$$\bar{c}_p = \frac{h_2 - h_1}{T_2 - T_1} \tag{6.16}$$

$$\bar{c}_v = \frac{u_2 - u_1}{T_2 - T_1} \tag{6.17}$$

For cases in which internal energy or enthalpy data are not available, equations have been developed either from empirical or spectroscopic data expressing the dependence of the specific heat on temeprature. These equations are commonly of the following forms, with A, B, and D as constants:

$$c = A + BT + DT^2 \tag{6.18a}$$

$$c = A' + \frac{B'}{T} + \frac{D'}{T^2} \tag{6.18b}$$

The mean or average specific heat of a substance can be defined as that value, when multiplied by the temperature interval, that will give the energy as heat interchanged during a process. Thus, if the specific heat varies as shown in Figure 6.5, the mean value occurs when areas *abcd* and *aefd* are equal. For gases whose specific heats are of the form given by Equation (6.18), and mean specific heats are, respectively,

$$\bar{c} = A + \frac{B}{2}(T_2 + T_1) + \frac{D}{3}(T_2^2 + T_2 T_1 + T_1^2) \tag{6.19a}$$

$$\bar{c} = A' + \frac{B' \ln(T_2/T_1)}{T_2 - T_1} + \frac{D'}{T_2 T_1} \tag{6.19b}$$

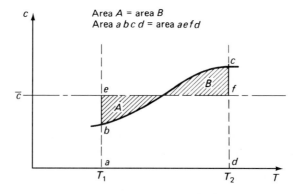

FIGURE 6.5 Mean specific heat.

---

* J. H. Keenan and J. Kaye, *Gas Tables* (New York: John Wiley & Sons, Inc., 1948).

250   THE IDEAL GAS

An alternative interpretation of the average specific heat can be illustrated graphically if reference is again made to Figure 6.5. The c and T scales are plotted linearly, and $\bar{c}$ expresses the condition that the equal positive and negative areas cancel each other. In other words, area A equals area B. Note that the mean specific heat does not necessarily occur at the arithmetic mean temperature unless c is constant or a linear function of T.

CALCULUS ENRICHMENT

The mathematical definition of the average value of a function can be written as

$$\bar{y} = \frac{\displaystyle\int_{x_1}^{x_2} y\, dx}{x_2 - x_1} \tag{a}$$

Using Equation (6.18a),

$$\bar{c} = \frac{\displaystyle\int_{T_1}^{T_2} (A + BT + DT^2)\, dT}{(T_2 - T_1)} \tag{b}$$

Integrating,

$$\bar{c} = \frac{A(T_2 - T_1) + \dfrac{B(T_2^2 - T_1^2)}{2} + \dfrac{D(T_2^3 - T_1^3)}{3}}{(T_2 - T_1)} \tag{c}$$

Factoring and dividing by $T_2 - T_1$,

$$\bar{c} = A + \frac{B}{2}(T_2 + T_1) + \frac{D}{3}(T_2^2 + T_2 T_1 + T_1^2) \tag{6.19a}$$

Using Equation (6.18b),

$$\bar{c} = \frac{\displaystyle\int_{T_1}^{T_2} \left(A' + \dfrac{B'}{T} + \dfrac{D'}{T^2}\right) dT}{(T_2 - T_1)} \tag{d}$$

Integrating,

$$\bar{c} = \frac{A'(T_2 - T_1) + B' \ln\left(\dfrac{T_2}{T_1}\right) - (T_2^{-1} - T_1^{-1})}{T_2 - T_1} \tag{e}$$

and

$$\bar{c} = A' + \frac{B' \ln (T_2/T_1)}{T_2 - T_1} - \frac{\left(\dfrac{1}{T_2} - \dfrac{1}{T_1}\right)}{T_2 - T_1} \tag{f}$$

Rearranging the last term in Equation (f),

$$\bar{c} = A' + \frac{B' \ln (T_2/T_1)}{T_2 - T_1} + \frac{1}{T_2 T_1} \qquad \textbf{(6.19b)}$$

ILLUSTRATIVE PROBLEM 6.9

For nitrogen between 540°R and 9000°R, an equation for the instantaneous specific heat at constant pressure is

$$\bar{c}_p = 0.338 - \frac{1.24 \times 10^2}{T} + \frac{4.15 \times 10^4}{T^2} \ \text{Btu/lb}_\text{m} \cdot °\text{R}$$

Determine the mean specific heat at constant pressure between 80°F and 500°F.

SOLUTION

This equation has the form of Equation (6.18b), with $A' = 0.338$, $B' = -1.24 \times 10^2$, and $D' = 4.15 \times 10^4$. Therefore, from Equation (6.19b), with temperatures in degrees Rankine,

$$\bar{c}_p = 0.338 - \frac{1.24 \times 10^2 \ln (960/540)}{960 - 540} + \frac{4.15 \times 10^4}{(540)(960)}$$
$$= 0.248 \ \text{Btu/lb}_\text{m} \cdot °\text{R}$$

ILLUSTRATIVE PROBLEM 6.10

Using the data for the properties of some gases at low pressure in Appendix 3, solve Illustrative Problem 6.9.

SOLUTION

The table in Appendix 3 does not give us the enthalpy data at 960°R and 540°R that we need. Interpolating yields

T	$\bar{h}$	T	$\bar{h}$
$_3\left(\begin{array}{c}537\\540\\600\end{array}\right._{63}$	3729.5	$_{60}\left(\begin{array}{c}900\\960\\1000\end{array}\right._{100}$	6268.1
	3750.4		6694.0
	4167.9		6977.9

$$\bar{h}_{540} = 3729.5 + \frac{3}{63}(4167.9 - 3729.5)$$

$$\bar{h}_{960} = 6268.1 + \frac{60}{100}(6977.9 - 6268.1)$$

Note that $\bar{h}$ is given for a mass of 1 lb mole. To obtain the enthalpy per pound, it is necessary to divide the values of $\bar{h}$ by the molecular weight, 28.

$$\bar{c} = \frac{h_2 - h_1}{T_2 - T_1} = \frac{6694.0 - 3750.4}{28(960 - 540)} = 0.250 \text{ Btu/lb}_m\cdot°\text{R}$$

With the more extensive *Gas Tables,* these interpolations are avoided. The *Gas Tables* provide a relatively easy and accurate method of obtaining average specific heats. Also, these tables have been computerized for ease of application.

ILLUSTRATIVE PROBLEM 6.11

The specific heat of air between 500°R and 2700°R is given by

$$c_p = 0.219 + 3.42 \times 10^{-5}T - 2.93 \times 10^{-9}T^2 \text{ Btu/lb}_m\cdot°\text{R}$$

Determine the mean specific heat at constant pressure for air between 80°F and 500°F.

SOLUTION

The equation for the instantaneous specific heat is of the form of Equation (6.18a), with $A = 0.219$, $B = 3.42 \times 10^{-5}$, and $D = 2.93 \times 10^{-9}$. Using these values and Equation (6.19a) yields

$$\bar{c}_p = 0.219 + \frac{3.42 \times 10^{-5}}{2}(960 + 540)$$
$$- \frac{2.93 \times 10^{-9}}{3}(960^2 + 960 \times 540 + 540^2)$$
$$= 0.243 \text{ Btu/lb}_m\cdot°\text{R}$$

For the ideal gas, certain unique and simple relations relating the specific heat at constant pressure to the specific heat at constant volume can be derived. Let us first recall the general definition of the term *enthalpy,*

$$h = u + \frac{pv}{J} \tag{6.20}$$

The change is enthalpy between any two states therefore is

$$h_2 - h_1 = (u_2 - u_1) + \frac{p_2v_2}{J} - \frac{p_1v_1}{J} \tag{6.21}$$

But $\qquad h_2 - h_1 = c_p(T_2 - T_1); \quad u_2 - u_1 = c_v(T_2 - T_1)$

and $\qquad pv = RT \tag{6.22}$

Substituting Equation (6.22) into Equation (6.21) gives

$$c_p(T_2 - T_1) = c_v(T_2 - T_1) + \frac{R}{J}(T_2 - T_1) \qquad (6.23)$$

Rearranging and simplifying Equation (6.23) yields the desired result:

$$\boxed{c_p - c_v = \frac{R}{J}} \qquad (6.24)$$

In SI units, the conversion factor J is not included. Therefore,

$$\boxed{c_p - c_v = R \quad \text{SI units}} \qquad (6.24a)$$

At this point, is often found convenient to express the results of Equation (6.24) in terms of the ratio of the specific heats. By defining

$$\boxed{k = \frac{c_p}{c_v}} \qquad (6.25)$$

and dividing Equation (6.24) by c_v, we obtain

$$\frac{c_p}{c_v} - \frac{c_v}{c_v} = \frac{R}{Jc_v} \qquad (6.26)$$

Using the definition of k given by Equation (6.25), we have

$$k - 1 = \frac{R}{Jc_v} \qquad (6.27)$$

Rearranging yields

$$\boxed{c_v = \frac{R}{J(k-1)}} \qquad (6.28)$$

In SI units,

$$\boxed{c_v = \frac{R}{k-1}} \qquad (6.28a)$$

Because $c_p = kc_v$,

$$\boxed{c_p = \frac{R}{J}\left(\frac{k}{k-1}\right)} \qquad (6.29)$$

In SI units,

$$c_p = R\left(\frac{k}{k-1}\right)$$

(6.29a)

ILLUSTRATIVE PROBLEM 6.12

Oxygen has a c_p of 0.24 But/lb$_m$ · °R at a given temperature. Determine c_v if its molecular weight is 32.

SOLUTION

The molecular weight of oxygen is 32. Therefore, R is $1545/32 = 48.28$. Because $c_p - c_v = R/J$,

$$c_v = c_p - \frac{R}{J}$$

$$= 0.24 - \frac{48.28}{778} = 0.178 \text{ Btu/lb}_m\cdot°R$$

ILLUSTRATIVE PROBLEM 6.13

If k of oxygen is 1.4, determine c_p and c_v in SI units.

SOLUTION

From Equation (6.28a),

$$c_v = \frac{R}{k-1} = \frac{8.314/32}{1.4-1} = 0.6495 \frac{\text{kJ}}{\text{kg}\cdot°K}$$

Because $c_p = kc_v$,

$$c_p = (1.4)(0.6495) = 0.9093 \frac{\text{kJ}}{\text{kg}\cdot°K}$$

ILLUSTRATIVE PROBLEM 6.14

A gas having an $R = 60$ ft lb$_f$/lb$_m$·°R undergoes a process in which $\Delta h = 500$ Btu/lb$_m$ and $\Delta u = 350$ Btu/lb$_m$. Determine k, c_v, and c_p for this gas.

SOLUTION

Because $\Delta h - c_p \Delta T$ and $\Delta u = c_v \Delta T$,

$$\frac{\Delta h}{\Delta u} = \frac{c_p \Delta T}{c_v \Delta T} = \frac{c_p}{c_v} = k$$

Therefore,

$$k = \frac{500}{300} = 1.429$$

From Equation (6.28),

$$c_v = \frac{R}{J(k-1)} = \frac{60}{(778)(1.429 - 1)}$$
$$= 0.180 \text{ Btu/lb}_m \cdot {}^\circ R$$

and

$$c_p = kc_v$$

yielding

$$c_p = (1.429)(0.180) = 0.257 \text{ Btu/lb}_m \cdot {}^\circ R$$

ILLUSTRATIVE PROBLEM 6.15

A closed, rigid container with a volume of 60 in³ contains 0.0116 lb$_m$ of a certain gas at 90 psia and 40°F (Figure 6.6). The gas is heated to 140°F and 108 psia, and the heat input is found to be 0.33 Btu. Assuming this to be a perfect gas, find the specific heat at constant pressure c_p.

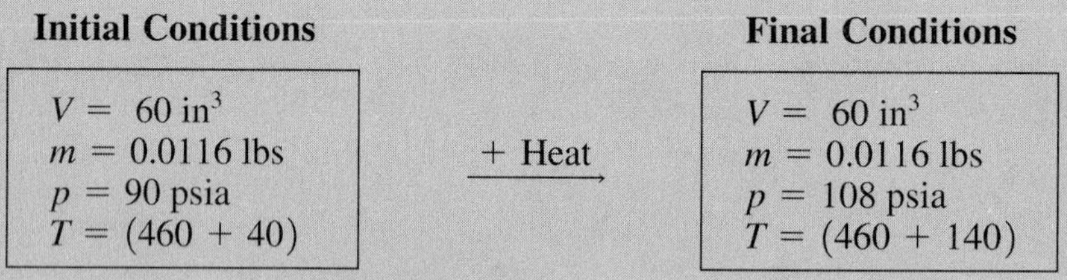

Initial Conditions   **Final Conditions**

$V = 60 \text{ in}^3$ $V = 60 \text{ in}^3$
$m = 0.0116 \text{ lbs}$ + Heat $m = 0.0116 \text{ lbs}$
$p = 90 \text{ psia}$ $p = 108 \text{ psia}$
$T = (460 + 40)$ $T = (460 + 140)$

FIGURE 6.6 Illustrative Problem 6.15.

SOLUTION

When solving this type of problem, it is necessary to note carefully the information given and to write the correct energy equation for the process. Because the process is carried out at constant volume, the heat added equals the change in internal energy. Because the change in internal energy per pound for the ideal gas is $c_v(T_2 - T_1)$, the total change in internal energy for m pounds must equal the heat added. Thus,

$$Q = m(u_2 - u_1) = mc_v(T_2 - T_1)$$

Using the data of the problem, $0.33 = 0.0116c_v(600 - 500)$, and

$$c_v = 0.284 \text{ Btu/lb}_m \cdot {}^\circ\text{R}$$

To obtain c_p, it is first necessary to obtain R. Enough information was given in the initial conditions of the problem to apply Equation (6.8) for R, $pV = mRT$,

$$(90)\,(144)\left(\frac{60}{1728}\right) = 0.0116(R)\,(460 + 40)$$

and
$$R = 77.6 \text{ ft lb}_f/\text{lb}_m \cdot {}^\circ\text{R}$$

Because

$$c_p - c_v = \frac{R}{J}$$

$$c_p = c_v + \frac{R}{J}$$

$$c_p = 0.284 + \frac{77.8}{778} = 0.384 \text{ Btu/lb}_m \cdot {}^\circ\text{R}$$

6.4   ENTROPY CHANGES OF THE IDEAL GAS

In the rest of this chapter, various nonflow processes of the ideal gas will be studied. It must be remembered that the energy quantities called work and heat depend on the path that the gas undergoes, whereas enthalpy, entropy, and internal energy are determined by the final and initial states of the fluid because they are properties.

Consider a gas that undergoes a change in state from A to B, as shown in Figure 6.7a, on pressure–volume coordinates. In this process (path A, B), the pressure, volume, and temperature change to the final values so that p_2, v_2, and T_2 are greater than p_1, v_1, and T_1.

To go from state A to state B, let us select the two separate reversible paths A, C and C, B. Path A, C is carried out at constant volume, and path C, B is carried out at constant pressure. These paths are shown in Figure 6.7a as dashed lines and are similarly shown on $T–s$ coordinates in Figure 6.7b. It is important to note that the properties at the initial and final states will determine the changes in enthalpy, entropy, and internal energy, but that the path selected between these states will determine the energy interchange as both heat and work.

The change in entropy for a gas undergoing the reversible paths A, C and C, B can readily be determined. For the constant-volume path, $q = \Delta u = c_v \Delta T$, and for the constant-pressure path, $q = \Delta h = c_p \Delta T$. Therefore, from the definition of the property entropy,

$$\Delta s = \frac{q}{T} = \left(\frac{c_v \Delta T}{T}\right)_v + \left(\frac{c_p \Delta T}{T}\right)_p \tag{6.30}$$

where the subscripts v and p refer to the constant-volume and constant-pressure processes, respectively.

To obtain the total change in entropy for a finite process between the temperature limits of T_2 and T_1, it is necessary to sum all the terms of Equation (6.30) over the entire temperature range. Let us first assume that the specific heats are independent of temperature. It is necessary only to sum

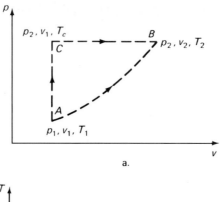

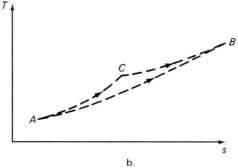

FIGURE 6.7 General gas process.

the $\Delta T/T$ values. This process has already been illustrated, but it is worth repeating at this point. Figure 6.8 shows a plot of $1/T$ as a function of T. By selecting a value of ΔT, as shown, the shaded area is $\Delta T/T$. Thus, the sum of the $\Delta T/T$ values is the area under the curve between the desired temperature limits. By the methods of the calculus, it can be shown that this area is $\ln T_2/T_1$.

By performing the summation required by Equation (6.30) and again noting that the specific heats are constant,

$$\Delta s = \left(c_v \ln \frac{T_c}{T_1} \right)_v + \left(c_p \ln \frac{T_2}{T_c} \right)_p \tag{6.31}$$

Equation (6.31) can be put into a more useful form by noting that for the constant-volume process, Equation (6.9) yields $T_c/T_1 = p_2/p_1$, and for the constant-pressure process, $T_2/T_c = v_2/v_1$. Thus,

$$\Delta s = c_v \ln \frac{p_2}{p_1} + c_p \ln \frac{v_2}{v_1} \tag{6.32}$$

By dividing both sides of Equation (6.32) by c_v and noting that $k = c_p/c_v$, we obtain

$$\frac{\Delta s}{c_v} = k \ln \frac{v_2}{v_1} + \ln \frac{p_2}{p_1} \tag{6.33}$$

258 THE IDEAL GAS

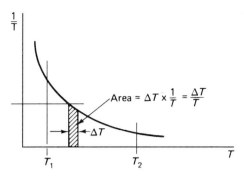

FIGURE 6.8 Evaluation of $\Delta T/T$.

Another form of Equation (6.32) can be derived by using Equations (6.24) and (6.33):

$$\Delta s = c_p \ln \frac{T_2}{T_1} - \frac{R}{J} \ln \frac{p_2}{p_1} \qquad \textbf{(6.34)}$$

Equation (6.34) can also be derived in the following alternative way; from Chapter 4,

$$T\Delta s = \Delta h - \frac{v\Delta p}{J}$$

Dividing through by T and noting that $\Delta h = c_p\,\Delta T$ and $v/T = R/p$,

$$\Delta s = c_p \frac{\Delta T}{T} - \frac{R}{J}\left(\frac{\Delta p}{p}\right)$$

Summing each term on the right side of this equation gives us Equation (6.34).

The utility of these considerations and equations will become apparent as the individual gas processes are studied. The student should note that Equations (6.31) through (6.34) are valid for *any process* of an *ideal gas* when the specific heats are *constant*. Note that in SI units, the J in the preceding equations is not necessary.

CALCULUS ENRICHMENT

In Chapter 4, we wrote the first of the Gibbs or $T\,ds$ equations as

$$T\,ds = du + p\,dv \qquad \textbf{(a)}$$

Assuming that we are dealing with an ideal gas,

$$du = c_v\,dT \qquad \textbf{(b)}$$

and

$$pv = RT \qquad \textbf{(c)}$$

Inserting Equations (b) and (c) into Equation (a) and solving for ds,

$$\int_{s_1}^{s_2} ds = \int \frac{du}{T} + \int \frac{p\,dv}{T} = \int_{T_1}^{T_2} c_v \frac{dT}{T} + \int_{v_1}^{v_2} R \frac{dv}{v} \qquad \textbf{(d)}$$

For constant specific heats, Equation (d) is integrated to yield

$$s_2 - s_1 = \Delta s = c_v \ln \frac{T_2}{T_1} + R \ln \frac{v_2}{v_1} \tag{e}$$

Equation (e) can be rearranged to yield Equation (6.32) using $c_p - c_v = R$. Thus,

$$\Delta s = c_v \ln \frac{T_2}{T_1} + (c_p - c_v) \ln \frac{v_2}{v_1} \tag{f}$$

and

$$\Delta s = c_v \ln \frac{T_2}{T_1} + c_p \ln \frac{v_2}{v_1} - c_v \ln \frac{v_2}{v_1} \tag{g}$$

Combining the c_v terms and using $\dfrac{p_1 v_1}{T_1} = \dfrac{p_2 v_2}{T_2}$ yields

$$\Delta s = c_v \ln \frac{p_2}{p_1} + c_p \ln \frac{v_2}{v_1} \tag{h}$$

which is Equation (6.32).

ILLUSTRATIVE PROBLEM 6.16

One pound of air (MW = 29) expands from 100 psia and 100°F to 15 psia and 0°F. Determine the change in entropy. Assume that c_p of air is 0.24 Btu/lb$_m$·°R and is constant.

SOLUTION

On the basis of the data given, we can use Equation (6.34) to solve this problem. Thus,

$$\Delta s = 0.24 \ln \frac{460}{460 + 100} - \frac{R}{J} \ln \frac{15}{100}$$

We can make these logarithms positive quantities by noting that the log $x = -\log 1/x$. Therefore,

$$-\Delta s = 0.24 \ln \frac{460 + 100}{460} - \frac{R}{J} \ln \frac{100}{15}$$

$$= 0.24 \ln \frac{560}{460} - \frac{1545/29}{778} \ln \frac{100}{15}$$

and

$$-\Delta s = 0.04721 - 0.1299$$

$$= -0.0827 \quad \text{or} \quad \Delta s = 0.0827 \text{ Btu/lb}_m \cdot °R$$

ILLUSTRATIVE PROBLEM 6.17

Solve Illustrative Problem 6.16 using Equation (6.32).

SOLUTION

Because c_p and R are given, let us first solve for c_v.

$$c_p = \frac{Rk}{J(k-1)}$$

$$0.24 = \frac{(1545/29)\,(k)}{778(k-1)}$$

Solving for k yields

$$k = 1.399 \quad \text{and} \quad c_v = \frac{c_p}{k} = \frac{0.24}{1.399} = 0.1716 \text{ Btu/lb}_m\cdot°R$$

Using Equation (6.32) gives us

$$\Delta s = c_v \ln \frac{p_2}{p_1} + c_p \ln \frac{v_2}{v_1}$$

But

$$\frac{v_2}{v_1} = \left(\frac{T_2}{T_1}\right)\frac{p_1}{p_2} = \left(\frac{460+0}{460+100}\right)\left(\frac{100}{15}\right) = 5.476$$

Thus,

$$\Delta s = 0.1716 \ln \frac{15}{100} + 0.24 \ln 5.476$$

$$= -0.3255 + 0.4081 = 0.0826 \text{ Btu/lb}_m\cdot°R$$

The agreement is very good.

ILLUSTRATIVE PROBLEM 6.18

Two kilograms of oxygen expand from 500 kPa and 100°C to 150 kPa and 0°C. Determine the change in entropy if $c_p = 0.9093$ kJ/kg·°K and remains constant.

SOLUTION

Using Equation (6.34) and dropping J gives us

$$\Delta s = 0.9093 \ln \frac{273}{273+100} - \frac{8.314}{32} \ln \frac{150}{500}$$

$$= 0.2901 \text{ kJ/kg}\cdot°K$$

For 2 kg,

$$\Delta S = 2(0.02901) = 0.05802 \text{ kJ/}°\text{K}$$

ILLUSTRATIVE PROBLEM 6.19

An ideal gas with constant specific heats undergoes a change during which its specific volume is halved and its entropy increases by an amount equal to one-fourth of its specific heat at constant volume. Assuming that k is 1.4, what was the increase in pressure for this process?

SOLUTION

From Equation (6.33) and the data given,

$$\frac{\Delta s}{c_v} = k \ln \frac{v_2}{v_1} + \ln \frac{p_2}{p_1}$$

$$\frac{1/4 c_v}{c_v} = 1.4 \ln \frac{1/2}{1} + \ln \frac{p_2}{p_1}$$

or

$$\frac{1}{4} - 1.4 \ln \frac{1}{2} = \ln \frac{p_2}{p_1}$$

But $-1.4 \ln \frac{1}{2} = -1.4(\ln 1 - \ln 2) = -1.4(0 - 0.693) = 1.4(0.693)$. Thus,

$$\frac{1}{4} + 1.4(0.693) = \ln \frac{p_2}{p_1}$$

$$\ln \frac{p_2}{p_1} = 1.22$$

and by taking antilogarithms,

$$\frac{p_2}{p_1} = 3.4 \quad \text{or} \quad p_2 = 3.4 p_1$$

6.5 NONFLOW GAS PROCESSES

In the following portions of this section, five nonflow gas processes are analyzed in detail. These processes are constant-volume, constant-pressure, isothermal, isentropic, and polytropic. The derivations in each case proceed from a consideration of the equation of state and the equation of the path. Paths shown as dashed lines indicate nonequilibrium paths.

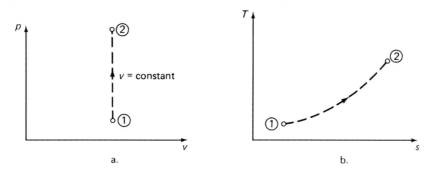

FIGURE 6.9 Constant-volume process.

6.5a Constant-Volume Process (Isometric Process)

The constant-volume process is best exemplified by a closed tank from which heat is either added or removed. In Figure 6.9, the indicated path is one in which heat is being added.

From Equation (6.9), the equation of the path can readily be written as

$$\frac{p_1}{T_1} = \frac{p_2}{T_2} \tag{6.35}$$

The applications of the first law yields the relationship between the heat added and the final state condition. Because there is no volume change, no work is done. Thus,

$$q = u_2 - u_1 = c_v(T_2 - T_1) \tag{6.36}$$

Also, $\qquad\qquad\qquad \mathbf{w} = 0 \tag{6.37}$

because there is no change in volume.

The change in entropy for this process is given directly by the defining equation for entropy or as the volume part of Equation (6.31). In either case,

$$\Delta s = c_v \ln \frac{T_2}{T_1} \tag{6.38}$$

Two further considerations complete all the necessary relations for the constant-volume process. The first of these states that the specific heat for this process is by definition c_v and that we have assumed it to be constant. The second consideration (whose utility will appear later) determines the exponent in the equation $pv^n =$ constant that will make this equation fit the constant-volume process. Rearranging this equation gives us

$$\left(\frac{p_1}{p_2}\right)\left(\frac{v_1}{v_2}\right)^n = 1 \quad \text{or} \quad \frac{v_1}{v_2} = \left(\frac{p_2}{p_1}\right)^{1/n} \tag{6.39}$$

Because $v_1/v_2 = 1$, it becomes necessary for $1/n$ to be zero or

$$n \to \infty \tag{6.40}$$

ILLUSTRATIVE PROBLEM 6.20

If $\frac{1}{2}$ lb of a gas is heated at constant volume from 70°F to 270°F, determine the change in entropy for this process. Assume that $c_v = 0.17$ Btu/lb$_m$·°R.

SOLUTION

For this process, $T_1 = 530$°R and $T_2 = 730$°R. Therefore,

$$\Delta s = c_v \ln \frac{T_2}{T_1} = 0.17 \ln \frac{730}{530}$$

$$\Delta s = 0.0544 \text{ Btu/lb}_m\text{·°R}$$

$$\Delta S = \left(\tfrac{1}{2}\right)(0.0544) = 0.0272 \text{ Btu/°R}$$

ILLUSTRATIVE PROBLEM 6.21

If 0.2 kg of air is heated at constant volume from 20°C to 100°C, determine the change in entropy for the process. Assume that c_v is constant and equal to 0.7186 kJ/kg·°K.

SOLUTION

$$\Delta s = c_v \ln \frac{T_2}{T_1} = 0.7186 \ln \frac{100 + 273}{20 + 273}$$

$$= 0.1735 \frac{\text{kJ}}{\text{kg·°K}}$$

For 0.2 kg,

$$\Delta S = (0.2)(0.1735) = 0.03470 \frac{\text{kJ}}{\text{°K}}$$

ILLUSTRATIVE PROBLEM 6.22

The entropy of 1 lb of gas changes by 0.0743 Btu/lb$_m$·°R when heated from 100°F to some higher temperature at constant volume. If $c_v = 0.219$ Btu/lb$_m$·°R, determine the higher temperature for this process.

SOLUTION

$$\Delta s = c_v \ln \frac{T_2}{T_1}$$

$$0.0743 = 0.219 \ln \frac{T_2}{100 + 460}$$

or
$$\ln \frac{T_2}{560} = 0.3393$$

Taking antilogarithms gives us

$$\frac{T_2}{560} = e^{0.3393}$$
$$= 1.404$$
$$T_2 = 786.2°R$$

ILLUSTRATIVE PROBLEM 6.23

If 1.5 kg of a gas is heated at constant volume to a final temperature of 425°C and the entropy increase is found to be 0.4386 kJ/°K, determine the initial temperature of the process. Use $c_v = 0.8216$ kJ/kg·°K.

SOLUTION

As before,

$$\Delta S = mc_v \ln \frac{T_2}{T_1}$$
$$0.4386 = 1.5 \times 0.8216 \ln \frac{273 + 425}{T_1}$$

or
$$\ln \frac{698}{T_1} = 0.35589$$

$$\frac{698}{T_1} = e^{0.35589} = 1.4275$$

and
$$T_1 = \frac{698}{1.4275} = 489°K = 216°C$$

6.5b   Constant-Pressure Process (Isobaric Process)

The constant-pressure process is a good approximation to many of the common physical processes with which we are familiar. The combustion of fuel in a boiler, the flow of fluids, the flow of air in ducts, and other processes can be used to illustrate constant pressure. Figure 6.10 shows a turbofan gas turbine engine in which the combustion process is carried out ideally at constant

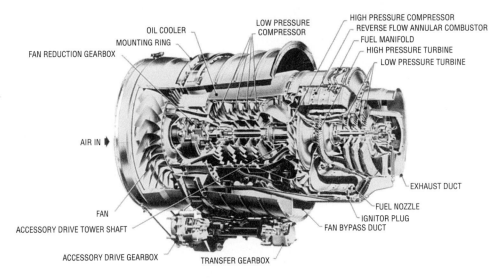

FAN REDUCTION GEARBOX
MOUNTING RING
OIL COOLER
LOW PRESSURE COMPRESSOR
HIGH PRESSURE COMPRESSOR
REVERSE FLOW ANNULAR COMBUSTOR
FUEL MANIFOLD
HIGH PRESSURE TURBINE
LOW PRESSURE TURBINE

AIR IN ►

EXHAUST DUCT
FUEL NOZZLE
IGNITOR PLUG
FAN BYPASS DUCT
FAN
ACCESSORY DRIVE TOWER SHAFT
ACCESSORY DRIVE GEARBOX
TRANSFER GEARBOX

FIGURE 6.10 Turbofan engine. The annular combustion chamber operates
at nearly constant pressure.
(Courtesy of Garrett Corp.)

pressure. Both the nonflow and flow processes yield to the following analysis: For an ideal gas, Equation (6.9) once again gives us the equation of the path for this process.

$$\frac{v_1}{v_2} = \frac{T_1}{T_2} \tag{6.41}$$

From the first law, the constant-pressure process has as its energy equation (neglecting kinetic and potential energy terms)

$$q = h_2 - h_1 \tag{6.42}$$

and consequently,

$$q = c_p(T_2 - T_1) \tag{6.43}$$

where c_p is the specific heat of the process.

During the constant-pressure process shown in Figure 6.11, work is removed as heat is added. The amount of work is the area under the pv curve, which is

$$p(v_2 - v_1) = p_2v_2 - p_1v_1 \tag{6.44}$$

Application of Equation (6.9) and reduction of the mechanical units of Equation (6.44) to thermal units yields

$$p_2v_2 - p_1v_1 = \frac{R}{J}(T_2 - T_1) \tag{6.45}$$

The exponent of pv^n that is applicable to this process is found as follows:

$$\left(\frac{p_2}{p_1}\right)\left(\frac{v_2}{v_1}\right)^n = 1 \quad \text{but} \quad \frac{p_2}{p_1} = 1$$

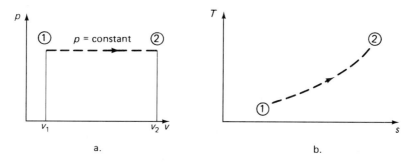

FIGURE 6.11   Constant-pressure process.

Therefore, $n = 0$ for a constant-pressure process.

Finally, the change in entropy for this process is

$$\Delta s = c_p \ln \frac{T_2}{T_1} \tag{6.46}$$

ILLUSTRATIVE PROBLEM 6.24

In a gas turbine cycle, heat is added to the working fluid (air) at constant pressure. Determine the heat transferred, the increase in entropy, and the flow work change per pound of air if the initial pressure is 100 psia and the temperature goes from 70°F to 400°F. The specific heat c_p for this process is constant and equals 0.24 Btu/lb$_\mathrm{m}$·°F.

SOLUTION

This process is shown graphically in Figure 6.11. From the energy equation for the constant-pressure process, the heat transferred is Δh. Therefore,

$$q = \Delta h = c_p(T_2 - T_1) = (0.24)(860 - 530)$$
$$= 79.2 \text{ Btu/lb} \quad \text{(into the system)}$$
$$\Delta s = c_p \ln \frac{T_2}{T_1} = 0.24 \ln \frac{860}{530}$$
$$= 0.116 \text{ Btu/lb}_\mathrm{m}\text{·°R}$$

The flow work change is

$$\frac{p_2 v_2}{J} - \frac{p_1 v_1}{J} = \frac{R}{J}(T_2 - T_1)$$
$$= \frac{(1545/29)(860 - 530)}{778}$$
$$= 22.6 \text{ Btu/lb}_\mathrm{m}$$

In addition to each of the assumptions made in all the processes being considered, it has further been tacitly assumed that these processes are carried out quasi-statically and without friction.

ILLUSTRATIVE PROBLEM 6.25

Heat is added to a turbofan engine such as the one shown in Figure 6.10 at constant pressure. If the air temperature is raised from 20°C to 500°C in the combustion chamber, which is operated at 1 MPa, determine the heat transferred and the increase in entropy per kilogram of air. Use $c_p = 1.0062$ kJ/kg·°K.

SOLUTION

We proceed as in Illustrative Problem 6.24.

$$q = \Delta h = c_p(T_2 - T_1) = (1.0062)(500 - 20) = 483.0 \text{ kJ/kg}$$

and
$$\Delta s = c_p \ln \frac{T_2}{T_1} = 1.0062 \ln \frac{500 + 273}{20 + 273}$$

$$= 0.9761 \text{ kJ/kg·°K}$$

6.5c   Constant-Temperature Process (Isothermal Process)

The expansion process shown in Figure 6.12 could conceivably be one in which a gas expands in a hot cylinder. Heat would be transferred from the hot cylinder walls to maintain the gas temperature constant. The path equation is

$$p_1v_1 = p_2v_2 = \text{constant} \tag{6.47}$$

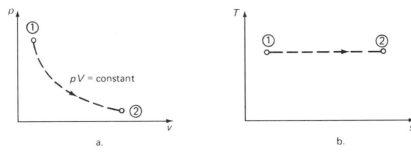

FIGURE 6.12   Isothermal process.

and from the first law, it can be deduced that both the changes in enthalpy and internal energy for this process are zero because the temperature is constant. Therefore, the work of the isothermal process must exactly equal the heat transferred. Thus,

$$q = \frac{w}{J} \tag{6.48}$$

or
$$q = w \quad \text{in SI units} \tag{6.48a}$$

Let us evaluate the change in entropy for the isothermal nonflow process from Equation (6.32) as follows:

$$\Delta s = c_p \ln \frac{v_2}{v_1} + c_v \ln \frac{p_2}{p_1} \tag{6.49}$$

Because $p_1 v_1 = p_2 v_2$,

$$\frac{v_1}{v_2} = \frac{p_2}{p_1} \tag{6.51}$$

$$\Delta s = c_p \ln \frac{v_2}{v_1} + c_v \ln \frac{v_1}{v_2} \tag{6.52}$$

However,

$$\ln \frac{v_1}{v_2} = -\ln \frac{v_2}{v_1} \tag{6.53}$$

Therefore,

$$\Delta s = c_p \ln \frac{v_2}{v_1} - c_v \ln \frac{v_2}{v_1} \tag{6.54}$$

Factoring yields

$$\Delta s = \left(\ln \frac{v_2}{v_1} \right) (c_p - c_v) \tag{6.55}$$

But $c_p - c_v = R/J$. Therefore,

$$\Delta s = \frac{R}{J} \ln \frac{v_2}{v_1} \tag{6.56}$$

Equation (6.56) can also be written as

$$\Delta s = \frac{pv}{TJ} \ln \frac{v_2}{v_1} = \frac{R}{J} \ln \frac{v_2}{v_1} \tag{6.57}$$

In SI units, Equation (6.57) is

$$\Delta s = \frac{pv}{T} \ln \frac{v_2}{v_1} = R \ln \frac{v_2}{v_1} \tag{6.57a}$$

We can now determine the energy interchange as heat and also the work by noting that

$$\Delta s = \frac{q}{T} \quad \text{or} \quad q = T\Delta s \tag{6.58}$$

But from Equation (6.48), $q = w/J$. This gives us

$$\frac{w}{J} = q = T\Delta s = \frac{RT}{J} \ln \frac{v_2}{v_1} = \frac{p_1 v_1}{J} \ln \frac{v_2}{v_1} = \frac{p_2 v_2}{J} \ln \frac{v_2}{v_1} \tag{6.59}$$

In SI units,

$$w = p_2 v_2 \ln \frac{v_2}{v_1} \tag{6.59a}$$

Because there is no change in temperature for this process and a finite quantity of energy as heat has crossed the system boundaries, the definition of specific heat requires that the specific heat of the isothermal process be infinite. Also, in order for both pv and pv^n to be the equation of the path, n must be unity.

ILLUSTRATIVE PROBLEM 6.26

If 0.1 lb of nitrogen is kept at a constant temperature of 200°F while its volume increases to twice its initial volume, determine the heat added and work out of the system.

SOLUTION

From Equation (6.59), we have

$$q = \frac{RT}{J} \ln \frac{v_2}{v_1}$$

$$= \frac{(1545/28)(460 + 200)}{778} \ln \frac{2}{1}$$

$$= 32.4 \text{ Btu/lb}_m$$

For 0.1 lb, $Q = 0.1 \times 32.4 = 3.24$ Btu (added to system). The work out of the system is equal to the heat added; thus,

$$\frac{W}{J} = 3.24 \text{ Btu} \quad \text{(out of system)}$$

ILLUSTRATIVE PROBLEM 6.27

If 1 kg of oxygen has its volume halved at a constant temperature of 50°C, determine the heat added and the work out of the system.

SOLUTION

$$q = RT \ln \frac{v_2}{v_1} = \frac{8.314}{32} (273 + 50) \ln \frac{1}{2}$$

$$= -58.17 \frac{\text{kJ}}{\text{kg}} \quad \text{(heat out of system)}$$

Therefore,

$$w = q = -58.17 \frac{\text{kJ}}{\text{kg}} \quad \text{(into system)}$$

ILLUSTRATIVE PROBLEM 6.28

Determine the change in entropy in Illustrative Problem 6.27.

SOLUTION

For constant temperature,

$$\Delta s = \frac{q}{T}$$

Therefore,

$$\Delta s = \frac{-58.17 \text{ kJ/kg}}{(50 + 273)^{\circ}\text{K}} = -0.1801 \text{ kJ/kg·}^{\circ}\text{K}$$

6.5d Constant-Entropy Process (Isentropic Process)

The turbofan engine and the gas turbine engine of Figures 6.10 and 6.2 produce useful work by the expansion of hot combustion gases in turbines. These processes are ideally isentropic. It will be recalled that the isentropic process is a reversible, adiabatic process. An alternative definition is a process carried out with no change in entropy. Equation (6.33) serves to define this path:

$$\frac{\Delta s}{c_v} = k \ln \frac{v_2}{v_1} + \ln \frac{p_2}{p_1} = 0 \tag{6.33}$$

or

$$\ln \left(\frac{v_2}{v_1} \right)^k = \ln \frac{p_1}{p_2} \tag{6.60}$$

Taking antilogarithms and rearranging yields

$$\boxed{p_1 v_1^k = p_2 v_2^k} \tag{6.61}$$

Equation (6.61) is the path equation for this process, and by using the equation of state [Equation (6.9)], it is possible to rewrite Equation (6.61) in terms of pressures and temperatures as

$$\frac{v_2}{v_1} = \left(\frac{p_1}{p_2}\right)^{1/k}$$

(6.62)

and $p_1 v_1 / T_1 = p_2 v_2 / T_2$ from Equation (6.9). Therefore,

$$\frac{p_1}{p_2}\left(\frac{T_2}{T_1}\right) = \left(\frac{p_1}{p_2}\right)^{1/k}$$

Rearranging gives us

$$\frac{T_1}{T_2} = \left(\frac{p_1}{p_2}\right)^{(k-1)/k}$$

(6.63)

Similarly, we obtain the p,v relation from Equation (6.61) as

$$\frac{p_1}{p_2} = \left(\frac{v_2}{v_1}\right)^{k} \quad \text{or} \quad p_1 v_1^k = p_2 v_2^k$$

(6.64)

From Equation (6.9), we have $p_1/p_2 = (v_2/v_1)(T_1/T_2)$. Substitution into Equation (6.64) yields

$$\frac{T_1}{T_2} = \left(\frac{v_1}{v_1}\right)^{k-1}$$

(6.65)

Equations (6.61), (6.63), and (6.65) define both the state and path of an ideal gas undergoing an isentropic change.

Because the energy interchange as heat is zero and there is a finite change in temperature, the specific heat for this process must be zero. Note that the process specific heat is defined by the path and differs from the specific heat at constant pressure or the specific heat at constant volume. These other specific heats serve to define the enthalpy and internal energy of the gas at the state conditions at the endpoints of the path. By the identification of pv^k with pv^n, the exponent n must equal the *isentropic exponent k* for the isentropic process.

The work of the nonflow process shown in Figure 6.13 can be evaluated by noticing that the work done in a nonflow process in the absence of heat transfer equals the change in internal energy. Therefore,

$$\text{work} = u_1 - u_2$$

$$\text{work} = c_v(T_1 - T_2)$$

But,

$$c_v = \frac{R}{J(k-1)}$$

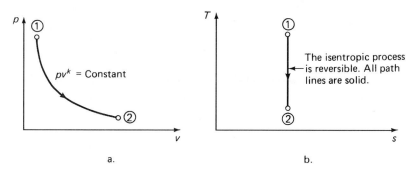

a.

b.

FIGURE 6.13 Isentropic process.

Therefore,

$$\frac{w}{J} = \frac{R(T_1 - T_2)}{J(k - 1)} \quad \text{or} \quad \frac{R(T_2 - T_1)}{J(1 - k)} \quad \text{Btu/lb}_m \tag{6.66}$$

Again, note that in SI units, the conversion factor J is not included.

CALCULUS ENRICHMENT

The work of the nonflow, isentropic process can be obtained by integration of the area under the p–v curve as shown in Figure A. Therefore,

$$w = \int_1^2 p \, dv \tag{a}$$

The path of the isentropic process is given by pv^k = constant. Inserting this into Equation (a),

$$w = \int_1^2 \frac{c}{v^k} \, dv \tag{b}$$

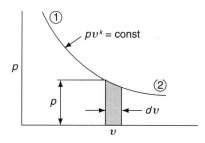

FIGURE A Work as an area.

Integrating,

$$w = c \left. \frac{v^{-k+1}}{-k+1} \right]_1^2 = \frac{c}{-k+1} \left[v_2^{-k+1} - v_1^{-k+1} \right] \qquad \textbf{(c)}$$

Because $c = pv^k = p_1 v_1^k = p_2 v_2^k$, Equation (c) becomes

$$w = \frac{p_2 v_2 - p_1 v_1}{-k+1} \qquad \textbf{(d)}$$

but $pv = RT$;

$$w = \frac{R(T_2 - T_1)}{1 - k} \qquad \textbf{(e)}$$

Equation (e) is the same as Equation (6.66) in appropriate units.

ILLUSTRATIVE PROBLEM 6.29

One pound of air expands isentropically from 5 atm absolute to 1 atm absolute. If the initial temperature is 1000°R, determine the final state and the work done by the air if k is 1.4 over this range of temperature.

SOLUTION

From Equation (6.63),

$$T_2 = T_1 \left(\frac{p_2}{p_1} \right)^{(k-1)/k}$$

$$= 1000 \left(\frac{1}{5} \right)^{(1.4-1)/1.4} = 631°R$$

From Equation (6.66),

$$\text{work} = \frac{R(T_2 - T_1)}{J(1 - k)} = \frac{(1545)(631 - 1000)}{(29)(778)(1 - 1.4)}$$

$$= 63 \text{ Btu/lb}_m \quad \text{(out)}$$

ILLUSTRATIVE PROBLEM 6.30

If the initial temperature in Illustrative Problem 6.29 is 500°C, determine the final state and the work done. Assume a mass of 1 kg.

SOLUTION

$$T_2 = T_1 \left(\frac{p_2}{p_1} \right)^{(k-1)/k}$$

$$= (500 + 273)\left(\frac{1}{5}\right)^{(1.4-1)/1.4}$$

$$= 488.06°\text{K} = 215.06°\text{C}$$

$$\text{work} = \frac{R(T_2 - T_1)}{1 - k} = \frac{8.314}{29}\left(\frac{215.06 - 500}{1 - 1.4}\right)$$

$$= 204.2\frac{\text{kJ}}{\text{kg}} \quad (\text{out})$$

ILLUSTRATIVE PROBLEM 6.31

A gas expands isentropically from 5 MPa to 1 MPa. If the initial temperature is 800°C and the final temperature is 500°C, determine k for this gas.

SOLUTION

Because

$$\frac{T_2}{T_1} = \left(\frac{p_2}{p_1}\right)^{(k-1)/k}$$

$$\frac{273 + 500}{273 + 800} = \left(\frac{1}{5}\right)^{(k-1)/k}$$

Taking logarithms of both sides of this equation yields

$$\ln\frac{273 + 500}{273 + 800} = \left(\frac{k-1}{k}\right)\ln\frac{1}{5}$$

or

$$\frac{k-1}{k} = \frac{\ln\dfrac{273 + 500}{273 + 800}}{\ln\dfrac{1}{5}} = 0.2038$$

Solving,

$$k = 1.256$$

6.5e Polytropic Process

Each of the processes discussed in this section can have its path equation written in terms of

$$\boxed{pv^n = \text{constant}} \tag{6.67}$$

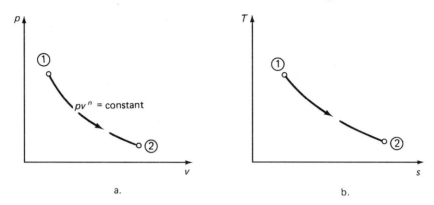

a.

b.

FIGURE 6.14 Polytropic process.

which is known as the polytropic equation and is used, in general, to characterize any mechanically reversible, nonadiabatic process. It is also used as an approximation for real processes. A diagram is shown in Figure 6.14. To recapitulate those processes already considered, see Table 6.3. If each of these processes is plotted on pv and TS coordinates (basically, the superposition of Figures 6.9 through 6.13), it is possible to show the entire spectrum of n values (see Figures 6.15 and 6.16).

TABLE 6.3

Process	n
Constant volume	∞
Constant pressure	0
Isothermal	1
Isentropic	k
Polytropic	$-\infty$ to $+\infty$

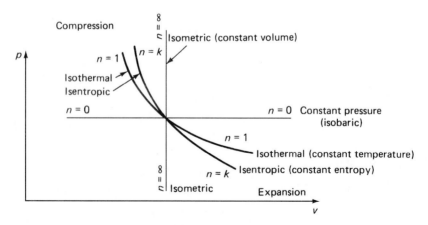

FIGURE 6.15 Polytropic processes.

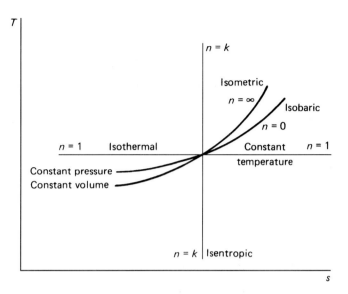

FIGURE 6.16 Polytropic processes.

By comparison, the pressure, temperature, and volume relations for the polytropic process corresponding to Equations (6.62), (6.63), and (6.65) can be written as follows:

$$\frac{v_2}{v_1} = \left(\frac{p_1}{p_2}\right)^{1/n} \tag{6.68}$$

$$\frac{p_1}{p_2} = \left(\frac{v_2}{v_1}\right)^{n} \quad \text{or} \quad p_1 v_1^n = p_2 v_2^n \tag{6.69}$$

$$\frac{T_1}{T_2} = \left(\frac{p_1}{p_2}\right)^{(n-1)/n} \tag{6.70}$$

$$\frac{T_1}{T_2} = \left(\frac{v_2}{v_1}\right)^{n-1} \tag{6.71}$$

The work of the polytropic process can be determined by the same method used to obtain the work of the isentropic process. The result is

$$\text{work} = \frac{p_2 v_2 - p_1 v_1}{1 - n} \tag{6.72}$$

and in terms of thermal units

$$\text{work} = \frac{R(T_2 - T_1)}{J(1 - n)} \tag{6.73}$$

To determine the specific heat of the polytropic process, we shall use the energy equation for the nonflow process:

$$q = (u_2 - u_1) + \frac{\text{w}}{J} \tag{6.74}$$

It is convenient to define the heat transferred by

$$c_n(T_2 - T_1) = q \tag{6.75}$$

where c_n is the polytropic specific heat. Substitution of Equations (6.73) and (6.75) into Equation (6.74) yields

$$c_n(T_2 - T_1) = c_v(T_2 - T_1) + \frac{R(T_2 - T_1)}{J(1 - n)} \tag{6.76}$$

$$c_n = c_v + \frac{R}{J(1 - n)} \tag{6.77}$$

and

$$c_n = \frac{Jc_v - nJc_v + R}{J(1 - n)} \tag{6.78}$$

Using Equation (6.29), we have

$$c_n = \frac{c_p - nc_v}{(1 - n)} \tag{6.79}$$

or

$$\boxed{c_n = c_v \left(\frac{k - n}{1 - n} \right)} \tag{6.80}$$

Note that c_n can be positive or negative depending on the values of n and k.

Substitution of c_n into Equation (6.75) gives the heat transferred in the polytropic process:

$$q = c_v \left(\frac{k - n}{1 - n} \right)(T_2 - T_1) \tag{6.81}$$

The change in entropy for the polytropic process can be arrived at by using Equation (6.30) and assuming that c_n is independent of temperature:

$$\Delta s = \frac{q}{T} = \frac{c_n \Delta T}{T} \tag{6.82}$$

Evaluating Equation (6.82) as done previously by summing the $\Delta T / T$ terms yields

$$\Delta s = c_n \ln \frac{T_2}{T_1} \tag{6.83}$$

ILLUSTRATIVE PROBLEM 6.32

For an internal combustion engine, the expansion process can be characterized by $pv^{1.3} = $ constant. If the ratio of specific heats for this gas is 1.4 and the specific heat at constant pressure is 0.24 Btu/lb$_m$·°R, determine the heat transferred as the gas expands from 1500°R to 600°R. Also evaluate the work done and the change in entropy for the process. R equals 53.3 for this gas.

SOLUTION

From

$$\frac{c_p}{c_v} = k \qquad c_v = \frac{0.24}{1.4} = 0.171$$

Therefore,

$$c_n = c_v\left(\frac{k - n}{1 - n}\right) = 0.171\left(\frac{1.4 - 1.3}{1 - 1.3}\right) = -0.0570 \text{ Btu/lb}_m\text{·°R}$$

The negative sign of c_n indicates that either the heat transfer for the process comes from the system or there is a negative temperature change while heat is transferred to the system.

The heat transferred is $c_n(T_2 - T_1)$. Therefore,

$$q = -0.0570\,(600 - 1500)$$
$$= 51.3 \text{ Btu/lb}_m \text{ (to the system)}$$

The work done can be found using Equation (6.73), giving us

$$\text{work} = \frac{R\,(T_2 - T_1)}{J(1 - n)} = \frac{53.3}{778}\left(\frac{600 - 1500}{1 - 1.3}\right)$$
$$= 205.5 \text{ Btu/lb}_m \text{ (from the system)}$$

The change in entropy Δs is given by

$$\Delta s = -0.0570 \ln \frac{600}{1500}$$
$$= -0.0570\left(-\ln \frac{1500}{600}\right) = 0.0522 \text{ Btu/lb}_m\text{·°R}$$

ILLUSTRATIVE PROBLEM 6.33

Determine the ratio of inlet to outlet pressure in Illustrative Problem 6.32.

SOLUTION

From Equation (6.70),

$$\frac{T_1}{T_2} = \left(\frac{p_1}{p_2}\right)^{(n-1)/n}$$

For this problem,

$$\frac{1500}{600} = \left(\frac{p_1}{p_2}\right)^{(1.3-1)/1.3}$$

$$2.5 = \left(\frac{p_1}{p_2}\right)^{0.3/1.3} = \left(\frac{p_1}{p_2}\right)^{0.2308}$$

Taking logarithms of both sides of this equation yields

$$\ln 2.5 = 0.2308 \ln \frac{p_1}{p_2}$$

$$\frac{0.9162}{0.2308} = \ln \frac{p_1}{p_2} = 3.97$$

and

$$\frac{p_1}{p_2} = 52.99$$

Keep in mind that when deriving the equations in this chapter, the specific heats were assumed to be constants, independent of temperature. The same assumption was made for k. When the variation of these quantities does not permit us to assume an essentially constant average value, the methods of the calculus must be employed. For these techniques, the student is referred to the references given in Appendix 1.

Table 6.4 summarizes the results of the derivations made for nonflow processes. In this table, the enthalpy and internal energies have also been tabulated, and it will be noted that these quantities depend only on the end states of the specified process and are independent of the path. This table has been included as a convenient reference. The student is cautioned against memorizing the table or using it without fully understanding the restrictions involved in each of the processes discussed.

The change in enthalpy and the change in internal energy for the ideal gas are useful quantities, and we can obtain these readily from the previous considerations in this chapter. Thus, for enthalpy we can write

$$h_2 - h_1 = c_p(T_2 - T_1) \tag{6.84}$$

From Equation (6.9), we have $T = pv/R$, and from Equation (6.29), we have $c_p = R/J(k - 1)$. Substitution of these quantities in Equation (6.84) gives us

$$h_2 - h_1 = \frac{R}{J}\left(\frac{k}{k - 1}\right)\left(\frac{p_2 v_2}{R} - \frac{p_1 v_1}{R}\right) \tag{6.85}$$

Simplifying yields

TABLE 6.4
Ideal Gas Relations (per Unit Mass of Gas)[a]

Process →	Constant volume $V =$ constant (isometric)	Constant pressure $p =$ constant (isobaric or isopiestic)	Isothermal $T =$ constant	Isentropic $s =$ constant	Polytropic $pv^n =$ constant
$\rightarrow$ p,v,T	$\dfrac{T_1}{T_2} = \dfrac{p_1}{p_2}$	$\dfrac{T_1}{T_2} = \dfrac{v_1}{v_2}$	$p_1v_1 = p_2v_2$	$p_1v_1^k = p_2v_2^k$ $\dfrac{T_1}{T_2} = \left(\dfrac{v_2}{v_1}\right)^{k-1}$ $\dfrac{T_1}{T_2} = \left(\dfrac{p_1}{p_2}\right)^{(k-1)/k}$	$p_1v_1^n = p_2v_2^n$ $\dfrac{T_1}{T_2} = \left(\dfrac{v_2}{v_1}\right)^{n-1}$ $\dfrac{T_1}{T_2} = \left(\dfrac{p_1}{p_2}\right)^{(n-1)/n}$
w	0	$p(v_2 - v_1)$	$p_1v_1 \ln \dfrac{v_2}{v_1}$	$\dfrac{p_2v_2 - p_1v_1}{1 - k}$	$\dfrac{p_2v_2 - p_1v_1}{1 - n}$
$u_2 - u_1$	$c_v(T_2 - T_1)$	$c_v(T_2 - T_1)$	0	$c_v(T_2 - T_1)$	$c_v(T_2 - T_1)$
q	$c_v(T_2 - T_1)$	$c_p(T_2 - T_1)$	$p_1v_1 \ln \dfrac{v_2}{v_1}$	0	$c_n(T_2 - T_1)$
n	∞	0	1	k	$-\infty$ to $+\infty$
c	c_v	c_p	∞	0	$c_n = c_v\left(\dfrac{k - n}{1 - n}\right)$
$h_2 - h_1$	$c_p(T_2 - T_1)$	$c_p(T_2 - T_1)$	0	$c_p(T_2 - T_1)$	$c_p(T_2 - T_1)$
$s_2 - s_1$	$c_v \ln \dfrac{T_2}{T_1}$	$c_p \ln \dfrac{T_2}{T_1}$	$R\left(\ln \dfrac{v_2}{v_1}\right)$	0	$c_n \ln \dfrac{T_2}{T_1}$

[a] The conversion factor J has been omitted from all equations. Care should be taken when using this table with English units.

$$h_2 - h_1 = \frac{k}{J(k - 1)}(p_2v_2 - p_1v_1) \tag{6.86}$$

or

$$h_2 - h_1 = \frac{k}{k - 1}(p_2v_2 - p_1v_1) \quad \text{SI units} \tag{6.86a}$$

Similarly, for internal energy,

$$u_2 - u_1 = c_v(T_2 - T_1)$$

and

$$u_2 - u_1 = \frac{1}{J(k - 1)}(p_2v_2 - p_1v_1) \tag{6.87}$$

or

$$u_2 - u_1 = \frac{1}{k - 1}(p_2v_2 - p_1v_1) \quad \text{SI units} \tag{6.87a}$$

As a check on Equations (6.86) and (6.87), note that the ratio of $h_2 - h_1$ to $u_2 - u_1$ is the ratio of c_p to c_v, which it should be.

6.6 THE GAS TABLES

The *Gas Tables* provide a convenient and most useful set of tables for the computation of gas processes. By designating Equation (6.9) as the equation of state, it has been possible to express the various thermodynamic properties of these gases in terms of their temperatures. For air at 32°F, the error in using this equation of state is of the order of 1% at 300 psia and only 0.1% at atmospheric pressure. In addition to using the equation of state of an ideal gas, the specific heats listed in these tables were based on spectroscopic data at zero pressure. The zero of entropy and enthalpy were chosen to be zero at zero degrees absolute. Use of the specific heats at zero pressure is equivalent to the assumption that the specific heat is a function of temperature only.

The outline of the *Gas Tables* can be easily developed by considering that the specific heats are temperature functions. The enthalpy values are listed as functions of the absolute temperature, and the change in enthalpy for a process is just the difference for the values tabulated at the final and initial temperatures. Because ideal gas relations are used,

$$u_2 - u_1 = (h_2 - h_1) - \frac{R}{J}(T_2 - T_1) \tag{6.88}$$

The general expression for the change in entropy of an ideal gas with a variable specific heat can be obtained, and the change in entropy per unit mass between states 1 and 2 is then

$$s_2 - s_1 = \phi_2 - \phi_1 - \frac{R}{J}\ln\frac{p_2}{p_1} \tag{6.89}$$

where ϕ is a function of temperature given in the tables for each gas at each temperature. For an isentropic change,

$$\ln\frac{p_2}{p_1} = \frac{J}{R}(\phi_2 - \phi_1) \tag{6.90}$$

CALCULUS ENRICHMENT

When the specific heat of an ideal gas is not constant, we need to return to the second Gibbs or $T\,ds$ equation, namely,

$$T\,ds = dh - v\,dp \tag{a}$$

Using $dh = c_p\,dT$ and $pv = RT$ for the ideal gas yields

$$ds = c_p\frac{dT}{T} - \frac{R}{p}\,dp \tag{b}$$

It is possible to remove the constant R from the integral of Equation (b). However, with $c_p = c_p(T)$, that is, a function of temperature, c_p cannot be removed from the integral. Therefore, we write,

$$\int_{s_1}^{s_2} ds = \int_{T_1}^{T_2} \frac{c_p \, dT}{T} - R \ln \frac{p_2}{p_1} \tag{c}$$

The integral in Equation (c) is a function only of temperature. In the *Gas Tables*, the tabulated function is

$$\int_{T_1}^{T_2} \frac{c_p \, dT}{T} = \phi_2 - \phi_1 \tag{d}$$

The change in entropy for an ideal gas with variable specific heats therefore is

$$\Delta s = s_2 - s_1 = \phi_2 - \phi_1 - R \ln \frac{p_2}{p_1} \tag{e}$$

In English units,

$$s_2 - s_1 = \phi_2 - \phi_1 - \frac{R}{J} \ln \frac{p_2}{p_1} \tag{f}$$

where Equation (f) is the same as Equation (6.89).

If we now use the first Gibbs or $T \, ds$ equation,

$$T \, ds = du + p \, dv \tag{g}$$

Dividing by T and using $du = c_v dT$,

$$\int_{s_1}^{s_2} ds = \int_{T_1}^{T_2} \frac{c_v \, dT}{T} + R \ln \frac{v_2}{v_1} \tag{h}$$

or

$$s_2 - s_1 = \int_{T_1}^{T_2} \frac{c_v \, dT}{T} + R \ln \frac{v_2}{v_1} \tag{i}$$

For an isentropic process,

$$-R \ln \left(\frac{v_2}{v_1} \right) + \int_{T_1}^{T_2} \frac{c_v \, dT}{T} \tag{j}$$

If we now define the relative volume, $v_r = \ln \left(\dfrac{v_2}{v_1} \right)$,

$$v_r = \frac{1}{R} \int_{T_1}^{T_2} \frac{c_v \, dT}{T} \tag{k}$$

The *Gas Tables* tabulate two other useful properties; the relative pressure p_r, and the relative volume v_r. If we define the relative pressure p_r as

$$\ln p_r = \frac{\phi}{R} \tag{6.90a}$$

then Equation (6.90) becomes

$$\ln\left(\frac{p_{r2}}{p_{r1}}\right) = (\phi_2 - \phi_1) \tag{6.90b}$$

Equating Equations (6.90) and (6.90b),

$$\ln\left(\frac{p_{r2}}{p_{r1}}\right) = \ln\left(\frac{p_2}{p_1}\right) \tag{6.90c}$$

or
$$\frac{p_1}{p_1} = \frac{p_{r2}}{p_{r1}} \quad \text{for an isentropic process} \tag{6.90d}$$

If we now use the definition of the relative volume v_r as

$$v_r = \ln\left(\frac{v_2}{v_1}\right) \tag{6.90e}$$

we can show that

$$\frac{v_{r2}}{v_{r1}} = \frac{v_2}{v_1} \quad \text{for an isentropic process.} \tag{6.90f}$$

The *Gas Tables* also have adopted the definition that

$$\boxed{v_r = \frac{RT}{p_r}} \tag{6.91}$$

with the units of R selected so that v_r is the specific volume in cubic feet per pound when the pressure is in psia for those tables based on a unit mass. When the mass is the pound-mole, v_r is the molal specific volume in cubic feet per pound-mole when the pressure is psia.

The following problems will serve to illustrate the use of the abridged air table, which is given in Appendix 3.

ILLUSTRATIVE PROBLEM 6.34

Determine the change in enthalpy, internal energy, and entropy when air is heated at constant pressure from 500°R to 1000°R.

SOLUTION

From the table at 1000°R:

$h = 240.98$ Btu/lb$_m$

$u = 172.43$ Btu/lb$_m$

$\phi = 0.75042$ Btu/lb$_m$·°R

From the table at 500°R:

$h = 119.48$ Btu/lb$_m$

$u = 85.20$ Btu/lb$_m$

$\phi = 0.58233$ Btu/lb$_m$·°R

The change in enthalpy is $h_2 - h_1 = 240.98 - 119.48 = 121.5$ Btu/lb$_m$. The change in internal energy is $u_2 - u_1 = 172.43 - 85.20 = 87.23$ Btu/lb$_m$. Because in the constant-pressure process $-R \ln(p_2/p_1)$ is zero,

$$\Delta s = \phi_2 - \phi_1 = 0.75042 - 0.58233 = 0.16809 \text{ Btu/lb}_m\text{·°R}$$

ILLUSTRATIVE PROBLEM 6.35

Solve Illustrative Problem 6.29 using the *Gas Tables.*

SOLUTION

In this problem, the air expands from 5 atm absolute to 1 atm absolute from an initial temperature of 1000°R. At 1000°R, $p_r = 12.298$ and $h = 240.98$ Btu/lb$_m$. The value of the final $p_r = 12.298/5 = 2.4596$. Interpolation in the air table yields the following:

T	p_r
620	2.249
	2.4596
640	2.514

$$620 + \frac{2.4596 - 2.249}{2.514 - 2.249} \times 20 = 635.9 \quad \text{or} \quad 636°R$$

The work done in an isentropic nonflow expansion is

$$\text{work} = u_1 - u_2 = 172.43 - 108.51 = 63.92 \text{ Btu/lb}_m$$

where the value of u_2 is obtained by interpolation at 636°R and the value of u_1 is read from the air table at 1000°R.

The fact that the results found by using the *Gas Tables* agrees well with those found from the ideal gas relations is not a coincidence. In all these problems, the pressures were low and the temperatures relatively high. Under these conditions, it would be expected that these gases would behave much like ideal gases. The principal differences are caused by the fact that the specific heats vary with temperature. The *Gas Tables* take this variation into account and reduce the work of computations considerably.

6.7 GAS FLOW PROCESSES

In recent years, with the advent of missiles, high-speed aircraft, and the flow of gases in such devices as gas turbines and rocket exhaust nozzles, the area of fluid mechanics denoted as compressible flow has taken on increased importance. Many devices and flow regimes can be treated to a reasonable approximation by considering the flow to be one-dimensional; that is, fluid properties are uniform over any cross-section. Figures 6.17 and 6.18 show modern, high-speed, high-performance military aircraft which can achieve speeds in excess of twice the local speed of sound. The speed of sound (*acoustic velocity*) in an ideal gas is given by

$$V_a = \sqrt{g_c kRT} \qquad (6.92)$$

where V_a is the speed of sound in feet per second, $g_c = 32.17 \ \text{lb}_m \cdot \text{ft}/\text{lb}_f \cdot \text{s}^2$, k is the ratio of c_p/c_v, and T is the absolute temperature in degrees Rankine. For air,

$$V_a = 49.0 \ \sqrt{T} \qquad \text{ft/s} \qquad (6.93)$$

where T is in degrees Rankine. In SI units,

$$V_a = \sqrt{kRT} = 20.05 \ \sqrt{T} \qquad \text{m/s} \qquad (6.93a)$$

where T is in Kelvin.

FIGURE 6.17 F-16 Air Force supersonic fighter. It is powered by the F100 turbofan engine, which has the highest thrust-to-weight ratio ever achieved in a production engine. *(Courtesy of Pratt and Whitney Aircraft Group, United Technologies Corp.)*

FIGURE 6.18 U.S. Navy F-14 Tomcat supersonic fighter. This aircraft has unique defensive capabilities. With its advanced weapons guidance system, the F-14 can fire Phoenix missiles at six different targets simultaneously at distances of over 100 miles.
(Courtesy of Grumman Aerospace Corp., Bethpage, N.Y.)

ILLUSTRATIVE PROBLEM 6.36

Determine the velocity of sound in air at 1000°F. Using the data in Appendix 3, determine the velocity of sound in hydrogen at this same temperature.

SOLUTION

The velocity of sound in air at 1000°F is

$$V_a = 49.0 \sqrt{(1000 + 460)} = 1872 \text{ ft/s}$$

Hydrogen has a specific heat ratio of 1.41 and $R = 766.53$. Therefore,

$$\frac{V_{a_{\text{hydrogen}}}}{V_{a_{\text{air}}}} = \sqrt{\frac{(Rk)_{\text{hydrogen}}}{(Rk)_{\text{air}}}} = \sqrt{\frac{766.53 \times 1.41}{53.36 \times 1.40}} = 3.804$$

$$V_{a_{\text{hydrogen}}} = 3.804 V_{a_{\text{air}}} = 3.804 \times 1872 = 7121 \text{ ft/s}$$

In the subsequent work in this chapter, it will be found that the *Mach number* is an important parameter. It is defined as the ratio of the velocity at a point in a fluid to the velocity of sound at that point at a given instant of time. Denoting the local velocity by V and the velocity of sound as V_a,

$$M = \frac{V}{V_a} \qquad (6.94)$$

and

$$M^2 = \frac{V^2}{g_c kRT} \qquad (6.95)$$

where Equation (6.95) is applicable only to an ideal gas. In SI, g_c is omitted.

ILLUSTRATIVE PROBLEM 6.37

Air is flowing in a duct at atmospheric pressure with a velocity of 1500 ft/s. If the air temperature is 200°F, what is the Mach number?

SOLUTION

The velocity of sound in air at 200°F is

$$V_a = 49.0 \sqrt{T} = 49.0 \sqrt{(200 + 460)} = 1259 \text{ ft/s}$$

The Mach number is

$$M = \frac{V}{V_a} = \frac{1500}{1259} = 1.191$$

6.7a Adiabatic Flow

When studying the steady-flow processes of gases, it is necessary to utilize the continuity equation, the equation of state, and the appropriate energy equation. Consider the special case of adiabatic flow without shaft work or elevation change. For this process, the first law of thermodyamics yields

$$h + \frac{V^2}{2g_cJ} = h_1 + \frac{V_1^2}{2g_cJ} = h^0 \tag{6.96}$$

Equation (6.96) defines the term *stagnation enthalpy* h^0. This terminology is best illustrated by referring to Figure 6.19. At section 0, it is assumed that the area can be considered infinite or that the velocity is essentially zero. Thus, from Equation (6.96), the stagnation enthalpy is $h + V^2/2g_cJ$; $h_1^0 = h_1 + V_1^2/2g_cJ$, and $h^0 = h_1^0 = h_2^0 = $ constant. The stagnation enthalpy (or total enthalpy) therefore is a constant by definition for the process in question if it is adiabatic, if no work is done on or by the fluid, and if the fluid does not change in elevation.

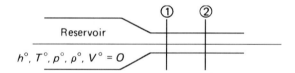

FIGURE 6.19 Stagnation conditions.

The introduction of the terminology of the preceding paragraph requires a brief explanation. By referring to Figure 6.20, it will be noted that an isentropic compression is shown in enthalpy–entropy coordinates by the line A,B. Because Equation (6.96) is applicable, it follows that the change in enthalpy indicated in the diagram must equal $V^2/2g_cJ$, because the velocity at 0 condition is zero. The final state (the 0 state) is commonly known as the *total* state, and the pressure, temperature, and density are known, respectively, as the total pressure, total temperature, and total density.

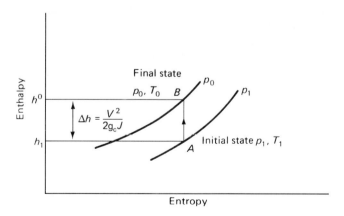

FIGURE 6.20 Isentropic stagnation.

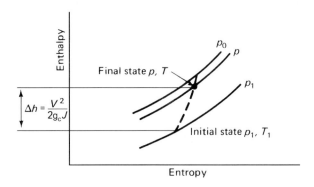

FIGURE 6.21 Adiabatic stagnation.

However, if the process is adiabatic but not isentropic, it may be shown as in Figure 6.21. The final enthalpy will be the same as that for the isentropic case, but the entropy will be greater and, consequently, the final pressure at the end of the actual process less. The end state of the non-isentropic (actual) process is commonly called the *stagnation state,* and the properties at this state are called *stagnation properties.* By noting that the final enthalpies shown in Figures 6.20 and 6.21 are equal, either the terminology of total or stagnation enthalpy for this state is correct. Again, for the ideal gas, certain relevant and important conclusions may be drawn from this discussion. The final temperature of each process, that is, the total temperature and the stagnation temperatures, will be equal. The final pressures will not, in general, be equal. To recapitulate, the total state corresponds to an isentropic stagnation process and the stagnation state to an adiabatic process. Caution should be exercised in the use of this terminology. The literature contains many examples of differing definitions for these properties, and all terms used should be defined carefully.

ILLUSTRATIVE PROBLEM 6.38

Assume that a fluid is flowing in a device and that, at some cross-section, the fluid velocity is 1000 ft/s. If the fluid is saturated steam having an enthalpy of 1204.4 Btu/lb$_m$, determine its total enthalpy.

SOLUTION

From the previous paragraph, we can simply write

$$h^0 - h = \frac{V^2}{2g_c J}$$

From this problem,

$$h^0 - h = \frac{(1000)^2}{2 \times 32.17 \times 778} = 20 \text{ Btu/lb}_m$$

and
$$h^0 = 1204.4 + 20 = 1224.4 \text{ Btu/lb}_m$$

It will be noted for this problem that if the initial velocity had been 100 ft/s, Δh would have been 0.2 Btu/lb$_m$, and for most practical purposes, the total properties and those of the flowing fluid would have been essentially the same. Thus, for low-velocity fluids, the difference in total and stream properties can be neglected.

6.7b Isentropic Flow of an Ideal Gas

For the isentropic flow of an ideal gas, it is necessary only to recall that ideal refers to the equation of state of the gas and isentropic to the equation of the path. For an ideal gas with constant specific heat, it is possible to write the enthalpy (measured from a base of absolute zero) as

$$ h^0 = c_p T^0 \tag{6.97} $$

Equation (6.96) can be written in the following form:

$$ V^2 = 2g_c J(h^0 - h_1) = 2g_c J c_p (T^0 - T_1) \tag{6.98} $$

Applying the equations of path and state to Equation (6.98) and simplifying yields

$$
\begin{aligned}
V_1 &= \left\{ 2g_c J c_p T^0 \left[1 - \left(\frac{p_1}{p_0} \right)^{(k-1)/k} \right] \right\}^{1/2} \\
&= \left\{ \frac{2g_c k}{k-1} \frac{p^0}{\rho^0} \left[1 - \left(\frac{p_1}{p_0} \right)^{(k-1)/k} \right] \right\}^{1/2}
\end{aligned}
\tag{6.99}
$$

6.7c Converging Nozzle

We apply Equation (6.99) to the throat of the section of minimum area tt (Figure 6.22) and let p_t be the pressure at that section. The mass of gas passing through the section tt per unit time is

$$
\begin{aligned}
\dot{m} &= \rho_t A_t V_t = A_t \rho^0 \left(\frac{p_t}{p^0} \right)^{1/k} (V_t) \\
&= A_t \left\{ \frac{2g_c k}{k-1} p^0 \rho^0 \left[\left(\frac{p_t}{p^0} \right)^{2/k} - \left(\frac{p_t}{p^0} \right)^{(k+1)/k} \right] \right\}^{1/2}
\end{aligned}
\tag{6.100}
$$

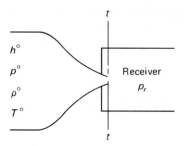

FIGURE 6.22 Converging nozzle.

The variation of $\dot{m}$ with p/p^0 is given by the curved line (partly dashed and partly solid) in Figure 6.23, which shows that $\dot{m}$ reaches a maximum value for a certain pressure ratio $p_t/p^0 = p_c/p^0$, where p_c is called the *critical pressure* and can be determined as follows: $\dot{m}$ is at its maximum when

$$\left(\frac{p_t}{p^0}\right)^{2/k} - \left(\frac{p_t}{p^0}\right)^{(k+1)/k}$$

is at a maximum.

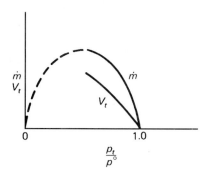

FIGURE 6.23 Mass flow variation in a nozzle.

By performing the required operations, using the methods of calculus, it is found that

$$\frac{p_t}{p^0} = \frac{p_c}{p^0} = \left(\frac{2}{k+1}\right)^{k/(k-1)} \tag{6.101}$$

CALCULUS ENRICHMENT

Let $(p_t/p^0) = x$. Then we wish to maximize.

$$y = (x)^{2/k} - x^{(k+1)/k} \tag{a}$$

---

292   THE IDEAL GAS

$$\frac{dv}{dx} = \frac{2}{k}(x)^{(2/k)-1} - \left(\frac{k+1}{k}\right)x^{[(k+1)/k]-1} \qquad \text{(b)}$$

For maximum flow, $dy/dx = 0$

$$0 = \frac{2}{k}(x)^{(2-k)/k} - \left(\frac{k+1}{k}\right)x^{1/k} \qquad \text{(c)}$$

Transposing and equating,

$$\frac{2}{k}x^{(2-k)/k} = \frac{k+1}{k}x^{1/k} \qquad \text{(d)}$$

and

$$\frac{\dfrac{2}{k}}{\dfrac{k+1}{k}} = \frac{x^{1/k}}{x^{(2-k)/k}} \qquad \text{(e)}$$

Therefore

$$\frac{2}{k+1} - x^{1/k-[(2-k)/k]} = x^{(1/k)-(2/k)+1} \qquad \text{(f)}$$

Then

$$\frac{2}{k+1} - x^{1-(1/k)} = x^{(k-1)/k} \qquad \text{(g)}$$

Finally

$$\left(\frac{2}{k+1}\right)^{k/(k-1)} = x \qquad \text{(h)}$$

Equation (h) is the same as Equation (6.101).

For air with $k = 1.4$, $p_c/p^0 = 0.53$. For superheated steam with $k = 1.3$, $p_t/p^0 = 0.546 = p_c p^0$. At the critical pressure ratio,

$$\boxed{\frac{T_t}{T_0} = \frac{2}{k+1}} \qquad \text{(6.102)}$$

$$\boxed{\frac{V_{a_t}}{V_{a_0}} = \sqrt{\frac{2}{k+1}}} \qquad \text{(6.103)}$$

$$\boxed{V_t = \sqrt{\frac{2g_c k}{k+1}(p^0/\rho^0)} = \sqrt{\frac{2}{k+1}}(V_{a_0}) = V_{a_t}} \qquad \text{(6.104)}$$

$$\boxed{\dot{m}_{max} = A_t\left(\frac{2}{k+1}\right)^{1/(k-1)}\sqrt{\frac{2g_c k}{k+1}(p^0/\rho^0)}} \qquad \text{(6.105)}$$

In general, also,

$$\left(\frac{A}{A_t}\right)^2 = \frac{k-1}{2}\left\{\frac{\left(\frac{2}{k+1}\right)^{(k+1)/(k-1)}}{\left(\frac{p}{p^0}\right)^{2/k}\left[1-\left(\frac{p}{p^0}\right)^{(k-1)/k}\right]}\right\} \qquad \textbf{(6.106)}$$

From the foregoing, we can see that for a given nozzle throat area there is a maximum mass rate of flow of fluid that can pass through the nozzle. When the pressure ratio p_t/p^0 is above the critical value, $\dot{m}$ increases with decreasing p_t. After p_t reaches p_c, there is no further increase in $\dot{m}$, and the flow is said to be choked. Under such conditions, the fluid leaving the nozzle decreases in pressure from p_c to p_t through irreversible flow processes. These flow conditions are shown in Figure 6.24. Because $V_t = V_{a_t}$, the throat velocity at the critical pressure ratio is the local velocity of sound.

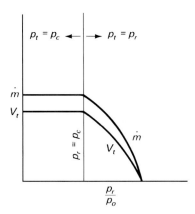

FIGURE 6.24 Choked flow.

6.7d Converging and Diverging Nozzles (de Laval Nozzle)

In the preceding discussion, it was shown for a converging nozzle that the maximum velocity the fluid will attain at the exit section is the local sonic velocity and that the minimum pressure corresponding to sonic velocity at the exit is the critical pressure. To obtain a higher velocity and lower pressure at the exit section of a nozzle, a divergent portion is added downstream of the throat section. The fluid continues to expand in the divergent portion, reaching V_{e_1} and p_{e_1} at the exit. The fluid could also be compressed in the divergent portion, reaching V_{e_2} and p_{e_2} at the exit. The process in the nozzle is determined by the pressure in the receiver. If $p_r = p_{e_1}$, the former occurs. If $p_r = p_{e_2}$, the latter occurs. If p_r lies between the two, the fluid would first follow the former and reach p_r at the nozzle exit, with a normal shock somewhere in the divergent portion of the nozzle or with compression waves in the receiver (see Figure 6.25).

The foregoing discussion is for a mass flow rate in the nozzle equal to the maximum value. If the weight flow is below this value, the velocity at the throat is subsonic, and at the exit section of the nozzle, $p = p_{e_3}$ and $V = V_{e_3}$.

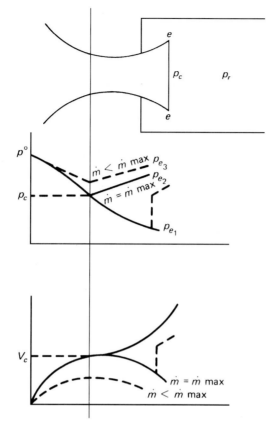

FIGURE 6.25 Flow conditions in a de Laval nozzle.

To facilitate the use of many of the equations already developed, they have been placed into nondimensional form in terms of only two variables, the Mach number and the ratio of specific heats. Tables 30 and 53 of *Gas Tables* list the results of numerical computations of these one-dimensional compressible flow functions. In these tables, the superscript * refers to conditions in which $M = 1$, and the subscript 0 refers to isentropic stagnation. With this notation in mind, the working nondimensional equations are

$$\boxed{T_0 = \text{constant} = T_0^*} \tag{6.107}$$

$$\boxed{T^* = \text{constant}} \tag{6.108}$$

$$\boxed{p^0 = \text{constant} = p_0^*} \tag{6.109}$$

$$\boxed{p^* = \text{constant}} \tag{6.110}$$

$$M^* = \frac{V}{V^*} = M\left\{\frac{k+1}{2[1+\frac{1}{2}(k-1)M^2]}\right\}^{1/2} \tag{6.111}$$

$$\frac{A}{A^*} = \frac{1}{M}\left\{\frac{2[1+\frac{1}{2}(k-1)M^2]}{k+1}\right\}^{(k+1)/2(k-1)} \tag{6.112}$$

$$\frac{T}{T^*} = \frac{k+1}{2[1+\frac{1}{2}(k-1)M^2]} \tag{6.113}$$

$$\frac{\rho}{\rho^*} = \left\{\frac{k+1}{2[1+\frac{1}{2}(k-1)M^2]}\right\}^{1(k-1)} \tag{6.114}$$

$$\frac{p}{p^*} = \left\{\frac{k+1}{2[1+\frac{1}{2}(k-1)M^2]}\right\}^{1/(k-1)} \tag{6.115}$$

Table 6.5 is abridged from Table 30 of the *Gas Tables,* and all the foregoing nondimensional equations are tabulated as functions of Mach number. The convenience of these equations as well as their use is illustrated by Illustrated Problems 6.39 and 6.40. The application of Table 6.5 is also demonstrated in these problems.

TABLE 6.5
One-Dimensional Isentropic Compressible Flow Functions for an Ideal Gas with Constant Specific Heat and Molecular Weight and $k = 1.4$

M	M^*	$\dfrac{A}{A^*}$	$\dfrac{p}{p_0}$	$\dfrac{\rho}{\rho_0}$	$\dfrac{T}{T_0}$
0	0	∞	1.00000	1.00000	1.00000
0.10	0.10943	5.8218	0.99303	0.99502	0.99800
0.20	0.21822	2.9635	0.97250	0.98027	0.99206
0.30	0.32572	2.0351	0.93947	0.95638	0.98232
0.40	0.43133	1.5901	0.89562	0.92428	0.96899
0.50	0.53452	1.3398	0.84302	0.88517	0.95238
0.60	0.63480	1.1882	0.78400	0.84045	0.93284
0.70	0.73179	1.09437	0.72092	0.79158	0.91075
0.80	0.82514	1.03823	0.65602	0.74000	0.88652
0.90	0.91460	1.00886	0.59126	0.68704	0.86058
1.00	1.00000	1.00000	0.52828	0.63394	0.83333
1.10	1.08124	1.00793	0.46835	0.58169	0.80515
1.20	1.1583	1.03044	0.41238	0.53114	0.77640
1.30	1.2311	1.06631	0.36092	0.48291	0.74738
1.40	1.2999	1.1149	0.31424	0.42742	0.71839
1.50	0.3646	1.1762	0.27240	0.39498	0.68965

1.60	0.4254	1.2502	0.23527	0.35573	0.66138
1.70	1.4825	1.3376	0.20259	0.31969	0.63372
1.80	1.5360	1.4390	0.17404	0.28682	0.60680
1.90	1.5861	1.5552	0.14924	0.25699	0.58072
2.00	1.6330	1.6875	0.12780	0.23005	0.55556
2.10	1.6769	1.8369	0.10935	0.20580	0.53135
2.20	1.7179	2.0050	0.09352	0.18405	0.50813
2.30	1.7563	2.1931	0.07997	0.16458	0.48591
2.40	1.7922	2.4031	0.06840	0.14720	0.46468
2.50	1.8258	2.6367	0.05853	0.13169	0.44444
2.60	1.8572	2.8960	0.05012	0.11787	0.42517
2.70	1.8865	3.1830	0.04295	0.10557	0.40684
2.80	1.9140	3.5001	0.03685	0.09462	0.38941
2.90	1.9398	3.8498	0.03165	0.08489	0.37286
3.00	1.9640	4.2346	0.02722	0.07623	0.35714
3.50	2.0642	6.7896	0.01311	0.04523	0.28986
4.00	2.1381	10.719	0.00658	0.02766	0.23810
4.50	2.1936	16.562	0.00346	0.01745	0.19802
5.00	2.2361	25.000	$189(10)^{-5}$	0.01134	0.16667
6.00	2.2953	53.180	$633(10)^{-6}$	0.00519	0.12195
7.00	2.3333	104.143	$242(10)^{-6}$	0.00261	0.09259
8.00	2.3591	190.109	$102(10)^{-6}$	0.00141	0.07246
9.00	2.3772	327.189	$474(10)^{-7}$	0.000815	0.05814
10.00	2.3904	535.938	$236(10)^{-7}$	0.000495	0.04762
∞	2.4495	∞	0	0	0

Source: Abridged from Table 30 in Joseph H. Keenan and Joseph Kaye, *Gas Tables* (New York: John Wiley & Sons, Inc., 1948).

ILLUSTRATIVE PROBLEM 6.39

An isentropic convergent nozzle is used to evacuate air from a test cell that is maintained at stagnation pressure and temperature of 300 psia and 800°R, respectively. The nozzle has inlet and outlet areas of 2.035 and 1 ft², respectively. Constant-pressure specific heat is 0.24, the specific heat ratio is 1.4, and the gas constant R in consistent units is 53.35. Calculate the pressure, temperature, velocity, Mach number, and weight flow at the inlet and pressure and velocity if the Mach number at the outlet is unity.

SOLUTION

Refer to Figure 6.26. For $k = 1.4$ and $M = 1$ (Table 6.5),

$$\frac{T^*}{T_0} = 0.8333$$

$$T^* = (800)(0.8333) = 666.6°R$$

and
$$V_{at} = \sqrt{g_c R T^*} = \sqrt{32.17 \times 1.4 \times 53.35 \times 666.6}$$
$$= 1266 \text{ ft/s} = V_2$$

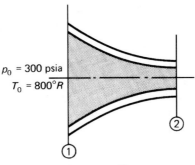

M = 1, therefore state ② is the * state

FIGURE 6.26 Illustrative Problem 6.52.

For

$$\frac{A}{A^*} = 2.035$$

the tables yield $M_1 = 0.3$ and

$$\frac{p^*}{p_0} = 0.52828 \qquad p^* = (300)(0.52828) = 158.5 \text{ psia}$$

Also,

$$\frac{T_1}{T_0} = 0.98232 \quad \text{and} \quad \frac{p_1}{p_0} = 0.93947$$

Therefore,

$$T_1 = (800)(0.98232) = 785.9°\text{R}$$

and $\qquad p_1 = (300)(0.93947) = 281.8 \text{ psia}$

From the inlet conditions derived,

$$V_{a_1} = \sqrt{g_c k R T_1} = \sqrt{32.17 \times 1.4 \times 53.35 \times 785.9}$$
$$= 1374 \text{ ft/s}$$
$$V_1 = M_1 V_{a_1} = (0.3)(1374) = 412 \text{ ft/s}$$

The specific volume at inlet is found from the equation of state for an ideal gas:

$$v = \frac{RT_1}{p_1} = \frac{53.35 \times 785.9}{281.8 \times 144} = 1.033 \text{ ft}^3/\text{lb}_\text{m}$$

and $\qquad \dot{m} = \rho A V = \dfrac{1}{1.033} \times 2.035 \times 412 = 812 \text{ lb}_\text{m}/\text{s}$

ILLUSTRATIVE PROBLEM 6.40

At a certain section of an air stream, the Mach number is 2.5, the stagnation temperature is 560°R, and the static pressure is 0.5 atm. Assuming that the flow is steady isentropic and follows one-dimensional theory, calculate the following at the point at which M is 2.5: (a) temperature, (b) stagnation pressure, (c) velocity, (d) specific volume, and (e) mass velocity.

SOLUTION

This problem will be solved by two methods (A and B).

A. *By equations:* Assume that $k = 1.4$ and $R = 53.3$:

(a) From Equations (6.112) and 6.113),

$$\frac{T}{T^*} = \frac{k+1}{2\left[1 + \frac{1}{2}(k-1)M^2\right]}$$

but from Equation (6.102),

$$\frac{T^*}{T_0} = \frac{2}{k+1}$$

Therefore,

$$\frac{T^*}{T_0} \times \frac{T}{T^*} = \frac{T}{T_0} = \frac{1}{\left[1 + \frac{1}{2}(k-1)M^2\right]}$$

and

$$T = \frac{560}{1 + \left(\frac{1}{2}\right)(1.4-1)(2.5)^2} = \frac{560}{2.25} = 249°R$$

(b)

$$\frac{p_0}{p} = \left(\frac{T_0}{T}\right)^{k/(k-1)} \qquad p_0 = (0.5)(14.7)\left(\frac{560}{249}\right)^{1.4/(1.4-1)}$$

$$= 125.4 \text{ psia}$$

(c)

$$V_a = \sqrt{g_c kRT} = \sqrt{32.17 \times 1.4 \times 53.3 \times 249}$$

$$V_a = 49.0\sqrt{249} = 773 \text{ ft/s}$$

$$V = MV_a = 2.5 \times 773 = 1932.5 \text{ ft/s}$$

(d) From the equation of state,

$$v = \frac{RT}{p} = \frac{53.3 \times 249}{(0.5)(14.7)(144)} = 12.54 \text{ ft}^3/\text{lb}_m$$

(e) Mass velocity is defined as the mass flow per unit area:

$$\frac{\dot{m}}{A} = \frac{AV}{vA} = \frac{V}{v} = \frac{1932.5}{12.54} = 154.1 \ \frac{\frac{lb_m}{s}}{ft^2}$$

B. *By the Gas Tables:* At $M = 2.5$, Table 6.5 gives

(a) $\dfrac{T}{T_0} = 0.44444$

(b) $\dfrac{p}{p_0} = 0.05853$

$$T = (560)(0.44444) = 249°R$$

$$p = \frac{0.5 \times 14.7}{0.05853} = 125.6 \ psia$$

(c) As before, 1932.5 ft/s.
(d) As before, 12.54 ft^3/lb_m.
(e) As before, 154.1 $\dfrac{\frac{lb}{m}}{\frac{s}{ft^2}}$.

6.7e Actual Nozzle Performance

In the preceding sections, certain idealizations were made regarding the character of the flow. Real fluids flowing in real devices deviate from these ideal conditions to some extent. There will always be frictional effects at the interface between the fluid and its containing walls; real fluids have viscosity and irreversible internal effects such as turbulence; temperature differences between the fluid and the walls give rise to nonadiabatic conditions; irreversible discontinuities (shocks) produce entropy increases; and so on. Although it is beyond the scope of this book to investigate each of these irreversible processes, certain overall performance parameters have been established and are commonly used to express the performance of real fluids flowing in real nozzles.

The first of these performance changes is the *nozzle efficiency.* This parameter may be defined as

$$\eta = \frac{\text{actual change in enthalpy}}{\text{isentropic change in enthalpy}} \tag{6.116}$$

or

$$\eta = \frac{\text{actual kinetic energy at nozzle exit}}{\text{kinetic energy at exit for an isentropic expansion}} \tag{6.117}$$

for an expansion to the same pressure at the exit of the nozzle. It is important to note that Equations (6.116) and (6.117) are strictly equivalent if the velocity entering the nozzle is zero or negligible. Figure 6.27 shows an isentropic expansion for state ① to state ②. Also shown is a poly-

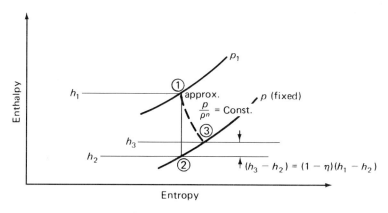

FIGURE 6.27   Nozzle coefficients.

tropic expansion characterized by a path equation $p/\rho^n = $ constant, where n is a function of the path. This approximation can be justified on the basis that the flow of a real fluid through a nozzle proceeds in such a rapid manner that it can be essentially adiabatic. Referring to Figure 6.27, we note that the nozzle efficiency (for negligible entrance velocity) is

$$\eta = \frac{h_1 - h_3}{h_1 - h_2} \quad \text{or} \quad \frac{T_1 - T_3}{T_1 - T_2} \tag{6.118}$$

for an ideal gas with constant specific heat. Equation (6.116) can also be written in terms of stagnation enthalpy by use of Equation (6.96):

$$\eta = \frac{h_1^0 - h_3}{h_1^0 - h_2}$$

$$\eta = \frac{h_1 + (V_1^2/2g_cJ) - h_3}{h_1 + (V_1^2/2g_cJ) - h_2} \tag{6.119}$$

When $V_1^2/2g_cJ$ is small, Equations (6.119) and (6.118) are essentially the same. Equation (6.119) should be used when this velocity term is not small and cannot be considered negligible compared to the other terms in these equations. Note that in SI units, the g_c and J are omitted.

For most nozzles, the efficiency varies from 90% upward, with larger nozzles having higher efficiencies. Fluids such as steam can show marked variations in efficiency depending on the pressure and temperature range of operation. Saturated steam will immediately expand into the wet region, and a two-phase fluid composed of steam and water will result. Superheated steam will behave more nearly as an ideal gas until saturation is approached. At this time, it will commence to behave as saturated steam and ultimately go into a two-phase flow of steam and water. Because of this two-phase flow, it is possible for a real nozzle to have a mass flow rate exceeding that for isentropic flow between the same pressure ranges. This leads to the second performance index, the coefficient of discharge:

$$C_D = \frac{\text{actual mass rate of flow}}{\text{mass rate of flow for an isentropic expansion}} \qquad \textbf{(6.120)}$$

C_D can exceed unity, but it is usually of the order of 95%.

The foregoing parameters do not completely specify the performance of real fluids in real nozzles. This is readily shown from Equation (6.120).

$$C_D = \frac{(\rho A V)_{\text{actual}}}{(\rho A V)_{\text{isentropic}}} = \frac{(\rho V)_{\text{actual}}}{(\rho V)_{\text{isentropic}}} \qquad \textbf{(6.121)}$$

Turbine designers are interested in the exit velocity of the nozzle as well as mass discharge. Taking the square root of both sides of Equation (6.118) yields

$$\sqrt{\eta} = C_v = \frac{\sqrt{h_1 - h_3}}{\sqrt{h_1 - h_2}} = \frac{\text{actual exit velocity}}{\text{isentropic exit velocity}} \qquad \textbf{(6.122)}$$

for the same exit pressure. C_v is called the velocity coefficient. By defining the efficiency, coefficient of discharge, and coefficient of velocity we have thus completely specified the exit conditions in the real nozzle. The calculation of these coefficients is complex and beyond the scope of this study. For further discussions of these coefficients and tabulated values, the interested student is referred to the references in Appendix 1.

6.8 REAL GASES

Many equations have been proposed to describe the pressure, volume, and temperature behavior of real gases where the simple equation of state for the ideal gas is inadequate. The earliest *equation of state,* based on elementary kinetic theory, was proposed by van der Waals. This equation is

$$\left(p + \frac{a}{v^2}\right)(v - b) = RT \qquad \textbf{(6.123)}$$

where v is the volume occupied by 1 lb mole of gas, R is the universal gas constant (1545), and a and b are constants for each gas. In the limit, as a and b go to zero, the van der Waals equation reduces to the ideal gas relation. This equation of state is most accurate when applied to gases that are removed from the saturated vapor state. It is possible to evaluate a and b in terms of the conditions at the critical state (the c state). When this is done, it will be found that the van der Waals equation of state can be rewritten in terms of the *reduced properties* p/p_c, T/T_c, and v/v_c. These ratios are known, respectively, as the *reduced pressure,* the *reduced temperature,* and the *reduced specific volume,* and are indicated by the symbols p_r, T_r, and v_r. Using this notation, we can show from the van der Waals equation that all gases have the same p, v, T relation when their reduced properties are the same. This is essentially the *law of corresponding states.* This law is not universally correct, and it has been modified to make it more generally applicable.

The utility of the law of corresponding states lies in the fact that a single curve (called a generalized compressibility chart) can be used to evaluate the p, v, T data for all gases. Most charts (made for various classes of gases) based on this principle use Z as the ordinate, where $Z = pv/RT$. A generalized compressibility chart is shown in Figure 6.28.

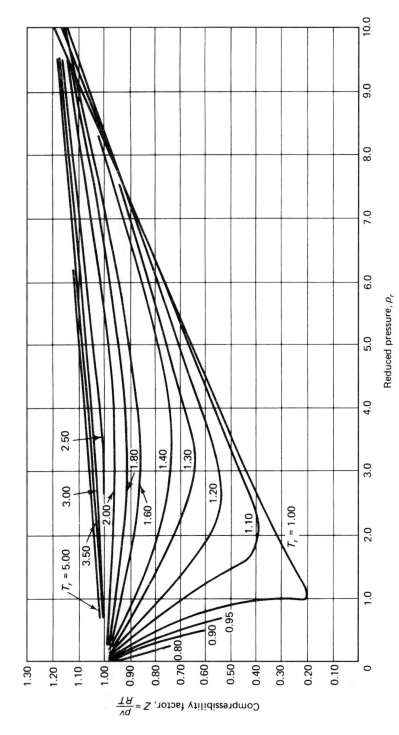

FIGURE 6.28 Compressibility factor versus reduced pressure for a series of reduced temperatures (low pressure range). *[From B. F. Dodge, Chemical Engineering Thermodynamics (New York: McGraw-Hill Book Co., 1944)].*

ILLUSTRATIVE PROBLEM 6.41

Methane (CH_4, MW = 16) has a critical temperature of 343°R and a critical pressure of 674 psia. Determine the specific volume of methane at 50°F and 500 psia, using both the ideal gas relation and Figure 6.28.

SOLUTION

$$p_r = \frac{p}{p_c} = \frac{500}{674} = 0.742$$

$$T_r = \frac{T}{T_c} = \frac{460 + 50}{343} = 1.487$$

Reading Figure 6.28 at these values gives Z equal to 0.93. Therefore,

$$Z = 0.93 = \frac{pv}{RT} \qquad v = 0.93\frac{RT}{p}$$

and

$$v = \frac{0.93(1545/16)(510)}{500 \times 144} = 0.635 \text{ ft}^3/\text{lb}_m$$

For the ideal gas,

$$v = \frac{RT}{p} = \frac{1545 \times 510}{16 \times 500 \times 144} = 0.683 \text{ ft}^3/\text{lb}_m$$

To obtain the other thermodynamic properties of the real gas, similar procedures based on the law of corresponding states have been developed. The interested reader is referred to the references in Appendix 1 for this information.

In general, the ideal gas relations yield more nearly correct results when the gas under consideration is at a pressure that is small compared to the critical pressure and when its temperature is much greater than the critical temperature. In other words, when $T_r \gg 1$ and $p_r \ll 1$, the ideal gas relations are usually applicable with negligible error.

6.9   REVIEW

Gases have been empirically found to have a simple equation that relates the properties, pressure, temperature, and specific volume. We called this relation the equation of state for an ideal gas and were able to express it mathematically in several equivalent forms. We also determined that the constant in the equation of state could be expressed as a constant divided by the molecular weight of the gas. For the ideal gas, the specific heats at constant pressure and constant temperature are functions only of the temeprature. We then discussed the concept of average specific heat. Because the specific heats are functions only of temperature, a simple relation was developed relating the specific heats and the universal gas constant. Several other useful equations were subsequently developed using the ratio of the two specific heats, k, as a useful parameter. A general relation for the change in entropy of the ideal gas was evaluated based on the general concepts that were discussed in Chapter 4. From this general relation, we then were able to write the equation of path for an isentropic process.

Once the foregoing was established, five nonflow processes were studied in detail: the constant-volume, constant-pressure, isothermal, isentropic, and polytropic processes. The relations developed were summarized in a convenient table. Much of the tedious work required to use the many equations developed can be minimized using the *Gas Tables*. The use of these tables was discussed and applied to several situations and compared to the solutions obtained using the equations developed earlier.

Our next discussion in this chapter involved the flow of gases at velocities near or greater than the speed of sound in the gas. This type of flow is called compressible flow and the parameter that characterizes it is called the Mach number, which is the ratio of the local velocity to the velocity of sound at the same location at the same time. After a general discussion, we then considered the isentropic flow of an ideal gas and subsequently studied the flow in nozzles in detail.

Finally, we briefly discussed some of the equations of state that have been proposed for real gases and to show how to calculate some of the properties of real gases.

The material covered in this chapter will probably be used either directly or as a basis for further study as frequently as the material in any chapter in this book. At first glance, it may appear to be formidable, because a large number of equations are developed. In actuality, for each of these situations, we have always invoked three basic concepts: an equation of state, an equation of path, and an energy equation. Rather than attempting to memorize formulas, each situation should be studied and the three basic concepts then applied. The definitions given throughout the chapter should be thoroughly understood as one proceeds in the chapter.

The material on gas flow and real gases is intended as a brief introduction to topics about which volumes have been written. This material will, however, provide the background for further studies in these fields.

KEY TERMS

Terms used for the first time in this chapter are:

acoustic velocity the speed of sound at a given location in a gas.

Boyle's law the path of a constant-temperature ideal gas process, $pv = $ constant.

Charles' laws $p_1/p_2 = T_1/T_2$ for a constant-volume process, $v_1/v_2 = T_1/T_2$ for a constant-pressure process.

critical pressure the pressure at which the mass flow rate in a nozzle is maximum.

de Laval nozzle a converging–diverging nozzle.

equation of state an equation relating p, v, and T for a gas.

gas constant (R) either 1545/MW ft lb_f $\text{lb}_m\cdot{}^\circ$R or 8.314/MW kJ/kg$\cdot{}^\circ$K.

Gas Tables J. H. Keenan and Joseph Kaye, *Gas Tables* (New York: John Wiley & Sons, Inc., 1948).

ideal gas a gas having the equation of state $pv = RT$.

isentropic exponent (k) $k = c_p/c_v$.

isobaric process a constant-pressure process.

isometric process a constant-volume process.

law of corresponding states all gases have the same p, v, T relation when their reduced properties are the same.

Mach number (M) the ratio of the velocity at a point in a fluid to the speed of sound at that point.

mole the amount of a substance whose mass is equal to its molecular weight in pounds mass or in kilograms.

molecular weight the relative weight of a molecule of gas using the atom of carbon as 12.000.

nozzle efficiency either the ratio of the actual change in enthalpy divided by the isentropic change in enthalpy or the ratio of the actual kinetic energy at the nozzle outlet divided by the kinetic energy at the exit for an isentropic expansion.

polytropic process a mechanically reversible, nonadiabatic process characterized by the equation of state, $pv^n = $ constant.

reduced properties the ratios of p/p_c, v/v_c, or T/T_c, where the subscript c refers to the critical state.

stagnation state the end state of a nonisentropic, adiabatic process.

EQUATIONS DEVELOPED IN THIS CHAPTER

ideal gas law	$pV = pmv = mRT$	**(6.8)**
ideal gas law	$pv = RT$	**(6.9)**
molar volume	$V_{\text{molar}} = \dfrac{1545T}{p}$ or $\dfrac{8.314T}{p}$ (in SI units)	**(6.12)**
specific heat of the ideal gas	$c_p = \dfrac{\Delta h}{\Delta T}$	**(6.15a)**
specific heat of the ideal gas	$c_v = \dfrac{\Delta u}{\Delta T}$	**(6.15b)**
property relation	$c_p - c_v = \dfrac{R}{J}$	**(6.24)**
property relation (SI)	$c_p - c_v = R$	**(6.24a)**
adiabatic exponent	$k = \dfrac{c_p}{c_v}$	**(6.25)**
property relation	$c_v = \dfrac{R}{J(k-1)}$	**(6.28)**
property relation (SI)	$c_v = \dfrac{R}{k-1}$	**(6.28a)**
property relation	$c_p = \dfrac{R}{J}\left(\dfrac{k}{k-1}\right)$	**(6.29)**
property relation (SI)	$c_p = R\left(\dfrac{k}{k-1}\right)$	**(6.29a)**
entropy change	$\Delta s = c_v \ln \dfrac{p_2}{p_1} + c_p \ln \dfrac{v_2}{v_1}$	**(6.32)**
entropy change	$\dfrac{\Delta s}{c_v} = k \ln \dfrac{v_2}{v_1} + \ln \dfrac{p_2}{p_1}$	**(6.61)**
isentropic path	$p_1 v_1^k = p_2 v_2^k$	**(6.61)**
isentropic path	$\dfrac{v_2}{v_1} = \left(\dfrac{p_1}{p_2}\right)^{1/k}$	**(6.62)**
isentropic path	$\dfrac{T_1}{T_2} = \left(\dfrac{p_1}{p_2}\right)^{(k-1)/k}$	**(6.63)**
isentropic path	$\dfrac{T_1}{T_2} = \left(\dfrac{v_2}{v_1}\right)^{k-1}$	**(6.65)**

| polytropic process | $pv^n = \text{constant}$ | (6.67) |

| polytropic process | $\dfrac{v_2}{v_1} = \left(\dfrac{p_1}{p_2}\right)^{1/n}$ | (6.68) |

| polytropic process | $\dfrac{T_1}{T_2} = \left(\dfrac{p_1}{p_2}\right)^{(n-1)/n}$ | (6.70) |

| polytropic process | $\dfrac{T_1}{T_2} = \left(\dfrac{v_2}{v_1}\right)^{n-1}$ | (6.71) |

| polytropic process | $c_n = c_v\left(\dfrac{k-n}{1-n}\right)$ | (6.80) |

| property relation | $u_2 - u_1 = (h_2 - h_1) - \dfrac{R}{J}(T_2 - T_1)$ | (6.88) |

| Gas Tables | $s_2 - s_1 = \phi_2 - \phi_1 - \dfrac{R}{J}\ln\dfrac{p_2}{p_1}$ | (6.89) |

| Gas Tables | $\ln\dfrac{p_2}{p_1} = \dfrac{J}{R}(\phi_2 - \phi_1)$ | (6.90) |

| Gas Tables | $v_r = \dfrac{RT}{p_r}$ | (6.91) |

| velocity of sound | $V_a = \sqrt{g_c k R T}$ | (6.92) |

| Mach number | $M = \dfrac{V}{V_a}$ | (6.94) |

| Mach number | $M^2 = \dfrac{V^2}{g_c k R T}$ | (6.95) |

| adiabatic flow | $h + \dfrac{V^2}{2 g_c J} = h_1 + \dfrac{V_1^2}{2 g_c J} = h^0$ | (6.96) |

| adiabatic flow | $h^0 = c_p T^0$ | (6.97) |

| adiabatic flow | $V^2 = 2 g_c J (h^0 - h_1) = 2 g_c J c_p (T^0 - T_1)$ | (6.98) |

adiabatic flow

$$V_1 = \left\{ 2 g_c J c_p T^0 \left[1 - \left(\frac{p_1}{p_0}\right)^{(k-1)/k} \right] \right\}^{1/2}$$

$$= \left\{ \frac{2 g_c k}{k-1}\frac{p^0}{\rho^0} \left[1 - \left(\frac{p_1}{p_0}\right)^{(k-1)/k} \right] \right\}^{1/2} \tag{6.99}$$

converging nozzle $\quad \dot{m} = \rho_t A_t V_t = A_t \rho^0 \left(\dfrac{p_t}{p^0}\right)^{1/k} (V_t)$

$$= A_t \left\{ \frac{2g_c k}{k-1} p^0 \rho^0 \left[\left(\frac{p_t}{p^0}\right)^{2/k} - \left(\frac{p_t}{p^0}\right)^{(k+1)/k} \right] \right\}^{1/2} \tag{6.100}$$

converging nozzle $\quad \dfrac{p_t}{p^0} = \dfrac{p_c}{p^0} = \left(\dfrac{2}{k+1}\right)^{k/(k-1)} \tag{6.101}$

converging nozzle $\quad \dfrac{T_t}{T_0} = \dfrac{2}{k+1} \tag{6.102}$

converging nozzle $\quad \dfrac{V_{a_t}}{V_{a_0}} = \sqrt{\dfrac{2}{k+1}} \tag{6.103}$

converging nozzle $\quad V_t = \sqrt{\dfrac{2g_c k}{k+1}(p^0/\rho^0)} = \sqrt{\dfrac{2}{k+1}}(v_{a_0}) = V_{a_t} \tag{6.104}$

converging nozzle $\quad \dot{m}_{\max} = A_t \left(\dfrac{2}{k+1}\right)^{1/(k-1)} \sqrt{\dfrac{2g_c k}{k+1}(p^0/\rho^0)} \tag{6.105}$

converging nozzle $\quad \left(\dfrac{A}{A_t}\right)^2 = \dfrac{k-1}{2} \left\{ \dfrac{\left(\dfrac{2}{k+1}\right)^{(k+1)/(k-1)}}{\left(\dfrac{p}{p^0}\right)^{2/k}\left[1 - \left(\dfrac{p}{p^0}\right)^{(k-1)/k}\right]} \right\} \tag{6.106}$

de Laval nozzle $\quad T_0 = \text{constant} = T_0^* \tag{6.107}$

de Laval nozzle $\quad T^* = \text{constant} \tag{6.108}$

de Laval nozzle $\quad p^0 = \text{constant} = p_0^* \tag{6.109}$

de Laval nozzle $\quad p^* = \text{constant} \tag{6.110}$

de Laval nozzle $\quad M^* = \dfrac{V}{V^*} = M\left\{ \dfrac{k+1}{2[1 + \frac{1}{2}(k-1)M^2]} \right\}^{1/2} \tag{6.111}$

de Laval nozzle $\quad \dfrac{A}{A^*} = \dfrac{1}{M}\left\{ \dfrac{2[1 + \frac{1}{2}(k-1)M^2]}{k+1} \right\}^{(k+1)/2(k-1)} \tag{6.112}$

de Laval nozzle $\quad \dfrac{T}{T^*} = \dfrac{k+1}{2[1 + \frac{1}{2}(k-1)M^2]} \tag{6.113}$

de Laval nozzle $\quad \dfrac{\rho}{\rho^*} = \left\{ \dfrac{k+1}{2[1 + \frac{1}{2}(k-1)M^2]} \right\}^{1/(k-1)} \tag{6.114}$

de Laval nozzle $\quad \dfrac{p}{p^*} = \left\{ \dfrac{k+1}{2[1 + \frac{1}{2}(k-1)M^2]} \right\}^{1/(k-1)} \tag{6.115}$

| nozzle performance | $\eta = \dfrac{\text{actual change in enthalpy}}{\text{isentropic change in enthalpy}}$ | **(6.116)** |

| nozzle performance | $\eta = \dfrac{h_1 - h_3}{h_1 - h_2}$ or $\dfrac{T_1 - T_3}{T_1 - T_2}$ | **(6.118)** |

| nozzle performance | $C_D = \dfrac{\text{actual mass rate of flow}}{(\text{mass rate of flow for an isentropic expansion})}$ | **(6.120)** |

| real gas | $\left(p + \dfrac{a}{v^2}\right)(v - b) = RT$ | **(6.123)** |

QUESTIONS

6.1 Define the term *ideal gas.*

6.2 How does an ideal gas differ from a perfect gas?

6.3 Define the term *mole.*

6.4 Write an expression for R in both systems of units. Clearly indicate the units in each case.

6.5 Air is a mechanical mixture of several gases. Which two gases make up the major part of air? What is the approximate molecular weight of air?

6.6 *True or false?* One mole of air at room pressure and temperature will occupy more volume than 1 mole of hydrogen at the same conditions.

6.7 The molecular weight of a gas in kilograms is that mass that contains 6.02×10^{26} molecules, where 6.02×10^{26} is called Avogadro's number. From this, estimate the mass of a nitrogen atom.

6.8 Define the specific heat at constant volume and the specific heat at constant pressure. For the ideal gas, what can be stated about the effects of temperature and pressure on these properties?

6.9 Illustrate what is meant by the term *mean specific heat.*

6.10 What is the lower limit of $k?$

6.11 A student conducts an experiment in a laboratory where the speed of sound is measured. Using Equation (6.92) show that this measurement can be used to obtain c_p and $c_v.$

6.12 Write the equation of path for a constant-volume process.

6.13 Using the p–v diagram show that the work of a constant-volume process is zero.

6.14 Define a *polytropic process.*

6.15 Write the equation of path for a constant-pressure process.

6.16 Using the p–v diagram show how the work of a constant-pressure process can be obtained.

6.17 Write the equation of path for an isothermal process.

6.18 Write the equation of path for an isentropic process.

6.19 How does a polytropic process differ from an isentropic process?

6.20. What is the equation of path for a polytropic process?

6.21 What single property governs the speed of sound in a given gas?

6.22 Define the term *Mach number.*

6.23 Differentiate between the total and stagnation states.

6.24 What is the maximum velocity that will exit from a converging nozzle?

6.25 How is the critical pressure related to the velocity of sound?

6.26 For a converging nozzle, what is the minimum pressure at the exit?

6.27 How would you increase the velocity and decrease the pressure at the exit of a nozzle?

6.28 How is a nozzle performance determined?

6.29 What is the range of the efficiency for a nozzle?

6.30 Does a larger nozzle have a different efficiency than a smaller nozzle?

6.31 In general, what conditions of pressure and temperature should be present for the ideal gas relations to be applicable with negligible error?

6.32 What is meant by the *law of corresponding states?*

PROBLEMS

Unless otherwise noted, use 29 for the molecular weight of air. Also, use $c_p = 0.24$ Btu/lb$_m$·°R (1.0061 kJ/kg·°K) and $c_v = 0.171$ Btu/lb$_m$·°R (0.7186 kJ/kg·°K).

Problems Involving the Ideal Gas Law

6.1 One pound of nitrogen initially at 14.7 psia and 32°F is compressed until its volume is halved and its pressure is increased by 50%. What is its final temperature?

6.2 One kilograms of nitrogen initially at 100 kPa and 0°C is expanded so that its volume is doubled while its pressure decreases to 0.3 times its initial value. What is the final temperature?

6.3 Basing your argument on the molecular weights of dry air and of water vapor, which would you expect to be heavier, dry air or moist air?

6.4 Two pounds of air is in a tank having a volume of 5 ft^3. What is the density of air in the tank?

6.5 Air is at 101.3 kPa and 20°C. Determine its specific volume.

6.6 Calculate the specific volume of air at 14.7 psia and 68°F.

6.7 Air at 40°C is in a tank at 1 MPa. If there is 1.5 kg of air in the tank, what is the volume of the tank?

6.8 What is the density of air at 72°F and 160 psia (MW = 29)?

6.9 Determine the density of air at 20°C and 100 kPa (MW = 29).

6.10 A closed tank having a volume of 5 ft^3 is filled with methane (CH$_4$, MW = 16) at 68°F. If the tank contains 6 lb$_m$, determine the pressure in the tank.

6.11 A closed cylinder having a volume of 0.14 m^3 is filled with a gas at 20°C. If the cylinder contains 2.75 kg of methane (MW = 16), determine the pressure in the cylinder.

6.12 A gas mixture occupies a volume of 1.4 m^3 at a temperature of 65°C and a pressure of 350 kPa. Determine R for the gas if the mass of gas in the container is 1.13 kg. Also, determine the molecular weight of the gas. Assume that the mixture behaves as a single ideal gas.

6.13 If the molecular weight of refrigerant HFC-134a is 102, determine its specific volume at 20 psia and 60°F.

6.14 If the molecular weight of refrigerant HFC-134a is 102, determine its specific volume at 140 kPa and 20°C.

6.15 A gas occupies 50 ft^3 at a temperature of 150°F and a pressure of 50 psia. Determine R for the gas if there is a mass of 2.5 lb in the tank.

6.16 A gas at 450 kPa occupies 1.5 m^3 at a temperature of 80°C. If there is a mass of 1.8 kg in the tank, determine R for this gas.

6.17   Three kilograms of CO_2 fills a tank having a volume of 0.2 m^3 at 20°C. What is the pressure in the tank?

6.18.   A tank of CO_2 is filled until the pressure is 5 MPa and the temperature is 100°C. If the volume of the tank is 1 m^3, how much mass is there in the tank?

6.19   A tank contains nitrogen at a pressure of 30 MPa and 250°K. If the tank has a volume of 100 L, determine the mass in the tank.

6.20   Determine the specific volume of CO_2 at a pressure of 190 psia and a temperature of 550°R.

6.21   Calculate the density and volume of 29 lb of air (1 lb mole) at 620°R and 14.7 psia.

6.22   A tank contains 1 kg of oxygen at 600 psia and 32°F. Determine the volume of the tank.

6.23   Ammonia vapor (NH_3) at 100 kPa and 100°C is stored in a tank. If the mass of the contents of the tank equals 10 kg, what is the volume of the tank?

6.24   Determine the temperature of 2 lb$_m$ of air that is in a tank whose volume is 5 ft^3 at a pressure of 50 psia.

*6.25   A "rule of thumb" has it that the pressure in a tire will change approximately 1 psi for every 10°F temperature change. Starting with the assumption that the tire is rigid and that the ambient temperature is 60°F, show how close this rule of thumb is for a tire that is initially at 30 psig. Also, show that the % change in pressure is $(\Delta/T_1)100$ where Δ is the temperature rise and T_1 is the initial absolute temperature.

6.26   An automobile tire contains air at 180 kPa and 0°C. After prolonged high-speed driving, the tire temperature is found to be 80°C. Assuming the tire to be rigid, determine its pressure when hot.

6.27   An automobile tire is inflated to 26 psig on a cold morning when the temperature is 32°F. After prolonged high-speed driving, the tire temperature is 140°F. If the tire is considered to be rigid, what is its hot pressure? Use Problem 6.25 as a check on your answer.

*6.28   A tank with a fixed volume contains O_2 at 30 psia and 60°F. Two pounds of O_2 is added to the contents so that the final pressure is 40 psia and the final temperature is 80°F. What is the volume of the tank?

Problems Involving Property Relations

6.29   An ideal gas has $R = 80$ ft lb$_f$/lb$_m$·°R and $k = 1.4$. Determine c_p and c_v for this gas.

6.30   An ideal gas has $k = 1.4$ and $R = 0.1890$ kJ/kg·°K. Determine c_p and c_v for this gas.

6.31   An ideal gas has MW equal to 29 and k equal to 1.3. Determine c_p and c_v of this gas in English units.

6.32   Solve Problem 6.31 in SI units.

6.33   An ideal gas has $c_p = 0.312$ Btu/lb$_m$·°R and $c_v = 0.240$ Btu/lb$_m$·°R. Determine R and the molecular weight of the gas.

6.34   An ideal gas has $c_p = 1.302$ kJ/kg·°K and $c_v = 0.986$ kJ/kg·°K. Determine R and the molecular weight of the gas.

6.35   If the specific heat at constant pressure of a gas having a molecular weight of 18 is 0.45 Btu/lb$_m$·°R, what is its specific heat at constant volume?

6.36   A gas has a specific heat at constant pressure of 0.34 Btu/lb$_m$·°R, and a specific heat at constant volume of 0.272 Btu/lb$_m$·°R. Determine the molecular weight of this gas.

6.37   If c_p of a gas is 5.2028 kJ/kg·°K and its molecular weight is 4, determine c_v.

6.38   If c_p of a gas is 1.0426 kJ/kg·°K and c_v is 0.7454 kJ/kg·°K. determine the molecular weight and k of the gas.

6.39   If c_p of a gas is 0.3 Btu/lb$_m$·°R, what is c_v if the gas has a molecular weight of 28?

6.40   An ideal gas has $c_p = 1.2$ kJ/kg·°K and $k = 1.32$. Determine the molecular weight, R, and c_v for the gas.

6.41   An ideal gas has $c_p = 0.286$ Btu/lb$_m$·°R and $k = 1.29$. Determine the molecular weight, R, and c_v for this gas.

Problems Involving Isothermal Processes

6.42   Nitrogen initially has a pressure of 55 psia and a temperature of 80°F. If its final pressure after an isothermal expansion is 18 psia, what is its final specific volume?

6.43   Oxygen is initially at 30°C and 350 kPa. What is its final specific volume after it undergoes an isothermal expansion to 120 kPa?

6.44   How much work is required to isothermally compress 1 kg of air from 200 kPa to 2 MPa? Assume that c_p is 1.0061 kJ/kg·°K, c_v is 0.7186 kJ/kg·°K, and the air is at 20°C (MW = 29).

6.45   Determine the change in entropy for Problem 6.43.

6.46   Determine the change in entropy for Problem 6.44.

6.47   Air is compressed from 20 psia to 120 psia at a constant temperature of 75°F. Calculate the work required for compression per lb$_m$.

6.48   How much work is required to isothermally compress 1 lb$_m$ of air from 25 psia and 125°F to 250 psia?

6.49   Air initially at 100°F is isothermally compressed from a pressure of 1 atm to a pressure of 5 atm. Determine the work done on 1 lb$_m$ of air and the heat removed during the process.

6.50   Air initially at 200°C is isothermally compressed from a pressure of 1 atm to a pressure of 5 atm. Determine the work done on 1 kg of air and the heat removed during the process.

6.51   Nitrogen (MW = 28) is isothermally compressed from 200°F and 40 psia to 1000 psia. Calculate the final specific volume, the change in entropy, and the heat transferred in this process.

6.52   One pound mass of air is expanded isothermally until its final volume is 50% greater than its initial volume. If the temperature of the air is 150°F, how much heat was supplied and what is the change in entropy of this process?

6.53   One kilogram of nitrogen (MW = 28) is expanded isothermally until its final volume is 100% greater than its initial volume. If the temperature of the nitrogen is 80°C, how much heat was supplied and what is the change in entropy for this process?

6.54   One pound of argon (MW = 39.90) is isothermally compressed from 100°F and 20 psia to 600 psia. Calculate the final specific volume, the change in entropy, and the heat transferred.

6.55   Heat is added to air at constant temperature from initial conditions of 80 psia and 250°F to a final pressure of 25 psia. Determine the heat added and the change in entropy per pound mass of air for this process (MW = 29).

6.56   Air is expanded at constant temperature from 200 psia and 2000°R to a final volume of 10 ft³. If the initial volume was 4 ft³, calculate the final pressure, the total work done, the total change in internal energy, and the total heat transfer for the process.

*6.57   A cylinder of nitrogen is 2 m high and has a diameter of 0.5 m. It is initially filled with gas until the pressure is 200 psia and the temperature is 70°F. Some of the gas is used in a welding process until the pressure is 50 psia while the temperature remains constant at 70°F. Calculate the mass of nitrogen that has been used.

*6.58   A cylinder is filled with air at 200°F and 150 psia. If 5 lb$_m$ is removed from the cylinder while the temperature is kept constant at 200°F and the pressure falls to 100 psia, determine the volume of the cylinder and the original mass in it.

6.59   Five pounds of air at 120°F and 100 psia undergo an isothermal process until the final pressure is 40 psia. Determine the work done, the change in internal energy, the heat removed, and the change in entropy for this process.

6.60 Two kilograms of a gas having $c_p = 1.205$ kJ/kg·°K and $c_v = 0.8607$ kJ/kg·°K undergo an isothermal process from initial conditions of 500 kPa and 200°C to a final pressure of 100 kPa. Determine the work done, the change in internal energy, the heat removed, and the change in entropy for the process.

6.61 A gas has $c_p = 0.3215$ Btu/lb$_m$·°R and $c_v = 0.2296$ Btu/lb$_m$·°R. Determine the work done, the heat removed, the change in internal energy, and the change in entropy for a pound mass that is initially at 85 psia and 100°F and is compressed to isothermally 255 psia.

*6.62 Ten pounds of carbon monoxide at 86°F occupies 100 ft^3 in a tank. An additional amount of carbon monoxide is added slowly to the tank, raising the pressure to 150 psia. Assuming the process is carried out isothermally, determine the mass of gas added.

*6.63 A gas has a molecular weight of 30 and is contained in a 1 m^3 tank at 350 kPa and 25°C. A second evacuated tank having a volume of 0.15 m^3 is attached to the first tank. A connecting valve is opened and the gas flows into the second tank until the pressures in both tanks are equal. If the second tank was initially at 25°C and the temperature remains constant, determine the final pressure in the tanks.

Problems Involving Constant-Volume Processes

6.64 A closed tank contains 8 ft^3 of air at 17 psia and 125°F. If 60 Btu is added to the air, determine the final pressure. Assume that $c_v = 0.17$ Btu/lb$_m$·°R.

6.65 What is the final pressure in a closed tank after 60 kJ is added to 0.3 m^3 of air that is at 150 kPa and 40°C? Assume that $c_p = 1.0061$ kJ/kg·°K and $c_v = 0.7186$ kJ/kg·°K.

6.66 If 0.1 lb$_m$ of a gas is heated at constant volume from 90°F to 350°F, determine the increase in entropy for the process. Assume that $c_v = 0.22$ Btu/lb$_m$·°R. Also, determine the ratio of the final pressure to the initial pressure for this process. Assume $c_p = 0.304$ Btu/lb$_m$·°R.

6.67 Determine the change in internal energy and enthalpy for the gas in Problem 6.66.

6.68 Air is heated at constant volume from 40°F to 1000°R. Calculate the change in entropy for this process.

6.69 Solve Problem 6.68 using the *Gas Tables*.

6.70 Calculate the change in internal energy for the gas in Problem 6.68.

6.71 Solve Problem 6.68 for the average specific heat at constant pressure for the process using the *Gas Tables*.

6.72 Solve Problem 6.68 for the average specific heat at constant volume for the process using the *Gas Tables*.

6.73 An air compressor is used to fill a rigid tank to 80 psia at a temperature of 70°F. Determine the temperature of the gas that will cause the relief valve to open if the valve is set at 150 psia.

6.74 If 0.05 kg of a gas is heated from 30°C to 175°C, what is the entropy change if the process is carried out at constant volume and $c_v = 0.90$ kJ/kg·°K? k for this gas is 1.3.

6.75 Determine the change in internal energy and enthalpy for the gas in Problem 6.74.

6.76 One hundred Btu is added to 1 lb$_m$ of air at constant volume whose initial conditions are 85°F and 35 psia. If $c_v = 0.19$ Btu/lb$_m$·°R, determine the final pressure and final temperature and the change in entropy for the process (MW = 29).

6.77 If 100 kJ is added to 0.5 kg of air at constant volume and the initial conditions are 30°C and 200 kPa, and $c_v = 0.7$ kJ/kg·°K, determine the final pressure, final temperature, and the change in entropy for the process (MW = 29).

6.78 One pound of a gas at 80°F is heated until its entropy increase is 0.15 Btu/lb$_m$·°R. If c_v is 0.21 Btu/lb$_m$·°R and the process is carried out at constant volume, determine the final gas temperature.

*6.79 A closed, rigid, well-insulated tank has a volume of air of 5 ft^3 that is initially at 150 psia and 50°F. The air is stirred by a paddle wheel until the pressure increases to 400 psia. Determine the final temperature and the paddle work input in Btu.

*6.80 A closed, rigid, well-insulated tank has a volume of air of 0.15 m^3 that is initially at 1 MPa and 10°C. The gas is stirred by a paddle wheel until the pressure increases to 2.8 MPa. Determine the final temperature and the paddle work input in kJ.

Processes Involving Constant Pressure

6.81 A process is carried out at a constant pressure of 350 kPa. If the volume changes from an initial value of 2.5 to 7.5 m^3, determine the work of the process.

6.82 Air is expanded from 1 to 8 ft^3 in a piston–cylinder apparatus at a constant pressure of 100 psia from an initial temperature of 100°F. Calculate the final air temperature and the work per lb$_m$ in the cylinder.

6.83 Air at 700 kPa is heated from 20°C to 120°C at constant pressure. Determine the heat required per kilogram.

6.84 Air at 100 psia is heated from 80°F to 200°F at constant pressure. Determine the heat required per pound mass.

6.85 Solve problem 6.84 using the *Gas Tables*.

6.86 Solve problem 6.84 for the average specific heat for the process using the *Gas Tables*.

6.87 Use the *Gas Tables* to determine the change in enthalpy of air when the temperature changes from 540°R to 1300°R.

6.88 Solve Problem 6.87 for the average specific heat for the process.

6.89 Air is heated at constant pressure from 40°F to 1240°F. Determine the change in entropy for this process.

6.90 Solve Problem 6.89 using the *Gas Tables*.

*6.91 If 4.5 lb$_m$ of oxygen is at 100°F, determine the change in density expressed as a percentage of the initial density when the temperature is changed to 250°F at a constant pressure of 100 psia.

*6.92 One kilogram of a gas is heated at a constant pressure of 425 kPa until the internal energy increases by 250 kJ/kg and the temperature increases by 100°C. If there is work done by the gas that equals 150 kJ/kg, determine the change in volume and c_p.

*6.93 If 2 kg of oxygen occupies 0.25 m^3 at 40°C, determine the change in density expressed as a percentage of the initial density when the temperature is changed to 125°C at a constant pressure of 0.7 MPa.

6.94 Compute the new volume in a cylinder if air is cooled at constant pressure until its initial temperature of 500°F is reduced to 100°F. The initial volume of the cylinder is 10 ft^3.

6.95 An ideal gas has $R = 80$ ft lb$_f$/lb$_m$·°R and $k = 1.4$. If the gas undergoes a nonflow, constant-pressure process from 1000°F to 500°F, determine the work done, the change in internal energy, and the change in enthalpy for the process. All answers should be given in Btu/lb$_m$.

6.96 Determine the heat removed from the cylinder of Problem 6.94. Assume that $c_p = 0.24$ Btu/lb$_m$·°R and $c_v = 0.171$ Btu/lb$_m$·°R.

6.97 An ideal gas (MW = 20) undergoes a nonflow, constant-pressure process from 500°C to 250°C. If k is 1.4, determine the work done, the change in internal energy, and the change in enthalpy for the process.

6.98 Determine the work done, the change in internal energy, and the heat that is supplied when air initially at 200 psia and 200°F expands in a cylinder at constant pressure from an initial volume of 20 ft^3 to triple its initial volume. Assume $c_p = 0.24$ Btu/lb$_m$·°R and $c_v = 0.171$ Btu/lb$_m$·°R.

6.99 The initial pressure in the combustor of a gas turbine is 80 psia. If this process is carried out at constant pressure, determine the increase in entropy and the flow work change per pound of air if the air temperature goes from 85°F to 550°F. Assume that $c_p = 0.24$ Btu/lb$_m$·°R.

6.100 Carbon dioxide (MW = 44) is heated at constant pressure from 14.7 psia and 90°F to 500°F. If $k = 1.4$, determine the heat transferred and the change in entropy for the process. Assume that c_p is constant.

6.101 Carbon dioxide is heated from 100 kPa and 35°C to 250°C at constant pressure. If k equals 1.3, determine the heat transferred and the change in entropy for the process if c_p is constant (MW = 44).

6.102 One pound of nitrogen occupies a volume of 5 ft³ at a temperature of 200°F. If the temperature is increased to 300°F at constant pressure, determine the work and the final volume.

*6.103 Ten cubic feet of air is compressed until the volume is 5 ft³. If the pressure is kept constant at 190 psia, determine the work, the change in internal energy, and the heat transfer for this process.

Problems Involving Isentropic and Polytropic Processes
Use Table A.13 for physical data (except for air) as necessary.

6.104 Argon is compressed isentropically from 80°F and 20 psia to 1200°F. Determine the final pressure.

6.105 Helium is compressed isentropically from 100 kPa and 0°C to 2 MPa. Calculate the final temperature.

6.106 Nitrogen is compressed isentropically from 14.7 psia and 70°F to 1400 psia. What is the final temperature of the nitrogen?

6.107 Air at 110 kPa and 20°C is isentropically compressed to 10 MPa. Determine the final temperature.

6.108 Air is expanded isentropically from a volume of 10 ft³ to a volume of 30 ft³. If the initial temperature is 200°F, determine the final temperature.

6.109 An ideal gas with $k = 1.4$ undergoes a reversible adiabatic expansion from 100 psia and 700°F to 350°F. Determine the final pressure of the gas.

6.110 An ideal gas expands isentropically from 120 psia and 200°F to 20 psia. Determine the final temperature for this nonflow process if $k = 1.4$.

6.111 Air is compressed isentropically from 1 atm absolute to 10 atm absolute. If the initial temperature is 70°F, determine the final temperature and the work required for this nonflow process if $k = 1.4$.

6.112 Solve Problem 6.110 using the *Gas Tables*.

6.113 Solve problem 6.111 using the *Gas Tables*.

*6.114 An ideal gas expands isentropically from 150 psia and 450°F to 25 psia and 100°F. Determine k for this process.

*6.115 An ideal gas expands isentropically from 600 kPa and 195°C to 110 kPa and 40°C. What is the k of this gas process?

*6.116 A gas expands isentropically from 90 psia and 380°F to 15 psia and 100°F. Determine k for this gas.

*6.117 Determine the final state and change in enthalpy and internal energy when air is expanded isentropically from 100 psia and 1000°R to 75 psia. Assume that $k = 1.4$. If all the work of this nonflow process could be recovered, how much work would be available?

6.118 Air undergoes a polytropic expansion from 2500°R to 1500°R. The final pressure of the air is 20 psia, and n for the process is 1.32. Determine the work output per pound mass of air.

*6.119 One pound mass of air is compressed from 20 to 60 psia polytropically from an initial temperature of 100°F. If $n = 1.26$, determine the final temperature and the heat transfer for the process.

*6.120 A gas initially at 170 psia and 270°F is expanded to 40 psia and 90°F. Determine n for this process.

*6.121 A process is carried out in which a gas is compressed from 14 to 28 psia, and its final volume is decreased to 60% of its initial volume. Determine n for this process.

*6.122 One pound mass of nitrogen undergoes a process during which it changes from an initial state of 100 psia and 400°F to a final state of 300 psia and 650°F. Determine n for this process.

*6.123 A gas that is initially at 1.16 MPa and 133°C is expanded to 275 kPa and 32°C. What is n for this process?

6.124 Air is compressed polytropically from 1 atm to 10 atm absolute. The initial temperature is 70°F and $n = 1.2$. Determine the work required for the compression if $c_v = 0.171$ Btu/lb$_m$·°R.

6.125 Determine the entropy change and heat transferred for the polytropic process described in Problem 6.124.

6.126 If n of a nonflow process is 1.2, how much work is required to compress 1 kg of air from 120 to 600 kPa if the air is initially at 30°C? Assume that k is 1.4 and MW is 29.

6.127 If n of a nonflow polytropic process is 1.18, determine the work required to compress 1 lb$_m$ of air from 20 to 100 psia if the initial air temperature is 80°F.

6.128 One pound mass of air undergoes a polytropic process with $n = 1.33$. The air is initially at 14.7 psia and 68°F. During the process 12 Btu/lb$_m$ is transferred from the air as heat. Determine the final temperature and pressure for the process.

6.129 A polytropic process for air has $n = 1.5$. If the initial state of the air is 80 psia and 900°F and the final state is at 20 psia, determine the final temperature and the work done per unit mass.

Problems Involving Average Specific Heat and Real Gases
6.130 An equation for the instantaneous specific heat of air is given as

$$c_p = 0.219 + \frac{0.342T}{10^4} - \frac{0.293T^2}{10^8} \qquad \text{where } T \text{ is in } °R$$

Determine the average specific heat of air between 340°F and 840°F.

6.131 Using the *Gas Tables,* solve Problem 6.130.

*6.132 The instantaneous specific heat of carbon dioxide is given by

$$c_p = 0.368 - \frac{148.4}{T} + \frac{3.2 \times 10^4}{T^2} \qquad \text{where } T \text{ is in } °R$$

Determine the average specific heat of carbon dioxide between 500 and 1000°R.

6.133 Carbon dioxide (MW = 44) has a critical pressure of 1071 psia and a critical temperature of 547.5°R. Determine the specific volume of this gas at atmospheric pressure (14.7 psia) and 1000°F, using Fig. 6.28. Compare it with the specific volume obtained from the ideal gas relation.

6.134 If the gas in Problem 6.135 is at 1470 psia, determine its specific volume and compare it to the value obtained from the ideal gas relation.

6.135 Carbon dioxide has a pressure of 3200 psia and a temperature of 1000°R. Calculate its specific volume using the generalized compressibility chart.

6.136 Use the generalized compressibility chart to determine the mass of nitrogen contained in a 500 ft^3 tank at 3000 psia and 400°R.

6.137 Use the generalized compressibility chart to determine the density of methane at 1500 psia and 0°F.

6.138 Determine the density of oxygen at 1500 psia and 400°F using the generalized compressibility chart.

Problems Involving Compressible Flow
6.139 Air at 50°F and 500 psia is flowing in a duct at 100 ft/s. What is its Mach number?

6.140 If air is slowed adiabatically from 500 to 300 ft/s, determine its temperature rise if $c_p = 0.24$ Btu/lb$_m$·°R.

*6.141 Air is slowed adiabatically from a velocity of 500 ft/s to another velocity. During the slowing-down process, the air temperature rises 10°F. If c_p of air is 0.24 Btu/lb$_m$·°R and constant, determine its final velocity.

6.142 Air is flowing at 500 ft/s at a temperature of 300°F and a pressure of 25 psia. Determine its isentropic stagnation temperature and pressure.

*6.143 Air flows in a nozzle. If the inlet velocity is negligible and the inlet pressure is 100 psia, determine the critical velocity in the nozzle if it is a converging–diverging nozzle. Assume that the inlet temperature is 100°F. What is the critical pressure?

*6.144 Air is flowing in a duct at a Mach number of 2.5. Assuming that all processes are isentropic and that the stagnation temperature is 500°F, determine the air temperature for which the Mach number is 1.2.

7

Mixtures of Ideal Gases

Learning Goals

After reading and studying the material in this chapter, you should be able to:

1. Apply the ideal gas relation to a gas mixture.

2. Understand and use the term *partial pressure*.

3. Apply Dalton's law to a gas mixture.

4. Define the *mole fraction* of a gas.

5. Arrive at the simple conclusion that the total number of moles of gas in a mixture is the sum of the moles of each of its constituents.

6. Use the fact that the mole fraction of a component of a gas mixture equals the partial pressure of that component.

7. Apply the ideal gas equation to both the components and the entire mixture.

8. Utilize the conclusion that the mole fraction is also equal to the volume fraction of a component of the mixture.

9. Convert from volume fraction to weight fraction and from weight fraction to volume fraction.

10. Recall that the thermodynamic properties of a mixture are each equal to the sum of the mass average of the individual properties of the components of the mixture.

11. Understand and use the terms that are used in conjunction with air–water vapor mixtures.

12. Determine the properties of air–water vapor mixtures.

13. Understand and use the psychrometric chart.

7.1 INTRODUCTION

Many mixtures of gases are of thermodynamic importance. For example, atmospheric air is principally a mixture of oxygen and nitrogen with some water vapor. The products of combustion from power plants and internal combustion engines are also mixtures of gases that are of interest. From the simple kinetic considerations of Chapter 1, it was possible to deduce an equation of state to relate the pressure, temperature, and volume of a gas. This particular equation, $pv = RT$, formed the basis of most of the work in Chapter 6. As we have remarked several times, it is surprising that so simple an equation can represent real gases with any degree of accuracy. The success of this equation for a single gas leads us to hope that mixtures of gases can be represented equally well by an equally simple equation of state. To some extent, this hope is realized for some real gaseous mixtures, but for others, it has not been possible to write a simple p,v,T relation in terms of the properties of the individual components of the mixture. By use of semiempirical methods, it has been found possible to correlate experimental p,v,T data for some gases, but no single method of correlation has yet been devised to correlate all gaseous mixtures satisfactorily.

In this chapter, we will treat gas mixtures as if they were pure substances to which the ideal gas relation can be applied. This will be taken to be true of the components of the mixture as well as to the mixture as a whole.

7.2   PRESSURE OF A MIXTURE

Consider that a mixture of several gases is contained in a volume V at a temperature T and that an absolute pressure gage placed in the volume would read a total mixture pressure p_m, as shown in Figure 7.1. In this gas mixture, the various component gas molecules are in constant random motion, and the gas mixture is assumed to be homogeneous. For such a mixture, *Dalton's law* (or Dalton's rule of additive pressures) is found to apply exactly if the ideal gas relation holds for the components and for the mixture. Dalton's law can be stated as follows:

1. Each gas behaves as if it alone occupied the volume.

2. Each gas behaves as if it exists at the temperature of the mixture and filled the entire volume by itself.

3. The pressure of the gas mixture is the sum of the pressures of each of its components when each behaves as described in statements 1 and 2.

Applying Dalton's law to the situation shown in Figure 7.1, where the mixture m and each of the components are all shown to occupy equal volumes, yields

$$p_m = p_a + p_b + p_c \qquad (7.1)$$

where p_m is the total pressure of the mixture and p_a, p_b, and p_c represent the pressures that each of the constituent gases would exert if they alone occupied the total volume at the temperature of the mixture. These pressures are called the *partial pressures* of each gas. By assuming that the

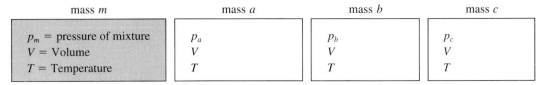

FIGURE 7.1   Dalton's law.

pressure–volume relationship for an ideal gas is applicable and denoting the molecular weight of the mixture as MW_m, we have

$$p_m V = m_m RT = m_m \left(\frac{1545T}{MW_m} \right)$$

(7.2)

where m_m is the mass of the mixture equal to the sum of the masses of the individual gases. The partial pressure of each gas can be substituted for p_m, the mass of each gas for m_m, and the molecular weight of each gas for MW_m in Equation (7.2). This yields a set of ideal gas equations for each component gas. Using each of these equations in Equation (7.1) yields

$$\frac{m_m}{MW_m} = \frac{m_a}{MW_a} + \frac{m_b}{MW_b} + \frac{m_c}{MW_c}$$

(7.2a)

Before continuing, it will be convenient to recall certain terms that were discussed briefly in Chapters 1 and 6. If the molecular weight of a substance is expressed in pounds, the resulting quantity is known as the pound molecular weight, a pound mole, or simply a mole. Thus, 32 lb_m of oxygen constitutes 1 lb mole of oxygen, and 28.02 lb_m of nitrogen is 1 lb mole of nitrogen. It has also been found that 1 mole of a substance, whether it be solid, liquid, or gas, contains the same number of molecules as 1 mole of any other substance. The number of molecules in 1 mole is known as the Avogadro number, or Avogadro's constant (6.02×10^{26}). Thus, by application of the ideal gas equation of state, it is found that at a given pressure and temperature, 1 mole of gas will occupy a fixed volume, regardless of the gas. This volume is known as the molar volume, and at 14.7 psia and 32°F, it is 358 ft^3. Table 7.1 is a convenient listing of the molecular weights of various gases.

TABLE 7.1
Molecular Weights of Gases

Gas	Chemical formula	Molecular weight
Acetylene	C_2H_2	26.02
Air		28.96
Ammonia	NH_3	17.024
Argon	A	39.90
Butane	C_4H_{10}	58.08
Carbon dioxide	CO_2	44.00
Carbon monoxide	CO	28.00
Dodecane	$C_{12}H_{26}$	170.3
Ethane	C_2H_6	30.05
Ethylene	C_2H_4	28.03
Helium	He	4.00
Hydrogen	H_2	2.016
Methane	CH_4	16.03
Nitrogen	N_2	28.02
Octane	C_8H_{18}	114.14
Oxygen	O_2	32.00
Propane	C_3H_8	44.06
Sulfur dioxide	SO_2	64.07
Water vapor	H_2O	18.01

It will now be noted that the ratio of m/MW in Equation (7.2a) is simply the number of moles of each component of the mixture. Denoting the moles of each constituent as n, with the appropriate subscripts,

$$n_m = n_a + n_b + n_c \qquad (7.2b)$$

In words, Equation (7.2b) states that the number of moles of gas in a mixture is equal to the sum of the moles of its constituent gases. It is also evident that the total mass of a gas mixture must equal the sum of the component masses. Referring to Figure 7.1, we have

$$m_m = m_a + m_b + m_c \qquad (7.3)$$

The ratio n_a/n_m is the *mole fraction* x_a of gas a, and the ratio m_a/m_m is the mass fraction of gas a. From Equations (7.1) and (7.2),

$$\frac{p_a}{p_m} = \frac{n_a}{n_m} = x_a$$

and

$$p_a = x_a p_m \qquad (7.4)$$

and from Equations (7.2a) and (7.3),

$$n_m(MW_m) = n_a(MW_a) + n_b(MW_b) + n_c(MW_c)$$

or

$$MW_m = x_a MW_a + x_b MW_b + x_c MW_c \qquad (7.5)$$

Because Equation (7.5) is based on the fact that the mixture and its components are ideal gases and that the mixture can also be treated as a single ideal gas, the gas constant for the mixture can be written as

$$R_m = \frac{1545}{MW_m} \qquad (7.6)$$

To sum up the foregoing:

1. The total pressure of a mixture of ideal gases is the sum of its partial pressures.

2. The mass of the mixture is the sum of the masses of the components of the mixture.

3. The total number of moles in the mixture equals the sum of the number of moles of the component gases of the mixture.

4. The partial pressure of a component equals the mole fraction of the component multiplied by the total pressure of the mixture.

5. The molecular weight of the mixture equals the sum of the products of mole fraction of each gas component multiplied by the molecular weight of each component gas.

6. The ideal gas equation of state is applicable to the components as well as to the entire mixture.

ILLUSTRATIVE PROBLEM 7.1

Dry air is a mixture of oxygen and nitrogen as the principal gases. Per pound of mixture, there is 0.2315 lb of oxygen and 0.7685 lb of nitrogen. Determine (a) the number of moles of each gas, the total number of moles, and the mole fraction of each component; (b) the partial pressure of each component if the air is at 14.7 psia; (c) the molecular weight of the air; and (d) the gas constant of the air.

SOLUTION

As the basis of this calculation, assume that we have 1 lb_m of mixture. Also, take the molecular weight of oxygen to be 32.00 and nitrogen to be 28.02 (from Table 7.1).

(a) $n_{O_2} = \dfrac{0.2315}{32} = 0.00723$ mole of oxygen/lb_m of mixture

$n_{N_2} = \dfrac{0.7685}{28.02} = 0.02743$ mole of nitrogen/lb_m of mixture

$n_m = n_{O_2} + n_{N_2} = 0.00723 + 0.02743$

$n_m = 0.03466$ mole/lb_m of mixture

$x_{O_2} = \dfrac{n_{O_2}}{n_m} = \dfrac{0.00723}{0.03466} = 0.209$

$x_{N_2} = \dfrac{n_{N_2}}{n_m} = \dfrac{0.02743}{0.03466} = 0.791$

(*Check:* $x_{O_2} = x_{N_2} = 0.209 + 0.791 = 1.000$)

(b) $p_{O_2} = x_O(14.7) = 0.209 \times 14.7 = 3.07$ psia

$p_{N_2} = x_{N_2}(14.7) = 0.791 \times 14.7 = 11.63$ psia

(c) $MW_m = x_{O_2}(32) + x_{N_2}(28.02)$

$= 0.209 \times 32 + 0.791 \times 28.02 = 28.85$

(d) $R_m = \dfrac{1545}{MW_m} = \dfrac{1545}{28.85} = 53.55$ or   53.6

ILLUSTRATIVE PROBLEM 7.2

Gaseous Freon-12 (CCl_2F_2) is used to slowly fill a tank that initially contained air at atmospheric pressure and temperature to a final pressure of 1000 psia and room temperature. What is the mole fraction of air in the final mixture? Also, find the molecular weight of the mixture and the partial pressure of the gases. Assume that the molecular weight of air is 29 and the molecular weight of Freon-12 is 120.9.

SOLUTION

The partial pressure of the air is by definition 14.7 psia. Therefore, the partial pressure of the Freon-12 is $1000 - 14.7 = 985.3$ psia. The mole fraction of air is $14.7/1000 = 0.0147$, and the mole fraction of Freon-12 is $985.3/1000 = 0.9853$.

The molecular weight of the mixture is

$$MW_m = 0.0147 \times 29 + 0.9853 \times 120.9 = 119.5$$

ILLUSTRATIVE PROBLEM 7.3

Ten pounds of air, 1 lb of carbon dioxide, and 5 lb of nitrogen are mixed at constant temperature until the mixture pressure is constant. The final mixture is at 100 psia. Determine the moles of each gas, the mole fraction of each gas, the partial pressure of each gas, the molecular weight of the mixture, and the gas constant of the mixture. Assume that the molecular weight of air is 29; of carbon dioxide, 44; and of nitrogen, 28.

SOLUTION

Let us consider as a basis for the calculation that we have 16 lb_m of mixture. Therefore,

$$n_{air} = \frac{10}{29} = 0.345 \qquad n_{CO_2} = \frac{1}{44} = 0.023 \qquad n_{N_2} = \frac{5}{28} = 0.179$$

The total number of moles is $0.345 + 0.023 + 0.179 = 0.547$. The mole fractions are

$$x_{air} = \frac{0.345}{0.547} = 0.631 \qquad x_{CO_2} = \frac{0.023}{0.547} = 0.042$$

$$x_{N_2} = \frac{0.179}{0.547} = 0.327$$

$$p_{air} = (0.631)(100) = 63.1 \text{ psia}$$
$$p_{CO_2} = (0.041)(100) = 4.1 \text{ psia}$$
$$p_{N_2} = (0.327)(100) = 32.7 \text{ psia}$$

The molecular weight of the mixture is found from

$$MW_m = (0.631)(29) + (0.041)(44) + (0.327)(28) = 29.3$$

and
$$R_m = \frac{1545}{29.3} = 52.7$$

7.3 VOLUME OF A MIXTURE

In arriving at Dalton's law in Section 7.2, it was assumed that each gas in a mixture expands to fill the entire volume of the container at the temperature of the mixture. Let us assume that three gases, *a, b,* and *c,* are placed in separate containers, each having different volumes but each at the same pressure and temperature. This condition is shown in Figure 7.2. If the three containers are brought together and the partitions are removed, the pressure and temperature of the mixture will remain constant, if these gases are ideal, and Dalton's law can be applied. The partial pressures will equal the initial pressure multiplied by the ratio of the initial to final volume. Therefore,

$$p_m = p_{oa}\frac{V_a}{V_m} + p_{ob}\frac{V_b}{V_m} + p_{oc}\frac{V_c}{V_m} \tag{7.7}$$

where p_{oa} is the initial pressure of *a* in its container, V_a the *volume of a before mixing,* and V_m the *volume of the mixture after mixing.* Because p_o of each gas is the same as the final pressure p_m,

$$1 = \frac{V_a}{V_m} + \frac{V_b}{V_m} + \frac{V_c}{V_m} \quad \text{or} \quad \boxed{V_m = V_a + V_b + V_c} \tag{7.8}$$

Equation (7.8) is an expression of *Amagat's law,* which states that the volume of a mixture is the sum of the volumes that each constituent gas would occupy if each were at the pressure and temperature of the mixture.

By applying the equation of state of an ideal gas to Equation (7.8), we obtain

$$\boxed{\frac{V_a}{V_m} = \frac{n_a}{n_m} = x_a} \tag{7.9}$$

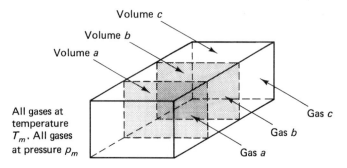

FIGURE 7.2 Law of additive volumes.

That is, *the mole fraction of a component also equals the volume fraction of that component in the mixture.* The volume fraction is defined as the ratio of the partial volume of a constituent to the total volume of the mixture. Amagat's law and Dalton's law are equivalent to each other if the gases and the mixture are ideal gases. The following two examples will serve to illustrate the concept of mixture volume.

ILLUSTRATIVE PROBLEM 7.4

Five moles of oxygen at 14.7 psia and 70°F are adiabatically mixed with 10 moles of hydrogen at 14.7 psia and 70°F. Determine the mixture volume, the partial volume of the constituents, the mole fraction of each constituent, and the molecular weight of the resulting mixture. The molecular weight of oxygen is 32.0 and of hydrogen, 2.016 (see Table 7.1).

SOLUTION

The total number of moles is $10 + 5 = 15$. Therefore, the mole fraction of each constituent is

$$x_{O_2} = \frac{5}{15} = \frac{1}{3} \qquad x_{H_2} = \frac{10}{15} = \frac{2}{3}$$

The molecular weight of the final mixture is

$$\left(\tfrac{1}{3}\right)(32) + \left(\tfrac{2}{3}\right)(2.016) = 12.01$$

The partial volume of the oxygen can be found as follows: per pound of oxygen,

$$(pv)_{O_2} = RT$$
$$14.7 \times 144 \times v_{O_2} = \frac{(1545)(460 + 70)}{32}$$

and
$$v_{O_2} = 12.09 \text{ ft}^3/\text{lb}_\text{m}$$

Because there are 5 moles of oxygen, each containing 32 lb_m,

$$V_{O_2} = 12.09 \times 5 \times 32 = 1934 \text{ ft}^3$$

For the hydrogen, we can simplify the procedure by noting that the fraction of the total volume occupied by the oxygen is the same as its mole fraction. Therefore,

$$V_m = (3)(1934) = 5802 \text{ ft}^3$$

and the hydrogen volume

$$V_{H_2} = 5802 - 1934 = 3868 \text{ ft}^3$$

We could obtain the partial volume of hydrogen by proceeding as we did for the oxygen. Thus,

$$(pv)_{H_2} = RT$$
$$14.7 \times 144 \times v_{H_2} = \frac{(1545)(460 + 70)}{2.016}$$

and
$$v_{H_2} = 191.88 \text{ ft}^3/\text{lb}_m$$

For 10 moles of hydrogen, each containing 2.016 lb_m, we have a total hydrogen volume of

$$V_{H_2} = 10 \times 2.016 \times 191.88 = 3868 \text{ ft}^3$$

which checks with the previous value. As an alternative to the foregoing, we could also use the fact that at 14.7 psia and 32°F a mole of *any* gas occupies a volume of 358 ft³. At 70°F and 14.7 psia, a mole occupies $358 \times [(460 + 70)/(460 + 32)] = 385.7$ ft³. Therefore, 5 moles of oxygen occupy

$$V_{O_2} = 5 \times 385.7 = 1928 \text{ ft}^3$$

and 10 moles of hydrogen occupy

$$V_{H_2} = 10 \times 385.7 = 3857 \text{ ft}^3$$

Both values are in good agreement with the previous calculations.

ILLUSTRATIVE PROBLEM 7.5

A mixture contains 10 lb of carbon dioxide and 5 lb of nitrogen. If the mixture is at 70°F and 100 psia, determine the mixture volume, the partial volume of each constituent, the partial pressure of each constituent, the mole fraction of each constituent, and the molecular weight of the mixutre (MW of CO_2 = 44.0; MW of N_2 = 28.02).

SOLUTION

Referring to Figure 7.3, we have for the CO_2,

$$n_{CO_2} = \frac{10}{44} = 0.227 \text{ mole}$$

and for N_2,

$$n_{N_2} = \frac{5}{28.02} = 0.178 \text{ mole}$$

The total moles in the mixture is therefore $0.227 + 0.178 = 0.405$ mole. Therefore,

Mixture		CO$_2$		N$_2$	
Mixture		M	= 10 lb$_m$	M	= 5 lb$_m$
Pressure	= p_m	Volume	= V_{CO_2}	Volume	= V_{N_2}
Temperature	= T	Temperature	= T	Temperature	= T
Volume	= V_m	Pressure	= p_m	Pressure	= p_m

FIGURE 7.3 Illustrative Problem 7.5.

$$x_{CO_2} = \frac{0.227}{0.405} = 0.560$$

and

$$x_{N_2} = \frac{0.178}{0.405} = 0.440$$

The molecular weight of the mixture is found from Equation (7.5) as

$$MW_m = (0.560)(44) + (0.440)(28.02) = 37.0$$

Because the mixture is 15 lb$_m$ (10 CO_2 + 5 N_2), the volume of the mixture is found from

$$p_m V_m = m_m R_m T_m \qquad V_m = \frac{\dfrac{15 \times 1545}{37.0} \times (460 + 70)}{100 \times 144}$$

$$= 23.05 \text{ ft}^3$$

The partial volume of carbon dioxide is the total volume multiplied by the mole fraction. Thus,

$$V_{CO_2} = (23.05)(0.560) = 12.91 \text{ ft}^3$$

and

$$V_{N_2} = (23.05)(0.440) = 10.14 \text{ ft}^3$$

The partial pressure of each constituent is proportional to its mole fraction. For these conditions,

$$p_{CO_2} = (100)(0.560) = 56.0 \text{ psia}$$

and

$$p_{N_2} = (100)(0.440) = 44.0 \text{ psia}$$

7.4   MIXTURE COMPOSITION

When a gas mixture is analyzed, it is possible to state its composition correctly in several different ways. For instance, the moles of each constituent gas could be given; alternatively, the weight or partial volumes of each gas could be stated. The manner of expressing the analysis is completely arbitrary. Therefore, it is necessary to be able to convert from one form of analysis to another.

If the moles (or mole fractions) are given, the volume fractions are immediately known. Conversely, if the volume fractions are known, the mole fractions are known. Because the mass of a gas is the product of the number of moles multiplied by the molecular weight, the conversion to a weight (gravimetric) analysis is readily achieved. The molecular weight of the mixture is also readily determined from Equation (7.5).

ILLUSTRATIVE PROBLEM 7.6

A gas mixture consists of the following volume percentages: 40% CO_2, 10% N_2, 40% O_2, and 10% H_2. Determine the mass fraction of each gas in the mixture, the molecular weight of the mixture, and the gas constant of the mixture.

SOLUTION

In solving this type of problem, we shall first assume a convenient volume for the gas mixture. Because we have percentages by volume for each constituent, we will assume that we have 100 volumes of gas mixture and set up Table 7.2. In the first column, we tabulate the gas, and in the second column, we tabulate the given volume fractions. Because the mole fraction equals the volume fraction, the values in column (3) are the same as those in column (2). The molecular weight is obtained from Table 7.1. Because the molecular weight of the mixture is the sum of the individual mole fractions multiplied by the respective molecular weights, the next column tabulates the product of mole fraction multiplied by molecular weight [column (3) multiplied by column (4)]. The sum of these entries is the molecular weight of the mixture, which for this case is 33.4. If we now consider that we have 33.4 lb_m of mixture and that the mass of each gas in the mixture is given by its product (x) (MW), the mass fraction of each constituent is the entry in column (5) divided by 33.4. The value of R for the mixture is simply $1545/33.4 = 46.3$. The use of a tabular solution helps to simplify the calculations and is also a timesaver when this type of repetitive calculation is involved.

TABLE 7.2
Illustrative Problem 7.6
Basis: 100 Volumes of Gas Mixture

Gas	(2) Volume fraction	(3) Mole fraction, x	(4) Molecular weight, MW	(5) (3) × (4) (x) (MW)	Mass fraction
CO_2	0.40	0.40	44.0	17.6	$17.6/33.4 = 0.527$
N_2	0.10	0.10	28.02	2.8	$2.8/33.4 = 0.084$
H_2	0.10	0.10	2.016	0.2	$0.2/33.4 = 0.006$
O_2	0.40	0.40	32.0	12.8	$12.8/33.4 = 0.383$
	1.00	1.00		$33.4 = MW_m$	$= 1.000$

The procedure when the weight (gravimetric) analysis is known and conversion to a volumetric basis is desired is similar to that given in the table. However, rather than assuming 100 volumes of gas, it is more convenient to use 100 lb_m of gas as the basis of the computations.

ILLUSTRATIVE PROBLEM 7.7

The result of Illustrative Problem 7.6 shows that the mass fractions of gases are $CO_2 = 0.527$, $N_2 = 0.084$, $H_2 = 0.006$, and $O_2 = 0.383$. Using these gases and mass fractions, determine the fraction by volume of each of the constituent gases of this mixture.

SOLUTION

We will again use a tabular solution, as we did for Illustrative Problem 7.6, but for this calculation, we will take as a basis 100 lb_m of mixture. The first three columns of Table 7.3 are similar to the items in Table 7.2 and are self-explanatory.

Dividing column (2) by column (3) gives us mass/molecular weight or moles of each constituent. The total number of moles in the mixture is the sum of column (4), and the molecular weight of the mixture is the mass of the mixture (100 lb_m) divided by the number of moles, 3, or 33.3. Column (5), the mole fraction, is the moles listed in column (4) divided by the total moles in the mixture, which for this problem is 3.0. Because mole fraction is also volume fraction, the last column, percent volume, is just the fifth column multiplied by 100. It should also be noted that the volume percent is also the pressure percent of the various constituents of the mixture. Because this problem is the inverse of Illustrative Problem 7.6, the results obtained are found to be in good agreement for both problems.

TABLE 7.3
Illustrative Problem 7.7
Basis: 100 Pounds of Mixture

Gas	(2) Mass lb_m	(3) Molecular weight, MW	(4) $\dfrac{\text{Mass}}{\text{Molecular weight}}$ = moles	(5) Mole fraction (moles/mole)	(6) Percent volume
CO_2	52.7	44	1.2	1.2/3 = 0.4	40
N_2	8.4	28.02	0.3	0.3/3 = 0.1	10
H_2	0.6	2.016	0.3	0.3/3 = 0.1	10
O_2	38.3	32	1.2	1.2/3 = 0.4	40
	100.0		Total moles = 3.0	1.0	100

$$MW_m = 100/3 = 33.3$$

ILLUSTRATIVE PROBLEM 7.8

Air consists of 23.18% O_2, 75.47% N_2, 1.30% A, and 0.05% CO_2 by weight. Determine the volumetric analysis.

SOLUTION

This problem is a weight-to-volume conversion similar to Illustrative Problem 7.7. We set up Table 7.4 in the same manner as Table 7.3 and use the same procedure to solve the problem.

TABLE 7.4
Illustrative Problem 7.8
Basis: 100 Pounds of Mixture

Gas	Mass (lb_m)	Molecular weight, MW	$\dfrac{\text{Mass}}{\text{Molecular weight}} = \text{moles}$	Moles fraction	Percent volume
O_2	23.18	32.00	0.724	0.210	21
N_2	75.47	28.02	2.693	0.780	78
A	1.30	39.90	0.033	0.010	1
CO_2	0.05	44.00	—	—	—
	100.00		3.45	1.00	100

$$MW_m = \frac{100}{3.45} = 28.99$$

7.5 THERMODYNAMIC PROPERTIES OF A GAS MIXTURE

When several gases are mixed, it is assumed that the gas molecules will disperse until the mixture is completely homogeneous. It is also assumed that the components of the mixture will reach the same average temperature as the mixture. In Chapter 6, it was noted that the internal energy and enthalpy of the ideal gas are solely functions of the gas temperature. With this in mind, it would seem reasonable to state that these thermodynamic properties of a gas mixture (internal energy and enthalpy) are simply the sum of the properties of the gases of which the mixture is composed. This statement, combined with Dalton's law, can be considered as the Gibbs–Dalton law of mixtures. As before, this law is valid for the ideal gas, and deviations will be found when the components of a mixture differ considerably from the ideal gas law.

As a consequence of the Gibbs–Dalton law, we can write the following property relations for a mixture m composed of three gases, a, b, and c, on the basis that the properties of the mixture equal the sum of the properties of each component of the mixture when each of the components is assumed to occupy the total volume at the temperature of the mixture and the gases behave as ideal gases. Thus, for the internal energy and enthalpy of a mixture, we write

$$U_m = m_a u_a + m_b u_b + m_c u_c \tag{7.10}$$

and
$$H_m = m_a h_a + m_b h_b + m_c h_c \tag{7.11}$$

where U_m and H_m are the total internal energy and total enthalpy of the mixture, respectively. Because the total value of these properties is the product of the mass and the specific value of each property,

$$U_m = m_m u_m = m_a u_a + m_b u_b + m_c u_c \tag{7.12}$$

and
$$H_m = m_m h_m = m_a h_a + m_b h_b + m_c h_c \tag{7.13}$$

From Equations (7.12) and (7.13), the specific internal energy and the specific enthalpy of the mixture are

$$u_m = \frac{m_a}{m_m} u_a + \frac{m_b}{m_m} u_b + \frac{m_c}{m_m} u_c \qquad (7.14)$$

and

$$h_m = \frac{m_a}{m_m} h_a + \frac{m_b}{m_m} h_b + \frac{m_c}{m_m} h_c \qquad (7.15)$$

where $m_m = m_a + m_b + m_c$.

From Equations (7.14) and (7.15), we can also obtain the specific heat at constant volume and the specific heat at constant pressure for the mixture, because both internal energy and enthalpy for the ideal gas are functions only of the temperature of the mixture. Therefore,

$$c_{vm} = \frac{m_a}{m_m} c_{va} + \frac{m_b}{m_m} c_{vb} + \frac{m_c}{m_m} c_{vc} \qquad (7.16)$$

and

$$c_{pm} = \frac{m_a}{m_m} c_{pa} + \frac{m_b}{m_m} c_{pb} + \frac{m_c}{m_m} c_{pc} \qquad (7.17)$$

It should be noted that Equations (7.14) through (7.17) state that the mixture properties in question are each equal to the sum of the mass average of the individual properties of the components of the mixture.

ILLUSTRATIVE PROBLEM 7.9

In a steady-flow process, 160 lb_m/hr of oxygen and 196 lb_m/hr of nitrogen are mixed adiabatically. Both gases are at atmospheric pressure, but the oxygen is at 500°F and the nitrogen is at 200°F. Assume c_p of oxygen to be constant and equal to 0.23 Btu/lb_m·°R and c_p of nitrogen also to be constant and equal to 0.25 Btu/lb_m·°R; determine the final temperature of the mixture (see Figure 7.4).

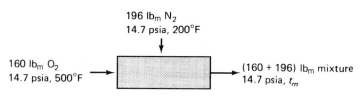

FIGURE 7.4   Illustrative Problem 7.9.

SOLUTION

The energy equation for the steady-flow, adiabatic mixing process gives us the requirement that the enthalpy of the mixture must equal the enthalpies of the components, because $\Delta h = q = 0$. An alternative statement of this requirement is that the gain in enthalpy of the nitrogen must equal the decrease in enthalpy of the oxygen. Using the latter statement, that the change in enthalpy of the oxygen must equal the change in enthalpy of the nitrogen, yields

$$(160)\,(0.23)\,(500 - t_m) = (196)\,(0.25)\,(t_m - 200)$$

where $mc_p\Delta t$ has been used for Δh. Solving yields

$$t_m = 328.7°F$$

Using the requirement that the enthalpy of the mixture must equal the sum of the enthalpies of the components yields an alternative solution to this problem. Let us assume that at 0°F, the enthalpy of each gas and of the mixture is zero. The enthalpy of the entering oxygen is $(160)\,(0.23)\,(500 - 0)$, and the enthalpy of the entering nitrogen is $(196)\,(0.25)\,(200 - 0)$. The enthalpy of the mixture is $(160 + 196)c_{pm}(t_m - 0)$. Therefore,

$$(160)\,(0.23)\,(500 - 0) + (196)\,(0.25)\,(200 - 0) = (160 + 196)c_{pm}(t_m - 0)$$

The required c_{pm} is obtained from Equation (7.17):

$$c_{pm} = \left(\frac{160}{160 + 196}\right)(0.23) + \left(\frac{196}{160 + 196}\right)(0.25)$$
$$= 2.41 \text{ Btu/lb}_m\cdot°R$$

Therefore,

$$t_m = \frac{(160)\,(0.23)\,(500 - 0) + (196)\,(0.25)\,(200 - 0)}{(160 + 196)\,(0.241)}$$
$$= 328.7°F$$

The use of 0°F as a base was arbitrary but convenient. Any base would yield the same results. Also, either method used to solve this problem yields the same mixture temperature. The specific heat data of Table A-13 are not applicable to this problem, because Table A-13 gives data at 77°F.

ILLUSTRATIVE PROBLEM 7.10

Assume that the process described in Illustrative Problem 7.9 is carried out as a nonflow mixing process. Also, assume that c_v of the oxygen is 0.164 Btu/lb$_m$·°R and the c_v of the nitrogen is 0.178 Btu/lb$_m$·°R. Determine the final temperature of the mixture.

SOLUTION

Because this is a nonflow process, the energy equation for this process requires the internal energy of the mixture to equal the sum of the internal energies of its components.

Alternatively, the decrease in internal energy of the oxygen must equal the increase in internal energy of the nitrogen. Using the latter statement gives us

$$(160)(0.164)(500 - t_m) = (196)(0.178)(t_m - 200)$$
$$t_m = 328.8°F$$

We obtained the same temperature for Illustrative Problems 7.9 and 7.10 because the ratio of c_p/c_v for both gases was deliberately taken to be the same. In effect, this constant was canceled from both sides of the equation.

In Chapter 6, we noted that the entropy of an ideal gas is in general a function of both pressure and temperature. For the mixture of ideal gases where each component of the mixture is taken to exist at its own partial pressure, we can write the following equation:

$$m_m s_m = m_a s_a + m_b s_b + m_c s_c \tag{7.18}$$

where the individual entropies s_a, s_b, and s_c are evaluated per pound at the mixture temperature and the component partial pressure.

ILLUSTRATIVE PROBLEM 7.11

Two ideal gases having constant specific heats each at the same pressure and temperature are adiabatically mixed by removing the partition shown in Figure 7.5. Assume gas a to have a molecular weight of MW_a and a mass of m_a lb. Assume gas b to have a molecular weight of MW_b and a mass of m_b lb. Determine the entropy change for this mixing process.

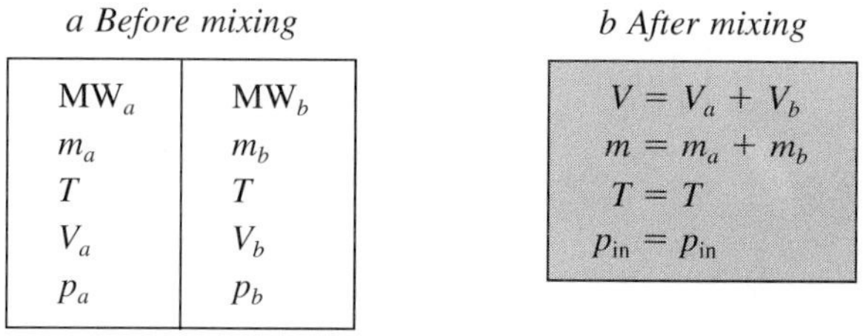

a Before mixing		b After mixing
MW_a $\quad$ MW_b		$V = V_a + V_b$
m_a $\quad$ m_b		$m = m_a + m_b$
T $\quad\quad$ T		$T = T$
V_a $\quad\quad$ V_b		$p_{in} = p_{in}$
p_a $\quad\quad$ p_b		

FIGURE 7.5   Illustrative Problem 7.11.

SOLUTION

The general equation applicable to the change in entropy of an ideal gas due to both a pressure and temperature change is

$$\Delta s = c_p \ln \frac{T_2}{T_1} - \frac{R}{J} \ln \frac{p_2}{p_1} \qquad \text{Btu/lb}_m \cdot °R \tag{a}$$

Because the process is carried out at constant temperature, the first term of Equation (a) applied to this process is zero. Also, the final partial pressure of gas a is p_a, and the final partial pressure of gas b is p_b. Denoting the total pressure of the mixture as p_m, we obtain

$$\Delta S_a = -m_a \frac{R_a}{J} \ln \frac{p_a}{p_m}$$

and

$$\Delta S_b = -m_b \frac{R_b}{J} \ln \frac{p_b}{p_m}$$

The total change in entropy for the mixture for $m_a + m_b$ lb$_m$ of mixture is

$$\Delta S = -m_a \frac{R_a}{J} \ln \frac{p_a}{p_m} - m_b \frac{R_b}{J} \ln \frac{p_b}{p_m}$$

However, $R = 1545/MW$. Therefore,

$$\Delta S = -\frac{m_a(1545)}{(MW_a)J} \ln \frac{p_a}{p_m} - \frac{m_b(1545)}{(MW_b)J} \ln \frac{p_b}{p_m}$$

We now note that $m_a/MW_a = n_a$, the number of moles of a; $m_b/MW_b = n_b$, the number of moles of b; $p_a/p_m = x_a$, the mole fraction of a; and $p_b/p_m = x_b$, the mole fraction of b:

$$\Delta S = -\frac{n_a(1545)}{J} \ln x_a - \frac{n_b(1545)}{J} \ln x_b \qquad \textbf{(b)}$$

Equation (b) is the desired result for the problem. However, it has an interesting conclusion in that the entropy increase due to the mixing process depends only on the moles of each gas present, not on the gas itself. Thus, the mixing of 1 mole of oxygen with 1 mole of hydrogen will yield the same entropy increase as the mixing of 1 mole of argon with 1 mole of Freon. One point that should be noted at this time is that the adiabatic mixing of *identical* gases initially at the same pressure and temperature does not result in an increase in entropy; Equation (b) is not applicable to this process. This has been called *Gibbs paradox,* and further discussion on it can be found in the references in Appendix 1.

ILLUSTRATIVE PROBLEM 7.12

Determine the change in entropy for the process described in Illustrative Problem 7.9.

SOLUTION

The change in entropy of the mixture is the sum of the changes in entropy of each component. For the oxygen, the temperature starts at 500°F (960°R) and decreases to 328.7°F. For the nitrogen, the temperature starts at 200°F (660°R) and increases to 328.7°F. Applying Equation (a) of Illustrative Problem 7.11 and noting that the process is carried out at constant pressure, which makes the pressure term zero (ln 1 = 0), *for the oxygen,*

$$\Delta s = c_p \ln \frac{T_2}{T_1} = 0.23 \ln \frac{328.7 + 460}{500 + 460}$$

$$= 0.23 \, (\ln 788.7 - \ln 960) = -0.045 \text{ Btu/lb}_m \cdot {}^\circ R$$

The total change in entropy of the oxygen is

$$\Delta S = (160)(-0.045) = -7.20 \text{ Btu/}{}^\circ R$$

For the nitrogen,

$$\Delta s = c_p \ln \frac{T_2}{T_1} = 0.25 \ln \frac{328.7 + 460}{200 + 460}$$

$$= 0.045 \text{ Btu/lb}_m \cdot {}^\circ R$$

The total change in entropy of the nitrogen is

$$\Delta S = (196)(0.045) = 8.82 \text{ Btu/}{}^\circ R$$

For the mixture, the total change in entropy is the sum of the changes of the constituent entropies. Therefore,

$$\Delta S = -7.20 + 8.82 = 1.62 \text{ Btu/}{}^\circ R$$

Per pound of mixture,

$$\Delta s_m = \frac{1.62}{196 + 160}$$

$$= 0.00455 \text{ Btu/lb}_m \cdot {}^\circ R \quad \text{(increase per pound mass of mixture)}$$

As an alternative solution, assume an arbitrary datum of 0°F (460°R). The initial entropy of the oxygen above this base is

$$\Delta s = c_p \ln \frac{T_2}{T_1} = 0.23 \ln \frac{500 + 460}{460} = 0.169 \text{ Btu/lb}_m \cdot {}^\circ R$$

The final entropy of the oxygen above this base is

$$\Delta s = c_p \ln \frac{T_2}{T_1} = 0.23 \ln \frac{328.7 + 460}{460} = 0.1240 \text{ Btu/lb}_m \cdot {}^\circ R$$

The entropy change of the oxygen therefore is

$$\Delta s = 0.124 - 0.169 = -0.045 \text{ Btu/lb}_m \cdot {}^\circ R$$

For the nitrogen,

$$\Delta s = c_p \ln \frac{T_2}{T_1} = 0.25 \ln \frac{660}{460}$$

$$= 0.090 \text{ Btu/lb}_m \cdot {}^\circ R \quad \text{(above a base of 0°F) initially}$$

$$\Delta s = c_p \ln \frac{T_2}{T_1} = 0.25 \ln \frac{460 + 328.7}{460}$$

$$= 0.135 \text{ Btu/lb}_m \cdot °R \quad \text{(above a base of } 0°F\text{)}$$

The entropy change of the nitrogen therefore is

$$\Delta s = 0.135 - 0.090 = 0.045 \text{ Btu/lb}_m \cdot °R$$

The remainder of the problem is as before. The advantage of this alternative method is that negative logarithms are avoided by choosing a reference temperature lower than any other temperature in the system.

7.6 AIR–WATER VAPOR MIXTURES*

A gas may be considered to be a vapor that is superheated. If the degree of superheat is high and the vapor is not close to critical conditions, it will usually behave as if it were an ideal gas. Although there are many exceptions to this generalization, it is found to be useful when treating air–water vapor mixtures. Air is a mixture of many gases, and dry air is defined as a mixture having the following composition by volume: oxygen, 20.95%; nitrogen, 78.09%; argon, 0.93%; carbon dioxide, 0.03%. In the treatment of air–water vapor mixtures, it is common to treat air as a single gas with a molecular weight of 28.966.

Some of the concepts involved in dealing with air–water vapor mixtures can best be illustrated by referring to Figure 7.6, a T–s diagram for water vapor. In general, atmospheric air contains superheated water vapor, which may typically be at state B in Figure 7.6.

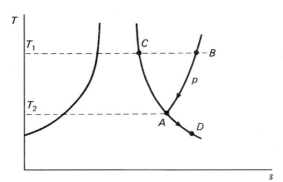

FIGURE 7.6 The water vapor portion of an air–water vapor mixture.

The *dry-bulb temperature* (T_1) is the temperature of the mixture of air and water vapor at rest, and it is measured by an instrument (thermometer) that is not affected by the amount of moisture in the air or by thermal radiation. If the mixture of air and water vapor that is originally at B is cooled at constant pressure, it will follow the path from B to A along the constant-pressure line. In practice, the *dew-point temperature* is the temperature (T_2) at which condensation will just begin when the moist air mixture under consideration is cooled at constant pressure. If the mixture is cooled further, the water vapor will follow along the saturation line to D, and some moisture will condense. If water is added to the mixture at constant temperature, it will follow the path from B to

---

* The material in this section has been based largely on the *ASHRAE Guide and Data Book* published by the American Society of Heating, Refrigerating, and Air-Conditioning Engineers, with permission.

C. Moist air is said to be saturated (*C*) when its condition is such that it can coexist in neutral equilibrium with an associated condensed moisture phase presenting a flat surface to it. The word *neutral* is used to exclude unstable states from the definition, and the flat surface is used to exclude surface-tension effects.

In addition to the quantities already defined, certain other terms describing air–water vapor mixtures require definition.

Relative Humidity (ϕ). The relative humidity of any mixture of air and water vapor is defined as the ratio of the mole fraction of water vapor in the mixture to the mole fraction of water vapor in saturated air at the same dry-bulb temperature and barometric pressure. It is also equal to the ratio of the partial pressure of the water vapor to the saturation pressure of water vapor at the temperature of the mixture.

Humidity Ratio (W). The humidity ratio of any mixture of air and water vapor is defined as the ratio of the mass of water vapor in the mixture to the mass of dry air in the mixture. The units of humidity ratio are pounds of water vapor per pound of dry air.

Specific Humidity. The term *specific humidity* has been and still is sometimes used for humidity ratio. They are synonymous.

Thermodynamic Wet-Bulb Temperature. The thermodynamic wet-bulb temperature is the temperature at which liquid water, by evaporating into air, can bring the air to saturation adiabatically at the same temperature. To illustrate the definition of the thermodynamic wet-bulb temperature, consider the idealized system shown in Figure 7.7. It is a steady-flow system in which no work is done between sections 1 and 2, and no energy as heat enters or leaves. Water enters the system at the thermodynamic wet-bulb temperature, and air leaves the system saturated at the thermodynamic wet-bulb temperature. The energy equation for this process is, per pound of dry air,

$$h_1 + h_w^*(W_2^* - W_1) = h_2^* \tag{7.19}$$

The * notation is used to denote properties at the thermodynamic wet-bulb temperature, and the subscript w is used to denote water. The temperature corresponding to h_2^* for given values of h_1 and W_1 is the defined thermodynamic wet-bulb temperature. The process that we have just considered is also known as the *adiabatic saturation process,* and the thermodynamic wet-bulb

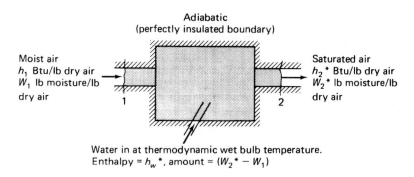

FIGURE 7.7 Adiabatic saturation process.

temperature is also known as the adiabatic saturation temperature. The usefulness of the foregoing discussion lies in the fact that the temperature of the saturated air–water vapor mixture leaving the duct is a function of the temperature, pressure, relative humidity of the entering mixture, and the exit pressure. Conversely, knowing the entering and exit pressures and temperatures, we can determine the relative humidity and humidity ratio of the entering air–water vapor mixture.

ILLUSTRATIVE PROBLEM 7.13

If the relative humidity of moist air is 50% when the mixture temperature is 80°F, determine the dew point of the air. Assume that the barometric pressure is 14.7 psia.

SOLUTION

Referring to Figure 7.6, it will be seen that the cooling of an air–water vapor mixture from B to A proceeds at constant pressure until the saturation curve is reached. The dew-point temperature of the air–water vapor mixture is equal to the saturation temperature, corresponding to the partial pressure of the water vapor. The relative humidity has been defined to be the ratio of the mole fraction of water vapor in the mixture to the mole fraction of water vapor in saturated air at the same dry-bulb temperature and barometric pressure. It will be recalled from Section 7.2 that the partial pressure of a component equals the mole fraction of the component multiplied by the total pressure of the mixture. Because the relative humidity is a mole fraction ratio, it is also the ratio of the partial pressure of the vapor in the air to the saturation pressure corresponding to the air temperature. At 80°F, the *Steam Tables* give us a saturation pressure of 0.5073 psia, and because the relative humidity is 50%, the vapor pressure of the water is $0.5 \times (0.5073) = 0.2537$ psia. Using the full *Steam Tables* or interpolating in the table in Appendix 3 yields a saturation temperature corresponding to 0.2537 psia as close to 60°F. This temperature is the desired dew-point temperature.

ILLUSTRATIVE PROBLEM 7.14

Air at 90°F contains 0.005 lb_m of water vapor per pound of dry air. Determine the partial pressure of each component, the relative humidity, and the dew-point temperature of the mixture if the mixture is at 14.7 psia.

SOLUTION

To solve this problem, it is necessary to determine the properties of the saturated mixture at 90°F. If the air is saturated at 90°F, the partial pressure of the water vapor is found directly from the *Steam Tables* as 0.6988 psia, and the specific volume of the water vapor is 467.7 ft^3/lb_m of vapor. At this point, let us assume that we have a container having a volume of 467.7 ft^3 and that it contains a saturated air–water vapor mixture at 90°F. Because the mixture is assumed to be saturated, there must be 1 lb_m of water vapor in the container, and the partial pressure of the dry air must be $14.7 - 0.6988 = 14.0$ psia. By applying the ideal gas equation to the air, we can determine the mass of air in the container. Thus,

$$v_{\text{dry air}} = \frac{RT}{p_{\text{dry air}}} = \frac{1545}{28.966} \times \frac{460 + 90}{14.0 \times 144}$$

$$= 14.55 \text{ ft}^3/\text{lb}_m \text{ dry air}$$

The mass of dry air in the 467.7-ft^3 container is $467.7/14.55 = 32.1$ lb$_m$. The saturated mixture therefore contains 1 lb$_m$ of water vapor per 32.1 lb$_m$ of dry air or 0.312 lb$_m$ water vapor/lb$_m$ dry air.

To obtain the relative humidity (ϕ), it is necessary to determine the mole fraction of water vapor for both the saturated mixture and the mixture in question. The saturated mixture contains 1 lb$_m$ of water vapor or $1/18.016$ moles $= 0.055$ mole of water vapor and $32.1/28.966 = 1.109$ moles of dry air. Therefore, for the saturated mixture, the ratio of moles of water vapor to moles of mixture is $0.0555/(1.109 + 0.055) = 0.0477$. For the actual mixture, the moles of water vapor per pound of dry air is $0.005/18.016 = 0.000278$, and 1 lb$_m$ of dry air is $1/28.966 = 0.0345$ mole. The moles of water vapor per mole of mixture at the conditions of the mixture is $0.000278/(0.0345 + 0.000278) = 0.00799$. From the definition of relative humidity, the relative humidity of this mixture is $0.00799/0.0477 = 0.168$ or 16.8%.

Because the mole ratio is also the ratio of the partial pressures for the ideal gas, ϕ can be expressed as the ratio of the partial pressure of the water vapor in the mixture to the partial pressure of the water vapor at saturation. Therefore, the partial pressure of the water vapor is $0.168 \times 0.6988 = 0.1173$ psia, and the partial pressure of the dry air in the mixture is $14.7 - 0.1173 = 14.58$ psia.

The dew-point temperature is the saturation temperature corresponding to the partial pressure of the water vapor in the mixture. The saturation temperature corresponding to 0.1173 psia is found from the *Steam Tables* to be close to 39°F.

The adiabatic saturation of air is a process that is useful to analytically study the thermodynamics of air–water vapor mixtures. If this process had to be used each time one wished to obtain the relative humidity of an air–water vapor mixture, it would be impractical. Instead, a relatively simple method is used to establish the partial pressure of the water vapor based upon a temperature known as the *wet-bulb temperature.* The concept of the wet-bulb temperature can be developed by reference to Figure 7.8. Two thermometers are placed in a moving airstream. The bulb of the wet-bulb thermometer is covered with a gauze that is immersed in clean water and acts to wick the water to the bulb. Air is continuously blown over the thermometer by a fan. At the wet-bulb thermometer, three actions occur. The first is that water will evaporate from the wick due to heat transfer to the air at the dry-bulb temperature; the second action is water evaporation from the wick due to radiant heat transfer from the surroundings; and the last action is a diffusion of water vapor from the wick to the surrounding air–water vapor mixture. Due to these actions, the temperature of the wet-bulb thermometer will be lower than the dry-bulb temperature. In practice, one method of obtaining the wet- and dry-bulb temperatures is by use of the *sling psychrometer* shown in Figure 7.9. As shown, two thermometers are mounted on a sling, with the wet-bulb thermometer having a gauze wick around its bulb which is saturated with water. The sling is whirled around,

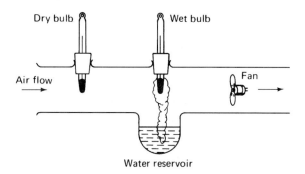

FIGURE 7.8 Steady-flow apparatus for measuring wet- and dry-bulb temperature.
[From G. J. Van Wylen, Thermodynamics *(New York: John Wiley & Sons, Inc., 1959),
p. 223, with permission.]*

which causes evaporation of water from the wick and lowers the temperature of the wet-bulb thermometer. When equilibrium is reached, both thermometers are read to obtain the wet- and dry-
bulb temperatures. The sling psychrometer is an inexpensive, portable device that is easy to use.
An alternative device, known as an aspiration psychrometer, is shown in Figure 7.10. In this device, a small fan is used to draw air over the wet- and dry-bulb thermometers as shown. While less
portable than the sling psychrometer, it gives continuous readings.

In principle, there is a difference between the wet-bulb temperature and the temperature of
adiabatic saturation (the thermodynamic wet-bulb temperature). The wet-bulb temperature is a function of both heat and mass diffusion rates, while the adiabatic saturation temperature is a function of
a thermodynamic equilibrium process. However, in practice, it has been found that for air–
water vapor mixtures at atmospheric pressures and temperatures, the wet-bulb and adiabatic saturation
temperatures are essentially the same. Thus, the observed readings of the wet-bulb thermometer can
be taken to be the same as the thermodynamic wet-bulb temperature defined by the adiabatic saturation process if the air velocity past the wick is approximately 1000 ft/min and if the wick is exposed
to radiation exchange with the surroundings at temperatures differing from the dry-bulb temperature
by no more than 20°F. It must be noted as a caution that the wet-bulb temperature and adiabatic saturation temperatures deviate considerably from each other for other than dry air–water vapor gas–vapor
mixtures and for air–water vapor mixtures at conditions other than usual atmospheric conditions.

Problems concerning air–water vapor mixtures can be considerably simplified by using the
ideal gas equations for each of the components. By denoting the subscripts a to be dry air, v to be
water vapor, vs to be saturated water vapor, s to be saturation, and m to mean mixture, the following relations can be derived:

Gas Constant of the Mixture

$$R_m = \left(\frac{m_a}{m_a + m_v}\right)R_a + \left(\frac{m_v}{m_a + m_v}\right)R_v \qquad (7.20)$$

By dividing the numerator and denominator of each of the terms in parentheses by m_a,

$$R_m = \left(\frac{1}{1 + \dfrac{m_v}{m_a}}\right)R_a + \left(\frac{m_v/m_a}{1 + m_v/m_a}\right)R_v$$

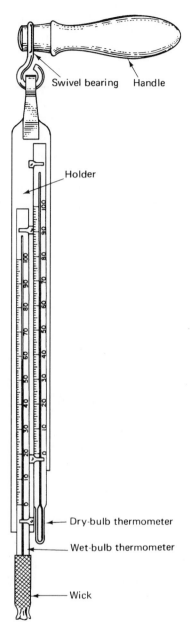

FIGURE 7.9 Sling psychrometer.

[From C. C. Dillio and E. P. Nye, Thermal Engineering *(Scranton, Pa.: International Textbook Co., 1959), p. 465, with permission.]*

But m_v/m_a is the humidity ratio W.

$$R_m = \left(\frac{1}{1+W}\right)R_a + \left(\frac{W}{1+W}\right)R_v \tag{7.21}$$

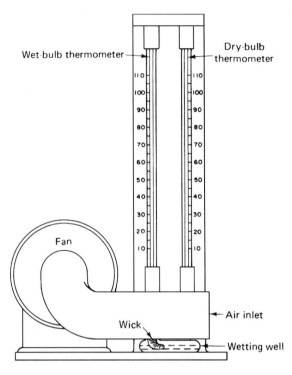

FIGURE 7.10 Aspiration psychrometer.
[From C. C. Dillio and E. P. Nye, Thermal Engineering *(Scranton, Pa.: International Textbook Co., 1959), p. 466, with permission.]*

Because W is a small number compared to unity, it is possible at a first approximation to use the gas constant of the mixture as the gas constant of dry air.

Relative Humidity. The relative humidity ϕ is

$$\phi = \frac{p_v}{p_{vs}} = \frac{v_{vs}}{v_v} = \frac{\rho_v}{\rho_{vs}} \qquad (7.22)$$

When the mixture is at the dew-point temperature, ϕ is 1.

Humidity Ratio.

$$W = \frac{m_v}{m_a} = \left(\frac{18.016}{28.966}\right)\frac{p_a}{p_a} = 0.622 \frac{p_v}{p_m - p_v} \qquad (7.23)$$

In Equation (7.23), 18.016 is the molecular weight of water vapor and 28.966 is the molecular weight of dry air.

Relative Humidity and Humidity Ratio. From Equations (7.22) and (7.23),

$$\boxed{\phi p_{vs} = p_v} \tag{7.24}$$

Therefore,

$$W = 0.622 \frac{p_v}{p_a} = 0.622\phi \frac{p_{vs}}{p_a} = \phi \frac{p_{vs}}{p_a}\left(\frac{R_a}{R_v}\right) = \phi \frac{\rho_{vs}}{\rho_a} = \phi \frac{v_a}{v_{vs}}$$

ILLUSTRATIVE PROBLEM 7.15

Solve Illustrative Problem 7.14 using Equations (7.20) through (7.24).

SOLUTION

W is given as 0.005. Therefore,

$$W = 0.005 = 0.622 \frac{p_v}{p_m - p_v}$$

with $p_m = 14.7$ psia. Thus, $p_v = (0.005 \times 14.7)/(0.622 + 0.005) = 0.117$ psia, and $p_a = 14.7 - 0.117 = 14.58$ psia. Because $\phi = p_v/p_{vs}$, it is necessary to obtain p_{vs} from the *Steam Tables* at 90°F. This is 0.6988 psia. Therefore, $\phi = 0.117/0.6988 = 0.167 = 16.7\%$. The dew-point temperature is the saturation temperature corresponding to 0.117 psia, which is found from the *Steam Tables* to be 39°F. The results of this problem and Illustrative Problem 7.14 are in good agreement.

ILLUSTRATIVE PROBLEM 7.16

Air at 90°F and 70% relative humidity is conditioned so that its final state is 80°F and 40% relative humidity. How much water was removed from the air? Assume that the barometer is at 14.7 psia.

SOLUTION

The amount of water vapor removed (per pound of dry air) is the difference between the humidity ratio (specific humidity) at inlet and outlet of the conditioning unit. We shall therefore evaluate W for both specified conditions. Because $\phi = p_v/p_{vs}$,

At 90°F: $p_{vs} = 0.6988$ psia

At 80°F: $p_{vs} = 0.5073$ psia

At 90°F: $p_v = 0.7 \times 0.6988 = 0.4892$ psia

At 80°F: $p_v = 0.4 \times 0.5073 = 0.2029$ psia

But $p_a = p_m - p_v$. Thus,

$$\text{At } 90°F: p_a = 14.7 - 0.4892 = 14.2108 \text{ psia}$$
$$\text{At } 80°F: p_a = 14.7 - 0.2029 = 14.4971 \text{ psia}$$

Because $W = 0.622 p_v/p_a$,

$$\text{At } 90°F: W = 0.622 \left(\frac{0.4892}{14.2108} \right)$$
$$= 0.0214 \text{ lb water/lb dry air}$$
$$\text{At } 80°F: W = 0.622 \left(\frac{0.2029}{14.4971} \right)$$
$$= 0.0087 \text{ lb water/lb dry air}$$

The amount of water removed per pound of dry air is

$$0.0214 - 0.0087 = 0.0127 \text{ lb water removed/lb dry air}$$

A unit of weight commonly used in air-conditioning work is the *grain*. By definition, there are 7000 grains per pound. Using this definition, the water removal per pound of dry air in Illustrative Problem 7.16 is $7000 \times 0.0127 = 88.9$ grains per pound of dry air.

7.7 THERMODYNAMIC PROPERTIES OF AIR–WATER VAPOR MIXTURES

The thermodynamic properties of air–water vapor mixtures can be determined from tabulated data available in the latest edition of the *ASHRAE Guide and Data Book,* published by the American Society of Heating, Refrigerating, and Air-Conditioning Engineers. These data are tabulated from $-160°F$ to $200°F$ for moist air at 29.921 in. Hg and are given for either perfectly dry or completely saturated air. By simple interpolation, the specific volume, enthalpy, or entropy of mixtures can be obtained. The interpolation formulas are given in the *Guide*.

For those designs or analyses that require accuracies that cannot be achieved by use of the ideal gas relations, these tables are invaluable. However, for those cases that can be dealt with by using ideal gas relations, it is possible to express the thermodynamic properties of air–water vapor mixtures in a relatively simple manner. It is convenient to express these properties per unit mass of dry air in the mixture, for example, Btu per pound of dry air.

Specific Heat of the Mixture. The specific heat of the mixture in Btu per pound mass per degree Fahrenheit (of dry air) is

$$\boxed{c_{pm} = c_{pa} + Wc_{pv}} \tag{7.25}$$

where $c_{pa} = 0.24$ Btu/$lb_m \cdot °R$ and $c_{pv} = 0.44$ Btu/$lb_m \cdot °R$ in the ordinary temperature range.

Enthalpy of the Mixture. The enthalpy of the mixture is given by

$$\boxed{h_m = h_a + Wh_v} \tag{7.26}$$

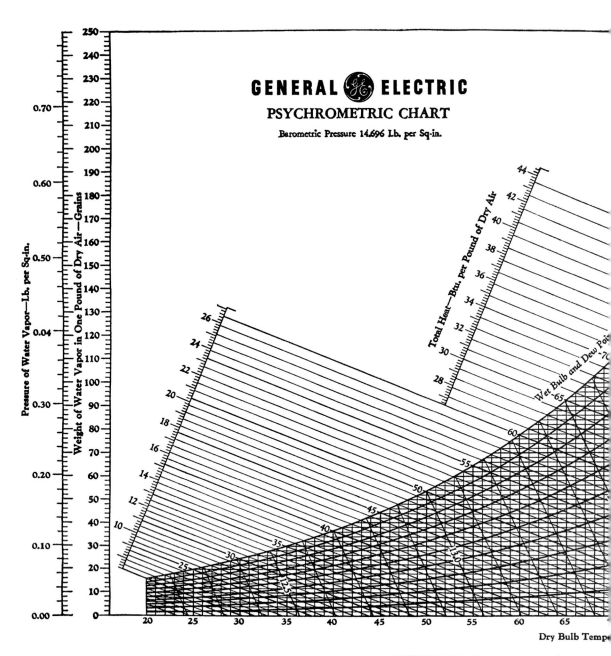

FIGURE 7.11 Psychrometric chart for ai

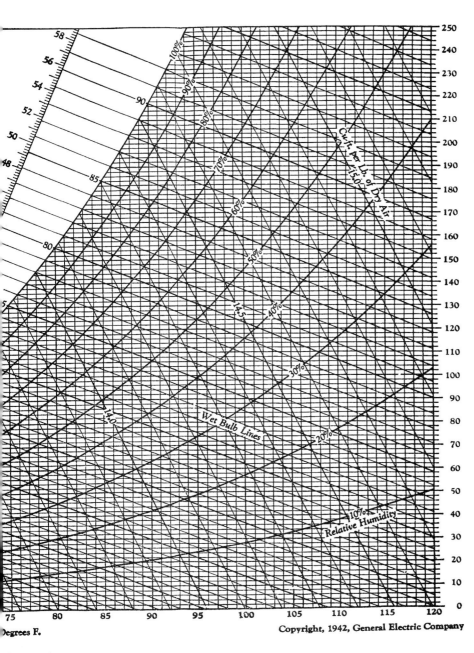

vapor mixtures.

It is possible to express h_m accurately to within 0.1% between 32°F and 100°F at 1 atm (29.921 in. Hg) by

$$h_m = (0.2402 + 0.44W)t + 1061W \tag{7.27}$$

where t is the dry-bulb temperature in degrees Fahrenheit.

Entropy of the Mixture. The entropy of the mixture can be given as

$$\boxed{s_m = s_a + Ws_v} \tag{7.28}$$

where s_a and s_v are evaluated at the respective partial pressures in the mixture.

7.8 THE PSYCHROMETRIC CHART

Just as the Mollier chart was found to be useful in plotting the thermodynamic properties of a fluid, a *psychrometric chart* is found to be equally useful when dealing with air–water vapor mixtures (see Figure 7.11). To understand the chart and its limitations fully, let us construct one in outline form. From the *Steam Tables,* we have the saturation data given in Table 7.5.

TABLE 7.5
Temperature and Pressure Data

Temperature (°F)	Pressure (psia)
32	0.08859
40	0.12166
50	0.17803
60	0.2563
70	0.3632
80	0.5073
90	0.6988
100	0.9503
110	1.2763
120	1.6945

If these data are plotted with the partial pressure of water vapor as the ordinate and the temperature (dry-bulb) as abscissa, the saturation curve is established as indicated in Figure 7.12 (100% relative humidity). From Equation (7.22), the relative humidity is the ratio of p_v/p_{vs}. Thus, by dividing the saturation pressures at each dry-bulb temperature into equal parts and connecting corresponding points, it is possible to construct lines of constant relative humidity. The outline of the 50% relative humidity line has been drawn on Figure 7.12.

Let us further assume that the chart will apply for a given barometric pressure, say, 29.921 in. Hg. From Equation (7.23),

$$W = 0.622 \frac{p_v}{p_m - p_v}$$

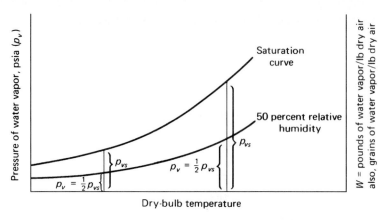

FIGURE 7.12 Skeleton psychrometric chart (showing constant ϕ).

Because the mixture pressure has been assumed to be constant, W becomes a simple linear function of p_v. Thus, the ordinate also becomes proportional to the weight of water vapor per pound of dry air. Because by definition there are 7000 grains/lb, we have plotted on the right ordinate of Figure 7.12 an auxiliary scale of W in grains of water vapor per pound of dry air.

From Equation (7.24), it is possible to evaluate the specific volume of the mixture per pound of dry air. Because ideal gas relations have been assumed, the specific volume of the air–water vapor mixture is a linear function of the humidity ratio W. Lines of constant specific volume are shown in Figure 7.13.

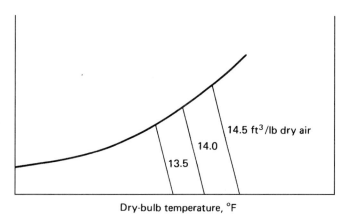

FIGURE 7.13 Skeleton psychrometric chart (with lines of constant v).

By using the ideal gas relation and Equation (7.27), the enthalpy of the mixture is also found to be linear with dry-bulb temperature and humidity ratio. Typical lines of constant enthalpy (total heat) are indicated in Figure 7.14. Figure 7.14 also shows the lines of constant "total heat" and wet-bulb temperatures as the same. Actually, wet-bulb temperature lines are straight, but they are not parallel. However, the error is small enough to be negligible. For a further discussion of the conclusions regarding the relation of total heat to enthalpy and so on, the reader is referred to the ASHRAE publications.

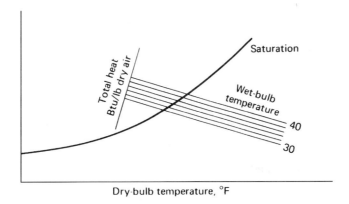

FIGURE 7.14 Skeleton psychrometric chart (with lines of constant *h*).

ILLUSTRATIVE PROBLEM 7.19

Solve Illustrative Problem 7.16 using the psychrometric chart.

SOLUTION

The initial conditions are 90°F and 70% relative humidity. Entering the chart at 90°F dry-bulb temperature and proceeding vertically to 70% relative humidity, we find the air to have 150 grains water vapor per pound of dry air. At the final condition of 80°F and 40% relative humidity, we read 61 grains of water/lb of dry air. The water removed is $150 - 61 = 89$ grains per pound of dry air or $89/7000 = 0.0127$ lb of water per pound of dry air removed.

Having constructed the chart, we can illustrate its usefulness best by showing typical processes on the chart. Remember, the chart in Figure 7.11 can only be used for atmospheric air at approximately 14.7 psia.

Dew-Point Determination—Heating and Cooling without Change in Moisture Content. From the definition of the dew point, it is apparent that a horizontal line on the chart (cooling at constant pressure of water vapor) extended to the saturation line will yield the dew-point temperature. This process of "sensible" heating or cooling (Figure 7.15) is distinguished by a change in dry-bulb temperature, relative humidity, wet-bulb temperature, total heat, and specific volume, and by no change in moisture content, dew-point temperature, and vapor pressure of the moisture in the air. Sensible heat is heat transfer due to a temperature difference without condensation or evaporation of the moisture in the air. Latent heat is heat transfer resulting from evaporation or condensation of the moisture in the air at saturation temperature.

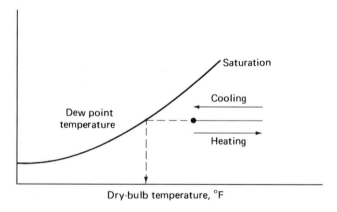

FIGURE 7.15 Sensible heating or cooling process.

Figure 7.16 shows an example of chilled-water coils using extended surfaces (fins) to increase the heat transfer of the unit. This type of construction is usually used in larger installations where cooling (or heating) is accomplished by flowing the heat-transfer medium inside the tubes and passing the

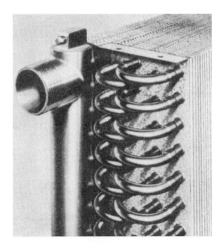

FIGURE 7.16 Air cooling and heating coils.

air over the external tube surface. When extended surfaces are used, they may be staggered or placed in line with the airflow. In those cases where frost may accumulate on the tube surface, plain tubes rather than extended surface elements are used. Coil performance is dependent on the way air flows over the outside of the tubes in relation to the way the heating or cooling fluid flows on the inside of the tubes. A discussion of the heat transfer in heat exchangers is given in Chapter 11.

ILLUSTRATIVE PROBLEM 7.20

How much heat is required to sensibly heat air having a dry-bulb temperature of 50°F and a relative humidity of 50% to 80°F?

SOLUTION

We first locate 50°F and 50% relative humidity on Figure 7.11. At this state, we read 26 grains of water per pound of dry air and a total heat of 16.1 Btu per pound of dry air. We now proceed horizontally to 80°F at a constant value of 26 grains of water per pound of dry air and read a total heat of 23.4 Btu per pound of dry air. The heat required therefore is 23.4 − 16.1 = 7.3 Btu per pound of dry air.

Humidifying of Air with No Change in Dry Bulb Temperature. As shown in Figure 7.17, this process is a vertical line on the psychrometric chart between the desired moisture limits. During this process, there is an increase in relative humidity, wet-bulb temperature, total heat, specific volume, moisture content, dew-point temperature, and vapor pressure of the moisture in the air. The addition of moisture to air is usually necessary for winter operation of air-conditioning systems. The process shown in Figure 7.17 can be accomplished by using heated spray water whose temperature is kept above the dry-bulb temperature of the air. In the limit, the air temperature approaches the final water temperature.

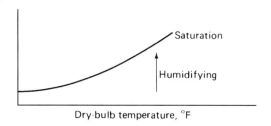

FIGURE 7.17 Humidification of air: constant dry-bulb temperature.

The control of the moisture content of air is one of the principle functions of air-conditioning systems. The addition of moisture is known as *humidification,* and the removal of moisture is known as *dehumidification.* The apparatus known as an air washer can be used to perform both processes as well as to clean the air. Figure 7.18 shows the principle of the spray-type air washer. In essence, an air washer consists of a casing with one or more spray bands through which air flows, and at the outlet, there is an eliminator to remove any entrained moisture. The operation of the spray washer can be summarized by noting that (1) if the final water temperature is held above the entering air dry-bulb temperature, the air is both heated and humidified; (2) if the final water temperature is kept below the entering dry-bulb temperature, the air will be cooled and humidified; and (3) if the final water temperature is held below the air's dew point, the air is both cooled and dehumidified.

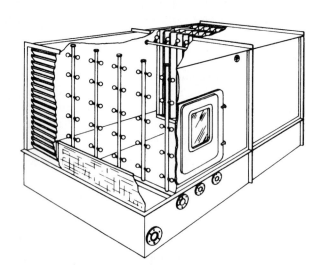

FIGURE 7.18 Spray-type air washer.
[From W. F. Stoecker, Refrigeration and Air Conditioning *(New York: McGraw-Hill, Inc., 1958), p. 291, with permission.]*

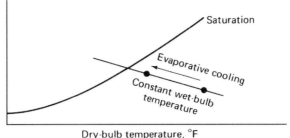

Dry-bulb temperature, °F

FIGURE 7.19 Evaporative cooling of air.

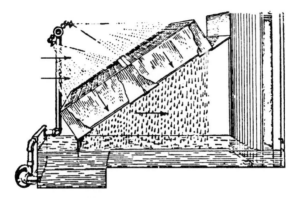

FIGURE 7.20 Capillary type of air washer.

Evaporative Cooling of Air.   If air is brought into contact with water that is at the wet-bulb temperature of the air, moisture will be added to the air at constant wet-bulb temperature. Because a line of constant wet-bulb temperature is also a line of constant total heat, the total heat for this process is constant. The heat "lost" by the incoming air serves to increase the moisture content of the final air, as given by Equation (7.27). Although both the wet-bulb temperature and total heat of the air remain constant, the dry-bulb temperature and the specific volume decrease as the relative humidity, the moisture content, and vapor pressure of the moisture in the air increase. This process is shown in Figure 7.19. As the process proceeds along the line of constant wet-bulb temperature, the dry-bulb temperature and dew point of the leaving air approach the wet-bulb temperature of the air. The spray-type air washer shown in Figure 7.18 can be used to accomplish this process. Because it is desired to obtain intimate contact between the air and water, another type of washer, shown in Figure 7.20, is often used. In this unit, the capillary action of glass filaments is used to thoroughly distribute water sprayed across the wetted surface. Air passing through the thoroughly soaked mass of glass fibers comes into intimate contact with a multiplicity of wetted surfaces. The excess water is collected in the bottom of the unit and recirculated. Evaporative cooling is one of the oldest methods of cooling known to man and is most successfully used in regions that are dry, such as in deserts. Portable evaporative coolers have been used in automobile and truck cabs in arid and desert regions for comfort control. For this method to be effective, there must be a large difference between the wet- and dry-bulb temperatures of the air, making it ineffective in humid, tropic areas.

ILLUSTRATIVE PROBLEM 7.21

In an evaporative cooling process, the exit air is found to be saturated at 50°F. If the entering air has a dry-bulb temperature of 80°F, determine the relative humidity of the entering air.

SOLUTION

Because the exit air is saturated, we find the exit condition on the saturation curve corresponding to a wet-bulb temperature of 50°F. The process is carried out at constant total enthalpy, which is along a line of constant wet-bulb temperature. Proceeding along the 50°F wet-bulb temperature line of Figure 7.11 diagonally to the right until it intersects with the vertical 80°F dry-bulb temperature line yields a relative humidity of approximately 4%.

Chemical Drying of Air. If air is passed over a drying agent that is not dissolved by the moisture extracted from the air and does not retain an appreciable amount of the heat of vaporization liberated when the water is condensed, the process can be said to be carried out along a wet-bulb temperature line. If the adsorber retains an appreciable amount of this heat, the process takes place along a line below the wet-bulb temperature line. When the adsorber is soluble in water (e.g., calcium chloride), the process line is above or below the wet-bulb temperature line depending on whether heat is liberated or adsorbed when the adsorber goes into solution. This process is shown graphically in Figure 7.21. Typical solid adsorbents are activated alumina, silica gel, activated carbon, and bauxite. These remove water from an airstream only if the vapor pressure of the water in the adsorbent is less than the partial pressure of the water vapor in the surrounding airstream. In other words, the process is primarily one of condensation. The latent heat of vaporization and heat of adsorption have to be taken away by the surroundings. Thus, the airstream, the adsorbent itself, and the adsorbent container all increase in temperature. Eventually, the sorbent temperature reaches a state of equilibrium, and the material has to be reactivated—heated to drive off the adsorbed moisture and cooled down for reuse.

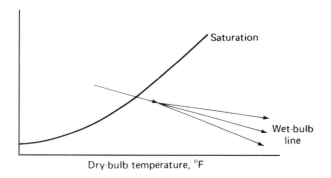

FIGURE 7.21 Chemical drying of air.

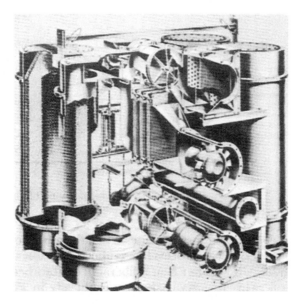

FIGURE 7.22 Two-bed solid dehumidifier.
(Reprinted with permission from Power, The Engineer's Reference Library,
copyright by McGraw-Hill, Inc., New York.)

Figure 7.22 shows a typical two-bed unit designed to operate continuously. Filtered air enters the unit and is ducted to one of the beds. While the first bed is working, the other is being reactivated by heating the sorbent (silica gel in this case) to approximately 300°F. Heating may be done by steam coils, electricity, gas, and so on.

Liquid sorbents such as the glycols, lithium chloride, calcium chloride, and lithium bromide functionally operate in a manner similar to the solid sorbents. These solutions must have concentrations such that their vapor pressure with respect to water is lower than that of the water vapor to be removed from the air. In the typical liquid-sorbent equipment, air passes through a tower into which the liquid sorbent is sprayed either in a fine mist or as a blanketing solution on the tower surfaces. Again, differences in partial pressure determine the absorption. Soon, a state of equilibrium is reached. The sorbent solution, the air, and the equipment increase in temperature as latent and chemical absorption heats are taken up.

The warm, weak sorbent solution is first cooled. Then part is diverted for concentration. This is done by heating it to a vapor pressure well above that of the air blown over it; water in solution goes off with the air. After cooling, the concentrated solution rejoins the main stream to maintain proper density, the principal factor for determining water absorption by liquid sorbents. Figure 7.23 shows a liquid sorbent unit in which the tower on the left fixes the moisture content of the conditioned air while in the one on the right the sorbent releases its water and is reconcentrated.

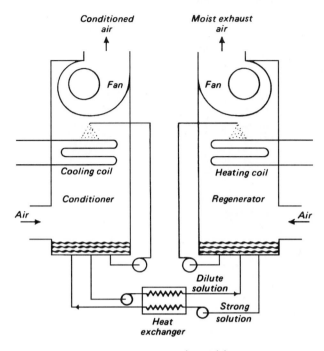

FIGURE 7.23 Dehumidifier.
[From W. F. Stoecker, Refrigeration and Air Conditioning *(New York: McGraw-Hill, Inc., 1958), p. 293, with permission.]*

Cooling and Dehumidifying of Air.   If the cooling surface temperature is below the initial dew-point temperature, this process can be portrayed as a straight line extending from the initial condition to the surface temperature on the saturation curve. The final condition of the air will depend on the total heat extracted from the air. During this process, the dry-bulb temperature, wet-bulb temperature, moisture content, specific volume, vapor pressure of the moisture, and total heat all decrease, as indicated in Figure 7.24. The coils shown in Figure 7.16 can also be used for cooling and dehumidifying air by keeping the coil temperature below the dew point of the entering air.

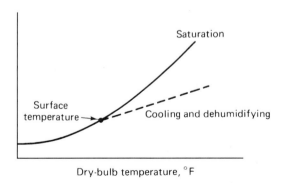

FIGURE 7.24 Cooling and dehumidifying process.

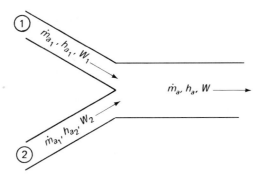

FIGURE 7.25 Adiabatic mixing of airstreams.

Adiabatic Mixing of Airstreams. The mixing of two airstreams at different initial conditions is quite commonly used in air-conditioning work. Let us assume that no condensation takes place during this process. Then, from Figure 7.25, a mass balance on the dry air is

$$\dot{m}_{a1} + \dot{m}_{a2} = \dot{m}_a \tag{7.29}$$

where $\dot{m}_{a1}$ is the mass flow rate of dry air in stream 1, $\dot{m}_{a2}$ is the mass flow rate of dry air in stream 2, and $\dot{m}_a$ is the mass flow rate of the combined streams. A mass balance on the water vapor yields

$$\dot{m}_{a1} W_1 + \dot{m}_{a2} W_2 = \dot{m}_a W \tag{7.30}$$

and
$$\dot{m}_{a1}h_{a1} + \dot{m}_{a2}h_{a2} = \dot{m}_a h_a$$

For an adiabatic mixing process in which velocity and elevation terms are negligible and in which no work enters or leaves the system, an energy balance yields

$$\dot{m}_{a1}h_{a1} + \dot{m}_{a2}h_{a2} = \dot{m}_a h_a \tag{7.31}$$

A combination of these equations yields

$$\boxed{\frac{h_a - h_{a2}}{h_a - h_{a1}} = \frac{W_2 - W}{W - W_1} = \frac{\dot{m}_{a1}}{\dot{m}_{a2}} = \frac{l_1}{l_2}} \tag{7.32}$$

where l_1 and l_2 are the corresponding line segment lengths shown in Figure. 7.26. Equation (7.32) can be interpreted to mean that the mixing of air at one condition with air at another condition can be

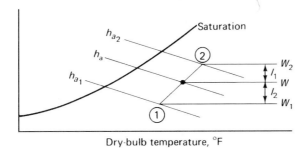

Dry-bulb temperature, °F

FIGURE 7.26 Adiabatic mixing of airstreams.

represented by a straight line connecting the initial and final conditions. The final mixture will lie along this line at a point, determined by the relative quantities of air being mixed, as shown in Figure 7.26.

ILLUSTRATIVE PROBLEM 7.22

Indoor air at 75°F dry bulb and 50% relative humidity is to be mixed with outdoor air at 90°F dry bulb and 60% relative humidity. If 4 parts of indoor air is mixed with 1 part of outdoor air (by weight), what is the final mixture composition?

SOLUTION

As noted from Figure 7.27, 1 lb of mixture, $\frac{4}{5}$ lb of indoor air, and $\frac{1}{5}$ lb of outdoor air are mixed per pound of mixture. We now locate the two end states on the psychrometric chart and connect them with a straight line. The line connecting the end states is divided into 5 equal parts. Using the results of Equation (7.32), we now proceed *from* the 75°F indoor air state 1 part *toward* the 90°F outdoor air state. This locates the state of the mixture, which is found to be a dry-bulb temperature of approximately 78°F, a wet-bulb temperature of approximately 66°F, and a relative humidity of approximately 54%.

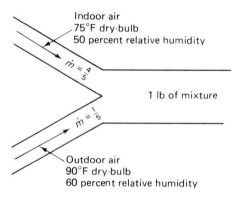

FIGURE 7.27   Illustrative Problem 7.22.

The Cooling Tower.   The cooling tower has been used in situations where the supply of water is limited or due to economic considerations. Recently, the effect on the environment due to the heat rejection to the cooling water of a power plant (thermal pollution) has become a factor in power-plant design. To minimize the environmental impact on rivers or other water sources, the cooling tower is used to cool the water discharged from the condensers, and this water is then recirculated. In effect, instead of using river or local water to carry off the heat rejected in the power plant, the atmosphere is used. The cooling tower is simply a device in which water is evaporatively cooled by air. In the natural-draft cooling tower, air is circulated through the tower in a horizontal direction while water is sprayed or trickled over wood filling. In the forced-draft cooling tower, a fan is used to positively circulate air countercurrent to the falling water. The fan can be located at the bottom of the tower, and this arrangement is known as a forced-draft tower. If the fan is located at the top to prevent recirculation of the hot moist air, we have an induced-draft tower. Figure 7.28

FIGURE 7.28 Large cooling tower installation.

shows a large bank of cooling towers used in a power installation. We can analyze the action in the cooling tower by referring to the schematic diagram of Figure 7.29. In this diagram, air flows from bottom to top of the tower, and water droplets flow counter to the airstream and are in intimate contact with the counterflowing airstream. Heat is transferred to the air, raising its wet-bulb temperature, its dry-bulb temperature, and its moisture content. The evaporation of a small portion of the water accounts for the water-cooling effect. The water that evaporates into the air stream must be replaced, and this replacement water is known as *makeup*. Referring to the symbols of Figure 7.29, we can write a heat balance for the tower by considering the air and water streams separately. *For the air,*

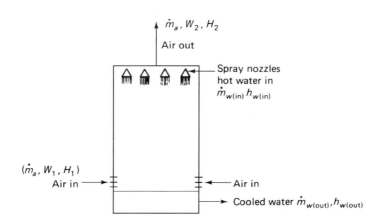

FIGURE 7.29 Schematic diagram of a cooling tower.

$$\text{heat pickup per lb dry air} = H_2 - H_1 \text{ where } H = \text{total heat/lb dry air}$$
$$\text{total heat pickup} = \dot{m}_a(H_2 - H_1)$$
$$\text{moisture pickup per lb dry air} = W_2 - W_1$$
$$\text{total moisture pickup} = \dot{m}_a(W_2 - W_1)$$

For the water,

$$\text{entering total enthalpy} = \dot{m}_{w(\text{in})}h_{w(\text{in})}$$
$$\text{leaving total enthalpy} = \dot{m}_{w(\text{out})}h_{w(\text{out})}$$
$$\text{total water enthalpy change} = \dot{m}_{w(\text{in})}h_{w(\text{in})} - \dot{m}_{w(\text{out})}h_{w(\text{out})}$$
$$\text{water evaporated} = \dot{m}_{w(\text{in})} - \dot{m}_{w(\text{out})}$$

Because the heat picked up by the air must equal the heat exchanged by the water, we have

$$\dot{m}_a(H_2 - H_1) = \dot{m}_{w(\text{in})}h_{w(\text{in})} - \dot{m}_{w(\text{out})}h_{w(\text{out})} \qquad (7.33)$$

The water picked up by the air must equal the water loss in the tower. Therefore,

$$\dot{m}_a(W_2 - W_1) = \dot{m}_{w(\text{in})} - \dot{m}_{w(\text{out})} \qquad (7.34)$$

Equations (7.33) and (7.34) provide the basic relations for the solution of cooling tower problems.

ILLUSTRATIVE PROBLEM 7.23

Hot water enters a cooling tower at the rate of 200,000 lb/hr at 100°F, and the water leaving is at 70°F. Air enters the tower at a dry-bulb temperature of 60°F with a 50% relative humidity and leaves at a dry-bulb temperature of 90°F with a relative humidity of 90%. Assuming that atmospheric pressure is 14.7 psia and the psychrometric chart (Figure 7.11) can be used, determine the amount of water lost per hour due to evaporation and the amount of air required per hour. The situation is summed up in Figure 7.30.

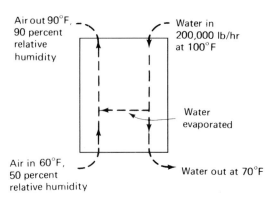

FIGURE 7.30 Illustrative Problem 7.23.

SOLUTION

Let us first obtain the necessary physical data required from the *Steam Tables* and the psychrometric chart. *For the water,*

$$h_{100°F} = 68.05 \text{ Btu/lb}_m$$
$$h_{70°F} = 38.09 \text{ Btu/lb}_m$$

For the air,

At inlet: $H = 20.4$ Btu/lb dry air

$W = 38.2$ grains/lb dry air (at 60°F D.B. and 50% R.H.)

At outlet: $H = 52.1$ Btu/lb dry air

$W = 194.0$ grains/lb dry air (at 90°F D.B. and 90% R.H.)

Per pound of dry air, the heat interchange is $52.1 - 20.4 = 31.7$ Btu per pound of dry air. Per pound of dry air, the moisture increase is $(194.0 - 38.2)/7000 = 0.0223$ lb per pound of dry air. From Equation (7.33),

$$\dot{m}_a(31.7) = (200,000)(68.05) - \dot{m}_{w(out)}(38.09)$$

From Equation (7.34),

$$\dot{m}_a(0.0223) = 200,000 - \dot{m}_{w(out)}$$

Solving the latter equation for $\dot{m}_{w(out)}$, we have

$$\dot{m}_{w(out)} = 200,000 - \dot{m}_a(0.0223)$$

Substituting this into the heat balance yields

$$\dot{m}_a(31.7) = (200,000)(68.05) - (200,000)(38.09) + \dot{m}_a(0.0223)(38.09)$$

Solving gives us

$$\dot{m}_a(31.7 - 0.85) = (200,000)(68.05 - 38.09)$$
$$= 5,992,000$$
$$\dot{m}_a = 194,230 \text{ lb}_m/\text{hr of dry air}$$
$$\text{water evaporated (lost)} = 194,230 \times 0.0223$$
$$= 4331 \text{ lb}_m/\text{hr}$$

Note that the water evaporated is slightly over 2% of the incoming water, and this is the makeup that has to be furnished to the tower.

7.9 AIR CONDITIONING

In the previous sections of this chapter, we have studied the properties of air–water vapor mixtures in some detail, the processes that these mixtures undergo, and the equipment in which the processes are carried out. The purpose of this section is to briefly study the factors that are involved in the design of air-conditioning systems that combine the elements that we have individually considered earlier in this chapter.

The term *air conditioning* as we will use it is meant to encompass the control of the properties of air in an enclosure such as a room or a building. If the object is to provide for the comfort of humans within an environment, it may be necessary to effectively control the dry-bulb temperature, the wet-bulb temperature, humidity, dust content, odors, bacteria, and toxic gases. The processes by which these objectives are attained are heating, cooling, ventilation, humidification, cleaning, and dehumidification. We have already considered the elements required for the control of dry-bulb temperature, wet-bulb temperature, and humidity. The removal of odors, gases, and dust is usually accomplished by passing the air through filter elements. In the *dry filter,* air is forced to flow through a screening material such as fiberglass, gauze, cellulose, or woven wool felt. Figure 7.31a shows a dry filter that is removable and cleanable. It uses wool felt arranged on a metal frame and can be vacuum cleaned, air blown, or dry cleaned. The dry filters serve best for relatively small airflows and light dust loadings. They are highly efficient and, when clean, offer relatively little resistance to airflow. Resistance builds up rapidly, however, and dust-holding capacity is relatively small. Change in resistance usually indicates when the filter should be replaced or cleaned.

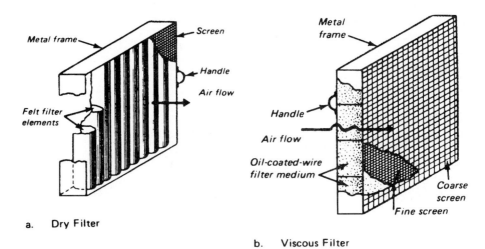

FIGURE 7.31 Dry and viscous filters.
(Reprinted with permission from Power, the Engineer's Reference Library, *copyright by McGraw-Hill, Inc., New York.)*

Other types of filters include viscous filters in which the filtering medium is coated with a sticky oil. Figure 7.31b shows this type of filter, which can either be thrown away or cleaned and reused. The electrostatic precipitator shown in Figure 7.32 utilizes the principle that a dust particle exposed to an electric field becomes charged and migrates to one of the electrodes. The design uses a viscous material to hold the attracted particles, and it is cleaned by shutting down the unit for washing and recoating. The electrostatic precipitator removes microscopic particles, smoke, and pollen that mechanical filters cannot remove. Odors are usually removed by being adsorbed by activated charcoal units placed in ducts or in the air washers shown in Figures 7.18 and 7.20.

Figure 7.33 shows schematically the elements of a year-round air-conditioning unit. This unit controls the basic items for both summer and winter use. In addition to control of temperature and humidity, air motion and distribution are controlled, and air purity is controlled using filters. There

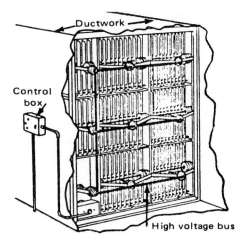

FIGURE 7.32 Electrostatic precipitator.
(Reprinted with permission from Power, The Engineer's Reference Library,
copyright by McGraw-Hill, Inc., New York.)

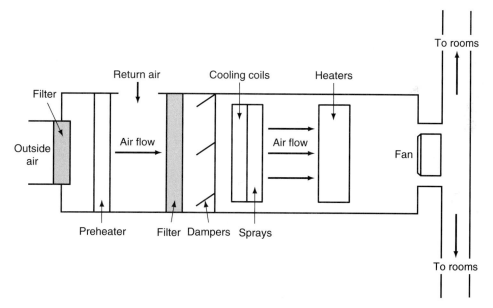

FIGURE 7.33 A year-round air-conditioning unit.

are many variations in the placement of the parts of such a system. Under certain conditions, depending on the condition of the entering air, one or more parts may not be used. As an example, in a moist climate, the spray would be eliminated. In general, in moderate climates, all of the elements shown would be used for all-year-round air conditioning.

It is necessary to maintain adequate ventilation in indoor spaces for both health and comfort. In recent years, the concern for energy conservation has caused the amount of outside air entering buildings to be reduced. Many of the modern buildings have materials and furnishings that out-gas volatile organic compounds (VOCs) that can be harmful to humans. Terms such as "sick building syndrome" have been used to indicate poor indoor air quality (IAQ).

When we consider human beings and their effect on the environment when they are in a relatively confined space, we must consider the bodily processes that occur. Oxygen is inhaled, and a moist mixture less rich in oxygen and richer in carbon dioxide is exhaled into the enclosure. In addition, heat is transferred from the human, and moisture is evaporated to the environment. Body odors, smoke, and organic material are also added to the air. For a person at rest, approximately 180 Btu/hr is given off as latent heat and 220 Btu/hr as sensible heat. For a person working, these values increase by approximately 25%. The term *latent heat* as used in air-conditioning work refers to the changes of liquid water to water vapor in humidification or the change of water vapor to liquid water in dehumidification. *Sensible heat* refers to the heating of dry air. One method of removing the heat, moisture, and odors that exist in a room is to ventilate the room. This process consists of supplying air to, or removing air from, the room. In ordinary residences, air infiltration usually provides sufficient ventilation.

Current ventilation standards are based upon the amount of air entering a room divided by the number of occupants in that room. The unit of ventilation is cubic feet per minute per person, or cfm/p. The American Society of Heating, Refrigerating, and Air-Conditioning Engineers (ASHRAE) has set standards for new construction. However, many states do not require that ventilation in older buildings be improved to meet current standards. This can present a major problem, because some of the current ventilation standards are as much as four times stricter than they were in the years 1975–1989, when they were relaxed to conserve energy. Table 7.6 shows some current ventilation standards for commercial uses.

TABLE 7.6

Some Current Ventilation Standards for Commercial Uses

Smoking lounges	60 cfm/person
Public restrooms	50 cfm/person
Hotel rooms	35 cfm/person
Offices and conference rooms	20 cfm/person
Classrooms and libraries	15 cfm/person
Hospital rooms	25 cfm/person
Beauty salons	25 cfm/person
Ballrooms, discos	25 cfm/person

Source: ASHRAE Standard 62-89, 1989.

In hot weather, the air-conditioning plant has to remove both sensible and latent heat (moisture). Some sources of the heat are as follows:

1. Heat liberated by occupants—sensible and latent.

2. Infiltration of outside air—sensible and latent.

3. Process heat (industrial, cooking, and so on)—sensible and latent.

4. Heat infiltration from walls and partitions—sensible and latent.

5. Solar heat absorption through walls, roofs, windows, and so on—sensible.

6. Heat transfer between the exterior and interior of the building due to temperature differences—sensible.

7. Air brought in for ventilation—sensible and latent.

The detailed calculation of these heat loads is given in the latest edition of the *ASHRAE Guide* and will not be considered further. The references in Appendix 1 should be consulted as well as the *ASHRAE Guide* for these details. Also, Chapter 11 on heat transfer should be studied prior to performing such calculations.

Figure 7.34 shows a complete heating and cooling plant used in a recent commercial industrial installation. All the necessary processes of heating, cooling, cleaning, and so on are performed in this central plant. From this plant, the treated air is ducted to the areas where it is used, and some of the "used" air is returned for recirculation. Fans or blowers are required to circulate the air through the heating and cooling coils, filters, washers, ducts, outlets, grilles, and diffusers. Both axial and centrifugal fans are used for providing air circulation, and the choice depends on the specific system requirements to which the fan is matched.

FIGURE 7.34 View of the district cooling and heating plant owned and operated by Houston Natural Gas Corp. at Nassau Bay, a commercial/residential project near Houston. *(Courtesy of Carrier Corp.)*

Securing proper distribution of air within a large room without creating drafts is difficult. Any air movement at a rate greater than 30 ft/min in a room of seated people may seem uncomfortable to many. If the temperature and relative humidity are within comfort zone limits, somewhat higher velocities can be tolerated. Where people are moving about, velocities as high as 120 ft/min can be used. Air can be withdrawn through grilles at up to 70 ft/min without discomfort.

Over and above the precautions taken to prevent undue drafts, it is desired to get even distribution of air despite the fact that incoming air may be different in temperature from the room air. Furthermore, some means of controlling air quantity at room inlets may be desirable. Thus, a considerable variety of grilles and diffusers is available. Figure 7.35 shows some of the air distribution grilles and diffusers used in many air-conditioning installations.

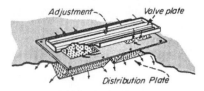

Grille has deflecting vanes so any desired air pattern can be set up

Valve plate and distribution space slow air down, smooth out the flow

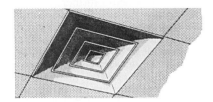

An adjustable damper permits varying the volume of entering air

Latest diffusers mount flush with ceiling, blend with the decorations

FIGURE 7.35 Air distribution grilles and diffusers.
(Reprinted with permission from Power, The Engineer's Reference Library, *copyright by McGraw-Hill, Inc., New York.)*

In the design of the air-conditioning system, the design outdoor dry-bulb and wet-bulb temperatures need to be known, and so do the dry- and wet-bulb temperature range in which humans can feel comfortable. For New York City, the design dry-bulb temperature is 95°F and the design wet-bulb temperature 75°F. In Mobile, Alabama, the corresponding values are 95°F and 80°F. Extensive tabulations of the design dry-bulb and design wet-bulb temperatures for many cities in the United States will be found in the *ASHRAE Guide*. Also given in the *Guide* is the range of dry- and wet-bulb temperatures known as the human comfort zone as an area on the psychrometric chart. For this chart, the *Guide* should be consulted, but for most people, the area enclosed by the 65°F and 85°F dry-bulb temperatures and the 30% and 70% relative humidity lines defines the human comfort zone.

The design of any air-conditioning system is a complex interaction of many considerations, and the *ASHRAE Guide* is an invaluable source of data for any design study.

7.10 REVIEW

As a logical extension of the material that we studied in Chapter 6, we have considered that a mixture of gases can be treated as a single gas having the equation of state of an ideal gas. With this as our starting point, we derived Dalton's law for the pressure of a gas mixture and Amagat's law for the volume of a gas mixture. From Dalton's law, we obtained the following conclusions:

1. The total pressure of a mixture of ideal gases is the sum of its partial pressures.

2. The mass of the mixture is the sum of the masses of the components of the mixture.

3. The total number of moles in the mixture equals the sum of the number of moles of the component gases of the mixture.

4. The partial pressure of a component equals the mole fraction of the component multiplied by the total pressure of the mixture.

5. The molecular weight of the mixture equals the sum of the products of mole fraction of each gas component multiplied by the molecular weight of each component gas.

6. The ideal gas equation of state is applicable to all components as well as to the entire mixture.

From Amagat's law, we obtained:

1. The mole fraction of a component equals the volume fraction of that component in the mixture.

2. The volume of a mixture is the sum of the volumes that each component gas would occupy if each were at the pressure and temperature of the mixture.

We then applied both of these laws to the determination of the composition of a gas. A gas analysis given on a weight basis was converted to a volumetric basis, and a gas analysis on a volumetric basis was converted to a weight basis. Having these analyses, it became a logical extension to ask for the thermodynamic properties of gas mixtures. By a relatively straightforward approach, we demonstrated that the mixture properties are each equal to the sum of the mass average of the individual properties of the components of the mixture.

Air–water vapor mixtures can be treated from fundamental considerations or by use of the psychrometric chart. The psychrometric chart is very useful, because various processes can be portrayed on it. The chart given is based on the ideal gas relations and a total pressure of 29.921 in. Hg and is usable for total pressures varying ±1 in. Hg from 29.921 in. After considering many processes on the psychrometric chart and the typical equipment used, we considered the cooling tower. The cooling tower is used to reject heat to the atmosphere while using a small amount of water. Consideration of the airstream and the water stream separately permitted us to write a heat balance for the cooling tower and to analyze fully its performance.

Our final brief discussion was on air conditioning. This topic was treated qualitatively; quantitative considerations, including the determination of heat loads, were not covered. After studying Chapter 11, the student will be in a better position to delve further into this subject.

KEY TERMS

Terms used for the first time in this chapter are:

air conditioning the control of the properties of air in an enclosure such as a room or a building.

Amagat's law the volume of a mixture is the sum of the volumes that each constituent gas would occupy if it were at the pressure and temperature of the mixture.

cooling tower a device that uses air to cool water evaporatively.

Dalton's law the pressure of a gas mixture is the sum of the partial pressures of each of its components.

dew-point temperature the temperature that a gas–vapor mixture must react at constant local mixture pressure to condense the vapor.

dry-bulb temperature the temperature of a mixture of air and water vapor as measured by an ordinary thermometer.

grain 1/7000 of a pound.

humidity ratio see *specific humidity.*

mole fraction the ratio of the number of moles of a constituent gas to the total number of moles in the mixture.

partial pressure the pressure that a constituent gas would exert if it alone occupied the total volume at the temperature of the mixture.

psychrometric chart a chart giving the thermodynamic properties of air–water vapor mixtures.

relative humidity the ratio of the partial pressure of vapor to the saturation pressure of the vapor at the same temperature in an air–water vapor mixture.

sling psychrometer two thermometers mounted on a board that can be slung (rotated); one ther-mometer, the wet-bulb thermometer, has a gauze wick wrapped around its bulb which is saturated with water.

specific humidity the ratio of the weight of vapor to the weight of the gas in a unit volume of an air–water vapor mixture.

wet-bulb temperature the equilibrium temperature registered by a thermometer wetted by a wick in a stream of an air–water vapor mixture. Also the temperature at which water, by evaporating into air, can bring the air to saturation adiabatically at the same temperature.

EQUATIONS DEVELOPED IN THIS CHAPTER

mixture pressure $\qquad p_m = p_a + p_b + p_c$ (7.1)

moles of gas $\qquad n_m = n_a + n_b + n_c$ (7.2b)

partial pressure $\qquad \dfrac{p_a}{p_m} = \dfrac{n_a}{n_m} = x_a$ (7.4)

mixture molecular weight $\qquad \mathrm{MW}_m = x_a \mathrm{MW}_a + x_b \mathrm{MW}_b + x_c \mathrm{MW}_c$ (7.5)

mixture gas constant $\qquad R_m = \dfrac{1545}{\mathrm{MW}_m}$ (7.6)

Amagat's law $\qquad V_m = V_a + V_b + V_c$ (7.8)

Amagat's law $\qquad \dfrac{V_a}{V_m} = \dfrac{n_a}{n_m} = x_a$ (7.9)

mixture internal energy $\qquad u_m = \dfrac{m_a}{m_m} u_a + \dfrac{m_b}{m_m} u_b + \dfrac{m_c}{m_m} u_c$ (7.14)

mixture enthalpy $\qquad h_m = \dfrac{m_a}{m_m} h_a + \dfrac{m_b}{m_m} h_b + \dfrac{m_c}{m_m} h_c$ (7.15)

specific heat $\qquad c_{vm} = \dfrac{m_a}{m_m} c_{va} + \dfrac{m_b}{m_m} c_{vb} + \dfrac{m_c}{m_m} c_{vc}$ (7.16)

specific heat $\qquad c_{pm} = \dfrac{m_a}{m_m} c_{pa} + \dfrac{m_b}{m_m} c_{pb} + \dfrac{m_c}{m_m} c_{pc}$ (7.17)

mixing entropy increase $\qquad \Delta S = -\dfrac{n_a(1545)}{J} \ln x_a - \dfrac{n_b(1545)}{J} \ln x_b$ **Illustrative Problem 7.11**

mixture gas constant $\qquad R_m = \left(\dfrac{1}{1+W}\right) R_a + \left(\dfrac{W}{1+W}\right) R_v$ (7.21)

relative humidity	$$\phi = \frac{p_v}{p_{vs}} = \frac{v_{vs}}{v_v} = \frac{\rho}{\rho_{vs}}$$	(7.22)

humidity ratio	$$W = 0.622 \frac{p_v}{p_m - p_v}$$	(7.23)

relative humidity	$\phi p_{vs} = p_v$	(7.24)
mixture specific heat	$c_{pm} = c_{pa} + Wc_{pv}$	(7.25)
mixture enthalpy	$h_m = h_a + Wh_v$	(7.26)
mixture entropy	$s_m = s_a + Ws_v$	(7.28)

adiabatic mixing	$$\frac{h_a - h_{a2}}{h_a - h_{a1}} = \frac{W_2 - W}{W - W_1} = \frac{\dot{m}_{a1}}{\dot{m}_{a2}} = \frac{l_1}{l_2}$$	(7.32)

cooling tower	$\dot{m}_a(H_2 - H_1) = \dot{m}_{w(\text{in})}h_{w(\text{in})} - \dot{m}_{w(\text{out})}h_{w(\text{out})}$	(7.33)
cooling tower	$\dot{m}_a(W_2 - W_1) = \dot{m}_{w(\text{in})} - \dot{m}_{w(\text{out})}$	(7.34)

QUESTIONS

7.1 State Dalton's law, and show how it leads to the statement that the total pressure of an ideal gas mixture is the sum of the partial pressures that each gas exerts.

7.2 The mole fraction of an ideal gas mixture is numerically equal to the fraction of at least three other items in the mixture. What are these items?

7.3 Would you use the mole fractions of the constituents to determine the specific heat of an ideal gas mixture?

7.4 In the determination of the enthalpy of an ideal gas mixture, is it necessary to know that the datum for each of the components be the same?

7.5 State Amagat's law in words.

7.6 The statement is made in the text that Amagat's law and Dalton's law are equivalent to each other if the gases and the mixture are ideal gases. Why is this so?

7.7 Are the gravimetric and volumetric analyses the same?

7.8 Give a reason why the entropy does not change when two identical ideal gases are mixed.

7.9 Define the dew point of an air–water vapor mixture.

7.10 Under what conditions will the wet-bulb temperature of an air–water vapor mixture equal its dry-bulb temperature?

7.11 What is meant by the term adiabatic saturation?

7.12 Describe the operation of a sling psychrometer.

7.13 What is the difference between relative humidity and humidity ratio?

7.14 Is a person's comfort a function of the humidity ratio, the relative humidity, both, or neither?

7.15 Sketch an outline of a psychrometric chart. Label each of the coordinates and scales on the chart with the customary set of units.

7.16 Illustrate the typical processes on the psychrometric chart.

7.17 Using the psychrometric chart, sketch those processes that can be used for dehumidification.

7.18 Why is evaporative cooling ineffective in humid tropical areas?

7.19 What happens to the chemical used for chemical drying of air? How is it restored to a usable condition?

7.20 When would you use a cooling tower?

7.21 How does the use of a cooling tower lessen the effect on the environment?

7.22 What is the principle on which the cooling tower is based?

7.23 What is the usual percentage of makeup in a cooling tower?

7.24 What is the difference between latent and sensible heat?

7.25 What air ventilation requirement is needed for a store?

7.26 What is the average relative humidity for the summer months where you live?

PROBLEMS

Use Table 7.1 for molecular weights of gases.

Problems Involving Gas Mixtures

7.1 Show that at 14.7 psia and 32°F, a mole of any ideal gas occupies a volume of 359 ft^3.

7.2 A gas mixture consists of 1 lb of methane (CH_4) and 1 lb of oxygen (O_2). Determine the number of moles of each gas, the total number of moles, and the mole fraction of each component. Also, determine the molecular weight of the mixture.

7.3 A gas mixture consists of 2 kg of methane (CH_4) and 2 kg of carbon dioxide (CO_2.) Determine the mole fraction of each component and the molecular weight of the mixture.

7.4 A certain gas mixture consists of 25% carbon dioxide, 55% nitrogen, and 20% helium by volume. Determine the gravimetric analysis of the mixture.

7.5 A gas mixture has a volumetric analysis of 35% CO_2, 35% N_2, and 30% O_2. If the mixture pressure is 14.7 psia and its temperature is 60°F, determine the partial pressure of each gas.

7.6 A mixture contains 1 kg of acetylene (C_2H_2) and 1 kg of butane (C_4H_{10}). If the total pressure of the mixture is 50 MPa, determine the partial pressure of each component.

7.7 What is the molecular weight of the mixture in Problem 7.6? Also, what is the gas constant of this mixture?

7.8 A mixture consists of 0.4 lb of CO_2, 0.2 lb of CH_4, and 0.4 lb of O_2. Determine the molecular weight and the gas constant of the mixture.

7.9 A mixture contains 1 lb$_m$ of CO_2 and 1 lb$_m$ of helium. If the mixture has a volume of 5 ft^3 at 90°F, determine its pressure.

7.10 Determine the molecular weight and volumetric analysis of the mixture in Problem 7.9.

7.11 A mixture of 60 lb of O_2, 20 lb of N_2, and 10 lb of H_2 is at 140°F and 17 psia. Determine the volume of the mixture.

7.12 A mixture of 20 kg of CO_2, 10 kg of O_2, and 15 kg of He is at 60°C and 0.2 MPa. Determine the volume of the mixture.

*7.13 Three pounds of carbon dioxide (CO_2) is mixed with 7 lb of an unknown gas. If the resulting mixture occupies a volume of 50 ft^3 when the mixture is at 50 psia and 200°F, determine the molecular weight of the unknown gas.

*7.14 Six kg of CO is mixed with 8 kg of an unknown gas. If the resulting mixture occupies a volume of 5 m^3 when the mixture is at 0.3 MPa and 150°C, determine the molecular weight of the unknown gas.

7.15 If a certain gas is CO (20% by weight), N_2 (50%), O_2 (10%), and CH_4 (20%), compute the partial pressure of each constituent if the total mixture pressure is 100 psia.

*7.16 A mixture contains 20 lb_m of N_2 and the rest CO_2. If the volume of the mixture is 100 ft^3 at 70 psia and 75°F, determine the average molecular weight of the mixture.

7.17 A gasoline engine's exhaust analysis is found to be CO_2 (10% by volume), CO (0.5%), N_2 (73%), H_2O (14.5%), and O_2 (2%). Determine the gravimetric analysis and the molecular weight of this mixture.

7.18 The volumetric analysis of a gas is CH_4 (86%), CO_2 (1%), N_2 (1%), and H_2 (12%). Determine the gravimetric analysis, the molecular weight, and the gas constant for the mixture.

*7.19 The density of a mixture of CO_2 and N_2 is 0.08 lb_m/ft^3 at 14.7 psia and 70°F. Determine the mass of CO_2 present in 1 lb_m of the mixture.

7.20 A mixture contains 10 lb of methane (CH_4) and 6 lb of oxygen (O_2). If the mixture is at 140°F and 75 psia, determine the mixture volume, the partial volume of each constituent, the partial pressure of each constituent, the mole fraction of each constituent, and the molecular weight of the mixture.

7.21 A tank has a volume of 5 ft^3. If it contains 2 lb of O_2 and 3 lb of CO_2, compute the partial pressures of each constituent and the total pressure of the mixture. The temperature of the mixture is 70°F.

7.22 If a tank has a volume of 2 m^3 and contains 1 kg of CO and 2 kg of He, compute the partial pressure of each constituent and the total mixture pressure when the temperature is 30°C.

7.23 It is desired to produce a gas mixture containing 50% by volume of CO_2 and O_2. If this mixture contains 4 lb of CO_2, how many pounds of O_2 is there in this mixture?

7.24 Determine the molecular weight and the gas constant of the mixture in Problem 7.23.

7.25 It is desired to produce a gas mixture containing 40% by volume of CO and 60% by volume of O_2. If there are 4 kg of CO in the mixture, how many kilograms of O_2 are there in this mixture?

7.26 What is the molecular weight and gas constant of the mixture in Problem 7.25?

*7.27 Two hundred cubic feet of hydrogen at 60 psia and room temperature is adiabatically mixed with 400 ft^3 of nitrogen at 15 psia and room temperature. If the temperature of the room is 70°F, determine the molecular weight, R, and the final pressure of the mixture.

*7.28 If 10 m^3 of hydrogen at 0.4 MPa and room temperature is adiabatically mixed with 20 m^3 of nitrogen at 0.1 MPa and room temperature, determine the molecular weight, R, and the final pressure of the mixture. Room temperature is 20°C.

7.29 A mixture consists of 7 moles of oxygen (O_2), 3 moles of methane (CH_4), and 1 mole of nitrogen (N_2). Determine the volumetric analysis, the weight of the mixture, the molecular weight of the mixture, and the gas constant of the mixture.

7.30 A gas mixture consists of the following volume percentages: 40% CO, 20% CH_4, 30% N_2, and 10% O_2. Determine the weight fraction of each gas, the molecular weight of the mixture, and the gas constant of the mixture.

7.31 A gas mixture has a volumetric analysis of 25% N_2, 35% O_2, and 40% CH_4. Determine the gravimetric analysis, the molecular weight of the mixture, and the gas constant of the mixture.

7.32 The volumetric analysis of a gas mixture is 40% N_2, 40% O_2, and 20% CO_2. The mixture is at 100 psia and 100°F. Determine the partial pressure of each gas, the molecular weight of the mixture, the gravimetric (weight) analysis, and the gas constant of the mixture.

7.33 A gas mixture of oxygen, nitrogen, and methane occupies a container, and each constituent has a partial pressure of 50, 20, and 65 psia, respectively. Determine the gravimetric analysis, the volumetric analysis, the molecular weight of the mixture, and the gas constant of the mixture.

*7.34 Determine the final temperature and pressure of a mixture of oxygen and nitrogen if 6 lb of oxygen at 70°F and 200 psia is mixed with 1 lb of nitrogen at 200°F and 100 psia. Use c_v of oxygen as 0.164 $Btu/lb_m \cdot °R$ and c_v of nitrogen as 1.79 $Btu/lb_m \cdot °R$. The process is adiabatic.

7.35 Determine c_v of the mixture in Problem 7.34.

*7.36 Determine the final temperature and pressure of a mixture of CO_2 and nitrogen if 3 kg of CO_2 at 20°C and 1.5 MPa is mixed with 1 kg of nitrogen at 100°C and 0.75 MPa. Use c_v of CO_2 as 0.656 kJ/kg·°K and c_v of nitrogen as 0.752 kJ/kg·°K. The process can be considered to be adiabatic.

7.37 Compute c_v of the mixture in Problem 7.36.

*7.38 Determine the final temperature and pressure of a mixture if it is made by mixing 10 ft³ of nitrogen at 120°F and 70 psia and 5 ft³ of oxygen at 60°F and 120 psia. Use c_v of oxygen as 0.164 Btu/lb$_m$·°R and c_v of nitrogen as 0.179 Btu/lb$_m$·°R. The process is adiabatic.

7.39 Determine c_v of the mixture in Problem 7.38.

7.40 One pound of nitrogen at 75°F and 14.7 psia is adiabatically mixed with 1 lb of helium at 80°F and 250 psia. Determine the final mixture pressure and temperature. For nitrogen, $c_v = 0.177$ Btu/lb$_m$·°R, and for helium, $c_v = 0.754$ Btu/lb$_m$·°R.

7.41 Determine the specific heat at constant volume for the mixture in Problem 7.40.

Problems Involving Air–Water Vapor Mixtures

7.42 An air–water vapor mixture at 100°F has a relative humidity of 40%. Calculate the partial pressure of the water vapor and its dew-point temperature.

7.43 Solve Problem 7.42 using the psychrometric chart.

7.44 A room contains air at 70°F with a humidity ratio of 0.008 lb$_m$ water vapor/lb$_m$ dry air. Determine the relative humidity and the dew point of the air.

7.45 Solve Problem 7.44 using the psychrometric chart.

7.46 An air–water vapor mixture at a dry-bulb temperature of 90°F and 14.7 psia is found to have a relative humidity of 50%. Calculate the dew point, the partial pressure of the water vapor, and the humidity ratio.

7.47 Solve Problem 7.46 using the psychrometric chart.

7.48 The air in a room is at 70°F. A bare pipe running through the room has a surface temperature of 55°F. At what relative humidity will condensation occur on the surface of the pipe?

7.49 Solve Problem 7.48 using the psychrometric chart.

*7.50 An air–water vapor mixture having a total pressure of 14.7 psia is found to have a humidity ratio of 0.010 lb of water vapor per pound of dry air. Calculate the relative humidity, the dew point, and the pressure of the water vapor if the dry-bulb temperature is 80°F.

7.51 Solve Problem 7.50 using the psychrometric chart.

7.52 The partial pressure of water vapor in an atmospheric pressure (14.7 psia) mixture is 0.4 psia. Calculate the relative humidity and the dew-point temperature if the dry-bulb temperature is 100°F.

7.53 Solve Problem 7.52 using the psychrometric chart.

7.54 Moist air at 14.7 psia has a dry-bulb temperature of 80°F and a wet-bulb temperature of 60°F. Determine the relative humidity, the enthalpy, and the specific humidity of the air using the psychrometric chart.

7.55 Moist air at 14.7 psia has a wet-bulb temperature of 68°F and a dew point of 50°F. Determine its dry-bulb temperature and relative humidity using the psychrometric chart.

7.56 Moist air at 14.7 psia has a dew-point temperature of 50°F and a dry-bulb temperature of 86°F. Determine the relative humidity and the wet-bulb temperature of the air using the psychrometric chart.

*7.57 A closed tank has a volume of 200 ft³. The tank contains atmospheric air at 70°F with a humidity ratio of 0.01 lb$_m$ water vapor/lb$_m$ air. Determine the mass of water that must be removed to obtain a relative humidity of 30%.

*7.58 Solve Problem 7.57 for a total pressure of 12 psia.

*7.59 Solve Problem 7.57 for a total pressure of 10 psia.

7.60 Atmospheric air at 14.7 psia and 80°F has a relative humidity of 30%. Calculate its dew point and its specific humidity.

7.61 Solve Problem 7.60 using the psychrometric chart.

*7.62 Atmospheric air is sensibly cooled at constant pressure from an initial dry-bulb temperature of 80°F and 50% relative humidity to a final dry-bulb temperature of 60°F. Calculate the heat removed per pound of dry air. Assume that the mixture is at 14.7 psia. Use Equation (7.27) for the enthalpy of the mixture.

7.63 Solve Problem 7.62 using the psychrometric chart.

*7.64 Air is dehumidified at constant dry-bulb temperature. Calculate the heat removed per pound of dry air if the initial conditions are 90°F dry-bulb temperature, 40% relative humidity, and the final relative humidity is 20%. The mixture is at 14.7 psia.

7.65 Solve Problem 7.64 using the psychrometric chart.

7.66 Air is cooled adiabatically from 70°F dry bulb and 57% relative humidity until 90% relative humidity is reached. Calculate the final dry-bulb temperature and the initial and final dew-point temperatures.

7.67 Solve Problem 7.66 using the psychrometric chart.

7.68 Atmospheric air at 10 psia and 70°F dry bulb has a relative humidity of 50%. If the air is sensibly heated at constant pressure to 80°F dry bulb, what is the final relative humidity?

7.69 Even though the psychrometric chart is not applicable at this pressure, determine the final relative humidity in Problem 7.68 using the chart and compare results.

7.70 Determine the humidity ratio and dew-point temperature for air at a total pressure of 30.921 in. Hg if it is at 80°F dry-bulb temperature and 60% relative humidity.

7.71 Solve Problem 7.70 using the psychrometric chart. Comment.

7.72 To condition it, air is passed over a coil at 50°F. If the initial air is at 85°F dry bulb and 70% relative humidity and a final 75°F dry-bulb air is specified, determine the final relative humidity, the amount of moisture removed per pound of dry air, and the amount of heat extracted per pound of dry air. Use the psychrometric chart.

*7.73 Determine whether it is possible to obtain air at a dry bulb of 60°F and 70% relative humidity by evaporative cooling and then sensible heating if the initial air is at 70°F dry bulb and 10% relative humidity.

7.74 In Problem 7.73, what is the least value of the initial relative humidity that will permit the process to be completed?

7.75 What final conditions will be obtained when 200 lb_m/min of air at 85°F and 60% relative humidity is mixed with 300 lb_m/min of air at 50°F and 20% relative humidity? (Temperatures are dry-bulb values.) Use the psychrometric chart.

*7.76 If the flow quantities in Problem 7.75 were given in cubic feet per minute, that is, 200 ft³/min and 300 ft³/min, determine the final mixture conditions.

*7.77 Water enters a cooling tower at 130°F and leaves at 100°F. Air at atmospheric pressure having an initial dry-bulb temperature of 60°F and a relative humidity of 60% leaves the tower saturated at 90°F. Calculate the weight of air required and the makeup water required (water lost by evaporation) if 150,000 lb_m/hr of water enters the tower.

*7.78 Water is cooled in a cooling tower from 110°F to 80°F. Air enters the tower at 80°F and a relative humidity of 40%, and it leaves at 95°F with a relative humidity of 90%. Determine the water entering that is cooled per pound of dry air and the makeup water required per pound of dry air.

*7.79 Air flows through a cooling tower to cool 30,000 lb$_m$/hr of water from 100°F to 80°F. Air enters the tower at 70°F with a relative humidity of 30% and leaves at 95°F with a relative humidity of 90%. Determine the makeup water required and the amount of air required.

*7.80 A cooling tower operates with 368 lb$_m$/s of 100°F water entering. The water is cooled to 70°F. Air enters at 60°F with 50% relative humidity and leaves at 90°F and 98% relative humidity. Determine the makeup water requirements and the weight of air required.

*7.81 A cooling tower cools 2400 lb$_m$/min of water from 85°F to 65°F. The air used for cooling flows at the rate of 4000 lb$_m$/min. The air enters the tower at a dry-bulb temperature of 65°F and a wet-bulb temperature of 55°F. If the air leaves the tower with 100% relative humidity, determine the temperature of the air leaving and the amount of makeup water.

8

Vapor Power Cycles

Learning Goals

After reading and studying the material in this chapter, you should be able to:

1. Understand the definition of the term cycle, and differentiate between gas and vapor cycles.

2. Recall that our conclusions regarding the Carnot cycle were independent of the working fluid used in the cycle.

3. Sketch and analyze the elements of the simple Rankine cycle.

4. Conclude that for the same pressures, the efficiency of the Rankine cycle is less than that of a Carnot cycle.

5. Define the type efficiency as the ratio of the ideal thermal efficiency of a given cycle divided by the efficiency of a Carnot cycle operating between the same maximum and minimum temperature limits.

6. Sketch the simple reheat cycle elements and T–s and h–s diagrams for this cycle.

7. Conclude that reheat does not greatly increase the efficiency of the Rankine cycle but does decrease the moisture content of the steam in the later stages of the turbine.

8. Understand that regeneration is a method of heating the feedwater with steam that has already done some work and that in the limit, regenerative cycle approaches the efficiency of the Carnot cycle operating between the same temperature limits.

9. Apply regeneration to the Rankine cycle, sketch the T–s diagram, and show how regeneration improves the Rankine cycle efficiency.

10. Qualitatively understand the construction and operation of the major pieces of equipment used in commercial steam-generating plants.

11. Understand the other cycles that have been proposed and/or constructed for the production of power, including direct energy devices.

12. Understand the use of cogeneration as a combination of power production and process or space use.

8.1 INTRODUCTION

A cycle has been defined as a series of thermodynamic processes during which the working fluid can be made to undergo changes involving energy transitions and subsequently is returned to its original state. The object of any practical cycle is to convert energy from one form to another, more useful form. For instance, the energy bound in a fossil fuel is released by the chemical process of combustion and, by undergoing appropriate thermodynamic processes, is made to yield useful work at the shaft of an engine. Similarly, the energy of the nuclear-fission process is made to yield useful work that ultimately appears as electrical energy.

In the following sections, several ideal cycles are discussed. They are "ideal" in the sense that they have been proposed as prototypes of practical cycles, and in the limit, their efficiencies approach a Carnot-cycle efficiency. These cycles usually bear the name of the person who either proposed them or developed them, and their composite study represents a large portion of applied thermodynamics. In practice, actual cycles deviate from the ideal because of unavoidable irreversibilities and for other practical reasons. The study of the ideal cycle, however, can and does yield invaluable results that are applicable to real cycles. Additionally, we discuss the actual devices (hardware) used in the practical realization of these ideal cycles. Along with these descriptions, we see where the actual cycles deviate from the "ideal" cycles and the attempts made to make actual cycles approach ideal performance.

Power cycles are often classified by the character of the working fluid in the cycle. The two general classes of cycle are the vapor cycle and the gas cycle. The vapor cycle differs from the gas cycle in two respects. In the gas cycle, there is no change of phase of the working substance, and the compression work of the gas cycles can therefore represent a large percentage of the useful work output of the cycle. In the vapor cycle, the working substance is condensed to a liquid at the lower temperature of the cycle. This liquid is pumped to the desired delivery pressure. Because the liquid is essentially incompressible, it would be expected that the pump work of the vapor cycle would represent a small percentage of the useful work output of the cycle. Also, in the vapor cycle, the working substance may contain moisture when it is expanded in a turbine. Because this is undesirable, modifications are made in the vapor cycle to alleviate this condition.

8.2 CARNOT CYCLE

The conclusions reached in the study of the Carnot cycle were independent of the working medium, and it is pertinent to review this cycle briefly, as applied to both vapors and gases. It will be recalled that a Carnot cycle consists of two reversible isothermal stages and two isentropic stages. Figure 8.1a shows a Carnot cycle in which the working fluid is indicated as being in the wet vapor region. The line *A, B* represents the isentropic compression of the fluid from the lower temperature (and pressure) to the upper temperature of the cycle. At the upper temperature, there is an isothermal and reversible reception of energy as heat from some reservoir. This transfer of the heat proceeds along path *B, C* with concurrent increase in the volume of the vapor. The fluid then expands isentropically along path *C, D* with energy extracted as work. As the fluid expands, the volume of the fluid increases. Finally, the wet fluid is condensed isothermally and reversibly at the lower temperature of the cycle. During the condensation process, the volume of the fluid decreases as heat is rejected to the sink of the system. The energy available is represented by area *A, B, C, D,* and the energy rejected is represented by area *A′, A, D, D′* on the *T–s* diagram of Figure 8.1a.

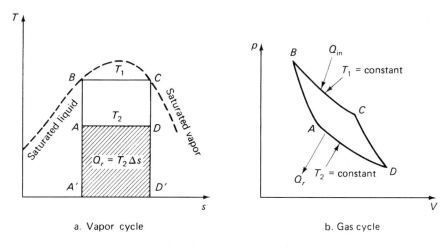

a. Vapor cycle   b. Gas cycle

FIGURE 8.1 Carnot cycle.

The efficiency of a Carnot cycle is solely a function of the upper and lower temperatures and is given by

$$\eta_c = \frac{T_1 - T_2}{T_1} \times 100 \tag{8.1}$$

Equation (8.1) is also the efficiency of any reversible heat engine operating between these constant temperature limits. All the heat is taken in or rejected at T_1 and T_2, respectively. Figure 8.1b shows the Carnot gas cycle plotted on pv coordinates. All the conclusions pertaining to the efficiency of the vapor cycle also pertain to the gas cycle, because we have already noted that the efficiency of a Carnot cycle is independent of the working fluid in the cycle.

  Unfortunately, there are certain practical limitations on the upper and lower temperatures. The upper temperature is limited by the strength of available materials and the lower by ambient conditions. Recent advances in the field of thermonuclear reactions have indicated that the upper temperature limitation may be removed by using magnetic fields to contain the working fluid. To date, the limits on the upper temperature of power cycles have depended on advances in the field of metallurgy.

8.3 THE RANKINE CYCLE

The Carnot cycle described in Section 8.2 cannot be used in a practical device for many reasons. Historically, the prototype of actual vapor cycles was the simple Rankine cycle. The elements of this cycle are shown in Figure 8.2; the T–s and h–s diagrams for the ideal *Rankine cycle* are illustrated in Figure 8.3.

  As indicated in the schematic of Figure 8.2, this cycle consists of four distinct processes. Starting with the feed pump, the liquid supplied to the boiler is first brought to the boiler pressure. In the ideal cycle, the liquid supplied to the pump is assumed to be saturated at the lowest pressure of the cycle. In an actual cycle, the liquid is usually slightly subcooled to prevent vapor bubbles from forming in the pump (which causes a process known as *cavitation,* which will subsequently

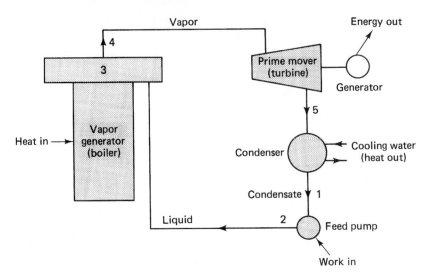

FIGURE 8.2 Elements of the simple Rankine cycle.

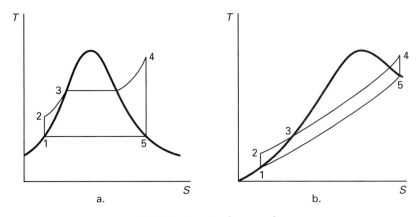

FIGURE 8.3 Rankine cycle.

damage the pump). For the ideal cycle, the compression process is taken to be isentropic, and the final state of the liquid supplied to the boiler is subcooled at the boiler pressure. This subcooled liquid is heated to saturation in the boiler, and it is subsequently vaporized to yield the steam for the prime mover in the cycle. The energy for the heating and vaporizing of the liquid is provided by the combustion of fuel in the boiler. If superheating of the vapor is desired, it is also accomplished in the boiler (also called a steam generator). The vapor leaves the steam generator and is expanded isentropically in a prime mover (e.g., a turbine or steam engine) to provide the work output of the cycle. After the expansion process is completed, the working substance is piped to the condenser, where it rejects heat to the cooling water.

Figure 8.4 gives an idea of the size and complexity of a modern electric generating station. The station shown is the Ravenswood station of New York City's Con Edison system. The three generators at the plant have a total electrical capability in excess of 1,700,000 kW. The plant is

FIGURE 8.4   Con Edison's Ravenswood Station.
(Courtesy of Con Edison, New York.)

located on the east bank of the East River, whose water is used as cooling water in the condensers. In the foreground can be seen the complex electrical substation equipment that ties this plant into the Con Edison system. The size of the plant can be visualized by comparing it to the automobiles parked near the buildings and to the six-story houses in the lower right of the photograph. Unit 3 of this station is a 1000-MW unit with a boiler capable of generating 6.5×10^6 lb$_m$/hr of steam at 2750 psia and 1000°F with a single reheat. This boiler is designed for a pressure of 2990 psia, and it will operate about 5% above its nominal pressure of 2400 psia. Feedwater enters the economizer at a temperature of 480°F, and steam enters the reheater at 650°F. The boiler is a twin-furnace unit and, outside of size, marks one important step forward: one furnace and one-half of the boiler are able to keep operating if the other half fails. Load reduction would be only 500 MW. Each half of the boiler has its own drum and circulating system and its own air and gas ductwork. Hot air for combustion is supplied by four regenerative air heaters. Fuel is heavy oil, and steam air heaters preheat it to about 150°F with the used air going to the main air heaters. Steam temperature and feedwater regulation, as well as combustion, are electronically controlled with solid-state parts; valves and dampers are pneumatically actuated.

Let us study the simple Rankine cycle shown in Figures 8.2 and 8.3 and consider some of the processes that occur.

Process 1–2

For steady-flow isentropic pumping of the feedwater it is assumed that the inlet condition to the pump corresponds to saturation of the lowest pressure of the cycle. The compression is assumed to be isentropic, and differences in potential and kinetic energies as inlet and outlet to the pump are negligible. For this ideal process the work in is

$$w = (h_2 - h_1)_s \qquad \text{Btu/lb}_m \tag{8.2}$$

or

$$w = (h_2 - h_1)_s \qquad \text{kJ/kg} \tag{8.2a}$$

where the subscript s denotes an isentropic compression. Because the water is essentially incompressible, it is possible to approximate Equation (8.2) as

$$w \simeq (p_2 - p_1)v_f / J \qquad \text{Btu/lb}_m \qquad (8.3)$$

or $\qquad\qquad w \simeq (p_2 - p_1)v_f \qquad \text{kJ/kg} \qquad (8.3a)$

where p_2 is the high pressure of the cycle, p_1 is the low pressure of the cycle (both in lb_f/ft^2 or kPa), and v_f is the specific volume of saturated liquid in ft^3/lb_m or m^3/kg at the inlet pressure to the pump.

Process 2–3

This process is the heating of the subcooled water to saturation by steam in the drum. We will consider that this occurs as part of the boiler process (2–4) and not evaluate it separately. However, it is considered as part of the boiler design by the designer.

Process 2–4

The liquid leaves the pump at the delivery pressure to the boiler. In the ideal cycle, pressure losses in the pipelines are negligible. Thus, the process of heating in the boiler and subsequent vaporization (and superheat) can be considered to be a steady-flow process carried out at constant pressure. For this process, neglecting differences in potential and kinetic energies at inlet and outlet of the boiler,

$$q = (h_4 - h_2) \qquad \text{Btu/lb}_m \text{ or kJ/kg} \qquad (8.4)$$

It should be noted that q is the energy required by the working fluid, not the energy released in the steam generator. Also, devices such as attemperators or desuperheaters, which are used to control the final steam temperature leaving the superheater, have to be accounted for in an energy balance of a steam generator.

Process 4–5

After leaving the boiler, the steam is piped to the turbine. Once again, friction losses will be neglected, as will differences in potential and kinetic energies at the inlet and outlet of the turbine. For the turbine, assuming an isentropic expansion,

$$w = (h_4 - h_5)_s \qquad \text{Btu/lb}_m \text{ or kJ/kg} \qquad (8.5)$$

The presence of moisture in the turbine can lead to mechanical difficulties such as excessive erosion of the turbine blades. This is highly undesirable, and to alleviate this condition, superheating of the steam is often used. This increases the efficiency of the unit (as deduced from the Carnot cycle) and decreases the amount of moisture in the turbine exhaust (see Figure 8.3).

Figure 8.5 illustrates an expansion of steam from 500 psia to a final pressure of $1\frac{1}{2}$ in. Hg. On $h - s$ coordinates, this ideal expansion is a vertical line. If the expansion is not isentropic, there is an increase in entropy, and the final state point must lie on the $1\frac{1}{2}$ in. Hg line at an increased value of s. As shown in Figure 8.5, the final enthalpy is required to be higher for the nonisentropic expansion than for the isentropic case. In other words, the change in enthalpy for the isentropic expansion is greater than the change in enthalpy for the nonisentropic expansion between the same pressure limits. The nonisentropic expansion has been shown as a dashed line. Actually, this line is not the path of the expansion, because the expansion is irreversible and nonequilibrium states

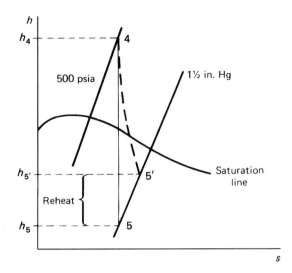

FIGURE 8.5 Rankine turbine expansion.

cannot be drawn on an $h–s$ diagram. The difference in enthalpy $(h_{5'} - h_5)$ is called *reheat,* because all the mechanical irreversibilities increase the final enthalpy, or "reheat" the fluid. To account for the nonideal effects in the turbine, the internal turbine efficiency is introduced as the ratio of the work actually done to the work that could be done ideally. The student should note that this procedure corresponds to the method used to define nozzle efficiencies in Chapter 6. Therefore, the internal turbine efficiency η_t is

$$\eta_t = \frac{\text{actual change in enthalpy}}{\text{isentropic change in enthalpy}} = \frac{h_4 - h_{5'}}{h_4 - h_5} \tag{8.5a}$$

Note that point 5 can be saturated vapor, wet steam, or superheated steam, depending on the location of point 4.

Process 5–1

After leaving the turbine, the fluid (which is usually wet steam) is piped to the condenser. Because the specific volume is high and we wish to keep the pressure losses in pipelines and equipment small, the piping and condenser are usually quite large at this part of the cycle. If pressure losses, kinetic terms, and potential terms are negligible and it is assumed that the equipment is well insulated, the heat rejected to the cooling water is

$$q_r = (h_5 - h_1) \qquad \text{Btu/lb}_\text{m} \text{ or kJ/kg} \tag{8.6}$$

This energy is rejected to the cooling water and raises the temperature of the cooling water. In the ideal cycle, the condensed steam is assumed to be saturated and is returned to the inlet of the circulating pump with no pressure drop in the piping. Actually, as we have noted, the condensate is usually slightly subcooled before leaving the condenser. A large surface condenser is shown in Figure 8.6. Note the size of the man in relation to the size of the equipment.

By definition, the thermal efficiency of any cycle is the useful (net) work out of the cycle divided by the thermal energy supplied (input) to the cycle. Therefore, the efficiency of the ideal Rankine cycle is

FIGURE 8.6 Large surface condenser.
(Courtesy of Foster Wheeler Corp.)

$$\eta_R = \frac{\text{turbine work out } - \text{ pump work}}{\text{heat supplied in boiler}}$$

(8.7)

$$= \frac{(h_4 - h_5)_s - (h_2 - h_1)_s}{h_4 - h_2}$$

where $h_4 - h_2 = h_4 - h_1 - (h_2 - h_1)$. If the pump work is small, as it usually is, it can be neglected, and the efficiency of the ideal Rankine cycle becomes

$$\eta'_R = \frac{(h_4 - h_5)_s}{h_4 - h_2}$$

(8.8)

Note that neglecting pump work is equivalent to saying that $h_2 = h_1$.

The engine that operates on the expansion of steam between the boiler and the condenser in the Rankine cycle is called the *Rankine engine*. If the expansion is isentropic, the maximum amount of work obtainable from the expansion is $(h_4 - h_5)_s$. If pump work is neglected, the efficiency of the Rankine cycle is the same as that of the Rankine engine.

ILLUSTRATIVE PROBLEM 8.1

A Rankine cycle operates with its upper temperature at 400°C and its upper pressure 3 MPa. If its lower pressure is 50 kPa, determine the thermal efficiency of the cycle (a) neglecting pump work and (b) including pump work.

SOLUTION

From the *Steam Tables* or Mollier chart in Appendix 3, we find that

$$h_1 = h_f \text{ at } 50 \text{ kPa} = 340.49 \text{ kJ/kg} \qquad h_4 = 3230.9 \text{ kJ/kg} \qquad h_5 = 2407.4 \text{ kJ/kg}$$

Here, point 5 is in the wet steam region.

(a) Neglecting pump work $(h_2 = h_1)$ gives

$$\eta'_R = \frac{3230.9 - 2407.4}{3230.9 - 340.49} \times 100 = 28.49\%$$

(b) Pump work $= (p_2 - p_1)v_f = (3000 - 50)(0.001030) = 3.04 \text{ kJ/kg}$

The efficiency of the cycle including pump work is

$$\eta_R = \frac{(3230.9 - 2407.4) - 3.04}{3230.9 - 340.49 - 3.04} \times 100 = 28.42\%$$

because $h_4 - h_2 = h_4 - h_1 - (h_2 - h_1)$.

ILLUSTRATIVE PROBLEM 8.2

Solve Illustrative Problem 8.1 using the computer disk to obtain the necessary properties.

SOLUTION

(a) For the conditions given in Illustrative Problem 8.1, the properties are found to be

$$h_2 = 340.54 \text{ kJ/kg} \qquad h_4 = 3230.7 \text{ kJ/kg} \qquad h_5 = 2407.2 \text{ kJ/kg}$$

$$\eta'_R = \left(\frac{3230.7 - 2407.2}{3230.7 - 340.54} \right) \times 100 = 28.49\%$$

(b) For the pump work, we do not need the approximation, because the computerized tables give us the necessary values directly. Assuming that the condensate leaving the condenser is saturated liquid gives us an enthalpy of 340.54 kJ/kg and an entropy of 1.0912 kJ/kg·°K. For an isentropic compression, the final condition is the boiler pressure of 3 MPa and an entropy of 1.0912 kJ/kg·°K. For these values, the program yields an enthalpy of 343.59 kJ/kg·°K. The isentropic pump work is equal to $(343.59 - 340.54) = 3.05$ kJ/kg. The efficiency of the cycle including pump work is

$$\eta_R = \frac{3230.7 - 2407.2 - 3.05}{3230.7 - 340.54 - 3.05} \times 100 = 28.42\%$$

Although individual values vary slightly, the final results in this problem agree with the results in Illustrative Problem 8.1.

The maximum and minimum pressures in a Rankine cycle are 400 and 14.696 psia. The steam is exhausted from the engine as saturated vapor. (a) Sketch the cycle on temperature–entropy and an enthalpy–entropy diagrams. (b) Find the thermal efficiency of the cycle.

SOLUTION

(a) Figure 8.3 with the cycle extending into the superheat region and expanding along $4 \to 5$ is the appropriate diagram for this process.

(b) This problem can be solved either by use of the Mollier chart or the *Steam Tables.* If the chart is used, 14.696 psia is first located on the saturated vapor line. Because the expansion, $4 \to 5$, is isentropic, a vertical line on the chart is the path of the process. The point corresponding to 4 in Figure 8.3 is found where this vertical line intersects 400 psia. At this point, the enthalpy is 1515 Btu/lb$_m$, and the corresponding temperature is approximately 980°F. Saturated vapor at 14.696 psia has an enthalpy of 1150.5 Btu/lb$_m$ (from the Mollier chart). The *Steam Tables* show that saturated liquid at 14.696 psia has an enthalpy of 180.15 Btu/lb$_m$. In terms of Figure 8.3, and neglecting pump work, we have

$$h_1 = h_2 = 180.15 \text{ Btu/lb}_m \qquad h_4 = 1515 \text{ Btu/lb}_m$$
$$h_5 = 1150.5 \text{ Btu/lb}_m$$

Neglecting pump work yields

$$\eta_R' = \frac{1515 - 1150.5}{1515 - 180.15} \times 100 = 27.3\%$$

The pump work is given approximately by Equation (5.5) as

$$\frac{(p_2 - p_1)v_f}{J} = \frac{(400 - 14.696)144}{778} \times 0.0167$$
$$= 1.19 \text{ Btu/lb}_m$$

The efficiency of the cycle including pump work is

$$\eta_R = \frac{(1515 - 1150.5) - (1.19)}{1515 - 180.15 - 1.19} \times 100 = 27.24\%$$

where the denominator is $h_4 - h_2 = h_4 - h_1 - (h_2 - h_1)$. Neglecting pump work is obviously justified in this case. An alternative solution is obtained by using the *Steam Tables:* at 14.696 psia and saturation, $s_g = 1.7567$; therefore, at 400 psia, $s = 1.7567$. From Table 3 (at 400 psia),

s	h	t
1.7632	1523.6	1000
1.7567	1514.2	982.4
1.7558	1512.9	980

The remainder of the problem proceeds along similar lines with essentially the same results. The use of the Mollier chart facilitates the solution of this problem.

ILLUSTRATIVE PROBLEM 8.4

Solve Illustrative Problem 8.3 using the computer generated property values.

SOLUTION

Refer to Figure 8.3. The desired quantities are obtained as follows:

at 14.696 psia, saturated vapor ($x = 1$), $s = 1.7566$ Btu/lb$_m$·°R, $h = 1150.4$ Btu/lb$_m$

at 14.696 psia, saturated liquid ($x = 0$), $s = 0.3122$ Btu/lb$_m$·°R, $h = 180.17$ Btu/lb$_m$

at 400 psia, $s = 1.7566$ Btu/lb$_m$·°R, $h = 1514.0$ Btu/lb$_m$, $t = 982.07$°F

at 400 psia, $s = 0.3122$ Btu/lb$_m$·°R, $h = 181.39$ Btu/lb$_m$

Note the agreement of these values with the ones obtained for Illustrative Problem 8.3. Also, note the temperature of 982.07°F compared to 982.4°F. Continuing,

$$\eta'_R = \frac{1514.0 - 1150.4}{1514.0 - 180.17} \times 100 = 27.26\%$$

The isentropic pump work $= (181.39 - 180.17) = 1.22$ Btu/lb$_m$. Therefore,

$$\eta_R = \frac{1514.0 - 1150.4 - 1.22}{1514.0 - 180.17 - 1.22} \times 100 = 27.19\%$$

The agreement with Illustrative Problem 8.3 is excellent. The effort required to obtain the necessary properties is greatly reduced by the use of the computer-generated values. However, as a word of caution, don't forget to draw a sketch of the cycle, label the points, and then use the computer program as needed.

Figure 8.7 shows a schematic of the Rankine cycle with the pertinent energy quantities shown for each of the principal units. Figure 8.8 shows the ideal Rankine cycle compared to a Carnot cycle on a $T-s$ plot. The plot clearly shows that for the same pressures, the efficiency of the Rankine cycle will be less than that of a Carnot cycle. This result is due to the slope of the saturation line. If the saturation line were vertical, then the thermodynamic efficiency of the Rankine and Carnot cycles would be the same. However, a vertical line on the $T-s$ diagram would require the liquid to have a specific heat of zero, which is not possible, but a fluid having a low specific heat compared to the value of h_{fg} would give us a higher Rankine cycle efficiency. We consider this point later when we consider the regenerative cycle.

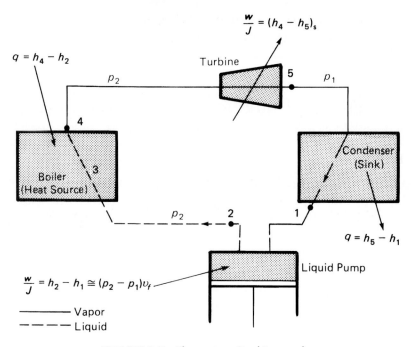

FIGURE 8.7 Elementary Rankine cycle.

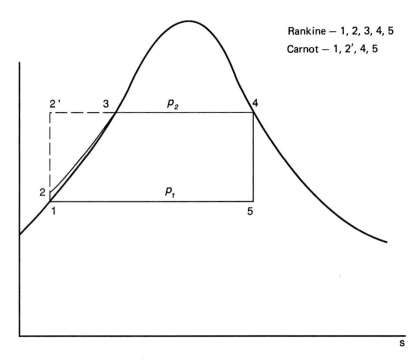

FIGURE 8.8 Carnot and Rankine cycle comparison.

8.4 RATING OF POWER-PLANT CYCLES

To evaluate the merit of the power-plant cycles, several yardsticks are employed. One of these, the type efficiency, is quite useful. The *type efficiency* is defined as the ratio of the ideal thermal efficiency of a given cycle divided by the efficiency of a Carnot cycle operating between the same maximum and minimum temperature limits. In essence, it measures the approach of a prototype ideal cycle to the Carnot cycle between the same maximum and minimum temperature limits.

ILLUSTRATIVE PROBLEM 8.5

What is the type efficiency of the ideal Rankine cycle of Illustrative Problem 8.3?

SOLUTION

The Carnot cycle would operate between 982.4°F and 212°F. Its efficiency is

$$\eta_c = \frac{T_1 - T_2}{T_1} \times 100$$

$$= \frac{(982.4 + 460) - (212 + 460)}{982.4 + 460} \times 100 = 53.4\%$$

The type efficiency would be

$$\frac{\eta_R'}{\eta_c'} \times 100 = \frac{27.3}{53.4} \times 100 = 51.1\%$$

In other words, at best, this cycle is only 51.1% as efficient as the Carnot cycle operating between the same temperature limits.

ILLUSTRATIVE PROBLEM 8.6

Calculate the type efficiency for the Rankine cycle of Illustrative Problem 8.1.

SOLUTION

For the upper temperature of the cycle, we have 400°C, and for 50 kPa, the *Steam Tables* give us a saturation temperature of 81.33°C. The efficiency of a Carnot cycle operating between the limits would be

$$\eta_c = \frac{T_1 - T_2}{T_1} \times 100$$

$$= \frac{(400 + 273) - (81.33 + 273)}{400 + 273} \times 100 = 47.4\%$$

The type efficiency is

$$\frac{\eta'_R}{\eta'_C} \times 100 = \frac{28.5}{47.4} \times 100 = 60.1\%$$

For actual cycles, the thermal efficiency is defined as the net work output of the system divided by the energy as heat supplied to the system. In evaluating the thermal efficiency, it is necessary to take care that the proper accounting of energies is carried out. This can best be illustrated by reference to Illustrative Problem 8.7.

ILLUSTRATIVE PROBLEM 8.7

In the Rankine cycle of Illustrative Problem 8.3, 50 Btu/lb_m of steam is lost by heat transfer from the turbine to the surroundings. What is the thermal efficiency of this cycle?

SOLUTION

From Illustrative Problem 8.3, it was found that the useful (ideal) work is 1515 − 1150.5 = 364.5 Btu/lb_m of steam if pump work is neglected. Because of the heat losses, 50 Btu/lb_m of the 364.5 Btu/lb_m becomes unavailable. Thus, 364.5 − 50 = 314.5 Btu/lb_m is available. The thermal efficiency of the cycle (neglecting pump work) is

$$\eta = \frac{314.5}{1515 - 180.15} \times 100 = 23.6\%$$

The heat rate and steam rate of a cycle are also figures of merit and have been used to describe cycle performance. The heat rate (actual or ideal) is defined as the amount of energy supplied per horsepower-hour or kilowatt-hour of net output of the cycle. Therefore, heat rate equals energy supplied per horsepower-hour (or kilowatt-hour) divided by η_R (or η'_R). Thus,

$$\text{heat rate} = \frac{2545}{\eta_R \ (\text{or } \eta'_R)} \quad \text{Btu/hp-hr} \tag{8.9a}$$

$$= \frac{3413}{\eta_R \ (\text{or } \eta'_R)} \quad \text{Btu/kWh} \tag{8.9b}$$

The denominator in Equation (8.9) can be the thermal efficiency of any cycle. It is not restricted to the Rankine cycle. It is left as an exercise for the student to verify the constants 2545 and 3413 used in Equation (8.9).

The steam rate of a cycle is defined as the ratio of the mass of steam supplied per net horsepower-hour or kilowatt-hour. Therefore,

$$\text{steam rate} = \frac{\text{pounds of steam supplied per hour}}{\text{net output hp-hr or kWh}} \qquad (8.10)$$

ILLUSTRATIVE PROBLEM 8.8

Determine the heat rate and the steam rate per kilowatt-hour of plant output in Illustrative Problem 8.3.

SOLUTION

Neglecting pump work, we have

$$\text{heat rate} = \frac{3413}{0.273} = 12,502 \text{ Btu/kWh}$$

Per pound of steam, $1515 - 1150.5 = 364.5$ Btu is delivered. Because 1 kWh = 3413 Btu, the steam rate becomes $3413/364.5 = 9.36$ lb$_m$ of steam per kilowatt-hour.

8.5 THE REHEAT CYCLE

The simple Rankine cycle suffers from the fact that heat is added while the temperature varies, and excess moisture during the expansion process is detrimental to the performance and life of the turbine. Superheating tends to help the turbine, but thermodynamically, it would not appear to yield much of an increase in the thermal efficiency of the cycle. To achieve a higher thermal efficiency and to help solve the turbine problem, the *reheat cycle* was proposed. The basis of this cycle is an attempt to approach Carnot-cycle efficiency by adding heat in increments at the highest possible temperature level. The steam is permitted to expand part of the way in the turbine and is then returned to the boiler, where it is heated again (reheated) and subsequently reexpanded through the turbine. Although there is no theoretical limit to the number of stages of reheat that can be employed in a cycle, two, or at most three, are used in practice. A schematic and a T–s diagram of this cycle are shown in Figure 8.9.

It will be noted that this cycle is similar to the Rankine cycle, with the addition of the constant pressure heating from $5 \rightarrow 6$ and the second isentropic expansion from $6 \rightarrow 7$. In an actual cycle, pressure and heat losses will occur in each part of the cycle, but for the present, it will be assumed that all processes are carried out to prevent these losses from occurring. Neglecting pump work, the energy supplied to the cycle per pound of fluid circulated is the sum of the heat supplied during the original heating plus that supplied during reheating. Thus, the heat in is $(h_4 - h_2) + (h_6 - h_5)$, which can also be written as $(h_4 - h_1) + (h_6 - h_5)$ when pump work is neglected. The work out of the cycle is due to the two expansions. Thus, the work out is $(h_4 - h_5)_s + (h_6 - h_7)_s$. The efficiency of the cycle is

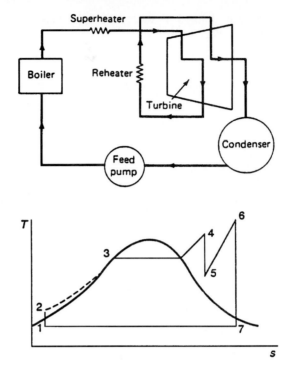

FIGURE 8.9 Simple reheat cycle.

$$\eta_{\text{reheat}} = \frac{(h_4 - h_5)_s + (h_6 - h_7)_s}{(h_4 - h_1) + (h_6 - h_5)} = \frac{(h_4 - h_7) + (h_6 - h_5)}{(h_4 - h_1) + (h_6 - h_5)} \qquad \textbf{(8.11)}$$

A reheat cycle is sometimes called an "ideal" reheat cycle if, in the first expansion, it is expanded isentropically to saturated vapor and, during the subsequent reheat, it is brought back to the same enthalpy that it had been before expansion, that is, $h_4 = h_6$.

ILLUSTRATIVE PROBLEM 8.9

Assume that the simple Rankine cycle described in Illustrative Problem 8.3 is made into a reheat cycle by first expanding the steam to 200 psia and then reheating it to the enthalpy that it had prior to the first expansion. If the final expansion is carried out to 14.696 psia, determine the efficiency of the reheat cycle and compare it to the simple cycle.

SOLUTION

The Mollier chart provides a convenient way of solving this problem. Expanding from 980°F, 400 psia, $s = 1.7567$ to 200 psia yields a final enthalpy of 1413 Btu/lb$_m$. Expanding from 200 psia and an enthalpy of 1515 Btu/lb$_m$ to 14.696 psia yields a final enthalpy of 1205 Btu/lb$_m$.

$$\eta_{\text{reheat}} = \frac{(1515 - 1205) + (1515 - 1413)}{(1515 - 180.15) + (1515 - 1413)} \times 100 = 28.7\%$$

It is apparent that for the conditions of this problem, the increase in efficiency is not very large. The final condition of the fluid after the second expansion is superheated steam at 14.696 psia. By condensing at this relatively high pressure condition, a large amount of heat is rejected to the condenser cooling water.

ILLUSTRATIVE PROBLEM 8.10

Solve Illustrative Problem 8.9 using the computerized properties.

SOLUTION

Some of the property data required was found in Illustrative Problem 8.4. In addition, we need

at 200 psia and $s = 1.7566$ Btu/lb$_m$·°R, $h = 1413.6$ Btu/lb$_m$

at 200 psia and $h = 1514.0$ Btu/lb$_m$, $s = 1.8320$ Btu/lb$_m$·°R

at 14.696 psia and $s = 1.8320$ Btu/lb$_m$·°R, $h = 1205.2$ Btu/lb$_m$

Using these data,

$$\eta_{\text{reheat}} = \frac{(1514.0 - 1205.2) + (1514.0 - 1413.6)}{(1514.0 - 180.17) + (1514.0 - 1413.6)} \times 100 = 28.53\%$$

Again, it is cautioned that a sketch of the problem on h–s or T–s coordinates should precede the use of the computerized *Steam Tables*.

8.6   THE REGENERATIVE CYCLE

As indicated in the preceding discussion, reheating is limited in its ability to improve the thermodynamic cycle efficiency and finds greatest usefulness in the reduction of moisture in the turbine. However, because the largest single loss of energy in a power plant occurs at the condenser in which heat is rejected to the coolant, it is pertinent to consider methods of reducing this rejected heat and improving cycle efficiency.

In both the ideal Rankine and reheat cycles, the condensate is returned to the boiler at the lowest temperature of the cycle. The fluid is heated to saturation by direct mixing in the steam drum of the boiler, by furnace radiation in the boiler tubes, or by gas convection heating by the flue gases in the economizer section of the unit. These methods depend on large temperature differences and are inherently irreversible. Rather than resorting to such procedures, which involve the internal arrangement of the boiler and its heat-transfer circuit, let us consider a method of feedwater heating that although idealized, leads to certain interesting and practical conclusions.

The most desirable method of heating the condensate would be by a continuous reversible method. Presuming that this is possible and imagining that the heating can be made to occur reversibly by an interchange within the turbine and in equilibrium with the expanding fluid, the T–s diagram will be given as shown in Figure 8.10. This diagram assumes saturated vapor at the initiation of expansion. Curve AB' is parallel to CD, because it was postulated that the heating is reversible. It will be noted that the increase in entropy during heating equals the decrease during the expansion and cooling of the vapor, and area $A'ABB'$ equals are $CD'D$. Thus, the cycle is equivalent to a Carnot cycle operating between the maximum and minimum temperatures T_1 and T_2. The efficiency is therefore given by Equation (8.1).

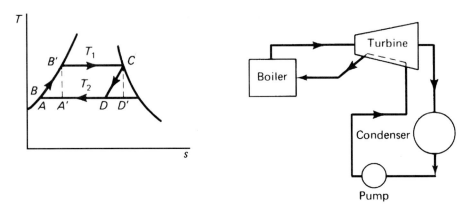

FIGURE 8.10   Ideal regenerative cycle.

It is interesting to note that by allowing the working fluid to deliver work and to transfer heat simultaneously in the manner indicated, the ideal-cycle efficiency approaches the Carnot efficiency as a limit.

In practice, this ideal cycle is approached by allowing the condensate from the feed pump to be heated in a separate heater (or heaters) by steam extracted from the turbine after it has partly expanded and done work. The extracted steam may mix directly with the condensate (as in an open heater) or may exchange heat indirectly and condense (as in a closed heater). A schematic of a practical cycle is shown in Figure 8.11.

The T–s diagram for a regenerative Rankine cycle using an open feedwater heater is shown in Figure 8.11a. Here, the feedwater heater takes regenerated steam from the turbine at an intermediate state 5 to be mixed with water pumped up from the condenser at state 8 to become hotter water at state 1. From there, a second feed pump delivers the liquid to the boiler at state 2. Superheated steam exits the boiler at state 4 and expands in the turbine to state 5, where a portion of the superheated steam is extracted and fed to the feedwater heater. The remaining steam completes its expansion in the turbine to state 6, where it enters the condenser to become liquid at state 7. From there, it is pumped up to the feedwater heater at state 8.

The T–s diagram for a Rankine cycle with two regenerative stages is shown in Figure 8.11b. Multiple regenerative stages are often used to get the maximum efficiency possible.

In large central station installations, from one to a dozen feedwater heaters are often used. These can attain lengths of more than 60 ft, diameters up to 7 ft, and have more than 30,000 ft^2 of surface. Figure 8.12 shows a straight condensing type of feedwater heater with steam entering at the center and flowing longitudinally, in a baffled path, on the outside of the tubes. Vents are provided to prevent the buildup of noncondensable gases in the heater and are especially important where

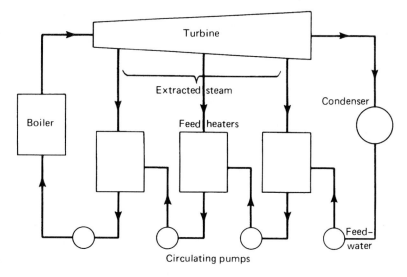

FIGURE 8.11   Regenerative cycle.

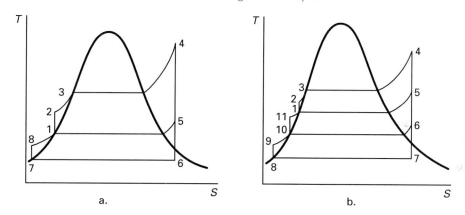

a.   b.

FIGURE 8.11a   Regenerative cycle with one feedwater heater.
FIGURE 8.11b   Regenerative cycle with two feedwater heaters.

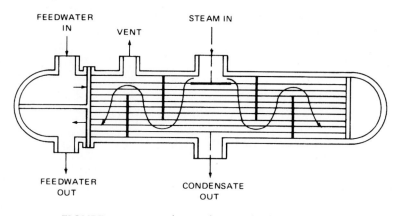

FIGURE 8.12   Straight condensing feedwater heater.

pressures are less than atmospheric pressure. Also, deaerators are often used in conjunction with feedwater heaters or as separate units to reduce the quantity of oxygen and other noncondensable gases in the feedwater to acceptable levels.

By combining the regenerative cycle with the reheat cycle, the modern regenerative–reheat cycle is obtained. An analysis of such a cycle involves writing a heat balance around each of the heaters to determine the quantities extracted. A subsequent heat balance around the cycle then establishes the cycle output and efficiency.

Figure 8.13 shows the heat balance for the large, modern central station shown in Figure 8.14. Note that there are seven heaters used and four turbine stages. Recently, the changes in fuel costs together with the problems in the nuclear industry have led to interest in the use of pulverized coal in conjunction with advanced steam conditions for use in power plants designed to be built for operation in the 1990s. A study of units having 800-MW nameplate capacity was conducted* for throttle conditions varying from 2400 psig to 4500 psig and temperatures up to 1050°F. This range of conditions represents the limits of present-day technology. Figure 8.15 shows the heat balance for the 4500-psig unit with double reheat using eight feed heaters. The turbine heat rate of 7283.4 Btu/kWh represents an overall plant efficiency of 46.9%. For the temperatures of the study (1050°F and 108.7°F corresponding to 2.5 in. Hg absolute), the type efficiency is found to be 75.2%, which is excellent.

---

* The author is indebted to H. Waxberg of Ebasco Services, Inc., who performed this study and made its results available.

FIGURE 8.14   Aerial view of the Charles Poletti Power Project in Astoria, New York.
(Courtesy of the Power Authority of the State of New York.)

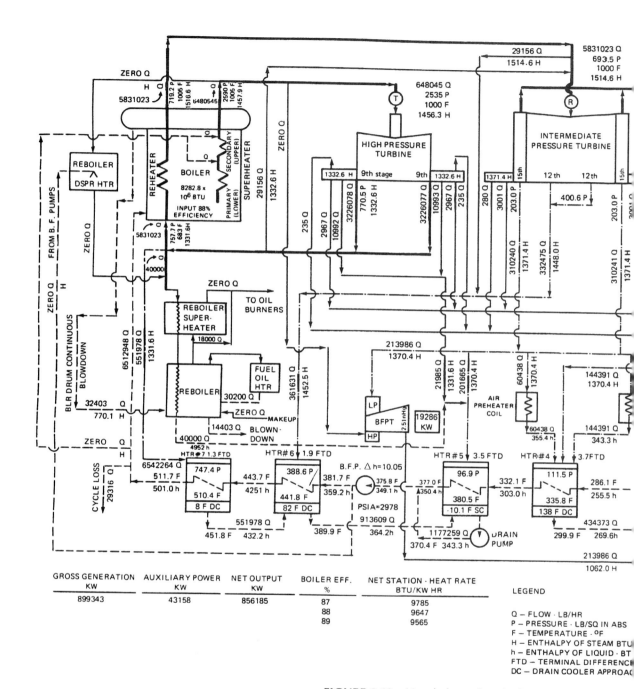

FIGURE 8.13 Heat balance for Charles Poletti Power Proje
(Courtesy of the Power Authority

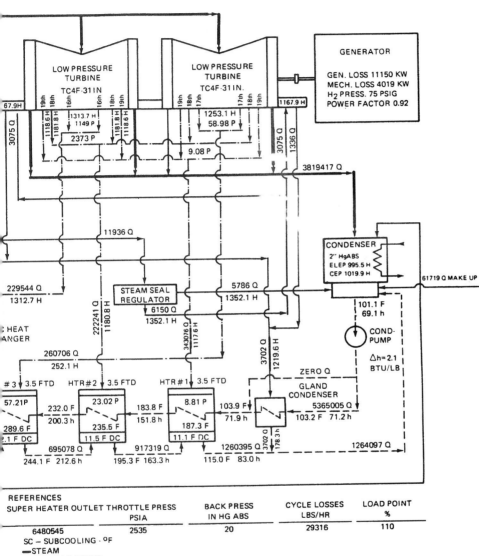

REFERENCES

SUPER HEATER OUTLET THROTTLE PRESS

	PSIA	BACK PRESS IN HG ABS	CYCLE LOSSES LBS/HR	LOAD POINT %
6480545	2535	20	29316	110

SC – SUBCOOLING · °F
— STEAM
— STEAM LEAKAGE
—·— EXTRACTION
——— CONDENSATE
T – THROTTLE VALVE
R – REHEAT INTERCEPT VALVE

...toria, New York, at 110% load point.
...ew York.)

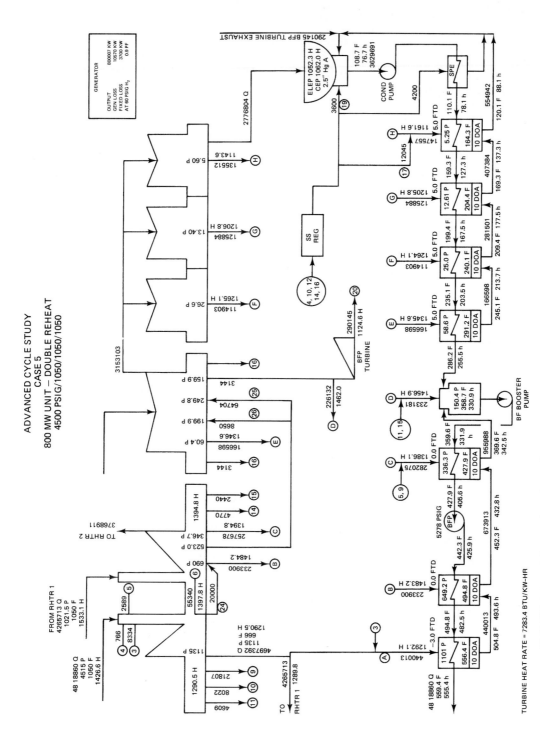

FIGURE 8.15 Advanced cycle study, case 5, 800-MW unit—double reheat 4500 psig/1050/1050/1050.

ILLUSTRATIVE PROBLEM 8.11

A Rankine cycle operates with an upper pressure of 1000 psia and an upper temperature of 1000°F. (a) If the condenser is at 1 psia, determine the efficiency of this cycle. (b) Compare it to a regenerative cycle operating under the same conditions with a single heater operating at 50 psia. Neglect pump work.

SOLUTION

(a) For the Rankine cycle, the Mollier chart gives

$$h_4 = 1505 \text{ Btu/lb}_m \qquad h_5 = 922 \text{ Btu/lb}_m$$

and at the condenser,

$$h_1 = 69.74 \text{ Btu/lb}_m$$

Then

$$\eta_R = \frac{1505 - 922}{1505 - 69.74} \times 100 = 40.6\%$$

(b) Figure 8.16 shows the regenerative cycle. After doing work (isentropically), W lbs of steam are bled from the turbine at 50 psia for each lb_m of steam leaving the steam generator, and $(1 - W)$ pound goes through the turbine and is condensed in the condenser to saturated liquid at 1 psia. This condensate is pumped to the heater, where it mixes with the extracted steam and leaves as saturated liquid at 50 psia. The required enthalpies are:

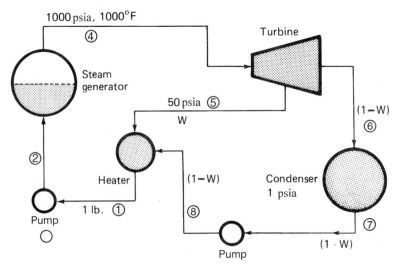

FIGURE 8.16 Illustrative Problem 8.11.

Leaving turbine: h_5 = 1168 Btu/lb$_m$ at 50 psia
Leaving condenser: h_7 = 69.74 Btu/lb$_m$ at 1 psia
                    = h_8 if pump work is neglected
Leaving heater: h_1 = 250.24 Btu/lb$_m$ at 50 psia
                    = h_2 if pump work is neglected (saturated liquid)

A heat balance around the heater (see Chapter 3.5f) gives

$$W(1168) + (1 - W)(69.74) = (1)(250.24)$$
$$W = 0.1644 \text{ lb}_m$$

The work output is

$$\text{work} = (1 - W)(1505 - 922) + W(1505 - 1168)$$
$$= (1 - 0.1644)(1505 - 922) + (0.1644)(1505 - 1168)$$
$$= 542.56 \text{ Btu/lb}_m$$

Heat into steam generator equals the enthalpy leaving minus the enthalpy of the saturated liquid entering at 50 psia:

$$q_{in} = 1505 - 250.24 = 1254.8 \text{ Btu/lb}_m$$

The efficiency is

$$\eta = \frac{\text{work}}{\text{heat in}} = \frac{542.56}{1254.8} \times 100 = 43.2\%$$

When compared to the simple Rankine cycle, it is seen that a single heater raises the efficiency by 2.6% for the same operating limits. This is a considerable efficiency increase.

It can also be shown that the efficiency of a regenerative cycle with one open heater is given by

$$\eta = \left[1 - \frac{(h_5 - h_1)(h_6 - h_7)}{(h_4 - h_1)(h_5 - h_7)} \right] \times 100 \qquad \textbf{(a)}$$

where the subscripts refer to those shown in Figure 8.16 with pump work neglected. Also,

$$W = \left(\frac{h_1 - h_7}{h_5 - h_7} \right) \qquad \textbf{(b)}$$

For this problem, $W = (250.24 - 69.74)/(1168 - 69.74) = 0.1644$ lb$_m$.

$$\eta = \left[1 - \frac{(1168 - 250.24)(922 - 69.74)}{(1505 - 250.24)(1168 - 69.74)} \right] \times 100 = 43.2\%$$

If a second open heater is used in Illustrative Problem 8.11 with extraction from the turbine at 100 psia, determine the efficiency of the cycle. Compare your results to Illustrative Problem 8.11. The T–s diagram for this problem is shown in Figure 8.11b.

SOLUTION

Figure 8.16(a) shows the cycle. For this cycle, W_2 pounds are extracted at 100 psia, and W_1 pounds are extracted at 50 psia for each pound produced by the steam generator. The enthalpies that are required are:

Leaving turbine:	922 Btu/lb$_m$ at 1 psia
Leaving condenser:	69.74 Btu/lb$_m$ at 1 psia (saturated liquid)
Leaving low pressure heater:	250.24 Btu/lb$_m$ at 50 psia (saturated liquid)
Leaving high pressure heater:	298.61 Btu/lb$_m$ at 100 psia
At low pressure extraction:	1168 Btu/lb$_m$ at 50 psia
At high pressure extraction:	1228.6 Btu/lb$_m$ at 100 psia
Entering turbine:	1505 Btu/lb$_m$

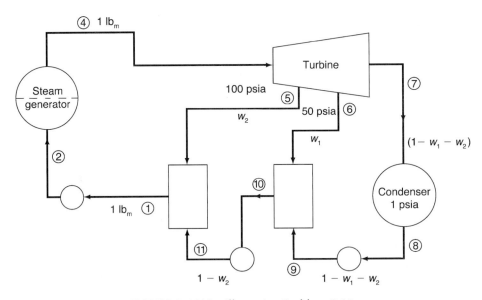

FIGURE 8.16(a)   Illustrative Problem 8.12.

A heat balance around the high pressure heater gives us

$$W_2(1228.6) + (1 - W_2)250.24 = 1(298.61)$$
$$W_2 = 0.049 \text{ lb}_m$$

A heat balance around the low-pressure heater yields

$$W_1(1168) + (1 - W_1 - W_2)69.74 = (1 - W_2)250.24$$
$$W_1 = 0.156 \text{ lb}_\text{m}$$

The work output is

$$\text{work} = (1505 - 1228.6)1 + (1 - W_2)(1228.6 - 1168)$$
$$+ (1 - W_1 - W_2)(1168 - 922)$$
$$= 529.6 \text{ Btu/lb}_\text{m}$$

Heat into the steam generator equals the enthalpy leaving minus the enthalpy of saturated liquid at 100 psia:

$$q_\text{in} = (1505 - 298.61) = 1206.4 \text{ Btu/lb}_\text{m}$$

The efficiency is

$$\eta = \frac{\text{work}}{\text{heat in}} = \frac{529.6}{1206.4} \times 100 = 43.9\%$$

The addition of a second heater increases the thermal efficiency of the cycle by 0.7% over the cycle with one heater.

It is possible to obtain analytical expressions for W_1, W_2, and η. This entails considerable algebraic manipulation. However, the results are useful for computer programming. In terms of Figure 8.16a,

$$W_2 = \frac{h_1 - h_{11}}{h_5 - h_{11}} \qquad \textbf{(a)}$$

$$W_1 = \left(\frac{h_5 - h_1}{h_6 - h_9}\right)\left(\frac{h_{10} - h_9}{h_5 - h_{10}}\right) \qquad \text{Neglecting pump work,} \qquad \textbf{(b)}$$

$$\eta = 1 - \left(\frac{h_7 - h_8}{h_4 - h_1}\right)\left(\frac{h_5 - h_1}{h_5 - h_{10}}\right)\left(\frac{h_6 - h_{10}}{h_6 - h_8}\right) \qquad \textbf{(c)}$$

For this problem, $h_8 = h_9$, $h_{10} = h_{11}$, and $h_1 = h_2$. Thus,

$$W_2 = \frac{298.61 - 250.24}{1228.6 - 250.24} = 0.049 \text{ lb}_\text{m}$$

$$W_1 = \left(\frac{1228.6 - 298.61}{1168 - 69.74}\right)\left(\frac{250.24 - 69.74}{1228.6 - 250.24}\right) = 0.156 \text{ lb}_\text{m}$$

$$\eta = 1 - \left(\frac{922 - 69.74}{1505 - 298.61}\right)\left(\frac{1228.6 - 298.61}{1228.6 - 250.24}\right)\left(\frac{1168 - 250.24}{1168 - 69.74}\right)$$
$$= 43.9\%$$

It is strongly suggested that a sketch similar to Figure 8.16a be used for all regenerative problems to avoid errors in terms.

As noted earlier, the use of several feedwater heaters in a regenerative cycle increases the efficiency of the cycle. Most power plants incorporate at least one open heater, one in which the fluid streams are mixed, operating at a pressure greater than atmospheric pressure to vent oxygen and other dissolved gases from the cycle in order to minimize corrosion. This process of removing dissolved gases is known as deaeration. Closed feedwater heaters are essentially shell-and-tube heat exchangers with the extracted steam condensing outside the tubes on the shell side of the heat exchanger and feedwater being heated while flowing through the tubes. The two fluid streams do not mix and are at different pressures. As shown in Figure 8.17, the condensate from a straight condensing, closed feedwater heater passes through a trap to a feedwater heater operating at a lower pressure or into the condenser. The trap is a valve that permits liquid only to pass to a lower pressure region. The liquid undergoes a throttling process in the trap as it flows to the lower pressure region. The operation of an open and closed heater in a regenerative cycle is shown in Illustrative Problem 8.13.

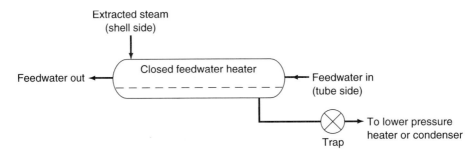

FIGURE 8.17   Straight condensing, closed feedwater heater with trap.

ILLUSTRATIVE PROBLEM 8.13

A regenerative cycle operates with an upper pressure of 1000 psia and an upper temperature of 1000°F. Determine the efficiency of the cycle if it has a single open heater at 50 psia and single closed heater at 100 psia. The condenser operates at 1 psia. Assume that the condensate from the closed heater is saturated liquid at 100 psia and that it is trapped (throttled) into the open feedwater heater. The enthalpy of the feedwater leaving the closed heater can be taken as that of saturated liquid at 310°F. Pump work can be neglected. Compare your results with Illustrative Problem 8.11.

SOLUTION

Figure 8.18 shows the regenerative cycle for this problem. Assume that 1 lb$_m$ of steam leaves the steam generator and that W_1 lb$_m$ is bled off to the closed heater at 100 psia and that W_2 lb$_m$ is bled off to the open heater at 50 psia. Also, assume that the feedwater leaving the closed heater is at 310°F, 18°F less than the saturation temperature corresponding to 100 psia. For calculation purposes, we will use h_f at 310°F for this

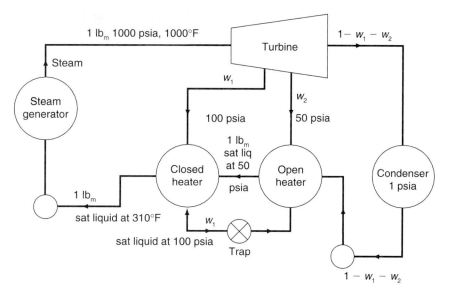

FIGURE 8.18 Illustrative Problem 8.13.

enthalpy. Using the Mollier diagram and the *Steam Tables,* we find the following values of enthalpy:

h to turbine $= 1505$ Btu/lb$_m$ (at 1000 psia and 1000°F)
h at first extraction $= 1228$ Btu/lb$_m$ (isentropically to 100 psia)
h at second extraction $= 1168$ Btu/lb$_m$ (isentropically to 50 psia)
h at turbine exit $= 922$ Btu/lb$_m$ (isentropically to 1 psia)
$h_f = 298.61$ Btu/lb$_m$ (at 100 psia)
$h_f = 250.24$ Btu/lb$_m$ (at 50 psia)
$h_f = 280.06$ Btu/lb$_m$ (at 310°F)
$h_f = 69.74$ Btu/lb$_m$ (at 1 psia)

A heat balance around the high pressure heater gives us

$$W_1(1228 - 298.61) = 1(280.06 - 250.24)$$

Solving, $$W_1 = 0.032 \text{ lb}_m$$

A heat balance around the open heater yields

$$W_2(1168) + (1 - 0.032 - W_2)69.74 + 0.032(298.61) = 1(250.24)$$

Solving for W_2, $W_2 = 0.158$ lb$_m$.

The work output of the cycle consists of the work that 1 lb$_m$ does in expanding isentropically to 100 psia, plus the work done by $(1 - W_1)$ lb$_m$ expanding isentropically from 100 to 50 psia, plus the work done by $(1 - W_1 - W_2)$ lb$_m$ expanding isentropically from

50 to 1 psia. Numerically, the work is $1(1505 - 1228) + (1 - 0.032)(1228 - 1168) + (1 - 0.032 - 0.158)(1168 - 922) = 534.34$ Btu/lb$_m$.

The heat input is $(1505 - 280.06) = 1224.94$ Btu/lb$_m$.

The efficiency of the cycle is

$$\eta = \frac{\text{work output}}{\text{heat input}} = \frac{534.34 \text{ Btu/lb}_m}{1224.94 \text{ Btu/lb}_m} \times 100 = 43.6\%$$

When compared to Illustrative Problem 8.11, we see that one open heater at 50 psia raised the efficiency of the simple Rankine cycle by 2.6%. The addition of an additional closed heater at 100 psia raises the efficiency by an additional 0.4%.

8.7 THE STEAM GENERATOR

At this time, we will consider the major pieces of equipment that are used in steam power plants. The first unit will be the boiler. The term *boiler* is often used to broadly describe the device known more properly as a steam-generating unit. The boiler section of a steam generator refers to the elements in which the change of state from water to steam takes place. The steam generator consists of the boiler section, the superheater, reheater, economizer, air heater, fuel system, air system, and ash-removal system. Thus, a complete steam generator is a combination of many elements, all integrated to yield an economical, efficient, reliable unit.

Let us examine the unit shown in Figure 8.19 and trace out all the elements that constitute this type of steam generator. The unit shown is known as a water-tube unit, in which steam is generated within the tubes and the combustion gases are outside the tubes. The path of the water can be traced out by noting that after leaving the feed pumps, it is usually pumped into the steam drum. The water is subcooled at the pump exit, and in the steam drum, it mixes with a steam–water mixture. This causes some of the steam to condense in the steam drum in order to heat the feed to saturation temperature. Depending on the particular design, it is possible for as much as 30% of the steam actually generated to be condensed to heat the feedwater in the steam drum. The steam drum acts as a collecting device and releasing point for the steam generated. In the water-tube boiler, water and steam flow in a relatively large number of externally heated paths. Circulation of the steam–water mixture in the tubes can be brought about in two ways. The first is illustrated in Figure 8.20 and is called natural circulation. In the circuit shown, steam is formed on the heated side. The steam–water mixture weighs less than the cooler mixture on the unheated downcomer side and is displaced by water circulating convectively in the circuit. The difference in force between the water leg in the downcomer and the steam–water mixture in the riser causes flow to occur. An equilibrium velocity is reached in the circuit when the force difference between the downcomer and riser equals the flow losses in the circuits. If the available pressure difference is greater than the flow resistance in the circuits, the flow will increase until a balance point is reached between available force and resistance. The steam–water mixture then goes to the steam drum, where the steam is released and the water is returned to the downcomer circuit. Unless adequate circulation is maintained, it is possible to have a stationary local steam pocket, which in turn leads to local overheating of the tube and, ultimately, to tube failure (burnout).

It will be recalled that at the critical point, there is no distinction between the vapor and liquid. Also, as the critical point is approached, the difference in density between the liquid and vapor decreases. Figure 8.21 shows this effect. To assure adequate circulation at all loads, some steam

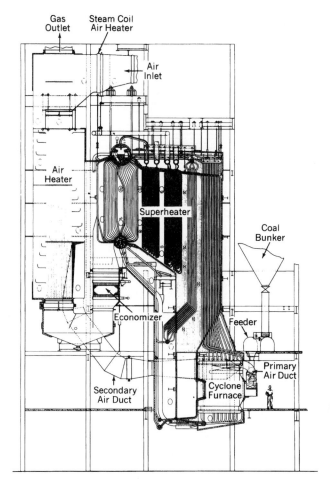

FIGURE 8.19 Modern water-tube steam generator. Design pressure, 1575 psia; steam temperature 900°F; maximum continuous steam output, 550,000 lb/hr.
(Courtesy of Babcock and Wilcox Co.)

generators use a pump unit to overcome circuit resistance. If this is done, it is usually done in high-pressure units where natural circulation forces are relatively small or in low-pressure units to give freedom of tube layout. Figure 8.22a shows a schematic of a forced circulation unit with a drum for steam separation. Figure 8.22b shows a schematic of a once-through forced circulation steam generator where all the water entering the tubes is vaporized and leaves the tubes as steam. As indicated, for pressures below the critical pressure, a separator may be used to remove any moisture in the steam.

The steam drum plays an important role in most boiler designs. The steam drum is the place where the cleaning and drying of the saturated steam is performed and the feedwater is distributed, and it provides protection for the unit. Large-diameter steam drums provide conservative steam release rates per square foot of separation surface, assuring high steam purity and stable water levels even with fluctuating loads. Figure 8.23 shows the interior of a steam drum in which

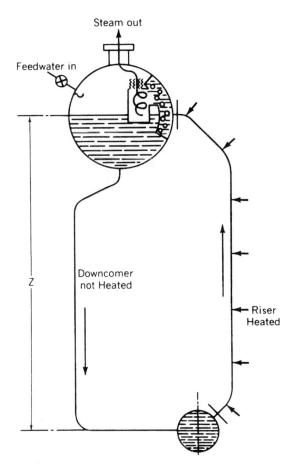

Steam out

Feedwater in

Downcomer
not Heated

Z

Riser
Heated

FIGURE 8.20 Schematic of natural circulation.
(Courtesy of Babcock and Wilcox Co.)

steam enters the drum through a primary separator, which removes carryover of water from the discharge of the generating tube and returns it to the drum water. The steam then passes through the closely spaced, Z-shaped steel fins of the water-cooled condenser tube elements, being, in effect, washed and scrubbed in its own condensate. In this design, sectionalized dryer cartons remove the remaining moisture from the steam before it enters the superheater. Steam drum diameters are of the order of 60 in. in many installations.

Clean steam is important, because modern turbines operate at rated capacity and high rates of speed for long periods. It is essential from the standpoint of safety to these prime movers, as well as to maintenance of their efficiency and capacity, that the blades remain clean. Deposits, however, are selective. Also, superheater tubing may overheat if solids are deposited in them. Because solids in steam are carried by the moisture associated with it, steam purification (as to solids, not gases) is primarily a matter of moisture removal. Modern power boilers deliver steam containing less than one part impurities per million. This is truly remarkable, because boiler-water impurity concentrations may run from a few to thousands of parts per million. It is interesting to note that steam is probably

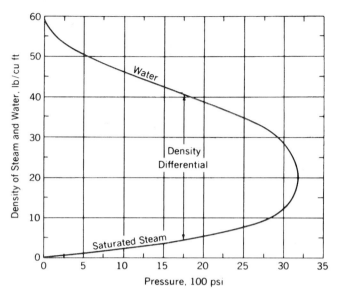

FIGURE 8.21 Specific weight of steam and water.
(Courtesy of Babcock and Wilcox Co.)

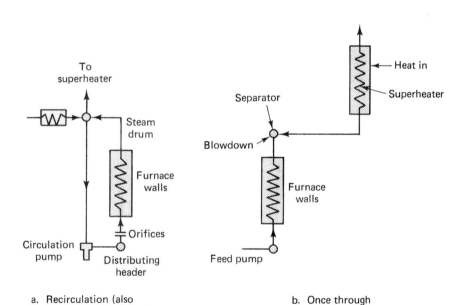

a. Recirculation (also controlled circulation)

b. Once through

FIGURE 8.22 Forced circulation steam generators.
(Courtesy of Combustion Engineering, Inc.)

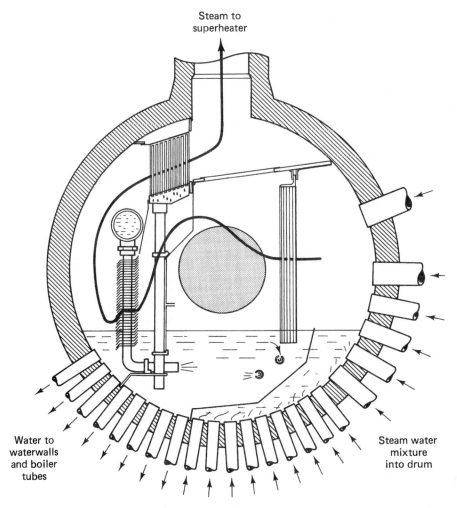

FIGURE 8.23 Steam drum internals.
(Courtesy of Riley Stoker Corp.)

the purest substance produced commercially—purer than 99.9999%. Commercially distilled water would be intolerably dirty as steam. While drum size has an important bearing on steam purity, internal baffling is even more effective. All such baffles are designed to use gravity, centrifugal force, and inertia to accomplish separation. Preliminary separation of water from steam is obtained by the manner in which the mixture is delivered from the tubes to the drum and by baffles that change the direction of flow. Because foam promotes carryover, baffles that direct the water toward a foam blanket to beat down minute bubbles have been used. A secondary separator may remove most of the moisture still remaining in the steam, after which the steam passes through a drier. Driers usually consist of several layers of screen or closely spaced undulated plates. They permit passage of steam without much resistance while offering a large area on which droplets may deposit.

For details of the design of the internal components of steam generators, the reader should consult *Steam Power* by the Babcock and Wilcox Company.

8.8 THE STEAM TURBINE

The steam leaves the steam generator and goes to the turbine where it expands to give useful work. There are many different types, sizes, and shapes of turbines depending on the specific application. Some idea of the size range can be obtained from Figure 8.24, which shows a small single-stage turbine and a large central station multistage turbine. Due to the large number of combinations possible, let us start our discussion with the blading and the basic principles used to convert the energy in the steam to useful work. The two classifications of blading used are *impulse blading* and *reaction blading*. The basic action that occurs in impulse blading is that the expansion of steam takes place in the nozzles of the turbine and not in the blading. In the simple single impulse stage shown in Figure 8.25a, the nozzles direct the steam into the moving blades (buckets) mounted on the rotor (see also Figure 8.24). In the single impulse stage, all the pressure drop is shown to occur in the nozzles. As the steam flows through the passages of the moving vanes, it is turned by the symmetrical moving blades in an axial direction, and at the same time, the steam does work on the blades. Thus, the final velocity of the steam is lower than at entry, and also, the enthalpy of the steam is lower. When two or more such stages are placed in series, this is known as *pressure compounding* (also known as Rateau staging). In a pressure-compounded impulse turbine, Figure 8.25b, the pressure drop is taken in equal steps in the alternate fixed nozzles. In Figure 8.25c, we have a velocity-compound impulse blading situation (also known as Curtis staging). In this arrangement, the complex expansion of the steam takes place in the fixed nozzles before the first stage. The steam gives up part of its kinetic energy in the first row of moving blades and is reversed in direction in the fixed blades. The fixed blading directs the steam into the second row of blades, where the remainder of the kinetic energy is absorbed.

FIGURE 8.24 Comparison of turbine sizes: 20-in. wheel diameter, single-stage turbine (courtesy of Worthington Corp.); large turbine during shop fabrication (courtesy of Westinghouse Electric Corp.).

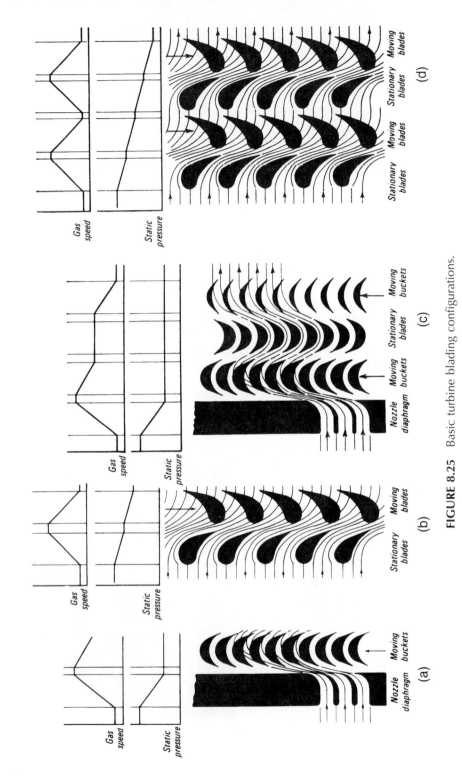

FIGURE 8.25 Basic turbine blading configurations.

In the reaction blading (also known as Parson's blading), a partial pressure drop occurs in the fixed blades, and the velocity of the steam is increased. The steam then flows between moving blades that are shaped to act as nozzles. A further pressure drop occurs in the moving blades, increasing the velocity of the steam relative to the blades. This expansion in the moving blades causes a reaction against the blades that does work in driving the rotor. Sufficient stages are employed to expand the steam down to the desired exhaust pressure. The right side of Figure 8.25d shows reaction stages being fed from velocity-compounded impulse staging. This is frequently done in multistage turbines where reaction staging is used in the latter stages due to their increased efficiency under these conditions.

There are two basic turbine types that depend on the exhaust-steam conditions. In the condensing type, the exhaust steam leaves at below atmospheric pressure, while in the noncondensing type, the steam is exhausted at pressures above atmospheric. In addition, extraction methods can be further categorized as straightflow, reheat, automatic, or nonautomatic. Figure 8.26 shows the common basic turbine types.

Figure 8.27 shows a straight, noncondensing steam turbine that is used for industrial applications. This turbine is used in applications requiring large quantities of single pressure steam for process or other use and where the unit can be operated in parallel with other units that can absorb load variations. This type of turbine is usually controlled by an exhaust pressure governor, which maintains the desired pressure of the process steam exhausted from the turbine. Noncondensing turbines are also used in topping arrangements, with steam exhausting to existing generating equipment to increase overall plant efficiency and output.

Figure 8.28 is also an industrial-type turbine, but the unit is a double automatic extraction condensing steam turbine. As shown, this turbine consists of high-, intermediate-, and low-pressure sections. Units of this type are used in applications demanding continuous steam at two pressure levels. During times of high process flow, the steam flowing to the condenser generates the electric power. An automatic electrohydraulic control maintains the extraction pressures even when the load on the unit fluctuates and extraction steam flows vary. Once the desired extraction pressure for the process is set, the turbine automatically holds that pressure. The steam flow is regulated by cam-operated, upper inlet control popper valves. These valves are designed to assure efficient handling and accurate governing of large volumes of steam.

The design of a large turbine for power generation may differ in detail from the units we have discussed thus far. Figure 8.29 shows a 3600-rpm, tandem-compound, four-flow, reheat steam turbine. This turbine is rated from 550,000 to 850,000 kW, with steam conditions to 3500 psig and steam at 1000°F. The 850,000-kW size has 33.5-in.-high last-stage buckets. The exhaust from the high-pressure portion goes to the reheater in the furnace, where the temperature is again raised to 1000°F. From the reheater, the steam returns to the double-flow intermediate section. The steam enters at the center of the reheat section and splits into two flow streams in opposite directions along the shaft of this section. After leaving the reheat section, the steam flows to the two low-pressure sections. Each low-pressure section is a double-flow section receiving steam at a central inlet. The steam divides and flows through low-pressure stages in opposite directions. These last stages are frequently followed by diffusers, which recover some of the kinetic energy from the steam before the steam flows to the condenser below. The latter stage blades are warped, because the diameter is large and the blade-velocity variation from tip to hub is also very large. Some typical turbine blades are shown in Figure 8.30.

Non-condensing

A straight non-condensing turbine is the economic choice when all exhaust steam can be used for process or heating. If high-pressure steam is available for the turbine and the lower pressure exhaust can be used to supply other prime movers or a process, a non-condensing turbine can be applied to increase power output for the same fuel consumption.

Condensing

A straight condensing turbine is used when electrical power generation is the only concern and it must be produced on a minimum amount of steam. To improve the plant's efficiency these units can be provided with multiple uncontrolled extraction openings for feedwater heating.

Condensing controlled extraction

A condensing controlled extraction turbine bleeds off part of the main steam flow at one (single extraction) or two (double extraction) points. It is applied where process steam is required at pressures below the inlet pressure.

Condensing mixed-pressure

A condensing mixed-pressure turbine automatically uses as much low-pressure steam as available, supplementing it with as much high-pressure steam as required to carry the load. This type is used when more power must be generated than is possible with the available low-pressure steam.

Mixed-pressure extraction

This unit operates either as a mixed-pressure or an extraction turbine, depending on steam balance. As the turbine generates power, it holds constant pressure on the low-pressure steam line, extracting or admitting steam as required.

Non-condensing controlled extraction

A non-condensing controlled extraction turbine bleeds part of the main steam flow at one (single extraction) or two (double extraction) points. It is applied where process steam is required at several different pressures with variable or intermittent flows.

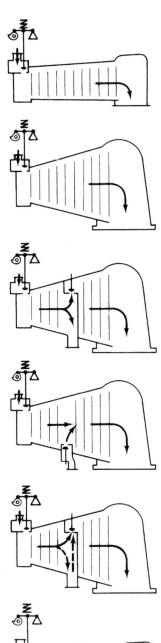

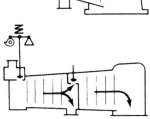

FIGURE 8.26 Basic turbine types.
(Courtesy of Worthington Corp.)

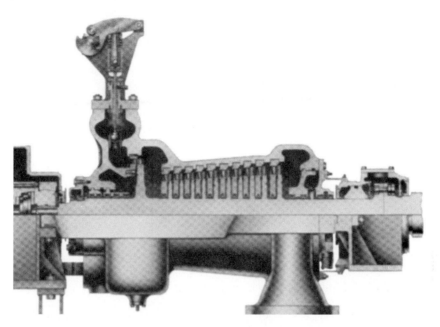

FIGURE 8.27   Straight, noncondensing steam turbine.
(Courtesy of General Electric Co.)

FIGURE 8.28   Double automatic, extraction condensing steam turbine.
(Courtesy of General Electric Co.)

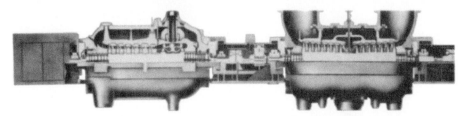

H-P Section   Double-Flow Reheat Section

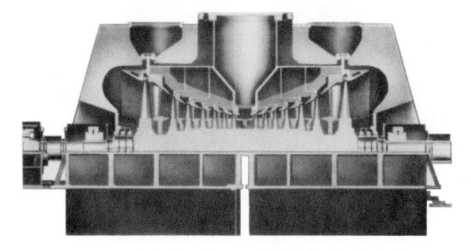

Typical Double-Flow Low-Pressure Section

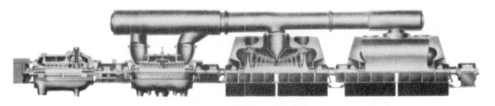

H-P Section   Reheat Section   L-P "A"   L-P "B"

FIGURE 8.29 Tandem-compound reheat steam turbine.
(Courtesy of General Electric Co.)

8.9 OTHER POWER-PLANT CYCLES

In the discussion so far, water has been the working fluid of the cycle because of its abundance, ease of handling, and nontoxicity. Actually, water has many disadvantages as a working fluid for a power-plant cycle. The vapor pressure of water is high; this necessitates thick piping and thick pressure vessels and gives rise to other design problems. From the standpoint of thermal efficiency, the high pressure is not necessary, because it is temperature that controls the efficiency of a cycle. Superheated steam is also a poor heat-transfer fluid and, from the equipment standpoint, requires larger heat-transfer surfaces and limits the heat inputs to the surface. Thermodynamically, it means

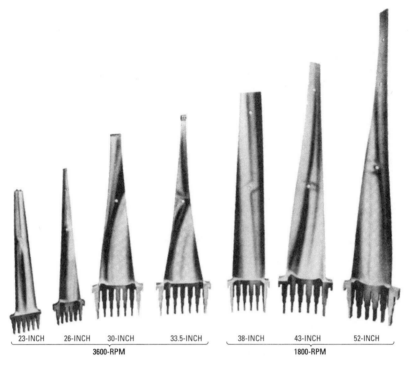

23-INCH 26-INCH 30-INCH 33.5-INCH 38-INCH 43-INCH 52-INCH

3600-RPM 1800-RPM

FIGURE 8.30 Typical turbine blading.
(Courtesy of General Electric Co.)

that large temperature differences must exist in parts of the equipment in order to transfer the heat. This obviously is not desirable.

Other qualities that make water undesirable are that it changes phase in the turbine; its lowest temperature in the cycle takes it below atmospheric pressure, requiring leaktight construction and pumps on the condenser; and it is corrosive and needs conditioning in order to be used in boilers. It also has a high specific heat of the vapor and a high specific heat of the liquid, which means that the temperature varies as heat is added (from the T–s diagram, it will be seen that a nearly vertical line for the water-heating part of the cycle makes it approach the Carnot cycle; a very low specific heat of the liquid is therefore desired).

In the binary vapor cycle, some of the advantages of water are utilized in conjunction with another fluid that has merit in other parts of the cycle. In the arrangement used in conventional power plants (as opposed to nuclear), mercury has been the high-temperature working fluid, and the heat rejected by the mercury part of the cycle has been the heat source for the steam part of the cycle. A schematic of such a cycle is shown in Figure 8.31.

A recent application of the *binary fluid cycle* is the OTEC (*ocean thermal energy conversion*) plant.* This plant uses the warm surface water of the ocean as a heat source and the cold water

---

* The author is indebted to Lockheed Missiles and Space Co., Inc., for having provided much of the material in this section on OTEC.

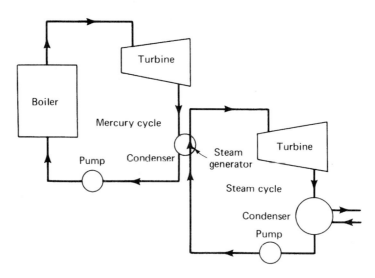

FIGURE 8.31 Binary fluid cycle.

from the ocean depths as a heat sink. The warm water vaporizes a liquid (1), such as ammonia, which, like steam, drives the turbine generator (2). The ammonia then is condensed to its liquid form (3) by the cold ocean water, and the closed cycle continues. Figure 8.32 shows this cycle schematically.

A demonstration plant, known as the Mini-OTEC, has been built and tested. The Mini-OTEC project objective was to design, build, and operate a closed-loop OTEC plant at sea to demonstrate technical feasibility. Tasks included site selection, design, fabrication, deployment, and operation.

The basic operating system in Mini-OTEC is a closed-loop ammonia cycle that powers a turbine-driven electric generator (50 kW). Warm surface water of about 80°F is used as a heat source to vaporize the ammonia. The vapor passes through the turbine and discharges into a condenser, where the ammonia condenses into a liquid to be pumped to the evaporator, repeating the cycle. Seawater from about 2200 ft at a temperature of 43°F is pumped through the condenser as a heat sink. Figure 8.33 shows an aerial view of Mini-OTEC on station and Figure 8.34 shows the power-plant schematic for this system. Mini-OTEC was made from off-the-shelf components with a plant layout dictated by the available barge. Plant operations began some 15 months after initiation of design and demonstrated that OTEC can extract energy from the ocean on a continuous basis, 24 hours a day.

Figure 8.35 shows an artist's conception of a large, central station OTEC plant, and Figure 8.36 shows a schematic of this plant. When fully developed, an OTEC plant of this type should be able to supply significant amounts of electric power at competitive prices. The U.S. Department of Energy, Solar Research Institute, is investigating the use of a thermoelectric OTEC concept. This system is simpler, safer, and potentially more reliable than ammonia closed-cycle designs.

8.10 NUCLEAR REACTOR POWER CYCLES

The advent of nuclear power for central stations, marine, and naval power plants has necessitated the study of the power production cycle from viewpoints that are novel compared to the technology

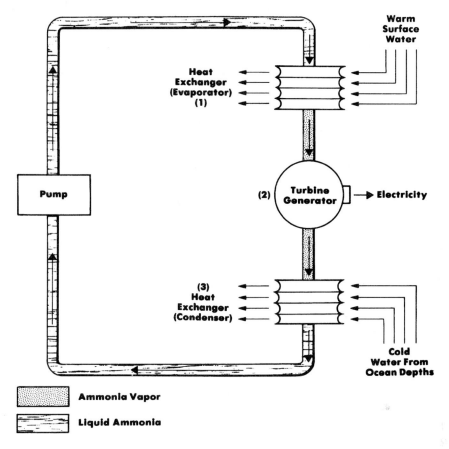

FIGURE 8.32   OTEC cycle.

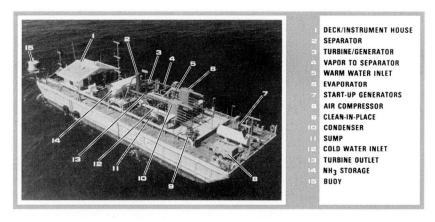

1	DECK/INSTRUMENT HOUSE
2	SEPARATOR
3	TURBINE/GENERATOR
4	VAPOR TO SEPARATOR
5	WARM WATER INLET
6	EVAPORATOR
7	START-UP GENERATORS
8	AIR COMPRESSOR
9	CLEAN-IN-PLACE
10	CONDENSER
11	SUMP
12	COLD WATER INLET
13	TURBINE OUTLET
14	NH_3 STORAGE
15	BUOY

FIGURE 8.33   Aerial view of Mini-OTEC on station.

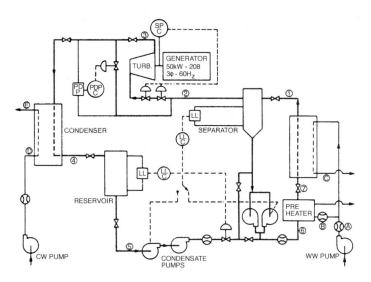

STATION	1	2	3	4	5	6	7		A	B	C	D	E
PRESSURE (PSIA)	130.1	130.0	92.0	92.3	92.5	136.5	132.3						
TEMPERATURE (°F)	70.7	70.5	51.6	48.0	48.0	49.3	70.6		80	80	75	43	50
ENTHALPY (BTU/LB)	404.1	629.0	609.8	95.7	95.7	96.9	121.3						
DENSITY (LB/FT³)	11.7	0.47	1.47	39.1	39.1	39.1	38.0						
QUALITY (%)	70.0	99.9	97.0	0.0	0.0	0.0	0.0						
FLOW (CPS)	0.62	10.85	3.47	58.5 GPM	58.5 GPM	83.8 GPM	83.8 GPM		2700 GPM	90 GPM	2610 GPM	2700 – 2900 GPM	
FLOW (LB/SEC)	7.3	5.1	5.1	5.1	5.1	7.3	7.3						

a. Power System Schematic

b. Turbine-Generator Specification

Inlet Pressure	130.0 psia
Inlet Temperature	70.5°F
Discharge Pressure	92.0 psia
Discharge Temperature	51.6°F
RPM	
High Speed	28,200
Low Speed	3,600
Generator Nameplate	50 kW
Overall Efficiency (T-G)	56%

FIGURE 8.34 Mini-OTEC demonstration plant.

that has been discussed earlier in this chapter. The extent of the use of nuclear plants for power production can be realized when the present and future installed capacities are studied. At present, 13% of the electrical energy produced in the U.S. is derived from central station nuclear power plants. The events that occurred in March 1979 at the Three Mile Island plant in Pennsylvania and the subsequent reappraisal of the safety problems by the Nuclear Regulatory Commission have caused the power industry to re-evaluate its position on the safe use of nuclear power. The reactor failure at

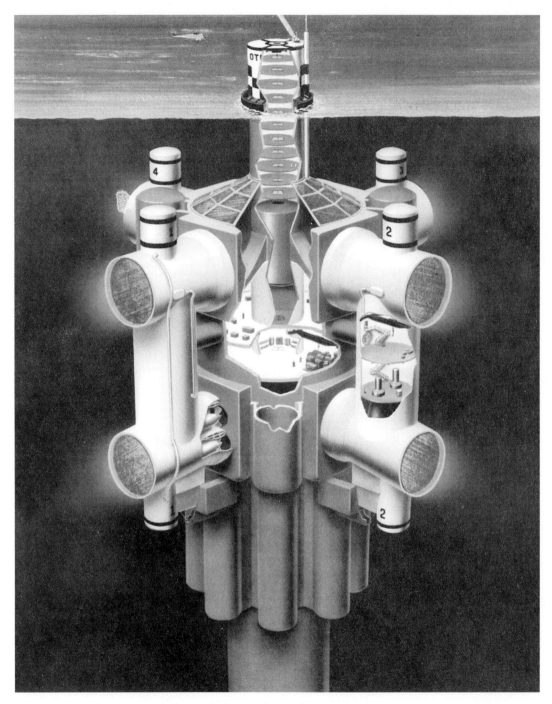

FIGURE 8.35   Artist's conception of a large OTEC plant.

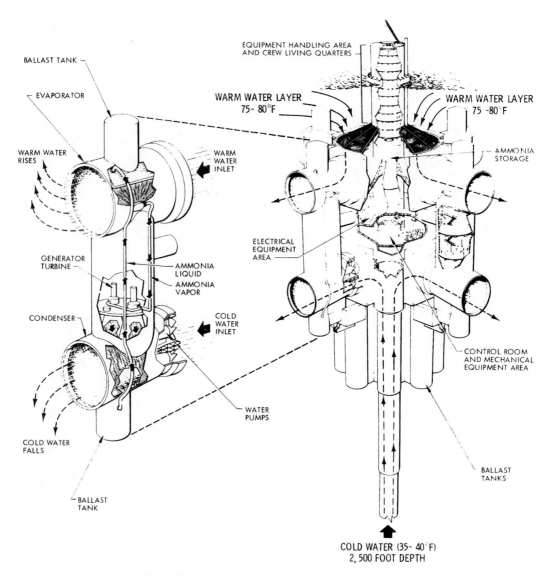

FIGURE 8.36 Schematic of large OTEC plant.

Chernobyl has also added to the public's concern regarding the safety of nuclear power plants. In addition to the safety question, the decreased use of electrical energy plus conservation methods has slowed down the construction of nuclear plants or caused the cancellation of proposed plants and those in various stages of construction. Also, the escalation of costs associated with the construction of these plants has made many of them economically unattractive, while alternative fuels such as coal have become economically competitive for use in central stations. For all these reasons, both economic and political, 114 commercial nuclear power plants have been canceled at some point in the planning or construction stages since 1972, and no new nuclear plants have been ordered in the United States since 1979. In addition, the Shoreham nuclear power plant in New York is the first completed nuclear plant to be canceled before producing commercial electricity. This plant, costing in excess of $5 billion, has been decommissioned and dismantled at a further cost of an estimated half-billion dollars. Approximately 15% of the power generation in the United States is from existing nuclear power plants, and more than 60% of the electrical energy generated in France, 65% in Belgium, and a large percentage in Japan are all produced from nuclear power plants.

All commercial power reactors in the United States generate electricity by first generating steam, which is then used to drive a turbine. Thus, we may consider the nuclear reactor and its fissionable material to be a fuel in a special container. To orient our discussion of the power-generating aspects of nuclear reactors, reference is made to Figure 8.37, which shows a simplified comparison of nuclear and conventional power systems. The fundamental process used for the production of power in a nuclear reactor is called the *fission process*. In this process, the nuclei of certain heavy atoms are bombarded by neutrons, causing the nuclei to split, in turn releasing heat, two or three additional neutrons, and fission fragments (as many as 80 different elements have been identified as fission fragments). The uranium isotopes ^{233}U and ^{235}U are fissionable, as is plutonium 239 (^{239}Pu). In addition, thorium 232 (^{232}Th) and uranium 238 (^{238}U) are fertile materials that may be transmuted to fissionable ^{233}U and ^{239}Pu. Due to the fission process, a large energy release occurs, which in engineering units would yield approximately 3.6×10^6 Btu (1×10^7 kWh) for each pound of material fissioned. From the standpoint of heat production, this is more than 2.5 million times greater than an equal weight of coal. For the fission process to be self-sustaining, each fission must produce enough neutrons to replace those lost due to leakage from the system, those captured in nonfissioning processes, those captured by fissioning nuclei in a process known as resonance capture, and also provide a neutron for the next fission process to occur. In general, approximately 2.5 neutrons must be produced per fission to have a self-sustaining chain reaction occur. When the reaction is self-sustaining, the reactor is said to be *critical*. A given reactor, having given fuel composition and geometry, has a given size for which the reaction will be self-sustaining, and the size is known as the *critical size*. It is important to note that once criticality is reached, the power output of a nuclear reactor is not related to the size of the reactor. The

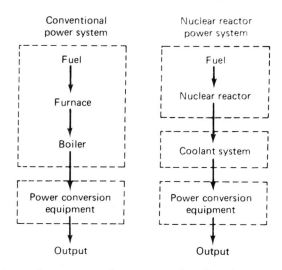

FIGURE 8.37 Comparison of conventional and nuclear power systems.

most important consideration is the ability to remove the heat produced from the system. This is quite different from conventional thermal systems where the heat output is directly related to the rate at which fuel is fired.

Associated with any nuclear reactor are problems of shielding personnel from the radioactive processes occurring within the reactor, designing of the system to cope with any leakage or accident, coolant stability, adequate heat removal, shielding of any circulating radioactive liquids, handling of radioactive fuel, chemical and metallurgical problems, and so on. Many of these problems have dictated solutions that often override any considerations of thermal efficiency in such a cycle. Without going into all these considerations for each of the reactor systems currently in use, we will briefly consider their basic power production cycles in the following paragraphs.

8.10a   Pressurized Water Reactor

The ideal reactor coolant is one that would have good heat-transfer properties while being a good moderator (slower down of neutrons) and also not be affected by the nuclear processes and reactions occurring in the reactor. Ordinary (light) water is suitable as a coolant and is used in many reactor systems as coolant-moderator. Because of questions concerning the stability of parallel coolant circuits when boiling occurs, many reactors have been proposed and built using nonboiling ordinary water as the coolant. These systems are known as *pressurized water reactors* (PWR). As shown schematically in Figure 8.38, primary water is circulated within the reactor to remove

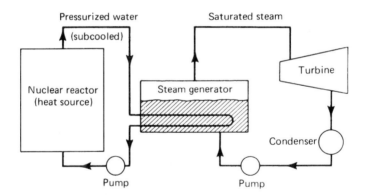

FIGURE 8.38   Pressurized-water reactor system.

the heat and act as moderator. The water from the reactor is circulated through a heat exchanger (steam generator), where it produces steam. The primary water returns to the reactor, and the secondary steam expands through the turbine to produce power. Due to the requirement that no steam be produced in the primary coolant circuit and the properties of water, even low primary water temperatures (under the critical temperature) require primary pressures of 1500 to 2000 psia. Some idea of the complexity of the reactor vessel for a PWR system can be obtained from Figure 8.39, which is a cross-section of a typical reactor vessel.

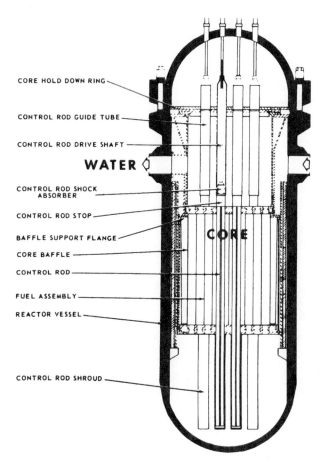

CORE HOLD DOWN RING

CONTROL ROD GUIDE TUBE

CONTROL ROD DRIVE SHAFT

WATER ◁

CONTROL ROD SHOCK
ABSORBER

CONTROL ROD STOP

BAFFLE SUPPORT FLANGE

CORE BAFFLE

CONTROL ROD

FUEL ASSEMBLY

REACTOR VESSEL

CONTROL ROD SHROUD

CORE

FIGURE 8.39 Typical PWR reactor vessel.

Because the steam produced in the heat exchanger is saturated, its temperature is lower than the average temperature of the primary water. If the steam goes directly to the turbine, problems occur that are associated with steam expanding into the wet region. Also, the thermal performance (efficiency) of such a low-temperature cycle is very poor. One way to alleviate the turbine problem and increase the overall plant thermal performance is to superheat the steam from the heat exchanger in a separate fossil-fueled superheater. One plant that utilizes this arrangement is the Indian Point Station of New York's Consolidated Edison System, which is shown in Figure 8.40. This plant has a design capacity of 2000 MW. The core of the reactor contains a small amount of highly enriched uranium and thorium. Some of the thorium is transmuted to ^{233}U, which supplements the original fuel loading. The reactor vessel for the plant is 41 ft $5\frac{1}{2}$ in. high, has an inside diameter of 9 ft 9 in., a wall thickness of $6\frac{15}{16}$ in. with the inside of the vessel clad with 0.109-in., type-304 stainless steel.

Because the primary water is radioactive to some extent, it is desired in this system to keep the primary water and steam circuits separate. This is accomplished for each of the reactor loops in a steam generator of the type shown in Figure 8.41. As shown, primary water circulates within the tubes, and steam and water circulate outside the tubes. The steam drum, risers, and downcomers provide the usual functions of these elements. The heat balance of a loop of this system is shown in Figure 8.42.

One further point should be noted: The entire reactor system and its associated heat-transfer loops are in a 160-ft-diameter steel containment sphere having a wall thickness of 1 in. and are enclosed in a concrete shielding structure.

FIGURE 8.40 Indian Point Station.
(Courtesy of Consolidated Edison Co.)

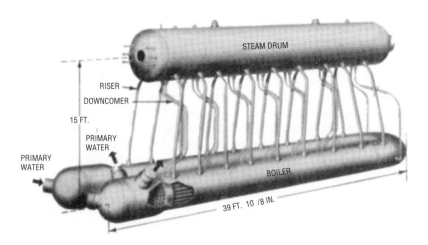

FIGURE 8.41 Indian Point steam generator.

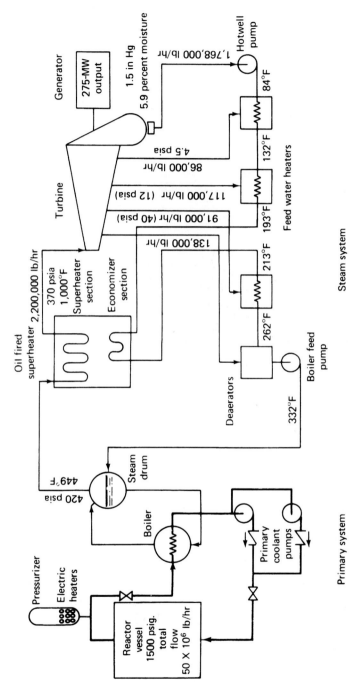

FIGURE 8.42 Heat balance of an Indian Point loop.

8.10b Boiling-Water Reactor

After a great deal of research, it was found possible to satisfactorily operate nuclear reactors with steam generation (boiling) occurring within the reactor. The *boiling-water reactor* (BWR) uses ordinary water as the coolant and moderator, and this allows operating pressures of about half those found in PWR systems. Also, the steam generator and its circulating system are eliminated.

The Oyster Creek unit shown in Figure 8.43 is such a single-cycle, 515-MW unit. Feedwater enters the reactor vessel at the top of the core. Here, it is mixed with the recirculated coolant

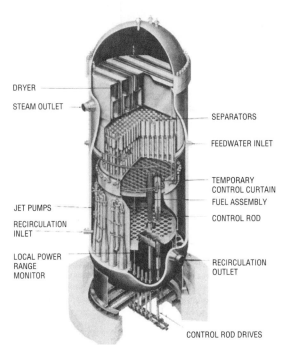

FIGURE 8.43 Oyster Creek boiling-water reactor.

and flows down between core and vessel walls. Recirculating flow, affording primary output control, is fed to external circulating pumps, which return it to the bottom of the vessel and up through the core. Steam generated within the core passes through steam separators and driers. Moisture content entering the turbine is about 0.1%. Considerable reduction in reactor vessel thickness is possible compared to the PWR. Bottom-entry control rods are an important feature of BWR vessels; space above the core is then free for steam separators and driers. Because scramming by gravity is no longer possible, a very reliable mechanical or piston-type control rod drive is essential. Radiation exposure in the vicinity of BWR is negligible. Pressure suppression containment, which traps radioactive steam and fission products, eliminates the familiar containment sphere, but the removal through the stack of radioactive carryover from the condenser can pose a problem.

Operational control of the BWR presents a problem in that the reactor is not self-controlling. A sudden increase in steam demand by the turbine reduces reactor pressure, which causes water to flash to steam in the reactor core. This reduces the reactivity of the reactor and steam output at the time that it should be increased. One way that this is overcome is to bypass excess steam past the turbine at part-load conditions. Any sudden load increase then closes the bypass valve, increasing steam flow to the turbine without affecting reactor pressure. Another method is to increase coolant flow by controlling the circulating pumps. This maintains reactor pressure by counteracting the tendency of core coolant to flash into steam.

Boiling-water reactors using nuclear superheat have also been built. The use of nuclear superheat has led to turbine steam conditions approaching those found in modern fossil-fueled plants. Without further discussion, it is noted that problems associated with nuclear superheat are problems with the fuel elements at the high reactor temperature, inadvertent flooding of the superheater, and steam carryover causing erosion and scaling of the fuel elements.

8.10c Liquid-Metal Fast Breeder Reactor

The basic fuel for light-water reactors is uranium 235, which is one of the few fissionable materials found in nature. As it is mined, natural uranium contains only about 0.7% fissionable ^{235}U. The balance is nonfissionable ^{238}U. To produce heat in a light-water reactor, the natural uranium is enriched by partially removing the ^{238}U from it until the concentration of ^{235}U reaches about 3%. The result is a small quantity of enriched uranium and a relatively large quantity of depleted uranium. The enriched uranium is made into fuel assemblies and the depleted uranium stockpiled, because it still contains a small percentage of ^{235}U and may have other uses. Domestic uranium will be able to carry the light-water reactor program well through the 1990s and past such time that the breeder reactor becomes a significant source for nuclear fuel. Yet, unless the fast breeder development program moves ahead, the date of commercial introduction will slip further ahead in the future, bringing with such slippage the possibility or probability of a uranium fuel shortage. Light-water reactors use ordinary water as both "moderator" and coolant, and this feature is the very reason that they cannot efficiently "breed," or produce more fuel than they consume. The moderator slows down the neutrons producing the chain reaction so that they are very efficient in splitting or fissioning the ^{235}U fuel. The slow neutrons, however, are not efficient at converting ^{238}U into ^{239}Pu.

In the fast breeder reactor, which uses no moderator, the fission process is carried forward by "fast" neutrons, which are very efficient in splitting the ^{239}Pu nucleus. The splitting of the ^{239}Pu nucleus in a fast reactor's fuel core emits more neutrons, which then can strike the ^{238}U atoms surrounding the core. This creates more ^{239}Pu atoms than the fission process is using up, thus creating the condition called *breeding*. In the typical fast breeder now under development, 1.3 new ^{239}Pu atoms are created for every one burned up or fissioned. The *breeding ratio* therefore is said to be 1.3.

The fast breeder reactor has a very compact fuel core, and the nuclear reactions produce great quantities of heat. To move the heat out of the reactor for use in power production, a very efficient coolant must be used. The coolant must also have desirable nuclear properties that will not slow down, or "moderate," fast neutrons. The coolant used is sodium. Being a metal, sodium has a good heat-transfer property. It melts and flows as a liquid at fairly low temperatures, and it will not interfere significantly with fast neutron reaction. Its vapor pressure is also low.

In the sodium-cooled fast breeder reactor shown schematically in Figure 8.44 [which is often called the *liquid-metal fast breeder reactor* (LMFBR)], the sodium coolant in the *primary loop* flows

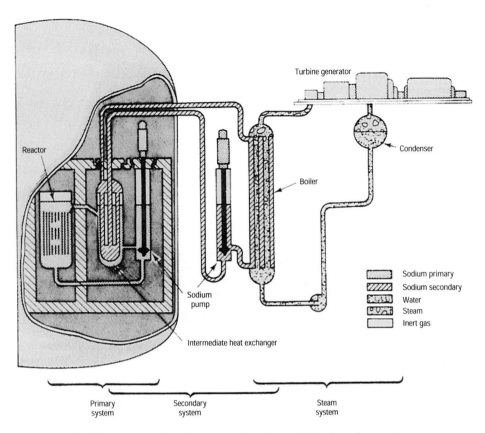

FIGURE 8.44 Flow diagram of liquid-metal fast breeder reactor.

through pipes into the bottom of the reactor and picks up heat as it moves up past the fuel assemblies. The primary sodium, which is radioactive, then is circulated through a heat exchanger, where it flows through tubes and gives up its heat to other sodium flowing outside the tubes in what is called the secondary loop. The heated sodium in the secondary loop then flows through a steam generator, where it gives up its heat to water flowing outside the tubes. The heat converts the water into steam, and the steam then drives a turbine generator to produce electricity. The secondary loop is used for two reasons. The first is for safety: Sodium and water react violently, and if a single loop were used, this high-pressure water could enter and react with primary sodium to cause a catastrophic reaction in the reactor coolant loop. The second reason for the use of a secondary system is the fact that the primary sodium is radioactive and could activate the water. Figure 8.45 shows the arrangement of the Enrico Fermi Atomic Power Plant, which is a fast breeder reactor using a sodium primary loop, a sodium secondary loop, and a separate steam generator.

Other power-producing nuclear reactor cycles are used. Among these are the gas-cooled reactor (GCR) that is popular in England and is the basis of the Philadelphia Electric 40-MW, high-temperature, gas-cooled reactor (HTGR) and the heavy-water reactor used in Canada (CANDU). We will not discuss these and other cycles, however, because they present no new principles from a power-production standpoint.

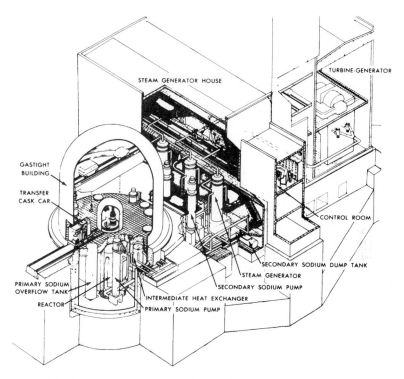

FIGURE 8.45 Enrico Fermi liquid-metal fast breeder reactor.

Based on data supplied by the International Atomic Energy Agency, there were 424 power reactors with a total capacity of 330,651 MW on the line throughout the world at the end of 1992. These reactors provide approximately 17% of the total power generated during the year. There are also 72 reactors under construction with a generating capacity of 59,720 MW. As noted earlier, in the United States, 15% of the total electrical energy is generated in nuclear power plants and in France, greater than 60% of the total energy is generated in nuclear power stations. Also, Finland produces more than 40% of its electricity from nuclear reactors.

While the nuclear reactor power-generation system is being developed rapidly in other countries of the world, there are several questions that have been raised concerning it. Due to low thermal performance, these plants reject more heat per kilowatt generated than do conventional power plants, which leads to problems concerning effects on the ecology (e.g., thermal pollution). In addition, the placing of a nuclear reactor in or near centers of population usually causes great public concern and, in some cases, has led to the abandoning of projects.

8.11 COGENERATION

Cogeneration is not a new idea but is a method of energy conservation that has been used for a long time. Basically, cogeneration plants use the same fuel source to produce both electricity and steam. The steam can be used for heating or in a process facility adjacent to the cogeneration facility, with the excess electricity sold to a local utility for its use. The utility benefits from cogeneration by gaining an additional generating source to help it meet its demand and also by purchasing the power at a rate lower than the rate at which it sells power. Whether the system is based on steam turbines, gas turbines, or a combination of both, cogeneration can achieve a significant increase in overall fuel efficiency over conventional power plants and steam generators. The advantage of cogeneration lies in the fact that both power and a transfer of heat for an intended use can be accomplished with a total fuel expenditure that is less than would be required to produce them individually.

The economic incentives that have made cogeneration more attractive in recent years include the following:

1. The Public Utility Regulatory Policy Act (PURPA) of 1978, which requires utilities to buy back excess power generated by cogeneration plants.

2. Cogeneration plants are easier and faster to install and do not require as much capital expenditure. Because components approach "off-the-shelf" availability, the lead time from inception to implementation is much shorter than for conventional plants.

3. A company can now be more independent of the local utility company by generating much of its own power at rates lower than the rates charged by the utility. Also, costly cutbacks in power by the company when the utility is having difficulty meeting demand need not occur.

4. A company can tailor its cogeneration facility to its own particular needs and project load growth and capital expenditures more accurately.

5. A cogeneration facility can easily add on additional capacity with a minimum of shutdown time and, by using available equipment, with a minimum of additional expenditure.

Let us consider the cogeneration system that is shown schematically in Figure 8.46. In this system, steam is extracted from the turbine and is used for heating or as process steam. The portion of the steam that goes to the condenser eventually combines with the extracted steam and is returned to the steam generator.

ILLUSTRATIVE PROBLEM 8.14

Assume that the regenerative cycle of Illustrative Problem 8.11 is operated with a heat exchanger replacing the heater similar to Figure 8.46. The heat exchanger will use the extracted steam to heat a building, and the condensate from the exchanger will be assumed to be saturated liquid at 50 psia. This liquid will mix with the condensate from the condenser and return as boiler feed. Assuming that 16% of the steam is extracted from the turbine and neglecting pressure drops and pump work, determine the conventional thermal efficiency of the cycle and the efficiency of energy utilization of the cycle. Compare your results with the results found in Illustrative Problem 8.11.

SOLUTION

Let us first determine the enthalpy values at the various points in the cycle:

1. At 1000 psia and 1000°F, $h = 1505$ Btu/lb$_m$.

2. At turbine inlet, $h = 1505$ Btu/lb$_m$.

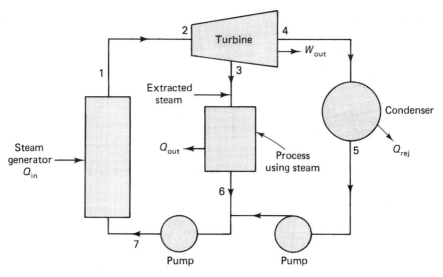

FIGURE 8.46 Cogeneration plant using process steam.

3. At turbine extraction of 50 psia, $h = 1168$ Btu/lb$_m$ (after isentropic expansion).

4. At condenser inlet, $h = 922$ Btu/lb$_m$ (after an isentropic expansion).

5. Saturated liquid at condenser pressure, $h = 69.74$ Btu/lb$_m$.

6. Saturated liquid at exchanger outlet, $h = 250.24$ Btu/lb$_m$ (50 psia).

7. Assuming that the condenser condensate mixes adiabatically with the liquid leaving the heat exchanger, $(0.16)(250.24) + (0.84)(69.74) = 98.62$ Btu/lb$_m$ based on 0.16 lb$_m$ extracted and 0.84 lb$_m$ going through the condenser.

Using these data, we have

$$\text{heat in at boiler} = 1505 \text{ Btu/lb}_m - 98.62 \text{ Btu/lb}_m$$
$$= 1406.38 \text{ Btu/lb}_m$$
$$\text{work out of turbine} = (0.84 \text{ lb}_m)(1505 - 922) \text{ Btu/lb}_m$$
$$+ (0.16)(1505 - 1168) \text{ Btu/lb}_m$$
$$= 543.64 \text{ Btu/lb}_m$$

The "conventional" thermal efficiency is defined as work out/heat in,

$$\eta = \frac{543.64 \text{ Btu/lb}_m}{1406.38 \text{ Btu/lb}_m} \times 100 = 38.66\%$$

If at this time we define the efficiency of energy utilization to be the ratio of the work out plus the useful heat out divided by the heat input to the cycle,

$$\eta_{\text{energy utilization}} = \frac{\mathbf{w} + q_{\text{out(useful)}}}{q_{\text{in}}} \times 100$$

and $q_{\text{out(useful)}}$ is $(0.16 \text{ lb}) (1168 - 250.24) \text{ Btu/lb}_m = 146.84 \text{ Btu/lb}_m$ of steam entering the turbine. Therefore,

$$\eta_{\text{energy utilization}} = \frac{(543.64 + 146.84) \text{ Btu/lb}_m}{1406.38 \text{ Btu/lb}_m} \times 100$$

$$\eta = 49.1\%$$

From these results, we see that when this cycle is compared with the regenerative cycle of Illustrative Problem 8.11, the "conventional" thermal efficiency is *decreased* by 4.5%. However, the efficiency based on the utilization of energy is *increased* by approximately 6%. Depending on the need, more or less steam can be bypassed to the heat exchanger or allowed to expand through the turbine.

8.12 DIRECT ENERGY CONVERSION

In all the devices and cycles that we have discussed thus far, the generation of electrical power from an energy source required the transfer of the energy from the source to a working fluid, which by undergoing circulation and other manipulation in a complex "prime mover" (e.g., turbine, engine, and so on) ultimately yielded the desired electrical output. Recent progress has been made in the development of direct energy conversion systems in which the energy of the fuel is converted directly to electrical energy without the use of a circulating fluid or any moving parts. At present, we will just qualitatively describe some of these systems and their limitations.

8.12a Thermoelectrical Converter

The first of these direct energy conversion devices, the *thermoelectric converter,* is based on the fact that a voltage is generated when two unlike conductors are connected at their ends with the ends kept at different temperatures. If the circuit is closed through a load resistance, current will flow through the load. This phenomenon has already been noted and is used to measure temperature in the thermocouple, which we described in Chapter 1. An associated phenomenon, known as the Peltier effect, is discussed in Chapter 10 in conjunction with the thermoelectric refrigerator. Figure 8.47 shows a simple thermoelectric energy converter. The desirable characteristics of the materials used in such a device are as follows:

1. Low internal electrical resistance to reduce internal heat generation resulting from current flow in the materials.

2. Low thermal conductivity (high thermal resistance) to reduce heat conduction from the heat source to the sink.

3. High values of open-circuit voltage. Most metals produce open-circuit voltages of the order of microvolts per degree of temperature difference.

The most suitable materials are the semiconductors, such as lead telluride, germanium–silicon alloys, and germanium telluride. The thermoelectric converter is basically a form of heat engine receiving heat from a source, rejecting heat to a sink, and converting heat to electrical work. Thus, it has as its theoretical upper limit the Carnot-cycle efficiency. Due to losses, the Carnot-cycle

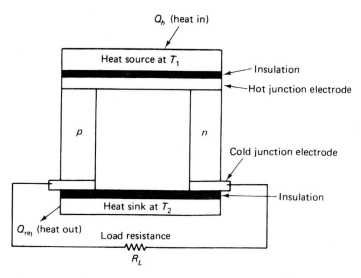

FIGURE 8.47 Thermoelectric converter.

limit is not even approached, and although in theory the thermoelectric converter is capable of operating at efficiencies greater than 10%, most of the devices built to date have shown efficiencies much lower than this figure. With development and using semiconductors, a thermal efficiency approaching 20% is a reasonable objective. Due to their reliability and lack of moving parts, they have been used in such diverse applications as converting waste heat from kerosene lamps to power radio receivers in rural areas and in conjunction with radioisotope heat sources to power long-life, unattended ocean buoys.

8.12b Fuel Cell

The *fuel cell* is an electrochemical device that resembles a car battery. As can be seen from Figure 8.48, it is unlike the battery in that it does not feed on itself. The fuel and the oxidant are consumed, and the electrodes stay intact. The components and structure of a fuel cell vary with the physical state of the reactants. Because gases, liquids, and solids have been used as fuel and oxidants, the structures and components of fuel cells have wide variations. Highly reactive hydrazine and almost inert carbon have drastically different requirements, just as fluorine and air must be handled differently, but even the same reactant pair (hydrogen and oxygen) is being utilized in a number of systems quite unlike each other. Regardless of the system, reactants must be admitted and heat and products removed approximately in proportion to the electric power demands. The ideal voltage output of a fuel cell at no load is approximately $1\frac{1}{4}$ V for a hydrogen–oxygen cell and depends to some extent on the reactants used.

Figure 8.49 shows a complete fuel-cell system. The system must provide for storage of at least one reactant (the fuel) or both reactants if air is not used as the oxidant. The system as shown must be capable of adjusting the flow of reactants according to demand while removing the products of the reaction and the heat generated. In some cases, such as water on a space vehicle, the product must also be stored. Where many cells are stacked to obtain the required voltage and power output, supply and removal must be uniform to all cells in the stack, the cells must perform uni-

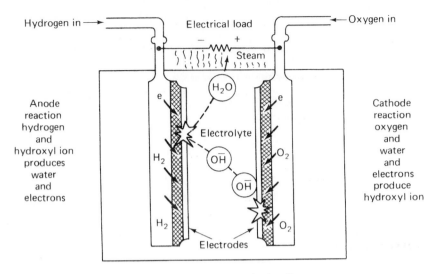

FIGURE 8.48 Basic fuel cell.

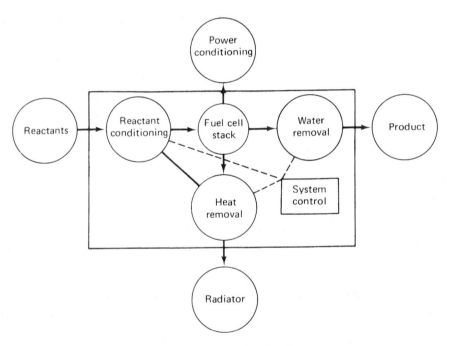

FIGURE 8.49 Complete fuel-cell system.

formly, and they must all have good external electrical connections. Figure 8.50 schematically shows a fuel-cell system that was used in the *Gemini* space vehicles. More than 30 cells were connected in series, in each of the three stacks, which were connected in parallel.

The fuel cell is not a heat engine, and it therefore is not limited by the Carnot efficiency. If we consider the function of the cell to be the conversion of chemical to electrical energy, then the maximum theoretical efficiency of a fuel cell is 100%.

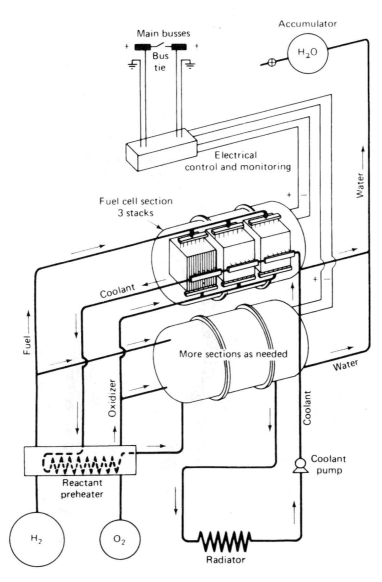

FIGURE 8.50 *Gemini* fuel-cell system.

8.12c Thermionic Converter

The *thermionic converter* is basically a high-temperature device that converts thermal energy to electricity by literally boiling off electrons from the hot cathode, which then travel to the cold anode. Figure 8.51 shows a schematic of this device in which the electrical circuit is completed by the external load, R_L. In principle, this device is a heat engine that uses electrons as the working fluid. Therefore, the upper limit of its efficiency is that of a Carnot engine operating between the temperature limits of the cathode and anode. Efficiencies of 10% have been achieved with cathode temperatures of 3200°F when the Carnot efficiency was 50%. Because electron emission is an exponential function of temperature, high cathode temperatures yield high power densities. Efficiencies of the order of 18% have been achieved in a thermionic converter with a cathode temperature of 4500°F.

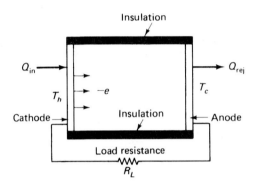

FIGURE 8.51   Schematic of a thermionic converter.

There are several problems associated with the operation of the thermionic converter. The first of these arises from the distribution of the electrons in the gap between electrodes. When the gap is maintained as nearly as possible as a "perfect" vacuum, it is necessary to keep the spacing of the electrodes as small as 0.001 in.; otherwise, the retarding potential (space-charge effects) becomes too large for satisfactory operation of the converter. These spacings are extremely difficult to maintain, especially with the large temperature differences between anode and cathode. To reduce the space-charge effect and, at the same time, maintain reasonable spacings (of the order of 0.040 in.), most present-day thermionic converters use cesium gas in the interelectrode space. Cesium is relatively easily ionized with positively charged ions, which tend to neutralize and negate the space-charge effect.

In addition to the space-charge problem, the thermionic converter performance is also degraded by heat radiated and conducted from the cathode to the anode, which is ultimately lost from the anode. Thus, to achieve high efficiencies in a thermionic converter, it is necessary to reduce the effects of space charge, to minimize radiation and conduction losses, and to keep the internal resistance of the converter to a minimum. It will be noted from the foregoing that mechanical design plays a vital role in the ultimate performance of a thermionic converter.

8.12d Magnetohydrodynamic Generator

When a conductor is moved through a magnetic field, an electromotive force results, and if the circuit is closed, a current will flow. This phenomenon is the basis of operation of all rotating electrical generators in which the rotor turns conductors in a magnetic field to produce electricity. The basic concept of the magnetohydrodynamic (MHD) converter is exactly this principle, with the conductor being a high-speed ionized gas. Figure 8.52 shows the replacement of the conducting wires of a conventional electric generator by a highly ionized gas called a *plasma*. The electrical properties of a plasma are determined by the density of free electrons in the gas. The ionization of the gas (and consequently its electrical conductivity) can be achieved in several ways (high gas temperatures, seeding of a gas with readily ionized atoms such as cesium, nuclear radiation of the gas, and others). When this ionized gas (plasma) flows at right angles to the magnetic field (Figure 8.52), a force is generated at right angles to both the velocity of flow and the magnetic field. This force drives the electrons and ions to the pickup electrodes, giving rise to a current when the electrodes are connected to an external load.

In the MHD cycle, the gas is first compressed, heat is added at constant pressure, and the gas is then accelerated in a nozzle before passing through the magnetic field. If we consider the overall cycle, such as the Rankine cycle or the Brayton cycle, which will be considered in Chapter 9,

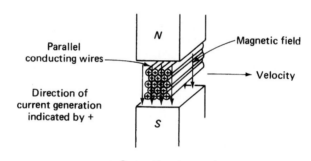

a. Conventional generator

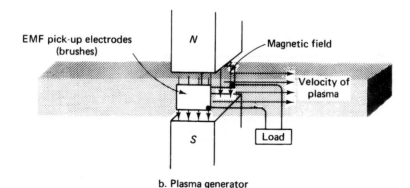

b. Plasma generator

FIGURE 8.52 Magnetohydrodynamic generator principle.

it will be seen that the MHD converter has basically replaced the turbine and generator. This device is obviously a heat engine device, and it is subject to the efficiency limitations of the Carnot cycle. Some of the factors that cause losses in the MHD converter are heat-transfer losses through the electrodes, internal electrical resistance losses in the plasma, energy losses in the magnets, and fluid friction losses associated with high-speed gas flow.

Due to the losses noted, it is found that the MHD converter is best suited for large central stations in which the rejected heat from the MHD converter is used as the input to a conventional power cycle. Using this binary type of cycle, it is estimated that practical efficiencies for the combined cycle approaching 50% can be achieved.

8.12e Solar Energy

The amount of solar energy potentially available on the earth's surface is many times greater than the present energy needs of humankind. However, in order to use this energy, enormous amounts of money must be expended. Useful amounts of energy have been generated by either *direct energy conversion* using photovoltaic devices (also called solar cells) or large arrays of reflectors used to concentrate the solar energy for use in a conventional steam power plant.

Figure 8.53 shows the principle of the photovoltaic effect. In the device shown, sunlight falls on the surface of a semiconductor and produces electrons by internal interactions. While silicon solar cells have been the dominant type of cell used, materials such as cadmium sulfide/cadmium telluride, germanium, selenium, and gallium arsenide have also been used for photovoltaic cells.

The largest photovoltaic operational power plant is a 1-MW plant located near Hesperia, California. It is constructed on a 20-acre tract of land and consists of arrays of cells (108), each 32 ft × 32 ft mounted on double-axis trackers. Each array has 256 modules rated at 40 W average

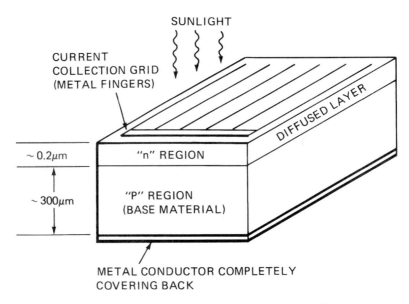

FIGURE 8.53 Structure of a photovoltaic cell.

output. The entire field is controlled by a minicomputer. Figure 8.54 shows this installation. Typical efficiencies of these direct conversion devices are from 10 to 12%.

In late 1994, a solar array panel developed by the Electric Power Research Institute achieved new levels of efficiency in converting sunlight directly to electrical energy. This array, which uses crystalline silicon as in conventional photovoltaic cells, collects the charge in a unique manner, enabling the cells to be manufactured in factories that make computer chips. Efficiencies greater than

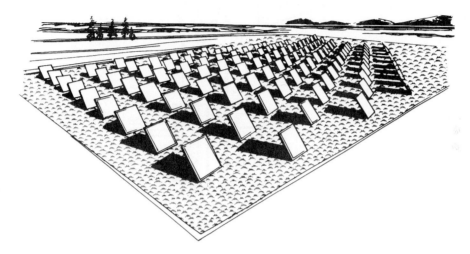

FIGURE 8.54 A 1-MW solar photovoltaic power station.

20% have been reported for laboratory models of this system, and a prototype has been constructed by Amonix of Torrance, California. This prototype, which is designed to generate 20 kW of electricity, covers 1,668 square feet. Tests of this prototype will be conducted at the Arizona Public Service Company's solar laboratory in Tempe, Arizona. It is estimated that if this system is successful, the total cost per kilowatt-hour will be 5.5 cents to 6 cents in commercial production.

Figure 8.55 shows a schematic of a solar steam power plant. In this system, the sun is tracked by a field of reflectors (heliostats) that reflect the sunlight to the boiler, which is situated on a tower. If the heliostat field is thought of as a reflector, then the boiler is located at the focal point of the reflector and the diffuse sunlight concentrated at the focal point, where steam is generated. Early in the 1980s, a 10-MW electric plant, called Solar One, started operation at Barstow, California. This plant, shown in Figure 8.56, produces steam used to drive turbines that deliver electric energy in synchronism with the network of Southern California Edison Co.

Other solar energy systems are used for hot water heating. These vary in size from the solar domestic hot water systems shown in Figure 8.57 to the 2863 ft^2 of collectors used by Monsanto Corp. to provide hot water and boiler feedwater as shown in Figure 8.58.

8.12f Wind Power

Since the beginning of recorded history, humans have used the wind to move ships on the seas and windmills to pump water and grind grain. In 1910, wind turbine generators were used to produce usable amounts of electricity in Denmark. In the United States, the first large experimental wind

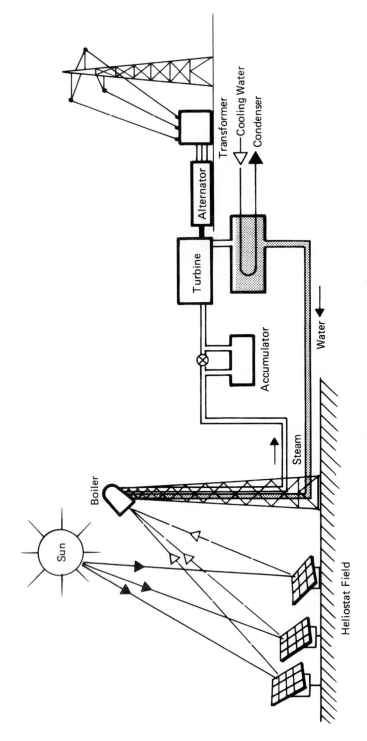

FIGURE 8.55 Solar-steam power plant.

FIGURE 8.56 Solar One steam power plant.

FIGURE 8.57 Domestic solar hot water heaters.

FIGURE 8.58 Large industrial solar collector installation.

turbine was built on Grandpa's Knob near Rutland, Vermont. Interest in wind power declined until the oil shortages of the 1970s occurred, because the total amount of convertible energy in the wind is tremendous and inexhaustible. Figure 8.59 shows the largest wind-energy system built, which is now operating in the Blue Ridge Mountains, atop Howard's Knob near Boone, North Carolina. This unit consists of a 200-ft-diameter, two-bladed rotor mounted on a 140-ft tower and is known as the MOD-1 unit. Its rated power is 2 MW at 25 mi/hr with a cut-in wind speed of 11 mi/hr and a cut-out wind speed of 35 mi/hr. The unit is designed to survive in a wind of 150 mi/hr, and the design rotor speed is 35 rpm.

Despite its environmental attractiveness as a nonpolluting power source, wind power probably could not provide a large fraction of the total U.S. energy requirements because of its variability. However, it can provide a much needed additional capacity for utilities and save large amounts of fuel by becoming a significant source of energy for smaller utilities, rural homes, and agricultural areas.

8.12g   Waste-to-Energy Resource Recovery

As a "rule of thumb," waste managers estimate that each person generates approximately 1 ton of refuse (garbage) a year. This includes household waste and waste produced by businesses. Waste-disposal in the past consisted of using dumps to get rid of this material. Due to environmental and financial considerations, dumps are being phased out and waste-to-energy resource recovery plants are being used to deal with the problem of refuse disposal. Figure 8.60 shows the Town of Hempstead waste-to-energy facility located in Long Island, New York. This $360 million facility uses a "mass burn technology" for clean burning and reliability. Waste is fed into the furnace on rolling drums, or grates. Computers that monitor the temperature and pollutants adjust the speed of the grates and the air mixture to obtain optimum combustion conditions. Next, "scrubbers" inject lime into the flue gas to neutralize acid rain components. Finally, a baghouse containing more than 1000 vacuum cleaner–like bags capture any ash particles.

From the refuse bunker, the waste is fed directly into the combustion chambers of three independent furnace/boiler units. Hot combustion gases pass through the waterwall boiler sections to generate steam. The steam drives a 72-MW turbine-generator, which produces electricity for in-plant use and with enough excess to keep the lights burning in 65,000 homes.

In addition to solving a difficult environmental problem, this plant is expected to earn more than $500 million over its lifetime from the sale of power.

8.12h   Geothermal Energy

The thermal energy that is stored in the earth's crust is known as *geothermal energy,* and this energy has been tapped to heat buildings and to generate electricity. A geothermal reservoir is formed when a layer of rock traps the heated water beneath it. While this water can be tapped with a well, it can also rise through porous rocks above the storage level to emerge from natural surface openings called *fumaroles.* The heated water will rise to the surface either in the form of geysers or hot springs. As the water rises, some of it will turn to steam due to the decrease in pressure. This steam is used to generate electricity using turbines. Low-temperature steam and hot water from the wells are used for local heating only.

There are two general types of geothermal energy systems: vapor, and hot water. In the vapor system, the earth's heating is sufficient to provide superheated steam at the well head. This

FIGURE 8.59 MOD-1, 2 MW wind power generator.

FIGURE 8.60 Town of Hempstead waste-to-energy facility, Long Island, New York. (Courtesy of American Ref-Fuel Company of Hempstead.)

steam can be piped directly to a turbine to produce power. A system of this type is in operation at the Geysers, in California, just north of San Francisco.

The most common geothermal system is the hot water system in which hot water is obtained from the wells. In one system, the hot water is then used to vaporize a working fluid that is piped to a turbine to produce power. In a second system, the flashed steam system, the water from the well is fed into a separator where it is throttled to a lower pressure to generate steam. The liquid, brine, is separated from the steam and discarded. The steam is then used to operate a turbine.

At present, geothermal energy accounts for a small percentage of the world's energy production. In the future, this percentage of energy production may increase. In order for this to occur, the problems associated with geothermal energy production must be overcome. Some of these problems are:

1. Low pressure and temperature exist at the well head.

2. The water has high concentrations of salt and minerals.

3. The discarded brine solution is harmful to the ecology of the earth's surface.

4. Most regions where geothermal energy is available are geologically unstable.

8.13   REVIEW

The stated purpose of this chapter is the study of the conversion of energy from one form to a more useful form. We started our study with a review of the Carnot cycle and then went on to the prototype Rankine, reheat, and regenerative steam cycles. These are "ideal" cycles in the sense that we ignored irreversible effects such as friction, pressure drops, heat losses, turbulence, and so on. Having studied these cycles, we then turned our attention to the equipment that is used for their practical realization.

Many other fluids and cycles using them have been proposed and constructed in addition to those cycles that use either water or gas as their working fluid. Binary fluid cycles using mercury or ammonia and water are examples of such alternates. In addition to the working fluids, the primary source of energy in a plant need not be from the combustion of a fuel. In the nuclear reactor, the fissionable fuel replaces the boiler, and the release of energy is used in a controlled manner as the primary energy source. The removal of the heat can be accomplished in many different ways, depending on the design of the system. Problems involving safety and disposal of the spent fuel have limited the use of nuclear energy in the United States.

The direct conversion of the energy of a source to electrical energy without the use of a circulating fluid or moving parts has come under intensive development in recent years. In this chapter, we looked briefly at the thermoelectric converter, fuel cell, thermionic converter, magnetohydrodynamic generation, solar energy, and wind power. Most of these sources are not suitable at present for the large-scale generation of power. However, they are a welcome supplement to our energy supplies and, with further development, will become increasingly economically useful.

Cogeneration has recently become a feasible method of energy utilization. By the simple expediency of using the low-temperature energy for process or heating requirements, it is possible to combine both electrical energy generation with heat requirements to obtain greater energy utilization than when each requirement is performed separately. While cogeneration does not introduce any new concepts, the savings in energy can be considerable. Most of the equipment needed to implement cogeneration can be obtained as standard "off-the-shelf" items.

KEY TERMS

Terms used for the first time in this chapter are:

binary fluid cycle a cycle using two fluids wherein one fluid rejects heat to the second fluid as a heat source.

boiling-water reactor (BWR) a nuclear reactor that uses ordinary water as the coolant-moderator with net steam generation occurring within the reactor.

breeding the creation of one or more neutrons for each neutron consumed in a nuclear reactor.

cogeneration the simultaneous generation of power and heating or process steam to increase the overall utilization of energy.

direct energy conversion the conversion of a fuel or energy source directly to electrical energy without the use of a circulating fluid or any moving parts.

fuel cell an electrochemical device in which the fuel and oxidant are consumed to produce a voltage across an electrical load.

geothermal energy the energy stored in the earth's crust.

liquid-metal fast breeder reactor (LMFBR) a nuclear reactor in which a liquid metal such as sodium is used as the coolant and in which there is no moderator.

ocean thermal energy conversion (OTEC) cycle a binary fluid cycle using the temperature gradients in the ocean to generate power.

pressurized water reactor (PWR) a nuclear reactor that uses ordinary water as both the moderator and coolant.

Rankine cycle a vapor cycle consisting of four processes: a constant-pressure vaporization, an isentropic expansion, a constant-pressure condensation, and a return to the initial conditions of the cycle.

regenerative cycle a cycle in which the condensate from the condenser is ideally continuously heated by an interchange with the steam as it expands in the turbine before it is returned to the boiler.

reheat cycle a Rankine cycle in which the steam is expanded in a turbine to an intermediate pressure, is returned to the steam generator, and then returned to the turbine to expand to condenser pressure.

solar energy conversion the use of a photovoltaic device to convert incident solar energy directly to electricity.

thermionic converter a high-temperature device in which thermal energy is converted to electricity by boiling off electrons from the hot cathode, which then travel to the cold anode.

thermoelectric converter a device that uses the fact that a voltage is generated when two unlike conductors are connected at their ends, with the ends kept at different temperatures.

type efficiency the ratio of the ideal efficiency of a given cycle divided by the efficiency of a Carnot cycle operating between the same maximum and minimum temperature limits.

wind power the use of the energy contained by the prevailing wind at a given location to operate a windmill to generate power.

EQUATIONS DEVELOPED IN THIS CHAPTER

Rankine cycle thermal efficiency

$$\eta_R = \frac{(h_4 - h_5)_s - (h_2 - h_1)_s}{h_4 - h_2} \tag{8.7}$$

heat rate

$$\text{heat rate} = \frac{2545}{\eta_R \text{ (or } \eta_R')} \text{ Btu/hp-hr} \tag{8.9a}$$

heat rate

$$\text{heat rate} = \frac{3413}{\eta_R' \text{ (or } \eta_R')} \text{ Btu/kWh} \tag{8.9b}$$

steam rate

$$\text{steam rate} = \frac{\text{pounds of steam supplied per hour}}{\text{net output (hp-hr or kWh)}} \tag{8.10}$$

QUESTIONS

8.1 In what two ways does a gas cycle differ from a vapor cycle?

8.2 How are the temperature limits of the Carnot cycle determined?

8.3 Cite some reasons why the practical realization of the Carnot cycle would be almost impossible.

8.4 How does the Rankine cycle differ from the Carnot cycle?

8.5 What is the basic reason that pump work usually is negligibly small? Would this be true if the steam at the outlet of the turbine was pumped directly back to the boiler without condensing it?

8.6 The reheat cycle causes two things to differ from the Rankine cycle. Indicate what these are.

8.7 What limits the number of reheats that are used in a given cycle?

8.8 The regenerative cycle has the potential of achieving the Carnot-cycle efficiency when operated between the same upper and lower temperatures. Basically, describe what is being done in the regenerative cycle that brings this about.

8.9 What factor limits the number of heaters used in the regenerative cycle?

8.10 Describe the operation of a binary fluid cycle.

8.11 What permits the OTEC cycle to function?

8.12 In what way does the nuclear reactor plant relate to the conventional steam generation plant?

8.13 How do the PWR, BWR, and LMFBR reactors differ?

8.14 What has made the concept of cogeneration attractive at this time?

8.15 Does the conventional thermal efficiency of a cogeneration plant increase, decrease, or remain the same as a conventional power plant operating between the same limits?

8.16 What does direct energy conversion attempt to do?

8.17 What makes the reversed thermoelectric converter attractive for use in space vehicles?

8.18 What is the limiting efficiency of a fuel cell?

8.19 Where would you expect to find a MHD power plant?

8.20 Solar energy conversion and wind power plants give us "something for nothing"; that is, with no energy input by humankind, we obtain energy output. Comment on this statement.

PROBLEMS

Wherever appropriate, sketch the cycle on $h-s$ or $T-s$ coordinates. Also, draw a sketch of the equipment showing the flow of the working fluid.

Problems Involving the Carnot Cycle

8.1 A Carnot engine operates between 1300°K and 300°K. Determine its thermal efficiency.

8.2 A Carnot engine operates between 60°F and 1200°F and delivers 400 Btu as work. Determine the heat rejected and the thermal efficiency of the engine.

8.3 A Carnot cycle is operated with its maximum temperature equal to 1000°R. If the cycle efficiency is such that 100 Btu/hr is rejected, what is the sink temperature if 500 Btu/hr enters at the upper temperature?

8.4 A Carnot cycle is operated with its maximum temperature 700°K. If the cycle efficiency is such that 300 kW is rejected, what is the sink temperature if 450 kW enters at the upper temperature?

8.5 A Carnot vapor cycle uses water as the working fluid. If the cycle operates within the wet region and upper pressure is 500 psia while lower pressure is 20 psia, determine the cycle efficiency.

*8.6 Two ideal reversible heat engines are operated in series, the first receiving heat at a temperature T_1 and the second rejecting heat at a temperature T_2. Prove that this arrangement is as efficient as a single reversible engine operating between the same minimum and maximum temperatures.

*8.7 Prove that the efficiency of a single Carnot engine operating between temperatures T_1 and T_2 has an efficiency of η_0, which is related to the efficiencies of two Carnot engines operated in series between the same temperature limits by

$$\eta_0 = \eta_1 + \eta_2 - \eta_1\eta_2$$

where η_1 is the efficiency of Carnot engine 1 operated between T_1 and an intermediate temperature T and η_2 is the efficiency of Carnot engine 2 that is operated in series with Carnot engine 1 between temperatures T and T_2.

8.8 A Carnot engine rejects 50 Btu of heat to a reservoir at 50°F. If this engine produces 80 Btu as work, determine the heat source temperature.

8.9 A Carnot cycle is operated between 2000°R and 400°R. If 500 kW enters as heat, how much heat is rejected, what is the thermal efficiency of the cycle, and how much work is obtained from the cycle?

8.10 A Carnot cycle is operated between 1000°K and 200°K. If 1 MJ/hr enters as heat, how much heat is rejected, what is the thermal efficiency of the cycle, and how much work is obtained from the cycle?

8.11 A Carnot cycle rejects 350 Btu/min while producing 5.5 hp of work. Determine the heat added, the lower temperature, and the efficiency of the cycle if the maximum temperature is 625°F.

Problems Involving the Rankine Cycle

8.12 A turbine manufacturer specifies that the maximum steam temperature in the turbine is to be 1000°F and that the last stage of the turbine is to have a moisture content not exceeding 10% by weight. If the expansion is carried out reversibly and adiabatically from 800 psia, what is the state of the steam as it leaves the turbine?

8.13 Solve Problem 8.12 using the Mollier chart.

8.14 Solve Problem 8.12 using the computerized *Steam Tables*.

8.15 In Problem 8.12, the steam is expanded irreversibly. If the actual $\Delta h = 90\%$ of the Δh that would be present if the process were carried out isentropically, what is the final moisture that would leave the turbine if the final pressure was that of Problem 8.12?

8.16 What is the internal efficiency of the turbine in Problem 8.15?

8.17 In a Rankine cycle steam enters the turbine at 900 psia and 1000°F and exhausts at 1 psia. Determine (a) the turbine work, (b) the heat supplied, and (c) the thermal efficiency of the cycle. Neglect pump work. Use the *Steam Tables* and Mollier chart.

8.18 Solve Problem 8.17 using the computerized *Steam Tables*.

8.19 Calculate the pump work in Problem 8.17 using the approximate formula and the computerized *Steam Tables*.

8.20 If the condenser pressure in a Rankine cycle is 1 psia and the maximum pressure in the cycle is 600 psia, calculate the efficiency of the ideal cycle if the steam at 600 psia is saturated vapor. Use the Mollier chart and neglect pump work.

8.21 Solve Problem 8.20 using the computerized *Steam Tables*.

8.22 Use the approximate formula to determine the pump work in Problem 8.20.

8.23 If the vapor in Problem 8.20 has 200°F superheat, calculate the efficiency of the ideal Rankine cycle.

8.24 What is the pump work in Problem 8.20? Use the computerized *Steam Tables*.

8.25 What is the thermal efficiency of the Rankine cycle in Problem 8.20 if pump work is included?

8.26 Repeat Problem 8.20 for an initial pressure of 400 psia.

8.27 Repeat Problem 8.23 for an initial pressure of 400 psia.

8.28 A Rankine cycle operates between pressures of 2 psia and 1500 psia. If the steam entering the turbine is superheated to 800°F, calculate the efficiency of the cycle and the work produced per pound of steam. Neglect pump work. Use the *Steam Tables*.

8.29 Determine the thermal efficiency of the Rankine cycle in Problem 8.28 if pump work is included. Use the computerized *Steam Tables*.

8.30 A Rankine cycle is operated with a turbine inlet pressure of 5.0 MPa and 300°C. Determine the efficiency of the cycle if the expansion is to 20 kPa. Neglect pump work.

8.31 If the expansion in Problem 8.30 is carried out to 50 kPa, determine the efficiency of the cycle.

8.32 If pump work is included in Problem 8.30, what is the cycle efficiency? Use the computerized *Steam Tables*.

8.33 If pump work is included in Problem 8.31, what is the efficiency of the cycle? Use the computerized *Steam Tables*.

8.34 A Rankine cycle is operated with a turbine inlet pressure of 600 psia and 600°F. Determine the efficiency if the expansion is to 1 psia. Use the Mollier diagram, and neglect pump work.

8.35 Determine the efficiency of the cycle in Problem 8.34 if the expansion is carried out to 10 psia. Compare results with those obtained in Problem 8.34.

8.36 If the turbine inlet pressure in a Rankine cycle is 5 MPa and the inlet temperature is 350°C, determine the efficiency of the cycle if pump work is neglected and the engine expansion is to 10 kPa.

8.37 If pump work is included in Problem 8.36, what is the efficiency of the cycle? Use the computerized *Steam Tables* for the pump work.

8.38 A Rankine cycle is operated with a turbine inlet pressure of 500 psia and a temperature of 620°F. If the engine expansion is to 2 psia, what is the efficiency of the cycle? Neglect pump work.

8.39 If pump work is included in Problem 8.38, what is the efficiency of the cycle?

8.40 Steam is expanded to 10% moisture at 2 psia in a Rankine cycle. If the initial pressure is 500 psia, what is the efficiency of the cycle? Neglect pump work.

8.41 If pump work is included in Problem 8.40, what is the efficiency of the cycle?

8.42 Steam is expanded to 12% moisture at 1 psia in a Rankine cycle. If the initial pressure is 400 psia, what is the efficiency of the cycle? Neglect pump work.

8.43 Include pump work in Problem 8.42 and determine the cycle efficiency. Use the computerized *Steam Tables* for the pump work and for the cycle.

8.44 Determine the type efficiency of the cycle of Problem 8.20.

8.45 Determine the type efficiency of the cycle of Problem 8.26.

8.46 Determine the type efficiency of the cycle of Problem 8.30.

8.47 Determine the type efficiency of the cycle of Problem 8.34.

8.48 Determine the type efficiency of the cycle of Problem 8.36.

8.49 Determine the type efficiency of the cycle of Problem 8.38.

8.50 Determine the type efficiency of the cycle of Problem 8.40.

8.51 Determine the heat rate in Btu/kWh for Problem 8.34.

8.52 Determine the heat rate in Btu/kWh for Problem 8.35.

8.53 Determine the ideal steam rate for Problem 8.34.

8.54 Determine the ideal steam rate for Problem 8.35.

8.55 If the specific heat of the condenser cooling water is assumed to be unity and it is desired to limit the condenser cooling water temperature change to 50°F in the condenser, determine the pounds of cooling water required per pound of steam for the ideal Rankine cycle operated from 600 psia and 600°F to 2 psia. The condensate is assumed to be saturated liquid.

8.56 Steam is supplied to the turbine of a Rankine cycle at 500 psia, and the turbine exhausts at 1 psia. Determine the turbine inlet temperature if there is 10% moisture in the exhaust steam. Use the Mollier chart.

8.57 An ideal Rankine cycle operates with a condenser pressure of 1 psia and with saturated vapor entering the turbine. What inlet pressure will produce a cycle efficiency of 28%? (*Hint:* Use the Mollier chart and a trial-and-error procedure.)

Problems Involving the Reheat Cycle

8.58 A reheat cycle is operated to decrease the amount of moisture in the final steam from a turbine. If the steam, initially at 1000 psia and 800°F, is permitted to expand only to saturation and is reheated to its initial enthalpy, what will the moisture content of the final steam be if it is expanded reversibly to 2 in. Hg? Use the Mollier chart.

8.59 Solve Problem 8.58 using the computerized *Steam Tables*.

8.60 An ideal reheat cycle operates with initial steam conditions of 1500 psia and 1000°F at the inlet to the turbine. The steam expands isentropically to 400 psia and is reheated at constant pressure to 1000°F. If the exhaust pressure of the cycle is 1 psia, determine the cycle efficiency. Use the Mollier chart.

8.61 Solve Problem 8.60 using the computerized *Steam Tables*.

8.62 Determine the heat rate in Problem 8.60.

8.63 If the steam in Problem 8.58 is reheated by the addition of one-half of the initial decrease in enthalpy, what will the moisture content of the final steam be after a reversible expansion to 2 in. Hg?

8.64 Steam in a reheat cycle is initially at 5 MPa and 500°C. It expands to saturation and is reheated to its initial enthalpy. If the final expansion is to 5 kPa, what is the moisture content of the steam leaving the turbine?

8.65 A reheat cycle is operated from 500 psia and 500°F to a final pressure of 1 in. Hg. If the amount of reheat added is $\frac{1}{2}\Delta h$ from the initial conditions to saturated vapor, determine the efficiency of the cycle (h_f at 1 in. Hg is 47.09 Btu/lb; temperature is close to 79°F.)

8.66 What is the type efficiency of the cycle in Problem 8.65?

8.67 Steam at 10 MPa and 500°C enters an ideal Rankine cycle that has one stage of reheat. The reheat is from saturated steam to the original temperature. If the final expansion is to 10 kPa, determine the efficiency of the cycle. Use the Mollier chart.

8.68 Solve Problem 8.67 using the computerized *Steam Tables*.

8.69 An ideal Rankine cycle with reheat operates with a turbine inlet pressure of 1300 psia and 1000°F. The condenser pressure is 1 psia. If the steam is first expanded to saturated vapor and then reheated to 900°F, determine the thermal efficiency of the cycle. Use the Mollier chart.

8.70 Solve Problem 8.69 using the computerized *Steam Tables*.

*8.71 Steam at 10 MPa and 600°C enters the turbine of an ideal Rankine cycle. The steam is reheated to 500°C. If saturated vapor enters the condenser at 5 kPa, determine the cycle efficiency. Use the Mollier chart.

*8.72 Solve Problem 8.71 using the computerized *Steam Tables*.

8.73 In a reheat cycle the steam is expanded from 700 psia and 600°F to 100 psia. It is subsequently reheated until the steam has the same degree of superheat as it had initially. If the final pressure of the cycle is 1 psia, determine its efficiency.

8.74 Solve Problem 8.73 using the computerized *Steam Tables*.

8.75 Solve Problem 8.73 including the pump work.

8.76 Solve Problem 8.74 including the pump work.

8.77 In an ideal Rankine cycle with reheat, steam enters the turbine at 10 MPa and 600°C. The steam is reheated to the initial temperature at a pressure of 1 MPa. If the condenser is operated at 5 kPa, determine the cycle efficiency. Use the Mollier chart. Neglect pump work.

8.78 Use the computerized *Steam Tables* to solve Problem 8.77.

8.79 Solve Problem 8.78 including pump work.

8.80 Starting with 500°F and 500 psia and assuming that the steam in a reheat cycle is reheated to an enthalpy equal to its initial enthalpy, determine the efficiency of this cycle if the final pressure is 1 psia and the initial expansion is carried out to 100 psia.

8.81 Assume that the final pressure in a reheat cycle is 1 psia and the first expansion is to saturated vapor. If the steam is reheated to its initial enthalpy, determine the efficiency of the cycle for an initial pressure of 500 psia and initial temperature of 600°F.

Problems Involving the Regenerative Cycle

8.82 Determine the efficiency of the cycle in Problem 8.20 if it operates as a regenerative cycle with one open heater at 100 psia. Compare your answers in both problems.

8.83 Solve Problem 8.82 using the computerized *Steam Tables*.

8.84 Use formulas (a) and (b) of Illustrative Problem 8.11 to solve Problem 8.82.

8.85 Determine the efficiency of the cycle in Problem 8.34 if it operates with one open feedwater heater at 50 psia. Compare your answers in both problems.

8.86 Solve Problem 8.85 using the computerized *Steam Tables*.

8.87 A regenerative cycle operates with a turbine inlet temperature of 600°C and a pressure of 10 MPa. The condenser is operated at 5 kPa. If there is one open heater in the cycle with extraction at 1 MPa, determine the efficiency of the cycle. Use the Mollier chart.

8.88 Solve Problem 8.87 using the computerized *Steam Tables*.

8.89 Use Equations (a) and (b) of Illustrative Problem 8.11 to solve Problem 8.88.

8.90 Steam enters the turbine of a regenerative cycle at 900 psia and 900°F. The condenser is operated at 1 psia. If extraction to the single open heater is at 200 psia, determine the thermal efficiency of the cycle. Use the Mollier chart.

8.91 Solve Problem 8.90 using the computerized *Steam Tables*.

8.92 Solve Problem 8.91 using Equations (a) and (b) of Illustrative Problem 8.11.

*8.93 Solve Problem 8.90 if a second open heater is used with extraction at 300 psia. Use the Mollier chart.

*8.94 Use the computerized *Steam Tables* to solve Problem 8.93.

8.95 Use Equations (a), (b), and (c) of Illustrative Problem 8.12 to solve Problem 8.94.

*8.96 A regenerative cycle operates with two open heaters, one at 50 psia and the other at 100 psia. The inlet to the turbine is superheated steam at 1000°F and 1000 psia, and the outlet of the turbine is at 1 psia. Using an ideal cycle, determine the efficiency and compare it to Illustrative Problem 8.11.

*8.97 An ideal regenerative cycle operates with reheat. The high-pressure turbine is supplied with steam at 2000 psia and 1000°F. The steam expands to 200 psia and is removed to a reheater where it is reheated back to 1000°F. Part of the steam leaving the high-pressure turbine is used in an open heater for feedwater heating. Steam is extracted from the low-pressure turbine at 50 psia and is used for feedwater heating in a second open feedwater heater. Determine the efficiency of the cycle if the condenser pressure is 1 psia. Use the Mollier chart.

*8.98 Solve Problem 8.97 using the computerized *Steam Tables*.

9

Gas Power Cycles

Learning Goals

After reading and studying the material in this chapter, you should be able to:

1. Differentiate between internal-combustion and external-combustion heat engines.

2. Describe the sequence of events in the four-stroke and two-stroke cycles.

3. State the assumptions made in air-standard cycle analyses.

4. Sketch the ideal Otto cycle on both $p-v$ and $T-s$ diagrams.

5. Show how an actual cycle differs from the ideal Otto cycle.

6. Derive the expression for the efficiency of the ideal air-standard Otto cycle, and show that it is a function only of the compression ratio.

7. Sketch the ideal Diesel cycle on both $p-v$ and $T-s$ diagrams.

8. Show that for the same compression ratio, the efficiency of the ideal Otto cycle is greater than the efficiency of the ideal Diesel cycle.

9. Understand why the Diesel cycle can be operated at higher compression ratios than the Otto cycle.

10. Sketch the ideal Brayton cycle on both $p-v$ and $T-s$ diagrams.

11. Show that for the same compression ratio, the Brayton and Otto cycles have the same thermal efficiency.

12. Understand why the dual cycle was proposed as the prototype for actual internal-combustion engines.

13. Show that the efficiency of the ideal dual-cycle engine lies between the efficiency of the Otto and Diesel cycles.

14. Understand that regeneration can be applied to internal-combustion cycles with the possibility of approaching the Carnot cycle efficiency, but that the practical realization of this goal is difficult due to the requirements and the complexity of the equipment needed.

9.1 INTRODUCTION

The power-plant cycles studied in Chapter 8 are known as external-combustion cycles, because heat is supplied from a source that is external to the engine and is rejected to a sink that is also external to the engine. The power plants that we use for propelling cars, boats, and so on are commonly known as *internal-combustion engines,* because the release of energy is by combustion within the engine. Mobile power plants have been developed to a high state of reliability and are commonplace in everyday life. Reliability is essentially a term that describes the level of development of the mechanical system. Thus, as the usual occurrence, we expect that a car will start easily and that a trip will be completed without mechanical failure. However, the size, weight, and cost of fuel for a given engine are all functions of the thermal efficiency of the unit. It should be noted that external-combustion engines have been studied more extensively in recent times with the aim of reducing air pollution, a problem in modern industrial societies.

In Chapter 8, our attention was focused on systems in which the release of energy by combustion occurred external to the device or engine that converted this energy to useful work. Several limitations occur when this is done. In the combustion device, temperatures are maintained as high as possible. This leads to problems associated with the strength of the heat-transfer surfaces (tubing) and requires large amounts of expensive, bulky insulation. The working fluid of the cycle is contained within tubular elements, and to keep thermal and pressure stresses within the strength capabilities of the tube material, the heat-transfer rates must be limited. This leads to the need for large heat-transfer surfaces requiring large units. Many of these objections can be overcome by releasing the energy in the fuel within the engine. In any device, it is usually true that the average temperature rather than the peak temperature of the working fluid will govern the strength limits of the material. In a periodic device (e.g., the automotive engine), the average temperature is kept reasonable by maintaining the duration of the peak temperature for a small fraction of the total cycle time.

Before considering the prototype ideal cycles of the internal-combustion engine, let us first describe the engine and then make those idealizations required for the analysis of the cycle. Figure 9.1 shows a 2.3-L, turbocharged, four-cylinder engine. These engines are mechanically complex and have hundreds of moving parts. To understand this engine, we will consider the four-stroke reciprocating automobile engine, which is the most commonly used. The four events that take place in the cycle are illustrated in Figure 9.2 and are as follows:

1. *Intake.* Consider the piston to be at the uppermost part of its stroke known as top dead center (t.d.c.). As the piston starts downward, it creates a partial vacuum in the cylinder. At this time, the intake valve is open and a mixture of air and vaporized gasoline is drawn into the cylinder from the intake manifold past the open intake valve.

2. *Compression.* When the piston reaches the end of its downward stroke, the piston is said to be at bottom dead center (b.d.c.). At this point, the intake valve closes and stays closed as the piston moves upward. Because the fuel–air mixture is totally confined within the cylinder, it is compressed as the piston moves upward. Both intake and exhaust valves remain closed.

3. *Power.* Ideally, when the piston again reaches top dead center and the fuel–air mixture is at its maximum compression, an electrical spark ignites the fuel–air mixture, causing combustion to occur. The large pressure force created by the combustion causes the piston to be driven downward. This driven-downward motion of the piston is the power stroke of

FIGURE 9.1 Modern 2.3-L, turbocharged, four-cylinder engine.
(Courtesy of Ford Motor Co.)

the cycle. In actual engines, ignition can occur before, at, or even after top dead center, based on details that we will not consider here. Also, both the intake and exhaust valves remain closed during the power stroke.

4. *Exhaust.* At the end of the power stroke, the gases have fully expanded and the exhaust valve opens. The upward motion of the piston causes the spent combustion gases to flow out of the cylinder past the open exhaust valve and into the exhaust manifold. At top dead center, the exhaust valve closes and the cycle starts again.

The four-stroke cycle just described produces one power stroke for each piston for two revolutions of the main shaft (crankshaft). The same sequence of events can also be accomplished in one revolution of the crankshaft to give one power stroke per revolution. This is the two-stroke engine cycle. The sequence of events is shown in Figure 9.3 and consists of the following for a fuel-injected engine:

1. *Air intake.* With transfer and exhaust ports open, air from the crankcase fills cylinder.

2. *Compression.* With all ports covered, the rising piston compresses air and creates suction in the crankcase. Injection starts.

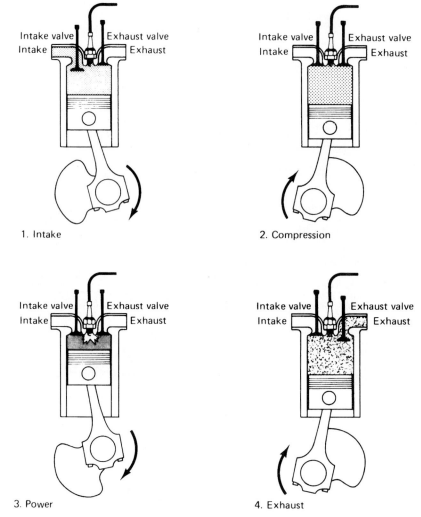

FIGURE 9.2 Four-stroke internal-combustion engine.
(From "How an Internal Engine Works," courtesy of Renewal Products Co., Fairless Hills, Pa.)

3. *Expansion.* The burning mixture expands, forcing the piston down. Air flows into the crankcase to be compressed as the piston descends.

4. *Exhaust.* The descending piston uncovers the exhaust port. Slight pressure builds up in the crankcase, which is enough to force air into the cylinder.

The distinguishing feature of the two-stroke engine is that *every* outward stroke of the piston is a power-producing stroke. It should be noted that for *a given output,* a definite air capacity is required, requiring the two-stroke engine to have an air input per unit time at least equal to the four-stroke equivalent engine.

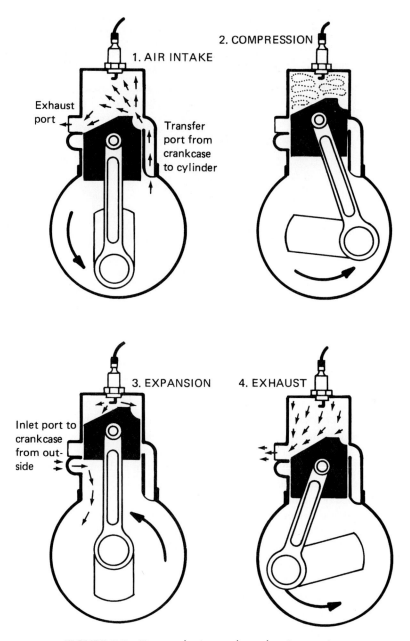

FIGURE 9.3 Two-stroke internal combustion engine.

Figure 9.4 shows a modern, compact four-cylinder engine. The cylinder block is usually an accurately machined housing for the cylinders, pistons, and connecting rods. All the other elements are supported by the cylinder block (or blocks) or bolted to it. At the top of the block is the cylinder head, which is also shown in Figure 9.4. To prevent melting of the metal parts, cooling must be provided. This can either be done by air cooling or water cooling. Water cooling is the com-

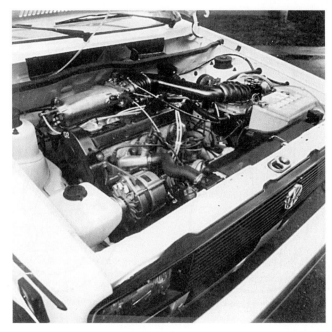

FIGURE 9.4   Modern four-cylinder, fuel-injected engine.
(Courtesy of Volkswagen of America.)

monest method used, and this water must circulate within the block and the heads. Thus, a series of matching openings is provided in the blocks and heads for water circulation.

The valves in the internal combustion engine are subjected to the severest conditions of all the elements of the engine. Operating as often as 2000 times per minute, they are required to be leakproof even though they are subjected to the direct pressure and temperature of the combustion occurring within the cylinders. Also, they must operate at the precisely correct moment in the cycle without valve bounce or chatter. The actuation of the valves is performed by one cam for each valve on the camshaft that is driven by the crankshaft. Because the four-cycle engine produces one power stroke per piston for two revolutions of the crankshaft, the camshaft is geared to rotate one revolution for each two revolutions of the crankshaft. The cam shape must be accurately designed to open the intake and exhaust valves precisely, hold them open for the necessary length of time, and close them all as dictated by the cycle.

Figure 9.5 shows a horizontally opposed, six-cylinder, 24-valve, 3.3-L, 230-hp engine used in a modern sports car. This configuration gives lower vibration and a higher torque at all engine speeds and a lower center of gravity for better roadability and handling. This engine uses programmed fuel injection, electronic direct ignition, and a computer-based electronic engine management system for optimum performance. Figure 9.6 shows the performance of this engine as a function of engine speed.

Many different arrangements of the cylinders and the manner in which they are grouped are possible. We have already seen and described the four-cylinder engine and the horizontally opposed, six-cylinder engine. In addition to the arrangements previously discussed, there are others, as shown in Figure 9.7, that are used in specific applications.

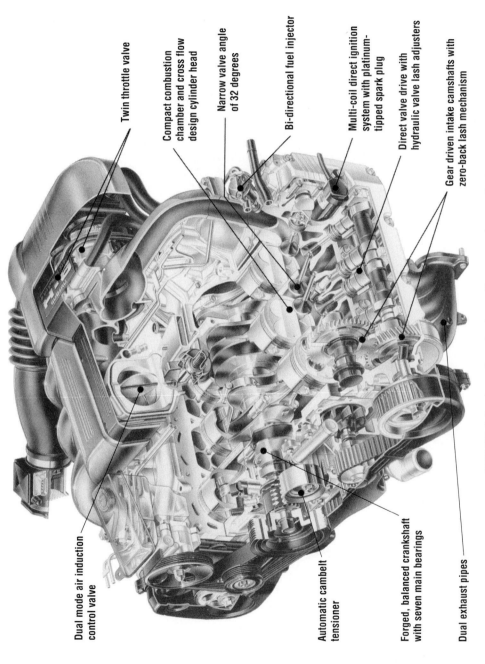

Twin throttle valve

Compact combustion chamber and cross flow design cylinder head

Narrow valve angle of 32 degrees

Bi-directional fuel injector

Multi-coil direct ignition system with platinum-tipped spark plug

Direct valve drive with hydraulic valve lash adjusters

Gear driven intake camshafts with zero-back lash mechanism

Dual mode air induction control valve

Automatic cambelt tensioner

Forged, balanced crankshaft with seven main bearings

Dual exhaust pipes

FIGURE 9.5 A modern, horizontally opposed, six-cylinder engine. *(Courtesy of Subaru of America, Inc.)*

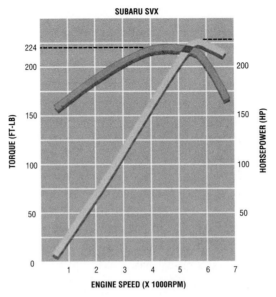

**6-cylinder, 24-valve engine delivers 230 hp
and 224 ft. lbs. torque.**

FIGURE 9.6 Performance of engine in Figure 9.5.
(Courtesy of Subaru of America, Inc.)

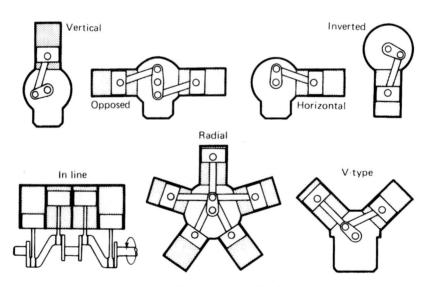

FIGURE 9.7 Internal-combustion cylinder arrangement.
(From "How an Internal Engine Works," courtesy of Renewal Products Co., Fairless Hills, Pa.)

To this point, our discussion of internal-combustion engines has considered only reciprocating engines. We will now consider the rotating-combustion engine, which is also known as the Wankel engine. The rotating-combustion engine was first conceived by a German inventor, Felix Wankel, in 1954. Since the early 1960s, intense development work on this engine has taken place throughout the world. At present, the Wankel has been produced in sizes ranging from $\frac{1}{2}$ to 400 hp and powers approximately 500,000 automobiles worldwide. To obtain an idea of the relative size and simplicity of the rotating-combustion engine compared to a conventional reciprocating automotive engine, one only has to note the following: A V-8 engine of 195 hp has 1029 parts, and 388 of these parts move. The engine weighs more than 600 lb and takes up 15 ft^3 in the engine compartment. In contrast, a 185-hp Wankel engine has only 633 parts, 154 of which move, and it only weighs 237 lb and occupies 5 ft^3. In addition, there are no valves in the Wankel engine.

The Wankel engine operates on the four-stroke Otto cycle, as indicated in Figure 9.8. As shown in the figure, there are two major moving assemblies in the Wankel engine: a three-sided rotor, and a main shaft. The rotor operates inside a chamber that is essentially an ellipse with sides pointed in. The mathematical name of the shape of the chamber is an epitrochoid. This shape is needed due to the motion of the rotor, which rotates on an eccentric axis off the main shaft in a wobbling motion. The cycle shown in Figure 9.8 starts when one of the apex points passes the intake port, causing the fuel–air mixture to flow into the housing in the volume between the rotor and housing. As the rotor turns, the volume between the rotor and chamber decreases, which effectively compresses the fuel–air mixture. The fuel–air mixture is ignited when its volume is at its minimum. The combustion of the fuel–air mixture causes a thrust on the rotor to keep it turning and, at the same time, causes the combustion gases to expand. At the end of expansion, the exhaust port is uncovered and the combustion gases leave through this port.

Thus far, we have considered only one face of the three-lobed rotor during the course of one revolution. It must be remembered that during this same revolution, the other two faces of the rotor are undergoing the same cycle. Thus, the torque output of the Wankel engine is positive for about

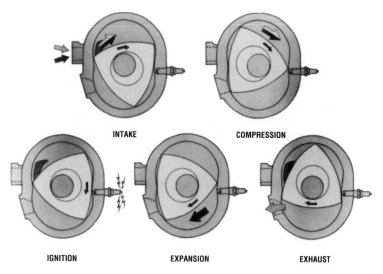

INTAKE COMPRESSION

IGNITION   EXPANSION   EXHAUST

FIGURE 9.8 Wankel engine combustion cycle.
(Courtesy of Curtiss-Wright Corp.)

two-thirds of one shaft revolution as compared to one-quarter of the two shaft revolutions required to complete a single cycle of a four-cycle reciprocating engine. This is due to the fact that there is a 1:3 rotor-to-crankshaft speed ratio in the Wankel engine, requiring three shaft revolutions for one complete rotor revolution to fire all three rotor flanks. A single-cylinder reciprocating engine fires once in two shaft revolutions, giving a three-cylinder reciprocating engine three power pulses in two shaft revolutions. On the basis of torque output, the Wankel single-rotor engine approaches the three-cylinder reciprocating engine. If we consider power, the single-rotor Wankel engine power output is twice the power output of the same displacement reciprocating four-cycle engine. This is due to the fact that for every mainshaft revolution there is one firing, similar to the two-cycle engine. Also, it must be noted that in the piston engine, the piston comes to a complete halt every time it reverses direction and that due to its linear motion, a connecting rod and crank are needed to convert its up-and-down motion into torque. As a result of these factors, the Wankel engine gives smoother torque output at all speeds without dynamic unbalance even when one rotor is used.

The advantages of the Wankel engine can be summarized as follows:

1. There are fewer moving parts: no valves, connecting rods, and so on.

2. There is a high power-to-weight ratio and a small volume.

3. The rotor center of gravity and rotational axis remain at a fixed eccentricity from the shaft centerline so that even a single-rotor engine can attain complete dynamic balance with simple shaft counterweights.

4. Smooth torque and power output exist at all speeds.

One of the Wankel engine's most serious engineering problems was the design of the apex seals that are used to seal off the three compartments from each other. Some of the initial engines built for automotive use seldom ran for more than 15,000 to 20,000 miles before the seals wore out. At present, it is claimed that seal lifetimes in automotive applications are in the 100,000-mile range, which is nearing the life of piston rings in conventional engines. Another problem that occurred during the initial development of the Wankel engine was its poor fuel economy, which was about half the miles per gallon of a comparative conventional automotive engine. The principal factors contributing to this poor mileage were the positioning of the spark plugs, insufficient distance between the inlet and exhaust ports, and the already-mentioned seal problem. Using a modified side entry port, repositioned spark plug, and the latest seals, current versions of the Wankel engine indicate that they are reaching the same fuel consumption levels *as comparably powered, conventional reciprocating engines.* The only commercially produced rotary engine was used in the Mazda RX-7 sports car. This car is no longer available in the United States but continues to be sold in other countries. A similar rotary engine car is expected to be introduced in the U.S. market in 2001.

9.2 AIR-STANDARD ANALYSIS OF THE OTTO CYCLE

The introduction of fuel into an internal-combustion engine gives rise to a variable mass in the cycle. Also, at the end of the cycle, the entire charge is discarded and a fresh charge is introduced to undergo a new cycle identical to the first. During the actual processes, both heat and work are interchanged at each portion of the cycle. The mass, specific heats, and state of the working fluid are all variable. Under these circumstances, the analysis of a cycle becomes quite difficult. To simplify the analysis, certain idealized cycles have been proposed as prototypes of the actual engine cycles. These models are considered for the rest of this chapter. In the analysis of each of the cycles, the following assumptions are made:

1. Each process is carried out reversibly. Friction, pressure differences, turbulence, and the like are neglected.

2. The working fluid is an ideal gas, and all relations already derived for the ideal gas are applicable.

3. The necessary energy is added or removed externally to achieve the desired state changes.

4. The working fluid is a gas with constant specific heats.

5. The amount of working fluid in the cycle is constant. In effect, the same air remains in the engine through each cycle.

It is apparent from these assumptions (and their implications) that the analysis of an engine cycle based on them is quite artificial. However, certain generalizations can be arrived at from this analysis, called the *air-standard analysis,* and as such, it can be quite useful.

The *Otto cycle* is the prototype for most internal-combustion engines commonly in use. Regardless of the number of strokes required to complete the cycle, it is conceived of as consisting of four separate and distinct processes. As shown in the *T–s* and *p–v* diagrams in Figure 9.9, the Otto cycle consists of (after induction of the gas) an isentropic compression followed by the reversible constant-volume addition of heat, then an isentropic expansion from which work is extracted, and finally a reversible constant-volume rejection of heat. Subsequently, the cycle is repeated. It will be noted that each step is an idealization of the events that we have previously described for the internal-combustion engine. Also, note that each portion of the cycle is a nonflow process.

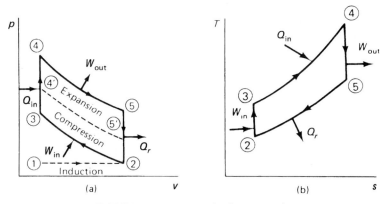

FIGURE 9.9   Air-standard Otto cycle.

It is instructive at this point to see how an actual cycle compares to the air-standard Otto cycle. Figure 9.10 shows a comparison of an actual and an Otto cycle engine on *p–v* coordinates. In the actual cycle, the air–fuel mixture is drawn into the cylinder from ① → ② with the cylinder pressure below atmospheric pressure. At a point before the piston reaches top dead center, or ③, the spark plug initiates ignition. Combustion is essentially complete at ④, with the piston starting down. Note that the peak pressure of the Otto cycle is not reached in the actual cycle. At

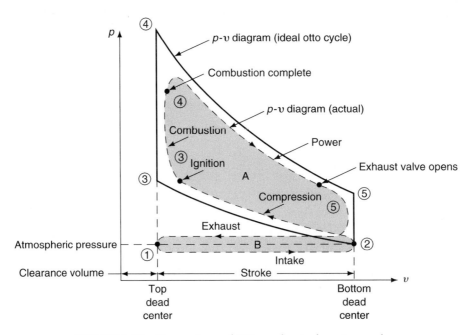

FIGURE 9.10 Comparison of Otto and actual engine cycles.

⑤, before bottom dead center, the exhaust valve opens. The pressure in the cylinder is always above atmospheric pressure during the exhaust process to overcome pressure drops in the exhaust system. When ① is reached, the process is repeated. The area noted as B (intake and exhaust) represents negative work, that is, work done *by* the engine, and the total indicated work in Figure 9.10 is equal to area A minus area B. In order to minimize area B, the intake and exhaust systems should be designed so as to minimize flow resistance.

Referring to Figure 9.9, we have the nonflow energy equation for the nonflow reception of heat (step ③ to ④) as

$$q_{in} = c_v(T_4 - T_5) \quad \text{Btu/lb}_m, \text{kJ/kg} \tag{9.1}$$

Similarly, the heat rejected during the constant-volume, reversible expansion (step ⑤ to ②) is

$$q_r = c_v(T_5 - T_2) \quad \text{Btu/lb}_m, \text{kJ/kg} \tag{9.2}$$

The net work in thermal units available from the cycle is the difference between q_{in} and q_r:

$$w = (q_{in} - q_r) \quad \text{Btu/lb}_m, \text{kJ/kg} \tag{9.3}$$

The efficiency of the cycle is

$$\eta_{Otto} = \frac{w}{q_{in}} = \frac{q_{in} - q_r}{q_{in}} = 1 - \frac{q_r}{q_{in}} \tag{9.4a}$$

and

$$\eta_{Otto} = 1 - \frac{c_v(T_5 - T_2)}{c_v(T_4 - T_3)} = 1 - \frac{T_5 - T_2}{T_4 - T_3} \tag{9.4b}$$

From Figure 9.9, it will be noted that the volumetric limits of the expansion and compression portions of the cycle are equal. From equations developed in Chapter 6,

$$\frac{T_4}{T_5} = \frac{T_3}{T_2} \quad \text{or} \quad \frac{T_2}{T_5} = \frac{T_3}{T_4} \tag{9.5}$$

By adding unity to both sides of Equation (9.5),

$$1 - \frac{T_2}{T_5} = 1 - \frac{T_3}{T_4} \quad \text{or} \quad \frac{T_5 - T_2}{T_5} = \frac{T_4 - T_3}{T_4} \tag{9.6}$$

By substituting Equation (9.6) into Equation (9.4b), we obtain

$$\eta_{\text{Otto}} = 1 - \frac{T_5}{T_4} = 1 - \frac{T_2}{T_3} = 1 - \left(\frac{v_3}{v_2}\right)^{k-1} \tag{9.7}$$

where we have used the equation of path for the isentropic compression to relate temperature and volumes from Table 6.4.

At this time, we will introduce a term called the *compression ratio*, r_c, which is defined as

$$r_c = \frac{v_2}{v_3} \tag{9.8}$$

Note that the compression ratio defined in this manner is the ratio of two volumes, not two pressures as is commonly thought. Using this definition of compression ratio, we have the efficiency of the Otto cycle as

$$\eta_{\text{Otto}} = \left[1 - \left(\frac{1}{r_c}\right)^{k-1} \right] \times 100 \tag{9.9}$$

It is emphasized that the term "compression ratio" is not the ratio of the peak-to-inlet pressure; it is the ratio of the volume before compression to the volume after compression.

Figure 9.11 is a plot of Equation (9.9) for $k = 1.4$. It is apparent that an increase in compression ratio yields an increased cycle efficiency. However, it will be noted that the efficiency increases at a decreasing rate, and an increase in compression ratio from 10 to 15 yields an increase in efficiency of only 6%. Until the mid-1960s, the compression ratio of the automotive engine was generally being increased in an effort to get more power out of a given engine. It was not uncommon to have engines with compression ratios in excess of 10, requiring highly leaded fuels to prevent preignition and detonation of the fuel–air mixture in the compression and ignition portions of the cycle. Environmental considerations, with the attendant desire to limit the pollutants from these engines, have led to engines having lower compression ratios. The use of pollution controls, fuel injection, multiple valves, and turbo-supercharging has enabled modern engines to achieve high power outputs while still being environmentally acceptable. The use of multipoint fuel injection, electronic ignition, and electronic engine management systems has made it possible to increase the compression ratio on modern engines. The engine shown in Figure 9.5 has a compression ratio of 10 and uses premium fuel.

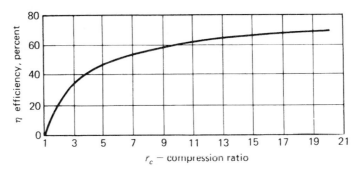

FIGURE 9.11 Efficiency of Otto cycle.

Returning to the air-standard Otto cycle, another point must be emphasized. The Otto air-standard cycle efficiency is determined only by the ratio of v_2 to v_3. Thus, the cycle operating between the limits of ②, ③, ④, and ⑤ in Figure 9.9 will have the same efficiency as the cycle operated between ②', ③', ④', and ⑤'. The heat in, work out, and heat rejected will be different for each cycle, but the efficiency will be the same. However, a Carnot cycle having as its maximum temperature the upper temperature of the Otto cycle and as its lower temperature the lowest temperature of the Otto cycle will always have a thermal efficiency that is greater than the Otto cycle. This can be determined from Figure 9.9 as follows: As heat is added to the Otto cycle, its maximum temperature T_4 is increased, but its efficiency is constant if v_2/v_3 is constant. The Carnot cycle efficiency increases as its peak cycle temperature is increased. In the limit, the least value of the efficiency of a Carnot cycle will be $1 - T_2/T_3$, which is the maximum efficiency of the Otto cycle.

ILLUSTRATIVE PROBLEM 9.1

Compute the efficiency of an air-standard Otto cycle if the compression ratio is 7 and $k = 1.4$. Compare this with a Carnot cycle operating between the same temperature limits if the lowest temperature of the cycle is 70°F and the peak temperature is 700°F, 1000°F, and 3000°F. Plot η versus temperature.

SOLUTION

The efficiency of the Otto cycle is a function only of the compression ratio. Therefore, for the entire range of this problem, η is constant and equal to

$$\eta_{\text{Otto}} = \left(1 - \frac{1}{7^{0.4}}\right) \times 100 = 54.1\%$$

For the Carnot cycle,

$$\eta_c = 1 - \frac{T_2}{T_4}$$

At 700°F,

$$\eta_C = \left(1 - \frac{530}{1160}\right) \times 100 = 54.3\%$$

At 1000°F,

$$\eta_C = \left(1 - \frac{530}{1460}\right) \times 100 = 63.7\%$$

At 3000°F,

$$\eta_C = \left(1 - \frac{530}{3460}\right) \times 100 = 84.7\%$$

These data are plotted in Figure 9.12.

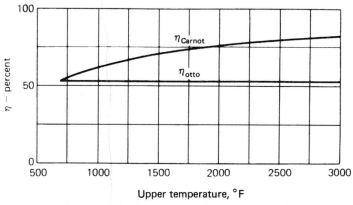

FIGURE 9.12   Solution of Illustrative Problem 9.1.

ILLUSTRATIVE PROBLEM 9.2

Assuming c_v to be 0.172 Btu/lb$_m$·°R, determine the heat added and the heat rejected per pound of fluid during the constant-volume portions of Illustrative Problem 9.1 for the case in which the peak temperature of the cycle is 1000°F. Also, compute the work output per pound of fluid. Determine the efficiency of the Otto cycle based on these values, and compare it with the results of Illustrative Problem 9.1.

SOLUTION

By referring to Figure 9.9 and using the relations developed in Chapter 6,

$$\frac{T_3}{T_2} = \left(\frac{v_2}{v_3}\right)^{k-1} = (7)^{0.4} = 2.18$$

$$T_3 = (2.18)(530) = 1155.4°R$$

$$q_{in} = c_v(T_4 - T_3) = (0.172)(1460 - 1155.4)$$
$$= 52.39 \text{ Btu/lb}_m$$

To determine T_5, further use is made of Table 6.4:

$$q_r = c_v(T_5 - T_2) \qquad \frac{T_5}{T_4} = \left(\frac{v_3}{v_2}\right)^{k-1} = \frac{1}{2.18}$$

Therefore,

$$T_5 = \frac{1460}{2.18} = 669.72$$

$$q_r = c_v(T_5 - T_2) = (0.172)(669.72 - 530) = 24.03 \text{ Btu/lb}_m$$

The net work out is

$$q_{in} - q_r = 52.39 - 24.03 = 28.36 \text{ Btu/lb}_m$$

Therefore,

$$\eta_{Otto} = \frac{28.36}{52.39} \times 100 = 54.1\%$$

This value of efficiency agrees with the results of Illustrative Problem 9.1.

ILLUSTRATIVE PROBLEM 9.3

An Otto cycle operates with a compression ratio of 8. If 50 kJ are added to the cycle which has as its lower pressure 150 kPa and as its lowest temperature 20°C, determine the peak temperature of the cycle if $c_v = 0.7186$ kJ/kg·°K and $k = 1.4$. Assume 1 kg of working fluid.

SOLUTION

Refer to Figure 9.9 and

$$\frac{T_3}{T_2} = \left(\frac{v_2}{v_3}\right)^{k-1} = 8^{0.4} = 2.297$$

Therefore,

$$T_3 = (2.297)(20 + 273) = 673.14°K$$

But

$$q_{in} = c_v(T_4 - T_3)$$

Therefore,

$$50 = 0.7186(T_4 - T_3)$$

$$69.58 = T_4 - T_3$$

and

$$T_4 = 742.72°K = 469.72°C$$

ILLUSTRATIVE PROBLEM 9.4

An Otto cycle operates with a compression ratio of 7. If 50 Btu/lb$_m$ is added to the cycle, which begins at 60°F and 14.7 psia, determine the pressure, temperature, and specific volume at each point of the cycle. Also, calculate the efficiency of the cycle. Assume that $c_p = 0.24$ Btu/lb$_m$·°R, $c_v = 0.171$ Btu/lb$_m$·°R, $R = 53.3$ ft lb$_f$/lb$_m$·°R, and $k = 1.4$.

SOLUTION

We will refer to Figure 9.9 and go through the cycle point to point. At ②, we need v_2. This will be found from the equation of state, $p_2 v_2 = RT_2$:

$$v_2 = \frac{53.3 \times 520}{14.7 \times 144} = 13.09 \text{ ft}^3/\text{lb}_m$$

Because the path from ② to ③ is isentropic,

$$p_3 v_3^k = p_2 v_2^k$$
$$p_3 = p_2 \left(\frac{v_2}{v_3}\right)^k = p_2 r_c^k = (14.7)(7)^{1.4} = 224.1 \text{ psia}$$
$$v_3 = \frac{v_2}{r_c} = \frac{13.09}{7} = 1.87 \text{ ft}^3/\text{lb}_m$$
$$T_3 = \frac{p_3 v_3}{R} = \frac{224.1 \times 144 \times 1.87}{53.3} = 1132.2°\text{R}$$

To obtain the values at ④, we note $v_3 = v_4$ and that $c_v(T_4 - T_3) = q_{in}$. Thus,

$$T_4 = T_3 + \frac{q}{c_v} = 1132.2 + \frac{50}{0.171} = 1424.6°\text{R}$$

For p_4, we use the equation of state to obtain

$$p_4 = \frac{RT_4}{v_4} = \frac{53.3 \times 1424.5}{144 \times 1.87} = 281.98 \text{ psia}$$

The last point has the same specific volume as ②, giving $v_5 = v_2 = 13.09$ ft^3/lb$_m$. The isentropic path equation gives

$$p_5 v_5^k = p_4 v_4^k \quad p_5 = p_4 \left(\frac{v_4}{v_5}\right)^k = p_4 \left(\frac{1}{r_c}\right)^k$$
$$p_5 = 281.96 \left(\frac{1}{7}\right)^{1.4} = 18.5 \text{ psia}$$

T_5 is found from the equation of state as

$$T_5 = \frac{p_5 v_5}{R} = \frac{18.49 \times 144 \times 13.09}{53.3} = 654.1°\text{R}$$

The efficiency can be expressed in terms of the temperature as

$$\eta = \frac{(T_4 - T_3) - (T_5 - T_2)}{T_4 - T_3} \times 100$$

$$= \frac{(1424.6 - 1132.2) - (654.1 - 520)}{1424.6 - 1132.2} \times 100$$

$$= 54.1\%$$

The *mean effective pressure* of a cycle (mep) is defined as the work out of a cycle divided by the volume swept out by the piston. In Figure 9.13, it is the shaded area (work) divided by $(v_2 - v_3)$. It becomes that number (the horizontal dashed line) which, when multiplied by the base, yields the same shaded area. As such, it is the mathematical mean ordinate of Figure 9.13. The area of the rectangle $[(\text{mep}) \times (v_2 - v_3)]$ equals the area enclosed by the cycle ②, ③, ④, ⑤. The value of the mean effective pressure can be obtained by integrating the area shown in Figure 9.13 and dividing by $(v_2 - v_3)$. However, it can also be obtained as follows:

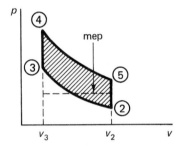

FIGURE 9.13 A p–v diagram of the Otto cycle.

The work out of the cycle equals $q_{in} - q_r$ in thermal units of Btu per pound:

$$q_{in} - q_r = c_v(T_4 - T_3) - c_v(T_5 - T_2)$$

In terms of pressure and volumes, we can show that

$$q_{in} - q_r = \frac{1}{k - 1}[v_3(p_4 - p_3) - v_2(p_5 - p_2)]$$

in mechanical units of ft lb_f/lb_m when pressure is in lb_f/ft^2 and specific volume in ft^3/lb_m. Dividing by $(v_2 - v_3)$, we obtain

$$\text{mep} = \frac{1}{k - 1}\left[\frac{v_3(p_4 - p_3) - v_2(p_5 - p_2)}{v_2 - v_3}\right] \tag{9.10}$$

or

$$\text{mep} = \frac{1}{k - 1}\left[\frac{(p_4 - r_c^k p_2)[1 - (1/r_c)^{k-1}]}{r_c - 1}\right] \tag{9.11}$$

The mean effective pressure is useful in evaluating the ability to produce power. Because the product of piston travel and piston area is piston volume (displacement), the relative power-producing capability of two pistons can be established by comparing the product of mep times displacement in each case.

ILLUSTRATIVE PROBLEM 9.5

Determine the power output of a cylinder having a cross-sectional area of A square inches, a length of stroke L inches, a mean effective pressure of p_m psi, and making N power strokes per minute.

SOLUTION

The force on the piston is $(p_m A)$ pounds. Because the mean effective pressure is equivalent to a constant force on the piston over the length of the piston stroke, the work done per power stroke is $(P_m A)L$. For N power strokes per minute,

$$ \mathrm{hp} = \frac{p_m LAN}{33{,}000} $$

Note that if p_m is in psi, A should be in square inches while L is in feet. Also, N is *not* the rpm of the engine; it is the number of power strokes per minute, which for a four-cycle engine is rpm/2.

ILLUSTRATIVE PROBLEM 9.6

It is common practice to denote the total volumetric displacement in liters. Using this notation to mean (LA), determine the power output of an engine having a mean effective pressure of p_m kPa and making N power strokes per minute.

SOLUTION

The basic solution is the same as for Illustrative Problem 9.5 with due respect to units. The displacement (LA) in cm^3/1000 needs to be converted to m^3.

$$ \mathrm{cm}^3 \times \frac{1}{(\mathrm{cm/m})^3} = \mathrm{cm}^3 \times \frac{1}{(100)^3} = \mathrm{cm}^3 \times 10^{-6} $$

or
$$ \mathrm{liters} \times 10^{-3} = \mathrm{m}^3 $$

One horsepower is equal to 746 W = 746 N·m/s. Per minute,

$$ 1\ \mathrm{hp} = 746 \times 60 = 44{,}760\ \frac{\mathrm{N \cdot m}}{\mathrm{min}} $$

Therefore,

$$\text{hp} = \frac{p_m \times 10^3 \times (LA)10^{-3} \times N}{44,760} = \frac{p_m LAN}{44,760}$$

where LA is expressed in liters, p_m in kPa, and N in power strokes per minute.

ILLUSTRATIVE PROBLEM 9.7

Using the results of Illustrative Problem 9.6, determine the horsepower of a 2.0-L engine that has a mep of 1 MPa operating at 4000 rpm. The engine is a four-cycle engine.

SOLUTION

$$\text{hp} = \frac{p_m LAN}{44,760} = \frac{1000 \text{ kPa} \times 2 \text{ L} \times \left(\dfrac{4000 \text{ rpm}}{2}\right)}{44,760 \dfrac{\text{N·m}}{\text{min·hp}}} = 89.4 \text{ hp}$$

ILLUSTRATIVE PROBLEM 9.8

Refer to Figure 9.14, and note that the difference in volume, $V_2 - V_3$, is commonly called the *displacement volume* and that V_3 is the *clearance volume*. Derive an expression for the compression ratio of an Otto engine in terms of the displacement volume and clearance volume. Also, derive an expression for the mep of an Otto engine in terms of the work, the compression ratio, and the initial specific volume.

SOLUTION

To simplify the analysis, let us call the clearance volume a fraction (c) of the displacement volume D. Thus, $V_3 = cD$, and $V_2 - V_3 = D$. By our definition of compression ratio,

$$r_c = \frac{v_2}{v_3} = \frac{V_2}{V_3} = \frac{D + cD}{cD} = \frac{1 + c}{c}$$

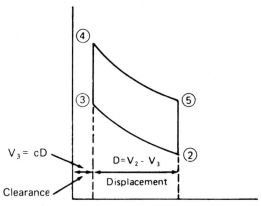

$V_3 = cD$

$D = V_2 - V_3$

②

Clearance

Displacement

FIGURE 9.14 Illustrative Problem 9.8.

The mean effective pressure $(\text{mep}) = \dfrac{\text{work}}{\text{displaced volume}} = \dfrac{w}{v_2 - v_3}$

$$\text{mep} = \frac{w}{v_2\left(1 - \dfrac{v_3}{v_2}\right)}$$

$$\boxed{\text{mep} = \frac{w}{v_2\left(1 - \dfrac{1}{r_c}\right)} = \frac{w}{(r_c - 1)v_3}}$$

ILLUSTRATIVE PROBLEM 9.9

An Otto engine has a clearance equal to 20% of its displacement. What is its compression ratio?

SOLUTION

From Illustrative Problem 9.8,

$$r_c = \frac{1 + c}{c} = \frac{1 + 0.2}{0.2} = \frac{6}{1}$$

ILLUSTRATIVE PROBLEM 9.10

Determine the mean effective pressure of a four-cycle, six-cylinder engine that has a 100-hp output at 4000 rpm. The cylinder bore is 3 in., and the stroke is 4 in.

SOLUTION

Based on the results of Illustrative Problem 9.5, we have

$$\text{hp} = \frac{p_m LAN}{33,000}$$

with $L = 4/12$ ft, $A = (\pi/4)(3)^2 \times 6$, and $N = 4000/2$ (because there are six cylinders and a four-cycle engine). Therefore,

$$p_m = \frac{100 \times 33,000}{\dfrac{4}{12} \times \dfrac{\pi(3)^2 \times 6}{4} \times \dfrac{4000}{2}} = 116.7 \text{ psia}$$

ILLUSTRATIVE PROBLEM 9.11

The engine shown in Figure 9.5 has an output of 230 hp at 5500 rpm. It is a six-cylinder engine with a displacement of 3.3 L. Determine the mean effective pressure of this engine at this engine speed.

SOLUTION

The displacement of 3.3 L can be converted to cubic inches as follows:

$$3.3 \text{ L} \times 1000 \frac{\text{cm}^3}{\text{L}} \times \left(\frac{\text{in.}}{2.54 \text{ cm}}\right)^3 = 201.4 \text{ in.}^3$$

From the results of Illustrative Problem 9.10, we have,

$$p_m = \frac{\text{hp} \times 33,000}{LAN} \quad \text{where } LA \text{ is the displacement}$$

$$p_m = \frac{230 \times 33,000 \times 12}{201.4 \times \dfrac{5500}{2}}$$

$$p_m = 164.4 \text{ psia}$$

9.3 DIESEL ENGINE (COMPRESSION IGNITION ENGINE)

In the Otto (or spark ignition) cycle, the fuel is mixed with the air prior to the compression stroke of the cycle. Ignition occurs due to an externally timed electrical spark. Because the fuel–air mixture is compressed, it is necessary to use volatile, readily vaporized fuels that can be distributed uniformly into the incoming air. Also, the compression of the fuel–air mixture can cause it to become prematurely ignited during the compression of the cycle, leading to the familiar phenomenon of "knock" in the engine. Recent emphasis on ecological considerations has placed severe limitations on the use of leaded fuels to prevent preignition and nonuniform burning of the fuel.

In the *Diesel engine,* the air is first compressed to a pressure and temperature sufficient to ignite the fuel, which is injected at the end of the compression stroke. Because there is no fuel present during the compression stroke, much higher compression ratios are used in the compression ignition engine than in the spark ignition engine. Figure 9.15 shows a cross-section of an automotive Diesel engine from which it will be seen that many of the mechanical features are the same as for the spark ignition engine. This engine has a displacement volume of 14 L, develops up to 525 hp, and uses glow plugs (resistance heaters) for starting. As will be noted from this figure, the entire spark ignition system is eliminated.

The operation of the Diesel engine shown in Figure 9.15 can be best illustrated by referring to Figure 9.16. The operation proceeds in the following manner:

FIGURE 9.15   A cutaway view of a 14-L Diesel engine.
(Courtesy of Cummins Engine Company.)

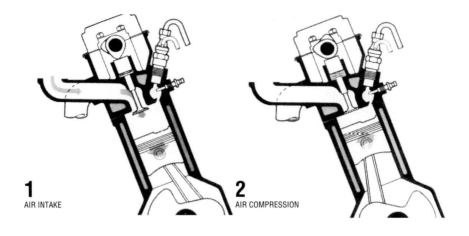

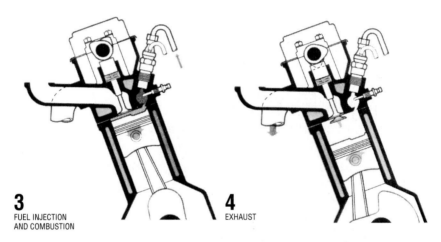

FIGURE 9.16 Operation of an automotive Diesel engine.

1. *Air intake.* Rotation of the crankshaft (bottom of drawing) drives a toothed belt that turns the overhead camshaft that opens the intake valve. As the piston moves downward, fresh air is sucked into the cylinder.

2. *Air compression.* As the piston moves upward, both the intake and exhaust valves are closed and compression of the trapped air begins. Because it is being compressed into a space 23.5-times smaller than its original volume, the air becomes very hot, reaching a maximum temperature of 1650°F (900°C)—far above the flashpoint for diesel fuel.

3. *Fuel injection and power.* As the piston reaches the top of its stroke and the temperature of the compressed air is at its maximum, a mist of fuel is injected into the spherical precham-

ber. The hot air ignites the fuel, and its burning causes the mixture to expand while the flame front spreads quickly from the prechamber to the cylinder. As the piston is now past its topmost point, it is driven downward by the force of the expanding gases produced by combustion, driving the crankshaft and the car.

4. *Exhaust.* As the energy of the fuel–air combustion is spent and the piston begins to move upward again, the exhaust valve opens, clearing the cylinder of burned gases, and as the piston starts downward, the air intake stroke (No. 1) begins again.

Before proceeding further, let us compare the features of the spark ignition and compression ignition engines:

Common Features

1. Both utilize the same mechanical patterns, that is, in line, V, and so on.

2. Both are similarly cooled and lubricated.

3. The valving and valve actions are similar.

4. Both use external starting systems.

5. Both are made in two- and four-cycle types.

Differing Features

1. Ignition in the spark ignition engine is initiated by an electrical spark at or near the completion of the compression stroke of the fuel–air mixture. In the compression ignition engine, air alone is compressed, and fuel ignition occurs when the fuel is injected into the cylinder at or near top dead center.

2. The prototype of the spark ignition engine is constant-volume combustion. In the compression ignition engine, constant-pressure combustion is the prototype.

3. Control of the spark ignition engine is accomplished by varying the quantity of fuel–air mixture while keeping a constant mixture composition. In the compression ignition engine, control is achieved by varying the fuel input to a constant mass of charging air.

4. Spark ignition engines use an electrical ignition system and a fuel injector. Compression ignition engines have no electrical ignition system, but they do have some form of high-pressure fuel injection system.

5. Spark ignition engines are restricted either to gases or readily vaporized fuels. Compression ignition engines do not have this limitation. The usual fuel is a refined crude petroleum oil. Surprisingly, both compression ignition and spark ignition engines are prone to detonation. The fuel qualities that suppress detonation in the spark ignition engine aggravate this condition in the compression ignition engine.

Although Diesel engines have been built in small sizes, such as for automotive use, most Diesels are large, relatively slow machines. An idea of the size of a Diesel-generator set that is used for a nuclear power plant for emergency power can be obtained from Figure 9.17. This unit is shipped as a single unit with the generator directly coupled to the Diesel engine.

FIGURE 9.17 Large industrial Diesel-generator set.
(Courtesy of Alco Power Inc.)

To increase the performance of Diesels, a turbine-type supercharger of the type shown in Figure 9.18 is frequently used. This is a self-contained unit composed of a gas turbine and a centrifugal blower mounted on a common shaft. The exhaust gases from the cylinders flow to the gas turbine where they drive the blower, which provides air for combustion and scavenging of exhaust gases from the cylinders. The use of a supercharger can increase the output of an engine as much as 50%. The injection of fuel into the cylinders of the compression ignition engine is one of the most exacting and difficult requirements for this type of engine. As an example, a 100-hp Diesel operating at 750 rpm requires the droplets of oil to be injected with the injection and burning completed in 0.01 s. The fuel system must have the following characteristics:

1. It must accurately measure the fuel.

2. Because the air is compressed to at least 500 psi, the fuel must be delivered under high pressure. The fuel system in some engines compresses the fuel to pressures up to 30,000 psi.

3. The timing of the fuel injection into the cylinder is critical.

4. The rate of fuel injection must be accurately controlled.

5. The fuel must be atomized to assure uniform combustion.

6. The fuel must be dispersed properly in the chamber.

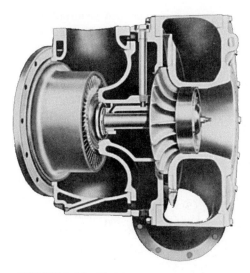

FIGURE 9.18 Turbine-type supercharger.
(Courtesy of Colt Industries, Fairbanks Morse Power Systems Division.)

In general, three types of fuel systems are used for Diesel engines:

1. *Unit system.* In this system, a single injection pump is used per cylinder. This pump provides the fuel pressurization, metering, and timing function. Each unit is usually mechanically operated off a cam shaft.

2. *Common rail system.* In this system, the fuel is kept under pressure at all times in a plenum chamber. Timed valves permit flow during the injection portion of the cycle, and the mechanically operated injection valve provides the timing and metering functions.

3. *Distributor system.* The metering of the fuel for all cylinders is carried out at low pressure by a single injection pump. Fuel is distributed through a distribution valve, which directs the metered fuel to the individual injection valves. The injection valves have mechanically operated plungers to raise the oil pressure to the required injection pressure. In this system, the injection valve pressurizes the fuel and also times the fuel injection. However, it does not meter the fuel.

Because of the exacting requirements for fuel injection in a Diesel engine, these engines run in a narrow range of speeds. Thus, Diesel engines used for transportation require more transmission gearing than comparable Otto cycle engines.

9.4 AIR-STANDARD ANALYSIS OF THE DIESEL CYCLE

All the assumptions made for the air-standard analysis of the Otto cycle regarding the working fluid and its properties apply to the present analysis of the idealized Diesel cycle.

The idealized air-standard Diesel cycle consists of four processes. The first is an isentropic compression of the air after it has been inducted into the cylinder. At the end of the compression process, fuel is injected and combustion is assumed to occur at constant pressure. Subsequent to the heat release by combustion, the gas is expanded isentropically to produce work, and finally, heat is rejected at constant volume. The gas is assumed to be recycled rather than rejected. Figure 9.19 shows the ideal Diesel cycle on both p–v and T–s coordinates.

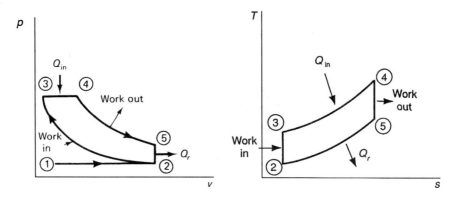

FIGURE 9.19 Diesel cycle..

Heat is received (reversibly) during the nonflow, constant-pressure process, ③ to ④. The energy equation for a constant-pressure (flow or nonflow) yields

$$q_{in} = c_p(T_4 - T_3) \quad \text{Btu/lb}_m; \text{kJ/kg} \tag{9.12}$$

The energy rejected during the constant-volume process is

$$q_r = c_v(T_5 - T_2) \quad \text{Btu/lb}_m; \text{kJ/kg} \tag{9.13}$$

The net work available from the cycle is

$$W = q_{in} - q_r = c_p(T_4 - T_3) - c_v(T_5 - T_2) \quad \text{Btu/lb}_m; \text{kJ/kg} \tag{9.14}$$

The efficiency thus becomes

$$\eta_{Diesel} = \frac{c_p(T_4 - T_3) - c_v(T_5 - T_2)}{c_p(T_4 - T_3)} = 1 - \frac{1}{k}\left(\frac{T_5 - T_2}{T_4 - T_3}\right) \tag{9.15}$$

At this point, it becomes conventional to introduce two terms and to define them as follows:

$$\text{compression ratio } r_c = \frac{v_2}{v_3} \tag{9.16}$$

and

$$\text{expansion ratio } r_e = \frac{v_5}{v_4} \tag{9.17}$$

Based on the nonflow processes discussed in Chapter 6, we can write

$$\frac{T_3}{T_2} = (r_c)^{k-1} \quad \text{and} \quad \frac{T_5}{T_4} = \frac{1}{(r_e)^{k-1}} \tag{9.18}$$

Because heat is received at constant pressure,

$$\frac{T_4}{T_3} = \frac{v_4}{v_3} \tag{9.19}$$

The ratio of

$$\frac{r_c}{r_e} = \frac{v_2/v_3}{v_5/v_4} = \frac{v_4 v_2}{v_3 v_5} \quad (\text{and because } v_2 = v_5) = \frac{v_4}{v_3} \tag{9.20}$$

This ratio v_4/v_3 is called the *cutoff ratio*. Therefore,

$$\frac{T_4}{T_3} = \frac{r_c}{r_e} = r_{\text{c.o.}} \tag{9.21}$$

By substituting Equations (9.16) through (9.21) into Equation (9.15) and rearranging, we obtain

$$\eta_{\text{Diesel}} = 1 - \frac{1}{k}\left[\frac{(r_c/r_e)^k - 1}{(r_c/r_e - 1)(r_c^{k-1})}\right] = 1 - \left(\frac{1}{r_c}\right)^{k-1}\left[\frac{(r_{\text{c.o.}})^k - 1}{k(r_{\text{c.o.}} - 1)}\right] \tag{9.22}$$

Equation (9.22) shows that the efficiency of the Diesel cycle is a function only of the two ratios r_c and r_e. The term $[(r_c/r_e)^k - 1]/k[(r_c/r_e - 1)]$ requires examination to determine whether the Diesel cycle is more or less efficient than the Otto cycle. Because v_4/v_3 is always greater than unity, it follows that this term is greater than unity. Therefore, by comparing Equation (9.22) with Equation (9.9), it can be concluded that the efficiency of the Otto cycle is greater than the efficiency of the Diesel cycle *for the same compression ratio*. We can also arrive at this conclusion by referring to Figure 9.20, which is a plot of the efficiency of the Diesel cycle as a function of the compression and cutoff ratios. When the cutoff ratio is unity, both cycles have equal efficiencies. For any other cutoff ratio, v_4/v_3, the efficiency of the Diesel is less than that of the Otto cycle.

However, as noted earlier in this section, the Diesel cycle can be operated at much higher compression ratios than the Otto cycle and therefore can have a higher efficiency than the Otto cycle. It should be further noted that as v_4/v_3 is increased, there is also a decrease in the efficiency of the Diesel cycle. Values between 2 and $2\frac{1}{2}$ are generally used as the upper limits of v_4/v_3. The mean effective pressure of the Diesel cycle is obtained in a similar manner to that used to obtain the mean effective pressure of the Otto cycle. The work of the Diesel cycle is given by Equation (9.14). Dividing it by the volume swept out by the piston per unit mass yields

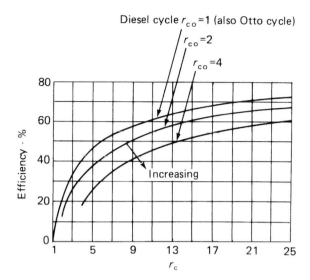

FIGURE 9.20 Efficiency of the Diesel cycle.

$$\text{mep}_{\text{Diesel}} = \left[\frac{c_p(T_4 - T_3) - c_v(T_5 - T_2)}{v_2 - v_3}\right]J \qquad \textbf{(9.23)}$$

By suitably rearranging Equation (9.23), the mean effective pressure of the Diesel cycle can be expressed in terms of pressures and volumes. Also, in SI units, J is not needed.

ILLUSTRATIVE PROBLEM 9.12

An air-standard Diesel engine has a compression ratio of 16 and a cutoff ratio of 2. Find the efficiency and temperature of the exhaust using $k = 1.4$ with the cycle starting at 14 psia and 100°F.

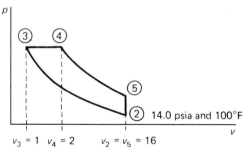

FIGURE 9.21 Illustrative Problem 9.14.

SOLUTION

From Figure 9.21, values of the volumes can be assigned as shown. The cutoff ratio v_4/v_3 is given as 2. Therefore, $v_4 = 2v_3 = 2$. From Equation (9.22),

$$\eta_{\text{Diesel}} = 1 - \left(\frac{1}{r_c}\right)^{k-1} \left[\frac{(v_4/v_3)^k - 1}{k(v_4/v_3 - 1)}\right]$$

For $k = 1.4$,

$$\eta_{\text{Diesel}} = 1 - \left(\frac{1}{16}\right)^{0.4} \left[\frac{(2)^{1.4} - 1}{(1.4)(2 - 1)}\right] = 0.614 = 61.4\%$$

From Equation (9.18),

$$\frac{T_3}{T_2} = r_c^{k-1} \quad \text{and} \quad \frac{T_5}{T_4} = \frac{1}{r_{\hat{e}}^{k-1}}$$

But

$$\frac{T_4}{T_3} = \frac{v_4}{v_3} = \frac{r_c}{r_e}$$

Substituting for T_3 and T_4 in terms of T_2, v_4, and v_3 yields

$$T_5 = T_2 \left(\frac{v_4}{v_3}\right)^k$$

Therefore,

$$T_5 = (100 + 460)(2)^{1.4} = 1478°R = 1018°F$$

ILLUSTRATIVE PROBLEM 9.13

In Illustrative Problem 9.12, determine the net work per pound of gas and the mean effective pressure. Assume that $c_v = 0.172 \text{ Btu/lb}_m \cdot °R$ and $c_p = 0.24 \text{ Btu/lb}_m \cdot °R$.

SOLUTION

From Illustrative Problem 9.12,

$$q_r = c_v(T_5 - T_2) = (0.172)(1018 - 100) = 157.9 \text{ Btu/lb}_m$$

Note that

$$\eta_{\text{Diesel}} = 1 - \frac{q_r}{q_{\text{in}}}$$

Therefore,

$$\frac{q_r}{1 - \eta} = q_{in}$$

$$q_{in} = \frac{157.9}{1 - 0.614} = 409.1 \text{ Btu/lb}_m$$

$$\text{net work out} = J(q_{in} - q_r)$$
$$= (778)(409.1 - 157.9)$$
$$= 195,433.6 \text{ ft lb}_f/\text{lb}_m$$

The mean effective pressure is net work divided by $(v_2 - v_3)$:

$$\text{mep} = \frac{195,433.6 \text{ ft lb}_f/\text{lb}_m}{(16 - 1)\dfrac{\text{ft}^3}{\text{lb}_m} \times \dfrac{144 \text{ in}^2}{\text{ft}^2}} = 90.5 \text{ psia}$$

9.5 BRAYTON CYCLE

The discussion of gas cycles has so far been limited to intermittent cycles. However, the gas turbine has recently become quite important, both from an aircraft propulsion standpoint and for the production of power in stationary power plants. Because the working processes occur continuously in a gas turbine engine (no "idle" strokes), it is capable of generating a higher power-output-to-weight ratio and so is ideal for use in aircraft. The emerging prominence of this device is due primarily to the mechanical development of its components and the metallurgical development of alloys that can be used at elevated temperatures. The prototype cycle for this device is the Brayton (or Joule) cycle. Its elements are shown in the p–v and T–s diagrams in Figure 9.22.

The gas is isentropically compressed along path ① to ②, and heat is added at constant pressure along path ② to ③. The gas then undergoes an isentropic expansion to its initial pressure, and heat is rejected at constant pressure. All the processes are reversible in the ideal cycle.

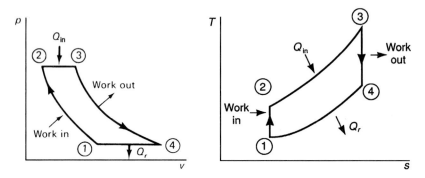

FIGURE 9.22 Ideal closed Brayton cycle.

Before we analyze the Brayton cycle, let us look at the hardware used in conjunction with this cycle. There are three basic elements: the compressor, the combustor, and the turbine in the simple cycle. Of these, the compressor and turbine play the key role in the power-generation cycle. The gas turbine cycle displayed in Figure 9.22 shows a working fluid that is returned to pass through the cycle after each sequence of events is completed. Such a cycle is known as a *closed cycle*. One application of a closed-cycle gas turbine plant is in conjunction with a nuclear reactor. The elements of this cycle are shown in Figure 9.23 with the nuclear reactor being the heat source.

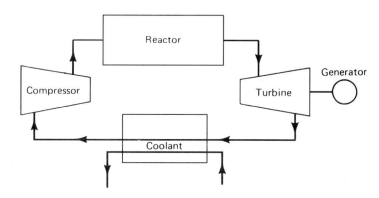

FIGURE 9.23   Elements of a closed-cycle gas turbine plant.

The *open-cycle* gas turbine system is a steady-flow system in which the working fluid is continuously discharged and fresh fuel and air are continuously added to maintain a steady flow of gases. The elements of this cycle are indicated in Figure 9.24. The cold air is compressed and flows to the combustion chamber where fuel is continuously supplied and burned to heat the fluid. The combustion gases are then expanded in the turbine to provide work. The successful operation of the gas turbine requires that the net output of the turbine be greater than the work required by the compressor. It also requires that the mechanical efficiencies of these components be high; otherwise, the losses in these pieces of equipment may decrease useful output of the system to a point at which it might become uneconomical. Figure 9.25 shows a cutaway view of an aircraft gas turbine engine. This unit operates on the open Brayton cycle and powers aircraft such as the vertical take-off V22 Osprey, which is shown in Figure 9.26. Another application of the open gas turbine

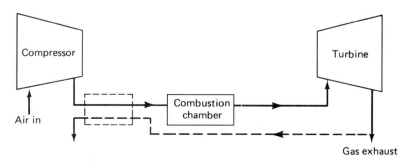

FIGURE 9.24   Open-cycle gas turbine plant.

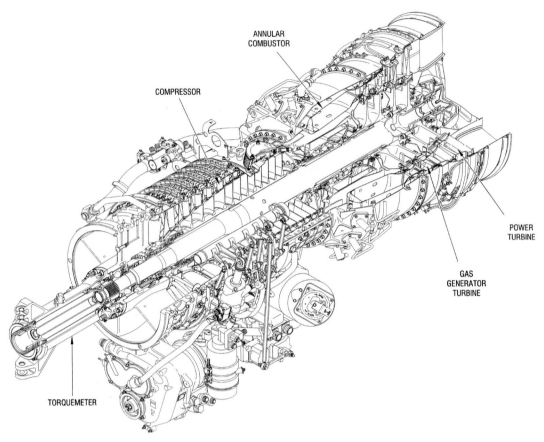

FIGURE 9.25 Aircraft gas turbine engine.
(Courtesy of Allison Engine Company, Inc., d.b.a. Rolls-Royce Allison.)

cycle is shown in Figure 9.27. This figure depicts the steam and gas turbine combined cycle plant which General Electric calls the STAG cycle (STAG is an acronym for *st*eam *a*nd *g*as turbine). The high-temperature exhaust gases from the open-cycle gas turbine generate steam in a forced circulation heat recovery steam generator.

The use of regeneration in vapor cycles has been shown to increase the efficiency of these cycles. Similarly, regeneration can be used to increase the efficiency of the Brayton cycle. Such an arrangement is indicated in Figure 9.24 by the dashed lines. The exhaust gas leaving the turbine is used to preheat the air after it leaves the compressor and before entering the combustion chamber. If the same quantity of fuel is used, a higher gas temperature is obtained at the inlet to the turbine. If a constant temperature to the turbine is desired, regeneration decreases the amount of fuel required to achieve this temperature.

The compression of the air in a gas turbine cycle is usually accomplished in two types of compressors; the centrifugal compressor, and the axial flow compressor. In the axial flow compressor, the blades have airfoil shapes and are arranged in concentric rings (stages) along the

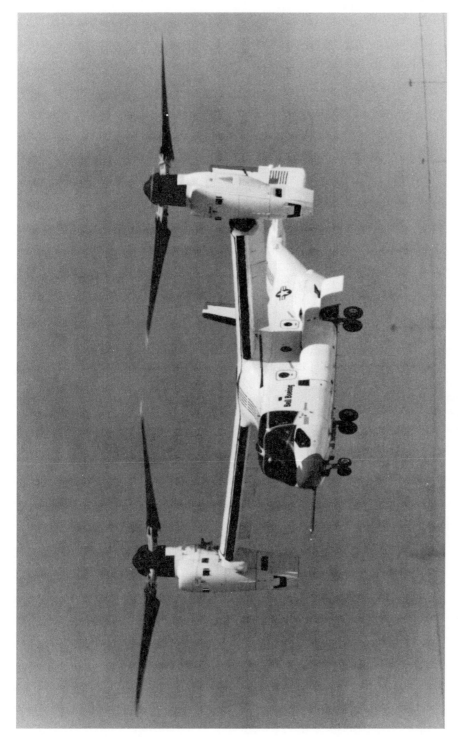

FIGURE 9.26 V22 Osprey.
(Courtesy of Allison Engine Company, Inc., d.b.a. Rolls-Royce Allison.)

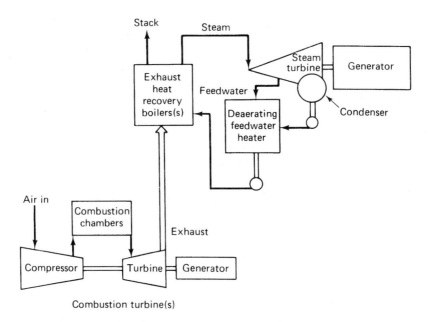

Stack   Steam

Exhaust heat recovery boilers(s)

Steam turbine

Generator

Feedwater

Condenser

Deaerating feedwater heater

Air in

Combustion chambers

Exhaust

Compressor   Turbine   Generator

Combustion turbine(s)

FIGURE 9.27 General Electric STAG combined cycle.

axis of the compressor shaft. In the usual design, each moving blade is followed by a stationary or fixed blade. Figure 9.28 shows the inlet to an axial flow unit and the compressor blading behind the inlet shroud. The moving blades impart a helical velocity to the air, increasing its velocity in each moving stage, while the stationary blades are shaped to act as diffusers, slowing the air but at the same time increasing its pressure. Because the specific volume of the air decreases as the pressure increases, blade heights decrease as the air progresses along the axis of the compressor. Also, blade angles are varied due to the differing air speeds and flow patterns

FIGURE 9.28 Axial flow gas turbine compressor.
(Courtesy of Pratt and Whitney Aircraft Group, United Technologies Corp.)

as the air proceeds along the compressor axis. Figure 9.29 shows the air path through the fixed and moving blades and also the rise in pressure along the axis of the compressor. It will be noted that each stage imparts approximately the same pressure rate as the preceding stage. Thus, if each stage produces a pressure ratio of 1.1:1 (that is, a 10 percent increase), 8 stages will produce a total pressure rise of $(1.1)^8 = 2.14$, and 16 stages will produce an overall pressure rise of $(1.1)^{16} = 4.6$. The action of the centrifugal compressor is shown schematically in Figure 9.29b. Air is taken in at the center (or eye) of the impeller. The rotation of the impeller causes the air to flow in a radial direction at high speed into the stationary diffuser. The air is slowed in the diffuser with an attendant increase in pressure. By varying the shape of the rotor blades and diffuser passages, it is possible to achieve almost any pressure characteristic as a function of rotor speed. Figure 9.30 shows a medium-sized, high-speed compressor having a first stage of axial compression and a second stage of centrifugal compression.

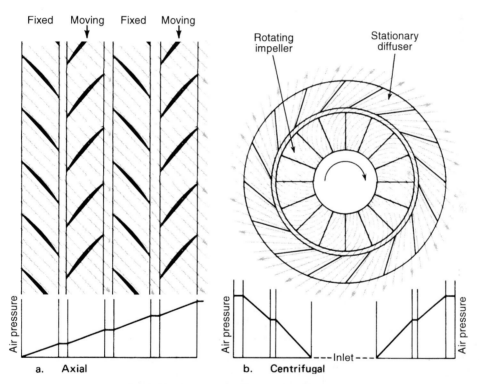

FIGURE 9.29 Basic flow in compressors.
(Reprinted with permission from Power, The Engineer's Reference Library, *copyright by McGraw-Hill, Inc., New York.)*

The gas turbine is similar to the steam turbine and utilizes both the impulse and reaction effects. Gas turbines have lower overall pressure drops than steam turbines, and consequently, they have fewer stages and less change in blade height from inlet to exhaust.

FIGURE 9.30 Combination axial flow–centrifugal compressor.
*(Reprinted with permission from Power, The Engineer's Reference Library,
copyright by McGraw-Hill, Inc., New York.)*

Figure 9.31 shows a single-cycle, single-shaft industrial gas turbine. Referring to the numbers in this figure, the following features are noted:

  2 Starting device: motor, steam turbine, or expansion (air) turbine.

  6 Radial inlet casing: provides uniform circumferential inlet flow to compressor.

  8 Compressor: axial flow.

  15 Fuel nozzles.

  16 Combustion chambers.

  19 Three-stage impulse turbine.

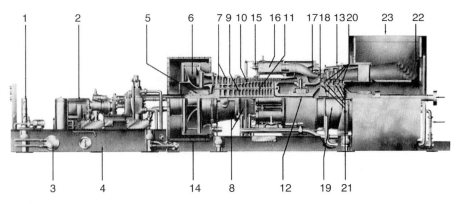

FIGURE 9.31 Single-shaft gas turbine.
(Courtesy of General Electric Co.)

Figure 9.32 shows a complete gas turbine power system. It consists of a two-stage, reaction-type turbine connected to a gas generator through a diffuser duct. The compressor of this unit is split into two sections, which are mechanically independent of each other. The first eight compressor stages (low-pressure compressor) are driven by the second and third turbine stages. Seven additional compressor stages (high-pressure compressor) are driven by the first turbine stage through a shaft that is independent of and concentric with the low-pressure drive shaft. The combustion section houses eight separate burner cans arranged circumferentially and interconnected by crossover tubes. The coupling between the gas generator and free turbine permits rapid response to varying power requirements. The air turbine requires a pressure of 45 psig for normal starting cycles.

The modern jet engine uses the power remaining from the turbine after it supplies the compressor to accelerate the air moving through it. This accelerated air exits the engine and, by momentum principles, provides the thrust (force) to propel the aircraft.

FIGURE 9.32 Gas turbine power unit.
(Courtesy of Pratt and Whitney Aircraft Group, United Technologies Corp.)

9.6 AIR-STANDARD BRAYTON CYCLE ANALYSIS

Let us now return to the ideal Brayton cycle shown in Figure 9.22 (repeated). The heat input in the constant-pressure, nonflow process per pound is

$$q_{in} = c_p(T_3 - T_2) \tag{9.24}$$

and for the constant-pressure, nonflow heat rejection part of the cycle ④ → ①,

$$q_r = c_p(T_4 - T_1)$$ (9.25)

The net work is

$$q_{in} - q_r = c_p(T_3 - T_2) - c_p(T_4 - T_1) \qquad \text{in thermal units}$$ (9.26)

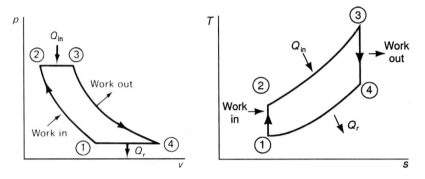

FIGURE 9.22 (*repeated*) Ideal closed Brayton cycle.

Thus,

$$\eta_{Brayton} = \frac{c_p(T_3 - T_2) - c_p(T_4 - T_1)}{c_p(T_3 - T_2)} = 1 - \frac{T_4 - T_1}{T_3 - T_2}$$ (9.27)

However,

$$\frac{T_4}{T_1} = \frac{T_3}{T_2} \quad \text{and} \quad \frac{T_4 - T_1}{T_1} = \frac{T_3 - T_2}{T_2}$$ (9.28)

Therefore,

$$\eta_{Brayton} = 1 - \frac{T_1}{T_2} = 1 - \left(\frac{v_2}{v_1}\right)^{k-1}$$ (9.29)

Denoting v_1/v_2 as r_c, we have

$$\boxed{\eta_{Brayton} = \left[1 - \left(\frac{1}{r_c}\right)^{k-1}\right] \times 100}$$ (9.30)

In terms of pressures (in which p_2/p_1 is called the pressure ratio),

$$\boxed{\eta_{Brayton} = \left[1 - \left(\frac{1}{p_2/p_1}\right)^{(k-1)/k}\right] \times 100}$$ (9.31)

Comparison of Equation (9.30) with Equation (9.9) yields the conclusion that the efficiency of the Brayton cycle is equal to the efficiency of the Otto cycle for the same compression ratio. Therefore, we can also conclude that the efficiency of the ideal Brayton cycle is solely a function of the compression ratio and not of the peak temperature of the cycle.

ILLUSTRATIVE PROBLEM 9.14

A Brayton cycle is operated with a compression ratio of 7. Assuming that $k = 1.4$ and $c_p = 0.24$ Btu/lb$_m$·°R, determine the heat in, the work out, the heat rejected, and the efficiency of the cycle if the peak temperature of the cycle is 1500°F and the initial conditions are 14.7 psia and 70°F. R for air is 53.3 ft lb$_f$/lb$_m$·°R.

SOLUTION

We can calculate the efficiency as

$$\eta_{\text{Brayton}} = 1 - \left(\frac{1}{r_c}\right)^{k-1} = 1 - \left(\frac{1}{7}\right)^{1.4-1} = 0.541 \quad \text{or} \quad 54.1\%$$

If we base our calculation on 1 lb$_m$ of gas and use subscripts that correspond to points ①, ②, ③, and ④ of Figure 9.22, we have

$$v_1 = \frac{RT_1}{p_1} = \frac{(53.3)(460 + 70)}{14.7 \times 144} = 13.35 \text{ ft}^3/\text{lb}_m$$

Because $r_c = 7$ and $v_2 = v_1/7$, then

$$v_2 = \frac{13.35}{7} = 1.91 \text{ ft}^3/\text{lb}_m$$

After the isentropic compression,

$$T_2(v_2)^{k-1} = T_1(v_1)^{k-1}$$

Thus,

$$T_2 = T_1\left(\frac{v_1}{v_2}\right)^{k-1} = (530)(7)^{(1.4-1)} = 1154°R = 694°F$$

T_3 is given at 1500°F. Therefore, the heat in is

$$q_{\text{in}} = c_p(T_3 - T_2) = (0.24)(1500 - 694) = 193.4 \text{ Btu/lb}_m$$

Because efficiency can be stated to be work out divided by heat in,

$$0.537 = \frac{w/J}{q_{\text{in}}} \quad \frac{w}{J} = (0.537)(193.4) = 103.9 \text{ Btu/lb}_m$$

The heat rejected is

$$q_{in} - \frac{W}{J} = 193.4 - 103.9 = 89.5 \text{ Btu/lb}_m$$

This problem can also be solved using the *Gas Tables*. In addition, a check is provided by solving for all the temperatures of the cycle and then solving for the efficiency.

9.7 THE DUAL COMBUSTION CYCLE (THE DUAL CYCLE)

A useful method of studying and comparing the performance of an engine experimentally is the indicator diagram. This diagram is a pressure–volume record of the events that are occurring within the cylinder of an operating engine. Indicator diagrams taken from both Otto and Diesel engines show a rounded top, as can be seen from Figure 9.10. For the ideal cycle to more closely represent the actual cycle events, it has been proposed that the ideal cycle have some combustion at constant volume and some at constant pressure. Such a cycle is called the dual combustion cycle, or simply the dual cycle. The *p–v* and *T–s* diagrams for the dual cycle are shown in Figure 9.33. Only the combustion or heat-addition portion of the cycle differs from either the Otto or Diesel cycles.

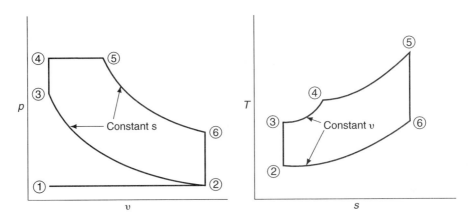

FIGURE 9.33 The dual combustion cycle.

Using the points indicated in Figure 9.33, we can write,

$$q_{in} = c_v(T_4 - T_3) + c_p(T_5 - T_4) \tag{9.32}$$

and

$$q_r = c_v(T_6 - T_2) \tag{9.33}$$

The efficiency is given in terms of q_{in} and q_r as

$$\eta_{\text{dual cycle}} = (q_{in} - q_r)/q_{in}$$

$$\eta_{\text{dual cycle}} = [c_v(T_4 - T_3) + c_p(T_5 - T_4) - c_v(T_6 - T_2)]/$$

$$[c_v(T_4 - T_3) + c_p(T_5 - T_4)] \tag{9.34}$$

Simplifying,

$$\eta_{\text{dual cycle}} = 1 - (T_6 - T_2)/[(T_4 - T_3) + k(T_5 - T_4)] \qquad \textbf{(9.35)}$$

If Equation (9.35) is examined, it is found that for a given compression ratio, the efficiency of the dual cycle is less than the efficiency of the Otto cycle but greater than the efficiency of the Diesel cycle.

9.8 STIRLING CYCLE AND ERICSSON CYCLE (REGENERATION)

The *Stirling cycle* has for some time been only of historical interest, because it has not been used as the prototype for practical power-producing systems. Recently, in this country and abroad, interest has been revived, and engines based on the Stirling cycle have been built and successfully operated. This development has been partially due to the possibility of using these engines for power production in space.

The elements of this cycle are shown in Figure 9.34. It consists of two isothermal and two constant-volume processes. During the isothermal compression, heat is rejected by cooling the air in the compressor. Heat is added during the constant-volume portion of the cycle (②→③). The gas is then isothermally expanded, and more heat is added. The gas finally rejects heat during the constant-volume position of the cycle (④→①). If all these paths were carried out reversibly, the efficiency of the Stirling cycle would equal that of the Carnot cycle between the same maximum and minimum temperature limits. This can be seen from the *T–s* diagram in Figure 9.34. Both the Stirling and the Ericsson cycles rely on a regenerator that reversibly takes the heat rejected in the constant volume or constant pressure process (④→①) and returns it to the engine in the other constant volume or constant pressure process (②→③). Paths ②→③ and ④→① are parallel, and the cycle can be "squared off" to yield the Carnot cycle shown as ②'—③—④—④'. The inability of actual devices to achieve this limiting efficiency arises principally from the heat-transfer processes and pressure losses in the engine and its associated piping.

Both the Stirling cycle and Ericsson cycle were originally developed as an effort to find a practical external heat-transfer engine using air as the working fluid. The *Ericsson engine* consists of two reversible constant-pressure processes and two reversible constant-temperature energy-addition and energy-rejection processes. The *T–s* diagram is similar to that of the Stirling cycle *T–s* diagram of Figure 9.34. The air-standard efficiency of the ideal Ericsson engine therefore is equal to the Carnot efficiency.

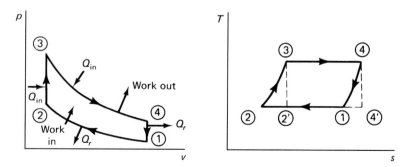

FIGURE 9.34 Stirling cycle.

The use of regeneration in the steam cycle or in the Stirling and Ericsson cycles is seen to cause the efficiencies of these cycles to approach the efficiency of the Carnot cycle. In the case of the Stirling and Ericsson cycles, the reduction of theory to actual working engines presents practical difficulties that cause the actual engine efficiencies to be below that of the Carnot cycle.

9.9   REVIEW

In Chapter 8, we considered vapor cycles in which the working fluid underwent a change in phase. In this chapter, we have concerned ourselves with cycles in which the working fluid is a gas and remains a gas throughout the cycle. Some of the cycles considered in this chapter have internal combustion; others have external combustion. Some have spark ignition; others have compression ignition. In order to simplify the calculations and to permit us to arrive at general conclusions, we utilized the air-standard cycle analysis. The air-standard cycle analysis assumes that air is the working fluid and that it behaves as an ideal gas. Each process of the cycle is assumed to be internally reversible, and the specific heats of the air are assumed to be constant throughout the cycle. In effect, the same air remains in the engine from cycle to cycle in the air-standard analysis.

Based on the air-standard analysis, the efficiency of the Otto cycle was determined to be a function only of the compression ratio. The efficiency of the Diesel cycle was found to be less than the efficiency of the Otto cycle for the same compression ratio. However, the compression ratio of Diesel engines is much greater than the compression ratio of the Otto cycle, because the fuel is injected at the end of the compression stroke rather than compressing the fuel–air mixture. The efficiency of the Brayton cycle was found to equal the efficiency of the Otto cycle for the same compression ratio. The dual cycle was proposed in an effort to come closer to actual indicator diagrams. In the dual cycle, heat is added at both constant volume and constant pressure. The efficiency of the dual cycle lies between the efficiency of the Otto and Diesel cycles for the same compression ratio.

The Stirling and Ericsson cycles both have thermal efficiencies that equal the thermal efficiency of the Carnot cycle. This performance is achieved through the use of regeneration, during which heat is transferred to an energy storage device during one portion of the cycle and is returned to the working fluid during another portion of the cycle. The hardware required to convert these theoretical cycles to working engines is complex and difficult to build. However, Stirling engine development has recently accelerated due to its potential space applications.

KEY TERMS

Terms used for the first time in this chapter are:

Brayton cycle an ideal four-cycle gas engine consisting of two isentropic and two constant-pressure processes.

compression ratio the ratio of the volume at the bottom of the stroke divided by the volume at the top of the stroke.

Diesel cycle an ideal four-cycle gas engine consisting of two isentropic, a constant-pressure, and a constant-volume process.

dual cycle an ideal gas engine cycle in which heat is added at both constant volume and constant pressure in order to come closer to actual indicator *p–v* diagrams.

Ericsson engine a regenerative internal-combustion engine consisting of two constant-temperature and two constant-pressure processes.

expansion ratio the ratio of the volume prior to expansion divided by the volume after compression in a Diesel cycle.

internal-combustion engine an engine in which the energy release occurs internally.

mean effective pressure (mep) defined as the work out of a cycle divided by the volume swept out by the piston.

Otto cycle an ideal four-cycle gas engine consisting of two isentropic and two constant-volume processes.

Stirling cycle a regenerative internal-combustion engine cycle consisting of two isothermal and two constant-volume processes.

EQUATIONS DEVELOPED IN THIS CHAPTER

compression ratio $\qquad r_c = \dfrac{v_2}{v_3}$ $\hfill$ **(9.8)**

efficiency of Otto cycle $\qquad \eta_{\text{Otto}} = \left[1 - \left(\dfrac{1}{r_c} \right)^{k-1} \right] \times 100$ $\hfill$ **(9.9)**

mean effective pressure $\qquad \text{mep} = \dfrac{\text{work out}}{\text{volume swept out by piston}}$ $\hfill$ **(definition)**

mean effective pressure $\qquad \text{mep} = \dfrac{1}{k-1} \left[\dfrac{v_3(p_4 - p_3) - v_2(p_5 - p_2)}{v_2 - v_3} \right]$ $\hfill$ **(9.10)**

mean effective pressure $\qquad \text{mep} = \dfrac{1}{k-1} \left[\dfrac{(p_4 - r_c^k p_2)[1 - (1/r_c)^{k-1}]}{r_c - 1} \right]$ $\hfill$ **(9.11)**

horsepower $\qquad \text{hp} = \dfrac{p_m LAN}{33{,}000}$ $\hfill$ **Illustrative Problem 9.5**

horsepower $\qquad \text{hp} = \dfrac{p_m LAN}{44{,}760}$ $\hfill$ **Illustrative Problem 9.6**

compression ratio $\qquad r_c = \dfrac{1 + c}{c}; \ \text{mep} = \dfrac{w}{v_2 \left(1 - \dfrac{1}{r_c} \right)} = \dfrac{w}{(r_c - 1)v_3}$ $\hfill$ **Illustrative Problem 9.8**

Diesel cycle $\qquad r_c = \dfrac{v_2}{v_3}$ $\hfill$ **(9.16)**

expansion ratio $\qquad r_e = \dfrac{v_5}{v_4}$ $\hfill$ **(9.17)**

cutoff ratio $\qquad \dfrac{r_c}{r_e} = \dfrac{v_4}{v_3}$ $\hfill$ **(9.20)**

cutoff ratio $\qquad r_{\text{c.o.}} = \dfrac{r_c}{r_e}$ $\hfill$ **(9.21)**

Diesel efficiency
$$\eta_{\text{Diesel}} = 1 - \frac{1}{k}\left[\frac{(r_c/r_e)^k - 1}{(r_c/r_e - 1)\,(r_c^{k-1})}\right]$$

$$= 1 - \left(\frac{1}{r_c}\right)^{k-1}\left[\frac{(r_{\text{c.o.}})^k - 1}{k(r_{\text{c.o.}} - 1)}\right] \quad \textbf{(9.22)}$$

Brayton efficiency
$$\eta_{\text{Brayton}} = \left[1 - \left(\frac{1}{r_c}\right)^{k-1}\right] \times 100 \quad \textbf{(9.30)}$$

Brayton efficiency
$$\eta_{\text{Brayton}} = \left[1 - \left(\frac{1}{p_2/p_1}\right)^{(k-1)/k}\right] \times 100 \quad \textbf{(9.31)}$$

QUESTIONS

9.1 In what two ways does a gas cycle differ from a vapor cycle?

9.2 List the four events in the four-stroke internal-combustion cycle.

9.3 How does the two-stroke cycle differ from the four-stroke cycle?

9.4 What is the ratio of the number of power strokes in the two-stroke cycle divided by the number of power strokes in the four-stroke cycle at a given value of engine revolutions per minute?

9.5 Describe the operation of the Wankel engine.

9.6 Describe the problems that the practical Wankel engine encountered and that have been overcome.

9.7 Describe the four events in the Otto cycle.

9.8 On what single factor does the efficiency of the Otto cycle depend?

9.9 What limits the practical realization of higher efficiencies in the Otto cycle?

9.10 How does the modern internal-combustion engine achieve higher power outputs without the use of higher compression ratios?

9.11 Of what utility is the concept of mean effective pressure?

9.12 How does the clearance volume affect the efficiency of the Otto cycle?

9.13 In what manner does the Diesel cycle differ from the Otto cycle?

9.14 How is the fuel introduced into the Diesel engine?

9.15 How does the efficiency of the Diesel engine compare to the efficiency of the Otto cycle for the same compression ratio?

9.16 How is it possible for a Diesel engine to operate at efficiencies greater than the efficiency of an Otto cycle?

9.17 How does the Diesel engine avoid the problem of preignition or detonation?

9.18 Describe the Brayton cycle.

9.19 How does the Brayton cycle efficiency compare to the efficiency of the Otto cycle for the same compression ratio?

9.20 How does the operation of the closed-cycle Brayton plant differ from the open-cycle Brayton plant?

9.21 Show how the dual cycle is a compromise between the Otto and Diesel cycles.

9.22 The Stirling and Ericsson cycles use a common concept in their operation. Describe this concept and its principle.

PROBLEMS

Wherever appropriate, sketch the cycle on p–v, h–s, or T–s coordinates. Use $c_p = 0.24$ Btu/lb$_m$·°R, $c_v = 0.171$ Btu/lb$_m$·°R, $c_p = 1.0061$ kJ/kg·°K, and $c_v = 0.7186$ kJ/kg·°K for air unless otherwise noted. Assume that these specific heats are constant and that the molecular weight of air is 29.

Problems Involving the Otto Cycle

9.1 An air-standard Otto cycle is operated with a compression ratio of 6. If k has values of 1.1, 1.2, 1.3, and 1.4, determine the efficiency of the cycle. Plot the efficiency as a function of k.

9.2 An air-standard Otto cycle operates with 520°R and 14.7 psia at the beginning of the compression portion of the cycle. If 300 Btu/lb$_m$ is added as heat and the maximum temperature of the cycle is 2900°R, determine the efficiency of the cycle.

9.3 The pressure at the beginning of the compression stroke in an air-standard Otto cycle is 15 psia. After compression the pressure is 140 psia. Determine the thermal efficiency of the cycle.

9.4 An air-standard Otto cycle has a clearance volume equal to 10% of its displacement volume. Calculate the compression ratio and the efficiency of the engine.

9.5 At the beginning of the compression stroke in an air-standard Otto cycle, the temperature is 68°F, the pressure is 15 psia, and the compression ratio is 7. If the heat rejected by the engine is 250 Btu/lb$_m$, determine the heat input to the cycle and the thermal efficiency of the cycle.

9.6 An air-standard Otto cycle has a compression ratio of 9. At the beginning of the compression process, the temperature is 20°C and the pressure is 100 kPa. The heat added is 500 kJ/kg. Determine the work output, the heat rejected, and the cycle efficiency.

9.7 At the beginning of the compression stroke of an air-standard Otto cycle, the volume is 0.1 ft³, the temperature is 90°F, and the pressure is 14 psia. At the end of compression, the pressure is 200 psia. Calculate the efficiency of the engine.

9.8 Determine the heat rejected, the net work, and the compression ratio in Problem 9.7 if 250 Btu/lb$_m$ is added as heat to the cycle.

9.9 An air-standard Otto cycle starts its compression stroke at 75°F. Assuming that the highest temperature that can be reached before the gasoline–air mixture will ignite without an electrical spark is 750°F, determine the maximum compression ratio that can be used in this engine.

9.10 An air-standard Otto cycle operates with 25% clearance. At the beginning of the compression stroke, the pressure is 14 psia and the temperature is 100°F. The maximum temperature in the cycle is 4040°F. Determine the efficiency of the cycle.

9.11 An air-standard Otto cycle with a compression ratio of 9 rejects 100 kJ/kg as heat. At the end of the compression process, the pressure is 2.2 MPa and the temperature is 720°K. Determine the heat input to the cycle, the work output, and the efficiency of the cycle.

9.12 Determine the mean effective pressure in the cycle of Problem 9.11.

9.13 An air-standard Otto cycle operates with the compression and expansion strokes being polytropic, with $n = 1.35$ for each stroke. Calculate the thermal efficiency of the cycle for a compression ratio of 9.

9.14 An air-standard Otto cycle operates with a compression ratio of 8. At the start of the compression stroke, the pressure is 15 psia and the temperature is 80°F. The maximum pressure in the cycle is 550 psia. Calculate the maximum temperature in the cycle and the thermal efficiency.

*9.15 An air-standard Otto cycle operates with a compression ratio of 9.5. At the start of compression, the conditions are 14.7 psia and 80°F. During the combustion portion of the cycle, 250 Btu/lb$_m$ is added as heat. Calculate p, v, and t at each point of the cycle and the thermal efficiency of the cycle.

*9.16 An air-standard Otto cycle operates with a minimum temperature of 300°K and a maximum temperature of 1700°K. The compression ratio of the cycle is 7. At the beginning of the compression process, the pressure is 105 kPa. Calculate p, v, and t at each point in the cycle, the mean effective pressure, and the thermal efficiency of the cycle.

*9.17 An air-standard Otto cycle has 2000 kJ/kg added as heat. The initial conditions are 20°C and 150 kPa. If the compression ratio is 7, determine the thermal efficiency of the cycle, the temperature and pressure at each point in the cycle, and its mean effective pressure.

*9.18 An air-standard Otto cycle with a compression ratio of 7 has air at 70°F and 15 psia at the start of the compression portion of the cycle. If 800 Btu/lb$_m$ is added as heat, determine the pressure and temperature in each portion of the cycle, the thermal efficiency of the cycle, and its mean effective pressure.

*9.19 An air-standard Otto cycle has 1400 kJ/kg added as heat. The initial conditions are 20°C and 100 kPa. If the compression ratio is 7, determine the thermal efficiency of the cycle, the temperature and pressure at each point in the cycle, and its mean effective pressure.

*9.20 An air-standard Otto cycle has 500 Btu/lb$_m$ added as heat. If the initial conditions are 70°F and 20 psia and the compression ratio is 7, determine the thermal efficiency of the cycle, the temperature and pressure at each point in the cycle, and its mean effective pressure.

*9.21 An air-standard Otto cycle has 300 Btu/lb$_m$ added as heat. If the initial conditions are 50°F and 15 psia and the compression ratio is 8, determine the thermal efficiency of the cycle, the temperature and pressure at each point in the cycle, and its mean effective pressure.

9.22 The mean effective pressure of an Otto cycle is 100 psia. If the displacement of an engine is 144 in^3, what horsepower can the engine deliver? This is a four-cycle engine operating at 3600 rpm.

9.23 The mean effective pressure of an Otto engine is 1 MPa. The displacement of the engine is 3.8 L. If the engine is operated as a four-cycle engine at 4000 rpm, what is its horsepower?

Problems Involving the Diesel Cycle

9.24 An air-standard Diesel cycle is operated with a compression ratio of 25. The inlet conditions to the cycle are 14.7 psia and 70°F. The gas is to be used after the expansion cycle to heat some process liquid in a heat exchanger. For this purpose, the temperature at the end of the expansion is to be 500°F. Determine the efficiency of the cycle.

9.25 Determine the work out of the engine in Problem 9.24. Is this an efficient way to operate this engine?

9.26 An air-standard Diesel cycle has a compression ratio of 22 and a cutoff ratio of 2.2. Determine the thermal efficiency of the cycle.

9.27 An air-standard Diesel cycle has a compression ratio of 18. At the beginning of compression, the pressure is 100 kPa and the temperature is 25°C. At the end of the combustion process, the temperature is 1300°C. Determine the thermal efficiency of the cycle and the mean effective pressure.

9.28 An air-standard Diesel cycle is operated with a compression ratio of 20. The ratio of $(v_4 - v_3)/(v_2 - v_3)$ is variable. Determine the efficiency of the cycle for values of the volume ratio of 0.01, 0.04, 0.08, and 0.1. Plot the results. Use $k = 1.4$.

9.29 An air-standard Diesel engine has 1000 kJ/kg added as heat. At the beginning of the compression, the temperature is 20°C and the pressure is 150 kPa. If the compression ratio is 20, determine the maximum pressure and temperature in the cycle.

9.30 An air-standard Diesel cycle is operated with a compression ratio of 20. The heat input is 300 Btu/lb$_m$. At the beginning of the compression stroke, the conditions are 15 psia and 70°F. Calculate the thermal efficiency of the engine and the mean effective pressure.

9.31 An air-standard Diesel cycle is operated with a compression ratio of 20. At the beginning of the compression stroke, the conditions are 15 psia and 80°F. The maximum temperature in the cycle is 3400°R. Calculate the thermal efficiency of this cycle.

*9.32 An air-standard Diesel engine has a temperature of 70°F at the beginning of compression and a temperature of 1400°F at the end of compression. If the cycle receives 500 Btu/lb$_m$ as heat, determine the compression ratio, the net work of the cycle in Btu/lb$_m$, and the thermal efficiency of the cycle.

*9.33 An air-standard Diesel cycle operates with a compression ratio of 18. The maximum and minimum temperatures in the cycle are 2000°K and 293°K, respectively. The maximum pressure in the cycle is 50 MPa. Calculate the p, v, and t at each point in the cycle, the thermal efficiency, and the mean effective pressure.

*9.34 Derive the following expression for the mean effective pressure in an air-standard Diesel:

$$\text{mep}_{\text{Diesel}} = p_1 \left[\frac{kr_c^k(v_3/v_2 - 1) - r_c[(v_3/v_2)^k - 1]}{(k-1)(r_c - 1)} \right]$$

*9.35 The net heat input to a Diesel cycle is 100 Btu/lb$_m$. If the initial conditions to the cycle are 14.7 psia and 60°F and the compression ratio is 20, determine the efficiency of the cycle, the net work out of the cycle, and its mean effective pressure. Assume that 1 lb$_m$ of air is used per cycle.

9.36 An air-standard Diesel engine has 500 Btu/lb$_m$ added as heat. At the beginning of the compression the temperature is 70°F and the pressure is 20 psia. If the compression ratio is 20, determine the maximum pressure and temperature in the cycle.

9.37 An air-standard Diesel engine has a compression ratio of 20. The cutoff ratio is 2.4. At the beginning of the compression stroke, the pressure is 100 kPa and the temperature is 293°K. Calculate the thermal efficiency and the heat added per kilogram of air.

Problems Involving the Brayton Cycle

9.38 A gas turbine cycle is operated with a pressure ratio of 5:1; that is, the pressure after compression is five times the pressure before compression. What is the efficiency of this cycle?

9.39 A Brayton cycle is operated to yield maximum work out of the cycle. If the compression ratio of the cycle is 4, the initial temperature to the cycle is 70°F, and the temperature at the start of the heat rejection portion of the cycle is 500°F, what is the peak temperature in the cycle?

9.40 An air-standard Brayton cycle has inlet air at 15 psia and 520°R. At the inlet to the turbine, the pressure is 150 psia and the temperature is 2500°R. Determine the heat input in Btu/lb$_m$, the outlet temperature from the turbine, and the thermal efficiency of the cycle.

9.41 An air-standard Brayton cycle receives air at 15 psia and 80°F. The pressure ratio is 6. If the upper temperature of the cycle is 2300°R, determine the thermal efficiency of the cycle.

9.42 An air-standard Brayton cycle is operated with a pressure ratio of 8 and a maximum temperature of 1400°K. Air enters the cycle at 30°C and 100 kPa. Determine the efficiency of the cycle.

9.43 An air-standard Brayton cycle operates with air entering at 14 psia and 80°F. At the turbine inlet, the temperature is 2200°F and the pressure is 140 psia. Determine the thermal efficiency of the cycle.

9.44 An air-standard Brayton cycle has an inlet temperature of 25°C and an inlet pressure of 110 kPa. The maximum pressure and temperature in the cycle are 1.1 MPa and 1450°K, respectively. Determine the heat input, the work output, and the thermal efficiency of the cycle.

*9.45 In an air-standard Brayton cycle, air enters the compressor at 15 psia and 60°F. The pressure leaving the compressor is 90 psia. The maximum temperature in the cycle is 2000°R. Determine p, v, and t at each point in the cycle and the thermal efficiency of the cycle.

*9.46 A Brayton cycle is operated to yield maximum work out of the cycle. If the compression ratio of the cycle is 4, the initial temperature of the cycle is 20°C, and the temperature at the start of the heat rejection portion of the cycle is 240°C, what is the peak temperature in the cycle?

9.47 A Brayton cycle has a compression ratio of 5. If the initial temperature is 70°F and the initial pressure is 15 psia and 1000 Btu/lb$_m$ is added as heat, what is the thermal efficiency of the cycle?

*9.48 A Brayton cycle has a compression ratio of 5. If the initial temperature is 20°C, the initial pressure is 100 kPa, and 1000 kJ/kg is added as heat, what is the temperature at each point in the cycle, and what is the thermal efficiency of the cycle?

9.49 A Brayton cycle is operated with a compression ratio of 8. The inlet conditions are 14.7 psia and 90°F. Determine the thermal efficiency of this cycle. If the compressor in this cycle is 90% efficient mechanically and the turbine is also 90% efficient mechanically, determine the overall efficiency of the cycle.

9.50 A turbine receives air at 1 MPa, 600°K, and discharges to a pressure of 150 kPa. The actual temperature at discharge is 400°K. What is the internal engine efficiency of the turbine?

9.51 A turbine receives air at 150 psia, 1000°R, and discharges to a pressure of 15 psia. The actual temperature at discharge is 600°R. What is the internal engine efficiency of the turbine?

Problems Involving Other Cycles

9.52 A Stirling cycle operates with a maximum temperature of 900°F and a minimum temperature of 40°F. Determine the efficiency of the cycle.

9.53 If the engine in Problem 9.52 is an Ericsson engine, determine its efficiency.

*9.54 A Stirling engine operates with temperatures of 900 and 70°F. The maximum pressure in the cycle is 700 psia, and the minimum pressure in the cycle is 14 psia. Determine the efficiency and work output of this engine per cycle.

*9.55 For an air-standard dual cycle, the conditions before compression are 15 psia and 100°F. After compression the specific volume is 1.04 ft^3/lb_m. If 150 Btu/lb_m is received as heat during the constant volume portion of the cycle and 200 Btu/lb_m is received as heat during the constant-pressure portion of the cycle, determine the efficiency of the cycle.

*9.56 A cycle operates on air for which the constant-volume and constant-pressure specific heats are 0.17 and 0.24 $Btu/lb_m \cdot °R$, respectively. All processes contained within the cycle are reversible and consist of (1) a constant-volume heat addition of 300 Btu/lb_m of air, which was initially at 15 psia and 70°F; (2) an adiabatic expansion to a temperature of 800°F; and (3) a constant-pressure heat rejection of a magnitude that will return the air to 15 psia and 70°F. Determine the net work of this cycle in Btu per pound of air. (*Note:* This cycle is not one of the cycles in this chapter. Draw a $p–v$ diagram first before trying to solve it.)

10

Refrigeration

Learning Goals

After reading and studying the material in this chapter, you should be able to:

1. Understand and use the fact that it is possible to reverse a power-producing cycle to remove heat from a place where it is undesirable and to reject it by the input of mechanical work.

2. Recall the elements of the Carnot cycle, and operate it as a reversed cycle.

3. Define the figure of merit for a refrigeration cycle, known as the coefficient of performance.

4. Conclude that the coefficient of performance of a reversed Carnot cycle is a function of the upper and lower temperatures of the cycle and is always greater than unity.

5. Use the defined ratings used for the ton of refrigeration, horsepower per ton of refrigeration, and kilowatt per ton of refrigeration.

6. Show how the reversed Rankine cycle operates using a sketch of the equipment and the T–s diagram.

7. Use the ammonia and HFC-134a tables to calculate the ideal performance of a refrigeration cycle.

8. Be aware of the many refrigerants that have been used and their advantages and disadvantages.

9. Explain that the elements of the gas refrigeration cycle amount to the reversal of the Brayton cycle, and draw the T–s and p–h diagrams for this cycle.

10. Explain the elements of the gas compression cycle and how it works.

11. Analyze the vacuum refrigeration system, and calculate the refrigerating effect of this system.

12. Show the thermoelectric effect and how it is used to produce a refrigerating effect.

13. Derive the performance index for a heat pump, and relate it to the performance index of a refrigeration cycle.

14. Show how the performance of a heat pump changes with a change in ambient temperatures.

15. Demonstrate that the use of well water or the earth as constant-temperature heat sources is more desirable than the use of atmospheric air.

16. Explain why the use of a heat pump for a given application depends on economic factors more than it does on thermodynamic considerations alone.

10.1 INTRODUCTION

In previous chapters, consideration was given to those cycles that (by the proper placing of elements) could yield useful work from a heat source. By the simple expedient of rearranging the sequence of events in these cycles, it is possible (in principle) to remove heat from a region of lower temperature and to deliver it to a region of higher temperature by the input of mechanical work. This removal of heat by the use of mechanical energy has been called the *refrigerating effect*. In more general terms, refrigeration can be defined as the art of maintaining a body at temperatures below its surroundings or, alternatively, as the removal of heat from a place in which it is undesirable to a place in which it is not.

The history of refrigeration can be traced back thousands of years, with natural ice providing the cooling effect desired. The field of refrigeration on a large scale was first developed in the nineteenth century, and in the mid-nineteenth century, the harvesting, storing, and shipping of natural ice became one of the leading industries of the New England states. By the end of the nineteenth century, mechanical refrigeration had become a practical reality, and the refrigeration industry as we know it today was in existence. Along with the use of industrial refrigeration for food preservation, chemical production, metallurgical applications, medicine, and so on, another facet of the refrigeration process appeared; the control of the temperature and humidity of the environment, which is commonly called air-conditioning. It is interesting to note that in 1904, a 450-ton air-conditioning system had been installed to air condition the New York Stock Exchange.

The Carnot cycle has served to establish the performance criteria of power cycles. In this chapter, the study of the reversed Carnot cycle will yield many of the thermodynamic limitations and performance criteria for refrigeration cycles.

10.2 THE REVERSED CARNOT CYCLE

It will be remembered that the Carnot cycle consists of four reversible processes: two isothermals, and two isentropics. In the direct cycle, in which the production of useful work is the primary objective, the elements are arranged so that the energy-flow diagram is as shown in Figure 10.1. Energy flows into the system from a reservoir at constant temperature T_1, work leaves the system through the agency of the prime mover, and heat is rejected to the receiver at constant temperature T_2. The $T-s$ diagram of the cycle is repeated in Figure 10.2. Because the ideal cycle can be treated independently of the working fluid, property lines are not shown on the diagram.

For the reversed cycle, the energy-flow diagram is as shown in Figure 10.3. Note that in this case, the work of the engine serves to take heat from the reservoir (sink at T_2) and rejects it to the source at T_1. The $T-s$ diagram for the reversed cycle is shown in Figure 10.4, and the areas are interpreted as various energy items corresponding to Figure 10.3. By considering these figures, certain general and important conclusions can be obtained for the reversed cycle. First, however, no-

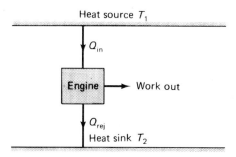

FIGURE 10.1 Direct Carnot cycle.

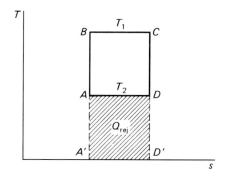

FIGURE 10.2 Direct Carnot cycle.

tice that the area *ABCD* of Figure 10.4 measures the net work supplied to the cycle, and the area *A′ADD′* on this same figure represents the heat removed from the colder region (at T_2). If the interest is in removing the greatest amount of heat from the sink, it is desirable to do so with the least possible energy input to the system. This mode of operation is equivalent to operating at the maximum efficiency.

To evaluate the maximum efficiency system, the same reasoning used for the direct cycle will serve to answer the question of the optimum efficiency conditions for operation of the reversed cycle. Consider a reversible engine operating as shown schematically in Figure 10.3. The amount of energy rejected at T_1 will equal the work in plus the heat removed from the lower temperature region at T_2. It has already been demonstrated that a reversible engine operating between two temperatures is the most efficient engine possible. Therefore, the energy rejected by the reversed cycle can be used in a reversible engine operating in a direct manner (Figure 10.1). The work out of the direct cycle will equal that required by the reversed cycle, and the energy rejected by the direct engine cycle will be equal to that removed by the reversed cycle. Let it now be assumed that the reversed cycle (Figure 10.3) is more efficient than a reversible engine operating in the same mode between the same temperature limits. For a given work input, this cycle will be capable of removing more energy from the sink at T_2 than the reversible engine could. However, the more efficient engine (reversed one) operating in a direct manner will, for a given work output between the same temperature limits, remove less energy from T_1 and reject more energy to T_2 than the equivalent energy amounts for the "more efficient than reversible engine." The inescapable conclusion of such a condition of combined operation is that without the use of energy or an external

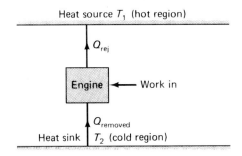

FIGURE 10.3   Reversed Carnot cycle.

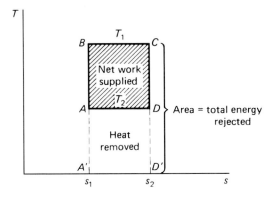

FIGURE 10.4   Reversed Carnot cycle.

agency, heat is removed continuously from a lower-temperature region and made to flow to a higher-temperature region. This is an obvious violation of the second law, and it must be concluded that the most efficient reversed engine cycle operating between two temperatures is a reversible engine; therefore, a reversed Carnot cycle represents the most efficient possible refrigerating cycle.

Based on the foregoing, let us now establish a figure of merit for the refrigeration cycle. For the direct power cycle, we were interested in the amount of work that was obtained from a given heat supply, and this ratio was termed the efficiency of the power cycle. For the reversed cycle, we are interested in the amount of work required to remove a given amount of heat from the low-temperature reservoir. This ratio of refrigeration effect to work input is known as the *coefficient of performance* (COP). The COP is a measure of the efficiency of the refrigeration cycle. However, it is not called efficiency, because it is usually greater than one. Thus,

$$\text{COP} = \frac{\text{refrigeration effect}}{\text{work input}} \tag{10.1}$$

Note that Equation (10.1) applies as well to rates of refrigeration effect per unit time and work per unit time (i.e., power) and also to refrigeration effect and work per unit mass of fluid flowing. Thus, the COP always equals the cooling rate divided by the input power.

For the reversed Carnot cycle shown in Figures 10.3 and 10.4, the heat removed from the reservoir at T_2 is $T_2(s_2 - s_1)$, and the work supplied is $(T_1 - T_2)(s_2 - s_1)$.

$$\text{COP}_{\text{Carnot refrigeration}} = \frac{T_2(s_2 - s_1)}{(T_1 - T_2)(s_2 - s_1)} = \frac{T_2}{T_1 - T_2}$$

$$(10.2)$$

Note that only absolute temperatures are to be used in Equation (10.2).

Note also that the COP for an ideal refrigeration cycle is always greater than unity. To summarize the foregoing:

1. No refrigeration cycle can have a higher COP than a reversible cycle operating between the same temperatures.

2. All reversible refrigeration cycles operating between the same temperature limits have the same COP.

3. The COP of a Carnot cycle, a reversible cycle, is a function only of the upper and lower temperatures of the cycle and increases as the difference between the upper and lower temperatures is decreased. These conclusions are independent of the working fluid of the cycle.

4. Equation (10.2) indicates that, for maximum COP, T_1 should be kept to a minimum. In most cases, either the atmosphere or some nearby body of water is the practical heat sink.

5. The value of T_2 has a more pronounced effect than T_1 on the COP of a Carnot cycle.

6. Any deviations of the actual cycle from the ideal processes predicated for the Carnot cycle lead to values of COP less than the ideal.

ILLUSTRATIVE PROBLEM 10.1

A Carnot refrigeration cycle is used to keep a freezer at 32°F. If the room to which the heat is rejected is at 70°F, calculate for a heat removal of 1000 Btu/min (a) the COP of the cycle, (b) the power required, and (c) the rate of heat rejected to the room. The data are shown in Figure 10.5.

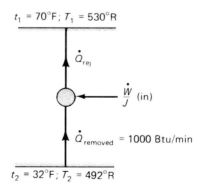

$t_1 = 70°F; T_1 = 530°R$

$\dot{Q}_{rej}$

$\dfrac{\dot{W}}{J}$ (in)

$\dot{Q}_{removed} = 1000$ Btu/min

$t_2 = 32°F; T_2 = 492°R$

FIGURE 10.5 Illustrative Problem 10.1.

SOLUTION

(a) The COP is given by Equation (10.2) for the reversed cycle. Thus,

$$COP = \frac{T_2}{T_1 - T_2} = \frac{492}{530 - 492} = 12.95$$

(b) Because the COP is defined as the ratio of the rate of the refrigeration effect to the power input, the power input is the rate of the refrigeration effect divided by the COP. Therefore,

$$\frac{\dot{W}}{J} = \frac{1000}{12.95} = 77.2 \, \text{Btu/min}$$

(c) $\dot{Q}_{rej}$ is the sum of the refrigeration effect plus the power into the cycle. Thus,

$$\dot{Q}_{rej} = 1000 + 77.2 = 1077.2 \, \text{Btu/min}$$

ILLUSTRATIVE PROBLEM 10.2

A Carnot refrigeration cycle is used to keep a freezer at $-5°C$. Heat is rejected to a room at 20°C. If 30 kW is removed as heat, determine (a) the COP of the cycle, (b) the power required, and (c) the rate of heat rejected to the room. Refer to Illustrative Problem 10.1 and Figure 10.5.

SOLUTION

Proceeding as in Illustrative Problem 10.1, we obtain:

$$\text{(a)} \quad COP = \frac{T_2}{T_1 - T_2} = \frac{268}{293 - 268} = 10.72$$

$$\text{(b)} \quad \dot{W} = \frac{30 \, \text{kW}}{10.72} = 2.8 \, \text{kW}$$

$$\text{(c)} \quad \dot{Q}_{rej} = 30 + 2.8 = 32.8 \, \text{kW}$$

10.3   DEFINED RATINGS

The use of mechanical means of producing refrigeration replaced the use of natural ice for this purpose. As a result, the capacity of a machine to extract heat in a refrigeration cycle was expressed in terms of equivalent ice-producing (or melting) capacity. The unit of refrigeration chosen was the amount of heat that is necessary to freeze 1 ton of water $(2000 \, lb_m)$ into ice in one day. Using approximately $144 \, \text{Btu/lb}_m$ as the latent heat of fusion of water, the unit of refrigeration is

$2000 \, \text{lb}_m \times 144 \, \text{Btu/lb}_m = 288{,}000 \, \text{Btu/day}$. This rate of steady heat removal is defined to be a *ton* of refrigeration. Thus,

$$1 \text{ ton of refrigeration} = 288{,}000 \text{ Btu/day}$$

For other times, we have 1 ton $= 12{,}000$ Btu/hr or 200 Btu/min.

From the definition of COP, it is evident that the product of COP and power input yields the rate at which heat is being removed, or

$$\text{rate of heat removal} = \text{COP} \times (\text{power input}) \, (\text{hp or kW}) \qquad \textbf{(10.3)}$$

Equation (10.3) leads to the following ratings:

$$\frac{\text{horsepower}}{\text{ton of refrigeration}} = \frac{200}{42.4 \times \text{COP}} = \frac{4.717}{\text{COP}} \qquad \textbf{(10.4)}$$

$$\frac{\text{kilowatts}}{\text{ton of refrigeration}} = \frac{200}{56.93 \times \text{COP}} = \frac{3.514}{\text{COP}} \qquad \textbf{(10.5)}$$

It is left as an exercise for the student to derive Equations (10.4) and (10.5). The following example will serve to illustrate the foregoing concepts.

ILLUSTRATIVE PROBLEM 10.3

A refrigeration cycle operates between 20°F and 70°F. (a) Determine the maximum COP for the cycle. (b) It is found that the actual COP is 2. How much more horsepower per ton of refrigeration effect is required by the actual cycle over the minimum possible requirement?

SOLUTION

Refer to Figure 10.3.

(a) $\text{COP}_{\text{ideal}} = \dfrac{460 + 20}{(460 + 70) - (460 + 20)} = 9.6$

(b) The minimum horsepower per ton of refrigeration [Equation (10.4)] is therefore $4.717/9.6 = 0.49$ hp/ton. The actual COP is stated to be 2; therefore, the actual horsepower per ton of refrigeration is $4.717/2 = 2.36$ hp/ton. The horsepower required by the actual cycle over the minimum is

$$\begin{array}{r} 2.36 \\ -0.49 \\ \hline 1.87 \text{ hp/ton} \end{array}$$

ILLUSTRATIVE PROBLEM 10.4

If the actual COP for Illustrative Problem 10.1 is 4.5, determine the power required. Compare this to the results found in Illustrative Problem 10.1.

SOLUTION

From Equation (10.4),

$$\frac{\text{horsepower}}{\text{ton}} = \frac{4.717}{4.5} = 1.048$$

In Illustrative Problem 10.1, 1000 Btu/min is the heat removed. This is (1000 Btu/min)/(200 Btu/min ton), or 5 tons of refrigeration. The horsepower required is $1.048 \times 5 = 5.24$ hp. In Illustrative Problem 10.1, we found that 77.2 Btu/min was required. In terms of horsepower,

$$77.2\,\frac{\text{Btu}}{\text{min}} \times 778\,\frac{\text{ft lb}_f}{\text{Btu}} \times \frac{1}{33,000\,\dfrac{\text{ft lb}_f}{\text{min hp}}} = 1.82\,\text{hp}$$

The ratio of the power required in each problem is the same as the inverse ratio of the COP values, that is, $(4.5/12.95)\,(5.24) = 1.82$, which checks with our results.

ILLUSTRATIVE PROBLEM 10.5

Using the conclusion of Illustrative Problem 10.4, determine the power required in Illustrative Problem 10.2 if the actual COP is 3.8.

SOLUTION

In Illustrative Problem 10.2, it was found that the COP was 10.72 and the power was 2.8 kW. Therefore,

$$\text{power} = 2.8 \times \frac{10.72}{3.8} = 7.9\,\text{kW}$$

10.4 REFRIGERATION CYCLES

To discuss the various refrigeration cycles that are in common use, we will first look at the "ideal" or prototype of each cycle to understand the basic thermodynamics involved and then consider how the practical realization of these cycles is accomplished.

At present, there are several common refrigeration cycles in use:

a. Vapor-compression cycle

b. Gas cycle refrigeration

c. Absorption refrigeration cycle

d. Vacuum refrigeration cycle

e. Thermoelectric refrigerator

10.4a Vapor-Compression Cycle

Just as it was found possible to reverse the Carnot cycle, it is equally possible in principle to reverse the Rankine cycle. An elementary vapor-compression cycle is shown in Figure 10.6, and a corresponding *T–s* diagram is shown in Figure 10.7.

The wet cycle consists of an expansion of the fluid from the saturation point to the wet region (path ① → ②). During this process (throttling), the enthalpy stays essentially constant (see Chapter 3). However, the pressure and temperature of the working fluid decrease, and the fluid becomes a vapor–liquid mixture at state ②. The cooled working fluid (*refrigerant*) then passes to the evaporator, and there (path ② → ③) heat enters from the region or fluid to be cooled. This part of the process is carried out at constant temperature and constant pressure (ideally), because the working fluid is in the wet region. The next part of the cycle (path ③ → ④) is a compression phase. If the compression proceeds from point ③ to point ④ on Figure 10.7, the refrigerant will start from the saturated vapor point and then proceed into the superheated vapor range. This path is called *dry compression*. An alternative path (③a → ④a) is shown where the refrigerant is

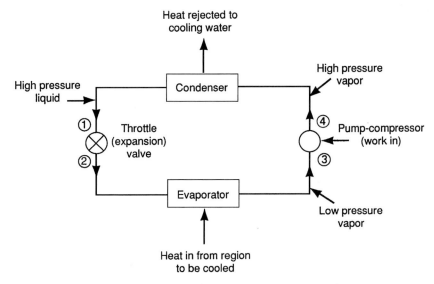

FIGURE 10.6 Simple vapor-compression cycle.

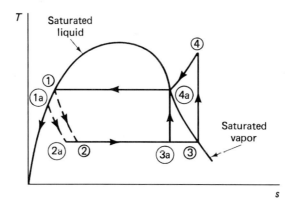

FIGURE 10.7 Simple vapor-compression cycle.

initially "wet" and is just brought to saturation conditions by the compression process. This path has been called *wet compression* for obvious reasons. From considerations of the cycle and the *T–s* diagram (Figure 10.7), it can be demonstrated that it is more efficient to operate the cycle with wet compression. However, most vapor-compression cycles operate with dry compression, because the loss in efficiency is small while the practical problems of the compressor design and operation are considerably eased. The last path of the cycle (path ④ → ①) rejects heat at essentially constant pressure until the saturated liquid line is reached. Once again, the *T–s* diagram shows an alternate path (①a → ②a). In practice, it would be difficult to just achieve point ① for all loads on the system, and some further cooling to point ①a would be expected.

Assuming that all processes are carried out without pressure or heat losses (other than specified) and neglecting kinetic energy and potential energy differences, it is possible to analyze the system as follows:

1. Path ① → ②, throttling:

$$\boxed{h_1 = h_2} \tag{10.6}$$

2. Path ② → ③, evaporator (cooling effect):

$$\boxed{q_{in} = h_3 - h_2 = h_3 - h_1} \tag{10.7}$$

3. Path ③ → ④, compressor:

$$\boxed{\text{work} = h_4 - h_3} \tag{10.8}$$

(For the refrigeration cycle, work in to the cycle is taken to be positive.)

4. Path $\textcircled{4} \rightarrow \textcircled{1}$, condenser (heat removal):

$$\boxed{q_{\text{rej}} = h_4 - h_1} \qquad \textbf{(10.9)}$$

From these quantities, it will be noted that the energy removed in the condenser must numerically equal the heat gain in the evaporator plus the work of the compressor. The refrigerating capability, however, is only the evaporator effect.

The mass of refrigerant circulated per ton of refrigeration can readily be found from the definition of the ton and the heat absorbed by the evaporator:

$$\frac{\text{mass of refrigerant}}{\text{minute}} = \frac{200 \times \text{tons}}{h_3 - h_2} = \frac{200 \times \text{tons}}{h_3 - h_1} \qquad \textbf{(10.10)}$$

The definition of the COP (in terms of this cycle) is the ratio of the heat absorbed in the evaporator to the net work into the cycle. In consistent heat units, this can be written as

$$\boxed{\text{COP} = \frac{h_3 - h_1}{h_4 - h_3}} \qquad \textbf{(10.11)}$$

For the ideal cycle, the work per ton of refrigeration will be given by

$$\boxed{\text{horsepower/ton} = 4.717 \left(\frac{h_4 - h_3}{h_3 - h_1} \right)} \qquad \textbf{(10.12)}$$

Note that this is the definition given in Equation (10.4).

ILLUSTRATIVE PROBLEM 10.6

An ammonia refrigerator plant is to operate between a saturated liquid at 120 psia at the condenser outlet and a saturated vapor at 15 psia at the evaporator outlet. If a capacity of 30 tons is desired, compute the following on the basis of an ideal cycle only: (a) COP; (b) work of compression, Btu per pound; (c) refrigerating effect, Btu per pound; (d) pounds per minute of ammonia required; and (e) ideal horsepower per ton of refrigeration.

SOLUTION

At 120 psia, the corresponding saturation temperature is 66°F. Referring to Figure 10.7, the enthalpies are

$h_1 = 116.0 \, \text{Btu/lb}_{\text{m}}$ (values of ammonia properties, from Appendix 3)
$h_2 = 116.0 \, \text{Btu/lb}_{\text{m}}$ (throttling gives us $h_1 = h_2$)
$h_3 = 602.4 \, \text{Btu/lb}_{\text{m}}$

From the consideration that $s_3 = s_4$, h_4 is found: at 15 psia, $s_3 = 1.3938$; therefore, by interpolation in the superheat tables at 120 psia, $t_4 = 237.4°F$ and $h_4 = 733.4\,Btu/lb_m$.

(a) $COP = \dfrac{602.4 - 116.0}{733.4 - 602.4} = 3.71$

(b) The work of compression is

$$h_4 - h_3 = 733.4 - 602.4 = 131.0\,Btu/lb_m$$

(c) The refrigerating effect is

$$h_3 - h_1 = 602.4 - 116.0 = 486.4\,Btu/lb_m$$

(d) The pounds per minute of ammonia required for circulation equals

$$\frac{200 \times 30}{602.4 - 116.0} = 12.34\,lb_m/min$$

(e) The ideal horsepower per ton of refrigeration equals

$$4.717\left(\frac{733.4 - 602.4}{602.4 - 116.0}\right) = 1.27\,hp/ton$$

ILLUSTRATIVE PROBLEM 10.7

Solve Illustrative Problem 10.6 if the refrigerant is Freon-12, with the high-pressure side at 110 psig. The Freon-12 enters the expansion valve as saturated liquid. It leaves the evaporator as a dry saturated vapor at $-20°F$.

SOLUTION

From Appendix 3, 110 psig corresponds to 96°F.

$$h_1 = 30.14\,Btu/lb_m$$
$$h_2 = 30.14\,Btu/lb_m$$
$$h_3 = 75.110\,Btu/lb_m$$

To determine h_4, we note that $s_3 = s_4$. At $-20°F$, $s_3 = 0.17102\,Btu/lb_m·°F$. By interpolation in the Freon-12 superheat table at these values, $h_4 = 89.293\,Btu/lb_m$.

(a) $COP = \dfrac{75.11 - 30.14}{89.293 - 75.11} = 3.17$

(b) Work of compression = $89.293 - 75.11 = 14.183\,Btu/lb_m$

(c) Refrigerating effect = $75.11 - 30.14 = 44.97\,Btu/lb_m$

(d) Pounds of Freon-12 required for circulation:

$$\frac{200 \times 30}{44.97} = 133.42$$

(e) Ideal horsepower per ton of refrigeration:

$$4.717\left(\frac{89.293 - 75.11}{75.11 - 30.14}\right) = 1.49 \, \text{hp/ton}$$

Before proceeding further with our discussion of refrigeration cycles, let us first look at some of the components used and their arrangement in practice. In a typical home refrigerator, the compressor is usually physically in the rear, near the bottom of the unit. The motor and compressor are usually located in a single housing with the electrical leads for the motor passing through the housing. While this is done to prevent leakage of refrigerant, it imposes the requirement that the refrigerant should be inert with respect to the insulation. One common refrigerant, Freon, cannot be used with natural rubber, because this class of refrigerant acts as a solvent with certain types of insulating materials and varnishes. Also, the electrical resistance of the refrigerant is of the utmost importance in hermetically sealed units such as those used in home refrigerators, where the motor windings are exposed to the refrigerant. Figure 10.8 shows a diagrammatic layout of a refrigerating system for a domestic-type refrigerator.

The condenser is physically arranged so that room air flows past the condenser by natural convection. The expansion valve is a long capillary tube, and the evaporator is shown around the outside of the freezing compartment inside the refrigerator. The pressures shown in Figure 10.8 are typical for the use of sulfur dioxide as the refrigerant. The compressor shown schematically is a reciprocating unit, but small, efficient, and economical rotary units are available for home refrigeration use.

Starting at point ③ of Figure 10.8, the vaporous refrigerant enters the compressor at low pressure and temperature. An isentropic compression (ideally) raises the pressure and temperature to p_2 and t_2. The saturation temperature corresponding to p_2 must be some value above atmospheric temperature or above the temperature of the cooling water that may be used in the condenser. Leaving the compressor in condition ④, the vapor enters the condenser, where it is condensed to a liquid at some temperature t_1. After the condenser, the liquid enters an expansion valve that separates the high- and low-pressure regions and passes through the valve in a throttling process with $h_1 = h_2$. The refrigerant then enters the evaporator (or freezing compartment) where it boils, because it is receiving heat from the refrigerator and its contents. The vapor from the evaporator enters the compressor and the cycle starts over.

The heart of any refrigeration system is the compressor. While the types of compressors used are similar to those used for air or other gases, the positive-displacement reciprocating compressor is the unit most widely used in industrial vapor-compression refrigeration installations. Centrifugal and gear-type positive-displacement compressors are also used. The gear-type compressor shows good volumetric efficiencies, but the centrifugal compressors are usually inefficient in smaller sizes and are used when the size of a reciprocating unit would be excessively large.

Figure 10.9 shows a typical vertical reciprocating compressor for refrigerant compression. It is a two-cylinder, single-acting type with safety head construction. The safety head construction is

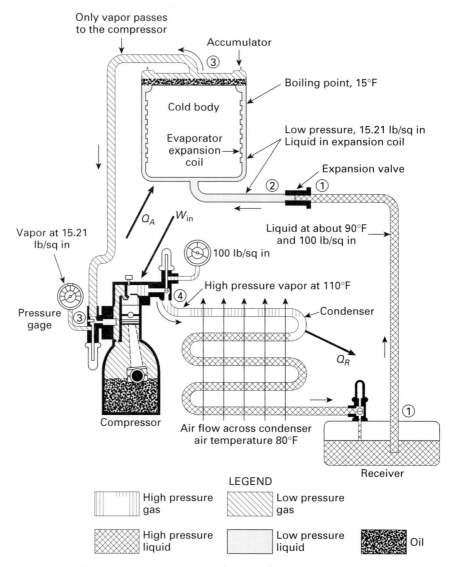

FIGURE 10.8 Diagrammatic layout of a refrigerating system.

[From V. M. Faires, Theory and Practice of Heat Engines (New York: Macmillan, Inc., 1948), p. 360, with permission.]

used in large compressors where there is danger of operation with wet compression or danger of operation with foreign material in the cylinder. The safety head is basically a second head placed at the end of the cylinder and held in position by heavy helical springs. If wet refrigerant or foreign material enters the cylinder space, the movement of the safety head relieves any excessive pressure buildup in the cylinder. The vapor from the evaporator is drawn into the crankcase and then passes upward through the suction valves mounted in the piston crown. Compressed vapor is forced past the discharge valves mounted in the safety head assembly. Multiple suction and dis-

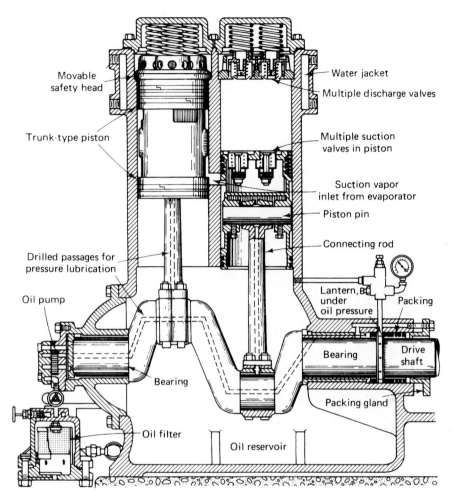

FIGURE 10.9 Vertical reciprocating refrigerant compressor.
[From C. C. Dillio and E. P. Nye, Thermal Engineering *(Scranton, Pa.: International Textbook Co., 1959), p. 438, with permission.]*

charge valves provide larger flow areas than for single valves and are often used to minimize the inevitable drop of pressure occurring at these valves. Reciprocating and rotary positive-displacement compressors in forms other than that shown are used in small domestic refrigerators (capacities in fractions of a ton). Multicylinder reciprocating compressors with *V* and *W* cylinder arrangements are commonly used for the range of capacities between 3 and 200 tons.

Figure 10.10 shows a large, two-stage, centrifugal compressor. Units of this type are used in large-capacity systems above 75 tons of refrigeration. This unit (Figure 10.10) is used in an air-conditioning system where brine is cooled and then circulated to provide cooling as needed. The condenser, which is water cooled, is shown in the upper portion of the figure, and the tube-type evaporator used for cooling the brine is shown in the lower-left portion of the unit. Single-unit machines of the type shown in Figure 10.10 have been built in capacities from 75 to 5000 tons.

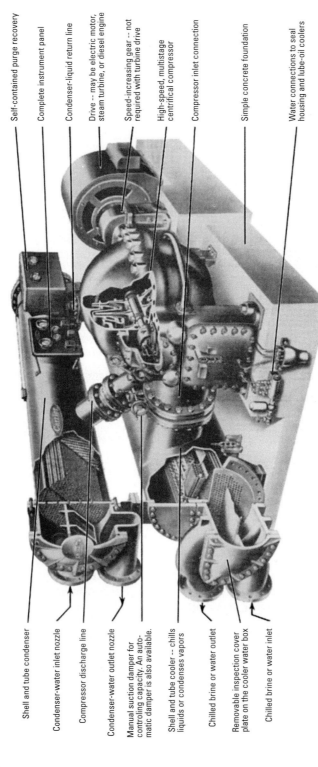

Self-contained purge recovery

Complete instrument panel

Condenser-liquid return line

Drive -- may be electric motor, steam turbine, or diesel engine

Speed-increasing gear -- not required with turbine drive

High-speed, multistage centrifical compressor

Compressor inlet connection

Simple concrete foundation

Water connections to seal housing and lube-oil coolers

Shell and tube condenser

Condenser-water inlet nozzle

Compressor discharge line

Condenser-water outlet nozzle

Manual suction damper for controling capacity. An automatic damper is also available.

Shell and tube cooler -- chills liquids or condenses vapors

Chilled brine or water outlet

Removable inspection cover plate on the cooler water box

Chilled brine or water inlet

FIGURE 10.10 Centrifugal refrigeration machine.

[From G. J. Van Wylen and R. E. Sonntag, Fundamentals of Classical Thermodynamics (New York: John Wiley & Sons, Inc., 1965), p. 10, with permission.]

The evaporator is another component of the refrigeration system that has to be considered. Many types of evaporators are used depending on the specific application. The direct expansion cooling coil shown in Figure 10.11 is fed just enough refrigerant by the expansion valve so that all the liquid is converted to vapor before the refrigerant reaches the outlet connection of the evaporator. This type of evaporator is widely used for air-conditioning ducts. The plate-type evaporator has continuous loops of steel tubing contained between evacuated plates.

Many considerations enter into the choice of refrigerant for a given application. Some of the refrigerants that are used commercially are ammonia, butane, CO_2, carrene, Freon, methyl chloride, sulfur dioxide, and propane. Based on the thermodynamics of the vapor-compression cycle that we have studied thus far, the following properties are desirable in a refrigerant:

1. The heat of vaporization of the refrigerant should be high. The higher h_{fg}, the greater the refrigerating effect per pound of fluid circulated.

2. The specific heat of the liquid should be low. The lower the specific heat of the liquid, the less heat it will pick up for a given change in temperature during either throttling or in flow through the piping and, consequently, the greater the refrigerating effect per pound of refrigerant.

3. The specific volume of the refrigerant should be low to minimize the work required per pound of refrigerant circulated.

4. The critical temperature of the refrigerant should be higher than the condensing temperature to prevent excessive power consumption.

In addition to the foregoing, it is necessary to consider toxicity, corrosiveness, dielectric strength for a hermetically sealed system, viscosity, thermal conductivity, explosiveness, effect on foods, stability, inertness, and cost. No one refrigerant has been found to meet all these requirements. Of the many possible refrigerants, ammonia, Freon, brine, methyl chloride, and sulfur dioxide are the most common that have been used.

Two environmental concerns have recently come to the attention of both industry and government, namely, the potential of a refrigerant to deplete ozone in the upper atmosphere and the potential of a refrigerant to persist in the upper atmosphere, trapping radiation emitted by the earth

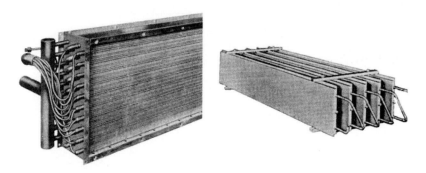

FIGURE 10.11 Evaporators.
(Reprinted with permission from Power, The Engineer's Reference Library, *copyright by McGraw-Hill, Inc., New York.)*

TABLE 10.1

Physical and Thermal Properties of Common Refrigerants

Refrigerant	Chemical formula	Boiling point (°F)	Freezing point (°F)	Critical points Temperature (°F)	Pressure (psia)	Specific gravity of liquid at atmos. pressure	Specific heat of liquid, ave. 5°F to 86°F
Ammonia	NH_3	−28.0	−107.9	271.4	1657.0	0.684	1.12
Azeotropic mixture Freon-12 and unsymmetrical difluoroethane (Carrene-7)	78.3% F-12, 26.2% G-100	−28.0	−254.0	221.1	631.0	—	—
Bromotrifluoromethane (Kulene-131)	CF_3Br	−73.6	−226.0	153.5	587.0	—	0.19
Carbon dioxide	CO_2	−109.3	−69.0	87.8	1069.9	1.560	0.77
Dichlorodifluoromethane (Freon-12, Genetron-12)	CCl_2F_2	−21.6	−252.0	233.6	596.9	1.480	0.23
Ethylene	C_2H_4	−155.0	−272.0	48.8	731.8	—	—
Isobutane	C_4H_{10}	−10.3	−229.0	272.7	537.0	0.549	0.62
Methyl chloride	CH_3Cl	−10.8	−144.0	289.4	968.7	1.002	0.38
Methylene chloride (Carrene-1)	CH_3Cl_2	105.2	−142.0	480.0	670.0	1.291	0.33
Monochlorodifluoromethane (Freon-12, Genetron-141)	$CHClF_2$	−41.4	−256.0	204.8	716.0	1.411	0.30
Sulfur dioxide	SO_2	14.0	−103.9	314.8	1141.5	1.357	0.34
Trichloromonofluoromethane (Freon-11, Genetron-11)	CCl_3F	74.7	−168.0	388.4	635.0	1.468	0.21
Trichlorotrifluoromethane (Freon-113)	$CCl_2F\text{-}CClF_2$	117.6	−31.0	417.4	495.0	1.559	0.21
HFC-134a (1,1,1,2-tetrafluoroethane)	CH_2FCF_3	−15.0	−14.9	213.9	588.9	1.376	0.33
Water	H_2O	212.0	32.0	706.1	3226.0	1.000	1.00

(the greenhouse effect). A major effort has been exerted to develop and use refrigerants having low ozone-depletion levels and low greenhouse-warming potential and on developing refrigerant recovery systems and reducing refrigerant leakage.

In September 1987, an international meeting was held in Montreal, Canada, to discuss the effects of chlorofluorocarbon (CFC) refrigerants on the atmosphere. In June 1990, representatives of 93 nations met in London and amended the 1987 Montreal Agreements. Basically, the London Amendments call for the complete phasing out of CFC refrigerants by the year 2000. Hydrochlorofluorocarbon (HCFC) refrigerants are to be phased out by the year 2020, with the absolute deadline being the year 2040. The refrigerant HFC (hydrofluorocarbon) contains no chlorine and poses no threat to the ozone layer. However, HFC-134a, a replacement for R-11 and R-12 refrigerants, has a global warming potential that is 420 times higher than carbon dioxide. Further study could lead to regulation on the use of HFC-134a as well as other HFC's.

Some of the major engineering problems that have arisen and have been overcome in the application of CFC replacement refrigerants are:

1. The immiscibility of HFC-134a with mineral oil lubricants previously used in vapor-compression systems.

2. The need for a new and different thermal expansion valve.

3. The need to use nylon as an interliner in the refrigerant hoses.

4. The need for a new desiccant to absorb water vapor.

5. The need for a new parallel-flow condenser to reduce system head pressure.

Table 10.1 summarizes some of the physical and thermal properties of commonly used refrigerants. Currently, HFC-134a is widely used in commercial systems, and Freon-22 is used in most residential units. HFC-134a tables are in the Appendix along with tables for Freon-12, which is still used in older systems. Ammonia is one of the most effective refrigerants on a per pound basis and so is also included in the tables in the Appendix. It is, however, highly toxic.

In the actual cycle, there are deviations from the ideal cycle due to compression inefficiencies, pressure drops in piping, temperature differences between the elements interchanging heat, and so on. Because heat is transferred in both the condenser and evaporator of the vapor-compression system, temperature differences exist between these units and the medium to which they are transferring heat. In the case of the evaporator, the coils must be at a lower temperature than the region being cooled. The temperature difference for the evaporator usually is of the order of 5°F, while in the condenser, the refrigeration fluid is of the order of 10°F higher than the coolant fluid. However, among these effects, those usually considered are the inefficiency in the compression part of the cycle and pressure differences in long piping systems. The other effects are usually considered to be either too difficult to analyze generally or small enough to be safely ignored.

ILLUSTRATIVE PROBLEM 10.8

An ideal refrigeration cycle uses Freon-12 as the working fluid. The temperature of the refrigerant in the condenser is 90°F, and in the evaporator, it is 10°F. Determine the COP of this ideal cycle.

SOLUTION

Using the Freon-12 tables in Appendix 3, we have

$$h_1 = 28.713 \text{ Btu/lb}_m$$
$$h_2 = h_1 = 28.713 \text{ Btu/lb}_m$$
$$h_3 = 78.335, s_3 = 0.16798 = s_4$$

From the superheat table, at 90°F and $s = 0.16798$, $h_4 = 87.192 \text{ Btu/lb}_m$.

$$\text{The heat extracted} = (78.335 - 28.713) = 49.622 \text{ Btu/lb}_m$$
$$\text{The work required} = (87.196 - 78.335) = 8.861 \text{ Btu/lb}_m$$

$$\text{Therefore, COP} = \frac{49.622}{8.861} = 5.60$$

ILLUSTRATIVE PROBLEM 10.9

Solve Illustrative Problem 10.8 if the refrigerant is HFC-134a.

SOLUTION

Using the HFC-134a tables in Appendix 3,

$$h_1 = 41.6 \text{ Btu/lb}_m$$
$$h_2 = h_1 = 41.6 \text{ Btu/lb}_m$$
$$h_3 = 104.6, s_3 = 0.2244, s_4 = s_3$$
$$h_4 = 116.0 \text{ Btu/lb}_m$$

$$\text{The heat extracted} = (104.6 - 41.6) = 63 \text{ Btu/lb}_m$$
$$\text{The work required} = (116.0 - 104.6) = 11.4 \text{ Btu/lb}_m$$

$$\text{Therefore COP} = \frac{63}{11.4} = 5.53$$

ILLUSTRATIVE PROBLEM 10.10

A Freon-12 vapor-compression refrigeration system has a compressor of 80% adiabatic efficiency. At inlet to the compressor, the refrigerant has a pressure of 5.3 psia and 40°F superheat; at discharge from the compressor the pressure is 100 psia.

Saturated liquid Freon at 100 psia leaves the condenser and passes through an expansion valve into the evaporative coils. The condenser is supplied with cooling water, which enters at 60° and leaves at 70°F. For a plant capacity of 5 tons, find (a) the horsepower required to drive the compressor if it has a mechanical efficiency of 100%, and (b) the required capacity in gallons per minute of cooling water to the pump.

SOLUTION

A T–s diagram for this problem is shown in Figure 10.12. This problem will require both the tables and p–h chart for Freon-12 given in Appendix 3.

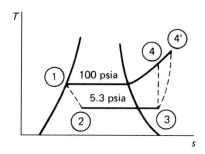

FIGURE 10.12   Illustrative Problem 10.10.

(a) By definition, the efficiency of the compressor is the ratio of the ideal compression work to actual compression work. Based on the points on Figure 10.12,

$$\eta = \frac{h_4 - h_3}{h_4' - h_3}$$

There is close correspondence between 5.3 psia and $-60°$F for saturated conditions. Therefore, state ③ is a superheated vapor at 5.3 psia and approximately $-20°$F, because the problem states that state ③ has a 40°F superheat. Interpolation in the Freon tables in Appendix 3 yields

$$T = -20°F$$

p	h	s
7.5	75.719	0.18371
5.3	76.885	0.18985
5.0	75.990	0.19069

$h_3 = 75.886\ \text{Btu}/\text{lb}_\text{m}$

At 100 psia and $s = 0.18985$.

t	s	h
170°F	0.18996	100.571
169.6°F	0.18985	100.5
160°F	0.18726	98.884

$h_4 = 100.5 \, \text{Btu/lb}_\text{m}$

The weight of refrigerant is given by

$$\frac{200(\text{tons})}{h_3 - h_1} = \frac{(200)(5)}{75.886 - h_1}$$

In the saturated tables, h_1 is

p	h
101.86	26.832
100 psia	26.542
98.87	26.365

and

$$\dot{m} = \frac{\text{mass flow}}{\text{min}} = \frac{(200)(5)}{75.886 - h_1} = \frac{(200)(5)}{75.886 - 26.542} = 20.3 \, \text{lb}_\text{m}/\text{min}$$

$$\text{total work of compression} = 20.3 \, (h_4' - h_3)$$

where

$$h_4' - h_3 = \frac{h_4 - h_3}{\eta} = \frac{100.5 - 75.886}{0.8} = 30.8 \, \text{Btu/lb}_\text{m}$$

Therefore,

$$\text{work} = \frac{(20.3)(30.8)(778)}{33,000} = 14.7 \, \text{hp}$$

(b) Assuming a specific heat of the water as unity, we obtain

$$\dot{m} = \frac{20.3 \, (h_4' - h_1)}{70 - 60}$$

From part (a),

$$h_4' - h_3 = 30.7$$
$$h_4' = 30.7 + 75.886$$
$$= 106.59 \, \text{Btu/lb}_\text{m}$$

Therefore,

$$\dot{m} = \frac{(20.3)\,(106.59 - 26.542)}{70 - 60}$$
$$= 162.5\ \text{lb}_m/\text{min of cooling water}$$
$$= 19.5\ \text{gal}/\text{min}$$

The student will note the effort involved in obtaining this answer when working directly from tables. Just as the Mollier chart simplifies calculations of power cycles, problems in refrigeration are simplified considerably by use of a modified Mollier chart. Such a chart uses pressure–enthalpy coordinates and usually has other curves of entropy and temperature on it. Figure 10.13 shows a skeleton chart using p–h coordinates of a refrigerant. The ordinate, pressure, is usually plotted on a logarithmic scale for convenience. In the superheat region, lines of constant volume, constant entropy, and constant temperature are displayed. Shown by dashed lines is an ideal simple vapor-compression cycle with end states corresponding to the events shown on the simplified schematic diagram of Figure 10.6. The following problem will serve to illustrate the use of the p–h diagram as well as the savings obtainable in both time and effort by using this modified Mollier chart.

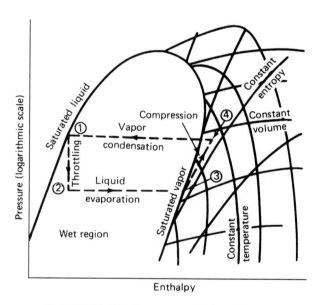

FIGURE 10.13   Pressure–enthalpy diagram.

ILLUSTRATIVE PROBLEM 10.11
Solve Illustrative Problem 10.10 using the p–h chart for Freon-12 in Appendix 3.

SOLUTION

(a) From Appendix 3, reading the p–h diagram directly, we have

$$h_3 = 76.2 \, \text{Btu/lb}_m$$
$$h_4 = 100.5 \, \text{Btu/lb}_m$$
$$\text{work of compression} = \frac{100.5 - 76.2}{0.8} = 30.4 \, \text{Btu/lb}_m$$

The enthalpy of saturated liquid at 100 psia is given at 26.1 Btu/lb$_m$. Proceeding as before yields

$$\frac{\text{mass flow}}{\text{min}} = \frac{(200)\,(5)}{76.2 - 26.1}$$
$$= 20 \, \text{lb}_m/\text{min}$$
$$\text{total ideal work of compression} = \frac{(20)\,(30.4)\,(778)}{33,000}$$
$$= 14.3 \, \text{hp}$$

(b) $h_4' = 76.2 + 30.4 = 106.8 \, \text{Btu/lb}_m$

$$\dot{m} = \frac{(20)\,(106.8 - 26.5)}{70 - 60} = 160.6 \, \text{lb}_m/\text{min} \quad \text{or} \quad 19.3 \, \text{gal/min}$$

The accuracy afforded by the diagram is sufficient for engineering applications.

10.4b Gas Cycle Refrigeration

Just as it was possible to reverse the Carnot and Rankine cycles to obtain a refrigerating effect, it should be possible to reverse any cycle to obtain a similar effect. Among the many gas cycles that have been proposed as prototypes for practical engines, the reversed Brayton cycle has been used to produce refrigeration. The earliest mechanical refrigeration system used air as the refrigerant due to its availability and safety. However, due to its high operating costs and low COP, it has largely become obsolete. However, the introduction of high-speed aircraft and missiles has required lightweight, high-refrigeration-capacity systems to cause a minimum reduction in payload. For example, a jet fighter traveling at 600 mph requires a refrigeration system having a capacity of 10 to 20 tons. In passenger aircraft, solar radiation, the occupants, and electrical and mechanical equipment all cause the heating of cabin air. In addition, the adiabatic stagnation of the air relative to the moving airplane (see Chapter 6) can cause temperature increases of the air of approximately 65°F at a relative air speed of 600 mph. Also, all modern aircraft operate with pressurized cabins, which requires compression of outside air to nearly atmospheric pressure, and in the process, the temperature of the compressed air increases. The gas cycle refrigeration system using air as the refrigerant has become generally used for this application due to the small size and weight of the components, which is of prime importance. The power to operate the system is of secondary importance in this application. In addition, the power source used for the air-cycle aircraft refrigeration system is the same as is used for pressurization of the cabin. Because pressurization is not needed at sea level but refrigeration is and, at high altitudes, pressurization is required but less refrigeration is needed, it is found that the power requirement for the combined pressurization and cooling is approximately constant.

Figure 10.14 shows an aircraft refrigeration unit, and Figure 10.15 shows the schematic diagram for this unit. In this unit, high-pressure bleed air, previously compressed by the jet engine compressor, passes into a surface-type heat exchanger. Using relatively cool air from outside the airplane as the coolant, the compressor bleed air temperature is lowered, in this case from 415°F to 172°F. The cooled bleed air enters the turbine, where it expands while producing shaft work to

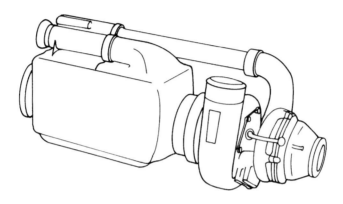

FIGURE 10.14 Aircraft refrigeration unit.
[From C. C. Dillio and E. P. Nye, Thermal Engineering (Scranton, Pa.: International Textbook Co., 1959), p. 428, with permission.]

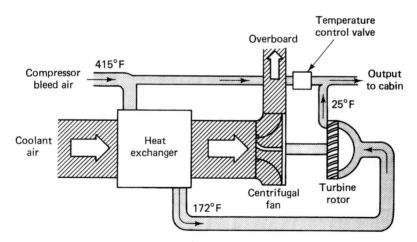

FIGURE 10.15 Schematic of an aircraft refrigeration unit.
[From C. C. Dillio and E. P. Nye, Thermal Engineering (Scranton, Pa.: International Textbook Co., 1959), p. 429, with permission.]

drive the centrifugal fan. The cooled, expanded air now at 25°F is mixed with compressor bleed air automatically (using a temperature control valve), and the tempered air is ducted to the cabin.

Let us now consider the reversed Brayton cycle shown schematically in Figure 10.16. This cycle is a closed cycle in that the working fluid (air) is continuously recycled. As shown in Figure 10.16, the cycle consists of an expander (turbine), a compressor (centrifugal fan), and two heat exchangers. Figure 10.17 shows the p–v and T–s diagrams for the cycle shown in Figure 10.16,

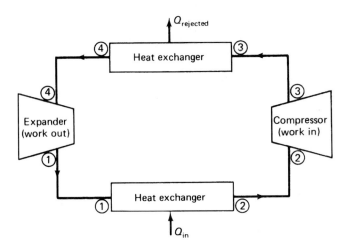

FIGURE 10.16   Schematic of simple, closed-cycle Brayton cycle.

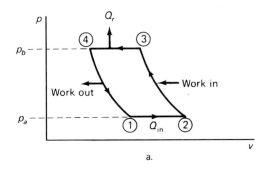

a.

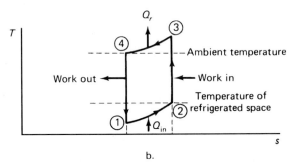

b.

FIGURE 10.17 T–s and p–v diagrams for reversed Brayton cycle.

with corresponding points (①, ②, ③, ④) noted on each of these figures. As noted in Chapter 9, the Brayton cycle consists of two isentropic and two constant-pressure (isopiestic or isobaric) portions.

Consider that a single charge of working fluid is in the device and that the cycle starts from the condition denoted as ① in Figures 10.16 and 10.17. At this point, the gas must be colder than the room (or second fluid) that is to be cooled. It flows at constant pressure through the required piping and absorbs heat from the area to be cooled. During this part of the cycle, the temperature of the cooled area will be greater than T_2. After being heated, the gas is compressed isentropically along path ② → ③, In general, T_3 will be greater than room temperature (or local cooling water temperature), and the gas therefore is capable of being cooled in the next part of the path. This is accomplished at constant pressure ③ → ④, in which T_4 is always greater than the temperature of available coolant. Finally, the gas is cooled to state ① by an isentropic expansion during which work is removed from the system. There is one basic point, however, that causes this cycle to deviate from the previously considered vapor cycle. The vapor cycle employs a throttling process to reduce the temperature of the working fluid. The same effect could be achieved by expanding the liquid in an engine, but this engine is not practical and would be prohibitively expensive. It will be recalled that a throttling process is carried out at constant enthalpy. For an ideal gas, a throttling process is carried out at constant temperature and therefore is incapable of lowering the temperature for the required refrigerating effect. Even in real gases, the throttling process does not provide much of a refrigerating effect. To obtain this effect when gas is used as the refrigerant, it is necessary to create the required low temperature level by letting the refrigerant do work by expanding in an engine.

Each of the nonflow processes constituting this gas cycle can be evaluated from the considerations given in Chapter 6. Thus, from the definition of COP and assuming an ideal gas with constant specific heat,

$$\boxed{\text{COP}_{\text{Brayton}} = \frac{\text{heat removed}}{\text{net work required}} = \frac{c_p\,(T_2 - T_1)}{_2w_3 - _4w_1}} \tag{10.13}$$

The denominator of Equation (10.13) represents the net area behind the ③ → ② and ④ → ① curves. Because all elements of the cycle are reversible, it follows that the net work of the cycle must equal the difference of the heat rejected by the cycle to the heat into the cycle. Therefore,

$$\boxed{\text{COP}_{\text{Brayton}} = \frac{c_p(T_2 - T_1)}{c_p(T_3 - T_4) - c_p(T_2 - T_1)} = \frac{1}{\dfrac{T_3 - T_4}{T_2 - T_1} - 1}} \tag{10.14a}$$

In terms of pressure,

$$\boxed{\text{COP}_{\text{Brayton}} = \frac{1}{(p_b/p_a)^{(k-1)/k} - 1}} \tag{10.14b}$$

Also,

$$\boxed{\text{COP}_{\text{Brayton}} = \frac{T_1}{T_4 - T_1} = \frac{T_2}{T_3 - T_2}} \qquad \textbf{(10.14c)}$$

By comparing the foregoing relations with a reversed Carnot cycle operating between the same limits, it will be noted that the index of performance (COP) of the reversed Brayton cycle is less than that of the Carnot cycle. The fundamental difference between the Brayton and Carnot cycles is that the Brayton does not operate at constant heat-rejection and constant heat-absorption temperatures whereas the Carnot does.

Because it may be of interest to know the work of the compressor and the turbine, it is necessary only to note that each of these processes represents a steady-flow isentropic process; thus,

$$\text{work of compressor per pound} = c_p(T_3 - T_2) \qquad \textbf{(10.15a)}$$

$$\text{work of expander per pound} = c_p(T_4 - T_1) \qquad \textbf{(10.15b)}$$

By proper substitution of the path equation, pv^k, these equations can be put into terms of pressures.

ILLUSTRATIVE PROBLEM 10.12

An ideal gas refrigeration cycle is operated so that the upper temperature of the cycle is 150°F, the lowest temperature of the cycle is −100°F, and the COP is 2.5. Determine the work of the expander, the work of the compressor, and the net work of the cycle per pound of air. Also, determine the airflow required per ton of refrigeration. Take $c_p = 0.24\,\text{Btu}/\text{lb}_\text{m}\cdot°\text{R}$.

SOLUTION

From Equation (10.14),

$$2.5 = \frac{T_1}{T_4 - T_1} \qquad 2.5(T_4 - T_1) = T_1 \qquad T_1 = 360°\text{R}$$

Therefore,

$$T_4 = \frac{3.5}{2.5}(T_1) = \frac{3.5}{2.5}(360) = 504°\text{R} \qquad (44°\text{F})$$

Also,

$$2.5 = \frac{T_2}{T_3 - T_2} \qquad (2.5)(150 + 460 - T_2) = T_2$$

$$T_2 = 436°\text{R} \qquad (-24°\text{F})$$

The work of the expander is

$$c_p(T_4 - T_1) = (0.24)(504 - 360) = 34.6\,\text{Btu}/\text{lb}_\text{m} \text{ of air}$$

The work of the compressor is

$$c_p(T_3 - T_2) = (0.24)(610 - 436) = 41.8 \, \text{Btu/lb}_m \text{ of air}$$

The net work required by the cycle is

$$41.8 - 34.6 = 7.2 \, \text{Btu/lb}_m \text{ (work in)}$$

Per ton of refrigeration, the required airflow is

$$\frac{200}{c_p(T_2 - T_1)} = \frac{200}{0.24(436 - 360)} = 10.96 \, \text{lb}_m/\text{min per ton}$$

In an actual cycle, the $\text{COP}_{\text{Brayton}}$ would be much lower because of the irreversible effects in each part of the cycle.

The air cycle described earlier for aircraft use is an example of an open cycle in which the working fluid is not caused to continually recycle. Instead, it is continuously discharged, and a fresh air charge is continuously introduced into the cycle at the entering conditions of the cycle. Figure 10.18 shows a simple, open air cycle used for refrigeration. The student should compare this cycle to the aircraft refrigeration cycle described earlier.

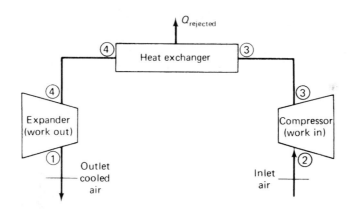

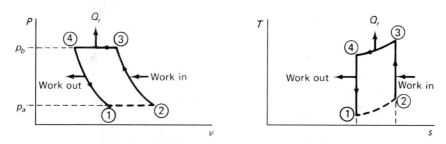

FIGURE 10.18 Open Brayton refrigeration cycle.

In addition to the simple Brayton refrigeration cycle, two other versions of this cycle are also used. The first of these is known as the *bootstrap system.* In this system, as shown in Figure 10.19, there are two heat exchangers with cooling air provided by the ram pressure of the aircraft forcing outside air through the exchangers. Air bled from the gas turbine engine compressor is first cooled in the initial heat exchanger, further compressed by the secondary compressor, then cooled further in the secondary heat exchanger, and finally expanded in the cooling turbine and delivered to the cabin of the airplane. The work delivered by the cooling turbine is used to drive the secondary compressor. The bootstrap system has the disadvantage that the airplane must be in flight to obtain the required ram air. To overcome this, a fan is provided to pull air over the secondary heat exchanger, giving a system having the features of both the simple and bootstrap system.

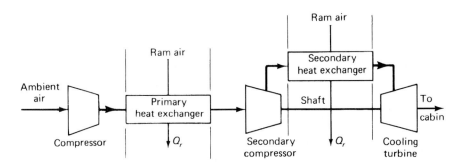

FIGURE 10.19 Bootstrap system.

The second version of the reversed Brayton cycle that is used is the regenerative system. This system is used when the discharge temperature of the simple cycle is too high, as may be the case for very high-speed aircraft. In this cycle, as shown in Figure 10.20, some of the turbine discharge air is diverted back to cool the air entering the turbine. This serves to cool the air entering the turbine to a temperature lower than that which would be obtained by using only ambient air.

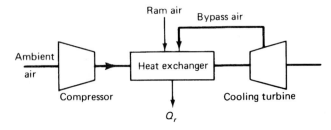

FIGURE 10.20 Regenerative system.

10.4c Absorption Refrigeration Cycle

One of the fundamental concepts of thermodynamics is the equivalence of various forms of energy. A good portion of this study has been devoted to the concept that heat and work are forms of energy in transition. The *absorption refrigeration* cycle takes advantage of this concept by essentially replacing the vapor-compression pump compressor of the simple vapor-compression cycle with an "equivalent" heat source. In practice, this cycle leads to certain complications in "plumbing," and it is usual to find that the vapor-compression system has a higher index of performance (COP) than the absorption system.

Figure 10.21 shows a simple absorption system in which the pump compressor of the vapor-compression system (shown dashed for reference on Figure 10.21) has been replaced by an absorber, a pump, and a vapor generator. The rest of the cycle is the same as the simple vapor-compression cycle.

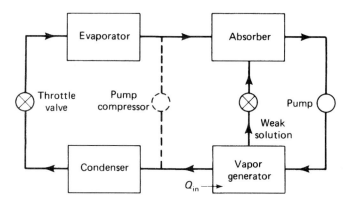

FIGURE 10.21 Simple absorption refrigeration cycle.

The most common absorption refrigeration cycle is one using ammonia as the refrigerant. In terms of the cycle shown in Figure 10.21, the ammonia solution in the vapor generator is heated to create ammonia vapor. The ammonia vapor liberated in the vapor generator proceeds along the indicated path to the condenser and then goes through the conventional part of the cycle. After leaving the evaporator, the ammonia enters the absorber. The weak solution in the vapor generator is mixed with the ammonia in the absorber, where the weak solution absorbs the ammonia and the resulting strong solution is pumped to the generator. In this cycle, the work of the circulating pump is very small for a given refrigerating effect, because the pump is pumping a liquid that has a small specific volume. To reduce the steam and cooling water requirements of the simple absorption system, a regenerative heat exchanger is placed between the absorber and generator as shown in Figure 10.22. The purpose of this heat exchanger is to transfer heat from the hot, weak solution to the cool, strong solution. A typical commercial absorption system is shown in Figure 10.23.

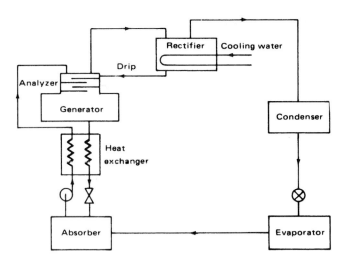

FIGURE 10.22 Simple regenerative absorption refrigeration cycle.
[From W. F. Stoecker, Refrigeration and Air Conditioning
(New York: McGraw-Hill Book Co., Inc., 1958), with permission.]

A domestic absorption-type refrigerator is shown (simplified) in Figure 10.24. Three fluids are used in this unit: ammonia as the refrigerant, water as the absorber, and hydrogen to maintain the total system pressure. In the generator, heat from a gas flame drives off the ammonia vapor, and the vapor carrying entrained liquid droplets flows to the separator. The separator serves to separate the liquid and vapor, with the vapor going to the condenser while the liquid drains into the absorber. The vapor is condensed in the condenser, and the condensate flows to the evaporator, where it receives heat and vaporizes.

In the absorber, the weak solution from the evaporator absorbs water vapor from the evaporator, and the mixture returns by gravity to the generator where the cycle recommences. Circulation within the system takes place due to the vapor-lift pump in which the density of the ammonia–ammonia vapor mixture is less than the density of the liquid in the absorber–generator lines. In the evaporator and absorber, hydrogen maintains the total pressure equal to the pressure of the ammonia and water in the condenser and generator. In the evaporator, the ammonia is evaporated at a low temperature due to its low vapor pressure, while in the condenser, where there is no hydrogen, condensation occurs at a temperature high enough to reject heat to the atmosphere. U-bend traps after the separator and condenser are used to maintain the seals that prevent hydrogen escape from the low side of the system.

Figure 10.25 shows an actual cycle in which several refinements have been made to the simple cycle of Figure 10.24. A liquid heat exchanger is used for the weak solution going to the absorber and the strong solution going to the generator. The analyzer and rectifier are added to remove the water vapor that may have formed in the generator. Thus, only ammonia vapor goes to the condenser. The condenser and evaporator each consist of two sections, permitting the condenser to extend below the top of the evaporator to segregate the freezing portion of the evaporator and to provide additional surface. A reserve hydrogen vessel has been added to give constant efficiency under variable room temperatures.

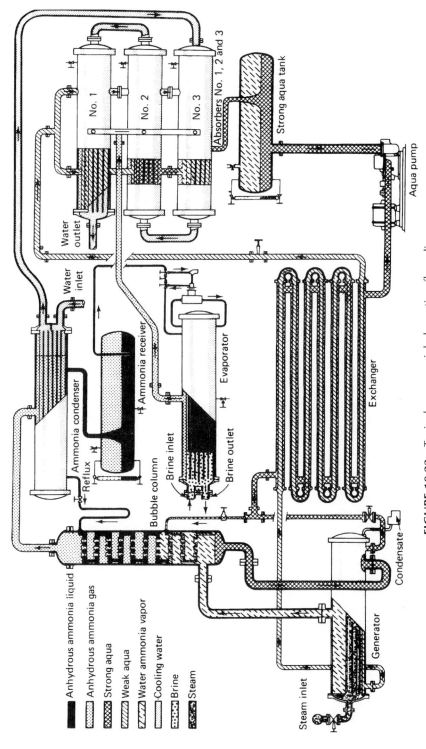

FIGURE 10.23 Typical commercial absorption flow diagram.

[From R. C. Jordan and G. B. Priester, Refrigeration and Air Conditioning, 2nd ed. (Englewood Cliffs, N.J.: Prentice-Hall, Inc., 1956), p. 372, with permission.]

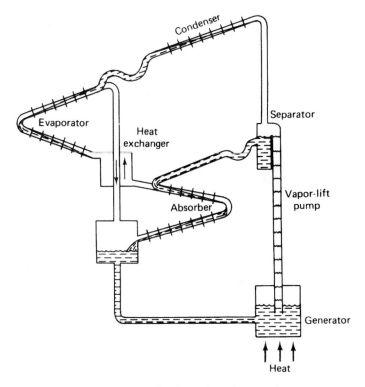

FIGURE 10.24 Simple Electrolux absorption system.
[From W. F. Stoecker, Refrigeration and Air Conditioning
(New York: McGraw-Hill Book Co., Inc., 1958), p. 178, with permission.]

Another development uses lithium bromide salt and water, with the water acting as the refrigerant. The boiling point of the salt, a solid at ordinary temperatures, is so high that it behaves like a nonvolatile substance. Thus, there is no vaporization of the absorbent in the generator and no carryover of the absorbent vapor to the condenser. The characteristics of the absorption cycle limit its usefulness to places where sources of heat and cooling water are plentiful and cheap.

Figure 10.26 shows an absorption refrigeration machine that uses steam or hot liquid with lithium bromide absorbent. This unit is typical of modern design, which has eliminated the earlier collection of tanks, pipes, and coils. Its condenser and generator are in a single vessel, and the evaporator and absorber are in the other vessel. Both are supported by the same set of legs, with solution and evaporator pumps underneath them.

10.4d Vacuum Refrigeration Cycle

To understand the physical processes involved in this refrigeration cycle, let us start with a simple experiment. Take a flask half-filled with water, say, at 70°F, and place it under a bell jar that can be evacuated. The partial pressure of the water vapor in the air in the flask will correspond to its initial temperature. As the pressure in the bell jar is lowered, no change can be seen in the flask. For all practical purposes, only the air is evacuated, and the water stays essentially isothermal. As

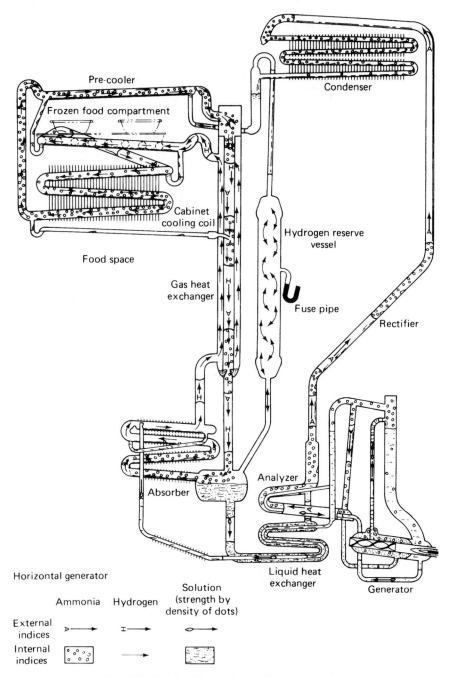

Pre-cooler

Frozen food compartment

Cabinet cooling coil

Food space

Gas heat exchanger

Condenser

Hydrogen reserve vessel

Fuse pipe

Rectifier

Analyzer

Absorber

Liquid heat exchanger

Generator

Horizontal generator

Ammonia   Hydrogen   Solution (strength by density of dots)

External indices

Internal indices

FIGURE 10.25   Absorption refrigerator cycle.

[From R. C. Jordan and G. B. Priester, Refrigeration and Air Conditioning, *2nd ed. (Englewood Cliffs, N.J.: Prentice-Hall, Inc., 1956), p. 379, with permission.]*

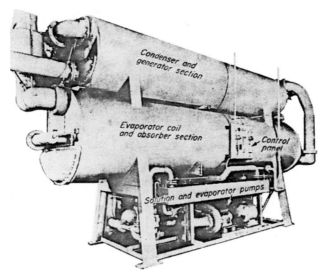

FIGURE 10.26 Absorption refrigeration unit.
(Reprinted with permission from Power, The Engineer's Reference Library,
copyright by McGraw-Hill, Inc., New York.)

the bell jar pressure reaches the saturation pressure (approximately 19 mm Hg absolute), the water will appear to "boil" violently. Actually, the violent agitation of the water is due to the air that was absorbed in the water *outgassing*. Once this violent motion has subsided, the continued pumping will lower the pressure below the initial saturation value. As this occurs, water vapor will be drawn off the surface of the liquid, and the temperature of the remaining fluid will be lowered until it corresponds to the pressure existing in the bell jar. The lowering of the temperature is caused by the decrease in internal energy of the remaining liquid necessitated by the energy required to vaporize some of the liquid at the lower pressure. During the period from the time the outgassing ceases until approximately 4.5 mm Hg absolute is reached, the water will be still and no visual change will be observed. At approximately 4.5 mm Hg absolute, the surface of the water will quite suddenly become coated with a layer of ice. Thus, by lowering the pressure, it has been possible to lower the temperature of the liquid below its surroundings. In this case, a change of phase from liquid to solid has been achieved.

A schematic of a steam-jet refrigeration system in which water is used as the refrigerant is shown in Figure 10.27. In this system, a steam ejector is used to establish and maintain the required vacuum, which causes a portion of the water to vaporize and cool the rest of the remaining water. The steam-jet ejector is used as the vacuum pump, drawing off both vaporized water and any entrained air and compressing them to a higher pressure. The compressed vapor and steam are then condensed. This condenser requires considerably more cooling water than the condenser for a conventional mechanical compression system, because it must remove the heat of the power steam as well as that liberated from the chilled water.

When the steam-jet ejector system is used, it is common to utilize low-pressure plant steam for reasons of economy. It is necessary to provide a condenser on the steam jet as well as an air

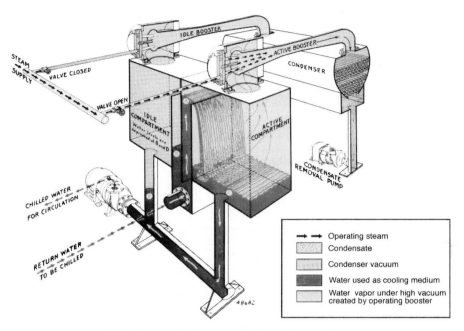

FIGURE 10.27 Steam-jet refrigeration system.

ejector to take care of outgassed air and any air that may leak into the subatmospheric parts of the system. The operation of the steam-jet ejector system is simple and does not require moving parts, but the jet is a poor compressor and requires large amounts of steam.

We can analyze the refrigeration process in the simple *vacuum refrigeration* system by referring to Figure 10.28. Because this is a steady-flow system, water is continuously supplied and with-

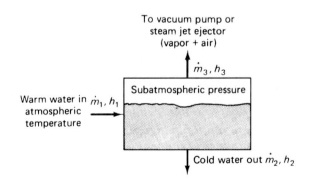

FIGURE 10.28 Analysis of vacuum refrigeration system.

drawn with the difference in the mass of water vapor going to the pumping system. Writing the mass balance for this system yields

$$\dot{m}_1 = \dot{m}_2 + \dot{m}_3 \qquad\qquad (10.16)$$

and

$$\dot{m}_1 h_1 = \dot{m}_2 h_2 + \dot{m}_3 h_3 \qquad\qquad (10.17)$$

The refrigerating effect (or heat removed from the system) can be obtained from the vapor as $\dot{m}_3(h_3 - h_1)$ or from the liquid as $\dot{m}_2(h_1 - h_2)$.

ILLUSTRATIVE PROBLEM 10.13

A vacuum refrigeration system is used to cool water from 90°F to 45°F. Calculate the mass of water vapor that must be removed per pound of entering water.

SOLUTION

Refer to Figure 10.29 and Equations (10.16) and (10.17). Assume that 1 lb_m of liquid enters the system and that the vapor leaves as saturated vapor at 45°F.

$$\dot{m}_1 h_1 = \dot{m}_2 h_2 + \dot{m}_3 h_3$$

$$1 \times 58.07 = (1 - \dot{m}_3)13.04 + \dot{m}_3 \times 1081.1$$

$$\dot{m}_3 = 0.04216 \ lb_m \ \text{vapor}/lb_m \ \text{incoming water}$$

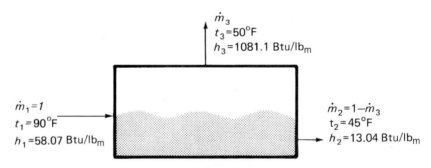

FIGURE 10.29 Illustrative Problem 10.13.

ILLUSTRATIVE PROBLEM 10.14

Determine the refrigeration effect in Illustrative Problem 10.13.

10.4e Thermoelectric Refrigerator

In 1822, Thomas J. Seebeck found that an electromotive force is generated between the junctions of dissimilar metals at different temperatures. This is the basis of the thermocouple as a temperature-measuring device (discussed in some detail in Chapter 1). For metals, the electromotive force is of the order of microvolts per degree of temperature difference, but recent advances in semiconductor materials have produced materials having much larger voltage outputs per degree of temperature difference. While such systems are relatively inefficient direct converters of heat to electricity, they are reliable and have no moving parts. Figure 10.30a shows a schematic of a thermoelectric converter.

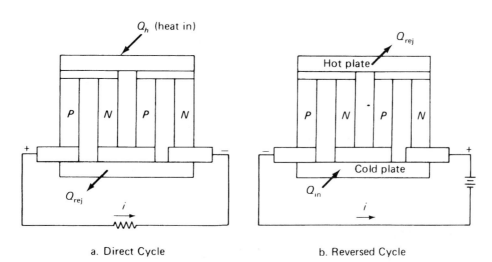

a. Direct Cycle   b. Reversed Cycle

FIGURE 10.30 Thermoelectric converters.

In 1834, Jean Peltier found that when a current flows around a loop of two dissimilar metals, one end warms up while the other end cools. In effect, the thermoelectric converter has been reversed in a manner similar to the reversal of a power cycle to obtain a refrigeration cycle. Thus, if one junction (the hot one) is kept in the ambient surroundings while the other (the cold one) is kept in the refrigerated space and a potential is applied, as shown schematically in Figure 10.30b, heat will be transferred from the space to be refrigerated to the cold junction. The hot junction will reject heat to the surroundings, because its temperature will rise above the ambient temperature.

At present, the thermoelectric refrigerator cannot compete with the more conventional refrigeration systems. However, in such space-limited applications as the removal of heat from electronic devices and assemblies, the thermoelectric refrigerator is in commercial use. More commercial applications will become economical as this refrigeration effect is developed further.

10.5 THE HEAT PUMP

Thus far, we have directed our attention to the heat removed from the region to be cooled and denoted this as the refrigeration effect. If instead we consider the energy rejected as heat from the refrigeration cycle, we have a device that has been called the *heat pump.* To understand the basic principles of operation of this device, let us once again consider the reversed Carnot cycle shown in Figures 10.3 (repeated) and 10.4 (repeated). During the process of cooling a given space, energy (heat) is rejected at some higher temperature. As can be seen from Figure 10.3 (repeated), the heat rejected is equal to the heat removed from the cold region plus the work into the system. For the heat to be rejected, say, as in a household refrigerator, the temperature of the fluid in the cycle must be above the temperature of the surroundings. As shown in Figure 10.4 (repeated), the total energy rejected is given by the rectangle $A'BCD'$, which is also equal to $T_1(s_2 - s_1)$. The energy (or work) supplied is equal to $(T_1 - T_2)(s_2 - s_1)$. If we now define the performance index for such a system as the ratio of the energy delivered (the heating effect or total energy rejected) to the net work supplied as the *coefficient of performance,* we have

$$\text{COP}_{\text{Carnot heat pump}} = \frac{T_1(s_2 - s_1)}{(T_1 - T_2)(s_2 - s_1)} = \frac{T_1}{T_1 - T_2} \qquad (10.18)$$

and

$$\text{COP}_{\text{heat pump}} = (1 + \text{COP}_{\text{refrig}})_{\text{Carnot cycle}} \qquad (10.18a)$$

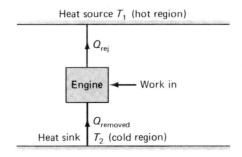

FIGURE 10.3 (*repeated*) Reversed Carnot cycle.

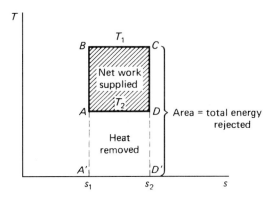

FIGURE 10.4 (*repeated*)   Reversed Carnot cycle.

In an actual machine, the necessity for finite temperature differences to transfer the heat, as well as other irreversibilities, will make the actual performance less than the optimum performance of a heat pump.

ILLUSTRATIVE PROBLEM 10.15

Determine the COP as a heat pump for the cycle described in Illustrative Problem 10.1.

SOLUTION

We can solve this problem by using the cycle temperatures, that is, a rejection temperature of 530°R and a cold temperature of 492°R. Thus,

$$\text{COP}_{\text{Carnot heat pump}} = \frac{T_1}{T_1 - T_2} = \frac{530}{530 - 492} = 13.95$$

The COP can also be obtained from the energy items solved for in Illustrative Problem 10.1. The power was found to be 77.2 Btu/min, and the total rate of heat rejection was 1077.2 Btu/min. Therefore, the COP = 1077.2/77.2 = 13.95, as before for the Carnot heat pump.

ILLUSTRATIVE PROBLEM 10.16

If the cycle described in Illustrative Problems 10.1 and 10.15 has its lower temperature reduced to 0°F, determine the work in, the heat rejected, and the COP as a heat pump.

SOLUTION

Let us first consider the cycle as a refrigeration cycle. The COP as a refrigeration cycle is $(0 + 460)/(530 - 460) = 6.57$. The power input would be $1000/6.57 = 152.2$ Btu/min, and the total rate of heat rejection would be $1000 + 152.2 = 1152.2$ Btu/min. The COP as a heat pump is $1152.2/152.2 = 7.57$. As a check, $6.57 + 1 = 7.57$, Eq. (10.18a).

Note, as a result of Illustrative Problems 10.15 and 10.16, that as the outside temperature decreased from 32°F to 0°F, the COP as a heat pump was almost halved and that the energy delivered to the higher temperature increased by 7% while the required work into the system doubled. Thus, when the demand is the greatest as the outside temperature drops, the ability of a given heat pump to meet this demand decreases. In northern latitudes, the heat pump is found to be uneconomical if it is used solely for heating, but it can and is used in southern latitudes where there are moderate heating loads and relatively long cooling needs. In such an installation, the same equipment is used for both heating and cooling, and whenever additional heating requirements are needed, electrical resistance elements in the air ducts are used to make up this additional heat requirement. Figure 10.31 shows the operation of the heat pump used for summer cooling and winter heating. Suitable valving is necessary to reverse the path of the refrigerant and to operate the system properly.

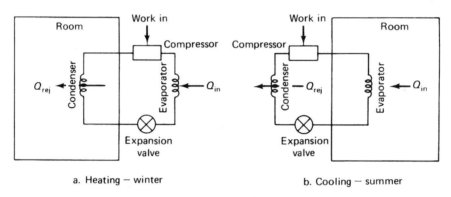

FIGURE 10.31 All-year heat pump operation.

The method by which a heat pump can be switched to operate in either a cooling or heating mode is illustrated in Figure 10.32. In the heating mode (top schematic), an expansion device drops the pressure and temperature of the refrigerant so that it may pass through the outside heat exchanger, which functions in this mode as an evaporator. There, the fluid's temperature is sufficiently low for it to pick up heat from the outside air and vaporize. The vapor is then passed through the compressor, where its temperature and pressure increase. Next, the refrigerant passes through the indoor heat exchanger, where it gives up its heat to the cooler room air and condenses. It then passes through the expansion valve to repeat the cycle.

In the cooling mode, activating the reversing valve and the appropriate expansion valve bypass reverses the flow of the refrigerant so that it completes the same cycle but with the outside heat-exchanger functioning as the condenser (heat given up to the outside air) and the inside heat exchanger functioning as the evaporator (picking up heat from the room air). Switching is accomplished using an interior thermostat.

In Figure 10.31, the atmosphere is indicated to be the energy source when the heat pump is operated on the heating cycle. Unfortunately, atmospheric air is a variable temperature fluid, and as noted earlier, as the temperature of the air decreases, requiring a greater heat output, the heat pump is less able to meet this demand. To overcome this problem, well water, and the earth itself, have been used rather than atmospheric air. If well water is used, it is necessary that the supply be

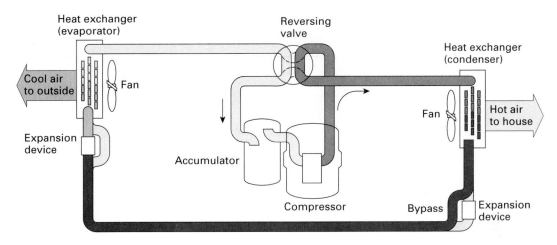

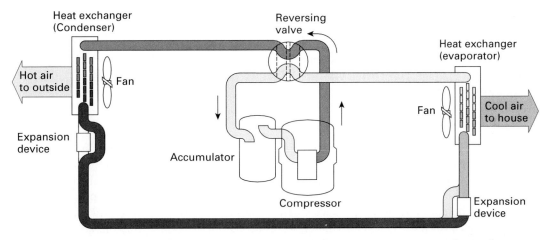

FIGURE 10.32 Basic heat-pump operation, air-to-air heat pump. (a) Heating. (b) Cooling. *(From EPRI,* Heat Pump Manual, *2nd ed., © 1997, p. 14, with permission.)*

adequate and that the water be returned underground to prevent the local water table from lowering. When the heated water is returned, it should be at some distance from where it is initially taken from the ground to prevent heating of the water source. If coils are buried in the earth, it is necessary to ensure good thermal contact between the coil and the earth. Figure 10.33 shows a packaged heat pump that is purchased as a complete unit; it is only necessary to connect the unit to a power source and add appropriate ducting to make it operative.

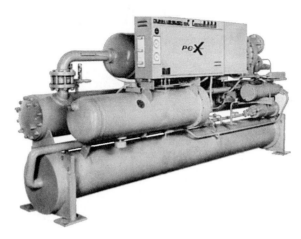

FIGURE 10.33 Packaged heat pump.
(Courtesy of Dunham-Bush, Inc.)

The application of the heat pump to a specific installation or for a specific use is an economic problem. In areas where low-cost electricity is available, the heat pump finds applications in the heating and cooling of both homes and industrial buildings. Also, industrial processes often require either heating or cooling, and the heat pump could be used for specific applications. One such application is where cooling of a process fluid is required, and the heat rejected by the heat pump could be used to heat the plant or office space. The heat pump should not be used without a prior, thorough investigation of the economics of the proposed installation being made after the engineering feasibility has been established.

10.6 REVIEW

Most of the elements in this chapter were studied earlier. However, instead of concentrating on the power-producing elements of the various cycles, our interest has been concentrated on keeping an area either cool or warm. Where we wished to keep an area cool, we called the system refrigeration, and where we wished to keep an area warm, we called the system a heat pump. For each of these desired effects, we were able to define a figure of merit called the coefficient of performance (COP). The cycles studied were the Carnot, Rankine, Brayton, absorption, and vacuum refrigeration. In addition to studying the cycles, we also looked at the refrigerants and the practical equipment used in them. Refrigeration is an accepted part of our daily lives and the lives of most of the people in the industrialized countries of the world. Its use for food preservation, comfort, and process control has become an indispensable part of our technology. The use of the heat pump, especially when it is combined with air conditioning, is also becoming more pervasive as time goes on.

In addition to the analytical tools that were already developed, we introduced the pressure–enthalpy diagram in this chapter. This is really a modified Mollier diagram that is very useful for the analysis of refrigeration cycles. For the interested student, it is suggested that the publications of the American Society of Heating, Refrigeration, and Air Conditioning Engineers (ASHRE) be consulted for further details and information on specific system variables.

KEY TERMS

Terms used for the first time in this chapter are:

absorption refrigeration the use of a heat source to replace the compressor in a vapor-compression system.

bootstrap system a Brayton air-refrigeration cycle with heat exchangers provided with cooling air by the ram pressure of the aircraft forcing outside air through the exchangers.

coefficient of performance (COP) (for heat pump) the ratio of the energy delivered to the work input.

coefficient of performance (COP) (for refrigeration) the ratio of the refrigeration effect to the work input.

dry compression compression of vapor only in a vapor-compression cycle.

heat pump a reversed power cycle whose object is to keep an area warm.

refrigerant the working fluid in a refrigeration cycle.

refrigerating effect the removal of heat from an area by the use of mechanical energy.

thermoelectric refrigerator the use of the reversed Seebeck–Peltier effects to keep a space at low temperature.

ton (of refrigeration) 288,000 Btu/day, or 12,000 Btu/hr.

vacuum refrigeration the lowering of the pressure below atmospheric pressure to lower the temperature of the refrigerant below that of the surroundings.

wet compression the compression of wet refrigerant to saturation conditions.

EQUATIONS DEVELOPED IN THIS CHAPTER

coefficient of performance

$$\text{COP} = \frac{\text{refrigeration effect}}{\text{work input}} \tag{10.1}$$

$\text{COP}_{\text{Carnot refrigeration}}$

$$\text{COP}_{\text{Carnot refrigeration}} = \frac{T_2}{T_1 - T_2} \tag{10.2}$$

definition

$$1 \text{ standard commercial ton} = 288,000 \text{ Btu/day} \tag{}$$

rating

$$\frac{\text{horsepower}}{\text{ton of refrigeration}} = \frac{4.717}{\text{COP}} \tag{10.4}$$

rating

$$\frac{\text{kilowatts}}{\text{ton of refrigeration}} = \frac{3.514}{\text{COP}} \tag{10.5}$$

throttling— vapor-compression cycle

$$h_1 = h_2 \tag{10.6}$$

cooling effect— vapor-compression cycle

$$q_{\text{in}} = h_3 - h_2 = h_3 - h_1 \tag{10.7}$$

compression— vapor-compression cycle

$$\text{work} = h_4 - h_3 \tag{10.8}$$

heat removal— vapor-compression cycle

$$q_{\text{rej}} = h_4 - h_1 \tag{10.9}$$

coefficient of performance (vapor-compression refrigeration)	$\text{COP} = \dfrac{h_3 - h_1}{h_4 - h_3}$	(10.11)

horsepower/ton (vapor-compression)	$\text{horsepower/ton} = 4.717\left(\dfrac{h_4 - h_3}{h_3 - h_1}\right)$	(10.12)

$\text{COP}_{\text{Brayton}}$	$\text{COP}_{\text{Brayton}} = \dfrac{\text{heat removed}}{\text{net work required}} = \dfrac{c_p(T_2 - T_1)}{{}_2W_3 - {}_4W_1}$	(10.13)

$\text{COP}_{\text{Brayton}}$	$\text{COP}_{\text{Brayton}} = \dfrac{c_p(T_2 - T_1)}{c_p(T_3 - T_4) - c_p(T_2 - T_1)}$	
	$= \dfrac{1}{\dfrac{T_3 - T_4}{T_2 - T_1} - 1}$	(10.14a)

$\text{COP}_{\text{Brayton}}$	$\text{COP}_{\text{Brayton}} = \dfrac{1}{(p_b/p_a)^{(k-1)/k} - 1}$	(10.14b)

$\text{COP}_{\text{Brayton}}$	$\text{COP}_{\text{Brayton}} = \dfrac{T_2}{T_3 - T_2}$	(10.14c)

mass balance— vacuum refrigeration	$\dot{m}_1 = \dot{m}_2 + \dot{m}_3$	(10.16)

heat balance— vacuum refrigeration	$\dot{m}_1 h_1 = \dot{m}_2 h_2 + \dot{m}_3 h_3$	(10.17)

$\text{COP}_{\text{heat pump}}$	$\text{COP}_{\text{Carnot heat pump}} = \dfrac{T_1}{T_1 - T_2}$	(10.18)

$\text{COP}_{\text{heat pump}}$	$\text{COP}_{\text{heat pump}} = (1 + \text{COP}_{\text{refrig}})_{\text{Carnot cycle}}$	(10.18a)

QUESTIONS

10.1 Why is it possible to reverse a Carnot cycle to obtain a refrigerating effect?

10.2 Is the COP an absolute parameter, or does it depend on the usage of cycle energy quantities?

10.3 List the five common refrigeration cycles that were discussed in this chapter.

10.4 Trace the path of the working fluid on a T–s diagram, indicating each step for each refrigeration cycle.

10.5 List the desirable properties as a refrigerant.

10.6 List some commonly used refrigerants.

10.7 Sketch on a T–s diagram the reversed Brayton cycle.

10.8 How does the bootstrap system differ from the simple Brayton cycle?

10.9 How does regeneration function to increase the efficiency of the simple Brayton cycle?

10.10 How is the pump replaced by an "equivalent heat source" in the absorption refrigeration cycle?

10.11 How does vacuum refrigeration work?

10.12 Thermoelectric refrigeration has been used for space applications. Why is this type of refrigeration used for this application?

10.13 Under what circumstances would you use the heat pump in northern climates?

PROBLEMS

Problems Involving the Carnot Cycle

10.1 A Carnot refrigeration cycle is used to keep a refrigerator at 36°F. During the summer, the room temperature in which the refrigerator is kept reaches 85°F. If 1500 Btu/min must be removed, determine (a) the COP of the cycle, (b) the power required; and (c) the rate of heat rejected to the room.

10.2 A freezer using a Carnot refrigeration cycle maintains its interior at −10°C. In the summer, it rejects heat to a room at 30°C. If 25 kW is removed as heat, determine (a) the COP of the cycle, (b) the power required, and (c) the rate of heat rejected to the room.

10.3 A refrigeration cycle operates between 15°F and 72°F. Determine the maximum COP for this cycle. If the actual COP is 3, how much more horsepower is required per ton of refrigeration by the actual cycle over the minimum possible requirement?

10.4 Derive Equation (10.4).

10.5 Derive Equation (10.5).

10.6 A reversed Carnot cycle operates between temperatures of 32°F and 86°F. Calculate the ratio of the cooling effect to the work input.

10.7 A Carnot refrigeration cycle has a COP of 4. It rejects heat to a room that is at 72°F. Determine the lowest temperature in the cycle.

10.8 A Carnot cycle is used to cool a building. If 20 kW is supplied as power and its COP is 4.2, determine the refrigerating effect in tons and T_1/T_2.

10.9 A Carnot refrigeration cycle takes heat from water at 33°F and discards it to a room at 72°F. Assume that the refrigeration effect is equal to 8 tons of refrigeration. (a) How much heat is discarded to the room? (b) How much power is required? (c) How much heat is removed from the water?

10.10 Determine the horsepower required for a Carnot refrigeration machine if its capacity is rated at 8 tons of refrigeration while operating between 10°F and 110°F.

10.11 If Illustrative Problem 10.6 had been a Carnot cycle operating between the same temperature limits, determine items (a), (b), and (c) required by this problem. Compare the results.

10.12 A reversed Carnot cycle is operated to remove 1000 Btu/hr from the low-temperature region of the cycle, and it rejects 1500 Btu/hr to the high-temperature region of the cycle. Calculate the COP for this application.

10.13 If the purpose of the unit in Problem 10.12 is to supply heat to the upper-temperature region, calculate its COP.

10.14 A Carnot refrigeration cycle operates with a COP of 4.5. The space to be cooled requires 15 MJ/hr to be removed as heat. What is the minimum size motor that would be needed for this application? State your answer in horsepower.

10.15 A refrigerator operating between 0°F and 100°F removes 1000 Btu/min. If the COP is three-fourths of a Carnot refrigerator operating between the same temperatures, determine the power input and the rate of heat rejected by the actual cycle and the Carnot cycle.

10.16 A refrigerator operates between −15°C and +40°C and removes 20 kW of heat from the unit. If the COP is 80% of a Carnot refrigerator operating between the same temperatures, determine the actual power and the rate of heat rejected.

*10.17 A Carnot refrigeration unit is used to cool a space to 40°F by the transfer of heat to the surroundings at 77°F. Assume that the cooling rate (the heat removed) is equivalent to 3 tons of refrigeration. If this same unit using the same power is now used to cool the space to 10°F, estimate the cooling rate required.

Problems Involving Vapor-Compression Cycles

10.18 A dry compression refrigeration cycle using ammonia as the working fluid operates between 80°F and 10°F. Determine the ideal COP for this cycle. Use the ammonia tables in Appendix 3.

10.19 Compare the result obtained in Problem 10.18 to an ideal Carnot cycle operating between the same temperature extremes.

10.20 Solve Problem 10.18 using the p–h diagram for ammonia given in Appendix 3.

10.21 If the compressor efficiency in Problem 10.18 is 80%, determine the COP of the cycle.

10.22 Compute the number of pounds of ammonia that must be circulated per minute in a 15-ton refrigeration system operating between pressures of 15 and 140 psia, with the ammonia entering the expansion valve at saturation temperature and leaving the evaporator as dry saturated vapor.

10.23 If an ideal Carnot refrigeration cycle operates between the temperature limits shown in Problem 10.22, determine the number of pounds of ammonia that would have to be circulated per minute. How does this compare to the results of Problem 10.22?

*10.24 An ammonia vapor compression system operates with a capacity of 100 tons. The condenser pressure is 200 psia, and the evaporator pressure is 34 psia. Determine the amount of ammonia that must be circulated per minute if the ammonia enters the expansion valve at saturation temperature and leaves the evaporator as dry saturated vapor.

10.25 An ammonia refrigeration system operates between a condenser temperature of 80°F and an evaporator temperature of −10°F. If saturated liquid leaves the condenser and dry saturated vapor leaves the evaporator, determine the pounds per minute of ammonia that is circulated in a 50-ton vapor-compression system.

10.26 A vapor-compression system uses Freon-12 as refrigerant. The vapor in the evaporator is at −10°F, and in the condenser, it is at 100°F. When the vapor leaves the evaporator, it is as a dry saturated vapor, and it enters the expansion valve as saturated liquid. Determine the refrigerating effect per pound of Freon circulated.

10.27 A 10-ton vapor-compression refrigeration system uses Freon-12 as refrigerant. The vapor in the evaporator is at −20°F and leaves as dry saturated vapor. The high-side pressure is 180 psia, with Freon-12 entering the expansion valve as a saturated liquid. Compute the Freon-12 flow.

10.28 Assuming the cycle in Problem 10.27 to be an ideal Carnot cycle, determine the Freon-12 flow. Compare your results to Problem 10.27.

*10.29 An ideal vapor-compression refrigeration cycle is used to produce 20 tons of refrigeration. If the system uses Freon-12 and has a condenser operating at 90°F and an evaporator operating at 38°F, determine the compressor power and the refrigerant flow. The vapor leaves the evaporator as a dry saturated vapor, and it enters the expansion valve as saturated liquid.

10.30 A vapor-compression system uses Freon-12. It is desired to calculate the refrigeration effect when the system is operated with dry saturated vapor leaving the evaporator at −6°F and saturated liquid entering the expansion valve at 90°F.

*10.31 A refrigeration system is operated to produce 40 tons of refrigeration. The evaporator is operated at −30°F, and the condenser is operated at 70°F. Using Freon-12 as the refrigerant, determine the theoretical COP and the weight of refrigerant circulated per minute. Dry saturated vapor leaves the evaporator and saturated liquid enters the expansion valve.

10.32 Solve Problem 10.26 if the working refrigerant is HFC-134a.

10.33 Solve Problem 10.27 if the working refrigerant is HFC-134a.

10.34 Solve Problem 10.28 if the working refrigerant is HFC-134a.

10.35 Solve Problem 10.29 if the working refrigerant is HFC-134a.

10.36 Solve Problem 10.30 if the working refrigerant is HFC-134a.

10.37 Solve Problem 10.31 if the working refrigerant is HFC-134a.

Problems Involving Gas Refrigeration

10.38 An ideal gas refrigeration cycle is operated so that the gas is expanded from 150 to 15 psia. Assuming that the ratio of specific heats (k) is 1.4, and the upper temperature of the cycle is 200°F, determine the COP and the temperature before the compression part of the cycle.

10.39 Solve for the COP in Problem 10.38 if the cycle is an ideal Carnot cycle operating between the high and low temperatures of the gas cycle. Compare this to Problem 10.38.

10.40 An ideal Brayton refrigeration cycle operates between pressures of 15 and 60 psia. Air enters the compressor at 70°F and leaves the turbine at −10°F. If the air flow is 20 lb_m/min, determine the COP and the horsepower required. Use $c_p = 0.24\,\text{Btu/lb}_\text{m}\cdot°\text{R}$ and $k = 1.4$.

10.41 An ideal Brayton refrigeration cycle uses air between pressures of 15 and 90 psia. The compressor inlet temperature is 78°F, and the turbine inlet temperature is 88°F. Determine the COP and the power required. Use $c_p = 0.24\,\text{Btu/lb}_\text{m}\cdot°\text{R}$ and $k = 1.4$.

10.42 If the upper temperature of a Brayton refrigeration cycle is 180°F when the lowest temperature is at −80°F and the COP is 2.7, determine the airflow per ton of refrigeration. Use $c_p = 0.24\,\text{Btu/lb}_\text{m}\cdot°\text{R}$.

*10.43 An air turbine is used to produce power and also to cool a space. The air enters the turbine at 200 psia and 140°F and is expanded isentropically to 20 psia. If the cooled space is kept at −10°F and the turbine output is 100 hp, determine the ideal refrigeration in tons. Assume that $c_p = 0.24\,\text{Btu/lb}_\text{m}\cdot°\text{R}$ and $k = 1.4$.

10.44 An ideal Brayton refrigeration cycle uses air between pressures of 15 and 75 psia. The compressor inlet temperature is 75°F, and the turbine inlet temperature is 95°F. The machine has a rated capacity of 6 tons. Determine the power required. Use $c_p = 0.24\,\text{Btu/lb}_\text{m}\cdot°\text{R}$ and $k = 1.4$.

10.45 An ideal Brayton refrigeration cycle uses air. The pressure ratio of the compressor is 5. At the compressor inlet, the pressure is 100 kPa and the temperature is 35°C. At the outlet of the turbine, the temperature is 0°C. Determine the work in the compressor, the work in the turbine, and the COP. Use $c_p = 1.0061\,\text{kJ/kg}\cdot°\text{K}$ and $k = 1.4$.

Problems Involving Vacuum Refrigeration

*10.46 A vacuum refrigeration system cools incoming water at 80°F to 50°F. Determine the mass of water vapor that must be removed per pound of entering water.

*10.47 A vacuum refrigeration system cools incoming water from 30°C ($h_f = 125.79$ and $h_g = 2556.3\,\text{kJ/kg}$) to 10°C ($h_f = 42.01$ and $h_g = 2519.8\,\text{kJ/kg}$). Determine the mass of water vapor that must be removed per kilogram of entering water.

10.48 A vacuum refrigeration cycle is operated at 0.12166 psia in the vacuum chamber. If the incoming water is at 70°F, determine per pound of incoming water (a) pounds of cooled water, (b) pounds of vapor drawn off, (c) and refrigerating effect.

*10.49 Solve Problem 10.48 if there is a heat leak into the system equal to 10 Btu/lb_m of incoming water.

*10.50 A vacuum refrigeration cycle is operated at 0.0010724 MPa in the vacuum chamber ($h_f = 33.60$ and $h_g = 2516.1$ kJ/kg). If the incoming water is at 20°C ($h_f = 83.96$ and $h_g = 2538.1$ kJ/kg) determine per kilogram of incoming water (a) kg of cooled water, (b) kg of vapor drawn off, and (c) refrigerating effect.

*10.51 Solve Problem 10.50 if there is a heat leak into the system equal to 6 W·h/kg of incoming water.

Problems Involving Heat Pumps

10.52 A Carnot cycle is used for both heating and cooling. If 5 hp is supplied as work and the COP for cooling is 4.0, determine the refrigerating effect and T_1/T_2.

10.53 A Carnot heat pump operates to keep a space cool by extracting 400 Btu/min from the space. The compressor required for this application uses 4 hp. Calculate the COP for this application.

10.54 If the purpose of the cycle in Problem 10.53 is to heat another space, calculate its COP.

10.55 If the cycle in Problem 10.10 was rated for 8 tons of equivalent heating when operating as a heat pump, determine the power required.

10.56 The device described in Illustrative Problem 10.3 is operated as a heat pump. Determine the maximum COP and the power input for the continuous supply of 1 ton of refrigeration.

10.57 A Carnot heat pump is required to maintain a house at 20°C when the outside temperature is at −10°C. Calculate the minimum horsepower required if the house loses 120 MJ/hr.

10.58 A Carnot heat pump is used to maintain a house at 68°F when the outside temperature is 0°F. If 80,000 Btu/hr is lost from the house, determine the horsepower required by the motor in the ideal cycle.

*10.59 Derive the expression

$$\text{COP}_{\text{heat pump}} = (1 + \text{COP}_{\text{refrig}})_{\text{Carnot cycle}}$$

10.60 A reversed Carnot cycle is used for building heating. Assuming that 25,000 Btu/min is required to keep the building at 72°F when the outside air temperature is 40°F, determine the power required and the COP for the cycle as a heat pump.

10.61 A Carnot heat pump is used to keep a house at 70°F when the outside temperature is −10°F. If it is determined that the house requires 1.5 MJ/min in order to maintain these conditions, determine the least horsepower motor required.

10.62 A reversed Carnot cycle heats a building to a constant inside temperature of 22°C when the outside air temperature is 5°C. If 500 kW is required by the building, determine the power (kW) required and the COP for this cycle used as a heat pump.

10.63 A plant is proposed in which a heat pump will provide the heat in the winter. It is estimated that the peak heating load will correspond to 50,000 Btu/hr. A well will be installed which can supply water with a constant temperature of 45°F. For comfort it is decided that the heated air will not exceed 85°F. If the cycle is operated as a Carnot heat pump, determine (a) the Carnot heat pump COP, (b) the compressor power, and (c) if an actual cycle achieves 80% of the Carnot COP, the compressor power.

*10.64 Two Carnot heat pumps operate in series, with the output of the first pump being the input of the second pump. If the lowest temperature in the cycle is 400°K and the highest temperature is 1100°K, determine the work input for each pump. Assume that both pumps have the same COP and that 500 kJ is removed from the lowest temperature sink.

11

Heat Transfer

Learning Goals

After reading and studying the material in this chapter, you should be able to:

1. Understand that there are three mechanisms of heat transfer—conduction, convection, and radiation—and that they can occur individually or in combination with each other.

2. Derive Fourier's law of heat conduction from some simple experiments.

3. Appreciate the fact that the thermal conductivity of solids is greater than the thermal conductivity of liquids and that, in turn, the thermal conductivity of liquids is greater than the thermal conductivity of gases.

4. Use the analogy that exists between Ohm's law and Fourier's law to solve conduction problems of thermal resistances in series and in parallel.

5. Calculate the heat transfer by conduction in cylinders.

6. Understand the need to allow for the problem of thermal contact resistance when two materials are brought into physical contact and heat flows across their interface.

7. Realize that convection heat transfer is due to the motion of a fluid relative to a body.

8. Use Newton's law of cooling to calculate the convection heat transfer; also, by comparing Fourier's equation with Newton's law of cooling, be able to write the resistance term for convection heat transfer involving the heat-transfer coefficient.

9. Understand why convection heat transfer is broken into two parts; natural convection, and forced convection.

10. Use the dimensionless Nusselt, Prandtl, and Grashof numbers to calculate natural convection.

11. Use the simplified forms for the calculation of natural convection for gases at atmospheric pressure.

12. Understand why we limit ourselves to the calculation of forced convection in turbulent flow inside tubes.

13. Realize that radiation heat transfer is a form of electromagnetic radiation and follows laws that are different from those controlling conduction and convection heat transfer.

14. Use the Stefan–Boltzmann law to calculate radiation heat transfer.

15. Apply the earlier portions of the chapter to calculate the performance of heat exchangers.

16. Understand that the temperatures of the fluids in a heat exchanger continuously change as the fluids go from one end of the unit to the other, leading to the concept of the logarithmic mean temperature difference.

17. Estimate the effect of fouling on the performance of heat exchangers.

18. Solve problems involving combined modes of heat transfer.

11.1 INTRODUCTION

In the earlier chapters of this book, we considered heat to be energy in transition, and there has been only one reference to the mechanisms that must exist for this type of energy transport to occur. The limitation that was found was from the second law of thermodynamics in the Clausius statement: "Heat cannot, of itself, pass from a lower temperature to a higher temperature." Thus, for heat transfer to occur, we can state that a temperature difference must exist between the bodies, with heat flowing from the body at the higher temperature to the body at the lower temperature. For the purposes of this book, we look at only steady-state heat transfer in the system being considered. The condition for steady state is that the temperatures in the system be independent of time, and as a consequence, the rate of heat transfer out of the system must equal the rate of heat transfer to the system.

The mechanisms of heat transfer that we consider in this chapter are *conduction, convection,* and *radiation*. In any industrial application, it is possible to encounter more than one of these mechanisms occurring simultaneously, and it is necessary to consider them in combination when designing or analyzing heat-transfer equipment. The formal categorization of heat transfer into separate and distinct mechanisms is a somewhat arbitrary but useful approach to complex technical problems. The usefulness of this approach will become apparent as our study progresses.

Figure 11.1 shows a modern steam generator in which specialized heat-transfer surfaces are used to achieve high efficiencies. Waterwalls within the unit receive their heat by radiation from the combustion of the fuel. This heat is conducted through the steel tubes composing the waterwalls and thence transferred to the water–steam mixture in the tubes. The superheater can use both radiation or convection heat transfer from the hot furnace gases to superheat the steam in the tubes. Thus, in the convection superheater, the flow of hot gas over the tubes causes heat to be transferred to the tubes, where it is then conducted through the metal wall and eventually is exchanged to the steam within the tubes. In the convection section, heat is transferred by the somewhat cooler gas to the tubing to vaporize water in the tubes. Additional convection heat-transfer surface is provided in the economizer and air heater to extract more heat from the now-cooler combustion gases to heat the entering feedwater and the incoming combustion air.

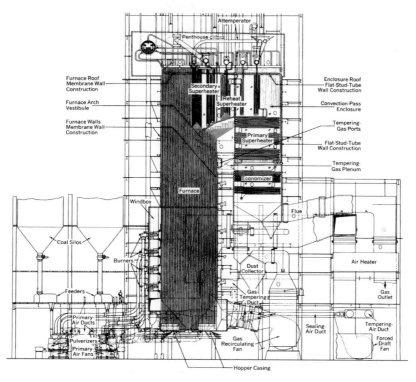

FIGURE 11.1 Modern steam generator. Note the size of the operator in the lower-left portion relative to the size of the unit. *(Courtesy of Babcock and Wilcox Co.)*

11.2 CONDUCTION

Consider the experiment shown in Figure 11.2, which consists of a uniform bar of cross-sectional area A perfectly insulated on all sides except at the ends; that is, heat can flow only in the x direction. When one end of the bar is maintained at the t_1 and the other at t_2, $\dot{Q}$ Btu/hr will be transferred steadily from station ① to station ②. If the cross-sectional area of the bar is doubled while all other conditions are kept constant, it will be found that $2\dot{Q}$ is now being transferred. In other words, the rate at which heat is being transferred is directly proportional to the cross-sectional area

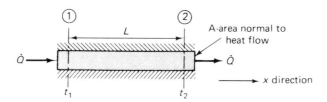

FIGURE 11.2 Conduction heat transfer.

of the bar normal to the direction of heat flow. Returning to the original bar, we now make the temperature difference $(t_1 - t_2)$ twice its original value, and again, we find the rate of heat transfer to be $2\dot{Q}$. Therefore, we conclude that the heat-transfer rate is directly proportional to the temperature difference between the ends of the bar. Finally, returning to the original conditions, we now make the bar twice as long $(2L)$, and this time, we find only half the amount of heat is being transferred, leading to the conclusion that the rate of heat transfer is inversely proportional to the length of the bar. If we combine all these events into a mathematical statement, we have

$$\dot{Q} \propto \frac{A(t_1 - t_2)}{L} \tag{11.1}$$

or

$$\dot{Q} = \frac{-kA(t_2 - t_1)}{L} \tag{11.2}$$

where the proportionality constant, k, is a property of the material called the thermal conductivity. The negative sign has been included in Equation (11.2) to indicate a positive heat flow in the increasing x direction, which is the direction of decreasing temperature. The conductivity, k, is usually found to be a function of temperature, but for moderate temperatures and temperature differences, it can be taken to be a constant. If we now rewrite Equation (11.2) in more general terms for one-dimensional conduction, we have

$$\dot{Q} = \frac{-kA \, \Delta t}{\Delta x} \tag{11.3}$$

Equation (11.3) is called *Fourier's law* of heat conduction in one dimension in honor of the noted French physicist, Joseph Fourier. In this equation, the heat transfer rate $\dot{Q}$ is expressed in usual English engineering units as Btu/hr, the normal area A is expressed in square feet, the temperature difference Δt is in °F, and the length Δx is in feet, giving us k in units of Btu/(hr) (ft^2) (°F/ft). This unit of k is often written as Btu/(hr·ft·°F). The nomenclature of Equation (11.3) is shown in Figure 11.3. Table 11.1 gives the thermal conductivities of some solids at room temperature, and Figure 11.4 shows the variation of thermal conductivity as a function of temperature for many ma-

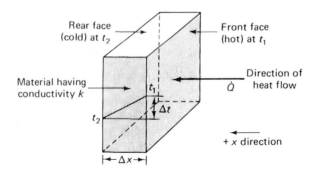

FIGURE 11.3   Nomenclature for Equation (11.3).

TABLE 11.1

Thermal Conductivities of Solids at Temperatures near 100°F

Material	Conductivity[a]	
	Btu/hr·ft·°F	W/m·°C
Cotton wool	0.01	0.017
Corkboard	0.025	0.043
Mineral wool	0.026	0.045
Balsa	0.040	0.069
Asbestos fiber	0.044	0.076
White pine	0.065	0.112
Fir	0.090	0.156
Gypsum plaster	0.30	0.519
Common brick	0.40	0.692
Concrete (average house construction)	0.80	1.385
Porcelain	0.95	1.644
Mild steel	26	45.0
Wrought iron	34.5	59.7
Yellow brass	52	90.0
Aluminum	118	204.2
Copper	220	381.0
Silver	242	419.0

Source: Reprinted with permission from F. P. Durham, *Thermodynamics,* 2nd ed. (Englewood Cliffs, N.J.: Prentice-Hall, Inc., 1959), p. 278.
[a] 1 Btu/hr·ft·°F = 1.7307 W/m·°C.

terials. A table of the thermal conductivity of some building and insulating equipment is given in Appendix 3, Table A.6. In SI units, k is W/m·°C.

The foregoing development was based on observable events in a hypothetical experiment. The conduction of heat can also be visualized as occurring as the transfer of energy by more active molecules at a higher temperature colliding with less active molecules at a lower temperature. Gases have longer molecular spacings than liquids and consistently exhibit much lower thermal conductivities than liquids. Due to the complex structure of solids, some have high values of k while others have low values of k. However, for pure crystalline metals, which are good electrical conductors, there is a large number of free electrons in the lattice structure that also makes them good thermal conductors. Table 11.2 gives the order of magnitude of the thermal conductivity k for various classes of materials, which may be used for rough estimates of conduction heat transfer.

ILLUSTRATIVE PROBLEM 11.1

A common brick wall 6 in. thick has one face maintained at 150°F and the other face is maintained at 80°F. Determine the heat transfer per square foot of wall.

(continues on p. 555)

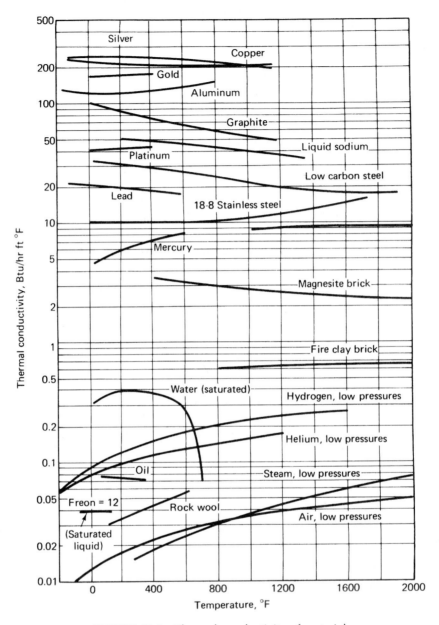

FIGURE 11.4 Thermal conductivity of material.

[Reprinted from W. M. Rohsenow and H. Y. Choi, Heat, Mass and Momentum Transfer (Englewood Cliffs, N.J.: Prentice-Hall, Inc., 1961), p. 508, with permission.]

TABLE 11.2

Order of Magnitude of k for Various Classes of Material

Material	k (Btu/hr·ft·°F)
Gases	0.005–0.02
Insulating materials	0.014–0.10
Wood	0.04–0.10
Liquids (nonmetallic)	0.05–0.40
Brick, concrete, stone, plaster	0.2–2.0
Refractory materials	0.50–10.0
Metals and alloys	10–240

Source: Reprinted with permission from A. I. Brown and S. M. Marco, *Introduction to Heat Transfer,* 3rd ed. (New York: McGraw-Hill Book Co., 1958), p. 24.

SOLUTION

From Table 11.1, the thermal conductivity of common brick near 100°F is 0.40 Btu/hr·ft·°F, and the wall is $\frac{6}{12}$ ft thick. Applying equation (11.3), we obtain

$$\frac{\dot{Q}}{A} = \frac{-k\,\Delta t}{\Delta x} = \frac{-(0.40)\,(80-150)}{6/12} = 56\ \text{Btu/hr·ft}^2$$

ILLUSTRATIVE PROBLEM 11.2

A brick wall 150 mm thick has one face maintained at 30°C while the other face is at 70°C. Determine the heat transfer per unit wall area.

SOLUTION

Using k of 0.692 W/m·°C, we obtain

$$\frac{\dot{Q}}{A} = \frac{-k\,\Delta t}{\Delta x} = \frac{-(0.692)\,(30-70)}{0.150} = 184.5\ \text{W/m}^2$$

For most cases of conduction heat transfer, it is satisfactory to select a value of k at the mean temperature of the process.

The Fourier equation has a direct analogue in *Ohm's law* for electrical circuits. This can be seen by rewriting Equation (11.3) in the following form:

$$\dot{Q} = \frac{\Delta t}{R_t} \tag{11.3a}$$

where $R_t = \Delta x/kA$ and is called the *thermal resistance*. Ohm's law for a direct current (dc) resistance can be expressed as

$$i = \frac{\Delta E}{R_e} \qquad (11.4)$$

where ΔE is the potential difference (in volts), R_e is the electrical resistance (in ohms), and i is the current (in amperes). Comparison of Ohm's law and the Fourier equation shows $\dot{Q}$ to be analogous to i, Δt to correspond to ΔE, and R_t to correspond to R_e. Table 11.3 shows the correspondence between these systems.

TABLE 11.3
Thermal-Electrical Analogy

Quantity	Thermal system	Electrical system
Potential	Temperature difference, °F	Voltage difference, volts
Flow	Heat transfer, Btu/hr	Current, amperes
Resistance	Resistance, hr·°F/Btu	Resistance, ohms

ILLUSTRATIVE PROBLEM 11.3

A simple resistive electrical circuit is to be used to determine the heat transfer through the wall of Illustrative Problem 11.1. If a 9-V battery is available and it is desired to have 1 A correspond to 100 Btu/hr·ft², what resistance is needed?

SOLUTION

Referring to Figure 11.5, we have for the thermal resistance $R_t = \Delta x/kA = (6/12)/0.4 \times 1 = 1.25$ hr·°F/Btu. If we now take the ratio of Equations (11.3a) and (11.4),

$$\frac{\dot{Q}}{i} = \frac{\Delta t/R_t}{\Delta E/R_e}$$

but the problem statement is that $\dot{Q}/i = 100$ (numerically). Therefore,

$$100 = \frac{\Delta t}{R_t} \bigg/ \frac{\Delta E}{R_e}$$

Solving for R_e yields

$$R_e = \frac{100\Delta E R_t}{\Delta t} = \frac{100 \times 9 \times 1.25}{70} = 16.07 \text{ ohms}$$

As a check, $i = \Delta E/R_e = 9/16 = 0.56$ A, which when multiplied by 100 yields $\dot{Q} = 56$ Btu/hr·ft².

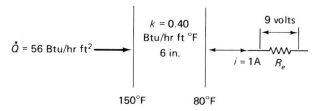

FIGURE 11.5 Illustrative Problem 11.5.

The analogy between the flow of heat and the flow of electricity is very useful in both visualizing and solving heat-transfer problems. The rules applying to electrical circuits can be used to solve thermal circuits that would otherwise be quite formidable. The techniques of solving dc circuits are applicable to solving steady-state thermal conduction problems. As an application of this type of technique, let us consider a wall composed of several thermal resistances in series, as shown in Figure 11.6a, and an analogous electrical circuit, as shown in Figure 11.6b. It will be recalled that for a series dc circuit, the overall resistance (R_{oe}) is the sum of the individual resistances:

$$R_{oe} = R_1 + R_2 + R_3$$ (11.5)

Therefore, by analogy, the overall resistance of the thermal circuit (R_{ot}) is

$$R_{ot} = R_1 + R_2 + R_3 = \frac{\Delta x_1}{k_1 A} + \frac{\Delta x_2}{k_2 A} + \frac{\Delta x_3}{k_3 A}$$ (11.6)

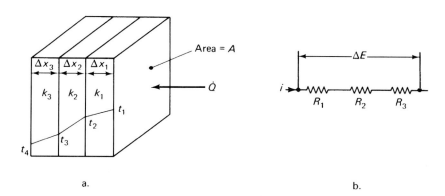

a.

b.

FIGURE 11.6 Thermal resistances in series.

ILLUSTRATIVE PROBLEM 11.4

A composite wall consists of 6 in. of ordinary brick ($k = 0.4$ Btu/hr·ft·°F), $\frac{1}{2}$ in. of concrete ($k = 0.8$ Btu/hr·ft·°F), and $\frac{1}{2}$ in. of plaster ($k = 0.3$ Btu/hr·ft·°F). If the inside of the wall is maintained at 70°F when the outside is 30°F, determine the heat transfer per square foot of wall. The overall situation is shown in Figures 11.6 and 11.7.

SOLUTION

For brick,

$$R = \frac{\Delta x}{kA} = \frac{6/12}{0.4 \times 1} = 1.25 \frac{\text{hr·°F}}{\text{Btu}}$$

For concrete,

$$R = \frac{\Delta x}{kA} = \frac{1/2}{12} \bigg/ (0.8 \times 1) = 0.052 \frac{\text{hr·°F}}{\text{Btu}}$$

For plaster,

$$R = \frac{\Delta x}{kA} = \frac{1/2}{12} \bigg/ (0.3 \times 1) = 0.039 \frac{\text{hr·°F}}{\text{Btu}}$$

The overall resistance $R_{ot} = R_1 + R_2 + R_3 = 1.25 + 0.053 + 0.139 = 1.442$ hr·°F/Btu, and the heat transfer $\dot{Q} = (70 - 30)/1.442 = 27.7$ Btu/hr·ft^2, because the area considered is 1 ft^2.

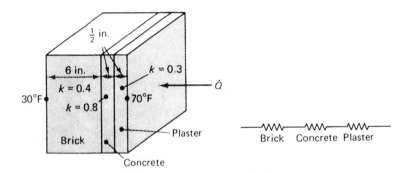

FIGURE 11.7 Illustrative Problem 11.4.

ILLUSTRATIVE PROBLEM 11.5

Calculate the temperature at the brick–concrete and concrete–plaster interfaces for the composite wall of Illustrative Problem 11.4.

SOLUTION

Because each resistance in a series circuit carries the same current (heat flow), it is only necessary to successively apply Ohm's law (Fourier's equation) to each element in turn. Thus,

$$\dot{Q} = \frac{\Delta t}{R} \quad \text{or} \quad \Delta t = R\dot{Q} \quad \text{with} \quad \dot{Q} = 27.8 \frac{\text{Btu}}{\text{hr·ft}^2}$$

from Illustrative Problem 11.3. Therefore, for the brick,

$$\Delta t = (1.25)(27.8) = 34.75°\text{F}$$

For the concrete,

$$\Delta t = (0.052)(27.7) = 1.44°\text{F}$$

For the plaster,

$$\Delta t = (0.139)(27.7) = \frac{3.85°\text{F}}{40.04°\text{F}}$$

Check:

$$\Delta t \,(\text{overall}) = 40.04$$

The interface temperatures are

$$\text{For brick–concrete:} \quad 30 + 34.75 = 64.75°\text{F}$$
$$\text{For concrete–plaster:} \quad 64.75 + 1.44 = 66.19°\text{F}$$

ILLUSTRATIVE PROBLEM 11.6

If the composite wall of Figure 11.7 consists of 150 mm of brick ($k = 0.692$ W/m·°C) 12 mm of concrete ($k = 1.385$ W/m·°C) and plaster ($k = 0.519$ W/m·°C) with wall temperatures of 0°C and 20°C, respectively, determine the heat transfer per square meter of wall.

SOLUTION

Proceeding as in Illustrative Problem 11.4, we have for the brick,

$$R = \frac{\Delta x}{kA} = \frac{0.150}{0.692 \times 1} = 0.217 \frac{°\text{C}}{\text{W}}$$

For the concrete,

$$R = \frac{\Delta x}{kA} = \frac{0.012}{1.385 \times 1} = 0.009 \frac{°\text{C}}{\text{W}}$$

For the plaster,

$$R = \frac{\Delta x}{kA} = \frac{0.012}{0.519 \times 1} = 0.023 \, \frac{°C}{W}$$

The total resistance is the sum of the individual resistances, which equals 0.249°C/W. The heat transfer therefore is $(20 - 0)/0.249 = 80.3 \, W/m^2$ for an area of 1 m².

ILLUSTRATIVE PROBLEM 11.7

Determine the temperature at each of the interfaces in Illustrative Problem 11.6.

SOLUTION

$$\Delta t = R\dot{Q}$$

For brick: $\Delta t = 80.3 \times 0.217 = 17.43°C$

For concrete: $\Delta t = 80.3 \times 0.009 = 0.72°C$

For plaster: $\Delta t = 80.3 \times 0.023 = \underline{1.85°C}$

As a check, overall $\Delta t = 20.00°C$

The temperatures at the interfaces are $(20 - 17.43) = 2.57°C$, $(2.57 - 0.72) = 1.85°C$, and $(1.85 - 1.85) = 0°C$.

The electrical analogy, when applied to the case of a composite plane wall with parallel sections, yields a considerable saving and is a distinct aid in visualizing the problem. Consider the wall shown in Figure 11.8a, which consists of sections side by side. Each wall section has a different area for heat transfer, their conductivities are different, but their thicknesses are equal and the front faces are kept at t_1 while the back faces are kept at t_2. This type of wall is called a parallel wall, and we shall solve it by using the parallel electrical circuit shown in Figure 11.8b. For

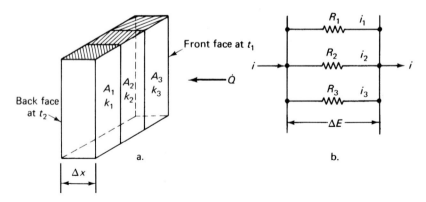

FIGURE 11.8 Parallel circuits.

the parallel electrical circuit, the potential difference is the same for all elements and the total current i is the sum of the branch currents. Thus,

$$i = i_1 + i_2 + i_3 \tag{11.7}$$

By analogy for the thermal circuit,

$$\boxed{\dot{Q} = \dot{Q}_1 + \dot{Q}_2 + \dot{Q}_3} \tag{11.8}$$

The overall resistance of the parallel electrical circuit is

$$R_{oe} = \frac{1}{(1/R_1) + (1/R_2) + (1/R_3)} \tag{11.9}$$

which gives us by analogy for the thermal circuit

$$\boxed{R_{ot} = \frac{1}{\dfrac{1}{\Delta x/k_1 A_1} + \dfrac{1}{\Delta x/k_2 A_2} + \dfrac{1}{\Delta x/k_3 A_3}}} \tag{11.10}$$

ILLUSTRATIVE PROBLEM 11.8

A wall 7 ft high and 6 ft wide is 4 in. thick. Two feet of the wall is made of fir ($k = 0.090$ Btu/hr·ft·°F), 2 ft is made of pine ($k = 0.065$ Btu/hr·ft·°F), and the final 2 ft is made of corkboard ($k = 0.025$ Btu/hr·ft·°F). Determine the heat loss from the wall if one face is maintained at 80°F while the other face is kept at 60°F.

SOLUTION

The problem is essentially the case shown in Figure 11.8. As the first step in the solution, we shall calculate the individual thermal resistances. Thus,

$$R_{fir} = \frac{\Delta x}{kA} = \frac{4/12}{0.090 \times 7 \times 2} = 0.265 \, \frac{\text{hr·°F}}{\text{Btu}}$$

$$\frac{1}{R} = 3.774$$

$$R_{pine} = \frac{\Delta x}{kA} = \frac{4/12}{0.065 \times 7 \times 2} = 0.366 \, \frac{\text{hr·°F}}{\text{Btu}}$$

$$\frac{1}{R} = 2.732$$

$$R_{corkboard} = \frac{\Delta x}{kA} = \frac{4/12}{0.025 \times 7 \times 2} = 0.952 \, \frac{\text{hr·°F}}{\text{Btu}}$$

$$\frac{1}{R} = 1.050$$

$$R_{\text{overall}} = \frac{1}{(1/R_1) + (1/R_2) + (1/R_3)}$$

$$= \frac{1}{3.774 + 2.732 + 1.050}$$

$$= \frac{1}{7.556} = 0.132$$

Therefore,

$$\dot{Q} = \frac{\Delta t}{R} = \frac{80 - 60}{0.132} = 151.5 \frac{\text{Btu}}{\text{hr}}$$

As a check,

$$\dot{Q}_{\text{fir}} = \frac{\Delta t}{R} = \frac{80 - 60}{0.265} = \frac{20}{0.265} = 75.5 \frac{\text{Btu}}{\text{hr}}$$

$$\dot{Q}_{\text{pine}} = \frac{\Delta t}{R} = \frac{80 - 60}{0.366} = \frac{20}{0.366} = 54.6 \frac{\text{Btu}}{\text{hr}}$$

$$\dot{Q}_{\text{corkboard}} = \frac{\Delta t}{R} = \frac{80 - 60}{0.952} = \frac{20}{0.952} = 21.0 \frac{\text{Btu}}{\text{hr}}$$

$$\dot{Q}_{\text{total}} = Q_1 + Q_2 + Q_3$$
$$= 75.5 + 54.6 + 21.0 = 151.1 \text{ Btu/hr}$$

Insulating materials, such as fiberglass batting for homes, are sold commercially with "R-values." These values relate directly to the thermal resistance R of 1 ft^2 of cross-sectional area. For a given material thickness Δx and its k value, the calculated R is what is referred to as the "R-value" for that thickness of insulation.

Thus far, we have been concerned with conduction through plane walls whose heat-transfer area remained constant. The conduction of heat through pipes represents a practical problem of considerable interest in which the heat-transfer area is constantly changing. To solve this problem, let us consider a long, uniform, hollow cylinder whose outer surface is maintained at a temperature t_o and whose inner surface is maintained at a temperature t_i. As is shown in Figure 11.9, we will denote the inner radius to be r_i, the outer radius to be r_o, and the radius at any portion of the cylinder to be r. The length of cylinder in question will be denoted as L. Let us now apply the Fourier equation to a small cylinder of thickness Δr, located at a radius r from the center. The surface area of this cylinder is $2\pi rL$, and its thickness is Δr. Therefore,

$$\dot{Q} = \frac{-kA \, \Delta t}{\Delta x} = \frac{-k2\pi rL \, \Delta t}{\Delta r} \tag{11.11}$$

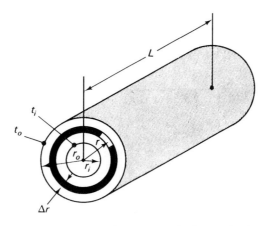

FIGURE 11.9   Conduction in a hollow cylinder.

Rearranging gives

$$\frac{\dot{Q}}{2\pi kL}\left(\frac{\Delta r}{r}\right) = -\Delta t \tag{11.12}$$

Because the heat transfer $\dot{Q}$ must go through all such elements, it is a constant, and we can rewrite Equation (11.12) as

$$\frac{\dot{Q}}{2\pi kL}\sum_{r_i}^{r_o}\frac{\Delta r}{r} = -\sum_{t_i}^{t_o}\Delta t \tag{11.13}$$

where the symbol Σ is used to denote the summation of the variable between the limits indicated. For the right side of the equation, we immediately note that the sum of all the Δt's is simply the overall temperature difference, and with the negative sign, the summation gives us $t_i - t_o$. The summation of $\Delta r/r$ is similar to the summation of $\Delta v/v$ or $\Delta t/t$, which we have already noted yields the natural logarithm (ln) of the argument between its limits. Therefore, Equation (11.13) can be summed to yield

$$t_i - t_o = \frac{\dot{Q}}{2\pi kL}\ln\left(\frac{r_o}{r_i}\right) \tag{11.14a}$$

$$\dot{Q} = \frac{(t_i - t_o)2\pi kL}{\ln(r_o/r_i)}\frac{\text{Btu}}{\text{hr}} \tag{11.14b}$$

and the thermal resistance

$$R_t = \frac{\ln\left(\dfrac{r_o}{r_i}\right)}{2\pi kL} \tag{11.14c}$$

CALCULUS ENRICHMENT

In 1822, Joseph Fourier put forth the fundamental equation governing the conduction of heat in a solid having constant thermal conductivity. For steady-state, one-dimensional heat conduction, this equation is

$$\dot{Q} = -kA\frac{dT}{dx} \tag{a}$$

where dT/dx is the temperature gradient, $\dot{Q}$ is the rate of heat flow, A is the area across which the heat is flowing, and k is a proportionality constant called the thermal conductivity. In the form of Equation (a), it is limited to one-dimensional steady flow, but it can be used to analyze the steady-state heat transfer of three shapes, namely, the plane wall, the hollow cylinder, and the hollow sphere due to the symmetry of these shapes.

For the plane wall with constant thermal conductivity, Equation (a) can be integrated as follows:

$$\int_0^L \dot{Q}\,dx = -\int_{T_h}^{T_c} kA\,dT \tag{b}$$

where T_h is the hot face temperature and T_c is the cold face temperature. Integrating Equation (b) yields

$$\dot{Q}L = kA(T_h - T_c) \tag{c}$$

or

$$\dot{Q} = \frac{kA}{L}(T_h - T_c) \tag{d}$$

Equation (d) is the same as Equation (11.3).

For the cylinder of Figure 11.9, symmetry allows us to use a one-dimensional solution. Rewriting Equation (a) for this condition in cylindrical coordinates gives us

$$\dot{Q} = -kA\frac{dT}{dr} \tag{e}$$

Because $A = 2\pi rL$,

$$\dot{Q} = -k2\pi rL\frac{dT}{dr} \tag{f}$$

Rearranging,

$$\dot{Q}\int_{r_i}^{r_o}\frac{dr}{r} = -2\pi kL\int_{T_i}^{T_o} dT \tag{g}$$

Integrating,

$$\dot{Q} = \frac{2\pi kL}{\ln\dfrac{r_o}{r_i}}(T_i - T_o) \tag{h}$$

Equation (h) is the same as Equation (11.14b).

For the sphere, with inside radius r_i, outside radius r_o, inside temperature T_i, and outside temperature T_o, we can again use Equation (a) by symmetry. Using $A = 4\pi r^2$,

$$\dot{Q} = -k4\pi r^2 \frac{dT}{dr} \tag{i}$$

Rearranging,

$$\dot{Q} \int_{r_i}^{r_o} \frac{dr}{r^2} = -4\pi k \int_{T_i}^{T_o} dT \tag{j}$$

Integrating and simplifying,

$$\dot{Q} = \frac{4\pi k r_o r_i}{(r_o - r_i)} (T_i - T_o) \tag{k}$$

The results that we have just obtained assumed that k was constant. If k varies linearly with temperature, as is often the case over limited temperature intervals, then

$$k = k_0 (1 + bT) \tag{l}$$

If we now define an average value of k to be used over the temperature interval, we find that

$$k_{avg} = k_0 \big[1 + (b/2)\big] (T_h + T_c) \tag{m}$$

However, the term $k_0\big[1 + (b/2)\big]$ is the arithmetic mean value of k in the temperature range T_h to T_c. The same result is obtained whether we consider the plane wall, the cylinder, or the sphere. We therefore conclude that for these cases, the average value of k over the temperature range is the appropriate value to use if k varies linearly with temperature.

ILLUSTRATIVE PROBLEM 11.9

A bare steel pipe having an outside diameter of 3.50 in. and an inside diameter of 3.00 in. is 5 ft long. Determine the heat loss from the pipe if the inside temperature is 240°F and its outside temperature is 120°F. Use a k of 26 Btu/hr·ft·°F for the steel.

SOLUTION

Because the ratio of r_o/r_i is the same as the ratio of the corresponding diameters, we can apply Equation (11.14b) directly:

$$\dot{Q} = \frac{2\pi(26)(5)(240 - 120)}{\ln(3.50/3.00)} = 635{,}858 \frac{\text{Btu}}{\text{hr}}$$

ILLUSTRATIVE PROBLEM 11.10

A bare steel pipe with an outside diameter of 90 mm and an inside diameter of 75 mm is 2 m long. The outside temperature is 40°C, while the inside temperature is 110°C. If k is 45 W/m·°C, determine the heat loss from the pipe.

SOLUTION

Using Equation (11.14b), we obtain

$$\dot{Q} = \frac{2\pi \times 45 \times 2(110 - 40)}{\ln(90/75)} = 217.1 \times 10^3 \, \text{W}$$

The results of Illustrative Problems 11.9 and 11.10 show the extremely large heat loss from a bare pipe. To decrease the loss, it is usual to insulate the pipe with a material having a low thermal conductivity. When this is done, we have a situation of two resistances in series, each having variable heat-transfer areas. Referring to Figure 11.10 and applying Equation (11.14b) to each of the cylinders yields the following: For the inner cylinder,

$$\dot{Q} = \frac{2\pi k_1 L (t_1 - t_2)}{\ln(r_2/r_1)}; \quad R_{ti} = \frac{\ln(r_2/r_1)}{2\pi k_1 L} \tag{11.14d}$$

For the outer cylinder,

$$\dot{Q} = \frac{2\pi k_2 L}{\ln(r_3/r_2)}(t_3 - t_2); \quad R_{to} = \frac{\ln(r_3/r_2)}{2\pi k_2 L} \tag{11.14e}$$

If at this point we apply the electrical analogy, we can write directly that the overall resistance is the sum of the individual resistances. The same result can be obtained by noting that $t_1 - t_3 = (t_1 - t_2) + (t_2 - t_3)$. Using Equations (11.14d) and (11.14e), we obtain $t_1 - t_2 = \dot{Q}R_{ti}$ and $t_2 - t_3 = \dot{Q}R_{to}$. Therefore, $t_1 - t_3 = \dot{Q}(R_{ti} + R_{to})$, or

$$\dot{Q} = \frac{(t_1 - t_3)2\pi L}{(1/k_1)\ln(r_2/r_1) + (1/k_2)\ln(r_3/r_2)} \tag{11.15}$$

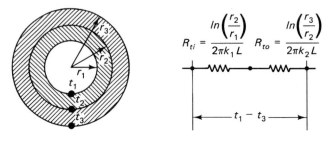

FIGURE 11.10 Compound cylinder.

ILLUSTRATIVE PROBLEM 11.11

If 1 in. of mineral wool ($k = 0.026$ Btu/hr·ft·°F) insulation is added to the pipe of Illustrative Problem 11.9 and the temperature on the outside of this insulation is found to be 85°F, determine the heat loss.

SOLUTION

Using Equation (11.15) and the data of Illustrative Problem 11.9 yields

$$\frac{1}{k_1} \ln \frac{r_2}{r_1} = \frac{1}{k_1} \ln \frac{D_2}{D_1} = \frac{1}{26} \times \ln \frac{3.500}{3.00} = 0.006$$

$$\frac{1}{k_2} \ln \frac{r_3}{r_2} = \frac{1}{k_2} \ln \frac{D_3}{D_2} = \frac{1}{0.026} \ln \frac{5.50}{3.50} = 17.384$$

Therefore,

$$\dot{Q} = \frac{(240 - 85)2\pi(5)}{0.006 + 17.384} = 280.0 \frac{\text{Btu}}{\text{hr}}$$

Two things are evident from the comparison of Illustrative Problems 11.9 through 11.11. The first is the decrease in heat loss of more than a thousandfold by the addition of a relatively small amount of insulation. The second is that the resistance of the steel could have been neglected, because it was only a minute fraction of the resistance of the insulation.

Before leaving the subject of conduction heat transfer, one topic must be mentioned: thermal contact resistance. When two materials are brought into physical contact and heat flows from one to the other across the interface, it is found that there is a discontinuous and sometimes very large temperature drop at the interface plane, as shown in Figure 11.11. This temperature drop is attributed to a thermal contact resistance, which can be thought of as being caused by two parallel heat paths at the interface: solid-to-solid conduction at discrete contact points and conduction through air entrapped at the gas interface. Due to the relatively poor thermal conductivity of air (or other gases) as compared to that of metals, the larger part of the joint thermal contact resistance is due to the entrapped air. The prediction of thermal contact resistance is a complex problem that as yet

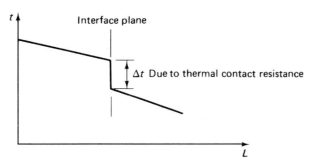

FIGURE 11.11 Thermal contact resistance.

has not been adequately solved, and the student is cautioned that in actual practice, allowance must be made for this effect when designing industrial equipment.

11.3 CONVECTION

Heat transfer by convection from a body involves the motion of a fluid (liquid or gas) relative to the body. If the motion is caused by the differences in density due to the difference in temperature at various locations in the fluid, it is known as natural convection. If the motion of the fluid is caused by an external agent such as a blower or fan, it is called forced convection. Heat transfer from a surface whose temperature is higher than the surrounding fluid occurs in a complex fashion. We can, however, visualize it as occurring in the following sequential order. First, the particles of fluid adjacent to the walls are heated by conduction from the wall, which increases their temperature. These "hot" particles will collide with colder particles, imparting some of their energy to these colder particles. This action will occur due to both particle motion as well as movement of the hotter fluid relative to the colder bulk fluid. To categorize the types of convective heat-transfer mechanisms, it is necessary to briefly discuss the mechanism of flow.

The term *laminar* or (*streamline*) *flow* is applied to a flow regime in which the flow is smooth and the fluid moves in layers or paths parallel to each other. When a fluid moves in laminar flow over a hotter surface, heat is transferred principally by molecular conduction within the fluid from layer to layer. This type of convection heat transfer yields low heat-transfer rates. In contrast to laminar flow, there is the flow regime known as *turbulent flow*. As the name implies, this type of flow is characterized by eddies that cause mixing of the layers of fluid until the layers are no longer distinguishable. The mixing of the fluid due to this turbulence causes an increase in the heat transfer, and consequently, the greater the turbulence, the greater the heat transfer rate.

The basic equation for convection heat transfer is known as *Newton's law of cooling* and is given by

$$\dot{Q} = hA(\Delta t) \tag{11.16}$$

where
$\dot{Q}$ = heat-transfer rate (Btu/hr)

A = heat-transfer area (ft^2)

Δt = temperature difference between the surface and the bulk of the fluid away from the surface (°F)

h = coefficient of heat transfer, film coefficient, thermal convective conductance, heat-transfer coefficient, or film heat-transfer factor (Btu/hr·ft^2·°F)

By a comparison of Equation (11.16) with Equation (11.3a), we can write the thermal resistance for convective heat transfer, R_c, as

$$R_c = \frac{1}{hA} \tag{11.17}$$

and it is treated in the same manner as we treated the resistance concept in conduction heat transfer. Some typical values of h are given in Table 11.4.

TABLE 11.4
Typical Values of h (Btu/hr·ft^2·°F)[a]

Gases (natural convection)	0.1–5
Flowing gases	2–50
Flowing liquids (nonmetallic)	30–1000
Flowing liquid metals	1000–50000
Boiling liquids	200–50000
Condensing vapors	500–50000

Source: W. M. Rohsenow and H. Y. Choi, *Heat, Mass, and Momentum Transfer* (Englewood Cliffs, N.J.: Prentice-Hall, Inc., 1961), p. 102, with permission.
[a] In SI units 1 Btu/hr·ft.^{2}·°F = 5.6786 W/m^2·°C.

ILLUSTRATIVE PROBLEM 11.12

Assume that the bare pipe in Illustrative Problem 11.9 has a heat-transfer coefficient on the outside due to natural convection to the surrounding air at 70°F of 0.9 Btu/hr·ft^2·°F. Determine the heat transfer from the pipe to the surrounding air.

SOLUTION

The resistance of the pipe is

$$R_{pipe} = \frac{\ln\left(\dfrac{r_2}{r_1}\right)}{2\pi kL} = \frac{\ln\left(\dfrac{3.50}{3.00}\right)}{2\pi(26)5} = 0.00019 \text{ hr} \cdot °F/Btu$$

$$R_{convection} = \frac{1}{hA} = \frac{1}{0.9 \times \pi(3.5)/12 \times 5} = 0.2425 \text{ hr} \cdot °F/Btu$$

$$R_{total} = 0.00019 + 0.2425 = 0.243 \text{ hr} \cdot °F/Btu$$

$$\dot{Q} = \frac{\Delta t}{R} = \frac{240 - 70}{0.243} = 699.6 \frac{Btu}{hr}$$

It is interesting to note that the outside film coefficient has acted in a manner similar to an insulator for this case (see Illustrative Problems 11.9 and 11.11) and that the resistance of the pipe wall is essentially negligible.

11.3a   Natural Convection*

The evaluation of the *heat-transfer coefficient h* is quite difficult, because it usually involves the interaction of complex physical phenomena. As noted earlier, *natural convection* heat transfer occurs due to density differences in the fluid caused by a body at a temperature different from the fluid exchanging heat to the fluid. These density differences cause a pumping action of the fluid relative to the body. Using the techniques of dimensional analysis, it can be shown that the parameters involved in natural convection heat transfer can be cast into the form

$$Nu = A(Gr)^a(Pr)^b \tag{11.18}$$

where

Nu = Nusselt number = hL/k or hD/k (dimensionless)

Pr = Prandtl number = $c_p\mu/k$ (dimensionless)

Gr = Grashof number = $g\beta(\Delta t)L^3\rho^2\mu^2$ (dimensionless)

A, a, b = constants depending on the system under consideration

β = coefficient of expansion

ρ = density

μ = viscosity

g = acceleration of gravity

D = diameter

L = length

c_p = specific heat at constant pressure

At this point, it should be noted that the character of the flow process must be considered. The boundary layer of the fluid will be either laminar or turbulent, and in turn, this will affect the constants in Equation (11.18). Based on experimentally determined values, it is found that when the product $(Gr)(Pr)$ exceeds 10^8, there is an increase in the heat-transfer coefficient, indicating a transition from a laminar boundary layer to a turbulent boundary layer.

Equation (11.18) can be recast into the following form for Pr near unity, which is the case for many gases.

$$Nu = C_L(Gr \times Pr)^{1/4} \tag{11.19}$$

$$Nu = C_T(Gr \times Pr)^{1/3} \tag{11.20}$$

Equation (11.19) applies to laminar flow, and C_L is the coefficient for laminar flow. Equation (11.20) applies to turbulent flow, and C_T is the coefficient for turbulent flow. The evaluation of these equations is, at best, tedious, and in some cases, it necessitates the use of an iterative process to obtain a solution. Fortunately, the properties of air, CO, N_2, and O_2 in the temperature range 100°F to 1500°F vary in such a manner that it is possible to lump all the properties that are individually temperature dependent into a single constant that is essentially temperature invariant to obtain the following simplified forms of Equations (11.19) and (11.20):

---

* A portion of this material in this section is from Irving Granet, "Natural Convection Heat Transfer Design Aids," *Design News,* March 1972, with permission.

$$h = C_L' \left(\frac{\Delta t}{L} \right)^{1/4} \qquad \textbf{(11.21a)}$$

$$h = C_T' (\Delta t)^{1/3} \qquad \textbf{(11.21b)}$$

Equation (11.21a) applies to laminar flow and Equation (11.21b) to turbulent flow, Δt is the temperature difference in °F between the surface and the bulk temperature of the gas, L in feet is a characteristic dimension (either a length or a diameter), h is the heat-transfer coefficient in Btu/hr·ft^2·°F, and C_L' and C_T' are the constants for laminar and turbulent flow, respectively.

If the references in Appendix 1 are examined, it will be found that there is as much as 100% difference among various authors for values of the coefficients C_L' and C_T'. A consistent set of data for C_L' and C_T' that yields conservative design values is given in Equations (11.22) and (11.23).

Vertical Plates

$$h = 0.29 \left(\frac{\Delta t}{L} \right)^{1/4} \qquad \text{for } 10^{-2} < (L^3 \Delta t) < 10^3 \text{ (laminar)} \qquad \textbf{(11.22)}$$

$$h = 0.21 (\Delta t)^{1/3} \qquad \text{for } 10^3 < (L^3 \Delta t) < 10^6 \text{ (turbulent)} \qquad \textbf{(11.23)}$$

Horizontal Pipes   Evidence exists to indicate that vertical pipes have higher heat-transfer coefficients than horizontal pipes, but this difference can be considered to be small and the equation given for horizontal pipes can be used for vertical pipes.

$$h = 0.25 \left(\frac{\Delta t}{D} \right)^{1/4} \qquad \text{for } 10^{-2} < (D^3 \Delta t) < 10^3 \text{ (laminar)} \qquad \textbf{(11.24)}$$

$$h = 0.18 (\Delta t)^{1/3} \qquad \text{for } 10^3 < (D^3 \Delta t) < 10^6 \text{ (turbulent)} \qquad \textbf{(11.25)}$$

Horizontal Square Plates

$$h = 0.27 \left(\frac{\Delta t}{L} \right)^{1/4} \qquad \text{for } 1.0 < (L^3 \Delta t) < 20 \text{ (laminar, hot side up)} \qquad \textbf{(11.26)}$$

$$h = 0.22 (\Delta t)^{1/3} \qquad \text{for } 20 < (L^3 \Delta t) < 30{,}000 \text{ (turbulent, hot side up)} \qquad \textbf{(11.27)}$$

$$h = 0.12 \left(\frac{\Delta t}{L} \right)^{1/4} \qquad \text{for } 0.3 < (L^3 \Delta t) < 30{,}000 \text{ (laminar, hot side down)} \qquad \textbf{(11.28)}$$

The foregoing is presented in graphical form in Figures 11.12 through 11.14. Figure 11.12 is used first to determine whether the character of the flow is laminar or turbulent. The necessary data for entering this chart are Δt and L (or D). Having determined the flow character, either Figure 11.13 or 11.14 is entered. Figure 11.13 is for the laminar region, and Figure 11.14 is for the turbulent region. The use of these charts is illustrated by the following examples. The student should note that Equations (11.22) through (11.28) are readily programmable on a digital computer. It is more instructive to see trends using Figures 11.12 through 11.14.

ILLUSTRATIVE PROBLEM 11.13

Determine the heat-transfer coefficient from a horizontal square plate with hot side facing down if the plate is 1 ft by 1 ft and $\Delta t = 100°F$.

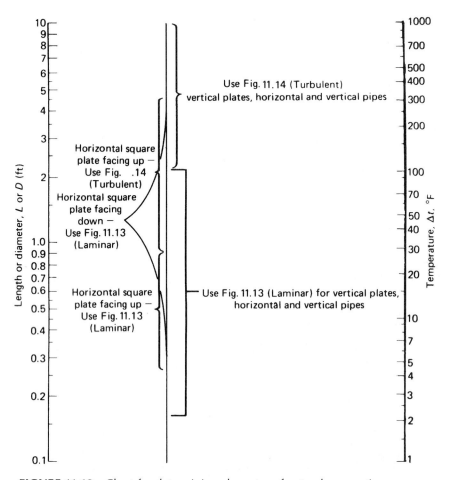

FIGURE 11.12 Chart for determining character of natural convection process.

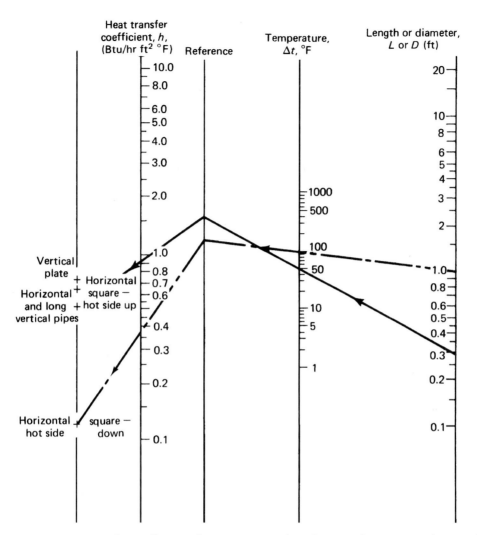

FIGURE 11.13 Heat-transfer coefficient of air, CO, N_2, and O_2 for natural convection, laminar flow.

SOLUTION

From Figure 11.12, the flow is found to be laminar and Figure 11.13 is to be used. Entering Figure 11.13 at $L = 1$ ft and $\Delta t = 100°F$ and connecting the reference scale to the point marked "horizontal square—hot side down" yields $h = 0.38$ Btu/hr·ft²·°F.

ILLUSTRATIVE PROBLEM 11.14

Determine the heat-transfer coefficient for a vertical plate if it is 10 ft high and $\Delta t = 2°F$.

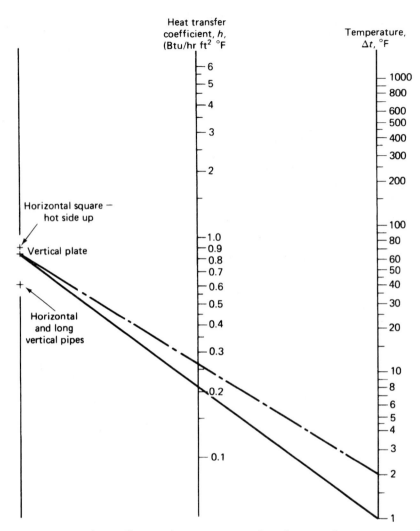

FIGURE 11.14 Heat-transfer coefficient of air, CO, N_2, and O_2 for natural convection, turbulent flow.

SOLUTION

From Figure 11.13, the flow is found to be turbulent. Entering Figure 11.14 at $\Delta t = 2$ and the point for a vertical plate yields $h = 0.26\ \text{Btu/hr·ft}^2\text{·°F}$.

ILLUSTRATIVE PROBLEM 11.15

A bare 3.50-in.-outside diameter pipe has an outside temperature of 120°F and is in a room at 70°F. Determine the heat loss from a section of 5 ft long (see Illustrative Problem 11.10) due to convection.

SOLUTION

From Figure 11.12 at $\Delta t = 50°F$, $D = 3.5/12 = 0.292$ ft, the flow is laminar, and Figure 11.13 is to be used. From Figure 11.13, $h = 0.9$ Btu/hr·ft²·°F, $\dot{Q} = hA\Delta t = 0.9 \times \pi(3.5)/12 \times 5(120 - 70) = 206.2$ Btu/hr.

ILLUSTRATIVE PROBLEM 11.16

A vertical wooden wall 8 ft high $\times$ 7 ft wide $\times$ 3 in. thick ($k = 0.07$ Btu/hr·ft²·°F) has warm air on one side at 80°F and cold air at 50°F on the other side. Determine the heat transfer through the wall and the wall temperature.

SOLUTION

This problem cannot be solved directly, because the individual film resistances are functions of unknown temperature differences. Therefore, as a first approximation, let us assume that h for both sides of the wall can be taken as $\frac{1}{2}$. The wall resistance is $\Delta x/kA$, which for 1 ft² is $(3/12)/0.07 = 3.57$. The overall series resistance therefore is $1/(1/2) + 1/(1/2) + 3.57 = 7.57$. Using 7.57, we can now obtain $\dot{Q}$ and the individual temperature differences. $\dot{Q} = (80 - 50)/7.57 = 3.96$ Btu/hr·ft², and Δt through the "hot" air film is $3.96/(1.2) = 7.92°F$. Through the wall, the Δt is $3.57 \times 3.96 = 14.1°F$, and through the "cold" air film, Δt is also $3.96/(1/2) = 7.92°F$. With these temperature differences, we can now enter Figures 11.12 and 11.14 to verify our approximation. From Figure 11.14, we find $h = 0.42$ Btu/hr·ft²·°F. Using $h = 0.42$, we have for the overall resistance $(1/0.42) + (1/0.42) + 3.57 = 8.33$. $\dot{Q} = (80 - 50)/8.33 = 3.6$ Btu/hr·ft². Δt through both air films is $3.6/0.42 = 8.57°F$ (say, 8.6°F), and through the wall, $\Delta t = 3.6 \times 3.57 = 12.85°F$. Entering Figure 11.14, we find that h stays essentially 0.42, and our solution is that the heat flow is 3.6 Btu/hr·ft², the "hot" side of the wall is at $80 - 8.6°F$ or 71.4°F, the "cold" side is at $50 + 8.6°F = 58.6°F$, and the temperature drop in the wall is $71.4 - 58.6 = 12.8°F$, which checks our wall Δt calculation.

Because natural convection results from density differences in the fluid, with the warmer fluid molecules tending to rise, the orientation of the hot surface affects the rate of heat transfer. Thus, the flow from a horizontal plate becomes turbulent more readily than that from a vertical plate, because the hot fluid molecules can more easily rise from the horizontal plate. Heat transfer from a plate with its hot side down has the lowest heat-transfer coefficient, because the hot fluid molecules have the most difficult time rising from this orientation. In addition, fluid flow over hot surfaces encounters frictional forces that can slow the fluid motion. This is more of a concern in forced convection, which is discussed in the following section. Thus, convection heat transfer is a more complex process than conduction heat transfer.

11.3b Forced Convection

Forced convection flow can be either laminar or turbulent, inside tubes or outside tubes, and involve changes of phase such as when a fluid is being boiled. Figure 11.15 shows a modern chemical plant where all these events occur. Due to the complexity and the number of cases that would

FIGURE 11.15 Large, modern chemical plant that utilizes large, tubular heat exchangers with specialized flow patterns to produce ethylene oxide and ethylene glycol.
(Courtesy of Foster Wheeler Corp.)

have to be studied to cover this topic, we shall limit ourselves to the situation where we have a liquid or gas flowing inside a tube in turbulent flow. For this condition, the heat-transfer coefficient can be calculated from the following type of equation:

$$Nu = C \, (Pr)^a \, (Re)^b \tag{11.29a}$$

where the nomenclature is the same as for Equation (11.18); that is, a, b, and C are constants. Nu is the Nusselt number, Pr is the Prandtl number, and Re is the Reynolds number, which is $DV\rho/\mu$ or DG/μ, where G is the mass flow rate per square foot of flow area, $\dot{m}/A$. It is to be remembered that these "numbers" are dimensionless groups, and consistent units must be used throughout. The form of Equation (11.29a) that is most generally used for turbulent flow (the Reynolds number must be greater than 2100) is

$$\boxed{\frac{hD}{k} = 0.023 \left(\frac{C_p\mu}{k}\right)^{1/3} \left(\frac{DV\rho}{\mu}\right)^{0.8}} \tag{11.29b}$$

with the properties of specific heat, viscosity, and thermal conductivity evaluated at the bulk temperature of the fluid. To facilitate the use of this equation for water and air flowing turbulently in tubes, Figures 11.16 through 11.20 have been developed.* Figures 11.16 and 11.17 give the viscosity of water and air and are used to check the Reynolds number to ensure that the flow is turbulent. Figures 11.18 and 11.19 yield the "basic" heat transfer coefficient h_1 as a function of the weight flow $W/1000$, where W is in pounds per hour. Note that in these figures, the bulk temperature of the fluid is used, not the temperature difference between the fluid and the surface. Finally, Figure 11.20 is a correction factor for the variation of the inside diameter from 1 in. The desired heat transfer coefficient h is then simply equal to $F \times h_1$.

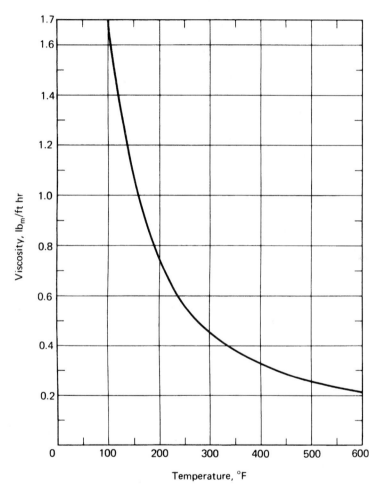

FIGURE 11.16 Viscosity of saturated water.
(Based on data from Steam Tables.)

* See Irving Granet, "Heat Transfer to Subsaturated Water, Dry Air, and Hydrogen in Turbulent Flow Inside a Tube," *Journal of the American Society of Naval Engineers*, November 1957, pp. 787–794.

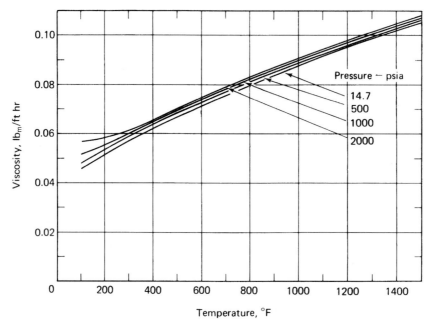

FIGURE 11.17   Viscosity of dry air.

ILLUSTRATIVE PROBLEM 11.17

Water flows in a tube that has a 1-in. outside diameter and a 0.065-in. wall (ID = 0.87 in.). If 20 lb/min of water at 400°F is flowing, calculate the heat-transfer coefficient on the inside of the tube. Figure 11.21 shows a high-temperature, forced-circulation hot water generator of this type.

SOLUTION

The first step is to check the Reynolds number. It will be recalled that the Reynolds number is given by $DV\rho/\mu$ and is dimensionless. Therefore, we can use D, diameter in feet; V, velocity in ft/hr; ρ, density in lb_m/ft^3, and μ, viscosity in $lb_m/ft\cdot hr$. Alternatively, the Reynolds number is given by DG/μ, where G is the mass flow rate per unit area ($lb_m/hr\cdot ft^2$). For this problem,

$$G = \frac{20 \times 60}{\dfrac{\pi(0.87)^2}{4 \times 144}} = 290{,}680 \ lb_m/hr\cdot ft^2$$

$$\mu = 0.33 \ lb_m/ft\cdot hr \quad \text{(from Fig. 11.16)}$$

Therefore, the Reynolds number is

$$\frac{DG}{\mu} = \frac{(0.87/12) \times 290{,}680}{0.33} = 63{,}862$$

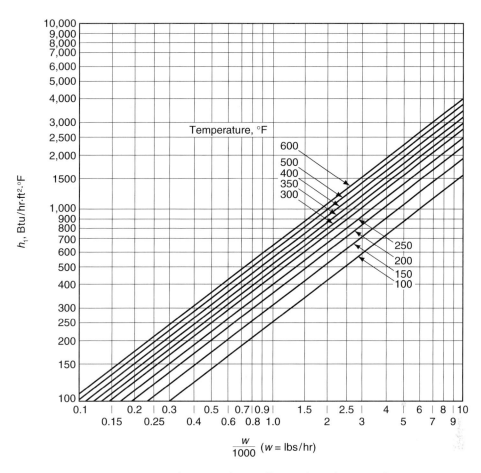

FIGURE 11.18 Basic heat-transfer coefficient for subsaturated water.

which is well into the turbulent flow regime. The next step is to enter Figure 11.18 at $W/1000$ of $20 \times 60/1000 = 1.2$ and 400°F to obtain $h_1 = 630$. From Figure 11.20, we obtain $F = 1.25$ for an inside diameter of 0.87 in. The final desired value of h therefore is equal to

$$F \times h_1 = 1.25 \times 630 = 788 \text{ Btu/hr·ft}^2\text{·°F}$$

ILLUSTRATIVE PROBLEM 11.18

If instead of water, atmospheric pressure air was flowing in the tube of Illustrative Problem 11.17, determine the inside film coefficient.

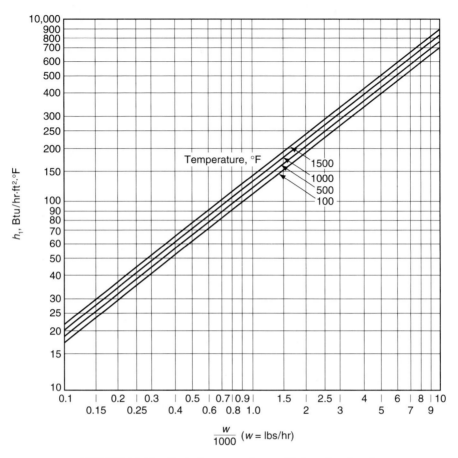

FIGURE 11.19 Basic heat-transfer coefficient for dry air.

SOLUTION

We first check the Reynolds number and note that G is the same as for Illustrative Problem 11.17. The viscosity of air at these conditions is obtained from Figure 11.17 as $0.062 \text{ lb}_m/\text{ft·hr}$. The Reynolds number, DG/μ, therefore is

$$\frac{DG}{\mu} = \frac{(0.87/12) \times 290{,}680}{0.062} = 339{,}908$$

which places the flow in the turbulent regime. Because $W/1000$ is the same as for Illustrative Problem 11.17 and equals 1.2, we now enter Figure 11.19 at 1.2 and 400°F to obtain $h_1 = 135$. Because the inside tube diameter is the same as before, $F = 1.25$. Therefore,

$$h = 135 \times 1.25 = 169 \text{ Btu/hr·ft}^2\text{·°F}$$

It is interesting to note that for equal mass flow rates, water yields a heat-transfer coefficient almost five times greater than air.

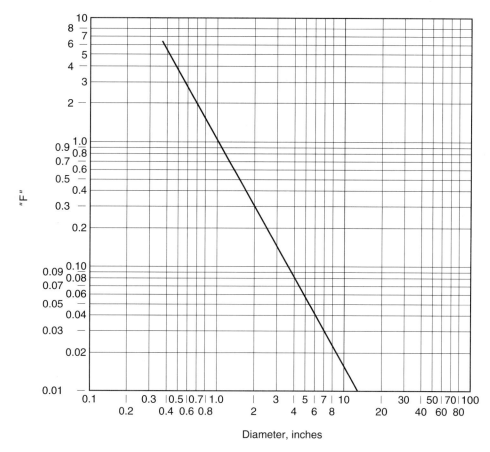

"F"

Diameter, inches

FIGURE 11.20 Factor F: multiplier of basic heat-transfer coefficient.

11.4 RADIATION

Radiant heat transfer differs from both conduction and convection in that a medium is not required to transfer the heat. Basically, radiant heat transfer is an electromagnetic phenomenon similar to the transmission of light, x-rays, and radio waves, and all bodies radiate heat. A net interchange of heat occurs when the absorption of radiant energy by one body exceeds the energy that it is radiating. A body that absorbs all the radiation that strikes it regardless of the wavelength of the radiation is said to be a *blackbody*. Real bodies reflect as well as absorb thermal radiation, and brightly polished metals are good reflectors of thermal radiation. The fraction of the incident heat that is reflected is known as the *reflectivity* of the body, the fraction absorbed is known as the *absorptivity,* and the effectiveness of the body as a thermal radiator at a given temperature is known as its *emissivity.* Thus, emissivity is also the ratio of the emission of heat at a given temperature to the emission of heat from a blackbody at the same temperature.

The radiation from a blackbody can be determined from the *Stefan–Boltzmann law,* which states that the radiation from a blackbody is proportional to the fourth power of the absolute temperature of the body. Thus,

FIGURE 11.21 High-temperature, forced-circulation hot water generator; partially assembled unit showing closely spaced tubes covering the walls; water flows inside tubes in circuits having equal pressure drops to ensure a hydraulically balanced system.

(Courtesy of Riley Stoker Corp.)

$$\dot{Q}_r = 0.173 \times 10^{-8} A T^4 \qquad\qquad \textbf{(11.30)}$$

where $\dot{Q}_r$ is the radiant heat transfer in Btu/hr, A is the radiating area in ft^2, and T is the absolute temperature in °R. The net interchange of heat by radiation between two bodies at different temperatures can be written as

$$\dot{Q}_r = \sigma F_e F_A A (T_1^4 - T_2^4) \qquad\qquad \textbf{(11.31)}$$

where σ = Stefan–Boltzmann constant = 0.173×10^{-8} Btu/hr·ft^2·°R^4 (in SI, 5.669×10^{-8} Watts/m^2·°K^4)

         F_e = emissivity factor to allow for the departure of the surfaces interchanging heat from complete blackness; F_e is a function of the surface emissivities and configurations

         F_A = geometric factor to allow for the average solid angle through which one surface "sees" the other

         A = area, ft^2 (in SI, m^2)

    T_1, T_2 = absolute temperatures, °R (in SI, °K)

Figure 11.22 shows a twin furnace unit where radiation plays a large role in the generation and superheating of steam. Table 11.5 and Figures 11.23 through 11.27 give the required values of F_e and F_A for most cases of practical interest. To evaluate F_e, it is necessary to know the emis-

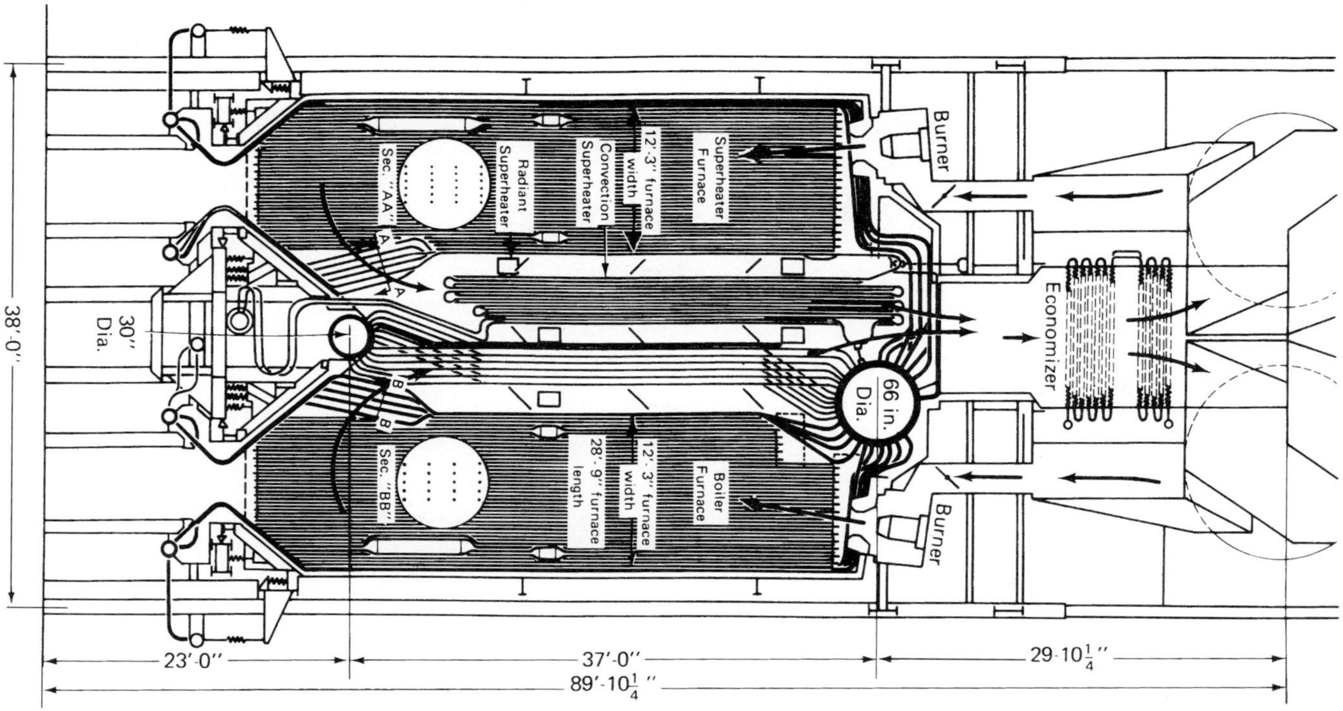
Burner
Burner
Economizer
66 in.
Dia.
Superheater
Furnace
Boiler
Furnace
12'-3" furnace
width
12'- 3" furnace
width
Convection
Superheater
28'- 9" furnace
length
Radiant
Superheater
Sec. "AA"
Sec. "BB"
A
A
B
B
30"
Dia.
38'-0"
23'-0"
37'-0"
29-10$\frac{1}{4}$"
89'-10$\frac{1}{4}$"

FIGURE 11.22 Large central station twin furnace unit. In this unit, steam generation is largely by radiation in the boiler furnace, and a large portion of the superheating is done by radiation in the superheating furnace.
(Courtesy of Foster Wheeler Corp.)

TABLE 11.5

Radiation between Solids: Factors for Use in Equation (11.31)

Surfaces between which radiation is being interchanged	Area, A	F_A	F_e
1. Infinite parallel planes.	A_1 or A_2	1	$\dfrac{1}{\dfrac{1}{\varepsilon_1}+\dfrac{1}{\varepsilon_2}-1}$
2. Completely enclosed body, small compared with enclosing body. (Subscript 1 refers to enclosed body.)	A_1	1	ε_1
3. Completely enclosed body, large compared with enclosing body. (Subscript 1 refers to enclosed body.)	A_1	1	$\dfrac{1}{\dfrac{1}{\varepsilon_1}+\dfrac{1}{\varepsilon_2}-1}$
4. Intermediate case between 2 and 3. (Incapable of exact treatment except for special shapes.) (Subscript 1 refers to enclosed body.)	A_1	1	$\varepsilon_1 > F_e > \dfrac{1}{\dfrac{1}{\varepsilon_1}+\dfrac{1}{\varepsilon_2}-1}$
5. Concentric spheres or infinite cylinders, special case of 4. (Subscript 1 refers to enclosed body.)	A_1	1	$\dfrac{1}{\dfrac{1}{\varepsilon_1}+\dfrac{A_1}{A_2}\left(\dfrac{1}{\varepsilon_2}-1\right)}$ [a]
6. Surface element dA and area A_2. There are various special cases of 6 with results presentable in graphical form. They follow cases 7, 8, and 9.	dA See special cases 7, 8, 9[b]		$\varepsilon_1\varepsilon_2$
7. Element dA and rectangular surface above and parallel to it, with one corner of rectangle contained in normal to dA.	dA	See Fig. 11.23	$\varepsilon_1\varepsilon_2$
8. Element dA and any rectangular surface above and parallel to it. Split rectangle into four having common corner at normal to dA and treat as in case 7.	dA	Sum of F_A's determined for each rectangle as in case 7	$\varepsilon_1\varepsilon_2$
9. Element dA and circular disk in plane parallel to plane of dA.	dA	Formula below[c]	$\varepsilon_1\varepsilon_2$
10. Two parallel and equal squares or disks of width or diameter D and distance between of L.	A_1 or A_2	Fig. 11.24, curves 1 and 2	$\varepsilon_1\varepsilon_2$

Case	Area	F_A	F_e
11. Same as case 10, except planes connected by nonconducting reradiating walls.	A_1 or A_2	Fig. 11.24, curve 3	$\varepsilon_1\varepsilon_2$
12. Two equal rectangles in parallel planes directly opposite each other and distance L between.	A_1 or A_2	$(F'_A F''_A)^{1/2}$ Fig. 11.25	$\varepsilon_1\varepsilon_2$ or $\dfrac{1}{\dfrac{1}{\varepsilon_1} + \dfrac{1}{\varepsilon_2} - 1}$
13. Two rectangles with common sides, in perpendicular planes.	A_1 or A_2	Fig. 11.26	$\varepsilon_1\varepsilon_2$
14. Radiation from a plane to a tube bank (1 or 2 rows) above and parallel to the plane.	A_1 or A_2	Fig. 11.27	$\varepsilon_1\varepsilon_2$

Source: A. I. Brown and S. M. Marco, *Introduction to Heat Transfer*, 3rd ed. (New York: McGraw-Hill Book Co., 1958), with permission.

[a] This form results from assumption of completely diffuse reflection. If reflection is completely specular (mirrorlike), then $F_e = 1/[(1/\varepsilon_1 + 1/\varepsilon_2) - 1]$.

[b] A complete treatment of this subject, including formulas for special complicated cases and the description of a mechanical device for solving problems in radiation, is given by H. C. Hottel, *Mech. Eng.*, 52(7), 699 (July 1930).

[c] Case 9, R = radius of disk + distance between planes; x = distance from dA to normal through center of disk + distance between planes.

$$F_A = \frac{1}{2}\left[1 - \frac{x^2 + 1 - R^2}{\sqrt{x^2 + 2(1 - R^2)x^2 + (1 + R^2)^2}}\right]$$

[d] $F'_A = F_A$ for squares equivalent to short side of rectangle (Fig. 11.24, curve 2).
$F''_A = F_A$ for squares equivalent to short side of rectangle (Fig. 11.24, curve 2).
$F_e = \varepsilon_1\varepsilon_2$ if the areas are small compared with L.
$F_e = 1/[(1/\varepsilon_1 + 1/\varepsilon_2) - 1]$ if the areas are large compared with L.

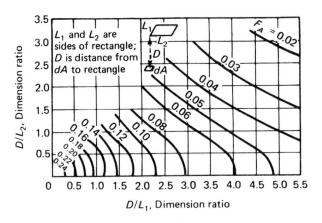

FIGURE 11.23 Radiation between a surface element and a rectangle above and parallel to the surface element.

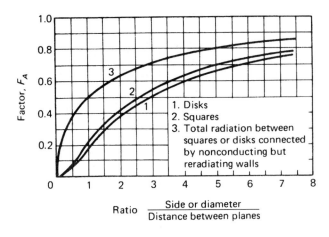

FIGURE 11.24 Direct radiation between equal disks or squares in parallel planes directly opposed.

sivities (ε) of the surface involved. A table of emissivity of various surfaces is given in Appendix 3 (Table A.21). In general, highly polished metals have low emissivities; the emissivity of most materials increases with temperature; most nonmetals have high emissivities; and the emissivity of a given surface will have wide variations depending on the conditions of the surface.

ILLUSTRATIVE PROBLEM 11.19

A bare steel pipe of outside diameter 3.50 in. runs through a room whose walls are at 70°F. Determine the heat loss by radiation from 5 ft of the pipe if its outside temperature is 120°F.

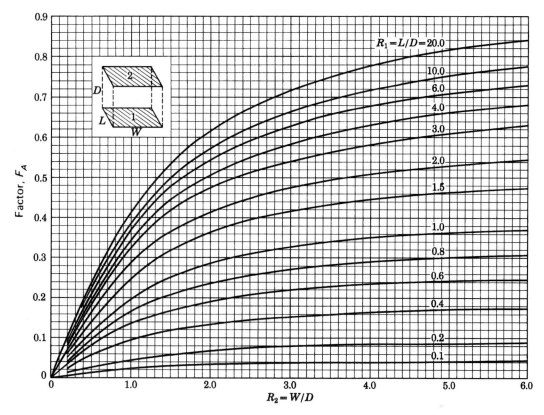

FIGURE 11.25 The radiation shape factor for parallel, directly opposed rectangles.

SOLUTION

From Table 11.5, case 2, $F_A = 1$, A is area of pipe, and F_e is the emissivity of the steel pipe. From Appendix 3, for oxidized steel, $\varepsilon = 0.79$. Therefore,

$$\dot{Q} = 0.173 \times 10^{-8} \times 0.79 \times 1 \times \frac{\pi(3.5)}{12} \times 5[(120 + 460)^4$$
$$- (70 + 460)^4] = 214.5 \text{ Btu}/\text{hr}$$

It should be noted that Equation (11.31) contains a $T_1^4 - T_2^4$ term, not a $(T_1 - T_2)^4$ term, a very common mistake.

In many situations where both radiation and convection occur simultaneously from a body, it is desirable to evaluate a combined heat-transfer coefficient for the process. To arrive at a heat-transfer coefficient for radiation, we will equate Equations (11.16) and (11.31) as follows:

$$\dot{Q}_r = h_r(T_1 - T_2)A = \sigma F_e F_A A(T_1^4 - T_2^4) \qquad \textbf{(11.32)}$$

and
$$h_r = \frac{\sigma F_e F_A (T_1^4 - T_2^4)}{T_1 - T_2} \qquad (11.33)$$

which can be rewritten as

$$h_r = F_e F_A \left[\frac{\sigma (T_1^4 - T_2^4)}{T_1 - T_2} \right] = F_e F_A h_r' \qquad (11.34)$$

For small temperature differences, the term in brackets, h_r', has been evaluated and plotted in terms of the upper temperature t_1 in degrees Fahrenheit and the temperature differences Δ in degrees Fahrenheit between the two bodies in Figure 11.28.

ILLUSTRATIVE PROBLEM 11.20

Determine the radiation heat-transfer coefficient for the pipe of Illustrative Problem 11.19.

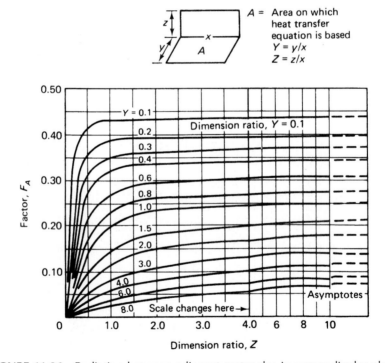

FIGURE 11.26 Radiation between adjacent rectangles in perpendicular planes.

SOLUTION

The upper temperature is given as 120°F and the temperature difference Δ is $120 - 70 = 50°F$. Using Figure 11.28, $h'_r = 1.18$ and $h_r = F_e F_A h'_r = 1 \times 0.79 \times 1.18 = 0.93$. As a check, using the results of Illustrative Problem 11.17

$$h_r = \frac{Q}{A\Delta t} = \frac{214.5}{[\pi(3.5/12] \times 5(120 - 70)}$$
$$= 0.94 \text{ Btu/hr·ft}^2·°F$$

ILLUSTRATIVE PROBLEM 11.21

Solve Illustrative Problem 11.15 taking into account both convection and radiation.

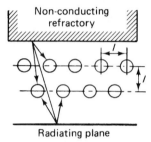

Non-conducting refractory

Radiating plane

Ordinate is fraction of heat radiated from the plane to an infinite number of rows of tubes or to a plane replacing the tubes

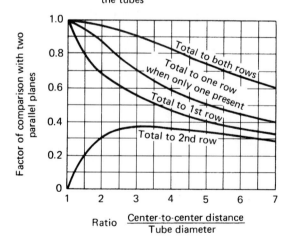

FIGURE 11.27   Radiation from a plane to one or two rows of tubes above and parallel to the plane.

SOLUTION

Because the conditions of Illustrative Problem 11.15 are the same as for Illustrative Problems 11.19 and 11.20 we can solve this problem in two ways to obtain a check. Thus, adding the results of these problems yields $\dot{Q}_{total} = 206.2 + 214.5 = 420.7$ Btu/hr. We can also approach this solution by obtaining a combined radiation and convection heat-transfer coefficient. Thus, $h_{combined} = 0.9 + 0.94 = 1.84$. $\dot{Q}_{total} = 184(\pi \times 3.5/12)5 \times (120 - 70) = 421.5$ Btu/hr. This procedure of obtaining combined or overall heat transfer coefficients is discussed further in Sections 11.5 and 11.6.

In solving radiation heat-transfer problems involving the fourth power of the absolute temperatures, the numbers can become excessively large. Because the radiation heat-transfer equa-

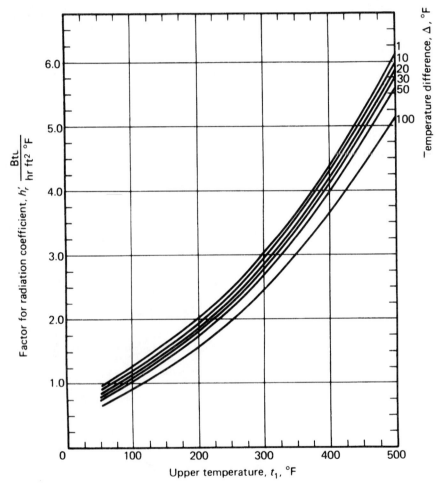

FIGURE 11.28 Factor for radiation coefficient.
(From I. Granet, "Coefficient of Radiation Heat Transfer," Design News, September 1970.)

tion involves the Stefan-Boltzmann constant with its factor of 10^{-8}, a useful device is to express the absolute temperature as a number times 10^2 (e.g., 1400 is 14×10^2). When this temperature is taken to the fourth power, a number times 10^8 results. This factor is conveniently cancelled by the 10^{-8} of the Stefan–Boltzmann constant, leaving the student with more manageable numbers.

Before proceeding, let us look at a general approach to obtaining F_e for two gray bodies interchanging radiation. This method is called the network method and can be adapted to determining F_e for systems in which more than two bodies are interchanging radiation.

When radiant energy is intercepted by a nonblackbody, that is, a gray body, some of the energy may be absorbed, some may be reflected, and some may be transmitted through the body, as in the case of glass. The fraction of the total incident radiation absorbed by the body is denoted by α, the *absorptivity* of the surface; the fraction of the incident radiation that is reflected is denoted by ρ, the *reflectivity* of the surface; and finally, the fraction of the incident radiation that is transmitted is denoted by τ, the *transmissivity* of the body. Based on these definitions, we can write,

$$\alpha + \rho + \tau = 1 \tag{11.35}$$

When a body is opaque, it does not transmit any radiation, giving us $\tau = 0$, which yields Equation (11.35a) for an opaque body:

$$\alpha + \rho = 1 \tag{11.35a}$$

We now denote the total incident energy per unit area to be the *irradiation*, $\dot{Q}_i$, having English units of Btu/hr·ft² and the total energy leaving the body as the *radiosity*, $\dot{Q}_r$ in consistent units. We further denote $\dot{Q}_b$ to be the amount of energy that a blackbody would emit at the same temperature as the gray surface being considered. Using these definitions, we can write

$$\dot{Q}_i = \rho \dot{Q}_i + \varepsilon \dot{Q}_b \tag{11.36}$$

Simplifying Equation (11.36),

$$\dot{Q}_i (1 - \rho) = \varepsilon \dot{Q}_b \tag{11.36a}$$

The net heat transferred to the body, $\dot{Q}_{net}$, is

$$\dot{Q}_{net} = A (\dot{Q}_r - \dot{Q}_i) \tag{11.37}$$

Combining Equation (11.36a) with Equation (11.37) yields

$$\dot{Q}_{net} = \frac{\dot{Q}_b - \dot{Q}_r}{\dfrac{(1 - \varepsilon)}{\varepsilon A}} \tag{11.38}$$

We can compare Equation (11.38) to Ohm's law, giving us $\dot{Q}_b - \dot{Q}_r$ as representing a potential, $(1 - \varepsilon)/\varepsilon A$ representing a resistance, and $\dot{Q}_{net}$ representing a current flow. Thus, we can represent a system of two surfaces that see each other and nothing else by the circuit diagram of Figure 11.29. Each gray surface has a resistance of $(1 - \varepsilon)/\varepsilon A$, and these resistances are connected by a resistance that is due to the geometry of the system, namely, $1/A_1 F_{1-2}$.

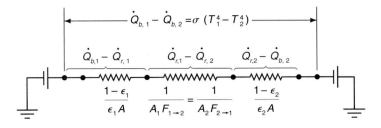

FIGURE 11.29 Electrical analogy for two gray bodies interchanging radiation; two infinite radiating planes that see each other and nothing else.

In Figure 11.29, we have three resistances in series across an overall potential difference equal to $\dot{Q}_{b,1} - \dot{Q}_{b,2} = \sigma(T_1^4 - T_2^4)$. Therefore,

$$\dot{Q}_{net} = \frac{\sigma(T_1^4 - T_2^4)}{\dfrac{(1 - \varepsilon_1)}{\varepsilon_1 A} + \dfrac{1}{A} + \dfrac{(1 - \varepsilon_2)}{\varepsilon_2 A}} \qquad (11.39)$$

If we factor the A in the denominator and rearrange terms, we obtain

$$\dot{Q}_{net} = \frac{\sigma(T_1^4 - T_2^4)A}{\dfrac{1}{\varepsilon_1} + \dfrac{1}{\varepsilon_2} - 1} \qquad (11.40)$$

Comparing Equation (11.40) with Equation (11.31) gives us the value of F_e for this case as

$$F_e = \frac{1}{\varepsilon_1} + \frac{1}{\varepsilon_2} - 1 \qquad (11.41)$$

which is the same as case 1 in Table 11.5.

The electric analogy can be applied to many situations that would be difficult to solve in any other manner. In many cases, the complex interactions lead to a group of simultaneous equations that can be written in condensed matrix form. A solution is then obtained by the process of inverting the matrix. The interested reader should consult the references for these procedures.

11.5 HEAT EXCHANGERS

When heat is transferred from one fluid to another in an industrial process without mixing, the fluids are separated and the heat transfer takes place in an apparatus known as a heat exchanger. A heat exchanger can be of varied shape and size and is usually designed to perform a specific function. The steam-generation plant uses heat exchangers as condensers, economizers, air heaters, feedwater heaters, reheaters, and so on. It is common to designate heat exchangers by their geometric shape and the relative directions of flow of the heat-transfer fluids. For example, Figure 11.30a shows a concentric-tube (or double-pipe) unit in which the fluids would be said to be flowing parallel to each other, and the unit would be called a parallel-flow, double-pipe unit. Other common types are shown in the remainder of Figure 11.30. Because of the extensive use of the shell-and-tube type of heat exchanger, a standard nomenclature has evolved for the parts of this class of unit. This nomenclature is shown in Figure 11.31.

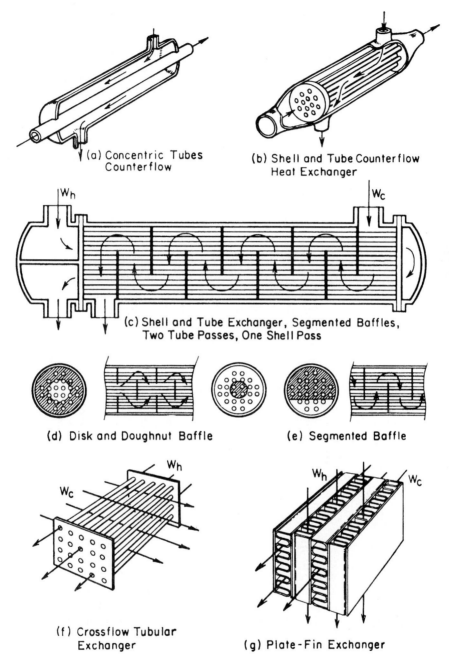

FIGURE 11.30 Representative types of heat exchangers.
[From W. M. Rohsenow and H. Y Choi, Heat, Mass and Momentum Transfer *(Englewood Cliffs, N.J.: Prentice-Hall, Inc., 1961), with permission.]*

The problem of calculating the heat transfer in these units differs from our previous discussion in that the temperature of one or both of the fluids varies continuously as the fluids proceed through the heat exchanger. This can be seen in Figure 11.32, where the fluid temperatures have been plotted as a function of the heat-transfer surface for the most common cases of parallel flow, counterflow, and for one fluid at constant temperature. The subscript h is used to denote the hot fluid, and the subscript c denotes the cold fluid. The sub-subscript 1 is used to denote the temperature at the entry of a fluid of the heat exchanger, and 2 denotes the temperature of the fluid at the exit of the heat exchanger. The direction of flow of each fluid through the exchanger is shown by the arrowheads on the temperature curves. The largest temperature difference between the fluids in the unit (at either inlet or outlet) is designated as θ_A, and the least temperature difference between the fluids (at either inlet or outlet) is designated as θ_B.

Newton's law of cooling [Equation (11.16)] can be written for the heat exchangers as

$$\dot{Q} = UA(\Delta t)_m \qquad (11.42)$$

where U is the overall conductance or overall heat-transfer coefficient having the same physical units as the convection coefficient h, Btu/hr·ft^2·°F; A is the heat transfer surface in square feet; and $(\Delta t)_m$ is an appropriate *mean temperature difference*. The *overall heat-transfer coefficient, U,* in Equation (11.42) is not usually constant for all locations in the heat exchanger, and its local value is a function of the local fluid temperatures. However, it is usual practice to evaluate the individual heat-transfer coefficients based on the arithmetic average fluid temperatures. By analogy

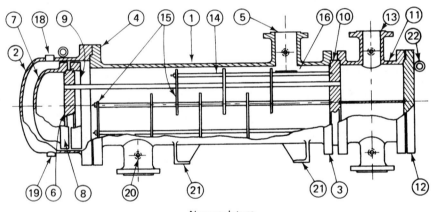

Nomenclature

1. Shell
2. Shell cover
3. Shell channel end flange
4. Shell cover end flange
5. Shell nozzle
6. Floating tubesheet
7. Floating head
8. Floating head flange
9. Floating head backing device
10. Stationary tubesheet
11. Channel
12. Channel cover
13. Channel nozzle
14. Tie rods and spacers
15. Transverse baffles or support plates
16. Impingement baffle
17. Longitudinal baffle
18. Vent connection
19. Drain connection
20. Test connection
21. Support saddles
22. Lifting ring

FIGURE 11.31   Single-pass, shell-and-tube heat exchangers.
(Used with permission from Standards of Tubular Exchanger Manufacturers Association.)

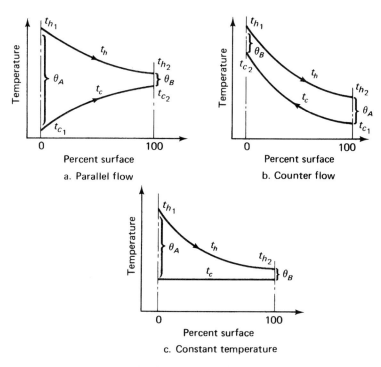

FIGURE 11.32 Fluid temperatures in heat exchangers.

to convection, we have $1/UA$ = resistance. The concept of the overall heat-transfer coefficient is best illustrated by Illustrative Problems 11.22 and 11.23.

ILLUSTRATIVE PROBLEM 11.22

Determine the overall heat-transfer coefficient for the composite wall of Illustrative Problem 11.4 if h on the hot side is 0.9 Btu/hr·ft²·°F and h on the cold side is 1.5 Btu/hr·ft²·°F. The temperatures given are to be the respective air temperatures (see Figure 11.7).

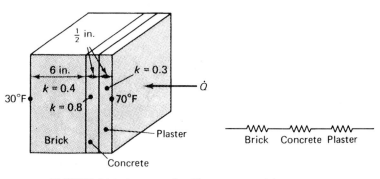

FIGURE 11.7 (*repeated*) Illustrative Problem 11.22.

SOLUTION

For a plane wall, the areas are all the same, and if we use 1 ft^2 of wall surface as the reference area,

$$\text{brick resistance} = \frac{\Delta x}{kA} = \frac{6/12}{0.4 \times 1} = 1.25$$

$$\text{concrete resistance} = \frac{\Delta x}{kA} = \frac{\frac{1}{2}/12}{0.8 \times 1} = 0.052$$

$$\text{plaster resistance} = \frac{\Delta x}{kA} = \frac{\frac{1}{2}/12}{0.3 \times 1} = 0.139$$

$$\text{"hot film" resistance} = \frac{1}{hA} = \frac{1}{0.9 \times 1} = 1.11$$

$$\text{"cold film" resistance} = \frac{1}{hA} = \frac{1}{1.5 \times 1} = 0.67$$

$$\text{total resistance} = \qquad\qquad 3.22$$

The overall conductance (or overall heat-transfer coefficient) $U = 1/(\text{overall resistance}) = 1/3.22 = 0.31$ Btu/hr·ft^2. In Illustrative Problem 11.21 the solution is straightforward, because the heat-transfer area is constant for all series resistances.

ILLUSTRATIVE PROBLEM 11.23

A steel pipe, $k = 26$ Btu/hr·ft^2·°F, having an outside diameter of 3.5 in., an inside diameter of 3.00 in., and 5 ft long, is covered with 1 in. of mineral wool, $k = 0.026$ Btu/hr·ft^2·°F. If the film coefficient on the inside of the pipe is 45 Btu/hr·ft^2·°F and on the outside is 0.9 Btu/hr·ft^2·°F, determine the overall heat transfer coefficient. (See Figure 11.33 and Illustrative Problems 11.9 and 11.11.)

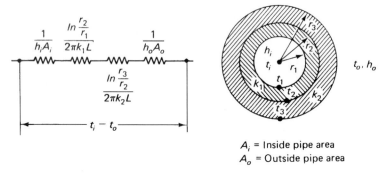

A_i = Inside pipe area
A_o = Outside pipe area

FIGURE 11.33 Illustrative Problem 11.23.

SOLUTION

Because the same amount of heat traverses each of the paths, we can write

$$\dot{Q} = \dot{Q}_i = A_i h_i (t_i - t_1) = 2\pi r_1 L h_i (t_i - t_1)$$

$$= \frac{2\pi k_1 L}{\ln(r_2/r_1)}(t_1 - t_2)$$

$$= \frac{2\pi k_2 L}{\ln(r_3/r_2)}(t_2 - t_3)$$

$$= A_o h_o (t_3 - t_o) = 2\pi r_3 L h_o (t_3 - t_o)$$

Solving for the temperature differences and adding them yields

$$\dot{Q} = \frac{t_i - t_o}{\dfrac{1}{2\pi r_1 L h_i} + \dfrac{\ln(r_2/r_1)}{2\pi k_1 L} + \dfrac{\ln(r_3/r_2)}{2\pi k_2 L} + \dfrac{1}{2\pi r_3 L h_o}} \qquad \text{(a)}$$

Although this equation looks formidable, we can simplify it and interpret it in terms of items discussed previously in this chapter. Thus,

$$\dot{Q} = \frac{(t_i - t_o)2\pi L}{\dfrac{1}{h_i r_1} + \dfrac{\ln(r_2/r_1)}{k_1} + \dfrac{\ln(r_3/r_2)}{k_2} + \dfrac{1}{h_o r_3}} \qquad \text{(b)}$$

Let us now define U_o as the overall heat-transfer coefficient based on the outside pipe surface A_o as

$$\dot{Q} = U_o A_o (t_i - t_o) \qquad \text{(c)}$$

If we multiply the numerator and denominator of Equation (b) by r_3, we obtain

$$\dot{Q} = \frac{(t_i - t_o)\,2\pi L r_3}{\dfrac{r_3}{h_i r_1} + \dfrac{r_3}{k_1}\ln\left(\dfrac{r_2}{r_1}\right) + \dfrac{r_3}{k_2}\ln\left(\dfrac{r_3}{r_2}\right) + \dfrac{r_3}{h_o r_3}} \qquad \text{(d)}$$

Comparison of Equations (c) and (d) yields

$$\boxed{U_o = \frac{1}{\dfrac{1}{h_i(r_1/r_3)} + \dfrac{r_3}{k_1}\ln\left(\dfrac{r_2}{r_1}\right) + \dfrac{r_3}{k_2}\ln\left(\dfrac{r_3}{r_2}\right) + \dfrac{1}{h_o}}} \qquad \text{(e)}$$

Note that U_o is the overall heat-transfer coefficient based on the outside tube surface. If we had multiplied Equation (b) by r_1/r_1 we would obtain the overall heat-transfer coefficient U_i based on the inside surface as

$$U_i = \cfrac{1}{\cfrac{1}{h_i} + \cfrac{r_1}{k_1}\ln\left(\cfrac{r_2}{r_1}\right) + \cfrac{r_1}{k_2}\ln\left(\cfrac{r_3}{r_2}\right) + \cfrac{1}{h_o(r_3/r_1)}}$$ (f)

In effect, we have required that $U_o A_o = U_i A_i$. When discussing an overall heat-transfer coefficient the reference area must be given. Proceeding with the numerical problems yields

$$\frac{1}{h_i} = \frac{1}{45} = 0.02222$$

$$\frac{r_1}{k_1}\ln\frac{r_2}{r_1} = \left[\frac{3.00}{2} \middle/ (26 \times 12)\right]\ln\frac{3.50}{3.00} = 0.00074$$

$$\frac{r_1}{k_2}\ln\frac{r_3}{r_2} = \left[\frac{3.00}{2} \middle/ (0.026 \times 12)\right]\ln\frac{5.50}{3.50} = 2.1730$$

$$\frac{1}{h_o(r_3/r_1)} = \frac{1}{0.9(5.50/3.00)} = 0.6061$$

$$\Sigma = 2.8021$$

Therefore,

$$U_i = \frac{1}{2.8021} = 0.357\,\frac{\text{Btu}}{\text{hr·ft}^2\text{·°F}}\ \text{(of inside area)}$$

Because $\quad U_o A_o = U_i A_i, \quad U_o = 0.357 \times A_i/A_o = 0.357 D_1/D_3 = (0.357)\,(3.00/5.50)$
$= 0.195\ \text{Btu}/(\text{hr·ft}^2\text{·°F})$ (of outside area).

The true mean temperature difference of two fluids exchanging heat in a heat exchanger cannot be determined by simply subtracting temperatures. Let us make the following assumptions in order to obtain the true temperature difference:

1. U is constant over the entire heat exchanger.

2. Both fluid flows are steady, that is, constant with time.

3. The specific heat of each fluid is constant over the entire heat exchanger.

4. Heat losses are negligible.

Then $(\Delta t)_m$ in Equation (11.42) is given for parallel flow, counterflow, and constant-temperature exchangers as

$$(\Delta t)_m = \frac{\theta_A - \theta_B}{\ln(\theta_A/\theta_B)}$$ (11.43)

where θ_A is the greatest temperature difference between the fluids (at either inlet or outlet) and θ_B is the least temperature difference between the fluids (at either inlet or outlet; see Figure 11.32). Thus, $(\Delta t)_m$ is known as the logarithmic mean temperature difference. Because heat losses are assumed to be negligible, the heat transferred from the hot fluid must equal that received by the cold fluid. Therefore,

$$\dot{m}_c(c_p)_c(t_{c_2} - t_{c_1}) = \dot{m}_h(c_p)_h(t_{h_1} - t_{h_2}) \qquad \textbf{(11.44)}$$

Some typical values of overall heat-transfer coefficients are given in Table 11.6 and are useful in making preliminary design calculations.

TABLE 11.6
Approximate Overall Coefficients for Preliminary Estimates

	Overall Coefficient, U	
Duty	$(Btu/hr \cdot ft^2 \cdot °F)^a$	$(W/m^2 \cdot °C)$
Steam to water		
Instantaneous heater	400–600	2270–3400
Storage-tank heater	175–300	990–1700
Steam to oil		
Heavy fuel	10–30	57–170
Light fuel	30–60	170–340
Light petroleum distillate	50–200	280–1130
Steam to aqueous solutions	100–600	570–3400
Steam to gases	5–50	28–280
Water to compressed air	10–30	57–170
Water to water, jacket water coolers	150–275	850–1560
Water to lubricating oil	20–60	110–340
Water to condensing oil vapors	40–100	220–570
Water to condensing alcohol	45–120	255–680
Water to condensing Freon-12	80–150	450–850
Water to condensing ammonia	150–250	850–1400
Water to organic solvents, alcohol	50–150	280–850
Water to boiling Freon-12	50–150	280–850
Water to gasoline	60–90	340–510
Water to gas, oil, or distillate	35–60	200–340
Water to brine	100–200	570–1130
Light organics to light organics	40–75	220–425
Medium organics to medium organics	20–60	110–340
Heavy organics to heavy organics	10–40	57–220
Heavy organics to light organics	10–60	57–340
Crude oil to gas oil	30–55	170–310

Source: Reproduced from F. Kreith, *Principles of Heat Transfer* (Scranton, Pa.: International Textbook Company, 1958), p. 463, with permission. See also F. Kreith and M. Bohn, *Principles of Heat Transfer* (West Publishing Co., 1993), p. 538.
^{a} 1 Btu/hr·ft^2·°F = 5.6786 W/m^2·°C.

CALCULUS ENRICHMENT

Let us consider the counterflow heat exchanger shown in Figure A. In this figure, the "hot" fluid, h, enters at the left and exits at the right while the "cold" fluid, c, enters at the right and exits at the left. A small section of heat exchanger having a heat transfer area dA is isolated as shown, with the temperature differences being dt_h and dt_c, respectively. Based on making the same assumptions noted earlier, a heat balance for each fluid yields

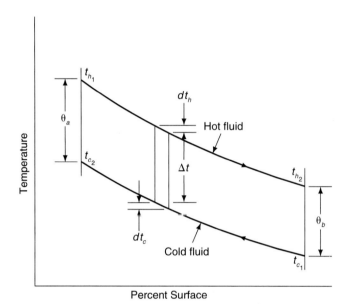

FIGURE A   Log mean temperature difference—derivation for counterflow.

$$d\dot{Q} = U\,dA(t_h - t_c) = \dot{m}_h c_{ph}\,dt_h \tag{a}$$

and
$$d\dot{Q} = U\,dA(t_h - t_c) = \dot{m}_c c_{pc}\,dt_c \tag{b}$$

For the entire heat exchanger,
$$\dot{Q} = UA(\Delta t)_m \tag{c}$$

where $(\Delta t)_m$ is the appropriate mean temperature difference. Also,
$$\dot{Q} = \dot{m}_h c_{ph}(t_{h_2} - t_{h_1}) = \dot{m}_c c_{pc}(t_{c_1} - t_{c_2}) \tag{d}$$

Equating the right sides of Equations (a) and (b) and integrating,
$$\frac{1}{\dot{m}_h c_{ph}} + \frac{1}{\dot{m}_c c_{pc}} = \frac{d(t_h - t_c)}{U\,dA(t_h - t_c)} \tag{e}$$

and from Equation (d),

$$\frac{1}{\dot{m}_h c_{ph}} + \frac{1}{\dot{m}_c c_{pc}} = \frac{(t_{h_1} - t_{h_2}) - (t_{c_2} - t_{c_1})}{UA(\Delta t)_m} \tag{f}$$

Combining Equations (e) and (f),

$$\frac{d(t_h - t_c)}{U\, dA(t_h - t_c)} = \frac{(t_{h_1} - t_{h_2}) - (t_{c_2} - t_{c_1})}{UA(\Delta t)_m} \tag{g}$$

Denoting $(t_h - t_c)$ as θ and rearranging terms yields

$$\int_{\theta_a}^{\theta_b} \frac{d\theta}{\theta} = \frac{\theta_a - \theta_b}{(\Delta t)_m} \tag{h}$$

Integrating Equation (h),

$$\ln \frac{\theta_a}{\theta_b} = \frac{\theta_a - \theta_b}{(\Delta t)_m} \tag{i}$$

and

$$(\Delta t)_m = \frac{\theta_a - \theta_b}{\ln \dfrac{\theta_a}{\theta_b}} \tag{j}$$

where $(\Delta t)_m$ is the log mean temperature difference.

Our derivation was for a counterflow heat exchanger. Using the same procedure, it can also be shown that Equation (j) is also applicable for parallel flow and for heat exchangers in which there is a change of phase in one of the fluids.

ILLUSTRATIVE PROBLEM 11.24

A counterflow heat exchanger is used to cool a flow of 400 lb/min of lubricating oil. Hot oil enters at 215°F and leaves at 125°F. The specific heat of the oil is 0.85 Btu/lb·°F, and the overall coefficient of heat transfer of the unit is 40 Btu/hr·ft²·°F (of outside tube surface). Water enters the unit at 60°F and leaves at 90°F. Determine the outside tube surface required.

SOLUTION

From Figure 11.34, $\theta_A = 215 - 90 = 125°F$, and $\theta_B = 125 - 60 = 65°F$. Therefore,

$$(\Delta t)_m = \frac{\theta_A - \theta_B}{\ln(\theta_A/\theta_B)} = \frac{125 - 65}{\ln(125/65)} = 92°F$$

From the oil data, the heat transfer $\dot{Q} = \dot{m} c_p\, \Delta t = 400 \times 60 \times 0.85(215 - 125) = 1{,}836{,}000$ Btu/hr, and from the heat transfer equations, $\dot{Q} = UA(\Delta t)_m = 1{,}836{,}000 = 40 \times A \times 92$. Therefore, $A = 499$ ft² of outside surface is required.

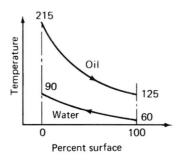

FIGURE 11.34 Illustrative Problem 11.24.

ILLUSTRATIVE PROBLEM 11.25

If the heat exchanger in Illustrative Problem 11.24 is operated in parallel flow, determine the outside tube surface required.

SOLUTION

From Figure 11.35, $\theta_A = 215 - 60 = 155°F$, and $\theta_B = 125 - 90 = 35°F$. Therefore,

$$(\Delta t)_m = \frac{\theta_A - \theta_B}{\ln(\theta_A/\theta_B)} = \frac{155 - 35}{\ln(155/35)} = 80.6°F$$

Because all other conditions are the same, $\dot{Q} = 1,836,000$ Btu/hr.

$$1,836,000 = 40 \times A \times 80.6$$

Therefore, $A = 569$ ft^2 of outside surface is required.

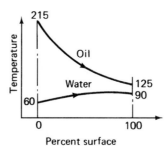

FIGURE 11.35 Illustrative Problem 11.25.

From the results of Illustrative Problems 11.24 and 11.25 we note that operation of a heat exchanger in parallel flow requires more surface for the same terminal temperatures than for a unit operated in counterflow. This conclusion is general, because $(\Delta t)_m$ between the same terminal temperature limits is always greater for counterflow than for parallel flow. Thus, where a choice exists, it is preferable to operate a unit in counterflow, because this type of operation will give the minimum surface requirements. Another advantage of counterflow is that it is possible to raise the exit temperature of the cooling fluid more closely to the inlet temperature of the hot fluid.

After a period of operation, it is found that heat exchangers cannot transfer as much heat as when they are first started and were clean. This is due to a buildup of scale, dirt, or oxide films and is known as *fouling*. This effect is taken into account in heat-exchanger design by introducing series resistances known as fouling factors in the design calculations. Table 11.7 gives typical values of fouling factors for various fluids.

ILLUSTRATIVE PROBLEM 11.26

Using Table 11.7 for the appropriate fouling factors, calculate the surface required for Illustrative Problem 11.25.

SOLUTION

For the oil side, a resistance (fouling factor) of 0.005 hr·°F·ft^2/Btu can be used, and for the water side, a fouling factor of 0.001 hr·°F·ft^2/Btu can be used. Due to the approximate nature of these resistances, we shall not correct them for inside or outside reference areas, and we will assume that they can be used directly with the value of U of 40 Btu/hr·ft^2·°F. The overall resistance and overall heat transfer coefficient are obtained as

$$
\begin{aligned}
\text{oil, } R &= & 0.005 \\
\text{water, } R &= & 0.001 \\
\text{clean unit, } R &= 1/40 & 0.025 \\
\hline
R_{\text{overall}} &= & 0.031
\end{aligned}
$$

$$
U_{\text{overall}} = \frac{1}{0.031} = 32.3 \, \frac{\text{Btu}}{\text{hr·ft}^2 \cdot {}^{\circ}\text{F}}
$$

Because all other parameters are the same, the surface required will vary inversely as U. Therefore, $A = 569 \times 40/32.3 = 705$ ft^2 or an increase in surface required of approximately 24% due to fouling. This obviously represents an important consideration in the design of industrial equipment.

In many heat exchangers, the flow paths of the fluids are not simply parallel or counterflow. In Figure 11.30c, we have an illustration of a shell-and-tube heat exchanger which is said to have one shell pass and two tube passes. In addition, the shell-side fluid passes over segmented baffles in a sinuous path. For this type of heat exchanger as well as others, correction factors have been developed to account for the deviation from parallel or counterflow. The correction factor is applied to the log mean temperature difference to obtain the true temperature difference. Thus,

True mean temperature difference $= F \times$ log mean temperature difference **(11.45)**

TABLE 11.7
Typical Fouling Factors

Types of fluid	Fouling resistance	
	(hr·°F·sq ft/Btu)[a]	(m²·°K/W)
Seawater below 125°F	0.0005	0.00009
Seawater above 125°F	0.001	0.0002
Treated boiler feedwater above 125°F	0.001	0.0002
East River water below 125°F	0.002–0.003	0.0004–0.0006
Fuel oil	0.005	0.0009
Quenching oil	0.004	0.0007
Alcohol vapors	0.0005	0.00009
Steam, nonoil-bearing	0.0005	0.00009
Industrial air	0.002	0.0004
Refrigerating liquid	0.001	0.0002

Source: Reproduced with permission from F. Kreith, *Principles of Heat Transfer* (Scranton, Pa.: International Textbook Co., Inc., 1958), p. 461. See also F. Kreith and M. Bohn, *Principles of Heat Transfer* (West Publishing Co., 1993), p. 510.
[a] Dividing the values by 5.6786 will yield $m^2 \cdot °C/W$.

Figures 11.36 and 11.37 show the correction factor for two arrangements of flow paths. To obtain the true mean temperature difference for any of these arrangements, the log mean temperature difference *for counterflow* is multiplied by the appropriate correction factor.

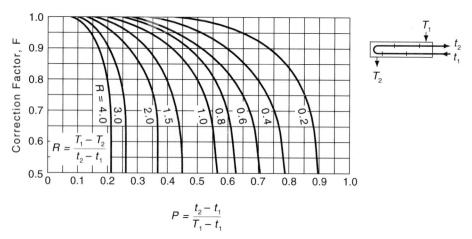

$$P = \frac{t_2 - t_1}{T_1 - t_1}$$

FIGURE 11.36 Correction factor for heat exchangers with 1 shell pass and 2, 4, or any multiple of 2 tube passes.

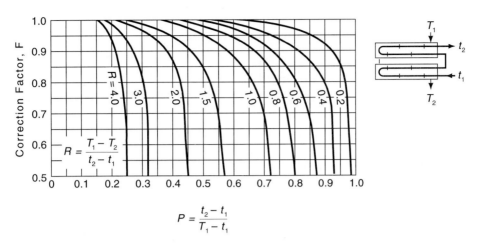

$$P = \frac{t_2 - t_1}{T_1 - t_1}$$

FIGURE 11.37 Correction factor for heat exchangers with 2 shell passes, and 4, 8, or any multiple of 4 tube passes.

ILLUSTRATIVE PROBLEM 11.27

A heat exchanger has 1 shell pass and 2 tube passes similar to Figure 11.30c. Oil flows in the tube side and is cooled from 280°F to 140°F. On the shell side, water is heated from 85°F to 115°F. Determine the true mean temperature difference.

SOLUTION

In order to use Figure 11.36, which is applicable to this problem, we need to calculate P and R. From the figure,

$$P = \frac{t_2 - t_1}{T_1 - t_1} = \frac{140 - 280}{85 - 280} = 0.72$$

$$R = \frac{T_1 - T_2}{t_2 - t_1} = \frac{85 - 115}{140 - 280} = 0.21$$

From Figure 11.36, $F = 0.91$. Thus,

The true mean temperature difference $= 0.91 \times$ LMTD counterflow

$$= 0.91 \left[\frac{(280 - 115) - (140 - 85)}{\ln \dfrac{(280 - 115)}{(140 - 85)}} \right]$$

$$= 91°F$$

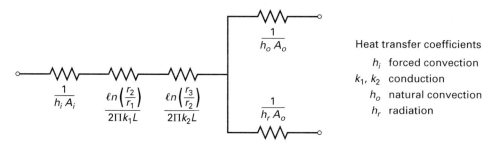

FIGURE 11.38 Electrical analogy for combined heat transfer from a hot fluid flowing in a pipe to still air in a large room.

11.6 COMBINED MODES OF HEAT TRANSFER

Heat-transfer situations often exhibit combined heat-transfer effects of conduction, convection, and radiation. For example, an insulated steel pipe could be used to carry flowing hot water in a large room of still air. The path taken by the heat as it flows from the hot water through the pipe wall and into the room would be as follows:

1. Forced convection from the hot water to the inside wall of the pipe.

2. Conduction through the pipe wall and its insulation.

3. Natural convection from the outer surface of the insulation to the room air in parallel with radiation to the room.

An equivalent resistance circuit is shown in Figure 11.38. An overall heat-transfer coefficient, U, would then be

$$U = \frac{1}{\dfrac{1}{h_i} + \dfrac{r_1}{k_1}\ln\left(\dfrac{r_2}{r_1}\right) + \dfrac{r_1}{k_2}\ln\left(\dfrac{r_3}{r_2}\right) + \left(\dfrac{1}{h_o + h_r}\right)\left(\dfrac{r_3}{r_1}\right)} \qquad \textbf{(11.46)}$$

Units of U are typically in Btu/hr·ft²·°F. Note that free convection and radiation occur simultaneously from the surface of the pipe's insulation. The total heat-transfer rate is thus the sum of the rates of convection and radiation.

11.7 REVIEW

In this chapter, we have surveyed the three mechanisms of heat transfer: conduction, convection, and radiation. We studied conduction through a simple plane wall and a wall consisting of several resistances. Using the electrical analogy where Ohm's law is compared to Fourier's law, it was found that the treatment of resistances in series and parallel could be handled readily. The same reasoning was used when conduction in a hollow cylinder was considered. In convection heat transfer, there is motion of a fluid relative to a body. It therefore becomes necessary to consider the character of the flow of the fluid. That is, is the flow laminar or turbulent, and is the flow due to natural or forced convection? Due to the complexity of the subject, we could only treat several common cases in a rather simplified manner. Radiation heat transfer differs from the other modes of heat transfer in that a medium is not required to transfer the heat. Using the concept of a blackbody and the Stefan–Boltzmann equation, we were able to calculate the radiation heat transfer and to combine it with the other modes of heat transfer. Finally, we studied heat exchangers and the application of Newton's law of cooling to the calculation of their performance.

The general subject of heat transfer is a vast field that encompasses many technical disciplines. In this chapter, we have been able to cover only briefly some of the basic concepts and topics of interest. In the past, empirical test data were used extensively, but in recent decades, great strides have been made in developing analytical and numerical procedures for the calculation of many heat-transfer problems. As examples from the space program alone, we have radiation to and from space vehicles, ablation, lightweight "super" insulation, heat transfer in high-speed flow, transpiration cooling, thermal contact resistances, heat transfer from plasmas, and compact heat exchangers. Many other examples could be cited from every branch of engineering and technology. The point to be made is that this is a truly fertile field with an almost infinite number of unsolved problems that represents great potential for original work. We hope that this chapter will serve as an introduction for the understanding and evaluation of industrial heat-transfer equipment as well as for more advanced work.

KEY TERMS

Terms used for the first time in this chapter are:

absorptivity (α) the fraction of the incident radiant heat that is absorbed by a body.

blackbody a body that absorbs all the radiation that strikes it.

coefficient of heat transfer (h) the factor of proportionality in Newton's law of cooling, $\dot{Q} = hA\,\Delta t$.

conduction the transfer of heat from one part of a body to another part of the same body without appreciable displacement of the particles of the body.

convection the transfer of heat due primarily to the motion of a fluid.

dimensionless numbers the arrangement of physical parameters that form dimensionless groups. Among these are the Nusselt, Prandtl, and Grashof numbers.

emissivity the effectiveness of a body as a thermal radiator at a given temperature.

forced convection the motion of a fluid or gas that is produced primarily by mechanical means.

Fourier's law the law of thermal conduction.

geometric factor (F_A) a factor in the Stefan–Boltzmann equation to allow for the average solid angle through which one surface "sees" the other.

heat-transfer coefficient see *coefficient of heat transfer.*

laminar flow a flow regime in which the fluid moves in layers or paths parallel to each other.

logarithmic mean temperature difference [$(\Delta t)_m$] the appropriate temperature difference to use for parallel flow, counterflow, and constant-temperature heat exchangers.

natural convection motion caused by the differences in density due to the difference in temperature at various locations in the fluid.

Ohm's law the basic electrical equation for direct current circuits. It is analogous to Fourier's law for heat conduction.

overall heat transfer coefficient (U) the proportionality term in Newton's law of cooling for heat exchangers, $\dot{Q} = UA(\Delta t)_m$.

radiation heat that is transferred from one body to another without the need for a medium to effect the transfer; an electromagnetic phenomenon similar to the transmission of light.

reflectivity (ρ) the fraction of the incident radiant heat that is reflected from a body, known as the reflectivity of the body.

Stefan–Boltzmann law an equation governing the radiant interchange of heat between two bodies.

thermal conductivity (k) the proportionality constant in Fourier's equation; also a physical property of a material. $\dot{Q} = -kA\,\Delta t/\Delta x$.

thermal resistance (R_t) as a consequence of the analogy with Ohm's law, the thermal resistance can be written as $R_t = \Delta x/kA$.

transmissivity (τ) the fraction of incident radiant heat that is transmitted.

turbulent flow the flow regime characterized by eddies that cause the mixing of the layers of fluid until no layers are distinguishable.

EQUATIONS DEVELOPED IN THIS CHAPTER

Fourier's equation
$$\dot{Q} = \frac{-kA\Delta t}{\Delta x} \qquad \textbf{(11.3)}$$

Ohm's law
$$\dot{Q} = \frac{\Delta t}{R_t} \qquad \textbf{(11.3a)}$$

Ohm's law
$$i = \frac{\Delta E}{R_e} \qquad \textbf{(11.4)}$$

series thermal circuit overall resistance
$$R_{oe} = R_1 + R_2 + R_3 \qquad \textbf{(11.5)}$$

series thermal circuit overall resistance
$$R_{ot} = R_1 + R_2 + R_3 = \frac{\Delta x_1}{k_1 A} + \frac{\Delta x_2}{k_2 A} + \frac{\Delta x_3}{k_3 A} \qquad \textbf{(11.6)}$$

parallel thermal circuit heat flow
$$\dot{Q} = \dot{Q}_1 + \dot{Q}_2 + \dot{Q}_3 \qquad \textbf{(11.8)}$$

parallel thermal circuit overall resistance
$$R_{ot} = \frac{1}{\dfrac{1}{\Delta x/k_1 A_1} + \dfrac{1}{\Delta x/k_2 A_2} + \dfrac{1}{\Delta x/k_3 A_3}} \qquad \textbf{(11.10)}$$

hollow cylinder
$$t_i - t_o = \frac{\dot{Q}}{2\pi kL} \ln\left(\frac{r_o}{r_i}\right) \qquad \textbf{(11.14a)}$$

hollow cylinder
$$\dot{Q} = \frac{(t_i - t_o)\,2\pi kL}{\ln(r_o/r_i)} \qquad \textbf{(11.14b)}$$

hollow cylinder
$$R_t = \frac{\ln\left(\dfrac{r_o}{r_i}\right)}{2\pi kL} \qquad \textbf{(11.14c)}$$

compound cylinder
$$\dot{Q} = \frac{(t_1 - t_3)2\pi L}{(1/k_1)\ln(r_2/r_1) + (1/k_2)\ln(r_3/r_2)} \qquad \textbf{(11.15)}$$

Newton's law of cooling
$$\dot{Q} = hA(\Delta t) \qquad \textbf{(11.16)}$$

convective thermal resistance
$$R_c = \frac{1}{hA} \qquad \textbf{(11.17)}$$

natural convection from gases—laminar flow	$h = C_{L'}\left(\dfrac{\Delta t}{L}\right)^{1/4}$	(11.21a)

natural convection from gases—laminar flow

$$h = C_{L'}\left(\frac{\Delta t}{L}\right)^{1/4} \tag{11.21a}$$

natural convection from gases—turbulent flow

$$h = C_{T'}(\Delta t)^{1/3} \tag{11.21b}$$

natural convection— vertical plates (laminar)

$$h = 0.29\left(\frac{\Delta t}{L}\right)^{1/4} \tag{11.22}$$

natural convection— vertical plates (turbulent)

$$h = 0.21(\Delta t)^{1/3} \tag{11.23}$$

natural convection— horizontal pipes (laminar)

$$h = 0.25\left(\frac{\Delta t}{D}\right)^{1/4} \tag{11.24}$$

natural convection— horizontal pipes (turbulent)

$$h = 0.18(\Delta t)^{1/3} \tag{11.25}$$

natural convection—horizontal square plates—hot side up (laminar)

$$h = 0.27\left(\frac{\Delta t}{L}\right)^{1/4} \tag{11.26}$$

natural convection— horizontal square plates— hot side up (turbulent)

$$h = 0.22(\Delta t)^{1/3} \tag{11.27}$$

natural convection— horizontal square plates— hot side down

$$h = 0.12\left(\frac{\Delta t}{L}\right)^{1/4} \tag{11.28}$$

forced convection inside tubes—turbulent flow

$$\frac{hD}{k} = 0.023\left(\frac{C_p\mu}{k}\right)^{1/3}\left(\frac{DV\rho}{\mu}\right)^{0.8} \tag{11.29b}$$

Stefan–Boltzmann equation

$$\dot{Q}_r = 0.173 \times 10^{-8} A T^4 \tag{11.30}$$

Stefan–Boltzmann equation

$$\dot{Q}_r = \sigma F_e F_A A(T_1^4 - T_2^4) \tag{11.31}$$

radiation heat-transfer coefficient

$$h_r = \frac{\sigma F_e F_A(T_1^4 - T_2^4)}{T_1 - T_2} \tag{11.33}$$

heat exchanger

$$\dot{Q} = UA(\Delta t)_m \tag{11.42}$$

overall heat-transfer coefficient for a cylinder

$$U_o = \frac{1}{\dfrac{1}{h_i(r_1/r_3)} + \dfrac{r_3}{k_1}\ln\left(\dfrac{r_2}{r_1}\right) + \dfrac{r_3}{k_2}\ln\left(\dfrac{r_3}{r_2}\right) + \dfrac{1}{h_o}}$$

Illustrative Problem 11.23, Equation (e)

overall heat-transfer
coefficient for a cylinder

$$U_i = \cfrac{1}{\cfrac{1}{h_i} + \cfrac{r_1}{k_1}\ln\left(\cfrac{r_2}{r_1}\right) + \cfrac{r_1}{k_2}\ln\left(\cfrac{r_3}{r_2}\right) + \cfrac{1}{h_o(r_3/r_1)}}$$

Illustrative Problem 11.23, Equation (f)

logarithmic mean
temperature difference

$$(\Delta t)_m = \frac{\theta_A - \theta_B}{\ln(\theta_A/\theta_B)} \qquad \textbf{(11.43)}$$

heat exchanger heat balance

$$\dot{m}_c(c_p)_c(t_{c_2} - t_{c_1}) = \dot{m}_h(c_p)_h(t_{h_1} - t_{h_2}) \qquad \textbf{(11.44)}$$

QUESTIONS

11.1   Explain why gases have a much lower thermal conductivity than liquids.

11.2   Compare the terms in Ohm's law and Fourier's law, and discuss their similarity.

11.3   Show how you can go from a series electrical circuit to a series thermal circuit.

11.4   Show how you can go from a parallel electrical circuit to a parallel thermal circuit.

11.5   In what way does heat transfer in a hollow cylinder differ from heat transfer in a plane wall?

11.6   What is the effect of thermal contact resistance in a series thermal circuit?

11.7   How does convection differ from conduction?

11.8   Discuss natural and forced convection. How do they differ?

11.9   What are the two flow regimes in convection heat transfer?

11.10 State Newton's law of cooling, and define the thermal resistance for convective heat transfer.

11.11 How is it possible to simplify the calculation of natural convection heat transfer for gases?

11.12 Which situation did we consider when we discussed forced convection?

11.13 How does radiant heat transfer differ from conduction and convection heat transfer?

11.14 What is the temperature dependence in radiation heat transfer?

11.15 How is combined convection and radiation handled?

11.16 How does the emissivity of metals vary with the condition of the surface?

11.17 How is Newton's law of cooling written for a heat exchanger?

11.18 What is the appropriate mean temperature difference in a heat exchanger?

11.19 How is the true mean temperature difference determined in heat exchangers with differing flow paths?

11.20 Would you expect that the condensation or boiling of water in a heat exchanger would yield high or low overall heat-transfer coefficients?

11.21 Describe the effect of fouling on the performance of a heat exchanger.

PROBLEMS

Unless indicated otherwise, use the data of Table 11.1 and Figure 11.4 for these problems.

Problems Involving Conduction Heat Transfer

11.1   A common brick wall is 4 in. thick. Determine the rate of heat transfer through the wall if one face is at 95°F while the other is at 32°F.

11.2 Compare the insulating qualities of the wall in Problem 11.1 with that of a wooden white pine wall 2 in. thick.

11.3 Compare the insulating qualities of a conducting air gap $\frac{1}{4}$ in. thick with the brick wall of Problem 11.1

11.4 A furnace wall has a thermal conductivity of 0.5 Btu/hr·ft·°F. If the wall is 9 in. thick with the hot face at 2000°F and the cold face at 470°F, determine the heat loss.

11.5 A concrete wall having a thermal conductivity of 1.385 W/m·°C is made 0.2 m thick. If the outer surface is kept at 0°C while the inner surface is kept at 21°C, determine the heat loss through the wall. Assume the wall is 12 m long and 2.5 m high.

11.6 A test is conducted to determine the thermal conductivity of a material. If the test specimen is 1 ft × 1 ft and 6 in. thick and steady-state surface temperatures are found to be 100°F and 92°F, respectively, while 60 Btu/hr is being applied, determine the thermal conductivity of the unknown material.

11.7 A test panel is made 3 ft × 3 ft × 1 in. thick. The panel is found to conduct 950 Btu in a test that lasts for $3\frac{1}{2}$ hr. If the surfaces of the panel are kept at 140°F and 110°F, respectively, determine the thermal conductivity of the panel.

11.8 A wall has a thermal conductivity of 0.4 W/m·°C. If the wall conducts 200 W/m² while one side is maintained at 100°C, determine the temperature of the colder surface. Assume the wall to be 2 cm thick.

11.9 A wall consists of a material that is 1 in. thick. If the surface area of the panel is 10 ft² and the k of the wall is 0.2 Btu/hr·ft·°F, determine the temperature of the "cold" side of the wall. The "hot" side of the wall is to be kept at 100°F, and there is a heat flow of 1000 Btu/hr through it.

11.10 A brick wall is 100 mm thick. How much heat is transferred through the wall if one face is held at 40°C while the other is held at 0°C?

11.11 A test is conducted in which the test specimen is 1 m² by 150 mm thick. Steady-state temperatures are found to be 35°C and 30°C, respectively. If 20 W is applied, determine the thermal conductivity of the material.

11.12 A furnace wall is 12 in. thick and is made of firebrick whose thermal conductivity can be taken to be 0.11 Btu/hr·ft·°F. If the heat loss is designed not to exceed 100 Btu/hr·ft², determine the outside wall temperature if the inside wall temperature is 1400°F.

11.13 A wall is designed to limit the heat transfer through it to 500 W/m². If it is of firebrick whose $k = 0.22$ W/m·°C and it is 500 mm thick, what is the design temperature differential across the wall?

11.14 A corkboard wall has an area of 14 ft² and allows a heat flow of 120 Btu/hr. If the surface temperatures are 70°F and 32°F, respectively, determine the thickness of the wall.

11.15 A glass window is 3 ft × 4 ft and $\frac{3}{8}$ in. thick. The glass has a thermal conductivity of 0.5 Btu/hr·ft·°F. When the inner surface temperature of the glass is 20°C, the outer surface is 10°C. Determine the heat transfer due to heat conduction only.

11.16 A portion of a wall is tested to determine its heat transfer characteristics. Assume that the wall is 6 in. thick and that 3500 Btu/hr is transferred through the wall. If a temperature differential of 60°F is maintained across the wall, determine the thermal conductivity of the wall if the area is 80 ft².

11.17 It is desired to limit the heat loss of a furnace wall to 500 Btu/hr·ft². The inner temperature of the wall is 2100°F, and the outer wall surface temperature is 470°F. If the material of the wall has a thermal conductivity of 0.5 Btu/hr·ft·°F, determine the necessary wall thickness.

11.18 A mild steel sheet is used to separate two gas streams that are at different temperatures. The steel sheet is 10 ft × 10 ft × $\frac{1}{16}$ in. thick, and the temperature of the faces of the sheet differ by 32°F. Calculate the heat flow through the wall.

11.19 A building has a wall composed of $\frac{3}{4}$ in. of white pine wood, 4 in. of mineral wool insulation, and $\frac{1}{2}$ in. of gypsum board. The value of k for each material can be taken from Table 11.1. Calculate the overall heat transfer coefficient for the wall.

11.20 If the temperature drop across the wall in Problem 11.16 is 40°F, determine the heat loss per square foot of wall.

11.21 A composite series wall has an inside surface temperature of 125°F and an outside surface temperature of 52°F. If it consists of 3 in. of mineral wool, $\frac{1}{2}$ in. of plaster on the outside, and $\frac{1}{2}$ in. of corkboard on the inside, determine the heat loss per square foot of wall.

11.22 Determine the interface temperatures in Problem 11.21.

11.23 A wall is 10 ft × 12 ft. In the wall, there is a window that is 4 ft × 4 ft. If the overall conductance of the wall is 0.28 Btu/hr·ft^2·°F and of the window is 1.13 Btu/hr·ft^2·°F, determine the total heat transfer if the outside temperature is 0°F and the inside temperature is 70°F.

11.24 If a wall is 3 m × 4 m and the outside temperature is −20°C while the room is maintained at 20°C, determine the heat transfer through the wall. The overall heat transfer coefficient of the wall is 0.07 W/m^2·°C. Assume also that a 1 m × 1 m window exists in the wall and that the window has an overall heat transfer coefficient of 0.2 W/m^2·°C.

11.25 A so-called "solid brick" construction consists of three courses of common brick, each 4 in. thick. Comment on the insulating qualities of this wall when compared to a hollow wall filled with 3 in. of mineral wool.

11.26 A bar of steel 1 in. in diameter and 6 in. long is cut in half, and the cut ends are pressed together. If the thermal contact resistance can be thought of as an equivalent air gap of 0.001 in. thick ($k = 0.01$ Btu/hr·ft·°F), determine the percent increase in resistance to heat transfer from one end of the bar to the other due to the cutting of the bar, that is, [(cut resistance − uncut resistance)/uncut resistance] × 100.

11.27 A wall is to be made of 4 in. of common brick, 2 in. of mineral wool, and $\frac{1}{2}$ in. of wallboard ($k = 0.04$ Btu/hr·ft·°F from Appendix 3). Determine the heat transfer from the wall if the inside wall temperature is at 74°F when the outside wall is at 0°F.

11.28 Determine the interface temperatures in Problem 11.27.

11.29 A wall consists of 100 mm of mineral wool, 12 mm of plaster, and 12 mm of corkboard. If a temperature difference of 30°C is maintained across the wall, what is the steady-state heat transfer?

11.30 What is the heat loss from a bare steel pipe ($k = 45$ W/m·°C) whose inside temperature is 190°C and whose outer temperature is 100°C? The inside diameter of the pipe is 25 mm, and the outside diameter is 40 mm. Its length is 2 m.

11.31 Solve Problem 11.30 for a copper pipe ($k = 390$ W/m·°C). Compare your results.

*11.32 The equation for the heat transfer through a cylindrical wall is given by Equation (11.14b). By comparison with a plane wall, show that the mean conducting area for a cylindrical shell is $2\pi L(r_2 - r_1)/\ln(r_2/r_1)$. If S_1 and S_2 are the inner and outer surface areas, show that the mean conducting area is also $(S_2 - S_1)/\ln(S_2/S_1)$.

*11.33 The heat transfer through a spherical shell of radii r_2 and r_1 is

$$Q = \frac{4\pi k r_1 r_2(t_1 - t_2)}{r_2 - r_1}$$

Show that the mean conducting area for a spherical shell is $4\pi r_1 r_2$ or $\sqrt{S_1 S_2}$, where S_1 and S_2 are the inner and outer spherical surface areas, respectively.

11.34 A steel pipe having a $4\frac{1}{2}$-in. outside diameter is insulated with a molded pipe covering of diatomaceous earth ($k = 0.036$ Btu/hr·ft·°F) $1\frac{1}{2}$ in. thick. Thermocouples embedded in the insulation read 212°F on the inside and 130°F on the outside. Find the heat loss per foot length of pipe.

11.35 Determine the heat loss from 50 ft of $3\frac{1}{2}$-in.-outside-diameter pipe covered with 1-in. insulation ($k = 0.03$ Btu/hr·ft·°F) if the inner temperature of the insulation is 350°F and the outer temperature is 100°F.

11.36 A large steel pipe having an 8-in. outside diameter carries superheated steam. If the inside insulation temperature is 800°F and the pipe is insulated with 3 in. of molded pipe covering ($k = 0.036$ Btu/hr·ft·°F), determine the heat loss per foot of pipe if the outside insulation temperature does not exceed 140°F.

11.37 If the steel pipe in Problem 11.36 has a wall thickness of $1\frac{1}{2}$ in. and $k_{steel} = 26$ Btu/hr·ft·°F, determine the heat loss per foot of pipe. Assume that the inside of the pipe is at 800°F and the outside of the insulation remains at 140°F. Compare the results with those for Problem 11.36.

11.38 A steel boiler tube is 0.18 in. thick and has an outside diameter of 4 in. If the inside surface is at 500°F and there is a heat loss of 3800 Btu/hr per foot length of pipe, determine the outside tube temperature.

11.39 If the boiler tube in Problem 11.38 has a scale thickness of 0.1 in. built up on the inside of the tube, determine the outside tube temperature. Assume that the inside of the scale is at 500°F and that the scale has a thermal conductivity of 1.5 Btu/hr·ft²·°F.

11.40 A 3-in.-thick brick wall separates 75°F air from the outside air at 25°F. What thickness of gypsum plaster overlay will keep the heat loss through the wall to 50 Btu/hr·ft²?

11.41 A composite wall 8 ft × 10 ft is made up of 2-in.-thick gypsum plaster and 1-in.-thick fir. When a temperature difference of 80°F exists across the wall, what thickness of mineral wool insulation must be added to limit heat transfer through the wall to 800 Btu/hr?

11.42 A 20-cm-thick aluminum plate is to be insulated to keep the heat loss through it down to 100 W/m² when a Δt of 40°C exists across it. What thickness of corkboard will accomplish this?

11.43 What is the heat loss rate per meter of a 10-cm-outside-diameter × 5-mm-thick copper pipe covered with 1 cm of asbestos when a 50°C temperature difference exists across the insulated pipe wall?

*11.44 What thickness of mineral wool insulation is required to limit the heat loss rate from a 6-in.-outside-diameter porcelain pipe, 1-in. thick, to 50 Btu/hr per ft of pipe when an overall temperature gradient of 100°F exists?

*11.45 In Problem 11.44, if the outside of the insulation is at 40°F, what is the temperature at the interface of the pipe and the insulation?

Problems Involving Convection Heat Transfer

11.46 A bare horizontal pipe with 6-in. outside diameter has its outer surface at 140°F in a room whose air temperature is 65°F. Determine the natural convection heat-transfer coefficient on the outside of the pipe.

11.47 If instead of a pipe the surface in Problem 11.46 was a 1 ft × 1 ft square plate with the hot side facing up, determine the convection heat-transfer coefficient.

11.48 Determine the heat loss to the room by convection if the pipe in Problem 11.46 is 10 ft long.

11.49 If the hot surface of the plate in Problem 11.47 faces down, determine the convection heat transfer coefficient.

11.50 Atmospheric air flows in a 1-in.-inside-diameter tube at the rate of 0.2 lb$_m$/s. If the average air temperature is 150°F, determine the film coefficient on the inside of the tube.

11.51 Air flows in a 1.5-in.-inside-diameter tube at the rate of 0.4 lb_m/s and has an average temperature of 500°F. If the air is at atmospheric pressure, determine the heat-transfer coefficient on the inside of the tube.

11.52 Solve Problem 11.50 if the average air temperature is raised to 250°F. Compare your results.

11.53 Solve Problem 11.51 for a 2.0-in.-inside-diameter tube. Compare the results of these two problems.

11.54 If water flows in a 1-in.-inside-diameter tube at the rate of 1 lb_m/s, determine the film coefficient on the inside of the tube. The average temperature is 150°F.

11.55 Solve Problem 11.54 for a 2-in.-inside-diameter tube. Compare the results of these problems.

11.56 Water having an average temperature of 200°F flows in a 2-in.-inside-diameter pipe at the rate of 1.5 lb_m/s. Determine the heat-transfer coefficient on the inside of the pipe.

11.57 If the flow rate in Problem 11.56 is increased to 2.0 lb_m/s, determine the heat-transfer coefficient of the inside of the pipe.

11.58 A square, hollow tube is made of steel and has a thermal conductivity of 26 Btu/hr·ft·°F. On one side, the film coefficient is 500 Btu/hr·ft²·°F, while on the other side, the film coefficient is 250 Btu/hr·ft²·°F. Estimate the overall heat-transfer coefficient if the tube is $\frac{1}{4}$-in. thick.

11.59 A bare vertical pipe, 5-in.-outside-diameter has its outer surface at 120°F in a 70°F room. Find the convective heat loss per foot of pipe.

11.60 A vertical plate, 3 ft × 3 ft, is at 180°F on one side; the other side is insulated. Find the convective heat-loss rate to 65°F still air.

11.61 In Problem 11.51, find the heat-transfer rate per foot of tube if the temperature of the inside of the tube is 350°F.

11.62 In Problem 11.56, find the heat-transfer rate if the pipe is 4 ft long and its inside surface is at 150°F.

Problems Involving Radiation Heat Transfer

11.63 Calculate the rate of heat radiation from a blackbody per square foot of radiating area if the temperature of the body is 1000°F, 100°F, and 0°F.

11.64 Two opposed, parallel infinite black planes are at 350°F and 450°F, respectively. Determine their net heat interchange.

11.65 If the upper temperature in Problem 11.64 is made 550°F, by what percentage is the net heat transfer increased over that in Problem 11.64?

11.66 Assume that the planes in Problem 11.64 are gray with emissivities of 0.8 and 0.7, respectively. Determine the net heat interchange, and compare your answer to Problem 11.64.

11.67 Two perfectly black squares, 10 ft × 10 ft, are spaced 20 ft apart. What is the net rate of heat exchange between them if their temperatures are 500°F and 200°F, respectively? The planes are parallel and directly opposite each other.

11.68 Solve Problem 11.67 if the planes have emissivities of 0.8 and 0.9, respectively.

11.69 Two perfectly black rectangles, 6 ft × 9 ft, are spaced 6 ft apart. Their temperatures are 800°F and 200°F, respectively. If the planes are parallel and directly opposite each other, determine the net heat interchange.

11.70 Solve Problem 11.69 if both planes have emissivities of 0.8.

11.71 Solve Problem 11.69 if one plane has an emissivity of 0.85 and the other has an emissivity of 0.75.

11.72 A $3\frac{1}{2}$-in.-outside-diameter oxidized iron pipe whose outer temperature is 175°F passes through a room at 75°F. If $\varepsilon_{pipe} = 0.8$, determine h_r, the radiation heat-transfer coefficient.

11.73 A body whose emissivity is 0.9 is placed in a large room whose walls are at 20°C. If the body is at 250°C, determine the heat radiated to the walls per square meter of body surface. Assume that the body is small compared to the dimensions of the room.

11.74 Calculate the radiant heat interchange between two square plates having 1-ft sides if they are separated by $\frac{1}{2}$ ft. Assume that one square has an emissivity of 0.8 and that the other has an emissivity of 0.6. The hotter surface is at 1000°F, and the colder surface is at 100°F.

11.75 A rough metal sphere having an emissivity of 0.85 and a 12-in. diameter is hung by a nonconducting wire in a large room. Inside the sphere is an electrical heater that maintains the surface at 200°F. If the room is at 70°F, determine the wattage of the heater.

11.76 A light bulb filament may be considered to be a blackbody. If such a filament has a diameter of 0.12 mm and a length of 4 cm, determine the filament temperature. Consider radiation only, and ignore the temperature of the room or the glass envelope. The bulb is rated at 60 W.

*11.77 Consider two parallel, infinite planes having the same emissivity. The planes are at temperatures T_1 and T_2, respectively. A third plane having the same emissivity is placed between the original planes and allowed to come to its equilibrium temperature T_3. Show that the presence of the third plane reduces the heat transfer between the original two planes to half the value obtained when no plane (shield) is present.

11.78 What is the net radiation heat-transfer rate between the floor of a furnace at 1000°F and one of its walls at 500°F? The emissivity of each is 0.8. The floor is 6 ft × 12 ft, and the wall is 6 ft × 9 ft with the 6-ft length common to both.

11.79 Two 3-ft-diameter disks hang vertically 9-in. apart in a 75°F room. The emissivity of each is 0.7. One disk is at 200°F, the other is at room temperature. Find the radiant heat-transfer rate between the two disks.

Problems Involving Heat Exchangers

11.80 A condenser operates with steam condensing at 1 psia (101.7°F). If the cooling water enters at 46°F and leaves at 90°F, determine the log mean temperature difference.

*11.81 The log mean temperature difference in a condenser operating at 1.5 psia (115.65°F) is 30°F. If the cooling water enters at 50°F, determine the exit temperature of the water.

*11.82 In an air heater operating as a counterflow heat exchanger, the air enters at 70°F and is heated to 200°F. The flue gas enters the heater at 620°F. The mass flow rate of the flue gas is 12% greater than the mass flow rate of the air. Assuming that the specific heats of the air and the flue gas are equal, determine the log mean temperature difference in the unit.

11.83 Solve Problem 11.82 for a parallel flow heat-transfer arrangement.

11.84 A counterflow heat exchanger operates with the hot fluid entering at 850°F and leaving at 650°F. The cold fluid enters at 150°F and leaves at 550°F. Calculate the log mean temperature difference.

11.85 Solve Problem 11.84 if the unit is operated as a parallel flow heat exchanger.

11.86 Solve Problem 11.84 for the true temperature difference if the unit is operated with one shell pass and two tube passes with the hot fluid on the shell side.

11.87 A condenser has condensed steam on its shell side at 81.7°F. If the cooling water enters at 60°F and leaves at 70°F, determine the logarithmic mean temperature difference in the unit.

11.88 If the overall heat-transfer coefficient in Problem 11.87 is 1000 Btu/hr·ft²·°F of outside tube surface, determine the heat transfer per square foot of exchanger outside tube surface.

11.89 A condenser in a power plant operates at 15°C. The cooling water enters at 5°C and leaves at 10°C. Determine the logarithmic mean temperature difference in the unit.

11.90 If the overall heat transfer coefficient in Problem 11.89 is 250 W/m²·°C of outside tube surface, determine the heat transfer per square meter of exchanger outside tube surface.

11.91 A counterflow heat exchanger cools oil from 175°F to 125°F. The cooling water enters at 65°F and leaves at 85°F. Calculate the logarithmic mean temperature difference in the unit.

11.92 If the heat exchanger in Problem 11.91 is operated as a parallel flow unit, determine the logarithmic mean temperature difference.

11.93 A counterflow heat exchanger operates with oil cooled from 90°C to 50°C. The cooling water enters at 15°C and leaves at 25°C. If the overall heat-transfer coefficient based on the outside tube surface is 200 W/m²·°C, determine the heat transfer per square meter of outside tube surface.

11.94 If the heat exchanger in Problem 11.93 is operated as a parallel flow unit, determine the heat transfer.

11.95 A condenser operates with steam condensing on the shell side at 27°C. Cooling water enters at 5°C and leaves at 10°C. If the overall heat transfer coefficient is 5000 W/m²·°C based on the outside tube surface, determine the heat transfer per square meter of outside tube surface.

11.96 A nuclear steam generator boils water (makes steam) at 500°F on the shell side of a steam generator. Water enters the tubes at 570°F and leaves at 525°F. Determine the logarithmic mean temperature difference.

11.97 A tank-type heat exchanger is used to generate steam at 275°C. Water enters the tubes at 325°C and leaves at 300°C. Determine the logarithmic mean temperature difference.

11.98 A heat exchanger is used to transfer 1×10^6 Btu/hr. At the inlet, the temperature difference between fluids is 75°F, and at the outlet, the temperature difference is 35°F. Determine the surface required if the overall heat-transfer coefficient is 175 Btu/hr·ft²·°F based on the outside area of the tubes.

11.99 A heat exchanger is used to transfer 300 kW. At the inlet, the temperature difference between the fluids is 40°C, and at the outlet, the difference is 20°C. Determine the surface required if the overall heat-transfer coefficient is 1000 W/m²·°C based on the outside area of the tubes.

11.100 How much heat is being exchanged in a heat exchanger having an overall heat-transfer coefficient of 60 Btu/hr·ft²·°F based on the outside area of the tubes, 625 ft² of outside tube area, and temperature differences between fluids at inlet of 62°F and at outlet of 36°F?

11.101 A heat exchanger has an overall heat-transfer coefficient of 550 W/m²·°C based on the outside area of the tubes. If there is 400 m² of outside tube area and the temperature differences between fluids at the inlet is 40°C and at the outlet is 22°C, determine the total heat transfer in the unit.

*11.102 A shell-and-tube heat exchanger has a film coefficient on the inside of the tube of 100 Btu/hr·ft²·°F, a steel tube wall thickness of 0.105 in. ($k = 26$ Btu/hr·ft·°F), a tube outside diameter of $\frac{3}{4}$ in., and a film coefficient on the shell side (outside of tube) of 50 Btu/hr·ft²·°F. Determine the overall film coefficient based on the inside area of the tube.

*11.103 Determine the overall film coefficient of the unit in Problem 11.102 based on the outside area of the tube.

11.104 A counterflow heat exchanger is used to cool 2000 lb$_m$/hr of oil from 150°F to 100°F. Assume the c_p of oil to be 0.5 Btu/lb·°F. Water ($c_p = 1$ Btu/lb·°F) enters at 55°F and leaves at 75°F. U, based on the outside area of the tubes, is 30 Btu/hr·ft²·°F. Determine the area required.

*11.105 A shell-and-tube heat exchanger is used to condense steam on the shell side at 81.7°F. Cooling water enters the tubes at 60°F and leaves at 70°F. The tubes have a 1-in. outside diameter, a 0.902-in. inside diameter, and a conductivity of 63 Btu/hr·ft·°F. The film coefficient on the inside of the tubes is 1200 Btu/hr·ft^2·°F and on the outside is 950 Btu/hr·ft^2·°F If the unit extracts 766 × 10^6 Btu/hr, determine the outside tube area.

11.106 In Problem 11.104, what should the cooling water flow rate be?

11.107 A heat exchanger must heat 12,000 lbs/hr of fluid whose c_p is 1.1 Btu/lb·°R from 75°F to 200°F on the tube side of the heat exchanger. Heat is supplied by condensing steam at 250°F on the shell side. The overall heat-transfer coefficient is 100 Btu/hr·ft^2·°F. What heat transfer surface area is required?

Problems Involving Combined Modes of Heat Transfer

11.108 Solve Problem 11.18 assuming air films exist on both sides of the wall and that the temperatures given are the temperatures of the air on each side of the wall, respectively. Calculate the heat loss through the wall if both surface heat-transfer coefficients are equal to 1.4 Btu/hr·ft^2·°F.

11.109 Solve Problem 11.15 assuming air film exists on both sides of the window and that the given temperatures are that of the air on each side, respectively. Determine the heat transfer through the window if the heat transfer coefficients on both sides can be taken to be 1.35 Btu/hr·ft^2·°F.

11.110 Solve Problem 11.21 if there are air films on both sides of the wall having heat transfer coefficients of 1.2 Btu/hr·ft^2·°F.

11.111 Determine the heat transfer in Problem 11.27 if there is an air film on each side of the wall, each having a heat transfer coefficient of 2 Btu/hr·ft^2·°F. Assume the temperatures to be the air temperatures on both sides of the wall.

11.112 If a film coefficient of 1 W/m^2·°C exists on the inside of the wall in Problem 11.29 and the temperature differential is across the film plus the composite wall, determine the heat transfer.

11.113 A wall, 7 ft × 7 ft, conducts 500 Btu/hr. If there is a film coefficient on the inside of the wall of 1 Btu/hr·ft^2·°F and the wall has a resistance equal to 1 in. of mineral wool, determine the temperature drop across each resistance.

11.114 A composite wall consists of $\frac{1}{2}$ in. plaster and 2 in. of fir. If the hot-side air is at 105°F, the cold-side air is at 10°F, the hot-side film coefficient is 2 Btu/hr·ft^2·°F, and the cold-side film coefficient is 4 Btu/hr·ft^2·°F, determine the heat transfer per square foot of wall.

11.115 Determine the temperature drops through each resistance in Problem 11.114.

11.116 If the pipe in Problem 11.36 has an outside heat transfer coefficient of 2.5 Btu/hr·ft^2·°F, and the room temperature is 140°F, determine the heat loss per foot of pipe. Neglect the steel-wall resistance.

11.117 If in Problem 11.72 there is a convection heat-transfer coefficient of 1.5 Btu/hr·ft^2·°F on the outside of the pipe, determine the combined heat-transfer coefficient and the heat loss per foot of pipe.

11.118 In Problem 11.79, find the *total* heat loss rate from the hotter disk if the room air is still. Consider the disk as a vertical plate with the diameter as the characteristic length.

11.119 Two 9 in. × 9 in. plates hang vertically separated by a 3-in. air space. One plate is at 350°F with an emissivity of 0.8; the other is at 100°F with an ε of 0.7. The air is still and at 200°F. Find the *total* heat loss rate from the hotter plate.

*11.120 In a large kitchen, a small griddle of 1 ft^2 is on a horizontal table and operates at 350°F. Its surface emissivity is 0.6. If the kitchen air is still at 70°F, how much heat does the griddle add to the room whose walls are at 75°F in its 2 hours of operation?

11.121 A 4-in.-outside-diameter horizontal pipe ($\varepsilon = 0.8$) whose surface is at 150°F is in a large room of still air at 70°F. What percentage of its heat loss is by convection? By radiation?

11.122 A 0.03-in. nichrome wire runs horizontally through a large enclosure containing 80°F air. The temperature of the wire can reach 2250°F. Find the maximum total heat loss per foot from the wire if the emissivity of the wire is 0.3. (*Hint:* consider the wire as a horizontal pipe.)

11.123 A 2 ft × 2 ft plate, $\varepsilon = 0.7$ and 150°F on one side with the other side insulated, is placed in a large room at 60°F. Find its *total* heat loss rate when it is (a) vertical; (b) horizontal, hot side up; and (c) horizontal, hot side down.

*11.124 A foot-long horizontal porcelain pipe, 5-in.-outside-diameter × 0.5-in. thick, is in still air at 70°F. Find the overall heat-transfer coefficient, including conduction and convection, based on the outer surface area when the surface is at 200°F.

*11.125 In Problem 11.124, if the emissivity of porcelain is 0.9, find the overall heat-transfer coefficient including radiation.

Appendix 1

References

Following are selected textbook references. Some specialized references are given later.

American Society of Heating, Refrigerating, and Air-Conditioning Engineers. *ASHRAE Guide and Data Book,* latest edition. New York: ASHRAE.

American Society of Mechanical Engineers. *ASME Steam Tables,* 6th ed. New York: ASME, 1993.

Anderson, E. E. *Thermodynamics.* Boston: PWS Publishing Co., 1994.

Babits, G. F. *Thermodynamics.* Boston: Allyn and Bacon, Inc., 1963.

Baker, H. D., E. A. Ryder, and H. N. Baker. *Temperature Measurement in Engineering.* New York: John Wiley & Sons, Inc., 1961.

Benedict, R. P. *Fundamentals of Temperature, Pressure, and Flow Measurements.* New York: John Wiley & Sons, Inc., 1969.

Black, W. M., and J. G. Hartley. *Thermodynamics.* New York: Harper & Row Publishers, Inc., 1985.

Brown, A. I., and S. M. Marco. *Introduction to Heat Transfer,* 3rd ed. New York: McGraw-Hill Book Company, 1958.

Burghardt, M. D., and J. A. Harbach. *Engineering Thermodynamics,* 4th ed. New York: Harper-Collins College Publishers, Inc., 1993.

Chapman, A. A. *Heat Transfer,* 4th ed. New York: Macmillan Publishing Co., Inc., 1974.

Clifford, G. E. *Modern Heating and Ventilating Systems Design.* Englewood Cliffs, N.J.: Regents/Prentice-Hall, 1993.

Dodge, B. F. *Chemical Engineering Thermodynamics.* New York: McGraw-Hill Book Company, 1944.

Faires, V. M. *Elementary Thermodynamics,* 3rd ed. New York: Macmillan Publishing Co., Inc., 1957.

Faires, V. M., and C. S. Simmang. *Thermodynamics,* 6th ed. New York: Macmillan Publishing Co., Inc., 1978.

Fellinger, R. C., and W. J. Cook. Introduction to Engineering Thermodynamics. Dubuque, Iowa: Wm. C. Brown Company, 1985.

Fermi, Enrico. *Thermodynamics.* New York: Dover Publications, Inc., 1956.

Glasstone, S. *Principles of Nuclear Reactor Engineering.* New York: Van Nostrand Reinhold Company, Inc., 1955.

Haberman, W. L., and J. E. A. John. *Engineering Thermodynamics.* Boston: Allyn and Bacon, 1980.

Hall, N. A. *Thermodynamics of Fluid Flow.* Englewood Cliffs, N.J.: Prentice-Hall, Inc., 1951.

Hertzfeld, M., ed. *Temperature—Its Measurement and Control in Industry,* Vol. II. New York: Van Nostrand Reinhold Company, Inc., 1962.

Holman, J. P. *Heat Transfer,* 4th ed. New York: McGraw-Hill Book Company, 1976.

Holman, J. P. *Thermodynamics,* 3rd ed. New York: McGraw-Hill Book Company, 1980.

Howell, J. R., and R. O. Buckius. *Fundamentals of Engineering Thermodynamics,* 2nd ed. New York: McGraw-Hill Book Company, 1987.

Huang, F. F. *Engineering Thermodynamics: Fundamentals and Applications.* New York: Macmillan Publishing Co., Inc., 1992.

Irvine, T. F., and J. P. Hartnett, *Steam and Gas Tables with Computer Equations.* New York: Hemisphere Publishing Corp., 1984.

Jordan, R. C., and G. B. Priester. *Refrigeration and Air Conditioning,* 2nd ed. Englewood Cliffs, N.J.: Prentice-Hall, Inc., 1956.

Keenan, J. H. *Thermodynamics.* New York: John Wiley & Sons, Inc., 1941.

Keenan, J. H., F. G. Keyes, P. G. Hill, and J. G. Moore. *Steam Tables—Thermodynamic Properties of Water Including Vapor, Liquid, and Solid Phases.* New York: John Wiley & Sons, Inc., 1969. Also by the same authors and under the same title (*SI Units*), 1978.

Kiefer, P. J., G. F. Kinney, and M. C. Stuart. *Principles of Engineering Thermodynamics.* New York: John Wiley & Sons, Inc., 1953.

Kreith, F., and M. S. Bohn. *Principles of Heat Transfer,* 5th ed. St. Paul, Minn.: West Publishing Co., 1993.

Liepmann, H. W., and A. E. Puckett. *Introduction to Aerodynamics of a Compressible Fluid.* New York: John Wiley & Sons, Inc., 1947.

Liepmann, H. W., and A. Roshko. *Elements of Gas Dynamics.* New York: John Wiley & Sons, Inc., 1957.

Liley, P. E. *2000 Solved Problems in Mechanical Engineering Thermodynamics.* New York: McGraw-Hill Book Co., 1989.

McAdams, W. H. *Heat Transmission,* 3rd ed. New York: McGraw-Hill Book Company, 1954.

McQuiston, F. C., and J. D. Parker. *Heating, Ventilating, and Air-Conditioning,* 4th ed. New York: John Wiley & Sons, Inc., 1994.

Moran, M. J., and H. N. Shapiro. *Fundamentals of Engineering Thermodynamics,* 2nd ed. New York: John Wiley & Sons, Inc., 1992.

Obert, E. F. *Concepts of Thermodynamics.* New York: McGraw-Hill Book Company, 1960.

Potter, M. C., and C. W. Sommerton. *Theory and Problems of Engineering Thermodynamics.* New York: McGraw-Hill, Inc., 1993.

Reynolds, W. C. *Thermodynamic Properties in SI (Graphs, Tables and Computational Equations for 40 Substances).* Stanford, Calif.: Department of Mechanical Engineering, Stanford University, 1979.

Reynolds, W. C., and H. C. Perkins. *Engineering Thermodynamics.* New York: McGraw-Hill Book Company, 1970.

Rohsenow, W. M., and H. Y. Choi. *Heat, Mass and Momentum Transfer.* Englewood Cliffs, N.J.: Prentice-Hall, Inc., 1961.

Rolle, K. C. *Thermodynamics and Heat Power,* 4th ed. New York: Macmillan Publishing Co., 1994.

Rotty, R. M. *Introduction to Gas Dynamics.* New York: John Wiley & Sons, Inc., 1962.

Severns, W. H., and J. F. Fellows. *Air Conditioning and Refrigeration.* New York: John Wiley & Sons, Inc., 1958.

Shapiro, A. H. *The Dynamics and Thermodynamics of Fluid Flow,* Vol. 1. New York: The Ronald Press Company, 1953.

Stoecker, W. F. *Refrigeration and Air Conditioning.* New York: McGraw-Hill Book Company, 1958.

Stoecker, W. F., and J. W. Jones. *Refrigeration and Air Conditioning,* 2nd ed. New York: McGraw-Hill Book Company, 1982.

Todd, J. P., and H. B. Ellis. *An Introduction to Thermodynamics for Engineering Technologists.* New York: John Wiley & Sons, Inc., 1981.

Van Wylen, G. J., and R. E. Sonntag. *Fundamentals of Classical Thermodynamics,* 2nd ed. New York: John Wiley & Sons, Inc., 1973; SI version, 1976.

Wolfe, ed. *Temperature, Its Measurement and Control in Science and Industry.* New York: Litton Educational Publishing Inc., 1955.

Wood, D. B. *Applications of Thermodynamics,* 2nd ed. Reading, Mass.: Addison-Wesley Publishing Co., Inc., 1982.

Zemansky, M. W. *Heat and Thermodynamics,* 4th ed. New York: McGraw-Hill Book Company, 1957.

Zuchrow, M. J. *Jet Propulsion and Gas Turbines.* New York: John Wiley & Sons, Inc., 1948.

Following are selected, specialized references.

Alexander, G. "The Little Engine That Could Be an Answer to Pollution," *New York Times Magazine,* October 3, 1971.

American Iron and Steel Institute. *AISI Metric Practice Guide: SI Units and Conversion Factors for the Steel Industry.* Washington, D.C.: AISI, 1975.

American Society of Mechanical Engineers. *ASME Orientation and Guide for Use of SI (Metric) Units,* 5th ed. New York: ASME, 1974.

American Society of Mechanical Engineers. *ASME Text Booklet: SI Units in Strength of Materials.* New York: ASME, 1975.

Cohen, Karl. *The Effect of Fast Breeder Reactors on Uranium Requirements.* General Electric Co., delivered at Governor's Conference on Uranium, Casper, Wyo., November 1, 1968.

Fagenbaum, J. "The Stirling: A "Hot" Engine for the 80's?" *Mechanical Engineering,* Vol. 105, No. 5, May 1983, pp. 18–29.

Ferris, E. A. *Modern Steam Generating Equipment.* New York: Combustion Engineering Inc., 1960.

Granet, I. "Heat Transfer Performance Curves," *Chemical Engineering,* March 1955, pp. 187–190.

Granet, I. "Heat Transfer to Subsaturated Water, Dry Air and Hydrogen in Turbulent Flow Inside a Tube," *Journal of the American Society of Naval Engineers,* November 1957, pp. 787–794.

Granet, I. "The Coefficient of Radiant Heat Transfer," *Design News,* September 29, 1970.

Granet, I. "Natural Convection Heat Transfer Aids," *Design News,* March 20, 1972.

Jones, C. *A Survey of Curtiss–Wright's 1958–1971 Rotary Combustion Engine Technological Developments,* Paper 720468. Society of Automotive Engineers National Automobile Engineering Meeting, Detroit, Mich., May 1972.

Kreith, F., and R. Bezdek. "Can Industry Afford Solar Energy?" *Mechanical Engineering,* Vol. 105, No. 3, March 1983, pp. 35–41.

Mechanical Engineering. "Energetics," a series of seven articles appearing June through December 1966.

Moebius Research, Inc. *Heat Pump Manual,* 2nd ed. Palo Alto, Calif.: Electric Power Research Institute, 1997.

Naef, F. E., and D. N. Burwell. *Mini-OTEC Results.* 7th Energy Technology Conference.

Parrish, J. R., G. M. Roy, and F. G. Bailey. *Nuclear Power Plant of TVA at Brown's Ferry.* Presented at American Power Conference, Chicago, April 25–27, 1967.

Power magazine. *The Engineer's Reference Library.* New York: McGraw-Hill Book Company.

Power magzine special reports: B. G. A. Skrotzki, "Steam Turbines," June 1962; B. G. A. Skrotzki, "Gas Turbines," December 1963; R. J. Bender, "Steam Generation," June 1964; R. K. Evans, "Nuclear Power Reactors," March 1965; R. G. Schweiger, "Heat Exchangers," June 1970.

Renwal Products Co. *How an Internal Combustion Engine Works.* Fairless Hills, Pa.: The Company, 1960.

Rinaldi, V. J. *Westinghouse Liquid Metal Fast Breeder Reactor,* Westinghouse Technical Paper 282–165A. Prepared by Westinghouse Advanced Reactors Division, Madison, Pa., May 1962.

Stultz, S. C., and J. B. Kitto, eds. *Steam: Its Generation and Use.* Barberton, Ohio: The Babcock and Wilcox Company, 1992.

U.S. Atomic Energy Commission. *Power Reactors.* Washington, D.C.: USAEC. Technical Information Service, May 1958.

Waxberg, H. *Economic Study of Advanced Steam Conditions.* New York: Ebasco Services, Inc., June 1981.

Appendix 2

Answers to Even-Numbered Problems

Chapter 1

1.2	9.78 cm
1.4	2.894 tons
1.6	2.95 ft
1.8	3900.9 g
1.10	256.54 cm
1.12	943.9 L/s
1.14	0.2103 L/s
1.16	0.00333 m^3/s
1.18	5.07×10^{-5} m^3/s
1.20	102,600 L/min
1.22	1579.9 gpm
1.24	0.00946 m^3/s
1.26	3650 L/s
1.28	135.92 L/s
1.30	0.0774 m^2
1.32	0.01416 m^3
1.34	0.003218 m^3
1.36	3.93 L
1.38	29.05 m^3
1.40	3.123 m^3
1.42	0.932 miles
1.44	39.77 mi/hr
1.46	5339.7 N; 2.54×10^{-5} m
1.48	68°F, 104°F, 140°F
1.50	4.44°C, 227°C, 87.8°C
1.52	160°C, 320°F
1.54	−160°F, −106.7°C
1.56	−130.15°ARB
1.58	490.5 N
1.60	(c)
1.62	979.063 N
1.64	10.198 kg

1.66	61.19 kg
1.68	0.39 lb_f
1.70	38.22 N
1.72	10 m/s^2
1.74	2 kg
1.76	2.4 N
1.78	25 lb_m/ft^3, 0.04 ft^3/lb_m
1.80	149.6 lb_f/ft^3, 149.6 lb_m/ft^3
1.82	56.16 lb_f/ft^3, 8829 N/m^3
1.84	2057.7 lb_f
1.86	7848 N/m^3, 800 kg/m^3, 0.00125 m^3/kg
1.88	1.26
1.90	10.37 lb_f/ft^3
1.92	8.14 N
1.94	8.679 lb_f
1.96	435.6 N
1.98	1.95 ft^3
1.100	40.7 psia
1.102	273.73 kPa
1.104	0.87 psig
1.106	250 kg/m^3, 0.25
1.108	39.5 psia, 7.0 ft
1.110	10.3 psia
1.112	2576.4 psfa
1.114	10.33 psf
1.116	18.72 psia

Chapter 2

2.2	300 ft lb_f, 600 ft lb_f; 380 ft lb_f, 600 ft lb_f
2.4	294.3 J
2.6	250 J

2.8	38.86 ft lb$_f$	**2.30**	234.38 kJ
2.10	9810 J, 6.26 m/s	**2.32**	Derivation
2.12	1000 ft lb$_f$, 80.2 ft/s	**2.34**	13,625 kW
2.14	7.67 m/s	**2.36**	50 J
2.16	288.86 ft lb$_f$, 250 ft lb$_f$, 38.86 ft lb$_f$	**2.38**	693.1 kJ/kg
2.18	4 kJ, 9 kJ	**2.40**	−1800 ft lb$_f$ (work in)
2.20	2000 N/m, 62.5 J	**2.42**	14,400 ft lb$_f$
2.22	10.4 ft lb$_f$	**2.44**	161.25 kJ
2.24	2500 N/m	**2.46**	100 kJ
2.26	3.37 ft	**2.48**	19.31 kJ
2.28	4330 J		

Chapter 3

3.2	215 kJ (into)	**3.46**	4.69 ft/s
3.4	70 Btu/lb$_m$ (increase)	**3.48**	4.59 ft/s, 47 ft/s
3.6	45 kJ/kg	**3.50**	6367.3 ft lb$_f$/s, 11.6 hp
3.8	−17.78 Btu/lb$_m$	**3.52**	−61.8 hp (in)
3.10	$c = 1$ kJ/kg·K, $\Delta V = 0.214$ m^3	**3.54**	100 J/kg
3.12	120 kJ	**3.56**	500 Btu/lb$_m$, not necessarily
3.14	−240 ft lb$_f$/lb$_m$	**3.58**	−350 kJ/kg
3.16	−8.84 Btu/lb$_m$	**3.60**	212.13 Btu/lb$_m$
3.18	171.5 Btu/lb$_m$	**3.62**	44.96 Btu/lb$_m$
3.20	0.005 kJ/kg·°K	**3.64**	−100 kJ/kg (in)
3.22	1 kJ/kg·K	**3.66**	79.5 Btu/lb$_m$
3.24	75 kJ/kg	**3.68**	2730 kW
3.26	149.4°C, 28.6 kJ/kg	**3.70**	188.05 kJ/kg
3.28	2 Btu/lb$_m$·°R	**3.72**	2451 ft/s
3.30	354.17 kJ/kg	**3.74**	4002 ft/s
3.32	−17.5 kJ (into), −24.5 kJ (decrease)	**3.76**	634.4 m/s
3.34	$x = 100, y = 600, z = -300,$ $t = -400$	**3.78**	1808 ft/s
		3.80	1793.9 ft/s
3.36	2.04 ft/s	**3.82**	2155.2 kJ/kg
3.38	1.8×10^6 kg/hr	**3.84**	1.86 ft^3/lb$_m$, 240.02 Btu/lb$_m$ (into system)
3.40	13.33 ft/s		
3.42	55 ft/s, 24.4 ft/s, 1872 lb$_m$/s	**3.86**	0.689 lb$_m$/lb$_m$
3.44	13.48 ft/s		

Chapter 4

4.2	845.84 Btu/min, 33.4%	**4.14**	34.2%, 38.5%, 41.1%, lower T_2
4.4	1370.6 MW	**4.16**	127.27 hp, 47.27 hp, 62.86%
4.6	requires an upper temperature of 5000°R!	**4.18**	115.17 Btu
		4.20	70.8%, $Q_r = 146$ Btu/min, work = 354 Btu/min
4.8	34.35%		
4.10	1,636,380 kJ/hr; 40.73 L/hr	**4.22**	retest
4.12	32.54%	**4.24**	Derivation

4.26 30.77%

4.28 3.37 h/day, $2.27/day

4.30 51.5%, 39.9 Btu/min

4.32 65.5%, 392.9 W

4.34 0.57 hp

4.36 7.88%

4.38 68.86%, 217.8 kJ, 67.8 kJ

4.40 42.1%, $T_1 = 932.9°R$

4.42 1428.6 Btu/hr

4.44 1.429 MJ

4.46 509.4 Btu/hr

4.48 0.82 hp

4.50 463.8 K

4.52 54.3 kW, 35.3 kW, 427.7°K

4.54 85.2°F; 6.74 Btu/min·°R

4.56 $\Delta s = 1.8528$ vs. 1.8526 Btu/lb$_m$·°R

4.58 0.312 Btu/lb$_m$·°R

4.60 0.1607 Btu/lb$_m$·°R -0.1607 Btu/lb$_m$·°R

4.62 0.4241 kJ/kg·°K

4.64 0.2267 Btu/lb$_m$·°R

4.66 150 Btu/lb$_m$

4.68 0.1552 Btu/lb$_m$·°R

4.70 1.2 MJ, 4 kJ/°K

4.72 0.6390 kJ/kg·°K

4.74 530°R

4.76 1105°R

4.78 0

4.80 58.6 kJ/kg, 63.6 kJ/kg

4.82 total $\Delta s = 0.026$ Btu/lb$_m$·°R

Chapter 5

5.2 at 1.0 MPa
$h_g = 2778.1$ kJ/kg
$s_g = 6.5865$ kJ/kg·°K
$v_g = 194.44 \times 10^{-3}$ m^3/kg
$u_g = 2583.6$ kJ/kg
at 1.1 MPa
$h_g = 2781.7$ kJ/kg
$s_g = 6.5536$ kJ/kg·°K
$v_g = 177.53 \times 10^{-3}$ m^3/kg
$u_g = 2586.4$ kJ/kg

5.4 at 350°F
$p = 134.53$ psia
$v_f = 0.017988$ ft^3/lb$_m$
$h_f = 321.80$ Btu/lb$_m$
at 500 °F
$p = 680$ psia
$v_f = 0.02043$ ft^3/lb$_m$
$h_f = 487.7$ Btu/lb$_m$

5.6 $h_{fg} = 826.8$ Btu/lb$_m$

5.8 897.44 Btu/lb$_m$, 897.5 Btu/lb$_m$

5.10 $h = 1143.35$ Btu/lb$_m$
$s = 1.54694$ Btu/lb$_m$·°R
$v = 4.213$ ft^3/lb$_m$
$u = 1065.41$ Btu/lb$_m$

5.12 82.3%

5.14 2.9261 ft^3/lb$_m$

5.16 819.71 Btu/lb$_m$

5.18 11.064 ft^3/lb$_m$

5.20 0.396, 41.85 psia

5.22 1.109 MPa

5.24 1098.58 Btu/lb$_m$, 467.13°F (wet)

5.26 2.004 ft^3/lb$_m$

5.28 541.7 psia

5.30 0.3066 m^3/kg

5.32 1327.3 Btu/lb$_m$

5.34 1333.2 Btu/lb$_m$

5.36 66.179 ft^3/lb$_m$, 1030.98 Btu/lb$_m$, 969.76 ft^3/lb$_m$

5.38 $h = 487.5$ Btu/lb$_m$
$v = 0.019766$ ft^3/lb$_m$
$u = 472.9$ Btu/lb$_m$
$s = 0.6758$ Btu/lb$_m$·°R

5.40 2.18 MPa

5.42 at 100 psia
$h_g = 1187.7$ Btu/lb$_m$
$s_g = 1.6034$ Btu/lb$_m$·°R
$v_g = 4.4329$ ft^3/lb$_m$
$u_g = 1105.7$ Btu/lb$_m$
at 1000 psia
$h_g = 1192.2$ Btu/lb$_m$
$s_g = 1.3902$ Btu/lb$_m$·°R
$v_g = 0.4459$ ft^3/lb$_m$
$u_g = 1109.7$ Btu/lb$_m$

5.44 at 1.0 MPa
$h_f = 762.88$ kJ/kg
$s_f = 2.1388$ kJ/kg·°K

$v_f = 0.001127$ m^3/kg

$u_f = 761.75$ kJ/kg

at 1.1 MPa

$\overline{h_f = 781.38}$ kJ/kg

$s_f = 2.1793$ kJ/kg

$v_f = 0.001133$ m^3/kg

$u_f = 780.14$ kJ/kg

5.46 $u = 3046.2$ kJ/kg

5.48 $s_{fg} = 1.6529$ Btu/lb$_m$·°R

5.50 $p = 5.6267$ kPa

$v_f = 0.001006$ m^3/kg

$s_f = 0.5050$ kJ/kg·°K

5.52 $h = 2582.2$ kJ/kg

$v = 0.1571$ m^3/kg

$s = 6.1107$ kJ/kg·°K

$u = 2406.2$ kJ/kg

5.54 0.3162

5.56 0.6723

5.58 0.07685 m^3/kg

5.60 $x = 0.0517, t = 78.186$°C

5.62 $x = 0.0909, 0.019612$ m^3,

0.00238 m^3, 2.3178 MPa

5.64 2.347 psia, 132.02°F, 0.01626 ft^3/lb$_m$

5.66 212.42°C, 2422.5 kJ/kg

5.68 1563.7 Btu/lb$_m$

5.70 0.5166 ft^3/lb$_m$

5.72 2835 kJ/kg

5.74 2811.7 kJ/kg

5.76 2245.9 kJ/kg

5.78 2620.1 kJ/kg

5.80 487.50 Btu/lb$_m$

5.82 1039.0°F, 1541.8 Btu/lb$_m$

5.84 6.47 kg

5.86 247.1 psia, 10.72 lb$_m$

5.88 8.581 MPa, 230.73 kg

5.90 52.78 lb$_m$ vapor

1063.4 lb$_m$ liquid

5.92 39.37 kg liquid

9.84 kg vapor

5.94 0.644

5.96 0.5157

5.98 235.1 Btu/lb$_m$

5.100 vapor = 57.52%, liquid = 42.49%

5.102 6238.7 Btu (removed)

5.104 557,600 Btu

5.106 1019.73 Btu/lb$_m$

5.108 316.9 Btu/lb$_m$

5.110 756.44°F; from chart approx. 755°F

5.112 312.07°F, moisture = 7.6%

5.114 moisture 2%

5.116 -130 Btu/lb$_m$

5.118 1058.08 Btu/lb$_m$

5.120 1179.88 Btu/lb$_m$

5.122 0.531 Btu/lb$_m$·°F

5.124 350 Btu/lb$_m$

5.126 891.31 Btu/lb$_m$

5.128 4030.6 kJ

5.130 412.87 psia

5.132 74.7 Btu/lb$_m$

5.134 31.15 Btu/lb$_m$

5.136 76.6 Btu/lb$_m$

5.138 4497 ft/s

5.140 1311.7 m/s

5.142 1354.3 m/s

5.144 6.46 kg

5.146 247.05 psia, 10.72 lb$_m$

5.148 8.583 kPa, 230.73 kg

5.150 52.80 lb$_m$ vapor

1063.3 lb$_m$ liquid

5.152 39.38 kg vapor

9.85 kg liquid

5.154 0.644

5.156 0.5156

5.158 235.4 Btu/lb$_m$

5.160 vapor = 57.51%, liquid = 42.49%

5.162 6259.5 Btu (removed)

5.164 567,285 Btu

5.166 1019.62 Btu/lb$_m$

5.168 317.0 Btu/lb$_m$

5.170 756.56°F

5.172 312.09°F, 7.64%

5.174 moisture = 2.05%

5.176 129.6 Btu/lb$_m$

5.178 1058.0 Btu/lb$_m$

5.180 1179.9 Btu/lb$_m$

5.182 0.531 Btu/lb$_m$·°R

5.184 353.2 Btu/lb$_m$

5.186 890.99 Btu/lb$_m$

5.188 4029.64 kJ/kg

5.190 412.87 psia

5.192 74.8 Btu/lb$_m$

5.194 31.1 Btu/lb$_m$

5.196 76.6 Btu/lb$_m$

5.198 4496.4 ft/s

5.200 1306.4 m/s

5.202 1354.1 m/s

Chapter 6

6.2 163.8°K

6.4 0.4 lb$_m$/ft^3

6.6 13.29 ft^3/lb$_m$

6.8 0.813 lb$_m$/ft^3

6.10 425 psia

6.12 1.283 kJ/kg·°K, 6.48

6.14 0.171 m^3/kg

6.16 1.06 kJ/kg·°K

6.18 70.94 kg

6.20 0.706 ft^3/lb$_m$

6.22 0.6062 ft^3

6.24 337.9°R

6.26 232.7 kPa

6.28 40.93 ft^3

6.30 $c_v = 0.4725$ kJ/kg·°K,
$c_p = 0.6615$ kJ/kg·°K

6.32 $c_v = 0.9556$ kJ/kg·°K,
$c_p = 1.2423$ kJ/kg·°K

6.34 $R = 0.316$ kJ/kg·°K, MW = 26.31

6.36 MW = 29.21

6.38 MW = 27.97, $k = 1.4$

6.40 MW = 28.57, $R = 0.291$ kJ/kg·°K,
$c_v = 0.9091$ kJ/kg·°K

6.42 11.5 ft^3/lb$_m$

6.44 −193.4 kJ/kg (in)

6.46 −0.6601 kJ/°K

6.48 −71,744 ft lb$_f$

6.50 218.2 kJ/kg

6.52 $q = 16.9$ Btu,
0.02777 Btu/°R

6.54 $v = 0.251$ ft^3/lb$_m$,
$\Delta s = -0.16928$ Btu/lb$_m$·°R,
$q = -94.8$ Btu/lb$_m$

6.56 80 psia, 105,557 ft lb$_f$, $\Delta u = 0$,
$Q = 135.7$ Btu

6.58 24.42 ft^3, 15.0 lb$_m$

6.60 524.2 kJ, 0, 524.2 kJ,
1.1082 kJ/kg·°K

6.62 61.70 lb$_m$

6.64 33.33 psia

6.66 0.00852 Btu/°R, 1.47

6.68 0.11853 Btu/lb$_m$·°R

6.70 85.5 Btu/lb$_m$

6.72 0.17446 Btu/lb$_m$·°R

6.74 0.01760 kJ/°K

6.76 68.8 psia, 611.3°F, 0.1283 Btu/°R

6.78 641.6°F

6.80 792.4°K, 676.8 kJ

6.82 4480°R, 208,696 ft lb$_f$/lb$_m$

6.84 28.8 Btu/lb$_m$

6.86 0.2405 Btu/lb$_m$·°R

6.88 0.2472 Btu/lb$_m$·°R

6.90 0.30525 Btu/lb$_m$·°R

6.92 +0.3529 m^3, 4.0 kJ/kg·°K

6.94 5.83 ft^3

6.96 −96 Btu/lb$_m$

6.98 work = 91.1 Btu/lb$_m$,
$\Delta u = 225.7$ Btu/lb$_m$,
$q = 316.0$ Btu/lb$_m$

6.100 $q = 64.8$ Btu/lb$_m$,
$\Delta s = 0.088$ Btu/lb$_m$·°R

6.102 work = 5517.9 ft lb$_f$,
$V_2 = 5.76$ ft^3

6.104 331.4 psia

6.106 3279.4°R

6.108 425.3°R

6.110 395.2°R

6.112 395.2°R

6.114 1.37

6.116 1.29

6.118 342.4 Btu/lb$_m$

6.120 1.24

6.122 1.30

6.124 −85.4 Btu/lb$_m$ (in)

6.126 −133.6 kJ/kg (in)

6.128 858.8°R, 104.4 psia

6.130 0.2516 Btu/lb$_m$·°R

6.132 0.226 Btu/lb$_m$ · °R

6.134 0.242 ft^3/lb (ideal gas),
0.244 ft^3/lb$_m$ (real)

6.136 10,194 lb_m
6.138 7.23 lb_m/ft^3
6.140 13.3°F

6.142 320.8°F, 27.5 psia
6.144 960.2°R

Chapter 7

7.2 moles CH_4 = 0.06238
moles O_2 = 0.03125
total moles = 0.09363
x_{CH_4} = 0.666
x_{O_2} = 0.334
MW = 21.36

7.4 CO_2 = 40.4%
N_2 = 56.6%
He = 2.9%

7.6 C_2H_2 = 34.53 MPa
C_4H_{10} = 15.47 MPa

7.8 MW = 29.38
R = 52.59 ft lb_f/lb_m·°R

7.10 R = 210.72 ft lb_f/lb_m·°R
CO_2 = 8.33%
He = 91.67%

7.12 62.6 m^3

7.14 MW = 37.68

7.16 MW = 34.62 ft lb_f/lb_m·°R

7.18 CH_4 = 93.5%
CO_2 = 2.98%
N_2 = 1.90%
H_2 = 1.60%
MW = 14.75
R = 104.75 ft lb_f/lb_m·°R

7.20 volume = 69.5 ft^3
CH_4 = 53.5 ft^3
O_2 = 16.2 ft^3
p_{CH_4} = 57.6 psia
p_{O_2} = 17.4 psia
x_{CH_4} = 0.768
x_{O_2} = 0.232
MW = 19.73

7.22 p_{CO} = 44.98 kPa
p_{He} = 629.80 kPa
p_{total} = 674.77 kPa

7.24 MW = 38.0
R = 40.66 ft lb_f/lb_m·°R

7.26 MW = 30.4
R = 0.2735 kJ/kg·°K

7.28 MW = 10.69
R = 0.778 kJ/kg·°K
p = 200 kPa

7.30 CO = 43.1%
CH_4 = 12.3%
N_2 = 32.3%
O_2 = 12.3%
MW = 26.02
R = 59.4 ft lb_f/lb_m·°R

7.32 p_{N_2} = 40 psia
p_{O_2} = 40 psia
p_{CO_2} = 20 psia
N_2 = 0.342
O_2 = 0.390
CO_2 = 0.268
MW = 32.81
R − 47.1 ft lb_f/lb_m·°R

7.34 t = 90°F
p = 167.4 psia

7.36 t = 42.1°C
p = 1.053 MPa

7.38 t = 89.9°F
p = 86.5 psia

7.40 p = 83.68 psia
t = 79.05°F

7.42 0.3801 psia, 71°F

7.44 51.4%, 51.5°F

7.46 69°F, 0.3494 psia,
105 grains/lb_m dry air

7.48 59.8%

7.50 45.9%, 57°F, 0.233 psia

7.52 42.1%, 73°F

7.54 30%, 26.5 Btu/lb_m dry air,
46 grains/lb_m dry air

7.56 30%, 64°F

7.58 0.0526 lb_m

7.60 46°F, 45.5 grains/lb_m dry air

7.62 4.9 Btu/lb_m dry air

7.64 6.69 Btu/lb_m dry air

7.66 62°F, 54°F, 59°F

7.68 35.8%

7.70 89 grains/lb$_m$ dry air, 65°F

7.72 81%, 21 grains/lb$_m$ dry air,
6.0 Btu/lb$_m$ dry air

7.74 20%

7.76 63°F, 55%, 48 grains/lb$_m$ dry air

7.78 0.9642 lb$_m$/lb$_m$ dry air,
0.0358 lb$_m$/lb$_m$ dry air

7.80 8.19 lb$_m$/s, 359.81 lb$_m$/s

Chapter 8

8.2 182.5 Btu, 68.67%

8.4 466.7°K

8.6 Derivation

8.8 1326°R

8.10 200 kJ/hr, 80%, 800 kJ/hr

8.12 $h = 1032$ Btu/lb$_m$, $p = 5.2$ psia

8.14 $h = 1031.3$ Btu/lb$_m$, $p = 5.167$ psia

8.16 90%

8.18 578.26 Btu/lb$_m$, 1439.175 Btu/lb$_m$,
40.2%

8.20 34.9%

8.22 1.79 Btu/lb$_m$

8.24 1.78 Btu/lb$_m$

8.26 32.9%

8.28 38.54%, 488.9 Btu/lb$_m$

8.30 32.91%

8.32 32.85%

8.34 35.4%

8.36 35.65%

8.38 32.8%

8.40 36.1%

8.42 36.86%

8.44 86%

8.46 78.50%

8.48 73%

8.50 59.7%

8.52 12,365 Btu/kWh

8.54 10.9 lb$_m$/kWh

8.56 1170°F

8.58 11.7%

8.60 43.65%

8.62 7819 Btu/kWh

8.64 2%

8.66 82.9%

8.68 41.47%

8.70 42.26%

8.72 43.80%

8.74 35.7%

8.76 35.5%

8.78 45.67%

8.80 33.7%

8.82 37.3%

8.84 37.26%

8.86 37.9%

8.88 46.6%

8.90 41.8%

8.92 41.8%

8.94 42.06%

8.96 43.9%, a gain of 0.7%

8.98 47.73%

Chapter 9

9.2 54.6%

9.4 11, 61.68%

9.6 292.4 kJ/kg, 207.6 kJ/kg, 58.48%

9.8 117 Btu/lb$_m$, 133 Btu/lb$_m$, 6.68

9.10 47.47%

9.12 187.6 kPa

9.14 2475.2°R, 56.47%

9.16 579 kPa, 54.08%

9.18 54.08%, 209 psia

9.20 54.08%, 173.9 psia

9.22 65.5 hp

9.24 69.7%

9.26 65.15%

9.28

Ratio	%
0.01	68.7
0.04	65.7
0.08	62.4
0.10	61.3

9.30 66.1%, 86.29 psia

9.32 23.07, 331.1 Btu/lb$_m$, 66.22%

9.34	Derivation	**9.46**	893.17°K
9.36	1325.78 psia, 3839°R	**9.48**	47.47%
9.38	36.86%	**9.50**	79.7%
9.40	359.04 Btu/lb$_m$, 1295°R, 48.21%	**9.52**	63.2%
9.42	44.8%	**9.54**	61%, 166.1 Btu/lb$_m$
9.44	880 kJ/kg, 424.2 kJ/kg, 48.21%	**9.56**	124.8 Btu/lb$_m$

Chapter 10

10.2	6.58, 3.8 kW, 28.8 kW	**10.36**	60.7 Btu/lb$_m$
10.4	Derivation	**10.38**	1.075, 342°R
10.6	9.11	**10.40**	2.06, 4.4 hp
10.8	23.9 tons	**10.42**	9.58 lb$_m$/min/ton
10.10	8.02 hp	**10.44**	16.5 hp
10.12	2	**10.46**	0.0282 lb$_m$/lb$_m$
10.14	25.13 hp	**10.48**	0.02808 lb$_m$/lb$_m$, 0.972 lb$_m$/lb$_m$,
10.16	5.33 kW, 25.33 kW		29.23 Btu/lb$_m$
10.18	5.74	**10.50**	0.020286 kg/kg, 0.979714 kg/kg,
10.20	5.68		49.34 kJ/kg
10.22	6.3 lb$_m$/min	**10.52**	848.3 Btu/min, $T_1/T_2 = 1.25$
10.24	43.26 lb$_m$/min	**10.54**	3.36
10.26	45.1 Btu/lb$_m$	**10.56**	10.62, 20.8 Btu/min
10.28	46.26 lb$_m$/min	**10.58**	4.05 hp
10.30	47.91 Btu/lb$_m$	**10.60**	16.63, 1505 Btu/min
10.32	56.6 Btu/lb$_m$	**10.62**	28.81 kW, 17.36
10.34	36.1 lb$_m$/min	**10.64**	329.1 kJ, 545.9 kJ

Chapter 11

11.2	pine wall = 3.08 × brick wall	**11.34**	36.3 Btu/hr·ft
11.4	$\dot{Q} = 1020$ Btu/hr	**11.36**	267 Btu/hr·ft
11.6	3.75 Btu/hr·ft·°F	**11.38**	502.2°F
11.8	90°C	**11.40**	1.35 in.
11.10	276.8 W/m^2	**11.42**	37.1 mm
11.12	491°F	**11.44**	1.11 in.
11.14	1.33 in.	**11.46**	0.9 Btu/hr·ft^2·°F
11.16	0.365 Btu/hr·ft^2·°F	**11.48**	1060 Btu/hr
11.18	15.98 × 10^6 Btu/hr	**11.50**	85 Btu/hr·ft^2·°F
11.20	2.87 Btu/hr·ft^2	**11.52**	86 Btu/hr·ft^2·°F
11.22	114.3°F at cork/mineral wool; 52.9°F	**11.54**	900 Btu/hr·ft^2·°F
	at mineral wool/plaster	**11.56**	435 Btu/hr·ft^2·°F
11.24	38.8 W	**11.58**	147 Btu/hr·ft^2·°F
11.26	43.3%	**11.60**	1056.96 Btu/hr
11.28	65°F at wallboard/mineral wool; 9.6°F	**11.62**	45,553 Btu/hr
	at mineral wool/brick	**11.64**	441.6 Btu/hr·ft^2
11.30	108.3 kW	**11.66**	263.1 Btu/hr·ft^2
11.32	Derivation	**11.68**	58.7 Btu/hr·ft^2

11.70 675 Btu/hr·ft^2

11.72 1.1 Btu/hr·ft^2·°F

11.74 1513 Btu/hr·ft^2

11.76 2894.3°K

11.78 41,229 Btu/hr

11.80 28.2°F

11.82 427°F

11.84 391.5°F

11.86 312.8°F

11.88 16,200 Btu/hr·ft^2

11.90 1800 W/m^2

11.92 69.2°F

11.94 9.1 kW/m^2

11.96 43.7°F

11.98 108.8°F

11.100 1.794 × 10^6 Btu/hr

11.102 40.5 Btu/hr·ft^2·°F

11.104 28.4 ft^2

11.106 2500 lb/hr

11.108 22.41 Btu/hr

11.110 5.6 Btu/hr·ft^2·°F

11.112 8.51 W/m^2

11.114 34.7 Btu/hr·ft^2

11.116 255.5 Btu/hr·ft

11.118 1568.9 Btu/hr

11.120 1510.43 Btu

11.122 349.54 Btu/hr·ft

11.124 1.308 Btu/hr·ft^2·°F

Appendix 3
Supplemental Tables

	Contents of Supplemental Tables	Page

632

TABLE A.1
Saturation: Temperature (Steam)*

Temp. Fahr. t	Press. Lbf./Sq. In. p	Specific Volume (ft³/lbₘ)		Internal Energy (Btu/lbₘ)			Enthalpy (Btu/lbₘ)			Entropy (Btu/lbₘ·°R)		
		Sat. Liquid v_f	Sat. Vapor v_g	Sat. Liquid u_f	Evap. u_{fg}	Sat. Vapor u_g	Sat. Liquid h_f	Evap. h_{fg}	Sat. Vapor h_g	Sat. Liquid s_f	Evap. s_{fg}	Sat. Vapor s_g
32	.08859	.016022	3305.	.01	1021.2	1021.2	.01	1075.4	1075.4	.00003	2.1870	2.1870
32.018	.08866	.016022	3302.	.00	1021.2	1021.2	.01	1075.4	1075.4	.00000	2.1869	2.1869
35	.09992	.016021	2948.	2.99	1019.2	1022.2	3.00	1073.7	1076.7	.00607	2.1704	2.1764
40	.12166	.016020	2445.	8.02	1015.8	1023.9	8.02	1070.9	1078.9	.01617	2.1430	2.1592
45	.14748	.016021	2037.	13.04	1012.5	1025.5	13.04	1068.1	1081.1	.02618	2.1162	2.1422
50	.17803	.016024	1704.2	18.06	1009.1	1027.2	18.06	1065.2	1083.3	.03607	2.0899	2.1259
60	.2563	.016035	1206.9	28.08	1002.4	1030.4	28.08	1059.6	1087.7	.05555	2.0388	2.0943
70	.3632	.016051	867.7	38.09	995.6	1033.7	38.09	1054.0	1092.0	.07463	1.9896	2.0642
80	.5073	.016073	632.8	48.08	988.9	1037.0	48.09	1048.3	1096.4	.09332	1.9423	2.0356
90	.6988	.016099	467.7	58.07	982.2	1040.2	58.07	1042.7	1100.7	.11165	1.8966	2.0083
100	.9503	.016130	350.0	68.04	975.4	1043.5	68.05	1037.0	1105.0	.12963	1.8526	1.9822
110	1.2763	.016166	265.1	78.02	968.7	1046.7	78.02	1031.3	1109.3	.14730	1.8101	1.9574
120	1.6945	.016205	203.0	87.99	961.9	1049.9	88.00	1025.5	1113.5	.16465	1.7690	1.9336
130	2.225	.016247	157.17	97.97	955.1	1053.0	97.98	1019.8	1117.8	.18172	1.7292	1.9109
140	2.892	.016293	122.88	107.95	948.2	1056.2	107.96	1014.0	1121.9	.19851	1.6907	1.8892
150	3.722	.016343	96.99	117.95	941.3	1059.3	117.96	1008.1	1126.1	.21503	1.6533	1.8684
160	4.745	.016395	77.23	127.94	934.4	1062.3	127.96	1002.2	1130.1	.23130	1.6171	1.8484
170	5.996	.016450	62.02	137.95	927.4	1065.4	137.97	996.2	1134.2	.24732	1.5819	1.8293
180	7.515	.016509	50.20	147.97	920.4	1068.3	147.99	990.2	1138.2	.26311	1.5478	1.8109
190	9.343	.016570	40.95	158.0	913.3	1071.3	158.03	984.1	1142.1	.27866	1.5146	1.7932
200	11.529	.016634	33.63	168.04	906.2	1074.2	168.07	977.9	1145.9	.29400	1.4822	1.7762
210	14.125	.016702	27.82	178.10	898.9	1077.0	178.14	971.6	1149.7	.30913	1.4508	1.7599
220	17.188	.016772	23.15	188.17	891.7	1079.8	188.22	965.3	1153.5	.32406	1.4201	1.7441
230	20.78	.016845	19.386	198.26	884.3	1082.6	198.32	958.8	1157.1	.33880	1.3901	1.7289
240	24.97	.016922	16.327	208.36	876.9	1085.3	208.44	952.3	1160.7	.35335	1.3609	1.7143
250	29.82	.017001	13.826	218.49	869.4	1087.9	218.59	945.6	1164.2	.36772	1.3324	1.7001
260	35.42	.017084	11.768	228.64	861.8	1090.5	228.76	938.8	1167.6	.38193	1.3044	1.6864

TABLE A.1 (cont'd.)

Temp. Fahr. t	Press. Lbf./Sq.In. p	Specific Volume (ft³/lb$_m$) Sat. Liquid v_f	Sat. Vapor v_g	Internal Energy (Btu/lb$_m$) Sat. Liquid u_f	Evap. u_{fg}	Sat. Vapor u_g	Enthalpy (Btu/lb$_m$) Sat. Liquid h_f	Evap. h_{fg}	Sat. Vapor h_g	Entropy (Btu/lb$_m$·°R) Sat. Liquid s_f	Evap. s_{fg}	Sat. Vapor s_g
270	41.85	.017170	10.066	238.82	854.1	1093.0	238.95	932.0	1170.9	.39597	1.2771	1.6731
280	49.18	.017259	8.650	249.02	846.3	1095.4	249.18	924.9	1174.1	.40986	1.2504	1.6602
290	57.53	.017352	7.467	259.25	838.5	1097.7	259.44	917.8	1177.2	.42360	1.2241	1.6477
300	66.98	.017448	6.472	269.52	830.5	1100.0	269.73	910.4	1180.2	.43720	1.1984	1.6356
310	77.64	.017548	5.632	279.81	822.3	1102.1	280.06	903.0	1183.0	.45067	1.1731	1.6238
320	89.60	.017652	4.919	290.14	814.1	1104.2	290.43	895.3	1185.8	.46400	1.1483	1.6123
330	103.00	.017760	4.312	300.51	805.7	1106.2	300.84	887.5	1188.4	.47722	1.1238	1.6010
340	117.93	.017872	3.792	310.91	797.1	1108.0	311.30	879.5	1190.8	.49031	1.0997	1.5901
350	134.53	.017988	3.346	321.35	788.4	1109.8	321.80	871.3	1193.1	.50329	1.0760	1.5793
360	152.92	.018108	2.961	331.84	779.6	1111.4	332.35	862.9	1195.2	.51617	1.0526	1.5688
370	173.23	.018233	2.628	342.37	770.6	1112.9	342.96	854.2	1197.2	.52894	1.0295	1.5585
380	195.60	.018363	2.339	352.95	761.4	1114.3	353.62	845.4	1199.0	.54163	1.0067	1.5483
390	220.2	.018498	2.087	363.58	752.0	1115.6	364.34	836.2	1200.6	.55422	.9841	1.5383
400	247.1	.018638	1.8661	374.27	742.4	1116.6	375.12	826.8	1202.0	.56672	.9617	1.5284
425	325.6	.019014	1.4249	401.24	717.4	1118.6	402.38	802.1	1204.5	.59767	.9066	1.5043
450	422.1	.019433	1.1011	428.6	690.9	1119.5	430.2	775.4	1205.6	.6282	.8523	1.4806
475	539.3	.019901	.8594	456.6	662.6	1119.2	458.5	746.4	1204.9	.6586	.7985	1.4571
500	680.0	.02043	.6761	485.1	632.3	1117.4	487.7	714.8	1202.5	.6888	.7448	1.4335
525	847.1	.02104	.5350	514.5	599.5	1113.9	517.8	680.0	1197.8	.7191	.6906	1.4007
550	1044.0	.02175	.4249	544.9	563.7	1108.6	549.1	641.6	1190.6	.7497	.6354	1.3851
575	1274.0	.02259	.3378	576.5	524.3	1100.8	581.9	598.6	1180.4	.7808	.5785	1.3593
600	1541.0	.02363	.2677	609.9	480.1	1090.0	616.7	549.7	1166.4	.8130	.5187	1.3317
625	1849.7	.02494	.2103	645.7	429.4	1075.1	654.2	492.9	1147.0	.8467	.4544	1.3010
650	2205.	.02673	.16206	685.0	368.7	1053.7	695.9	423.9	1119.8	.8831	.3820	1.2651
675	2616.	.02951	.11952	731.0	289.3	1020.3	745.3	332.9	1078.2	.9252	.2934	1.2186
700	3090.	.03666	.07438	801.7	145.9	947.7	822.7	167.5	990.2	.9902	.1444	1.1346
705.44	3204.	.05053	.05053	872.6	0	872.6	902.5	0	902.5	1.0580	0	1.0580

TABLE A.2
Saturation Pressures (Steam)

Press. Lbf. Sq. In. p	Temp. Fahr. t	Specific Volume (ft³/lbm) Sat. Liquid v_f	Sat. Vapor v_g	Internal Energy (Btu/lbm) Sat. Liquid u_f	Evap. u_{fg}	Sat. Vapor u_g	Enthalpy (Btu/lbm) Sat. Liquid h_f	Evap. h_{fg}	Sat. Vapor h_g	Entropy (Btu/lbm°R) Sat. Liquid s_f	Evap. s_{fg}	Sat. Vapor s_g
.50	79.56	.016071	641.5	47.64	989.2	1036.9	47.65	1048.6	1096.2	.09250	1.9443	2.0368
1.0	101.70	.016136	333.6	69.74	974.3	1044.0	69.74	1036.0	1105.8	.13266	1.8453	1.9779
1.5	115.65	.016187	227.7	83.65	964.8	1048.5	83.65	1028.0	1111.7	.15714	1.7867	1.9438
2.0	126.04	.016230	173.75	94.02	957.8	1051.8	94.02	1022.1	1116.1	.17499	1.7448	1.9198
3.0	141.43	.016300	118.72	109.38	947.2	1056.6	109.39	1013.1	1122.5	.20089	1.6852	1.8861
4.0	152.93	.016358	90.64	120.88	939.3	1060.2	120.89	1006.4	1127.3	.21983	1.6426	1.8624
5.0	162.21	.016407	73.53	130.15	932.9	1063.0	130.17	1000.9	1131.0	.23486	1.6093	1.8441
7.5	179.91	.016508	50.30	147.88	920.4	1068.3	147.90	990.2	1138.1	.26297	1.5481	1.8110
10	193.19	.016590	38.42	161.20	911.0	1072.2	161.23	982.1	1143.3	.28358	1.5041	1.7877
14.696	211.99	.016715	26.80	180.10	897.5	1077.6	180.15	970.4	1150.5	.31212	1.4446	1.7567
15	213.03	.016723	26.29	181.14	896.8	1077.9	181.19	969.7	1150.9	.31367	1.4414	1.7551
20	227.96	.016830	20.09	196.19	885.8	1082.0	196.26	960.1	1156.4	.33580	1.3962	1.7320
25	240.08	.016922	16.306	208.44	876.9	1085.3	208.52	952.2	1160.7	.35345	1.3607	1.7142
30	250.34	.017004	13.748	218.84	869.2	1088.0	218.93	945.4	1164.3	.36821	1.3314	1.6996
35	259.30	.017078	11.900	227.93	862.4	1090.3	228.04	939.3	1167.4	.38093	1.3064	1.6873
40	267.26	.017146	10.501	236.03	856.2	1092.3	236.16	933.8	1170.0	.39214	1.2845	1.6767
45	274.46	.017209	9.403	243.37	850.7	1094.0	243.51	928.8	1172.3	.40218	1.2651	1.6673
50	281.03	.017269	8.518	250.08	845.5	1095.6	250.24	924.2	1174.4	.41129	1.2476	1.6589
55	287.10	.017325	7.789	256.28	840.8	1097.0	256.46	919.9	1176.3	.41963	1.2317	1.6513
60	292.73	.017378	7.177	262.06	836.3	1098.3	262.25	915.8	1178.0	.42733	1.2170	1.6444
65	298.00	.017429	6.657	267.46	832.1	1099.5	267.67	911.9	1179.6	.43450	1.2035	1.6380
70	302.93	.017478	6.209	272.56	828.1	1100.6	272.79	908.3	1181.0	.44120	1.1909	1.6321
75	307.63	.017524	5.818	277.37	824.3	1101.6	277.61	904.8	1182.4	.44749	1.1790	1.6265
80	312.07	.017570	5.474	281.95	820.6	1102.6	282.21	901.4	1183.6	.45344	1.1679	1.6214
85	316.29	.017613	5.170	286.30	817.1	1103.5	286.58	898.2	1184.8	.45907	1.1574	1.6165
90	320.31	.017655	4.898	290.46	813.8	1104.3	290.76	895.1	1185.9	.46442	1.1475	1.6119
95	324.16	.017696	4.654	294.45	810.6	1105.0	294.76	892.1	1186.9	.46952	1.1380	1.6076
100	327.86	.017736	4.434	298.28	807.5	1105.8	298.61	889.2	1187.8	.47439	1.1290	1.6034
105	331.41	.017775	4.234	301.97	804.5	1106.5	302.31	886.4	1188.7	.47906	1.1204	1.5995
110	334.82	.017813	4.051	305.52	801.6	1107.1	305.88	883.7	1189.6	.48355	1.1122	1.5957
115	338.12	.017850	3.884	308.95	798.8	1107.7	309.33	881.0	1190.4	.48786	1.1042	1.5921
120	341.30	.017886	3.730	312.27	796.0	1108.3	312.67	878.5	1191.1	.49201	1.0966	1.5886
125	344.39	.017922	3.588	315.49	793.3	1108.8	315.90	875.9	1191.8	.49602	1.0893	1.5853

Press. Lbf. Sq. In. p	Temp. Fahr. t	Specific Volume (ft³/lb$_m$)		Internal Energy (Btu/lb$_m$)			Enthalpy (Btu/lb$_m$)			Entropy (Btu/lb$_m$·°R)		
		Sat. Liquid v_f	Sat. Vapor v_g	Sat. Liquid u_f	Evap. u_{fg}	Sat. Vapor u_g	Sat. Liquid h_f	Evap. h_{fg}	Sat. Vapor h_g	Sat. Liquid s_f	Evap. s_{fg}	Sat. Vapor s_g
130	347.37	.017957	3.457	318.61	790.7	1109.4	319.04	873.5	1192.5	.49989	1.0822	1.5821
135	350.27	.017991	3.335	321.64	788.2	1109.8	322.08	871.1	1193.2	.50364	1.0754	1.5790
140	353.08	.018024	3.221	324.58	785.7	1110.3	325.05	868.7	1193.8	.50727	1.0688	1.5761
145	355.82	.018057	3.115	327.45	783.3	1110.8	327.93	866.4	1194.4	.51079	1.0624	1.5732
150	358.48	.018089	3.016	330.24	781.0	1111.2	330.75	864.2	1194.9	.51422	1.0562	1.5704
160	363.60	.018152	2.836	335.63	776.4	1112.0	336.16	859.8	1196.0	.52078	1.0443	1.5651
170	368.47	.018214	2.676	340.76	772.0	1112.7	341.33	855.6	1196.9	.52700	1.0330	1.5600
180	373.13	.018273	2.533	345.68	767.7	1113.4	346.29	851.5	1197.8	.53292	1.0223	1.5553
190	377.59	.018331	2.405	350.39	763.6	1114.0	351.04	847.5	1198.6	.53857	1.0122	1.5507
200	381.86	.018387	2.289	354.9	759.6	1114.6	355.6	843.7	1199.3	.5440	1.0025	1.5464
225	391.87	.018523	2.043	365.6	750.2	1115.8	366.3	834.5	1200.8	.5566	.9799	1.5365
250	401.04	.018653	1.8448	375.4	741.4	1116.7	376.2	825.8	1202.1	.5680	.9594	1.5274
275	409.52	.018777	1.6813	384.5	733.0	1117.5	385.4	817.6	1203.1	.5786	.9406	1.5192
300	417.43	.018896	1.5442	393.0	725.1	1118.2	394.1	809.8	1203.9	.5883	.9232	1.5115
350	431.82	.019124	1.3267	408.7	710.3	1119.0	409.9	795.0	1204.9	.6060	.8917	1.4978
400	444.70	.019340	1.1620	422.8	696.7	1119.5	424.2	781.2	1205.5	.6218	.8638	1.4856
450	456.39	.019547	1.0326	435.7	683.9	1119.6	437.4	768.2	1205.6	.6360	.8385	1.4746
500	467.13	.019748	.9283	447.7	671.7	1119.4	449.5	755.8	1205.3	.6490	.8154	1.4645
550	477.07	.019943	.8423	458.9	660.2	1119.1	460.9	743.9	1204.8	.6611	.7941	1.4551
600	486.33	.02013	.7702	469.4	649.1	1118.6	471.7	732.4	1204.1	.6723	.7742	1.4464
700	503.23	.02051	.6558	488.9	628.2	1117.0	491.5	710.5	1202.0	.6927	.7378	1.4305
800	518.36	.02087	.5691	506.6	608.4	1115.0	509.7	689.6	1199.3	.7110	.7050	1.4160
900	532.12	.02123	.5009	523.0	589.6	1112.6	526.6	669.5	1196.0	.7277	.6750	1.4027
1000	544.75	.02159	.4459	538.4	571.5	1109.9	542.4	650.0	1192.4	.7432	.6471	1.3903
1250	572.56	.02250	.3454	573.4	528.3	1101.7	578.6	603.0	1181.6	.7778	.5841	1.3619
1500	596.39	.02346	.2769	605.0	486.9	1091.8	611.5	557.2	1168.7	.8082	.5276	1.3359
1750	617.31	.02450	.2268	634.4	445.9	1080.2	642.3	511.4	1153.7	.8361	.4748	1.3109
2000	649.20	.02565	.18813	662.4	404.2	1066.6	671.1	464.4	1136.3	.8623	.4238	1.2861
2250	652.90	.02698	.15692	689.9	360.7	1050.6	701.1	414.8	1115.9	.8876	.3728	1.2604
2500	668.31	.02860	.13059	717.7	313.4	1031.0	730.9	360.5	1091.4	.9131	.3196	1.2327
2750	682.46	.03077	.10717	747.3	258.6	1005.9	763.0	297.4	1060.4	.9401	.2604	1.2005
3000	695.52	.03431	.08404	783.4	185.4	968.8	802.5	213.0	1015.5	.9732	.1843	1.1575
3203.6	705.44	.05053	.05053	872.6	0	872.6	902.5	0	902.5	1.0580	0	1.0580

TABLE A.3

Properties of Superheated Steam

p (t Sat.)		14.696 (211.99)				20 (227.96)				30 (250.34)		
Vapor (psia)												
t	v	u	h	s	v	u	h	s	v	u	h	s
Sat.	26.80	1077.6	1150.5	1.7567	20.09	1082.0	1156.4	1.7320	13.748	1088.0	1164.3	1.6996
150	24.10	1054.5	1120.0	1.7090	17.532	1052.0	1116.9	1.6710	11.460	1047.3	1111.0	1.6185
160	24.54	1058.2	1125.0	1.7171	17.870	1056.0	1122.1	1.6795	11.701	1051.6	1116.5	1.6275
170	24.98	1062.0	1129.9	1.7251	18.204	1059.9	1127.2	1.6877	11.938	1055.8	1122.0	1.6364
180	25.42	1065.7	1134.9	1.7328	18.535	1063.8	1132.4	1.6957	12.172	1059.9	1127.5	1.6449
190	25.85	1069.5	1139.8	1.7405	18.864	1067.6	1137.4	1.7036	12.403	1064.0	1132.9	1.6533
200	26.29	1073.2	1144.7	1.7479	19.191	1071.4	1142.5	1.7113	12.631	1068.1	1138.2	1.6615
210	26.72	1076.9	1149.5	1.7553	19.515	1075.2	1147.5	1.7188	12.857	1072.1	1143.5	1.6694
220	27.15	1080.6	1154.4	1.7624	19.837	1079.0	1152.4	1.7262	13.081	1076.1	1148.7	1.6771
230	27.57	1084.2	1159.2	1.7695	20.157	1082.8	1157.4	1.7335	13.303	1080.0	1153.9	1.6847
240	28.00	1087.9	1164.0	1.7764	20.475	1086.5	1162.3	1.7405	13.523	1084.0	1159.0	1.6921
250	28.42	1091.5	1168.8	1.7832	20.79	1090.3	1167.2	1.7475	13.741	1087.9	1164.1	1.6994
260	28.85	1095.2	1173.6	1.7899	21.11	1094.0	1172.1	1.7543	13.958	1091.7	1169.2	1.7064
270	29.27	1098.8	1178.4	1.7965	21.42	1097.7	1177.0	1.7610	14.173	1095.6	1174.2	1.7134
280	29.69	1102.4	1183.1	1.8030	21.73	1101.4	1181.8	1.7676	14.387	1099.4	1179.2	1.7202
290	30.11	1106.0	1187.9	1.8094	22.05	1105.0	1186.6	1.7741	14.600	1103.2	1184.2	1.7269
300	30.52	1109.6	1192.6	1.8157	22.36	1108.7	1191.5	1.7805	14.812	1106.9	1189.2	1.7334
310	30.94	1113.2	1197.4	1.8219	22.67	1112.4	1196.3	1.7868	15.023	1110.7	1194.1	1.7399
320	31.36	1116.8	1202.1	1.8280	22.98	1116.0	1201.0	1.7930	15.233	1114.4	1199.0	1.7462
330	31.77	1120.4	1206.8	1.8340	23.28	1119.7	1205.8	1.7991	15.442	1118.2	1203.9	1.7525
340	32.19	1124.0	1211.6	1.8400	23.59	1123.3	1210.6	1.8051	15.651	1121.9	1208.8	1.7586
350	32.60	1127.6	1216.3	1.8458	23.90	1126.9	1215.4	1.8110	15.859	1125.6	1213.6	1.7646
360	33.02	1131.2	1221.0	1.8516	24.21	1130.6	1220.1	1.8168	16.067	1129.3	1218.5	1.7706
370	33.43	1134.8	1225.7	1.8574	24.51	1134.2	1224.9	1.8226	16.273	1133.0	1223.3	1.7765
380	33.84	1138.4	1230.5	1.8630	24.82	1137.8	1229.7	1.8283	16.480	1136.7	1228.1	1.7822
390	34.26	1142.0	1235.2	1.8686	25.12	1141.4	1234.4	1.8340	16.686	1140.3	1233.0	1.7880
400	34.67	1145.6	1239.9	1.8741	25.43	1145.1	1239.2	1.8395	16.891	1144.0	1237.8	1.7936
420	35.49	1152.8	1249.3	1.8850	26.03	1152.3	1248.7	1.8504	17.301	1151.4	1247.4	1.8047
440	36.31	1160.1	1258.8	1.8956	26.64	1159.6	1258.2	1.8611	17.709	1158.7	1257.0	1.8155
460	37.13	1167.3	1268.3	1.9060	27.25	1166.9	1267.7	1.8716	18.116	1166.1	1266.6	1.8260
480	37.95	1174.6	1277.8	1.9162	27.85	1174.2	1277.2	1.8819	18.523	1173.4	1276.2	1.8364

TABLE A.3 (cont'd)

Vapor (psia)

p (t Sat.)	14.696 (211.99)				20 (227.96)				30 (250.34)			
t	v	u	h	s	v	u	h	s	v	u	h	s
500	38.77	1181.8	1287.3	1.9263	28.46	1181.5	1286.8	1.8919	18.928	1180.8	1285.9	1.8465
520	39.59	1189.1	1296.8	1.9361	29.06	1188.8	1296.3	1.9018	19.333	1188.2	1295.5	1.8564
540	40.41	1196.5	1306.4	1.9457	29.66	1196.1	1305.9	1.9114	19.737	1195.5	1305.1	1.8661
560	41.22	1203.8	1315.9	1.9552	30.26	1203.5	1315.5	1.9210	20.140	1203.0	1314.8	1.8757
580	42.04	1211.2	1325.5	1.9645	30.87	1210.9	1325.2	1.9303	20.543	1210.4	1324.4	1.8851
600	42.86	1218.6	1335.2	1.9737	31.47	1218.4	1334.8	1.9395	20.95	1217.8	1334.1	1.8943
620	43.67	1226.1	1344.8	1.9827	32.07	1225.8	1344.5	1.9485	21.35	1225.3	1343.8	1.9034
640	44.49	1233.5	1354.5	1.9916	32.67	1233.3	1354.2	1.9575	21.75	1232.8	1353.6	1.9123
660	45.30	1241.0	1364.2	2.0004	33.27	1240.8	1363.9	1.9662	22.15	1240.4	1363.3	1.9211
680	46.12	1248.6	1374.0	2.0090	33.87	1248.3	1373.7	1.9749	22.55	1247.9	1373.1	1.9298
700	46.93	1256.1	1383.8	2.0175	34.47	1255.9	1383.5	1.9834	22.95	1255.5	1383.0	1.9384
720	47.75	1263.7	1393.6	2.0259	35.07	1263.5	1393.3	1.9918	23.35	1263.2	1392.8	1.9468
740	48.56	1271.4	1403.4	2.0342	35.66	1271.2	1403.2	2.0001	23.75	1270.8	1402.7	1.9551
760	49.37	1279.0	1413.3	2.0424	36.26	1278.8	1413.0	2.0082	24.15	1278.5	1412.6	1.9633
780	50.19	1286.7	1423.2	2.0504	36.86	1286.5	1423.0	2.0163	24.55	1286.2	1422.5	1.9714
800	51.00	1294.4	1433.1	2.0584	37.46	1294.3	1432.9	2.0243	24.95	1294.0	1432.5	1.9793
850	53.03	1313.9	1458.1	2.0778	38.96	1313.8	1457.9	2.0438	25.95	1313.5	1457.6	1.9988
900	55.07	1333.6	1483.4	2.0967	40.45	1333.5	1483.2	2.0627	26.95	1333.2	1482.8	1.0178
950	57.10	1353.5	1508.8	2.1151	41.94	1353.4	1508.6	2.0810	27.95	1353.2	1508.3	2.0362
1000	59.13	1373.7	1534.5	2.1330	43.44	1373.5	1534.3	2.0989	28.95	1373.3	1534.0	2.0541
1100	63.19	1414.6	1586.4	2.1674	46.42	1414.5	1586.3	2.1334	30.94	1414.3	1586.1	2.0886
1200	67.25	1456.5	1639.3	2.2003	49.41	1456.4	1639.2	2.1663	32.93	1456.2	1639.1	2.1215
1300	71.30	1499.3	1693.2	2.2318	52.39	1499.2	1693.1	2.1978	34.92	1499.1	1692.9	2.1530
1400	75.36	1543.0	1747.9	2.2621	55.37	1542.9	1747.9	2.2281	36.91	1542.8	1747.7	2.1833
1500	79.42	1587.6	1803.6	2.2912	58.35	1587.6	1803.5	2.2572	38.90	1587.5	1803.4	2.2125
1600	83.47	1633.2	1860.2	2.3194	61.33	1633.2	1860.1	2.2854	40.88	1633.1	1860.0	2.2407
1800	91.58	1727.0	1976.1	2.3731	67.29	1727.0	1976.1	2.3391	44.86	1726.9	1976.0	2.2944
2000	99.69	1824.4	2095.5	2.4237	73.25	1824.3	2095.4	2.3897	48.83	1824.2	2095.3	2.3450
2200	107.80	1924.8	2218.0	2.4716	79.21	1924.8	2218.0	2.4376	52.81	1924.7	2217.9	2.3929
2400	115.91	2028.1	2343.4	2.5170	85.17	2028.1	2343.3	2.4830	56.78	2028.0	2343.3	2.4383

p (t Sat.)	40 (267.26)				50 (281.03)				60 (292.73)			
t	v	u	h	s	v	u	h	s	v	u	h	s
Sat.	10.501	1092.3	1170.0	1.6767	8.518	1095.6	1174.4	1.6589	7.177	1098.3	1178.0	1.6444
200	9.346	1064.6	1133.8	1.6243	7.370	1060.9	1129.1	1.5940	6.047	1057.1	1124.2	1.5680
210	9.523	1068.8	1139.3	1.6327	7.519	1065.4	1135.0	1.6029	6.178	1061.9	1130.5	1.5774
220	9.699	1073.0	1144.8	1.6409	7.665	1069.9	1140.8	1.6115	6.307	1066.6	1136.6	1.5864
230	9.872	1077.2	1150.3	1.6488	7.810	1074.2	1146.5	1.6198	6.432	1071.2	1142.6	1.5952
240	10.043	1081.3	1155.6	1.6565	7.952	1078.6	1152.1	1.6279	6.556	1075.7	1148.5	1.6036
250	10.212	1085.4	1161.0	1.6641	8.092	1082.8	1157.7	1.6358	6.677	1080.1	1154.3	1.6118
260	10.380	1089.4	1166.2	1.6714	8.231	1087.0	1163.1	1.6434	6.797	1084.5	1160.0	1.6198
270	10.546	1093.4	1171.4	1.6786	8.368	1091.1	1168.5	1.6509	6.915	1088.8	1165.6	1.6275
280	10.711	1097.3	1176.6	1.6857	8.504	1095.2	1173.9	1.6582	7.031	1093.0	1171.1	1.6351
290	10.875	1101.2	1181.7	1.6926	8.639	1099.2	1179.2	1.6653	7.146	1097.2	1176.5	1.6424
300	11.038	1105.1	1186.8	1.6993	8.772	1103.2	1184.4	1.6722	7.260	1101.3	1181.9	1.6496
310	11.200	1109.0	1191.9	1.7059	8.904	1107.2	1189.6	1.6790	7.373	1105.4	1187.3	1.6565
320	11.360	1112.8	1196.9	1.7124	9.036	1111.2	1194.8	1.6857	7.485	1109.5	1192.6	1.6634
330	11.520	1116.6	1201.9	1.7188	9.166	1115.1	1199.9	1.6922	7.596	1113.5	1197.8	1.6700
340	11.680	1120.4	1206.9	1.7251	9.296	1119.0	1205.0	1.6986	7.706	1117.4	1203.0	1.6766
350	11.838	1124.2	1211.8	1.7312	9.425	1122.8	1210.0	1.7049	7.815	1121.4	1208.2	1.6830
360	11.996	1128.0	1216.8	1.7373	9.553	1126.7	1215.1	1.7110	7.924	1125.3	1213.3	1.6893
370	12.153	1131.7	1221.7	1.7432	9.681	1130.5	1220.1	1.7171	8.032	1129.2	1218.4	1.6955
380	12.310	1135.5	1226.6	1.7491	9.808	1134.3	1225.0	1.7231	8.139	1133.1	1223.5	1.7015
390	12.467	1139.2	1231.5	1.7549	9.935	1138.1	1230.0	1.7290	8.246	1136.9	1228.5	1.7075
400	12.623	1143.0	1236.4	1.7606	10.061	1141.9	1235.0	1.7348	8.353	1140.8	1233.5	1.7134
420	12.933	1150.4	1246.1	1.7718	10.312	1149.4	1244.8	1.7461	8.565	1148.4	1243.5	1.7249
440	13.243	1157.8	1255.8	1.7828	10.562	1156.9	1254.6	1.7572	8.775	1156.0	1253.4	1.7360
460	13.551	1165.2	1265.5	1.7934	10.811	1164.4	1264.4	1.7679	8.984	1163.6	1263.3	1.7469
480	13.858	1172.7	1275.2	1.8038	11.059	1171.9	1274.2	1.7784	9.192	1171.1	1273.2	1.7575
500	14.164	1180.1	1284.9	1.8140	11.305	1179.4	1284.0	1.7887	9.399	1178.6	1283.0	1.7678
520	14.469	1187.5	1294.6	1.8240	11.551	1186.8	1293.7	1.7988	9.606	1186.2	1292.8	1.7780
540	14.774	1194.9	1304.3	1.8338	11.796	1194.3	1303.5	1.8086	9.811	1193.7	1302.6	1.7879
560	15.078	1202.4	1314.0	1.8434	12.041	1201.8	1313.2	1.8183	10.016	1201.2	1312.4	1.7976
580	15.382	1209.9	1323.7	1.8529	12.285	1209.3	1323.0	1.8277	10.221	1208.8	1322.2	1.8071

TABLE A.3 (cont'd.)

p (t Sat.)		40 (267.26)				50 (281.03)				60 (292.73)			
t	v	u	h	s	v	u	h	s	v	u	h	s	
600	15.685	1217.3	1333.4	1.8621	12.529	1216.8	1332.8	1.8371	10.425	1216.3	1332.1	1.8165	
620	15.988	1224.8	1343.2	1.8713	12.772	1224.4	1342.5	1.8462	10.628	1223.9	1341.9	1.8257	
640	16.291	1232.4	1353.0	1.8802	13.015	1231.9	1352.4	1.8552	10.832	1231.5	1351.7	1.8347	
660	16.593	1239.9	1362.8	1.8891	13.258	1239.5	1362.2	1.8641	11.035	1239.1	1361.6	1.8436	
680	16.895	1247.5	1372.6	1.8977	13.500	1247.1	1372.0	1.8728	11.237	1246.7	1371.5	1.8523	
700	17.196	1255.1	1382.4	1.9063	13.742	1254.8	1381.9	1.8814	11.440	1254.4	1381.4	1.8609	
720	17.498	1262.8	1392.3	1.9147	13.984	1262.4	1391.8	1.8898	11.642	1262.0	1391.3	1.8694	
740	17.799	1270.5	1402.2	1.9231	14.226	1270.1	1401.7	1.8982	11.844	1269.7	1401.2	1.8778	
760	18.100	1278.2	1412.1	1.9313	14.467	1277.8	1411.7	1.9064	12.045	1277.5	1411.2	1.8860	
780	18.401	1285.9	1422.1	1.9394	14.708	1285.6	1421.7	1.9145	12.247	1285.2	1421.2	1.8942	
800	18.701	1293.7	1432.1	1.9474	14.949	1293.3	1431.7	1.9225	12.448	1293.0	1431.2	1.9022	
850	19.452	1313.2	1457.2	1.9669	15.551	1312.9	1456.8	1.9421	12.951	1312.7	1456.4	1.9218	
900	20.202	1333.0	1482.5	1.9859	16.152	1332.7	1482.2	1.9611	13.452	1332.5	1481.8	1.9408	
950	20.951	1352.9	1508.0	2.0043	16.753	1352.7	1507.7	1.9796	13.954	1352.5	1507.4	1.9593	
1000	21.700	1373.1	1533.8	2.0223	17.352	1372.9	1533.5	1.9975	14.454	1372.7	1533.2	1.9773	
1100	23.20	1414.2	1585.9	2.0568	18.551	1414.0	1585.6	2.0321	15.454	1413.8	1585.4	2.0119	
1200	24.69	1456.1	1638.9	2.0897	19.747	1456.0	1638.7	2.0650	16.452	1455.8	1638.5	2.0448	
1300	26.18	1498.9	1692.8	2.1212	20.943	1498.8	1692.6	2.0966	17.449	1498.7	1692.4	2.0764	
1400	27.68	1542.7	1747.6	2.1515	22.138	1542.6	1747.4	2.1269	18.445	1542.5	1747.3	2.1067	
1500	29.17	1587.4	1803.3	2.1807	23.332	1587.3	1803.2	2.1561	19.441	1587.2	1803.0	2.1359	
1600	30.66	1633.0	1859.9	2.2089	24.53	1632.9	1859.8	2.1843	20.44	1632.8	1859.7	2.1641	
1800	33.64	1726.9	1975.9	2.2626	26.91	1726.8	1975.8	2.2380	22.43	1726.7	1975.7	2.2179	
2000	36.62	1824.2	2095.3	2.3132	29.30	1824.1	2095.2	2.2886	24.41	1824.0	2095.1	2.2685	
2200	39.61	1924.7	2217.8	2.3611	31.68	1924.6	2217.8	2.3365	26.40	1924.5	2217.7	2.3164	
2400	42.59	2028.0	2343.2	2.4066	34.07	2027.9	2343.1	2.3820	28.39	2027.8	2343.1	2.3618	

p (t Sat.)	80 (312.07)				100 (327.86)				140 (353.08)				180 (373.13)			
t	v	u	h	s	v	u	h	s	v	u	h	s	v	u	h	s
Sat.	5.474	1102.6	1183.6	1.6214	4.434	1105.8	1187.8	1.6034	3.221	1110.3	1193.8	1.5761	2.533	1113.4	1197.8	1.5553
250	*4.902*	*1074.5*	*1147.1*	*1.5720*												
260	*4.998*	*1079.3*	*1153.3*	*1.5806*												
270	*5.093*	*1083.9*	*1159.3*	*1.5890*												
280	*5.186*	*1088.5*	*1165.2*	*1.5970*												
290	*5.277*	*1092.9*	*1171.1*	*1.6049*												
300	*5.367*	*1097.4*	*1176.8*	*1.6125*	*4.228*	*1093.1*	*1171.4*	*1.5822*								
310	*5.456*	*1101.7*	*1182.5*	*1.6199*	*4.303*	*1097.7*	*1177.4*	*1.5900*	*2.978*	*1089.2*	*1166.4*	*1.5414*				
320	5.544	1106.0	1188.0	1.6271	*4.377*	*1102.3*	*1183.3*	*1.5976*	*3.037*	*1094.3*	*1173.0*	*1.5499*				
330	5.631	1110.2	1193.5	1.6341	4.449	1106.7	1189.1	1.6050	*3.094*	*1099.3*	*1179.4*	*1.5581*				
340	5.717	1114.3	1199.0	1.6409	4.521	1111.1	1194.8	1.6121	*3.150*	*1104.1*	*1185.7*	*1.5661*				
350	5.802	1118.5	1204.3	1.6476	4.592	1115.4	1200.4	1.6191	3.205	1108.9	1191.9	1.5737	*2.429*	*1101.7*	*1182.6*	*1.5368*
360	5.886	1122.5	1209.7	1.6541	4.662	1119.7	1205.9	1.6259	3.259	1113.5	1198.0	1.5812	*2.475*	*1106.9*	*1189.3*	*1.5450*
370	5.970	1126.6	1215.0	1.6605	4.731	1123.9	1211.4	1.6326	3.312	1118.1	1203.9	1.5884	*2.520*	*1111.9*	*1195.8*	*1.5529*
380	6.053	1130.6	1220.2	1.6668	4.799	1128.0	1216.8	1.6391	3.364	1122.6	1209.7	1.5954	2.563	1116.7	1202.1	1.5605
390	6.135	1134.6	1225.4	1.6730	4.867	1132.1	1222.2	1.6455	3.415	1127.0	1215.5	1.6022	2.606	1121.5	1208.3	1.5678
400	6.217	1138.5	1230.6	1.6790	4.934	1136.2	1227.5	1.6517	3.466	1131.4	1221.2	1.6088	2.648	1126.2	1214.2	
410	6.299	1142.5	1235.7	1.6850	5.001	1140.3	1232.8	1.6578	3.517	1135.7	1226.8	1.6153	2.690	1130.8	1220.4	1.5818
420	6.380	1146.4	1240.8	1.6908	5.068	1144.3	1238.1	1.6638	3.567	1139.9	1232.3	1.6217	2.731	1135.3	1226.3	1.5886
430	6.460	1150.3	1245.9	1.6966	5.134	1148.3	1243.3	1.6697	3.616	1144.1	1237.8	1.6279	2.771	1139.8	1232.1	1.5951
440	6.541	1154.2	1251.0	1.7022	5.199	1152.3	1248.5	1.6755	3.665	1148.3	1243.3	1.6339	2.811	1144.2	1237.8	1.6015
450	6.621	1158.0	1256.0	1.7078	5.265	1156.2	1253.6	1.6812	3.713	1152.4	1248.6	1.6399	2.850	1148.5	1243.4	1.6078
460	6.700	1161.9	1261.1	1.7133	5.330	1160.1	1258.8	1.6868	3.762	1156.5	1254.0	1.6458	2.889	1152.8	1249.0	1.6139
470	6.780	1165.7	1266.1	1.7188	5.394	1164.1	1263.9	1.6923	3.810	1160.6	1259.3	1.6515	2.928	1157.0	1254.6	1.6199
480	6.859	1169.6	1271.1	1.7241	5.459	1168.0	1269.0	1.6978	3.857	1164.7	1264.6	1.6572	2.966	1161.3	1260.1	1.6257
490	6.938	1173.4	1276.1	1.7294	5.523	1171.8	1274.0	1.7032	3.904	1168.7	1269.9	1.6627	3.005	1165.4	1265.5	1.6315
500	7.017	1177.2	1281.1	1.7346	5.587	1175.7	1279.1	1.7085	3.952	1172.7	1275.1	1.6682	3.042	1169.6	1270.9	1.6372
510													3.080	1173.7	1276.3	1.6427
520	7.173	1184.8	1291.0	1.7449	5.714	1183.5	1289.2	1.7189	4.045	1180.7	1285.5	1.6789	3.117	1177.8	1281.6	1.6482
530													3.154	1181.9	1286.9	1.6536
540	7.329	1192.4	1300.9	1.7549	5.840	1191.2	1299.2	1.7290	4.138	1188.6	1295.8	1.6893	3.191	1185.9	1292.2	1.6589
550													3.228	1190.0	1297.5	1.6642
560	7.485	1200.1	1310.9	1.7647	5.966	1198.9	1309.3	1.7390	4.230	1196.5	1306.0	1.6995	3.264	1194.0	1302.7	1.6693
570													3.301	1198.0	1307.9	1.6744
580	7.640	1207.7	1320.8	1.7743	6.091	1206.6	1319.3	1.7487	4.321	1204.3	1316.2	1.7094	3.337	1202.0	1313.1	1.6795
590													3.373	1206.0	1318.3	1.6844

p (t Sat.) t	80 (312.07) v	u	h	s	100 (327.86) v	u	h	s	140 (353.08) v	u	h	s	180 (373.13) v	u	h	s
600	7.794	1215.3	1330.7	1.7838	6.216	1214.2	1329.3	1.7582	4.412	1212.1	1326.4	1.7191	3.409	1210.0	1323.5	1.6893
620	7.948	1222.9	1340.6	1.7930	6.340	1221.9	1339.3	1.7675	4.502	1219.9	1336.6	1.7286	3.481	1217.9	1333.9	1.6990
640	8.102	1230.5	1350.5	1.8021	6.464	1229.6	1349.2	1.7767	4.592	1227.7	1346.7	1.7379	3.552	1225.8	1344.1	1.7084
660	8.255	1238.2	1360.4	1.8111	6.588	1237.3	1359.2	1.7857	4.682	1235.6	1356.8	1.7470	3.623	1233.8	1354.4	1.7177
680	8.408	1245.9	1370.4	1.8199	6.711	1245.0	1369.2	1.7946	4.771	1243.4	1367.0	1.7560	3.693	1241.7	1364.7	1.7268
700	8.561	1253.6	1380.3	1.8285	6.834	1252.8	1379.2	1.8033	4.860	1251.2	1377.1	1.7648	3.763	1249.6	1374.9	1.7357
720	8.714	1261.0	1390.3	1.8371	6.957	1260.5	1389.3	1.8118	4.949	1259.0	1387.2	1.7735	3.833	1257.5	1385.2	1.7445
740	8.866	1269.0	1400.3	1.8455	7.079	1268.3	1399.3	1.8203	5.037	1266.9	1397.4	1.7820	3.903	1265.4	1395.4	1.7531
760	9.018	1276.8	1410.3	1.8538	7.201	1276.1	1409.4	1.8286	5.125	1274.7	1407.5	1.7904	3.972	1273.4	1405.7	1.7615
780	9.170	1284.6	1420.3	1.8619	7.324	1283.9	1419.5	1.8368	5.214	1282.6	1417.7	1.7986	4.041	1281.3	1415.9	1.7699
800	9.321	1292.4	1430.4	1.8700	7.445	1291.8	1429.6	1.8449	5.301	1290.5	1427.9	1.8068	4.110	1289.3	1426.2	1.7781
820					7.567	1299.7	1439.7	1.8529	5.389	1298.5	1438.1	1.8148	4.179	1297.3	1436.5	1.7862
840					7.689	1307.6	1449.9	1.8607	5.477	1306.4	1448.3	1.8228	4.248	1305.3	1446.7	1.7942
850	9.700	1312.1	1455.7	1.8897												
860					7.810	1315.5	1460.0	1.8685	5.564	1314.4	1458.5	1.8306	4.316	1313.3	1457.1	1.8020
880					7.932	1323.5	1470.2	1.8762	5.651	1322.4	1468.8	1.8383	4.385	1321.3	1467.4	1.8098
900	10.078	1332.0	1481.2	1.9087	8.053	1331.5	1480.5	1.8838	5.739	1330.4	1479.1	1.8459	4.453	1329.4	1477.7	1.8175
920					8.174	1339.5	1490.7	1.8913	5.826	1338.5	1489.4	1.8535	4.521	1337.5	1488.1	1.8250
940					8.295	1347.5	1501.0	1.8987	5.913	1346.6	1499.8	1.8609	4.589	1345.6	1498.5	1.8325
950	10.455	1352.0	1506.8	1.9273												
960					8.416	1355.6	1511.3	1.9060	6.000	1354.7	1510.1	1.8683	4.657	1353.8	1508.9	1.8399
980					8.537	1363.7	1521.7	1.9132	6.086	1362.9	1520.5	1.8755	4.725	1362.0	1519.4	1.8472
1000	10.831	1372.3	1532.6	1.9453	8.657	1371.9	1532.1	1.9204	6.173	1371.0	1531.0	1.8827	4.793	1370.2	1529.8	1.8545
1020					8.778	1380.1	1542.5	1.9275	6.260	1379.3	1541.4	1.8898	4.861	1378.4	1540.3	1.8616
1040					8.899	1388.3	1552.9	1.9345	6.346	1387.5	1551.9	1.8969	4.928	1386.7	1550.9	1.8687
1060					9.019	1396.5	1563.4	1.9414	6.433	1395.8	1562.4	1.9039	4.996	1395.0	1561.4	1.8757
1080					9.140	1404.8	1573.9	1.9483	6.519	1404.1	1573.0	1.9107	5.063	1403.4	1572.0	1.8826
1100	11.583	1413.5	1584.9	1.9799	9.260	1413.1	1584.5	1.9551	6.605	1412.4	1583.6	1.9176	5.131	1411.7	1582.6	1.8894
1200	12.333	1455.5	1638.1	2.0130	9.861	1455.2	1637.7	1.9882	7.036	1454.6	1636.9	1.9507	5.467	1454.0	1636.1	1.9227
1300	13.081	1498.4	1692.1	2.0446	10.461	1498.2	1691.8	2.0198	7.466	1497.7	1691.1	1.9824	5.802	1497.2	1690.4	1.9544
1400	13.830	1542.3	1747.0	2.0749	11.060	1542.0	1746.7	2.0502	7.895	1541.6	1746.1	2.0129	6.137	1541.2	1745.6	1.9849
1500	14.577	1587.0	1802.8	2.1041	11.659	1586.8	1802.5	2.0794	8.324	1586.4	1802.0	2.0421	6.471	1586.0	1801.5	2.0142
1600	15.324	1632.6	1859.5	2.1323	12.257	1632.4	1859.3	2.1076	8.752	1632.1	1858.8	2.0704	6.804	1631.7	1858.4	2.0425
1800	16.818	1726.4	1975.5	2.1861	13.452	1726.4	1975.3	2.1614	9.607	1726.1	1975.0	2.1242	7.470	1725.8	1974.6	2.0964
2000	18.310	1823.9	2094.9	2.2367	14.647	1823.7	2094.8	2.2121	10.461	1823.5	2094.5	2.1749	8.135	1823.2	2094.2	2.1470
2200	19.802	1924.4	2217.5	2.2846	15.842	1924.3	2217.4	2.2600	11.315	1924.0	2217.1	2.2228	8.800	1923.7	2216.8	2.1950
2400	21.294	2027.7	2342.9	2.3301	17.036	2027.5	2342.8	2.3054	12.169	2027.3	2342.5	2.2682	9.465	2027.0	2342.2	2.2404

p (t Sat.)	200 (381.86)				300 (417.43)				350 (431.82)				400 (444.70)			
t	v	u	h	s	v	u	h	s	v	u	h	s	v	u	h	s
Sat.	2.289	1114.6	1199.3	1.5464	1.5442	1118.2	1203.9	1.5115	1.3267	1119.0	1204.9	1.4978	1.1620	1119.5	1205.5	1.4856
400	2.361	1123.5	1210.8	1.5600	*1.4915*	*1108.2*	*1191.0*	*1.4967*								
410	2.399	1128.2	1217.0	1.5672	*1.5221*	*1114.0*	*1198.5*	*1.5054*								
420	2.437	1132.9	1223.1	1.5741	1.5517	1119.6	1205.7	1.5136	*1.2950*	*1112.0*	*1195.9*	*1.4875*				
430	2.475	1137.5	1229.1	1.5809	1.5805	1125.0	1212.7	1.5216	*1.3219*	*1117.9*	*1203.6*	*1.4962*	*1.1257*	*1110.3*	*1193.6*	*1.4723*
440	2.511	1142.0	1234.9	1.5874	1.6086	1130.3	1219.6	1.5292	1.3480	1123.7	1211.0	1.5045	*1.1506*	*1116.6*	*1201.7*	*1.4814*
450	2.548	1146.4	1240.7	1.5938	1.6361	1135.4	1226.2	1.5365	1.3733	1129.2	1218.2	1.5125	1.1745	1122.6	1209.6	1.4901
460	2.584	1150.8	1246.5	1.6001	1.6630	1140.4	1232.7	1.5436	1.3979	1134.6	1225.2	1.5201	1.1977	1128.5	1217.1	1.4984
470	2.619	1155.2	1252.1	1.6062	1.6894	1145.3	1239.1	1.5505	1.4220	1139.9	1232.0	1.5275	1.2202	1134.1	1224.4	1.5063
480	2.654	1159.5	1257.7	1.6122	1.7154	1150.1	1245.3	1.5572	1.4455	1145.0	1238.6	1.5346	1.2421	1139.6	1231.5	1.5139
490	2.689	1163.7	1263.3	1.6181	1.7410	1154.8	1251.5	1.5637	1.4686	1150.0	1245.1	1.5415	1.2634	1144.9	1238.4	1.5212
500	2.724	1168.0	1268.8	1.6239	1.7662	1159.5	1257.5	1.5701	1.4913	1154.9	1251.5	1.5482	1.2843	1150.1	1245.2	1.5282
510	2.758	1172.2	1274.2	1.6295	1.7910	1164.1	1263.5	1.5763	1.5136	1159.7	1257.8	1.5546	1.3048	1155.2	1251.8	1.5351
520	2.792	1176.3	1279.7	1.6351	1.8156	1168.6	1269.4	1.5823	1.5356	1164.5	1263.9	1.5610	1.3249	1160.2	1258.2	1.5417
530	2.826	1180.5	1285.0	1.6405	1.8399	1173.1	1275.2	1.5882	1.5572	1169.1	1270.0	1.5671	1.3447	1165.0	1264.6	1.5481
540	2.860	1184.6	1290.4	1.6459	1.8640	1177.5	1281.0	1.5940	1.5786	1173.7	1276.0	1.5731	1.3642	1169.8	1270.8	1.5544
550	2.893	1188.7	1295.7	1.6512	1.8878	1181.9	1286.7	1.5997	1.5998	1178.3	1281.9	1.5790	1.3833	1174.6	1277.0	1.5605
560	2.926	1192.7	1301.0	1.6565	1.9114	1186.2	1292.3	1.6052	1.6207	1182.8	1287.7	1.5848	1.4023	1179.2	1283.0	1.5665
570	2.960	1196.8	1306.3	1.6616	1.9348	1190.5	1297.9	1.6107	1.6414	1187.2	1293.5	1.5904	1.4210	1183.8	1289.0	1.5723
580	2.993	1200.8	1311.6	1.6667	1.9580	1194.8	1303.5	1.6161	1.6619	1191.6	1299.3	1.5960	1.4395	1188.4	1294.9	1.5781
590	3.025	1204.9	1316.8	1.6717	1.9811	1199.0	1309.0	1.6214	1.6823	1196.0	1304.9	1.6014	1.4579	1192.9	1300.8	1.5837
600	3.058	1208.9	1322.1	1.6767	2.004	1203.2	1314.5	1.6266	1.7025	1200.3	1310.6	1.6068	1.4760	1197.3	1306.6	1.5892
620	3.123	1216.9	1332.5	1.6864	2.049	1211.6	1325.4	1.6368	1.7424	1208.9	1321.8	1.6172	1.5118	1206.1	1318.0	1.5999
640	3.188	1224.9	1342.9	1.6959	2.094	1220.0	1336.2	1.6467	1.7818	1217.4	1332.8	1.6274	1.5471	1214.8	1329.3	1.6103
660	3.252	1232.8	1353.2	1.7053	2.139	1228.2	1347.0	1.6564	1.8207	1225.8	1343.8	1.6372	1.5819	1223.4	1340.5	1.6203
680	3.316	1240.8	1363.5	1.7144	2.183	1236.4	1357.6	1.6658	1.8593	1234.2	1354.6	1.6469	1.6163	1231.9	1351.6	1.6301
700	3.379	1248.8	1373.8	1.7234	2.227	1244.6	1368.3	1.6751	1.8975	1242.5	1365.4	1.6562	1.6503	1240.4	1362.5	1.6397
720	3.442	1256.7	1384.1	1.7322	2.270	1252.8	1378.9	1.6841	1.9354	1250.8	1376.2	1.6654	1.6840	1248.8	1373.4	1.6490
740	3.505	1264.7	1394.4	1.7408	2.314	1261.0	1389.4	1.6930	1.9731	1259.1	1386.9	1.6744	1.7175	1257.2	1384.3	1.6581
760	3.568	1272.7	1404.7	1.7493	2.357	1269.1	1400.0	1.7017	2.0104	1267.3	1397.5	1.6832	1.7506	1265.5	1395.1	1.6670
780	3.631	1280.6	1415.0	1.7577	2.400	1277.3	1410.5	1.7103	2.0476	1275.6	1408.2	1.6919	1.7836	1273.8	1405.9	1.6758

TABLE A.3 (cont'd.)

p (t Sat.)	200 (381.86)				300 (417.43)				350 (431.82)				400 (444.70)			
t	v	u	h	s	v	u	h	s	v	u	h	s	v	u	h	s
800	3.693	1288.6	1425.3	1.7660	2.442	1285.4	1421.0	1.7187	2.085	1283.8	1418.8	1.7004	1.8163	1282.1	1416.6	1.6844
820	3.755	1296.6	1435.6	1.7741	2.485	1293.6	1431.5	1.7270	2.121	1292.0	1429.4	1.7088	1.8489	1290.5	1427.3	1.6928
840	3.818	1304.7	1446.0	1.7821	2.527	1301.7	1442.0	1.7351	2.158	1300.3	1440.0	1.7170	1.8813	1298.8	1438.0	1.7011
860	3.879	1312.7	1456.3	1.7900	2.569	1309.9	1452.5	1.7432	2.194	1308.5	1450.6	1.7251	1.9135	1307.1	1448.7	1.7093
880	3.941	1320.8	1466.7	1.7978	2.611	1318.1	1463.1	1.7511	2.231	1316.7	1461.2	1.7331	1.9456	1315.4	1459.4	1.7173
900	4.003	1328.9	1477.1	1.8055	2.653	1326.3	1473.6	1.7589	2.267	1325.0	1471.8	1.7409	1.9776	1323.7	1470.1	1.7252
920	4.064	1337.0	1487.5	1.8131	2.695	1334.5	1484.1	1.7666	2.303	1333.3	1482.5	1.7487	2.0094	1332.0	1480.8	1.7330
940	4.126	1345.2	1497.9	1.8206	2.736	1342.8	1494.7	1.7742	2.339	1341.6	1493.1	1.7563	2.0411	1340.4	1491.5	1.7407
960	4.187	1353.3	1508.3	1.8280	2.778	1351.1	1505.3	1.7817	2.375	1349.9	1503.7	1.7639	2.0727	1348.7	1502.2	1.7483
980	4.249	1361.6	1518.8	1.8353	2.819	1359.3	1515.8	1.7891	2.411	1358.2	1514.4	1.7713	2.1043	1357.1	1512.9	1.7558
1000	4.310	1369.8	1529.3	1.8425	2.860	1367.7	1526.5	1.7964	2.446	1366.6	1525.0	1.7787	2.136	1365.5	1523.6	1.7632
1020	4.371	1378.0	1539.8	1.8497	2.902	1376.0	1537.1	1.8036	2.482	1375.0	1535.7	1.7859	2.167	1373.9	1534.3	1.7705
1040	4.432	1386.3	1550.3	1.8568	2.943	1384.4	1547.7	1.8108	2.517	1383.4	1546.4	1.7931	2.198	1382.4	1545.1	1.7777
1060	4.493	1394.6	1560.9	1.8638	2.984	1392.7	1558.4	1.8178	2.553	1391.8	1557.1	1.8002	2.229	1390.8	1555.9	1.7849
1080	4.554	1403.0	1571.5	1.8707	3.025	1401.2	1569.1	1.8248	2.588	1400.2	1567.9	1.8072	2.261	1399.3	1566.6	1.7919
1100	4.615	1411.4	1582.2	1.8776	3.066	1409.6	1579.8	1.8317	2.624	1408.7	1578.6	1.8142	2.292	1407.8	1577.4	1.7989
1200	4.918	1453.7	1635.7	1.9109	3.270	1452.2	1633.8	1.8653	2.799	1451.5	1632.8	1.8478	2.446	1450.7	1631.8	1.8327
1300	5.220	1496.9	1690.1	1.9427	3.473	1495.6	1688.4	1.8973	2.974	1495.0	1687.6	1.8799	2.599	1494.3	1686.8	1.8648
1400	5.521	1540.9	1745.3	1.9732	3.675	1539.8	1743.8	1.9279	3.148	1539.3	1743.1	1.9106	2.752	1538.7	1742.4	1.8956
1500	5.822	1585.8	1801.3	2.0025	3.877	1584.8	1800.0	1.9573	3.321	1584.3	1799.4	1.9401	2.904	1583.8	1798.8	1.9251
1600	6.123	1631.6	1858.2	2.0308	4.078	1630.7	1857.0	1.9857	3.494	1630.2	1856.5	1.9685	3.055	1629.8	1855.9	1.9535
1800	6.722	1725.6	1974.4	2.0847	4.479	1724.9	1973.5	2.0396	3.838	1724.5	1973.1	2.0225	3.357	1724.1	1972.6	2.0076
2000	7.321	1823.0	2094.0	2.1354	4.879	1822.3	2093.2	2.0904	4.182	1822.0	2092.8	2.0733	3.658	1821.6	2092.4	2.0584
2200	7.920	1923.6	2216.7	2.1833	5.280	1922.9	2216.0	2.1384	4.525	1922.5	2215.6	2.1212	3.959	1922.2	2215.2	2.1064
2400	8.518	2026.8	2342.1	2.2288	5.679	2026.1	2341.4	2.1838	4.868	2025.8	2341.1	2.1667	4.260	2025.4	2340.8	2.1519

t (Sat.)	450 (456.39) v	u	h	s	500 (467.13) v	u	h	s	550 (477.07) v	u	h	s	600 (486.33) v	u	h	s
Sat.	1.0326	1119.6	1205.6	1.4746	.9283	1119.4	1205.3	1.4645	.8423	1119.1	1204.8	1.4551	.7702	1118.6	1204.1	1.4464
450	*1.0183*	*1115.5*	*1200.3*	*1.4687*												
460	1.0405	1121.8	1208.5	1.4777	*.9133*	*1114.7*	*1199.2*	*1.4578*								
470	1.0620	1128.0	1216.4	1.4863	.9342	1121.3	1207.8	1.4671	*.8283*	*1114.1*	*1198.4*	*1.4483*				
480	1.0828	1133.8	1224.0	1.4944	.9543	1127.7	1216.0	1.4759	.8480	1121.1	1207.4	1.4579	*.7582*	*1113.9*	*1198.1*	*1.4401*
490	1.1029	1139.5	1231.4	1.5022	.9736	1133.8	1223.9	1.4843	.8669	1127.7	1215.9	1.4669	.7769	1121.1	1207.4	1.4500
500	1.1226	1145.1	1238.5	1.5097	.9924	1139.7	1231.5	1.4923	.8850	1134.0	1224.1	1.4755	.7947	1128.0	1216.2	1.4592
510	1.1417	1150.4	1245.5	1.5170	1.0106	1145.4	1238.9	1.4999	.9056	1140.1	1232.0	1.4836	.8118	1134.5	1224.6	1.4679
520	1.1605	1155.7	1252.3	1.5239	1.0283	1150.9	1246.1	1.5073	.9196	1146.0	1239.6	1.4914	.8283	1140.7	1232.7	1.4762
530	1.1788	1160.8	1258.9	1.5307	1.0456	1156.3	1253.1	1.5144	.9361	1151.7	1246.9	1.4989	.8443	1146.8	1240.5	1.4841
540	1.1969	1165.8	1265.5	1.5372	1.0625	1161.6	1259.9	1.5212	.9522	1157.2	1254.1	1.5061	.8598	1152.6	1248.0	1.4917
550	1.2146	1170.7	1271.9	1.5436	1.0792	1166.7	1266.6	1.5279	.9679	1162.6	1261.1	1.5131	.8749	1158.2	1255.4	1.4990
560	1.2320	1175.6	1278.2	1.5498	1.0955	1171.8	1273.1	1.5343	.9834	1167.8	1267.9	1.5198	.8896	1163.7	1262.5	1.5060
570	1.2492	1180.3	1284.4	1.5559	1.1115	1176.7	1279.5	1.5406	.9985	1173.0	1274.6	1.5263	.9040	1169.1	1269.5	1.5128
580	1.2662	1185.0	1290.5	1.5618	1.1273	1181.6	1285.9	1.5467	1.0133	1178.0	1281.1	1.5327	.9181	1174.3	1276.3	1.5194
590	1.2830	1189.7	1296.5	1.5675	1.1429	1186.4	1292.1	1.5527	1.0280	1183.0	1287.6	1.5388	.9320	1179.5	1282.9	1.5253
600	1.2996	1194.3	1302.5	1.5732	1.1583	1191.1	1298.3	1.5585	1.0424	1187.9	1293.9	1.5448	.9456	1184.5	1289.5	1.5320
610	1.3160	1198.8	1308.4	1.5788	1.1735	1195.8	1304.3	1.5642	1.0566	1192.7	1300.2	1.5507	.9590	1189.5	1295.9	1.5381
620	1.3323	1203.3	1314.2	1.5842	1.1885	1200.4	1310.4	1.5698	1.0706	1197.4	1306.4	1.5565	.9722	1194.4	1302.3	1.5440
630	1.3484	1207.8	1320.0	1.5896	1.2033	1205.0	1316.3	1.5753	1.0845	1202.1	1312.5	1.5621	.9853	1199.2	1308.6	1.5497
640	1.3644	1212.2	1325.8	1.5948	1.2181	1209.5	1322.2	1.5807	1.0982	1206.7	1318.5	1.5676	.9982	1203.9	1314.8	1.5554
650	1.3803	1216.6	1331.5	1.6000	1.2327	1214.0	1328.0	1.5860	1.1118	1211.3	1324.5	1.5730	1.0109	1208.6	1320.9	1.5609
660	1.3960	1221.0	1337.2	1.6051	1.2472	1218.4	1333.8	1.5912	1.1252	1215.9	1330.4	1.5783	1.0235	1213.3	1326.9	1.5664
670	1.4116	1225.3	1342.9	1.6101	1.2615	1222.9	1339.6	1.5963	1.1386	1220.4	1336.3	1.5836	1.0360	1217.9	1332.9	1.5717
680	1.4272	1229.6	1348.5	1.6151	1.2758	1227.3	1345.3	1.6014	1.1518	1224.9	1342.1	1.5887	1.0483	1222.5	1338.9	1.5769
690	1.4426	1233.9	1354.1	1.6199	1.2899	1231.7	1351.0	1.6063	1.1649	1229.4	1347.9	1.5938	1.0606	1227.0	1344.8	1.5821
700	1.4580	1238.2	1359.6	1.6248	1.3040	1236.0	1356.7	1.6112	1.1779	1233.8	1353.7	1.5987	1.0727	1231.5	1350.6	1.5872
720	1.4884	1246.7	1370.7	1.6342	1.3319	1244.7	1367.9	1.6208	1.2037	1242.6	1365.1	1.6085	1.0968	1240.4	1362.2	1.5971
740	1.5186	1255.2	1381.7	1.6435	1.3594	1253.3	1379.1	1.6302	1.2291	1251.3	1376.4	1.6180	1.1205	1249.3	1373.7	1.6067
760	1.5485	1263.7	1392.6	1.6525	1.3867	1261.8	1390.1	1.6394	1.2543	1260.0	1387.6	1.6273	1.1439	1258.1	1385.1	1.6161
780	1.5782	1272.1	1403.5	1.6614	1.4138	1270.3	1401.1	1.6483	1.2793	1268.6	1398.8	1.6364	1.1671	1266.8	1396.3	1.6253

TABLE A.3 (cont'd.)

p (t Sat.) t	450 (456.39) v	u	h	s	500 (467.13) v	u	h	s	550 (477.07) v	u	h	s	600 (486.33) v	u	h	s
800	1.6077	1280.5	1414.4	1.6701	1.4407	1278.8	1412.1	1.6571	1.3040	1277.1	1409.8	1.6452	1.1900	1275.4	1407.6	1.6343
820	1.6369	1288.9	1425.2	1.6786	1.4673	1287.3	1423.0	1.6657	1.3285	1285.7	1420.9	1.6539	1.2128	1284.1	1418.7	1.6430
840	1.6660	1297.3	1436.0	1.6870	1.4938	1295.7	1433.9	1.6742	1.3528	1294.2	1431.9	1.6625	1.2353	1292.7	1429.8	1.6517
860	1.6950	1305.6	1446.8	1.6952	1.5201	1304.2	1444.8	1.6825	1.3770	1302.7	1442.9	1.6708	1.2577	1301.2	1440.9	1.6601
880	1.7238	1314.0	1457.5	1.7033	1.5463	1312.6	1455.7	1.6906	1.4010	1311.2	1453.8	1.6791	1.2800	1309.8	1451.9	1.6684
900	1.7524	1322.4	1468.3	1.7113	1.5723	1321.0	1466.5	1.6987	1.4249	1319.7	1464.7	1.6872	1.3021	1318.4	1462.9	1.6766
920	1.7810	1330.8	1479.1	1.7191	1.5982	1329.5	1477.4	1.7066	1.4487	1328.2	1475.6	1.6951	1.3240	1326.9	1473.9	1.6846
940	1.8094	1339.2	1489.8	1.7269	1.6240	1337.9	1488.2	1.7144	1.4723	1336.7	1486.6	1.7030	1.3459	1335.5	1484.9	1.6925
960	1.8377	1347.6	1500.6	1.7345	1.6497	1346.4	1499.0	1.7221	1.4958	1345.2	1497.5	1.7107	1.3676	1344.0	1495.9	1.7003
980	1.8660	1356.0	1511.4	1.7420	1.6753	1354.9	1509.9	1.7296	1.5193	1353.7	1508.4	1.7183	1.3893	1352.6	1506.8	1.7079
1000	1.8941	1364.4	1522.2	1.7495	1.7008	1363.3	1520.7	1.7371	1.5426	1362.3	1519.3	1.7259	1.4108	1361.2	1517.8	1.7155
1020	1.9221	1372.9	1533.0	1.7568	1.7262	1371.8	1531.6	1.7445	1.5659	1370.8	1530.2	1.7333	1.4322	1369.7	1528.9	1.7230
1040	1.9501	1381.4	1543.8	1.7641	1.7515	1380.4	1542.4	1.7518	1.5890	1379.3	1541.1	1.7406	1.4536	1378.3	1539.7	1.7303
1060	1.9780	1389.9	1554.6	1.7712	1.7768	1388.9	1553.3	1.7590	1.6121	1387.9	1552.0	1.7478	1.4749	1386.9	1550.7	1.7376
1080	2.0058	1398.4	1565.4	1.7783	1.8020	1397.4	1564.2	1.7661	1.6352	1396.5	1562.9	1.7550	1.4961	1395.6	1561.7	1.7448
1100	2.034	1406.9	1576.3	1.7853	1.8271	1406.0	1575.1	1.7731	1.6581	1405.1	1573.9	1.7620	1.5173	1404.2	1572.7	1.7519
1150	2.103	1428.4	1603.5	1.8025	1.8896	1427.5	1602.4	1.7904	1.7152	1426.7	1601.3	1.7793	1.5699	1425.9	1600.2	1.7692
1200	2.172	1450.0	1630.8	1.8192	1.9518	1449.2	1629.8	1.8072	1.7720	1448.5	1628.8	1.7962	1.6222	1447.7	1627.8	1.7861
1250	2.240	1471.8	1658.3	1.8355	2.0137	1471.1	1657.4	1.8235	1.8285	1470.3	1656.5	1.8126	1.6742	1469.6	1655.5	1.8026
1300	2.308	1493.7	1685.9	1.8515	2.075	1493.1	1685.1	1.8395	1.8848	1492.4	1684.2	1.8286	1.7260	1491.7	1683.4	1.8186
1350	2.376	1515.8	1713.7	1.8670	2.137	1515.2	1712.9	1.8551	1.9409	1514.6	1712.2	1.8443	1.7775	1514.0	1711.4	1.8343
1400	2.444	1538.1	1741.7	1.8823	2.198	1537.6	1741.0	1.8704	1.9967	1537.0	1740.2	1.8596	1.8289	1536.5	1739.5	1.8497
1450	2.512	1560.6	1769.8	1.8972	2.259	1560.1	1769.2	1.8853	2.0525	1559.6	1768.5	1.8746	1.8801	1559.1	1767.5	1.8647
1500	2.580	1583.3	1798.2	1.9119	2.320	1582.8	1797.5	1.9000	2.1080	1582.3	1796.9	1.8892	1.9312	1581.9	1796.3	1.8794
1600	2.715	1629.3	1855.4	1.9403	2.442	1628.9	1854.8	1.9285	2.219	1628.4	1854.3	1.9178	2.033	1628.0	1853.7	1.9080
1800	2.983	1723.3	1972.1	1.9944	2.684	1723.3	1971.7	1.9827	2.440	1722.9	1971.2	1.9720	2.236	1722.6	1970.8	1.9622
2000	3.251	1821.3	2092.0	2.0453	2.926	1820.9	2091.6	2.0335	2.660	1820.6	2091.2	2.0229	2.438	1820.2	2090.8	2.0131
2200	3.519	1921.8	2214.9	2.0933	3.167	1921.5	2214.5	2.0815	2.879	1921.1	2214.2	2.0709	2.639	1920.8	2213.8	2.0612
2400	3.787	2025.1	2340.4	2.1388	3.408	2024.7	2340.1	2.1270	3.098	2024.4	2339.7	2.1164	2.840	2024.0	2339.4	2.1067

TABLE A.3 (cont'd.)

p (t Sat.)	800 (518.36)				1000 (544.75)			
t	v	u	h	s	v	u	h	s
Sat.	.5691	1115.0	1199.3	1.4160	.4459	1109.9	1192.4	1.3903
500								
510	.5554	1107.9	1190.1	1.4066				
520	.5717	1116.4	1201.0	1.4178				
530	.5870	1124.3	1211.2	1.4282				
540	.6015	1131.8	1220.8	1.4378	.4389	1105.2	1186.4	1.3844
550	.6154	1138.8	1229.9	1.4469	.4534	1114.8	1198.7	1.3966
560	.6287	1145.6	1238.6	1.4555	.4669	1123.6	1210.0	1.4077
570	.6415	1152.0	1247.0	1.4637	.4795	1131.7	1220.5	1.4179
580	.6539	1158.3	1255.1	1.4714	.4915	1139.4	1230.4	1.4275
590	.6659	1164.3	1262.9	1.4789	.5030	1146.7	1239.8	1.4365
600	.6776	1170.1	1270.4	1.4861	.5140	1153.7	1248.8	1.4450
610	.6890	1175.8	1277.8	1.4930	.5245	1160.3	1257.4	1.4531
620	.7002	1181.3	1285.0	1.4997	.5348	1166.7	1265.7	1.4609
630	.7111	1186.7	1292.0	1.5062	.5447	1172.9	1273.7	1.4682
640	.7218	1192.0	1298.9	1.5125	.5543	1178.9	1281.5	1.4754
650	.7324	1197.2	1305.6	1.5186	.5637	1184.7	1289.1	1.4822
660	.7428	1202.3	1312.3	1.5245	.5730	1190.4	1296.5	1.4889
670	.7530	1207.4	1318.8	1.5304	.5820	1196.0	1303.7	1.4953
680	.7631	1212.3	1325.3	1.5361	.5908	1201.4	1310.8	1.5015
690	.7730	1217.2	1331.7	1.5416	.5995	1206.8	1317.7	1.5076
700	.7829	1222.1	1338.0	1.5471	.6080	1212.0	1324.6	1.5135
720	.8023	1231.6	1350.4	1.5577	.6247	1222.3	1337.9	1.5249
740	.8212	1241.0	1362.6	1.5680	.6410	1232.3	1350.9	1.5359
760	.8399	1250.3	1374.6	1.5779	.6569	1242.1	1363.7	1.5464
780	.8583	1259.4	1386.5	1.5875	.6725	1251.7	1376.2	1.5566
800	.8764	1268.5	1398.2	1.5969	.6878	1261.2	1388.5	1.5664
820	.8943	1277.4	1409.8	1.6061	.7028	1270.6	1400.6	1.5760
840	.9120	1286.4	1421.4	1.6150	.7176	1279.9	1412.7	1.5853
860	.9295	1295.2	1432.8	1.6238	.7323	1289.1	1424.6	1.5944
880	.9468	1304.1	1444.2	1.6324	.7467	1298.2	1436.4	1.6033
900	.9640	1312.9	1455.6	1.6408	.7610	1307.3	1448.1	1.6120
920	.9811	1321.7	1466.9	1.6490	.7752	1316.3	1459.7	1.6205
940	.9980	1330.4	1478.2	1.6572	.7892	1325.3	1471.3	1.6288
960	1.0149	1339.2	1489.4	1.6651	.8031	1334.3	1482.9	1.6370
980	1.0316	1348.0	1500.7	1.6730	.8169	1343.2	1494.4	1.6451
1000	1.0482	1356.7	1511.9	1.6807	.8305	1352.2	1505.9	1.6530
1020	1.0647	1365.5	1523.1	1.6883	.8441	1361.1	1517.3	1.6608
1040	1.0812	1374.2	1534.3	1.6959	.8576	1370.0	1528.7	1.6684
1060	1.0975	1383.0	1545.5	1.7033	.8710	1378.9	1540.1	1.6760
1080	1.1138	1391.7	1556.6	1.7106	.8844	1387.9	1551.5	1.6834
1100	1.1300	1400.5	1567.8	1.7178	.8976	1396.8	1562.9	1.6908
1120	1.1462	1409.3	1579.0	1.7249	.9108	1405.7	1574.3	1.6980
1140	1.1623	1418.1	1590.2	1.7319	.9240	1414.6	1585.6	1.7052
1160	1.1783	1426.9	1601.4	1.7389	.9370	1423.6	1597.0	1.7122
1180	1.1943	1435.8	1612.6	1.7458	.9500	1432.5	1608.3	1.7192
1200	1.2102	1444.6	1623.8	1.7526	.9630	1441.5	1619.7	1.7261
1220	1.2261	1453.5	1635.9	1.7593	.9759	1450.4	1631.0	1.7329
1240	1.2420	1462.3	1646.2	1.7659	.9888	1459.4	1642.4	1.7396
1260	1.2578	1471.3	1657.4	1.7725	1.0016	1468.4	1653.8	1.7462
1280	1.2735	1480.2	1668.7	1.7790	1.0144	1477.4	1665.1	1.7528
1300	1.2892	1489.1	1680.0	1.7854	1.0272	1486.5	1676.5	1.7593
1350	1.3284	1511.6	1708.2	1.8013	1.0589	1509.1	1705.1	1.7753
1400	1.3674	1534.2	1736.6	1.8167	1.0905	1531.9	1733.7	1.7909
1450	1.4062	1556.9	1765.1	1.8319	1.1218	1554.8	1762.4	1.8061
1500	1.4448	1579.9	1793.7	1.8467	1.1531	1577.8	1791.2	1.8210
1600	1.5218	1626.2	1851.5	1.8754	1.2152	1624.4	1849.3	1.8499
1800	1.6749	1721.0	1969.0	1.9298	1.3384	1719.5	1967.2	1.9046
2000	1.8271	1818.8	2089.3	1.9808	1.4608	1817.4	2087.7	1.9557
2200	1.9789	1919.4	2212.4	2.0290	1.5828	1918.1	2211.0	2.0038
2400	2.1305	2022.7	2338.1	2.0745	1.7046	2021.3	2336.7	2.0494

TABLE A.4
Properties of Compressed Liquid (Steam)

Liquid (psia)

p (t Sat.)	0				500(467.13)				1000(544.75)			
t	v	u	h	s	v	u	h	s	v	u	h	s
Sat.					.019748	447.70	449.53	.64904	.021591	538.39	542.38	.74320
32	.016022	0.01	0.01	.00003	.015994	.00	1.49	.00000	.015967	.03	2.99	.00005
50	.016024	18.06	18.06	.03607	.015998	18.02	19.50	.03599	.015972	17.99	20.94	.03592
100	.016130	68.05	68.05	.12963	.016106	67.87	69.36	.12932	.016082	67.70	70.68	.12901
150	.016344	117.95	117.95	.21504	.016318	117.66	119.17	.21457	.016293	117.38	120.40	.21410
200	.016635	168.05	168.05	.29402	.016608	167.65	169.19	.29341	.016580	167.26	170.32	.29281
250	.017003	218.52	218.52	.36777	.016972	217.99	219.56	.36702	.016941	217.47	220.61	.36628
300	.017453	269.61	269.61	.43732	.017416	268.92	270.53	.43641	.017379	268.24	271.46	.43352
350	.018000	321.59	321.59	.50359	.017954	320.71	322.37	.50249	.017909	319.83	323.15	.50140
400	.018668	374.85	374.85	.56740	.018608	373.68	375.40	.56604	.018550	372.55	375.98	.56472
450	.019503	429.96	429.96	.62970	.019420	428.40	430.19	.63798	.019340	426.89	430.47	.62632
500	.02060	488.1	488.1	.6919	.02048	485.9	487.8	.6896	.02036	483.8	487.5	.6874
510	.02087	500.3	500.3	.7046	.02073	497.9	499.8	.7021	.02060	495.6	499.4	.6997
520	.02116	512.7	512.7	.7173	.02100	510.1	512.0	.7146	.02086	507.6	511.5	.7121
530	.02148	525.5	525.5	.7303	.02130	522.6	524.5	.7273	.02114	519.9	523.8	.7245
540	.02182	538.6	538.6	.7434	.02162	535.3	537.3	.7402	.02144	532.4	536.3	.7372
550	.02221	552.1	552.1	.7569	.02198	548.4	550.5	.7532	.02177	545.1	549.2	.7499
560	.02265	566.1	566.1	.7707	.02237	562.0	564.0	.7666	.02213	558.3	562.4	.7630
570	.02315	580.8	580.8	.7851	.02281	576.0	578.1	.7804	.02253	571.8	576.0	.7763
580					.02332	590.8	592.9	.7946	.02298	585.9	590.1	.7899
590					.02392	606.4	608.6	.8096	.02349	600.6	604.9	.8041
600									.02409	616.2	620.6	.8189
610									.02482	632.9	637.5	.8348

p	4000				18,000				20,000			
t	v	u	h	s	v	u	h	s	v	u	h	s
32	.015807	.10	11.80	.00005	.015188	0.54	50.05	.00372	.015116	.65	55.30	.00446
50	.015821	17.76	29.47	.03534	.015227	16.32	67.04	.03021	.015154	16.14	72.23	.02936
100	.015942	66.72	78.52	.12714	.015372	62.83	114.03	.11817	.015298	62.37	118.99	.11688
150	.016150	115.77	127.73	.21136	.015572	109.64	161.51	.19941	.015497	108.91	166.26	.19778
200	.016425	165.02	177.18	.28931	.015813	156.62	209.29	.27474	.015736	155.62	213.86	.27281
250	.016765	214.52	226.93	.36200	.016096	203.57	257.19	.34472	.016015	202.28	261.55	.34250
300	.017174	264.43	277.15	.43038	.016422	250.58	305.28	.41021	.016333	248.96	309.41	.40766
350	.017659	314.98	328.05	.49526	.016791	297.80	353.73	.47197	.016693	295.83	357.61	.46911
400	.018235	366.35	379.85	.55734	.017207	345.32	402.63	.53057	.017096	342.97	406.24	.52738
450	.018924	418.83	432.84	.61725	.017669	393.15	452.00	.58639	.017541	390.39	455.31	.58286
500	.019766	472.9	487.5	.6758	.018183	441.3	501.9	.6397	.018031	438.1	504.8	.6358
520	.020161	495.2	510.1	.6990	.018404	460.6	521.9	.6605	.018242	457.2	524.7	.6564
540	.020600	517.9	533.1	.7223	.018636	480.0	542.1	.6808	.018461	476.4	544.7	.6766
560	.021091	541.2	556.8	.7457	.018879	499.5	562.4	.7009	.018689	495.6	564.8	.6965
580	.021648	565.2	581.2	.7694	.019134	519.0	582.8	.7207	.018928	514.9	585.0	.7160
600	.02229	590.0	606.5	.7936	.01940	538.7	603.3	.7403	.01918	534.3	605.2	.7354
620	.02304	616.0	633.0	.8183	.01968	558.4	623.9	.7596	.01944	553.7	625.6	.7544
640	.02394	643.3	661.1	.8441	.01998	578.2	644.8	.7787	.01972	573.2	646.1	.7732
660	.02506	672.7	691.2	.8712	.02030	598.1	665.7	.7976	.02001	592.7	666.8	.7918
680	.02653	704.9	724.5	.9007	.02064	618.2	686.9	.8163	.02031	612.4	687.6	.8102
700	.02867	742.1	763.4	.9345	.02099	638.4	708.3	.8349	.02063	632.1	708.5	.8285
710	.03026	764.3	786.7	.9545	.02118	648.5	719.1	.8442	.02080	642.1	719.1	.8375

TABLE A.1 (SI)

Saturation: Temperature (Steam)

Temp. °C T	Press. kPa P	Specific Volume (m³/kg) Sat. Liquid v_f	Sat. Vapor v_g	Internal Energy (kJ/kg) Sat. Liquid u_f	Evap. u_{fg}	Sat. Vapor u_g	Enthalpy (kJ/kg) Sat. Liquid h_f	Evap. h_{fg}	Sat. Vapor h_g	Entropy (kJ/kg·°k) Sat. Liquid s_f	Evap. s_{fg}	Sat. Vapor s_g
0.01	0.6113	0.001 000	206.14	.00	2375.3	2375.3	.01	2501.3	2501.4	.0000	9.1562	9.1562
5	0.8721	0.001 000	147.12	20.97	2361.3	2382.3	20.98	2489.6	2510.6	.0761	8.9496	9.0257
10	1.2276	0.001 000	106.38	42.00	2347.2	2389.2	42.01	2477.7	2519.8	.1510	8.7498	8.9008
15	1.7051	0.001 001	77.93	62.99	2333.1	2396.1	62.99	2465.9	2528.9	.2245	8.5569	8.7814
20	2.339	0.001 002	57.79	83.95	2319.0	2402.9	83.96	2454.1	2538.1	.2966	8.3706	8.6672
25	3.169	0.001 003	43.36	104.88	2304.9	2409.8	104.89	2442.3	2547.2	.3674	8.1905	8.5580
30	4.246	0.001 004	32.89	125.78	2290.8	2416.6	125.79	2430.5	2556.3	.4369	8.0164	8.4533
35	5.628	0.001 006	25.22	146.67	2276.7	2423.4	146.68	2418.6	2565.3	.5053	7.8478	8.3531
40	7.384	0.001 008	19.52	167.56	2262.6	2430.1	167.57	2406.7	2574.3	.5725	7.6845	8.2570
45	9.593	0.001 010	15.26	188.44	2248.4	2436.8	188.45	2394.8	2583.2	.6387	7.5261	8.1648
50	12.349	0.001 012	12.03	209.32	2234.2	2443.5	209.33	2382.7	2592.1	.7038	7.3725	8.0763
55	15.758	0.001 015	9.568	230.21	2219.9	2450.1	230.23	2370.7	2600.9	.7679	7.2234	7.9913
60	19.940	0.001 017	7.671	251.11	2205.5	2456.6	251.13	2358.5	2609.6	.8312	7.0784	7.9096
65	25.03	0.001 020	6.197	272.02	2191.1	2463.1	272.06	2346.2	2618.3	.8935	6.9375	7.8310
70	31.19	0.001 023	5.042	292.95	2176.6	2469.6	292.98	2333.8	2626.8	.9549	6.8004	7.7553
75	38.58	0.001 026	4.131	313.90	2162.0	2475.9	313.93	2321.4	2635.3	1.0155	6.6669	7.6824
80	47.39	0.001 029	3.407	334.86	2147.4	2482.2	334.91	2308.8	2643.7	1.0753	6.5369	7.6122
85	57.83	0.001 033	2.828	355.84	2132.6	2488.4	355.90	2296.0	2651.9	1.1343	6.4102	7.5445
90	70.14	0.001 036	2.361	376.85	2117.7	2494.5	376.92	2283.2	2660.1	1.1925	6.2866	7.4791
95	84.55	0.001 040	1.982	397.88	2102.7	2500.6	397.96	2270.2	2668.1	1.2500	6.1659	7.4159

TABLE A.1 (SI) (cont'd.)

Temp. °C T	Press. kPa P	Specific Volume (m³/kg)		Internal Energy (kJ/kg)			Enthalpy (kJ/kg)			Entropy (kJ/kg · °k)		
		Sat. Liquid v_f	Sat. Vapor v_g	Sat. Liquid u_f	Evap. u_{fg}	Sat. Vapor u_g	Sat. Liquid h_f	Evap. h_{fg}	Sat. Vapor h_g	Sat. Liquid s_f	Evap. s_{fg}	Sat. Vapor s_g
	MPa											
100	0.101 35	0.001 044	1.6729	418.94	2087.6	2506.5	419.04	2257.0	2676.1	1.3069	6.0480	7.3549
105	0.120 82	0.001 048	1.4194	440.02	2072.3	2512.4	440.15	2243.7	2683.8	1.3630	5.9328	7.2958
110	0.143 27	0.001 052	1.2102	461.14	2057.0	2518.1	461.30	2230.2	2691.5	1.4185	5.8202	7.2387
115	0.169 06	0.001 056	1.0366	482.30	2041.4	2523.7	482.48	2216.5	2699.0	1.4734	5.7100	7.1833
120	0.198 53	0.001 060	0.8919	503.50	2025.8	2529.3	503.71	2202.6	2706.3	1.5276	5.6020	7.1296
125	0.2321	0.001 065	0.7706	524.74	2009.9	2534.6	524.99	2188.5	2713.5	1.5813	5.4962	7.0775
130	0.2701	0.001 070	0.6685	546.02	1993.9	2539.9	546.31	2174.2	2720.5	1.6344	5.3925	7.0269
135	0.3130	0.001 075	0.5822	567.35	1977.7	2545.0	567.69	2159.6	2727.3	1.6870	5.2907	6.9777
140	0.3613	0.001 080	0.5089	588.74	1961.3	2550.0	589.13	2144.7	2733.9	1.7391	5.1908	6.9299
145	0.4154	0.001 085	0.4463	610.18	1944.7	2554.9	610.63	2129.6	2740.3	1.7907	5.0926	6.8833
150	0.4758	0.001 091	0.3928	631.68	1927.9	2559.5	632.20	2114.3	2746.5	1.8418	4.9960	6.8379
155	0.5431	0.001 096	0.3468	653.24	1910.8	2564.1	653.84	2098.6	2752.4	1.8925	4.9010	6.7935
160	0.6178	0.001 102	0.3071	674.87	1893.5	2568.4	675.55	2082.6	2758.1	1.9427	4.8075	6.7502
165	0.7005	0.001 108	0.2727	696.56	1876.0	2572.5	697.34	2066.2	2763.5	1.9925	4.7153	6.7078
170	0.7917	0.001 114	0.2428	718.33	1858.1	2576.5	719.21	2049.5	2768.7	2.0419	4.6244	6.6663
175	0.8920	0.001 121	0.2168	740.17	1840.0	2580.2	741.17	2032.4	2773.6	2.0909	4.5347	6.6256
180	1.0021	0.001 127	0.194 05	762.09	1821.6	2583.7	763.22	2015.0	2778.2	2.1396	4.4461	6.5857
185	1.1227	0.001 134	0.174 09	784.10	1802.9	2587.0	785.37	1997.1	2782.4	2.1879	4.3586	6.5465
190	1.2544	0.001 141	0.156 54	806.19	1783.8	2590.0	807.62	1978.8	2786.4	2.2359	4.2720	6.5079
195	1.3978	0.001 149	0.141 05	828.37	1764.4	2592.8	829.98	1960.0	2790.0	2.2835	4.1863	6.4698
200	1.5538	0.001 157	0.127 36	850.65	1744.7	2595.3	852.45	1940.7	2793.2	2.3309	4.1014	6.4323
205	1.7230	0.001 164	0.115 21	873.04	1724.5	2597.5	875.04	1921.0	2796.0	2.3780	4.0172	6.3952
210	1.9062	0.001 173	0.104 41	895.53	1703.9	2599.5	897.76	1900.7	2798.5	2.4248	3.9337	6.3585
215	2.104	0.001 181	0.094 79	918.14	1682.9	2601.1	920.62	1879.9	2800.5	2.4714	3.8507	6.3221
220	2.318	0.001 190	0.086 19	940.87	1661.5	2602.4	943.62	1858.5	2802.1	2.5178	3.7683	6.2861
225	2.548	0.001 199	0.078 49	963.73	1639.6	2603.3	966.78	1836.5	2803.3	2.5639	3.6863	6.2503
230	2.795	0.001 209	0.071 58	986.74	1617.2	2603.9	990.12	1813.8	2804.0	2.6099	3.6047	6.2146
235	3.060	0.001 219	0.065 37	1009.89	1594.2	2604.1	1013.62	1790.5	2804.2	2.6558	3.5233	6.1791
240	3.344	0.001 229	0.059 76	1033.21	1570.8	2604.0	1037.32	1766.5	2803.8	2.7015	3.4422	6.1437
245	3.648	0.001 240	0.054 71	1056.71	1546.7	2603.4	1061.23	1741.7	2803.0	2.7472	3.3612	6.1083

TABLE A.1 (SI) (cont'd.)

Temp. °C T	Press. MPa P	Specific Volume (m³/kg) Sat. Liquid v_f	Sat. Vapor v_g	Internal Energy (kJ/kg) Sat. Liquid u_f	Evap. u_{fg}	Sat. Vapor u_g	Enthalpy (kJ/kg) Sat. Liquid h_f	Evap. h_{fg}	Sat. Vapor h_g	Entropy (kJ/kg·°K) Sat. Liquid s_f	Evap. s_{fg}	Sat. Vapor s_g
250	3.973	0.001 251	0.050 13	1080.39	1522.0	2602.4	1085.36	1716.2	2801.5	2.7927	3.2802	6.0730
255	4.319	0.001 263	0.045 98	1104.28	1496.7	2600.9	1109.73	1689.8	2799.5	2.8383	3.1992	6.0375
260	4.688	0.001 276	0.042 21	1128.39	1470.6	2599.0	1134.37	1662.5	2796.9	2.8838	3.1181	6.0019
265	5.081	0.001 289	0.038 77	1152.74	1443.9	2596.6	1159.28	1634.4	2793.6	2.9294	3.0368	5.9662
270	5.499	0.001 302	0.035 64	1177.36	1416.3	2593.7	1184.51	1605.2	2789.7	2.9751	2.9551	5.9301
275	5.942	0.001 317	0.032 79	1202.25	1387.9	2590.2	1210.07	1574.9	2785.0	3.0208	2.8730	5.8938
280	6.412	0.001 332	0.030 17	1227.46	1358.7	2586.1	1235.99	1543.6	2779.6	3.0668	2.7903	5.8571
285	6.909	0.001 348	0.027 77	1253.00	1328.4	2581.4	1262.31	1511.0	2773.3	3.1130	2.7070	5.8199
290	7.436	0.001 366	0.025 57	1278.92	1297.1	2576.0	1289.07	1477.1	2766.2	3.1594	2.6227	5.7821
295	7.993	0.001 384	0.023 54	1305.2	1264.7	2569.9	1316.3	1441.8	2758.1	3.2062	2.5375	5.7437
300	8.581	0.001 404	0.021 67	1332.0	1231.0	2563.0	1344.0	1404.9	2749.0	3.2534	2.4511	5.7045
305	9.202	0.001 425	0.019 948	1359.3	1195.9	2555.2	1372.4	1366.4	2738.7	3.3010	2.3633	5.6643
310	9.856	0.001 447	0.018 350	1387.1	1159.4	2546.4	1401.3	1326.0	2727.3	3.3493	2.2737	5.6230
315	10.547	0.001 472	0.016 867	1415.5	1121.1	2536.6	1431.0	1283.5	2714.5	3.3982	2.1821	5.5804
320	11.274	0.001 499	0.015 488	1444.6	1080.9	2525.5	1461.5	1238.6	2700.1	3.4480	2.0882	5.5362
330	12.845	0.001 561	0.012 996	1505.3	993.7	2498.9	1525.3	1140.6	2665.9	3.5507	1.8909	5.4417
340	14.586	0.001 638	0.010 797	1570.3	894.3	2464.6	1594.2	1027.9	2622.0	3.6594	1.6763	5.3357
350	16.513	0.001 740	0.008 813	1641.9	776.6	2418.4	1670.6	893.4	2563.9	3.7777	1.4335	5.2112
360	18.651	0.001 893	0.006 945	1725.2	626.3	2351.5	1760.5	720.5	2481.0	3.9147	1.1379	5.0526
370	21.03	0.002 213	0.004 925	1844.0	384.5	2228.5	1890.5	441.6	2332.1	4.1106	.6865	4.7971
374.14	22.09	0.003 155	0.003 155	2029.6	0	2029.6	2099.3	0	2099.3	4.4298	0	4.4298

TABLE A.2 (SI)
Saturation Pressures (Steam)

Press. kPa P	Temp. °C T	Specific Volume (m³/kg) Sat. Liquid v_f	Sat. Vapor v_g	Internal Energy (kJ/kg) Sat. Liquid u_f	Evap. u_{fg}	Sat. Vapor u_g	Enthalpy (kJ/kg) Sat. Liquid h_f	Evap. h_{fg}	Sat. Vapor h_g	Entropy (kJ/kg · °K) Sat. Liquid s_f	Evap. s_{fg}	Sat. Vapor s_g
0.6113	0.01	0.001 000	206.14	.00	2375.3	2375.3	.01	2501.3	2501.4	.0000	9.1562	9.1562
1.0	6.98	0.001 000	129.21	29.30	2355.7	2385.0	29.30	2484.9	2514.2	.1059	8.8697	8.9756
1.5	13.03	0.001 001	87.98	54.71	2338.6	2393.3	54.71	2470.6	2525.3	.1957	8.6322	8.8279
2.0	17.50	0.001 001	67.00	73.48	2326.0	2399.5	73.48	2460.0	2533.5	.2607	8.4629	8.7237
2.5	21.08	0.001 002	54.25	88.48	2315.9	2404.4	88.49	2451.6	2540.0	.3120	8.3311	8.6432
3.0	24.08	0.001 003	45.67	101.04	2307.5	2408.5	101.05	2444.5	2545.5	.3545	8.2231	8.5776
4.0	28.96	0.001 004	34.80	121.45	2293.7	2415.2	121.46	2432.9	2554.4	.4226	8.0520	8.4746
5.0	32.88	0.001 005	28.19	137.81	2282.7	2420.5	137.82	2423.7	2561.5	.4764	7.9187	8.3951
7.5	40.29	0.001 008	19.24	168.78	2261.7	2430.5	168.79	2406.0	2574.8	.5764	7.6750	8.2515
10	45.81	0.001 010	14.67	191.82	2246.1	2437.9	191.83	2392.8	2584.7	.6493	7.5009	8.1502
15	53.97	0.001 014	10.02	225.92	2222.8	2448.7	225.94	2373.1	2599.1	.7549	7.2536	8.0085
20	60.06	0.001 017	7.649	251.38	2205.4	2456.7	251.40	2358.3	2609.7	.8320	7.0766	7.9085
25	64.97	0.001 020	6.204	271.90	2191.2	2463.1	271.93	2346.3	2618.2	.8931	6.9383	7.8314
30	69.10	0.001 022	5.229	289.20	2179.2	2468.4	289.23	2336.1	2625.3	.9439	6.8247	7.7686
40	75.87	0.001 027	3.993	317.53	2159.5	2477.0	317.58	2319.2	2636.8	1.0259	6.6441	7.6700
50	81.33	0.001 030	3.240	340.44	2143.4	2483.9	340.49	2305.4	2645.9	1.0910	6.5029	7.5939
75	91.78	0.001 037	2.217	384.31	2112.4	2496.7	384.39	2278.6	2663.0	1.2130	6.2434	7.4564
MPa												
0.100	99.63	0.001 043	1.6940	417.36	2088.7	2506.1	417.46	2258.0	2675.5	1.3026	6.0568	7.3594
0.125	105.99	0.001 048	1.3749	444.19	2069.3	2513.5	444.32	2241.0	2685.4	1.3740	5.9104	7.2844
0.150	111.37	0.001 053	1.1593	466.94	2052.7	2519.7	467.11	2226.5	2693.6	1.4336	5.7897	7.2233
0.175	116.06	0.001 057	1.0036	486.80	2038.1	2524.9	486.99	2213.6	2700.6	1.4849	5.6868	7.1717
0.200	120.23	0.001 061	0.8857	504.49	2025.0	2529.5	504.70	2201.9	2706.7	1.5301	5.5970	7.1271
0.225	124.00	0.001 064	0.7933	520.47	2013.1	2533.6	520.72	2191.3	2712.1	1.5706	5.5173	7.0878

TABLE A.2 (SI) (cont'd.)

Press. MPa P	Temp. °C T	Specific Volume		Internal Energy			Enthalpy			Entropy		
		Sat. Liquid v_f	Sat. Vapor v_g	Sat. Liquid u_f	Evap. u_{fg}	Sat. Vapor u_g	Sat. Liquid h_f	Evap. h_{fg}	Sat. Vapor h_g	Sat. Liquid s_f	Evap. s_{fg}	Sat. Vapor s_g
0.250	127.44	0.001 067	0.7187	535.10	2002.1	2537.2	535.37	2181.5	2716.9	1.6072	5.4455	7.0527
0.275	130.60	0.001 070	0.6573	548.59	1991.9	2540.5	548.89	2172.4	2721.3	1.6408	5.3801	7.0209
0.300	133.55	0.001 073	0.6058	561.15	1982.4	2543.6	561.47	2163.8	2725.3	1.6718	5.3201	6.9919
0.325	136.30	0.001 076	0.5620	572.90	1973.5	2546.4	573.25	2155.8	2729.0	1.7006	5.2646	6.9652
0.350	138.88	0.001 079	0.5243	583.95	1965.0	2548.9	584.33	2148.1	2732.4	1.7275	5.2130	6.9405
0.375	141.32	0.001 081	0.4914	594.40	1956.9	2551.3	594.81	2140.8	2735.6	1.7528	5.1647	6.9175
0.40	143.63	0.001 084	0.4625	604.31	1949.3	2553.6	604.74	2133.8	2738.6	1.7766	5.1193	6.8959
0.45	147.93	0.001 088	0.4140	622.77	1934.9	2557.6	623.25	2120.7	2743.9	1.8207	5.0359	6.8565
0.50	151.86	0.001 093	0.3749	639.68	1921.6	2561.2	640.23	2108.5	2748.7	1.8607	4.9606	6.8213
0.55	155.48	0.001 097	0.3427	655.32	1909.2	2564.5	655.93	2097.0	2753.0	1.8973	4.8920	6.7893
0.60	158.85	0.001 101	0.3157	669.90	1897.5	2567.4	670.56	2086.3	2756.8	1.9312	4.8288	6.7600
0.65	162.01	0.001 104	0.2927	683.56	1886.5	2570.1	684.28	2076.0	2760.3	1.9627	4.7703	6.7331
0.70	164.97	0.001 108	0.2729	696.44	1876.1	2572.5	697.22	2066.3	2763.5	1.9922	4.7158	6.7080
0.75	167.78	0.001 112	0.2556	708.64	1866.1	2574.7	709.47	2057.0	2766.4	2.0200	4.6647	6.6847
0.80	170.43	0.001 115	0.2404	720.22	1856.6	2576.8	721.11	2048.0	2769.1	2.0462	4.6166	6.6628
0.85	172.96	0.001 118	0.2270	731.27	1847.4	2578.7	732.22	2039.4	2771.6	2.0710	4.5711	6.6421
0.90	175.38	0.001 121	0.2150	741.83	1838.6	2580.5	742.83	2031.1	2773.9	2.0946	4.5280	6.6226
0.95	177.69	0.001 124	0.2042	751.95	1830.2	2582.1	753.02	2023.1	2776.1	2.1172	4.4869	6.6041
1.00	179.91	0.001 127	0.194 44	761.68	1822.0	2583.6	762.81	2015.3	2778.1	2.1387	4.4478	6.5865
1.10	184.09	0.001 133	0.177 53	780.09	1806.3	2586.4	781.34	2000.4	2781.7	2.1792	4.3744	6.5536
1.20	187.99	0.001 139	0.163 33	797.29	1791.5	2588.8	798.65	1986.2	2784.8	2.2166	4.3067	6.5233
1.30	191.64	0.001 144	0.151 25	813.44	1777.5	2591.0	814.93	1972.7	2787.6	2.2515	4.2438	6.4953
1.40	195.07	0.001 149	0.140 84	828.70	1764.1	2592.8	830.30	1959.7	2790.0	2.2842	4.1850	6.4693

TABLE A.2 (SI) (cont'd.)

Press. kPa P	Temp. °C T	Specific Volume (m³/kg) Sat. Liquid v_f	Sat. Vapor v_g	Internal Energy (kJ/kg) Sat. Liquid u_f	Evap. u_{fg}	Sat. Vapor u_g	Enthalpy (kJ/kg) Sat. Liquid h_f	Evap. h_{fg}	Sat. Vapor h_g	Entropy (kJ/kg·°k) Sat. Liquid s_f	Evap. s_{fg}	Sat. Vapor s_g
1.50	198.32	0.001 154	0.131 77	843.16	1751.3	2594.5	844.89	1947.3	2792.2	2.3150	4.1298	6.4448
1.75	205.76	0.001 166	0.113 49	876.46	1721.4	2597.8	878.50	1917.9	2796.4	2.3851	4.0044	6.3896
2.00	212.42	0.001 177	0.099 63	906.44	1693.8	2600.3	908.79	1890.7	2799.5	2.4474	3.8935	6.3409
2.25	218.45	0.001 187	0.088 75	933.83	1668.2	2602.0	936.49	1865.2	2801.7	2.5035	3.7937	6.2972
2.5	223.99	0.001 197	0.079 98	959.11	1644.0	2603.1	962.11	1841.0	2803.1	2.5547	3.7028	6.2575
3.0	233.90	0.001 217	0.066 68	1004.78	1599.3	2604.1	1008.42	1795.7	2804.2	2.6457	3.5412	6.1869
3.5	242.60	0.001 235	0.057 07	1045.43	1558.3	2603.7	1049.75	1753.7	2803.4	2.7253	3.4000	6.1253
4	250.40	0.001 252	0.049 78	1082.31	1520.0	2602.3	1087.31	1714.1	2801.4	2.7964	3.2737	6.0701
5	263.99	0.001 286	0.039 44	1147.81	1449.3	2597.1	1154.23	1640.1	2794.3	2.9202	3.0532	5.9734
6	275.64	0.001 319	0.032 44	1205.44	1384.3	2589.7	1213.35	1571.0	2784.3	3.0267	2.8625	5.8892
7	285.88	0.001 351	0.027 37	1257.55	1323.0	2580.5	1267.00	1505.1	2772.1	3.1211	2.6922	5.8133
8	295.06	0.001 384	0.023 52	1305.57	1264.2	2569.8	1316.64	1441.3	2758.0	3.2068	2.5364	5.7432
9	303.40	0.001 418	0.020 48	1350.51	1207.3	2557.8	1363.26	1378.9	2742.1	3.2858	2.3915	5.6772
10	311.06	0.001 452	0.018 026	1393.04	1151.4	2544.4	1407.56	1317.1	2724.7	3.3596	2.2544	5.6141
11	318.15	0.001 489	0.015 987	1433.7	1096.0	2529.8	1450.1	1255.5	2705.6	3.4295	2.1233	5.5527
12	324.75	0.001 527	0.014 263	1473.0	1040.7	2513.7	1491.3	1193.6	2684.9	3.4962	1.9962	5.4924
13	330.93	0.001 567	0.012 780	1511.1	985.0	2496.1	1531.5	1130.7	2662.2	3.5606	1.8718	5.4323
14	336.75	0.001 611	0.011 485	1548.6	928.2	2476.8	1571.1	1066.5	2637.6	3.6232	1.7485	5.3717
15	342.24	0.001 658	0.010 337	1585.6	869.8	2455.5	1610.5	1000.0	2610.5	3.6848	1.6249	5.3098
16	347.44	0.001 711	0.009 306	1622.7	809.0	2431.7	1650.1	930.6	2580.6	3.7461	1.4994	5.2455
17	352.37	0.001 770	0.008 364	1660.2	744.8	2405.0	1690.3	856.9	2547.2	3.8079	1.3698	5.1777
18	357.06	0.001 840	0.007 489	1698.9	675.4	2374.3	1732.0	777.1	2509.1	3.8715	1.2329	5.1044
19	361.54	0.001 924	0.006 657	1739.9	598.1	2338.1	1776.5	688.0	2464.5	3.9388	1.0839	5.0228
20	365.81	0.002 036	0.005 834	1785.6	507.5	2293.0	1826.3	583.4	2409.7	4.0139	.9130	4.9269
21	369.89	0.002 207	0.004 952	1842.1	388.5	2230.6	1888.4	446.2	2334.6	4.1075	.6938	4.8013
22	373.80	0.002 742	0.003 568	1961.9	125.2	2087.1	2022.2	143.4	2165.6	4.3110	.2216	4.5327
22.09	374.14	0.003 155	0.003 155	2029.6	0	2029.6	2099.3	0	2099.3	4.4298	0	4.4298

TABLE A.3 (SI)
Properties of Superheated Steam

T	P = .010 MPa (45.81) v	u	h	s	P = .050 MPa (81.33) v	u	h	s	P = .10 MPa (99.63) v	u	h	s
Sat.	14.674	2437.9	2584.7	8.1502	3.240	2483.9	2645.9	7.5939	1.6940	2506.1	2675.5	7.3594
50	14.869	2443.9	2592.6	8.1749								
100	17.196	2515.5	2687.5	8.4479	3.418	2511.6	2682.5	7.6947	1.6958	2506.7	2676.2	7.3614
150	19.512	2587.9	2783.0	8.6882	3.889	2585.6	2780.1	7.9401	1.9364	2582.8	2776.4	7.6134
200	21.825	2661.3	2879.5	8.9038	4.356	2659.9	2877.7	8.1580	2.172	2658.1	2875.3	7.8343
250	24.136	2736.0	2977.3	9.1002	4.820	2735.0	2976.0	8.3556	2.406	2733.7	2974.3	8.0333
300	26.445	2812.1	3076.5	9.2813	5.284	2811.3	3075.5	8.5373	2.639	2810.4	3074.3	8.2158
400	31.063	2968.9	3279.6	9.6077	6.209	2968.5	3278.9	8.8642	3.103	2967.9	3278.2	8.5435
500	35.679	3132.3	3489.1	9.8978	7.134	3132.0	3488.7	9.1546	3.565	3131.6	3488.1	8.8342
600	40.295	3302.5	3705.4	10.1608	8.057	3302.2	3705.1	9.4178	4.028	3301.9	3704.7	9.0976
700	44.911	3479.6	3928.7	10.4028	8.981	3479.4	3928.5	9.6599	4.490	3479.2	3928.2	9.3398
800	49.526	3663.8	4159.0	10.6281	9.904	3663.6	4158.9	9.8852	4.952	3663.5	4158.6	9.5652
900	54.141	3855.0	4396.4	10.8396	10.828	3854.9	4396.3	10.0967	5.414	3854.8	4396.1	9.7767
1000	58.757	4053.0	4640.6	11.0393	11.751	4052.9	4640.5	10.2964	5.875	4052.8	4640.3	9.9764
1100	63.372	4257.5	4891.2	11.2287	12.674	4257.4	4891.1	10.4859	6.337	4257.3	4891.0	10.1659
1200	67.987	4467.9	5147.8	11.4091	13.597	4467.8	5147.7	10.6662	6.799	4467.7	5147.6	10.3463
1300	72.602	4683.7	5409.7	11.5811	14.521	4683.6	5409.6	10.8382	7.260	4683.5	5409.5	10.5183

T	P = .20 MPa (120.23) v	u	h	s	P = .30 MPa (133.55) v	u	h	s	P = .40 MPa (143.63) v	u	h	s
Sat.	.8857	2529.5	2706.7	7.1272	.6058	2543.6	2725.3	6.9919	.4625	2553.6	2738.6	6.8959
150	.9596	2576.9	2768.8	7.2795	.6339	2570.8	2761.0	7.0778	.4708	2564.5	2752.8	6.9299
200	1.0803	2654.4	2870.5	7.5066	.7163	2650.7	2865.6	7.3115	.5342	2646.8	2860.5	7.1706
250	1.1988	2731.2	2971.0	7.7086	.7964	2728.7	2967.6	7.5166	.5951	2726.1	2964.2	7.3789
300	1.3162	2808.6	3071.8	7.8926	.8753	2806.7	3069.3	7.7022	.6548	2804.8	3066.8	7.5662
400	1.5493	2966.7	3276.6	8.2218	1.0315	2965.6	3275.0	8.0330	.7726	2964.4	3273.4	7.8985

TABLE A.3 (SI) (cont'd.)

T	$P = .20$ MPa (120.23)				$P = .30$ MPa (133.55)				$P = .40$ MPa (143.63)			
	v	u	h	s	v	u	h	s	v	u	h	s
500	1.7814	3130.8	3487.1	8.5133	1.1867	3130.0	3486.0	8.3251	.8893	3129.2	3484.9	8.1913
600	2.013	3301.4	3704.0	8.7770	1.3414	3300.8	3703.2	8.5892	1.0055	3300.2	3702.4	8.4558
700	2.244	3478.8	3927.6	9.0194	1.4957	3478.4	3927.1	8.8319	1.1215	3477.9	3926.5	8.6987
800	2.475	3663.1	4158.2	9.2449	1.6499	3662.9	4157.8	9.0576	1.2372	3662.4	4157.3	8.9244
900	2.706	3854.5	4395.8	9.4566	1.8041	3854.2	4395.4	9.2692	1.3529	3853.9	4395.1	9.1362
1000	2.937	4052.5	4640.0	9.6563	1.9581	4052.3	4639.7	9.4690	1.4685	4052.0	4639.4	9.3360
1100	3.168	4257.0	4890.7	9.8458	2.1121	4256.8	4890.4	9.6585	1.5840	4256.5	4890.2	9.5256
1200	3.399	4467.5	5147.3	10.0262	2.2661	4467.2	5147.1	9.8389	1.6996	4467.0	5146.8	9.7060
1300	3.630	4683.2	5409.3	10.1982	2.4201	4683.0	5409.0	10.0110	1.8151	4682.8	5408.8	9.8780

T	$P = .50$ MPa (151.86)				$P = .60$ MPa (158.85)				$P = .80$ MPa (170.43)			
	v	u	h	s	v	u	h	s	v	u	h	s
Sat.	.3749	2561.2	2748.7	6.8213	.3157	2567.4	2756.8	6.7600	.2404	2576.8	2769.1	6.6628
200	.4249	2642.9	2855.4	7.0592	.3520	2638.9	2850.1	6.9665	.2608	2630.6	2839.3	6.8158
250	.4744	2723.5	2960.7	7.2709	.3938	2720.9	2957.2	7.1816	.2931	2715.5	2950.0	7.0384
300	.5226	2802.9	3064.2	7.4599	.4344	2801.0	3061.6	7.3724	.3241	2797.2	3056.5	7.2328
350	.5701	2882.6	3167.7	7.6329	.4742	2881.2	3165.7	7.5464	.3544	2878.2	3161.7	7.4089
400	.6173	2963.2	3271.9	7.7938	.5137	2962.1	3270.3	7.7079	.3843	2959.7	3267.1	7.5716
500	.7109	3128.4	3483.9	8.0873	.5920	3127.6	3482.8	8.0021	.4433	3126.0	3480.6	7.8673
600	.8041	3299.6	3701.7	8.3522	.6697	3299.1	3700.9	8.2674	.5018	3297.9	3699.4	8.1333
700	.8969	3477.5	3925.9	8.5952	.7472	3477.0	3925.3	8.5107	.5601	3476.2	3924.2	8.3770
800	.9896	3662.1	4156.9	8.8211	.8245	3661.8	4156.5	8.7367	.6181	3661.1	4155.6	8.6033
900	1.0822	3853.6	4394.7	9.0329	.9017	3853.4	4394.4	8.9486	.6761	3852.8	4393.7	8.8153
1000	1.1747	4051.8	4639.1	9.2328	.9788	4051.5	4638.8	9.1485	.7340	4051.0	4638.2	9.0153
1100	1.2672	4256.3	4889.9	9.4224	1.0559	4256.1	4889.6	9.3381	.7919	4255.6	4889.1	9.2050
1200	1.3596	4466.8	5146.6	9.6029	1.1330	4466.5	5146.3	9.5185	.8497	4466.1	5145.9	9.3855
1300	1.4521	4682.5	5408.6	9.7749	1.2101	4682.3	5408.3	9.6906	.9076	4681.8	5407.9	9.5575

TABLE A.3 (SI) (cont'd.)

T	v	u	h	s	v	u	h	s	v	u	h	s
	P = 1.00 MPa (179.91)				P = 1.20 MPa (187.99)				P = 1.40 MPa (195.07)			
Sat.	.194 44	2583.6	2778.1	6.5865	.163 33	2588.8	2784.8	6.5233	.140 84	2592.8	2790.0	6.4693
200	.2060	2621.9	2827.9	6.6940	.169 30	2612.8	2815.9	6.5898	.143 02	2603.1	2803.3	6.4975
250	.2327	2709.9	2942.6	6.9247	.192 34	2704.2	2935.0	6.8294	.163 50	2698.3	2927.2	6.7467
300	.2579	2793.2	3051.2	7.1229	.2138	2789.2	3045.8	7.0317	.182 28	2785.2	3040.4	6.9534
350	.2825	2875.2	3157.7	7.3011	.2345	2872.2	3153.6	7.2121	.2003	2869.2	3149.5	7.1360
400	.3066	2957.3	3263.9	7.4651	.2548	2954.9	3260.7	7.3774	.2178	2952.5	3257.5	7.3026
500	.3541	3124.4	3478.5	7.7622	.2946	3122.8	3476.3	7.6759	.2521	3121.1	3474.1	7.6027
600	.4011	3296.8	3697.9	8.0290	.3339	3295.6	3696.3	7.9435	.2860	3294.4	3694.8	7.8710
700	.4478	3475.3	3923.1	8.2731	.3729	3474.4	3922.0	8.1881	.3195	3473.6	3920.8	8.1160
800	.4943	3660.4	4154.7	8.4996	.4118	3659.7	4153.8	8.4148	.3528	3659.0	4153.0	8.3431
900	.5407	3852.2	4392.9	8.7118	.4505	3851.6	4392.2	8.6272	.3861	3851.1	4391.5	8.5556
1000	.5871	4050.5	4637.6	8.9119	.4892	4050.0	4637.0	8.8274	.4192	4049.5	4636.4	8.7559
1100	.6335	4255.1	4888.6	9.1017	.5278	4254.6	4888.0	9.0172	.4524	4254.1	4887.5	8.9457
1200	.6798	4465.6	5145.4	9.2822	.5665	4465.1	5144.9	9.1977	.4855	4464.7	5144.4	9.1262
1300	.7261	4681.3	5407.4	9.4543	.6051	4680.9	5407.0	9.3698	.5186	4680.4	5406.5	9.2984

T	v	u	h	s	v	u	h	s	v	u	h	s
	P = 1.60 MPa (201.41)				P = 1.80 MPa (207.15)				P = 2.00 MPa (212.42)			
Sat.	.123 80	2596.0	2794.0	6.4218	.110 42	2598.4	2797.1	6.3794	.099 63	2600.3	2799.5	6.3409
225	.132 87	2644.7	2857.3	6.5518	.116 73	2636.6	2846.7	6.4808	.103 77	2628.3	2835.8	6.4147
250	.141 84	2692.3	2919.2	6.6732	.124 97	2686.0	2911.0	6.6066	.111 44	2679.6	2902.5	6.5453
300	.158 62	2781.1	3034.8	6.8844	.140 21	2776.9	3029.2	6.8226	.125 47	2772.6	3023.5	6.7664
350	.174 56	2866.1	3145.4	7.0694	.154 57	2863.0	3141.2	7.0100	.138 57	2859.8	3137.0	6.9563
400	.190 05	2950.1	3254.2	7.2374	.168 47	2947.7	3250.9	7.1794	.151 20	2945.2	3247.6	7.1271
500	.2203	3119.5	3472.0	7.5390	.195 50	3117.9	3469.8	7.4825	.175 68	3116.2	3467.6	7.4317
600	.2500	3293.3	3693.2	7.8080	.2220	3292.1	3691.7	7.7523	.199 60	3290.9	3690.1	7.7024
700	.2794	3472.7	3919.7	8.0535	.2482	3471.8	3918.5	7.9983	.2232	3470.9	3917.4	7.9487

TABLE A.3 (SI) (cont'd.)

T	v	u	h	s	v	u	h	s	v	u	h	s
	P = 1.60 MPa (201.41)				P = 1.80 MPa (207.15)				P = 2.00 MPa (212.42)			
800	.3086	3658.3	4152.1	8.2808	.2742	3657.6	4151.2	8.2258	.2467	3657.0	4150.3	8.1765
900	.3377	3850.5	4390.8	8.4935	.3001	3849.9	4390.1	8.4386	.2700	3849.3	4389.4	8.3895
1000	.3668	4049.0	4635.8	8.6938	.3260	4048.5	4635.2	8.6391	.2933	4048.0	4634.6	8.5901
1100	.3958	4253.7	4887.0	8.8837	.3518	4253.2	4886.4	8.8290	.3166	4252.7	4885.9	8.7800
1200	.4248	4464.2	5143.9	9.0643	.3776	4463.7	5143.4	9.0096	.3398	4463.3	5142.9	8.9607
1300	.4538	4679.9	5406.0	9.2364	.4034	4679.5	5405.6	9.1818	.3631	4679.0	5405.1	9.1329

T	v	u	h	s	v	u	h	s	v	u	h	s
	P = 2.50 MPa (223.99)				P = 3.00 MPa (233.90)				P = 3.50 MPa (242.60)			
Sat.	.079 98	2603.1	2803.1	6.2575	.066 68	2604.1	2804.2	6.1869	.057 07	2603.7	2803.4	6.1253
225	.080 27	2605.6	2806.3	6.2639								
250	.087 00	2662.6	2880.1	6.4085	.070 58	2644.0	2855.8	6.2872	.058 72	2623.7	2829.2	6.1749
300	.098 90	2761.6	3008.8	6.6438	.081 14	2750.1	2993.5	6.5390	.068 42	2738.0	2977.5	6.4461
350	.109 76	2851.9	3126.3	6.8403	.090 53	2843.7	3115.3	6.7428	.076 78	2835.3	3104.0	6.6579
400	.120 10	2939.1	3239.3	7.0148	.099 36	2932.8	3230.9	6.9212	.084 53	2926.4	3222.3	6.8405
450	.130 14	3025.5	3350.8	7.1746	.107 87	3020.4	3344.0	7.0834	.091 96	3015.3	3337.2	7.0052
500	.139 98	3112.1	3462.1	7.3234	.116 19	3108.0	3456.5	7.2338	.099 18	3103.0	3450.9	7.1572
600	.159 30	3288.0	3686.3	7.5960	.132 43	3285.0	3682.3	7.5085	.113 24	3282.1	3678.4	7.4339
700	.178 32	3468.7	3914.5	7.8435	.148 38	3466.5	3911.7	7.7571	.126 99	3464.3	3908.8	7.6837
800	.197 16	3655.3	4148.2	8.0720	.164 14	3653.5	4145.9	7.9862	.140 56	3651.8	4143.7	7.9134
900	.215 90	3847.9	4387.6	8.2853	.179 80	3846.5	4385.9	8.1999	.154 02	3845.0	4384.1	8.1276
1000	.2346	4046.7	4633.1	8.4861	.195 41	4045.4	4631.6	8.4009	.167 43	4044.1	4630.1	8.3288
1100	.2532	4251.5	4884.6	8.6762	.210 98	4250.3	4883.3	8.5912	.180 80	4249.2	4881.9	8.5192
1200	.2718	4462.1	5141.7	8.8569	.226 52	4460.9	5140.5	8.7720	.194 15	4459.8	5139.3	8.7000
1300	.2905	4677.8	5404.0	9.0291	.242 06	4676.6	5402.8	8.9442	.207 49	4675.5	5401.7	8.8723

TABLE A.3 (SI) (cont'd.)

T	v	u	h	s	v	u	h	s	v	u	h	s
	P = 4.0 MPa (250.40)				P = 4.5 MPa (257.49)				P = 5.0 MPa (263.99)			
Sat.	.049 78	2602.3	2801.4	6.0701	.044 06	2600.1	2798.3	6.0198	.039 44	2597.1	2794.3	5.9734
275	.054 57	2667.9	2886.2	6.2285	.047 30	2650.3	2863.2	6.1401	.041 41	2631.3	2838.3	6.0544
300	.058 84	2725.3	2960.7	6.3615	.051 35	2712.0	2943.1	6.2828	.045 32	2698.0	2924.5	6.2084
350	.066 45	2826.7	3092.5	6.5821	.058 40	2817.8	3080.6	6.5131	.051 94	2808.7	3068.4	6.4493
400	.073 41	2919.9	3213.6	6.7690	.064 75	2913.3	3204.7	6.7047	.057 81	2906.6	3195.7	6.6459
450	.080 02	3010.2	3330.3	6.9363	.070 74	3005.0	3323.3	6.8746	.063 30	2999.7	3316.2	6.8186
500	.086 43	3099.5	3445.3	7.0901	.076 51	3095.3	3439.6	7.0301	.068 57	3091.0	3433.8	6.9759
600	.098 85	3279.1	3674.4	7.3688	.087 65	3276.0	3670.5	7.3110	.078 69	3273.0	3666.5	7.2589
700	.110 95	3462.1	3905.9	7.6198	.098 47	3459.9	3903.0	7.5631	.088 49	3457.6	3900.1	7.5122
800	.122 87	3650.0	4141.5	7.8502	.109 11	3648.3	4139.3	7.7942	.098 11	3646.6	4137.1	7.7440
900	.134 69	3843.6	4382.3	8.0647	.119 65	3842.2	4380.6	8.0091	.107 62	3840.7	4378.8	7.9593
1000	.146 45	4042.9	4628.7	8.2662	.130 13	4041.6	4627.2	8.2108	.117 07	4040.4	4625.7	8.1612
1100	.158 17	4248.0	4880.6	8.4567	.140 56	4246.8	4879.3	8.4015	.126 48	4245.6	4878.0	8.3520
1200	.169 87	4458.6	5138.1	8.6376	.150 98	4457.5	5136.9	8.5825	.135 87	4456.3	5135.7	8.5331
1300	.181 56	4674.3	5400.5	8.8100	.161 39	4673.1	5399.4	8.7549	.145 26	4672.0	5398.2	8.7055
	P = 6.0 MPa (275.64)				P = 7.0 MPa (285.88)				P = 8.0 MPa (295.06)			
Sat.	.032 44	2589.7	2784.3	5.8892	.027 37	2580.5	2772.1	5.8133	.023 52	2569.8	2758.0	5.7432
300	.036 16	2667.2	2884.2	6.0674	.029 47	2632.2	2838.4	5.9305	.024 26	2590.9	2785.0	5.7906
350	.042 23	2789.6	3043.0	6.3335	.035 24	2769.4	3016.0	6.2283	.029 95	2747.7	2987.3	6.1301
400	.047 39	2892.9	3177.2	6.5408	.039 93	2878.6	3158.1	6.4478	.034 32	2863.8	3138.3	6.3634
450	.052 14	2988.9	3301.8	6.7193	.044 16	2978.0	3287.1	6.6327	.038 17	2966.7	3272.0	6.5551
500	.056 65	3082.2	3422.2	6.8803	.048 14	3073.4	3410.3	6.7975	.041 75	3064.3	3398.3	6.7240
550	.061 01	3174.6	3540.6	7.0288	.051 95	3167.2	3530.9	6.9486	.045 16	3159.8	3521.0	6.8778
600	.065 25	3266.9	3658.4	7.1677	.055 65	3260.7	3650.3	7.0894	.048 45	3254.4	3642.0	7.0206

TABLE A.3 (SI) (cont'd.)

T	P = 6.0 MPa (275.64)				P = 7.0 MPa (285.88)				P = 8.0 MPa (295.06)			
	v	u	h	s	v	u	h	s	v	u	h	s
700	.073 52	3453.1	3894.2	7.4234	.062 83	3448.5	3888.3	7.3476	.054 81	3443.9	3882.4	7.2812
800	.081 60	3643.1	4132.7	7.6566	.069 81	3639.5	4128.2	7.5822	.060 97	3636.0	4123.8	7.5173
900	.089 58	3837.8	4375.3	7.8727	.076 69	3835.0	4371.8	7.7991	.067 02	3832.1	4368.3	7.7351
1000	.097 49	4037.8	4622.7	8.0751	.083 50	4035.3	4619.8	8.0020	.073 01	4032.8	4616.9	7.9384
1100	.105 36	4243.3	4875.4	8.2661	.090 27	4240.9	4872.8	8.1933	.078 96	4238.6	4870.3	8.1300
1200	.113 21	4454.0	5133.3	8.4474	.097 03	4451.7	5130.9	8.3747	.084 89	4449.5	5128.5	8.3115
1300	.121 06	4669.6	5396.6	8.6199	.103 77	4667.3	5393.7	8.5473	.090 80	4665.0	5391.5	8.4842

T	P = 9.0 MPa (303.40)				P = 10.0 MPa (311.06)				P = 12.5 MPa (327.89)			
	v	u	h	s	v	u	h	s	v	u	h	s
Sat.	.020 48	2557.8	2742.1	5.6772	.018 026	2544.4	2724.7	5.6141	.013 495	2505.1	2673.8	5.4624
325	.023 27	2646.6	2856.0	5.8712	.019 861	2610.4	2809.1	5.7568				
350	.025 80	2724.4	2956.6	6.0361	.022 42	2699.2	2923.4	5.9443	.016 126	2624.6	2826.2	5.7118
400	.029 93	2848.4	3117.8	6.2854	.026 41	2832.4	3096.5	6.2120	.020 00	2789.3	3039.3	6.0417
450	.033 50	2955.2	3256.6	6.4844	.029 75	2943.4	3240.9	6.4190	.022 99	2912.5	3199.8	6.2719
500	.036 77	3055.2	3386.1	6.6576	.032 79	3045.8	3373.7	6.5966	.025 60	3021.7	3341.8	6.4618
550	.039 87	3152.2	3511.0	6.8142	.035 64	3144.6	3500.9	6.7561	.028 01	3125.0	3475.2	6.6290
600	.042 85	3248.1	3633.7	6.9589	.038 37	3241.7	3625.3	6.9029	.030 29	3225.4	3604.0	6.7810
650	.045 74	3343.6	3755.3	7.0943	.041 01	3338.2	3748.2	7.0398	.032 48	3324.4	3730.4	6.9218
700	.048 57	3439.3	3876.5	7.2221	.043 58	3434.7	3870.5	7.1687	.034 60	3422.9	3855.3	7.0536
800	.054 09	3632.5	4119.3	7.4596	.048 59	3628.9	4114.8	7.4077	.038 69	3620.0	4103.6	7.2965
900	.059 50	3829.2	4364.8	7.6783	.053 49	3826.3	4361.2	7.6272	.042 67	3819.1	4352.5	7.5182
1000	.064 85	4030.3	4614.0	7.8821	.058 32	4027.8	4611.0	7.8315	.046 58	4021.6	4603.8	7.7237
1100	.070 16	4236.3	4867.7	8.0740	.063 12	4234.0	4865.1	8.0237	.050 45	4228.2	4858.8	7.9165
1200	.075 44	4447.2	5126.2	8.2556	.067 89	4444.9	5123.8	8.2055	.054 30	4439.3	5118.0	8.0987
1300	.080 72	4662.7	5389.2	8.4284	.072 65	4660.5	5387.0	8.3783	.058 13	4654.8	5381.4	8.2717

TABLE A.3 (SI) (cont'd.)

T	v	u	h	s	v	u	h	s	v	u	h	s
	P = 15.0 MPa (342.24)				P = 17.5 MPa (354.75)				P = 20.0 MPa (365.81)			
Sat.	.010 337	2455.5	2610.5	5.3098	.007 920	2390.2	2528.8	5.1419	.005 834	2293.0	2409.7	4.9269
350	.011 470	2520.4	2692.4	5.4421								
400	.015 649	2740.7	2975.5	5.8811	.012 447	2685.0	2902.9	5.7213	.009 942	2619.3	2818.1	5.5540
450	.018 445	2879.5	3156.2	6.1404	.015 174	2844.2	3109.7	6.0184	.012 695	2806.2	3060.1	5.9017
500	.020 80	2996.6	3308.6	6.3443	.017 358	2970.3	3274.1	6.2383	.014 768	2942.9	3238.2	6.1401
550	.022 93	3104.7	3448.6	6.5199	.019 288	3083.9	3421.4	6.4230	.016 555	3062.4	3393.5	6.3348
600	.024 91	3208.6	3582.3	6.6776	.021 06	3191.5	3560.1	6.5866	.018 178	3174.0	3537.6	6.5048
650	.026 80	3310.3	3712.3	6.8224	.022 74	3296.0	3693.9	6.7357	.019 693	3281.4	3675.3	6.6582
700	.028 61	3410.9	3840.1	6.9572	.024 34	3398.7	3824.6	6.8736	.021 13	3386.4	3809.0	6.7993
800	.032 10	3610.9	4092.4	7.2040	.027 38	3601.8	4081.1	7.1244	.023 85	3592.7	4069.7	7.0544
900	.035 46	3811.9	4343.8	7.4279	.030 31	3804.7	4335.1	7.3507	.026 45	3797.5	4326.4	7.2830
1000	.038 75	4015.4	4596.6	7.6348	.033 16	4009.3	4589.5	7.5589	.028 97	4003.1	4582.5	7.4925
1100	.042 00	4222.6	4852.6	7.8283	.035 97	4216.9	4846.4	7.7531	.031 45	4211.3	4840.2	7.6874
1200	.045 23	4433.8	5112.3	8.0108	.038 76	4428.3	5106.6	7.9360	.033 91	4422.8	5101.0	7.8707
1300	.048 45	4649.1	5376.0	8.1840	.041 54	4643.5	5370.5	8.1093	.036 36	4638.0	5365.1	8.0442

T	v	u	h	s	v	u	h	s	v	u	h	s
	P = 25.0 MPa				P = 30.0 MPa				P = 35.0 MPa			
375	.001 973 1	1798.7	1848.0	4.0320	.001 789 2	1737.8	1791.5	3.9305	.001 700 3	1702.9	1762.4	3.8722
400	.006 004	2430.1	2580.2	5.1418	.002 790	2067.4	2151.1	4.4728	.002 100	1914.1	1987.6	4.2126
425	.007 881	2609.2	2806.3	5.4723	.005 303	2455.1	2614.2	5.1504	.003 428	2253.4	2373.4	4.7747
450	.009 162	2720.7	2949.7	5.6744	.006 735	2619.3	2821.4	5.4424	.004 961	2498.7	2672.4	5.1962
500	.011 123	2884.3	3162.4	5.9592	.008 678	2820.7	3081.1	5.7905	.006 927	2751.9	2994.4	5.6282
550	.012 724	3017.5	3335.6	6.1765	.010 168	2970.3	3275.4	6.0342	.008 345	2921.0	3213.0	5.9026
600	.014 137	3137.9	3491.4	6.3602	.011 446	3100.5	3443.9	6.2331	.009 527	3062.0	3395.5	6.1179
650	.015 433	3251.6	3637.4	6.5229	.012 596	3221.0	3598.9	6.4058	.010 575	3189.8	3559.9	6.3010

TABLE A.3 (SI) (cont'd.)

T	v	P = 25.0 MPa u	h	s	v	P = 30.0 MPa u	h	s	v	P = 35.0 MPa u	h	s
700	.016 646	3361.3	3777.5	6.6707	.013 661	3335.8	3745.6	6.5606	.011 533	3309.8	3713.5	6.4631
800	.018 912	3574.3	4047.1	6.9345	.015 623	3555.5	4024.2	6.8332	.013 278	3536.7	4001.5	6.7450
900	.021 045	3783.0	4309.1	7.1680	.017 448	3768.5	4291.9	7.0718	.014 883	3754.0	4274.9	6.9886
1000	.023 10	3990.9	4568.5	7.3802	.019 196	3978.8	4554.7	7.2867	.016 410	3966.7	4541.1	7.2064
1100	.025 12	4200.2	4828.2	7.5765	.020 903	4189.2	4816.3	7.4845	.017 895	4178.3	4804.6	7.4057
1200	.027 11	4412.0	5089.9	7.7605	.022 589	4401.3	5079.0	7.6692	.019 360	4390.7	5068.3	7.5910
1300	.029 10	4626.9	5354.4	7.9342	.024 266	4616.0	5344.0	7.8432	.020 815	4605.1	5333.6	7.7653

T	v	P = 40.0 MPa u	h	s	v	P = 50.0 MPa u	h	s	v	P = 60.0 MPa u	h	s
375	.001 640 7	1677.1	1742.8	3.8290	.001 559 4	1638.6	1716.6	3.7639	.001 502 8	1609.4	1699.5	3.7141
400	.001 907 7	1854.6	1930.9	4.1135	.001 730 9	1788.1	1874.6	4.0031	.001 633 5	1745.4	1843.4	3.9318
425	.002 532	2096.9	2198.1	4.5029	.002 007	1959.7	2060.0	4.2734	.001 816 5	1892.7	2001.7	4.1626
450	.003 693	2365.1	2512.8	4.9459	.002 486	2159.6	2284.0	4.5884	.002 085	2053.9	2179.0	4.4121
500	.005 622	2678.4	2903.3	5.4700	.003 892	2525.5	2720.1	5.1726	.002 956	2390.6	2567.9	4.9321
550	.006 984	2869.7	3149.1	5.7785	.005 118	2763.6	3019.5	5.5485	.003 956	2658.8	2896.2	5.3441
600	.008 094	3022.6	3346.4	6.0114	.006 112	2942.0	3247.6	5.8178	.004 834	2861.1	3151.2	5.6452
650	.009 063	3158.0	3520.6	6.2054	.006 966	3093.5	3441.8	6.0342	.005 595	3028.8	3364.5	5.8829
700	.009 941	3283.6	3681.2	6.3750	.007 727	3230.5	3616.8	6.2189	.006 272	3177.2	3553.5	6.0824
800	.011 523	3517.8	3978.7	6.6662	.009 076	3479.8	3933.6	6.5290	.007 459	3441.5	3889.1	6.4109
900	.012 962	3739.4	4257.9	6.9150	.010 283	3710.3	4224.4	6.7882	.008 508	3681.0	4191.5	6.6805
1000	.014 324	3954.6	4527.6	7.1356	.011 411	3930.5	4501.1	7.0146	.009 480	3906.4	4475.2	6.9127
1100	.015 642	4167.4	4793.1	7.3364	.012 496	4145.7	4770.5	7.2184	.010 409	4124.1	4748.6	7.1195
1200	.016 940	4380.1	5057.7	7.5224	.013 561	4359.1	5037.2	7.4058	.011 317	4338.2	5017.2	7.3083
1300	.018 229	4594.3	5323.5	7.6969	.014 616	4572.8	5303.6	7.5808	.012 215	4551.4	5284.3	7.4837

TABLE A.4 (SI)
Properties of Compressed Liquid (Steam)

T	P = 5 MPa (263.99)				P = 10 MPa (311.06)				P = 15 MPa (342.24)			
	v	u	h	s	v	u	h	s	v	u	h	s
Sat.	.001 285 9	1147.8	1154.2	2.9202	.001 452 4	1393.0	1407.6	3.3596	.001 658 1	1585.6	1610.5	3.6848
0	.000 997 7	.04	5.04	.0001	.000 995 2	.09	10.04	.0002	.000 992 8	.15	15.05	.0004
20	.000 999 5	83.65	88.65	.2956	.000 997 2	83.36	93.33	.2945	.000 995 0	83.06	97.99	.2934
40	.001 005 6	166.95	171.97	.5705	.001 003 4	166.35	176.38	.5686	.001 001 3	165.76	180.78	.5666
60	.001 014 9	250.23	255.30	.8285	.001 012 7	249.36	259.49	.8258	.001 010 5	248.51	263.67	.8232
80	.001 026 8	333.72	338.85	1.0720	.001 024 5	332.59	342.83	1.0688	.001 022 2	331.48	346.81	1.0656
100	.001 041 0	417.52	422.72	1.3030	.001 038 5	416.12	426.50	1.2992	.001 036 1	414.74	430.28	1.2955
120	.001 057 6	501.80	507.09	1.5233	.001 054 9	500.08	510.64	1.5189	.001 052 2	498.40	514.19	1.5145
140	.001 076 8	586.76	592.15	1.7343	.001 073 7	584.68	595.42	1.7292	.001 070 7	582.66	598.72	1.7242
160	.001 098 8	672.62	678.12	1.9375	.001 095 3	670.13	681.08	1.9317	.001 091 8	667.71	684.09	1.9260
180	.001 124 0	759.63	765.25	2.1341	.001 119 9	756.65	767.84	2.1275	.001 115 9	753.76	770.50	2.1210
200	.001 153 0	848.1	853.9	2.3255	.001 148 0	844.5	856.0	2.3178	.001 143 3	841.0	858.2	2.3104
220	.001 186 6	938.4	944.4	2.5128	.001 180 5	934.1	945.9	2.5039	.001 174 8	929.9	947.5	2.4953
240	.001 226 4	1031.4	1037.5	2.6979	.001 218 7	1026.0	1038.1	2.6872	.001 211 4	1020.8	1039.0	2.6771
260	.001 274 9	1127.9	1134.3	2.8830	.001 264 5	1121.1	1133.7	2.8699	.001 255 0	1114.6	1133.4	2.8576
280					.001 321 6	1220.9	1234.1	3.0548	.001 308 4	1212.5	1232.1	3.0393
300					.001 397 2	1328.4	1342.3	3.2469	.001 377 0	1316.6	1337.3	3.2260
320									.001 472 4	1431.1	1453.2	3.4247
340									.001 631 1	1567.5	1591.9	3.6546

TABLE A.4 (SI) (cont'd.)

T	P = 20 MPa (365.81)				P = 30 MPa				P = 50 MPa			
	v	u	h	s	v	u	h	s	v	u	h	s
Sat.	.002 036	1785.6	1826.3	4.0139								
0	.000 990 4	.19	20.01	.0004	.000 985 6	.25	29.82	.0001	.000 976 6	.20	49.03	.0014
20	.000 992 8	82.77	102.62	.2923	.000 988 6	82.17	111.84	.2899	.000 980 4	81.00	130.02	.2848
40	.000 999 2	165.17	185.16	.5646	.000 995 1	164.04	193.89	.5607	.000 987 2	161.86	211.21	.5527
60	.001 008 4	247.68	267.85	.8206	.001 004 2	246.06	276.19	.8154	.000 996 2	242.98	292.79	.8052
80	.001 019 9	330.40	350.80	1.0624	.001 015 6	328.30	358.77	1.0561	.001 007 3	324.34	374.70	1.0440
100	.001 033 7	413.39	434.06	1.2917	.001 029 0	410.78	441.66	1.2844	.001 020 1	405.88	456.89	1.2703
120	.001 049 6	496.76	517.76	1.5102	.001 044 5	493.59	524.93	1.5018	.001 034 8	487.65	539.39	1.4857
140	.001 067 8	580.69	602.04	1.7193	.001 062 1	576.88	608.75	1.7098	.001 051 5	569.77	622.35	1.6915
160	.001 088 5	665.35	687.12	1.9204	.001 082 1	660.82	693.28	1.9096	.001 070 3	652.41	705.92	1.8891
180	.001 112 0	750.95	773.20	2.1147	.001 104 7	745.59	778.73	2.1024	.001 091 2	735.69	790.25	2.0794
200	.001 138 8	837.7	860.5	2.3031	.001 130 2	831.4	865.3	2.2893	.001 114 6	819.7	875.5	2.2634
220	.001 169 3	925.9	949.3	2.4870	.001 159 0	918.3	953.1	2.4711	.001 140 8	904.7	961.7	2.4419
240	.001 204 6	1016.0	1040.0	2.6674	.001 192 0	1006.9	1042.6	2.6490	.001 170 2	990.7	1049.2	2.6158
260	.001 246 2	1108.6	1133.5	2.8459	.001 230 3	1097.4	1134.3	2.8243	.001 203 4	1078.1	1138.2	2.7860
280	.001 296 5	1204.7	1230.6	3.0248	.001 275 5	1190.7	1229.0	2.9986	.001 241 5	1167.2	1229.3	2.9537
300	.001 359 6	1306.1	1333.3	3.2071	.001 330 4	1287.9	1327.8	3.1741	.001 286 0	1258.7	1323.0	3.1200
320	.001 443 7	1415.7	1444.6	3.3979	.001 399 7	1390.7	1432.7	3.3539	.001 338 8	1353.3	1420.2	3.2868
340	.001 568 4	1539.7	1571.0	3.6075	.001 492 0	1501.7	1546.5	3.5426	.001 403 2	1452.0	1522.1	3.4557
360	.001 822 6	1702.8	1739.3	3.8772	.001 626 5	1626.5	1675.4	3.7494	.001 483 8	1556.0	1630.2	3.6291
380					.001 869 1	1781.4	1837.5	4.0012	.001 588 4	1667.2	1746.6	3.8101

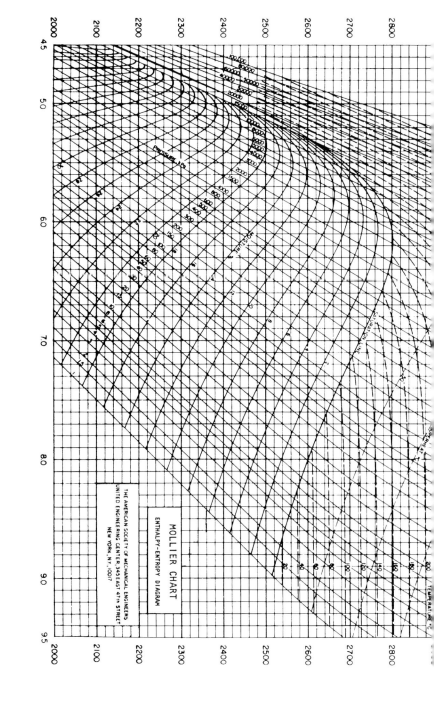

MOLLIER CHART

ENTHALPY–ENTROPY DIAGRAM

THE AMERICAN SOCIETY OF MECHANICAL ENGINEERS
UNITED ENGINEERING CENTER, 345 EAST 47th STREET
NEW YORK, N.Y., 10017

Enthalpy, h, kJ/kg

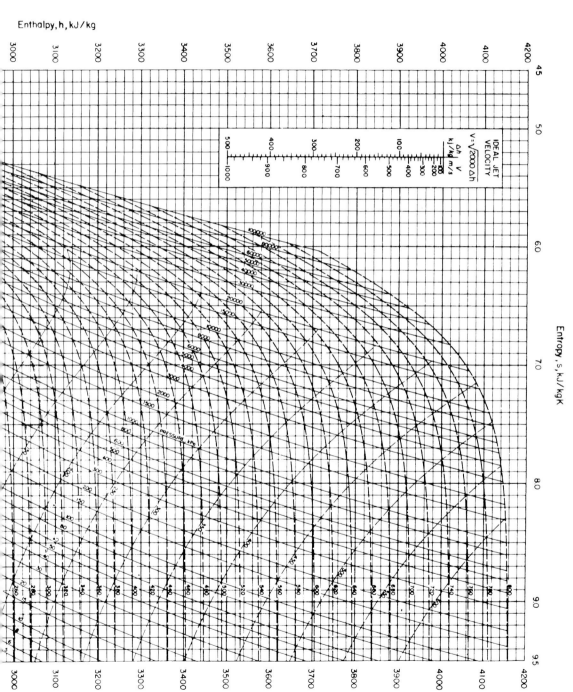

Entropy, s, kJ/kgK

TABLE A.5 (SI) Enthalpy—Entropy Diagram for Steam, SI Units

Source: American Society of Mechanical Engineers.

TABLE A.6
Thermal Conductivities of Some Building and Insulating Materials
$(k = \text{Btu/hr} \cdot \text{ft}^2 \cdot {}^\circ\text{F/ft})$

Material	Apparent density ρ_1 lb$_m$/ft^3 at room temperature	°F	k
Aerogel, silica, opacified	8.5	248	0.013
		554	0.026
Asbestos-cement boards	120	68	0.43
Asbestos sheets	55.5	124	0.096
Asbestos slate	112	32	0.087
	112	140	0.114
Aluminum foil, 7 air spaces per 2.5 in.	0.2	100	0.025
		351	0.038
Asphalt	132	68	0.43
Bricks:			
Alumina (92–99% Al$_2$O$_3$ by weight) fused	.....	801	1.8
Alumina (64–65% Al$_2$O$_3$ by weight)	.....	2399	2.7
(See also Bricks, fire clay)	115	1472	0.62
Building brickwork	.....	68	0.4
Carbon	96.7	.....	3.0
Chrome brick (32% Cr$_2$O$_3$ by weight)	200	392	0.67
	200	1202	0.85
	200	2399	1.0
Diatomaceous earth, molded and fired	38	399	0.14
	38	1600	0.18
Diatomaceous earth, high burn, large pores	37	392	0.13
	37	1832	0.34
Fire clay, Missouri	.....	392	0.58
		1112	0.85
		1832	0.95
		2552	1.02
Kaolin insulating brick	27	932	0.15
	27	2102	0.26
Kaolin insulating firebrick	19	392	0.050
	19	1400	0.113
Magnesite (86.8% MgO, 6.3% Fe$_2$O$_3$, 3% CaO, 2.6% SiO$_2$ by weight)	158	399	2.2
	158	1202	1.6
	158	2192	1.1
Calcium carbonate, natural	162	86	1.3
White marble	.....	.....	1.7
Chalk	96	.....	0.4
Calcium sulfate (4H$_2$O), artificial	84.6	104	0.22
Plaster, artificial	132	167	0.43
Building	77.9	77	0.25
Cardboard, corrugated	.....	.....	0.037
Celluloid	87.3	86	0.12
Charcoal flakes	11.9	176	0.043
	15	176	0.051
Coke, petroleum	.....	212	3.4
		932	2.9
Coke, powdered	.....	32–212	0.11
Concrete, cinder	.....	.....	0.20
1:4 dry	.....	.....	0.44
Stone	.....	.....	0.54
Cotton wool	5	86	0.024
Cork board	10	86	0.025
Cork, ground	9.4	86	0.025
Diatomaceous earth power, coarse	20.0	100	0.036
	20.0	1600	0.082
Fine	17.2	399	0.040
	17.2	1600	0.074
Molded pipe covering	26.0	399	0.051
	26.0	1600	0.088

Source: G. F. Babits, *Thermodynamics*, Allyn & Bacon, Inc., Boston, 1963, pp. 231-234, with permission.

TABLE A.6 (cont'd.)

Material	Apparent density ρ_1 lb$_m$/ft^3 at room temperature	°F	k
Dolomite	167	122	1.0
Ebonite	. . . . .	. . . . .	0.10
Enamel, Silicate	38	. . . . .	0.5–0.75
Felt, wool	20.6	86	0.03
Fiber insulating board	14.8	70	0.028
Glass	. . . . .	. . . . .	0.2–0.73
Borosilicate type	139	86–167	0.63
Soda glass	. . . . .	. . . . .	0.3–0.44
Window glass	. . . . .	. . . . .	0.3–0.61
Granite	. . . . .	. . . . .	1.0–2.3
Graphite, longitudinal	. . . . .	68	95
Gypsum, molded and dry	78	68	0.25
Hair felt, perpendicular to fibers	17	86	0.021
Ice	57.5	32	1.3
Kapok	0.88	68	0.020
Lampblack	10	104	0.038
Leather sole	62.4	. . . . .	0.092
Limestone (15.3 vol. %H_2O)	103	75	0.54
Magnesia, powdered	49.7	117	0.35
Magnesia, light carbonate	19	70	0.034
Magnesium oxide, compressed	49.9	68	0.32
Marble	. . . . .	. . . . .	1.2–1.7
Mica, perpendicular to planes	. . . . .	122	0.25
Mineral wool	9.4	86	0.0225
	19.7	86	0.024
Paper	. . . . .	. . . . .	0.075
Paraffin wax	. . . . .	32	0.14
Porcelain	. . . . .	392	0.88
Portland cement (see Concrete)	. . . . .	194	0.17
Pumice stone	. . . . .	70–151	0.14
Pyroxylin plastics	. . . . .	. . . . .	0.075
Rubber, hard	74.8	32	0.087
Para	. . . . .	70	0.109
Soft	. . . . .	70	0.075–0.092
Sand, dry	94.6	68	0.19
Sandstone	140	104	1.06
Sawdust	12	70	0.03
Slag, blast furnace	. . . . .	75–261	0.064
Slag wool	12	86	0.022
Slate	. . . . .	201	0.86
Snow	34.7	32	0.27
Sulfur, monoclinic	. . . . .	212	0.09–0.097
Rhombic	. . . . .	70	0.16
Wallboard, insulating type	14.8	70	0.028
Wallboard, stiff pasteboard	43	86	0.04
Wood shavings	8.8	86	0.034
Wood, across grain			
Balsa	7–8	86	0.025–0.03
Oak	51.5	59	0.12
Maple	44.7	122	0.11
Pine, white	34.0	59	0.087
Teak	40.0	59	0.10
White fir	28.1	140	0.062
Wood, parallel to grain			
Pine	34.4	70	0.20
Wool, animal	6.9	86	0.021

TABLE A.7
Ammonia: Properties of Liquid and Saturated Vapor

Temp.	Pressure		Volume	Density	Enthalpy from −40°F			Entropy from −40°F	
°F	Abs. lb/in.^{2}	Gage lb/in.^{2}	Vapor ft^3/lb	Liquid lb/ft^3	Liquid Btu/lb	Vapor Btu/lb	Latent Btu/lb	Liquid Btu/lb °F	Vapor Btu/lb °F
t	p	p_d	v_g	$1/v_f$	h_f	h_g	h_{fg}	s_f	s_g
−100	1.24	*27.4	182.90	45.51	−61.5	571.4	632.9	−0.1579	1.6025
− 90	1.86	*26.1	124.28	45.12	−51.4	575.9	627.3	−0.1309	1.5667
− 80	2.74	*24.3	86.54	44.73	−41.3	580.1	621.4	−0.1036	1.5336
− 70	3.94	*21.9	61.65	44.32	−31.1	584.4	615.5	−0.0771	1.5026
− 65	4.69	*20.4	52.34	44.11	−26.0	586.6	612.6	−0.0642	1.4833
− 60	5.55	*18.6	44.73	43.91	−20.9	588.8	609.7	−0.0514	1.4747
− 55	6.54	*16.6	38.38	43.70	−15.7	591.0	606.7	−0.0382	1.4614
− 50	7.67	*14.3	33.08	43.49	−10.5	593.2	603.7	−0.0254	1.4487
− 45	8.95	*11.7	28.62	43.28	− 5.3	595.7	600.7	−0.0128	1.4363
− 40	10.41	*8.7	24.86	43.08	0.0	597.6	597.6	0.0000	1.4242
− 38	11.04	*7.4	23.53	42.99	2.1	598.3	596.2	.0051	1.4193
− 36	11.71	*6.1	22.27	42.90	4.3	599.1	594.8	.0101	1.4144
− 34	12.41	*4.7	21.10	42.82	6.4	599.9	593.5	.0151	1.4096
− 32	13.14	*3.2	20.00	42.73	8.5	600.6	592.1	.0201	1.4048
− 30	13.90	*1.6	18.97	42.65	10.7	601.4	590.7	0.0250	1.4001
− 28	14.71	0.0	18.00	42.57	12.8	602.1	589.3	.0300	1.3955
− 26	15.55	0.8	17.09	42.48	14.9	602.8	587.9	.0350	1.3909
− 24	16.42	1.7	16.24	42.40	17.1	603.6	586.5	.0399	1.3863
− 22	17.34	2.6	15.43	42.31	19.2	604.3	585.1	.0448	1.3818
− 20	18.30	3.6	14.68	42.22	21.4	605.0	583.6	0.0497	1.3774
− 18	19.30	4.6	13.97	43.13	23.5	605.7	582.2	.0545	1.3729
− 16	20.34	5.6	13.29	42.04	25.6	606.4	580.8	.0594	1.3686
− 14	21.43	6.7	12.66	41.96	27.8	607.1	579.3	.0642	1.3643
− 12	22.56	7.9	12.06	41.87	30.0	607.8	577.8	.0690	1.3600
− 10	23.74	9.0	11.50	41.78	32.1	608.5	576.4	0.0738	1.3558
− 8	24.97	10.3	10.97	41.69	34.3	609.2	574.9	.0786	1.3516
− 6	26.26	11.6	10.47	41.60	36.4	609.8	573.4	.0833	1.3474
− 4	27.59	12.9	9.991	41.52	38.6	610.5	571.9	.0880	1.3433
− 2	28.98	14.3	9.541	41.43	40.7	611.1	570.4	.0928	1.3393
0	30.42	15.7	9.116	41.34	42.9	611.8	568.9	0.0975	1.3352
2	31.92	17.2	8.714	41.25	45.1	612.4	567.3	.1022	1.3312
4	33.47	18.8	8.333	41.16	47.2	613.0	565.8	.1069	1.3273
6	35.09	20.4	7.971	41.07	49.4	613.6	564.2	.1115	1.3234
8	36.77	22.1	7.629	40.98	51.6	614.3	562.7	.1162	1.3195
10	38.51	23.8	7.304	40.89	53.8	614.9	561.1	0.1208	1.3157
12	40.31	25.6	6.996	40.80	56.0	615.5	559.5	.1254	1.3118
14	42.18	27.5	6.703	40.71	58.2	616.1	557.9	.1300	1.3081
16	44.12	29.4	6.425	40.61	60.3	616.6	556.3	.1346	1.3043
18	46.13	31.4	6.161	40.52	62.5	617.2	554.7	.1392	1.3006
20	48.21	33.5	5.910	40.43	64.7	617.8	553.1	0.1437	1.2969
22	50.36	35.7	5.671	40.34	66.9	618.3	551.4	.1483	1.2933
24	52.59	37.9	5.443	40.25	69.1	618.9	549.8	.1528	1.2897
26	54.90	40.2	5.227	40.15	71.3	619.4	548.1	.1573	1.2861
28	57.28	42.6	5.021	40.06	73.5	619.9	546.4	.1618	1.2825
30	59.74	45.0	4.825	39.96	75.7	620.5	544.8	0.1663	1.2790
32	62.29	47.6	4.637	39.86	77.9	621.0	543.1	.1708	1.2755
34	64.91	50.2	4.459	39.77	80.1	621.5	541.4	.1753	1.2721
36	67.63	52.9	4.289	39.67	82.3	622.0	539.7	.1797	1.2686
38	70.43	55.7	4.126	39.50	84.6	622.5	537.9	.1841	1.2652

Source: Abstracted, by permission, from *Tables of Thermodynamic Properties of Ammonia*, U.S. Department of Commerce, Bureau of Standards Circular 142, 1945.

* Inches of mercury below one atmosphere.

Temp.	Pressure		Volume	Density	Enthalpy from −40°F			Entropy from −40°F	
°F	Abs. lb/in.²	Gage lb/in.²	Vapor ft³/lb	Liquid lb/ft³	Liquid Btu/lb	Vapor Btu/lb	Latent Btu/lb	Liquid Btu/lb °F	Vapor Btu/lb °F
t	p	p_d	v_g	$1/v_f$	h_f	h_g	h_{fg}	s_f	s_g
40	73.32	58.6	3.971	39.49	86.8	623.0	536.2	0.1885	1.2618
42	76.31	61.6	3.823	39.39	89.0	623.4	534.4	.1930	1.2585
44	79.38	64.7	3.682	39.29	91.2	623.9	532.7	.1974	1.2552
46	82.55	67.9	3.547	39.19	93.5	624.4	530.9	.2018	1.2519
48	85.82	71.1	3.418	39.10	95.7	624.8	529.1	.2062	1.2486
50	89.19	74.5	3.294	39.00	97.9	625.2	527.3	0.2105	1.2453
52	92.66	78.0	3.176	38.90	100.2	625.7	525.5	.2149	1.2421
54	96.23	81.5	3.063	38.80	102.4	626.1	523.7	.2192	1.2389
56	99.91	85.2	2.954	38.70	104.7	626.5	521.8	.2236	1.2357
58	103.7	89.0	2.851	38.60	106.9	626.9	520.0	.2279	1.2325
60	107.6	92.9	2.751	38.50	109.2	627.3	518.1	0.2322	1.2294
62	111.6	96.9	2.656	38.40	111.5	627.7	516.2	.2365	1.2262
64	115.7	101.0	2.565	38.30	113.7	628.0	514.3	.2408	1.2231
66	120.0	105.3	2.477	38.20	116.0	628.4	512.4	.2451	1.2201
68	124.3	109.6	2.393	38.10	118.3	628.8	510.5	.2494	1.2170
70	128.8	114.1	2.312	38.00	120.5	629.1	508.6	0.2537	1.2140
72	133.4	118.7	2.235	37.90	122.8	629.4	506.6	.2579	1.2110
74	138.1	123.4	2.161	37.79	125.1	629.8	504.7	.2622	1.2080
76	143.0	128.3	2.089	37.69	127.4	630.1	502.7	.2664	1.2050
78	147.9	133.2	2.021	37.58	129.7	630.4	500.7	.2706	1.2020
80	153.0	138.3	1.955	37.48	132.0	630.7	498.7	0.2749	1.1991
82	158.3	143.6	1.892	37.37	134.3	631.0	496.7	.2791	1.1962
84	163.7	149.0	1.831	37.26	136.6	631.3	494.7	.2833	1.1933
86	169.2	154.5	1.772	37.16	138.9	631.5	492.6	.2875	1.1904
88	174.8	160.1	1.716	37.05	141.2	631.8	490.6	.2917	1.1875
90	180.6	165.9	1.661	36.95	143.5	632.0	488.5	0.2958	1.1846
92	186.6	171.9	1.609	36.84	145.8	632.2	486.4	.3000	1.1818
94	192.7	178.0	1.559	36.73	148.2	632.5	484.3	.3041	1.1789
96	198.9	184.2	1.510	36.62	150.5	632.6	482.1	.3083	1.1761
98	205.3	190.6	1.464	36.51	152.9	632.9	480.0	.3125	1.1733
100	211.9	197.2	1.419	36.40	155.2	633.0	477.8	0.3166	1.1705
102	218.6	203.9	1.375	36.29	157.6	633.2	475.6	.3207	1.1677
104	225.4	210.7	1.334	36.18	159.9	633.4	473.5	.3248	1.1649
106	232.5	217.8	1.293	36.06	162.3	633.5	471.2	.3289	1.1621
108	239.7	225.0	1.254	35.95	164.6	633.6	469.0	.3330	1.1593
110	247.0	232.3	1.217	35.84	167.0	633.7	466.7	0.3372	1.1566
112	254.5	239.8	1.180	35.72	169.4	633.8	464.4	.3413	1.1538
114	262.2	247.5	1.145	35.61	171.8	633.9	462.1	.3453	1.1510
116	270.1	255.4	1.112	35.49	174.2	634.0	459.8	.3495	1.1483
118	278.2	263.5	1.079	35.38	176.6	634.0	457.4	.3535	1.1455
120	286.4	271.7	1.047	35.26	179.0	634.0	455.0	0.3576	1.1427
122	294.8	280.1	1.017	35.14	181.4	634.0	452.6	.3618	1.1400
124	303.4	288.7	0.987	35.02	183.9	634.0	450.1	.3659	1.1372

TABLE A.8
Ammonia: Properties of Superheated Vapor

Temp. °F	Absolute Pressure in lb/in.² (Saturation Temperature in italics)											
	5 −63.11			10 −41.34			15 −27.20			20 −16.64		
t	v	h	s	v	h	s	v	h	s	v	h	s
Sat.	49.31	588.3	1.4857	25.81	597.1	1.4276	17.67	602.4	1.3988	13.50	606.2	1.3700
−50	51.05	595.2	1.5025									
−40	52.36	600.3	.5149									
−30	53.67	605.4	.5269	26.58	603.2	1.4420						
−20	54.97	610.4	.5385	27.26	608.5	.4542	18.01	606.4	1.4031			
−10	56.26	615.4	.5498	27.92	613.7	.4659	18.47	611.9	.4154	13.74	610.0	1.3784
0	57.55	620.4	1.5608	28.58	618.9	1.4773	18.92	617.2	1.4272	14.09	615.5	1.3907
10	58.84	625.4	.5716	29.24	624.0	.4884	19.37	622.5	.4386	14.44	621.0	.4025
20	60.12	630.4	.5821	29.90	629.1	.4992	19.82	627.8	.4497	14.78	626.4	.4138
30	61.41	635.4	.5925	30.55	634.2	.5097	20.26	633.0	.4604	15.11	631.7	.4248
40	62.69	640.4	.6026	31.20	639.3	.5200	20.70	638.2	.4709	15.45	637.0	.4356
50	63.96	645.5	1.6125	31.85	644.4	1.5301	21.14	643.4	1.4812	15.78	642.3	1.4460
60	65.24	650.5	.6223	32.49	649.5	.5400	21.58	648.5	.4912	16.12	647.5	.4562
70	66.51	655.5	.6319	33.14	654.6	.5497	22.01	653.7	.5011	16.45	652.8	.4662
80	67.79	660.6	.6413	33.78	659.7	.5593	22.44	658.9	.5108	16.78	658.0	.4760
90	69.06	665.6	.6506	34.42	664.8	.5687	22.88	664.0	.5203	17.10	663.2	.4856
100	70.33	670.7	1.6598	35.07	670.0	1.5779	23.31	669.2	1.5296	17.43	668.5	1.4950
110	71.60	675.8	.6689	35.71	675.1	.5870	23.74	674.4	.5388	17.76	673.7	.5042
120	72.87	680.9	.6778	36.35	680.3	.5960	24.17	679.6	.5478	18.08	678.9	.5133
130	74.14	686.1	.6865	36.99	685.4	.6049	24.60	684.8	.5567	18.41	684.2	.5223
140	75.41	691.2	.6952	37.62	690.6	.6136	25.03	690.0	.5655	18.73	689.4	.5312
150	76.68	696.4	1.7038	38.26	695.8	1.6222	25.46	695.3	1.5742	19.05	694.7	1.5399
160	77.95	701.6	.7122	38.90	701.1	.6307	25.88	700.5	.5827	19.37	700.0	.5485
170	79.21	706.8	.7206	39.54	706.3	.6391	26.31	705.8	.5911	19.70	705.3	.5569
180	80.48	712.1	.7289	40.17	711.6	.6474	26.74	711.1	.5995	20.02	710.6	.5653
190	......	......	......	40.81	716.9	.6556	27.16	716.4	.6077	20.34	715.9	.5736
200	......	......	......	41.45	722.2	1.6637	27.59	721.7	1.6158	20.66	721.2	1.5817
220	......	......	......	......	......	......	28.44	732.4	.6318	21.30	732.0	.5978
240	......	......	......	......	......	......	......	......	......	21.94	742.8	.6135

Temp. °F	25 −7.96			30 −0.57			35 5.89			40 11.66		
t	v	h	s	v	h	s	v	h	s	v	h	s
Sat.	10.96	609.1	1.3515	9.236	611.6	1.3364	7.991	613.6	1.3236	7.047	615.4	1.3125
0	11.19	613.8	1.3616									
10	11.47	619.4	.3738	9.492	617.8	1.3497	8.078	616.1	1.3289			
20	11.75	625.0	.3855	9.731	623.5	.3618	8.287	622.0	.3413	7.203	620.4	1.3231
30	12.03	630.4	.3967	9.966	629.1	.3733	8.493	627.7	.3532	7.387	626.3	.3353
40	12.30	635.8	.4077	10.20	634.6	.3845	8.695	633.4	.3646	7.568	632.1	.3470
50	12.57	641.2	1.4183	10.43	640.1	1.3953	8.895	638.9	1.3756	7.746	637.8	1.3583
60	12.84	646.5	.4287	10.65	645.5	.4059	9.093	644.4	.3863	7.922	643.4	.3692
70	13.11	651.8	.4388	10.88	650.9	.4161	9.289	649.9	.3967	8.096	648.9	.3797
80	13.37	657.1	.4487	11.10	656.2	.4261	9.484	655.3	.4069	8.268	654.4	.3900
90	13.64	662.4	.4584	11.33	661.6	.4359	9.677	660.7	.4168	8.439	659.9	.4000
100	13.90	667.7	1.4679	11.55	666.9	1.4456	9.869	666.1	1.4265	8.609	665.3	1.4098
110	14.17	673.0	.4772	11.77	672.2	.4550	10.06	671.5	.4360	8.777	670.7	.4194
120	14.43	678.2	.4864	11.99	677.5	.4642	10.25	676.8	.4453	8.945	676.1	.4288
130	14.69	683.5	.4954	12.21	682.9	.4733	10.44	682.2	.4545	9.112	681.5	.4381
140	14.95	688.8	.5043	12.43	688.2	.4823	10.63	687.6	.4635	9.278	686.9	.4471
150	15.21	694.1	1.5131	12.65	693.5	1.4911	10.82	692.9	1.4724	9.444	692.3	1.4561
160	15.47	699.4	.5217	12.87	698.8	.4998	11.00	698.3	.4811	9.609	697.7	.4648
170	15.73	704.7	.5303	13.08	704.2	.5083	11.19	703.7	.4897	9.774	703.1	.4735
180	15.99	710.1	.5387	13.30	709.6	.5168	11.38	709.1	.4982	9.938	708.5	.4820
190	16.25	715.4	.5470	13.52	714.9	.5251	11.56	714.5	.5066	10.10	714.0	.4904
200	16.50	720.8	1.5552	13.73	720.3	1.5334	11.75	719.9	1.5148	10.27	719.4	1.4987
220	17.02	731.6	.5713	14.16	731.1	.5495	12.12	730.7	.5311	10.59	730.3	.5150
240	17.53	742.5	.5870	14.59	742.0	.5653	12.49	741.7	.5469	10.92	741.3	.5309
260	18.04	753.4	.6025	15.02	753.0	.5808	12.86	752.7	.5624	11.24	752.3	.5465
280	......	......	......	15.45	764.1	.5960	13.23	763.7	.5776	11.56	763.4	.5617
300	......	......	......	......	......	......	......	......	......	11.88	774.6	1.5766

Source: Abstracted, by permission, from *Tables of Thermodynamic Properties of Ammonia*, U.S. Department of Commerce, Bureau of Standards Circular 142, 1945.

TABLE A.8 (cont'd.)

Temp. °F	50 21.67			60 30.21			70 37.70			80 44.40		
t	v	h	s	v	h	s	v	h	s	v	h	s
Sat.	5.710	618.2	1.2939	4.805	620.5	1.2787	4.151	622.4	1.2658	8.655	624.0	1.2545
30	5.838	623.4	1.3046									
40	5.988	629.5	.3169	4.933	626.8	1.2913	4.177	623.9	1.2688			
50	6.135	635.4	1.3286	5.060	632.9	1.3035	4.290	630.4	1.2816	3.712	627.7	1.2619
60	6.280	641.2	.3399	5.184	639.0	.3152	4.401	636.6	.2937	3.812	634.3	.2745
70	6.423	646.9	.3508	5.307	644.9	.3265	4.509	642.7	.3054	3.909	640.6	.2866
80	6.564	652.6	.3613	5.428	650.7	.3373	4.615	648.7	.3166	4.005	646.7	.2981
90	6.704	658.2	.3716	5.547	656.4	.3479	4.719	654.6	.3274	4.098	652.8	.3092
100	6.843	663.7	1.3816	5.665	662.1	1.3581	4.822	660.4	1.3378	4.190	658.7	1.3199
110	6.980	669.2	.3914	5.781	667.7	.3681	4.924	666.1	.3480	4.281	664.6	.3303
120	7.117	674.7	.4009	5.897	673.3	.3778	5.025	671.8	.3579	4.371	670.4	.3404
130	7.252	680.2	.4103	6.012	678.9	.3873	5.125	677.5	.3676	4.460	676.1	.3502
140	7.387	685.7	.4195	6.126	684.4	.3966	5.224	683.1	.3770	4.548	681.8	.3598
150	7.521	691.1	1.4286	6.239	689.9	1.4058	5.323	688.7	1.3863	4.635	687.5	1.3692
160	7.655	696.6	.4374	6.352	695.5	.4148	5.420	694.3	.3954	4.722	693.2	.3784
170	7.788	702.1	.4462	6.464	701.0	.4236	5.518	699.9	.4043	4.808	698.8	.3874
180	7.921	707.5	.4548	6.576	706.5	.4323	5.615	705.5	.4131	4.893	704.4	.3963
190	8.053	713.0	.4633	6.687	712.0	.4409	5.711	711.0	.4217	4.978	710.0	.4050
200	8.185	718.5	1.4716	6.798	717.5	1.4493	5.807	716.6	1.4302	5.063	715.6	1.4136
210	8.317	724.0	.4799	6.909	723.1	.4576	5.902	722.2	.4386	5.147	721.3	.4220
220	8.448	729.4	.4880	7.019	728.6	.4658	5.998	727.7	.4469	5.231	726.9	.4304
240	8.710	740.5	.5040	7.238	739.7	.4819	6.187	738.9	.4631	5.398	738.1	.4467
260	8.970	751.6	.5197	7.457	750.9	.4976	6.376	750.1	.4789	5.565	749.4	.4626
280	9.230	762.7	1.5350	7.675	762.1	1.5130	6.563	761.4	1.4943	5.730	760.7	1.4781
300	9.489	774.0	.5500	7.892	773.3	.5281	6.750	772.7	.5095	5.894	772.1	.4933

Temp. °F	80 50.47			100 56.05			120 66.02			140 74.79		
t	v	h	s	v	h	s	v	h	s	v	h	s
Sat.	3.266	625.3	1.2445	2.952	626.5	1.2356	2.476	628.4	1.2201	2.132	629.9	1.2068
50												
60	3.353	631.8	1.2571	2.985	629.3	1.2409						
70	3.442	638.3	.2695	3.068	636.0	.2539	2.505	631.3	1.2255			
80	3.529	644.7	.2814	3.149	642.6	.2661	2.576	638.3	.2386	2.166	633.8	1.2140
90	3.614	650.9	.2928	3.227	649.0	.2778	2.645	645.0	.2510	2.228	640.9	.2272
100	3.698	657.0	1.3038	3.304	655.2	1.2891	2.712	651.6	1.2628	2.288	647.8	1.2396
110	3.780	663.0	.3144	3.380	661.3	.2999	2.778	658.0	.2741	2.347	654.5	.2515
120	3.862	668.9	.3247	3.454	667.3	.3104	2.842	664.2	.2850	2.404	661.1	.2628
130	3.942	674.7	.3347	3.527	673.3	.3206	2.905	670.4	.2956	2.460	667.4	.2738
140	4.021	680.5	.3444	3.600	679.2	.3305	2.967	676.5	.3058	2.515	673.7	.2843
150	4.100	686.3	1.3539	3.672	685.0	1.3401	3.029	682.5	1.3157	2.569	679.9	1.2945
160	4.178	692.0	.3633	3.743	690.8	.3495	3.089	688.4	.3254	2.622	686.0	.3045
170	4.255	697.7	.3724	3.813	696.6	.3588	3.149	694.3	.3348	2.675	692.0	.3141
180	4.332	703.4	.3813	3.883	702.3	.3678	3.209	700.2	.3441	2.727	698.0	.3236
190	4.408	709.0	.3901	3.952	708.0	.3767	3.268	706.0	.3531	2.779	704.0	.3328
200	4.484	714.7	1.3988	4.021	713.7	1.3854	3.326	711.8	1.3620	2.830	709.9	1.3418
210	4.560	720.4	.4073	4.090	719.4	.3940	3.385	717.6	.3707	2.880	715.8	.3507
220	4.635	726.0	.4157	4.158	725.1	.4024	3.442	723.4	.3793	2.931	721.6	.3594
230	4.710	731.7	.4239	4.226	730.8	.4108	3.500	729.2	.3877	2.981	727.5	.3679
240	4.785	737.3	.4321	4.294	736.5	.4190	3.557	734.9	.3960	3.030	733.3	.3763
250	4.859	743.0	1.4401	4.361	742.2	1.4271	3.614	740.7	1.4042	3.080	739.2	1.3846
260	4.933	748.7	.4481	4.428	747.9	.4350	3.671	746.5	.4123	3.129	745.0	.3928
280	5.081	760.0	.4637	4.562	759.4	.4507	3.783	758.0	.4281	3.227	756.7	.4088
300	5.228	771.5	.4789	4.695	770.8	.4660	3.895	769.6	.4435	3.323	768.3	.4243

TABLE A.8 (cont'd.)

Temp. °F	160 82.64			180 89.78			200 96.34			220 102.42		
t	v	h	s	v	h	s	v	h	s	v	h	s
Sat.	1.872	631.1	1.1952	1.667	632.0	1.1850	1.502	632.7	1.1756	1.367	633.2	1.1671
90	1.914	636.6	1.2055	1.668	632.2	1.1853						
100	1.969	643.9	1.2186	1.720	639.9	1.1992						
110	2.023	651.0	.2311	1.770	647.3	.2123	1.567	643.4	1.1947	1.400	639.4	1.1781
120	2.075	657.8	.2429	1.818	654.4	.2247	1.612	650.9	.2077	1.443	647.3	.1917
130	2.125	664.4	.2542	1.865	661.3	.2364	1.656	658.1	.2200	1.485	654.8	.2045
140	2.175	670.9	.2652	1.910	668.0	.2477	1.698	665.0	.2317	1.525	662.0	.2167
150	2.224	677.2	1.2757	1.955	674.6	1.2586	1.740	671.8	1.2429	1.564	669.0	1.2281
160	2.272	683.5	.2859	1.999	681.0	.2691	1.780	678.4	.2537	1.601	675.8	.2394
170	2.319	689.7	.2958	2.042	687.3	.2792	1.820	684.9	.2641	1.638	682.5	.2501
180	2.365	695.8	.3054	2.084	693.6	.2891	1.859	691.3	.2742	1.675	689.1	.2604
190	2.411	701.9	.3148	2.126	699.8	.2987	1.897	697.7	.2840	1.710	695.5	.2704
200	2.457	707.9	1.3240	2.167	705.9	1.3081	1.935	703.9	1.2935	1.745	701.9	1.2801
210	2.502	713.9	.3331	2.208	712.0	.3172	1.972	710.1	.3029	1.780	708.2	.2896
220	2.547	719.9	.3419	2.248	718.1	.3262	2.009	716.3	.3120	1.814	714.4	.2989
230	2.591	725.8	.3506	2.288	724.1	.3350	2.046	722.4	.3209	1.848	720.6	.3079
240	2.635	731.7	.3591	2.328	730.1	.3436	2.082	728.4	.3296	1.881	726.8	.3168
250	2.679	737.6	1.3675	2.367	736.1	1.3521	2.118	734.5	1.3382	1.914	732.9	1.3255
260	2.723	743.5	.3757	2.407	742.0	.3605	2.154	740.5	.3467	1.947	739.0	.3340
270	2.766	749.4	.3838	2.446	748.0	.3687	2.189	746.5	.3550	1.980	745.1	.3424
280	2.809	755.3	.3919	2.484	753.9	.3768	2.225	752.5	.3631	2.012	751.1	.3507
290	2.852	761.2	.3998	2.523	759.9	.3847	2.260	758.5	.3712	2.044	757.2	.3588
300	2.895	767.1	1.4076	2.561	765.8	1.3926	2.295	764.5	1.3791	2.076	763.2	1.3668
320	2.980	778.9	.4229	2.637	777.7	.4081	2.364	776.5	.3947	2.140	775.3	.3825
340	3.064	790.7	.4379	2.713	789.6	.4231	2.432	788.5	.4099	2.203	787.4	.3978
360	......	......	......	......	......	......	2.500	800.5	.4247	2.265	799.5	.4127
380	......	......	......	......	......	......	2.568	812.5	.4392	2.327	811.6	.4273

Temp. °F	240 108.09			260 113.42			280 118.45			300 123.21		
t	v	h	s	v	h	s	v	h	s	v	h	s
Sat.	1.253	633.6	1.1592	1.155	633.9	1.1518	1.072	634.0	1.1449	0.999	634.0	1.1383
110	1.261	635.3	1.1621									
120	1.302	643.5	.1764	1.182	639.5	1.1617	1.078	635.4	1.1473			
130	1.342	651.3	.1898	1.220	647.8	.1757	1.115	644.0	.1621	1.023	640.1	1.1487
140	1.380	658.8	.2025	1.257	655.6	.1889	1.151	652.2	.1759	1.058	648.7	.1632
150	1.416	666.1	1.2145	1.292	663.1	1.2014	1.184	660.1	1.1888	1.091	656.9	1.1767
160	1.452	673.1	.2259	1.326	670.4	.2132	1.217	667.6	.2011	1.123	664.7	.1894
170	1.487	680.0	.2369	1.359	677.5	.2245	1.249	674.9	.2127	1.153	672.2	.2014
180	1.521	686.7	.2475	1.391	684.4	.2354	1.279	681.9	.2239	1.183	679.5	.2129
190	1.554	693.3	.2577	1.422	691.1	.2458	1.309	688.9	.2346	1.211	686.5	.2239
200	1.587	699.8	1.2677	1.453	697.7	1.2560	1.339	695.6	1.2449	1.239	693.5	1.2344
210	1.619	706.2	.2773	1.484	704.3	.2658	1.367	702.3	.2550	1.267	700.3	.2447
220	1.651	712.6	.2867	1.514	710.7	.2754	1.396	708.8	.2647	1.294	706.9	.2546
230	1.683	718.9	.2959	1.543	717.1	.2847	1.424	715.3	.2742	1.320	713.5	.2642
240	1.714	725.1	.3049	1.572	723.4	.2938	1.451	721.8	.2834	1.346	720.0	.2736
250	1.745	731.3	1.3137	1.601	729.7	1.3027	1.478	728.1	1.2924	1.372	726.5	1.2827
260	1.775	737.6	.3224	1.630	736.0	.3115	1.505	734.4	.3013	1.397	732.9	.2917
270	1.805	743.6	.3308	1.658	742.2	.3200	1.532	740.7	.3099	1.422	739.2	.3004
280	1.835	749.8	.3392	1.686	748.4	.3285	1.558	747.0	.3184	1.447	745.5	.3090
290	1.865	755.9	.3474	1.714	754.5	.3367	1.584	753.2	.3268	1.472	751.8	.3175
300	1.895	762.0	1.3554	1.741	760.7	1.3449	1.610	759.4	1.3350	1.496	758.1	1.3257
320	1.954	774.1	.3712	1.796	772.9	.3608	1.661	771.7	.3511	1.544	770.5	.3419
340	2.012	786.3	.3866	1.850	785.2	.3763	1.712	784.0	.3667	1.592	782.9	.3576
360	2.069	798.4	.4016	1.904	797.4	.3914	1.762	796.3	.3819	1.639	795.3	.3729
380	2.126	810.6	.4163	1.957	809.6	.4062	1.811	808.7	.3967	1.686	807.7	.3878
400	......	......	......	2.009	821.9	1.4206	1.861	821.0	1.4112	1.732	820.1	1.4024

TABLE A.10
Dichlorodifluoromethane (Freon-12): Properties of Liquid and Saturated Vapor

Temp.	Pressure		Volume		Density		Enthalpy from −40°F			Entropy from −40°F	
°F t	Abs. lb/in^2 p	Gage lb/in^2 p_d	Liquid ft^3/lb v_f	Vapor ft^3/lb v_g	Liquid lb/ft^3 $1/v_f$	Vapor lb/ft^3 $1/v_g$	Liquid Btu/lb h_f	Latent Btu/lb h_{fg}	Vapor Btu/lb h_g	Liquid Btu/lb$_m$°R s_f	Vapor Btu/lb$_m$°R s_g
−152	0.13799	29.64024*	0.0095673	197.58	104.52	0.0050614	−23.106	83.734	60.628	−0.063944	0.20818
−150	.15359	29.60849*	.0095822	178.65	104.36	.0055976	−22.697	83.534	60.837	−0.062619	.20711
−145	.19933	29.51537*	.0096198	139.83	103.95	.0071517	−21.674	83.039	61.365	−0.059344	.20452
−140	.25623	29.39951*	.0096579	110.46	103.54	.0090533	−20.652	82.548	61.896	−0.056123	.20208
−135	.32641	29.25663*	.0096966	88.023	103.13	.011361	−19.631	82.061	62.430	−0.052952	.19978
−130	.41224	29.08186*	.0097359	70.730	102.71	.014138	−18.609	81.577	62.968	−0.049830	.19760
−125	0.51641	28.86978*	0.0097758	57.283	102.29	0.017457	−17.587	81.096	63.509	−0.046754	0.19554
−120	.64190	28.61429*	.0098163	46.741	101.87	.021395	−16.565	80.617	64.052	−0.043723	.19359
−115	.79200	28.30869*	.0098574	38.410	101.45	.026035	−15.541	80.139	64.598	−0.040734	.19176
−110	.97034	27.94558*	.0098992	31.777	101.02	.031470	−14.518	79.663	65.145	−0.037786	.19002
−105	1.1809	27.5169*	.0099416	26.458	100.59	.037796	−13.492	79.188	65.696	−0.034877	.18838
−100	1.4280	27.0138*	.0099847	22.164	100.15	0.045119	−12.466	78.714	66.248	−0.032005	0.18683
− 95	1.7163	26.4268*	.010029	18.674	99.715	.053550	−11.438	78.239	66.801	−0.029169	.18536
− 90	2.0509	25.7456*	.010073	15.821	99.274	.063207	−10.409	77.764	67.355	−0.026367	.18398
− 85	2.4371	24.9593*	.010118	13.474	98.830	.074216	− 9.3782	77.289	67.911	−0.023599	.18267
− 80	2.8807	24.0560*	.010164	11.533	98.382	.086708	− 8.3451	76.812	68.467	−0.020862	.18143
− 75	3.3879	23.0234*	0.010211	9.9184	97.930	0.10082	− 7.3101	76.333	69.023	−0.018156	0.18027
− 70	3.9651	21.8482*	.010259	8.5687	97.475	.11670	− 6.2730	75.853	69.580	−0.015481	.17916
− 65	4.6193	20.5164*	.010308	7.4347	97.016	.13451	− 5.2336	75.371	70.137	−0.012834	.17812
− 60	5.3575	19.0133*	.010357	6.4774	96.553	.15438	− 4.1919	74.885	70.693	−0.010214	.17714
− 55	6.1874	17.3237*	.010407	5.6656	96.086	.17650	− 3.1477	74.397	71.249	−0.007622	.17621
− 50	7.1168	15.4313*	0.010459	4.9742	95.616	0.20104	− 2.1011	73.906	71.805	−0.005056	0.17533
− 45	8.1540	13.3196*	.010511	4.3828	95.141	.22816	− 1.0519	73.411	72.359	−0.002516	.17451
− 40	9.3076	10.9709*	0.010564	3.8750	94.661	0.25806	0	72.913	72.913	0	0.17373
− 38	9.8035	9.9611*	.010586	3.6922	94.469	.27084	0.4215	72.712	73.134	0.001000	.17343
− 36	10.320	8.909*	.010607	3.5198	94.275	.28411	.8434	72.511	73.354	.001995	.17313
− 34	10.858	7.814*	.010629	3.3571	94.081	.29788	1.2659	72.309	73.575	.002988	.17285
− 32	11.417	6.675*	.010651	3.2035	93.886	.31216	1.6887	72.106	73.795	.003976	.17257
− 30	11.999	5.490*	0.010674	3.0585	93.690	0.32696	2.1120	71.903	74.015	0.004961	0.17229
− 28	12.604	4.259*	.010696	2.9214	93.493	0.34231	2.5358	71.698	74.234	.005942	.17203
− 26	13.233	2.979*	.010719	2.7917	93.296	.35820	2.9601	71.494	74.454	.006919	.17177
− 24	13.886	1.649*	.010741	2.6691	93.098	.37466	3.3848	71.288	74.673	.007894	.17151
− 22	14.564	0.270*	.010764	2.5529	92.899	.39171	3.8100	71.081	74.891	.008864	.17126
− 20	15.267	0.571	0.010788	2.4429	92.699	0.40934	4.2357	70.874	75.110	0.009831	0.17102
− 18	15.996	1.300	.010811	2.3387	92.499	.42758	4.6618	70.666	75.328	.010795	.17078
− 16	16.753	2.057	.010834	2.2399	92.298	.44645	5.0885	70.456	75.545	.011755	.17055
− 14	17.536	2.840	.010858	2.1461	92.096	.46595	5.5157	70.246	75.762	.012712	.17032
− 12	18.348	3.652	.010882	2.0572	91.893	.48611	5.9434	70.036	75.979	.013666	.17010
− 10	19.189	4.493	0.010906	1.9727	91.689	0.50693	6.3716	69.824	76.196	0.014617	0.16989
− 8	20.059	5.363	.010931	1.8924	91.485	.52843	6.8003	69.611	76.411	.015564	.16967
− 6	20.960	6.264	.010955	1.8161	91.280	.55063	7.2296	69.397	76.627	.016508	.16947
− 4	21.891	7.195	.010980	1.7436	91.074	.57354	7.6594	69.183	76.842	.017449	.16927
− 2	22.854	8.158	.011005	1.6745	90.867	.59718	8.0898	68.967	77.057	.018388	.16907
0	23.849	9.153	0.011030	1.6089	90.659	0.62156	8.5207	68.750	77.271	0.019323	0.16888
2	24.878	10.182	.011056	1.5463	90.450	.64670	8.9522	68.533	77.485	.020255	.16869
4	25.939	11.243	.011082	1.4867	90.240	.67263	9.3843	68.314	77.698	.021184	.16851
5†	26.483	11.787	.011094	1.4580	90.135	.68588	9.6005	68.204	77.805	.021647	.16842
6	27.036	12.340	.011107	1.4299	90.030	.69934	9.8169	68.094	77.911	.022110	.16833
8	28.167	13.471	.011134	1.3758	89.818	.72687	10.250	67.873	78.123	.023033	.16815
10	29.335	14.639	0.011160	1.3241	89.606	0.75523	10.684	67.651	78.335	0.023954	0.16798
12	30.539	15.843	.011187	1.2748	89.392	.78443	11.118	67.428	78.546	.024871	.16782
14	31.780	17.084	.011214	1.2278	89.178	.81449	11.554	67.203	78.757	.025786	.16765
16	33.060	18.364	.011241	1.1828	88.962	.84544	11.989	66.977	78.966	.026699	.16750
18	34.378	19.682	.011268	1.1399	88.746	.87729	12.426	66.750	79.176	.027608	.16734
20	35.736	21.040	0.011296	1.0988	88.529	0.91006	12.863	66.522	79.385	0.028515	0.16719
22	37.135	22.439	.011324	1.0596	88.310	.94377	13.300	66.293	79.593	.029420	.16704
24	38.574	23.878	.011352	1.0220	88.091	.97843	13.739	66.061	79.800	.030322	.16690
26	40.056	25.360	.011380	0.98612	87.870	1.0141	14.178	65.829	80.007	.031221	.16676
28	41.580	26.884	.011409	.95173	87.649	1.0507	14.618	65.596	80.214	.032118	.16662

Source: R. C. Jordan and G. B. Priester, *Refrigeration and Air Conditioning*, 2nd ed., Prentice-Hall, Inc., Englewood Cliffs, N.J., 1956. Courtesy of E. I. Dupont de Nemours and Co.

* Inches of mercury below one atmosphere.

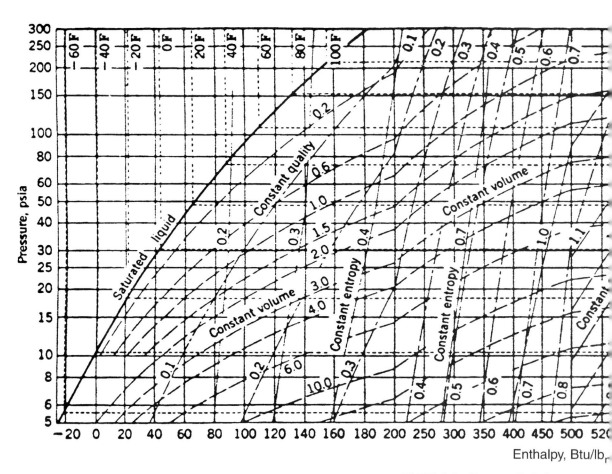

TABLE A.9   Pressure–Enthalpy Diagram

Source: Data from *Tables of Thermodynamic Properties of Ammonia*, Bureau

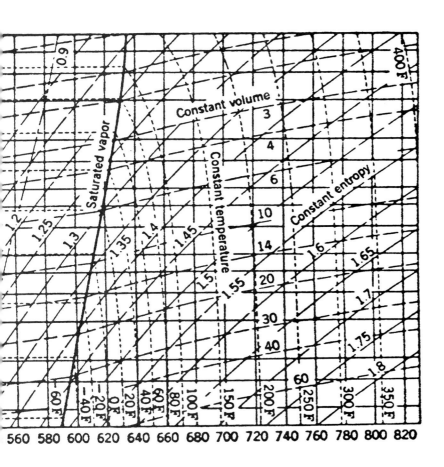

560 580 600 620 640 660 680 700 720 740 760 780 800 820

TABLE A.10 (cont'd.)

Temp.	Pressure		Volume		Density		Enthalpy from −40°F			Entropy from −40°F	
°F	Abs. lb/in²	Gage lb/in²	Liquid ft³/lb	Vapor ft³/lb	Liquid lb/ft³	Vapor lb/ft³	Liquid Btu/lb	Latent Btu/lb	Vapor Btu/lb	Liquid Btu/lb_m·°R	Vapor Btu/lb_m·°R
t	p	p_d	v_f	v_g	1/v_f	1/v_g	h_f	h_{fg}	h_g	s_f	s_g
30	43.148	28.452	0.011438	0.91880	87.426	1.0884	15.058	65.361	80.419	0.033013	.16648
32	44.760	30.064	.011468	.88725	87.202	1.1271	15.500	65.124	80.624	.033905	.16635
34	46.417	31.721	.011497	.85702	86.977	1.1668	15.942	64.886	80.828	.034796	.16622
36	48.120	33.424	.011527	.82803	86.751	1.2077	16.384	64.647	81.031	.035683	.16610
38	49.870	35.174	.011557	.80023	86.524	1.2496	16.828	64.406	81.234	.036569	.16598
40	51.667	36.971	.011588	0.77357	86.296	1.2927	17.273	64.163	81.436	.037453	.16586
42	53.513	38.817	.011619	.74798	86.066	1.3369	17.718	63.919	81.637	.038334	.16574
44	55.407	40.711	.011650	.72341	85.836	1.3823	18.164	63.673	81.837	.039213	.16562
46	57.352	42.656	.011682	.69982	85.604	1.4289	18.611	63.426	82.037	.040091	.16551
48	59.347	44.651	.011714	.67715	85.371	1.4768	19.059	63.177	82.236	.040966	.16540
50	61.394	46.698	0.011746	0.65537	85.136	1.5258	19.507	62.926	82.433	0.041839	0.16530
52	63.494	48.798	.011779	.63444	84.900	1.5762	19.957	62.673	82.630	.042711	.16519
54	65.646	50.950	.011811	.61431	84.663	1.6278	20.408	62.418	82.826	.043581	.16509
56	67.853	53.157	.011845	.59495	84.425	1.6808	20.859	62.162	83.021	.04449	.16499
58	70.115	55.419	.011879	.57632	84.185	1.7352	21.312	61.903	83.215	.045316	.16489
60	72.433	57.737	0.011913	0.55839	83.944	1.7909	21.766	61.643	83.409	0.046180	0.16479
62	74.807	60.111	.011947	.54112	83.701	1.8480	22.221	61.380	83.601	.047044	.16470
64	77.239	62.543	.011982	.52450	83.457	1.9066	22.676	61.116	83.792	.047905	.16460
66	79.729	65.033	.012017	.50848	83.212	1.9666	23.133	60.849	83.982	.048765	.16451
68	82.279	67.583	.012053	.49305	82.965	2.0282	23.591	60.580	84.171	.049624	.16442
70	84.888	70.192	0.012089	0.47818	82.717	2.0913	24.050	60.309	84.359	0.050482	0.16434
72	87.559	72.863	.012126	.46383	82.467	2.1559	24.511	60.035	84.546	.051338	.16425
74	90.292	75.596	.012163	.45000	82.215	2.2222	24.973	59.759	84.732	.052193	.16417
76	93.087	78.391	.012201	.43666	81.962	2.2901	25.435	59.481	84.916	.053047	.16408
78	95.946	81.250	.012239	.42378	81.707	2.3597	25.899	59.201	85.100	.053900	.16400
80	98.870	84.174	0.012277	0.41135	81.450	2.4310	26.365	58.917	85.282	0.054751	0.16392
82	101.86	87.16	.012316	.39935	81.192	2.5041	26.832	58.631	85.463	.055602	.16384
84	104.92	90.22	.012356	.38776	80.932	2.5789	27.300	58.343	85.643	.056452	.16376
86	108.04	93.34	.012396	.37657	80.671	2.6556	27.769	58.052	85.821	.057301	.16368
88	111.23	96.53	.012437	.36575	80.407	2.7341	28.241	57.757	85.998	.058149	.16360
90	114.49	99.79	0.012478	0.35529	80.142	2.8146	28.713	57.461	86.174	0.058997	0.16353
92	117.82	103.12	.012520	.34518	79.874	2.8970	29.187	57.161	86.348	.059844	.16345
94	121.22	106.52	.012562	.33540	79.605	2.9815	29.663	56.858	86.521	.06090	.16338
96	124.70	110.00	.012605	.32594	79.334	3.0680	30.140	56.551	86.691	.061536	.16330
98	128.24	113.54	.012649	.31679	79.061	3.1566	30.619	56.242	86.861	.062381	.16323
100	131.86	117.16	0.012693	0.30794	78.785	3.2474	31.100	55.929	87.029	0.063227	0.16315
102	135.56	120.86	.012738	.29937	78.508	3.3404	31.583	55.613	87.196	.064072	.16308
104	139.33	124.63	.012783	.29106	78.228	3.4357	32.067	55.293	87.360	.064916	.16301
106	143.18	128.48	.012829	.28303	77.946	3.5333	32.553	54.970	87.523	.065761	.16293
108	147.11	132.41	.012876	.27524	77.662	3.6332	33.041	54.643	87.684	.066606	.16286
110	151.11	136.41	0.012924	0.26769	77.376	3.7357	33.531	54.313	87.844	0.067451	0.16279
112	155.19	140.49	.012972	.26037	77.087	3.8406	34.023	53.978	88.001	.068296	.16271
114	159.36	144.66	.013022	.25328	76.795	3.9482	34.517	53.639	88.156	.069141	.16264
116	163.61	148.91	.013072	.24641	76.501	4.0584	35.014	53.296	88.310	.069987	.16256
118	167.94	153.24	.013123	.23974	76.205	4.1713	35.512	52.949	88.461	.070833	.16249
120	172.35	157.65	0.013174	0.23326	75.906	4.2870	36.013	52.597	88.610	0.071680	0.16241
122	176.85	162.15	.013227	.22698	75.604	4.4056	36.516	52.241	88.757	.072528	.16234
124	181.43	166.73	.013280	.22089	75.299	4.5272	37.021	51.881	88.902	.073376	.16226
126	186.10	171.40	.013335	.21497	74.991	4.6518	37.529	51.515	89.044	.074225	.16218
128	190.86	176.16	.013390	.20922	74.680	4.7796	38.040	51.144	89.184	.075075	.16210
130	195.71	181.01	0.013447	0.20364	75.367	4.9107	38.553	50.768	89.321	0.075927	0.16202
132	200.64	185.94	.013504	.19821	74.050	5.0451	39.069	50.387	89.456	.076779	.16194
134	205.67	190.97	.013563	.19294	73.729	5.1829	39.588	50.000	89.588	.077633	.16185
136	210.79	196.09	.013623	.18782	73.406	5.3244	40.110	49.608	89.718	.078489	.16177
138	216.01	201.31	.013684	.18283	73.079	5.4695	40.634	49.210	89.844	.079346	.16168
140	221.32	206.62	0.013746	0.17799	72.748	5.6184	41.162	48.805	89.967	0.080205	0.16159
142	226.72	212.02	.013810	.17327	72.413	5.7713	41.693	48.394	90.087	.081065	.16150
144	232.22	217.52	.013874	.16868	72.075	5.9283	42.227	47.977	90.204	.081928	.16140
146	237.82	223.12	.013941	.16422	71.732	6.0895	42.765	47.553	90.318	.082794	.16130
148	243.51	228.81	.014008	.15987	71.386	6.2551	43.306	47.122	90.428	.083661	.16120

TABLE A.11
Dichlorodifluoromethane (Freon-12): Properties of Superheated Vapor

Temp. °F	Abs. Pressure 0.14 lb/in.³ Gage Pressure 29.64 in. vac. (Sat. Temp. −151.7°F)			Abs. Pressure 0.20 lb/in.³ Gage Pressure 29.51 in. vac. (Sat. Temp. −144.9°F)			Abs. Pressure 0.40 lb/in.³ Gage Pressure 29.11 in. vac. (Sat. Temp. −130.7°F)			Abs. Pressure 0.60 lb/in.³ Gage Pressure 28.70 in. vac. (Sat. Temp. −121.6°F)		
t	v	h	s	v	h	s	v	h	s	v	h	s
Sat.	(194.91)	(60.656)	(0.20804)	(39.38)	(61.372)	(0.20449)	(72.756)	(62.897)	(0.19788)	(49.786)	(63.881)	(0.19419)
−150	196.01	60.840	0.20864	.....	.....	.....	.....	.....	.....	.....	.....	.....
−140	202.37	61.916	.21205	141.58	61.906	0.20617	.....	.....	.....	.....	.....	.....
−130	208.73	63.012	.21543	146.04	63.002	.20955	72.903	62.970	0.19810	.....	.....	.....
−120	215.09	64.127	.21876	150.50	64.118	.21288	75.139	64.088	.20144	50.020	64.058	0.19472
−110	221.44	65.261	.22205	154.95	65.253	.21618	77.374	65.225	.20474	51.515	65.197	.19802
−100	227.80	66.414	.22530	159.40	66.406	0.21943	79.607	66.381	0.20799	53.009	66.355	0.20128
−90	234.15	67.585	.22851	163.85	67.578	.22264	81.839	67.554	.21121	54.502	67.530	.20451
−80	240.50	68.774	.23169	168.30	68.768	.22582	84.070	68.746	.21439	55.993	68.723	.20769
−70	246.85	69.981	.23482	172.75	69.975	.22896	86.299	69.954	.21753	57.483	69.934	.21084
−60	253.20	71.206	.23793	177.19	71.200	.23206	88.527	71.181	.22064	58.972	71.161	.21395
−50	259.54	72.447	0.24099	181.64	72.442	0.23513	90.755	72.424	0.22371	60.460	72.406	0.21702
−40	265.89	73.706	.24403	186.09	73.701	.23816	92.982	73.684	.22675	61.947	73.667	.22006
−30	272.23	74.980	.24703	190.53	74.976	.24117	95.207	74.960	.22976	63.433	74.944	.22307
−20	278.58	76.271	.25000	194.97	76.267	.24414	97.433	76.252	.23273	64.919	76.238	.22605
−10	284.92	77.578	.25294	199.42	77.574	.24708	99.657	77.560	.23567	66.404	77.546	.22899
0	291.27	78.901	0.25585	203.86	78.897	0.24998	101.88	78.884	0.23858	67.889	78.871	0.23190
10	297.61	80.239	.25873	208.30	80.235	.25286	104.11	80.223	.24146	69.373	80.210	.23479
20	303.95	81.591	.26158	212.74	81.588	.25571	106.33	81.576	.24431	70.857	81.565	.23764
30	310.30	82.959	.26440	217.18	82.955	.25854	108.55	82.945	.24714	72.340	82.934	.24046
40	316.64	84.341	.26719	221.62	84.338	.26133	110.77	84.327	.24993	73.823	84.317	.24326
50	322.98	85.737	0.26996	226.07	85.734	0.26410	113.00	85.724	0.25270	75.306	85.714	0.24603
60	329.32	87.147	.27270	230.51	87.144	.26684	115.22	87.135	.25544	76.789	87.126	.24877
70	335.66	88.570	.27541	234.95	88.567	.26955	117.44	88.559	.25816	78.271	88.550	.25149
80	342.01	90.007	.27810	239.39	90.004	.27224	119.66	89.996	.26084	79.753	89.988	.25417
90	348.35	91.457	.28076	243.82	91.454	.27490	121.88	91.447	.26351	81.235	91.439	.25684
100	354.69	92.920	0.28340	248.26	92.917	0.27754	124.10	92.910	0.26614	82.716	92.902	0.25948
110	361.03	94.395	.28601	252.70	94.393	.28015	126.32	94.386	.26876	84.197	94.378	.26209
120	367.37	95.882	.28860	257.14	95.880	.28274	128.54	95.874	.27135	85.679	95.867	.26468
130	373.71	97.382	.29116	261.58	97.380	.28530	130.77	97.374	.27391	87.160	97.367	.26725
140	380.05	98.893	.29370	266.02	98.891	.28784	132.99	98.885	.27645	88.641	98.879	.26979

Source: R. C. Jordan and G. B. Priester, *Refrigeration and Air Conditioning*, 2nd ed., Prentice-Hall, Inc., Englewood Cliffs, N.J. 1956. Courtesy of E. I. Dupont de Nemours and Co.

TABLE A.11 (cont'd.)

Temp. °F	Abs. Pressure 0.80 lb/in.² Gage Pressure 28.29 in. vac. (Sat. Temp. −114.8°F)			Abs. Pressure 1.00 lb/in.² Gage Pressure 27.88 in. vac. (Sat. Temp. −109.3°F)			Abs. Pressure 2.0 lb/in.² Gage Pressure 25.85 in. vac. (Sat. Temp. −90.7°F)			Abs. Pressure 3.0 lb/in.² Gage Pressure 23.81 in. vac. (Sat. Temp. −78.8°F)		
t	v	h	s	v	h	s	v	h	s	v	h	s
Sat.	(38.051)	(64.624)	(0.19167)	(30.896)	(65.229)	(0.18977)	(16.195)	(67.276)	(0.18417)	(11.106)	(68.604)	(0.18114)
−110	38.586	65.170	0.19324	.....	.....	.....	.....	.....	.....	.....	.....	.....
−100	39.710	66.329	0.19651	31.730	66.303	0.19279	.....	.....	.....	.....	.....	.....
−90	40.833	67.506	.19974	32.631	67.482	.19602	16.228	67.361	0.18440	.....	.....	.....
−80	41.954	68.701	.20292	33.531	68.679	.19922	16.684	68.567	.18762	11.375	69.683	0.18394
−70	43.074	69.913	.20608	34.429	68.892	.20237	17.139	69.788	.19079	11.681	70.928	.18709
−60	44.194	71.142	.20919	35.327	71.123	.20549	17.593	71.026	.19393	11.986	72.188	.19021
−50	45.312	72.388	.21227	36.223	72.370	.20857	18.046	72.279	.19703	12.290	73.463	.19328
−40	46.429	73.650	.21531	37.119	73.633	.21162	18.498	73.549	.20009	12.594	74.754	.19632
−30	47.546	74.928	.21832	38.014	74.913	.21463	18.949	74.834	.20311	12.897	76.059	.19932
−20	48.662	76.223	.22130	38.908	76.208	.21761	19.400	76.134	.20611	13.199	77.379	.20229
−10	49.778	77.533	.22424	39.802	77.519	.22056	19.850	77.449	.20906	13.500	78.714	.20523
0	50.893	78.858	.22716	40.695	78.845	.22347	20.299	78.780	.21199	13.802	80.063	.20813
10	52.007	80.198	.23004	41.587	80.186	.22636	20.748	80.125	.21488	14.102	81.426	.21100
20	53.121	81.553	.23290	42.480	81.542	.22922	21.197	81.484	.21775	14.403	82.803	.21384
30	54.235	82.923	.23572	43.372	82.912	.23204	21.645	82.858	.22058	14.703	84.194	.21665
40	55.348	84.307	.23852	44.263	84.297	.23484	22.093	84.245	.22339	15.002	85.598	.21944
50	56.461	85.705	.24129	45.154	85.695	.23761	22.540	85.646	.22616	15.302	87.015	.22219
60	57.574	87.116	.24403	46.045	87.107	.24036	22.988	87.061	.22891	15.601	88.445	.22492
70	58.686	88.541	.24675	46.936	88.533	.24307	23.435	88.489	.23163	15.900	89.889	.22762
80	59.799	89.980	.24944	47.826	89.971	.24576	23.881	89.930	.23433	16.198	91.345	.23029
90	60.911	91.431	.25210	48.716	91.423	.24843	24.328	91.384	.23700	16.497	92.813	.23293
100	62.023	92.895	.25474	49.606	92.887	.25107	24.774	92.850	.23964	16.795	94.293	.23556
110	63.134	94.371	.25736	50.496	94.364	.25368	25.220	94.329	.24226	17.093	95.785	.23815
120	64.246	95.860	.25995	51.386	95.853	.25628	25.666	95.819	.24485	17.391	97.289	.24073
130	65.357	97.361	.26251	52.275	97.354	.25884	26.112	97.322	.24742	17.689	98.805	.24327
140	66.468	98.873	.26506	53.165	98.867	.26139	26.558	98.836	.24997	17.987	100.332	.24580
150	67.579	100.397	.26758	54.054	100.391	.26391	27.003	100.361	.25249	18.284	101.869	.24830
160	68.690	101.932	.27007	54.943	101.926	.26640	27.449	101.898	.25499	18.582	103.418	.25078
170	69.801	103.478	.27255	55.832	103.472	.26888	27.894	103.445	.25747	18.879	104.977	.25324
180	70.912	105.035	.27500	56.721	105.030	.27133	28.340	105.003	.25992	19.176	106.547	.25567
190	.....	.....	.....	.....	.....	.....	28.785	106.572	.26236	19.473	108.127	.25808
200	.....	.....	.....	.....	.....	.....	29.230	108.151	.26477	.....	.....	.....

TABLE A.11 (cont'd.)

Temp. °F	Abs. Pressure 5.0 lb/in.² Gage Pressure 19.74 in. vac. (Sat. Temp. −62.4°F)			Abs. Pressure 7.5 lb/in.² Gage Pressure 14.65 in. vac. (Sat. Temp. −48.1°F)			Abs. Pressure 10.0 lb/in.² Gage Pressure 0.56 in. vac. (Sat. Temp. −37.2°F)			Abs. Pressure 15 lb/in.² Gage Pressure 0.3 lb/in.³ (Sat. Temp. −20.8°F)		
t	v	h	s	v	h	s	v	h	s	v	h	s
Sat.	(6.9069)	(70.432)	(0.17759)	(4.7374)	(72.017)	(0.17501)	(3.6246)	(73.219)	(0.17331)	(2.4835)	(75.208)	(0.17111)
−60	6.9509	70.729	0.17834									
−50	7.1378	72.003	.18149									
−40	7.3239	73.291	.18459	4.8401	73.073	0.17755						
−30	7.5092	74.593	.18766	4.9664	74.390	.18065	3.6945	74.183	.17557			
−20	7.6938	75.909	.19069	5.0919	75.719	.18371	3.7906	75.526	.17866	2.4885	75.131	0.17134
−10	7.8777	77.239	.19368	5.2169	77.061	.18673	3.8861	76.880	.18171	2.5546	76.512	.17445
0	8.0611	78.582	0.19663	5.3412	78.415	0.18971	3.9809	78.246	.18471	2.6201	77.902	0.17751
10	8.2441	79.939	.19955	5.4650	79.782	.19265	4.0753	79.624	.18768	2.6850	79.302	.18052
20	8.4265	81.309	.20244	5.5884	81.162	.19556	4.1691	81.014	.19061	2.7494	80.712	.18349
30	8.6086	82.693	.20529	5.7114	82.555	.19843	4.2626	82.415	.19350	2.8134	82.131	.18642
40	8.7903	84.090	.20812	5.8340	83.959	.20127	4.3556	83.828	.19635	2.8770	83.561	.18931
50	8.9717	85.500	0.21091	5.9562	85.377	0.20408	4.4484	85.252	0.19918	2.9402	85.001	0.19216
60	9.1528	86.922	.21367	6.0782	86.806	.20685	4.5408	86.689	.20197	3.0031	86.451	.19498
70	9.3336	88.358	.21641	6.1999	88.247	.20960	4.6329	88.136	.20473	3.0657	87.912	.19776
80	9.5142	89.806	.21912	6.3213	89.701	.21232	4.7248	89.596	.20746	3.1281	89.383	.20051
90	9.6945	91.266	.22180	6.4425	91.166	.21501	4.8165	91.067	.21016	3.1902	90.865	.20324
100	9.8747	92.738	0.22445	6.5636	92.643	0.21767	4.9079	92.548	0.21283	3.2521	92.357	0.20593
110	10.055	94.222	.22708	6.6844	94.132	.22031	4.9992	94.042	.21547	3.3139	93.860	.20859
120	10.234	95.717	.22968	6.8051	95.632	.22292	5.0903	95.546	.21809	3.3754	95.373	.21122
130	10.414	97.224	.23226	6.9256	97.143	.22550	5.1812	97.061	.22068	3.4368	96.896	.21382
140	10.594	98.743	.23481	7.0459	98.665	.22806	5.2720	98.586	.22325	3.4981	98.429	.21640
150	10.773	100.272	0.23734	7.1662	100.198	0.23060	5.3627	100.123	0.22579	3.5592	99.972	0.21895
160	10.952	101.812	.23985	7.2863	101.741	.23311	5.4533	101.669	.22830	3.6202	101.525	.22148
170	11.131	103.363	.24233	7.4063	103.295	.23560	5.5437	103.226	.23080	3.6811	103.088	.22398
180	11.311	104.925	.24479	7.5262	104.859	.23806	5.6341	104.793	.23326	3.7419	104.661	.22646
190	11.489	106.497	.24723	7.6461	106.434	.24050	5.7243	106.370	.23571	3.8025	106.243	.22891
200	11.668	108.079	0.24964	7.7658	108.018	0.24292	5.8145	107.957	0.23813	3.8632	107.835	0.23135
210	11.847	109.670	.25204	7.8855	109.612	.24532	5.9046	109.553	.24054	3.9237	109.436	.23375
220	12.026	111.272	.25441	8.0051	111.215	.24770	5.9946	111.159	.24291	3.9841	111.046	.23614
230	12.205	112.883	.25677	8.1246	112.828	.25005	6.0846	112.774	.24527	4.0445	112.665	.23850
240				8.2441	114.451	.25239	6.1745	114.398	.24761	4.1049	114.292	.24085
250							6.2643	116.031	0.24993	4.1651	115.929	0.24317

TABLE A.11 (cont'd.)

Temp. °F	Abs. Pressure 20 lb/in.² Gage Pressure 5.3 lb/in.² (Sat. Temp. −8.1°F)			Abs. Pressure 26 lb/in.² Gage Pressure 11.3 lb/in.² (Sat. Temp. 4.1°F)			Abs. Pressure 32 lb/in.² Gage Pressure 17.3 lb/in.² (Sat. Temp. 14.4°F)			Abs. Pressure 40 lb/in.² Gage Pressure 25.3 lb/in.² (Sat. Temp. 25.9°F)		
t	v	h	s	v	h	s	v	h	s	v	h	s
Sat.	(1.8977)	(76.397)	(0.16969)	(1.4835)	(77.710)	(0.16850)	(1.2198)	(78.793)	(0.16763)	(0.98743)	(80.000)	(0.16676)
0	1.9390	77.550	0.17222	.....	.....	.....	.....	.....	.....	.....	.....	.....
10	1.9893	78.973	.17528	1.5071	78.566	0.17033	.....	.....	.....	.....	.....	.....
20	2.0391	80.403	.17829	1.5468	80.024	.17340	1.2387	79.634	0.16939	.....	.....	.....
30	2.0884	81.842	.18126	1.5861	81.487	.17642	1.2717	81.123	.17246	0.99865	80.622	0.16804
40	2.1373	83.289	.18419	1.6248	82.956	.17939	1.3042	82.616	.17548	1.0258	82.148	.17112
50	2.1858	84.745	0.18707	1.6632	84.432	0.18232	1.3363	84.113	0.17845	1.0526	83.676	0.17415
60	2.2340	86.210	.18992	1.7013	85.916	.18520	1.3681	85.616	.18137	1.0789	85.206	.17712
70	2.2819	87.684	.19273	1.7390	87.407	.18804	1.3995	87.124	.18424	1.1049	86.739	.18005
80	2.3295	89.168	.19550	1.7765	88.906	.19084	1.4306	88.639	.18707	1.1306	88.277	.18292
90	2.3769	90.661	.19824	1.8137	90.413	.19361	1.4615	90.161	.18987	1.1560	89.819	.18575
100	2.4241	92.164	0.20095	1.8507	91.929	0.19634	1.4921	91.690	0.19263	1.1812	91.367	0.18854
110	2.4711	93.676	.20363	1.8874	93.453	.19904	1.5225	93.227	.19535	1.2061	92.920	.19129
120	2.5179	95.198	.20628	1.9240	94.986	.20171	1.5528	94.771	.19803	1.2309	94.480	.19401
130	2.5645	96.729	.20890	1.9605	96.527	.20435	1.5828	96.323	.20069	1.2554	96.047	.19669
140	2.6110	98.270	.21149	1.9967	98.078	.20695	1.6127	97.883	.20331	1.2798	97.620	.19933
150	2.6573	99.820	0.21405	2.0329	99.637	0.20953	1.6425	99.451	0.20590	1.3041	99.200	0.20195
160	2.7036	101.380	.21659	2.0689	101.204	.21208	1.6721	101.027	.20847	1.3282	100.788	.20453
170	2.7497	102.949	.21910	2.1048	102.781	.21461	1.7017	102.611	.21100	1.3522	102.383	.20708
180	2.7957	104.528	.22159	2.1406	104.367	.21710	1.7311	104.204	.21351	1.3761	103.985	.20961
190	2.8416	106.115	.22405	2.1763	105.961	.21958	1.7604	105.805	.21600	1.3999	105.595	.21210
200	2.8874	107.712	0.22649	2.2119	107.563	0.22202	1.7896	107.414	0.21845	1.4236	107.212	0.21457
210	2.9332	109.317	.22891	2.2474	109.174	.22445	1.8187	109.031	.22089	1.4472	108.837	.21702
220	2.9789	110.932	.23130	2.2828	110.794	.22685	1.8478	110.656	.22330	1.4707	110.469	.21944
230	3.0245	112.555	.23367	2.3182	112.422	.22923	1.8768	112.289	.22568	1.4942	112.109	.22183
240	3.0700	114.186	.23602	2.3535	114.058	.23158	1.9057	113.930	.22804	1.5176	113.757	.22420
250	3.1155	115.826	0.23835	2.3888	115.703	0.23391	1.9346	115.579	0.23038	1.5409	115.412	0.22655
260	3.1609	117.475	.24065	2.4240	117.355	.23623	1.9634	117.235	.23270	1.5642	117.074	.22888
270	3.2063	119.131	.24294	2.4592	119.016	.23852	1.9922	118.900	.23500	1.5874	118.744	.23118
280	3.2517	120.796	.24520	2.4943	120.684	.24079	2.0209	120.572	.23727	1.6106	120.421	.23347
290	3.2970	122.469	.24745	2.5293	122.360	.24304	2.0495	122.251	.23953	1.6337	122.105	.23573

TABLE A.11 (cont'd.)

Temp. °F	Abs. Pressure 50 lb/in.² Gage Pressure 35.3 lb/in.² (Sat. Temp. 38.2°F)			Abs. Pressure 60 lb/in.² Gage Pressure 45.3 lb/in.² (Sat. Temp. 48.6°F)			Abs. Pressure 80 lb/in.² Gage Pressure 65.3 lb/in.² (Sat. Temp. 66.2°F)			Abs. Pressure 100 lb/in.² Gage Pressure 85.3 lb/in.² (Sat. Temp. 80.8°F)		
t	v	h	s	v	h	s	v	h	s	v	h	s
Sat.	(0.79824)	(81.249)	(0.16597)	(0.67005)	(82.299)	(0.16537)	(0.50680)	(84.003)	(0.16450)	(0.40674)	(85.351)	(0.16389)
40	0.80248	81.540	0.16655	.....	.....	.....	.....	.....	.....	.....	.....	.....
50	0.82502	83.109	0.16966	0.67272	82.518	0.16580	.....	.....	.....	.....	.....	.....
60	.84713	84.676	.17271	.69210	84.126	.16892	0.51269	84.640	0.16571	.....	.....	.....
70	.86886	86.243	.17569	.71105	85.729	.17198	.52795	86.316	.16885	.....	.....	.....
80	.89025	87.811	.17862	.72964	87.330	.17497	.54281	87.981	.17190	0.41876	86.964	0.16685
90	.91134	89.380	.18151	.74790	88.929	.17791	.55734	89.640	.17489	.43138	88.694	.16996
100	.93216	90.953	.18434	.76588	90.528	.18079	.57158	91.294	.17782	.44365	90.410	.17300
110	.95275	92.529	.18713	.78360	92.128	.18362	.58556	92.945	.18070	.45562	92.116	.17597
120	.97313	94.110	.18988	.80110	93.731	.18641	.59931	94.594	.18352	.46733	93.814	.17888
130	.99332	95.695	.19259	.81840	95.336	.18916	.61286	96.242	.18629	.47881	95.507	.18172
140	1.0133	97.286	.19527	.83551	96.945	.19186	.62623	97.891	.18902	.49009	97.197	.18452
150	1.0332	98.882	.19791	.85247	98.558	.19453	.63943	99.542	.19170	.50118	98.884	.18726
160	1.0529	100.485	.20051	.86928	100.176	.19716	.65250	101.195	.19435	.51212	100.571	.18996
170	1.0725	102.093	.20309	.88596	101.799	.19976	.66543	102.851	.19696	.52291	102.257	.19262
180	1.0920	103.708	.20563	.90252	103.427	.20233	.67824	104.511	.19953	.53358	103.944	.19524
190	1.1114	105.330	.20815	.91896	105.060	.20486	.69095	106.174	.20207	.54413	105.633	.19782
200	1.1307	106.958	.21064	.93531	106.700	.20736	.70356	107.841	.20458	.55457	107.324	.20036
210	1.1499	108.593	.21310	.95157	108.345	.20984	.71609	109.513	.20706	.56492	109.018	.20287
220	1.1690	110.235	.21553	.96775	109.997	.21229	.72853	111.190	.20951	.57519	110.714	.20535
230	1.1880	111.883	.21794	.98385	111.655	.21471	.74090	112.872	.21193	.58538	112.415	.20780
240	1.2070	113.539	.22032	.99988	113.319	.21710	.75320	114.559	.21432	.59549	114.119	.21022
250	1.2259	115.202	.22268	1.0159	114.989	.21947	.76544	116.251	.21669	.60554	115.828	.21261
260	1.2447	116.871	.22502	1.0318	116.666	.22182	.77762	117.949	.21903	.61553	117.540	.21497
270	1.2636	118.547	.22733	1.0476	118.350	.22414	.78975	119.652	.22135	.62546	119.258	.21731
280	1.2823	120.231	.22962	1.0634	120.039	.22644	.80183	121.361	.22364	.63534	120.980	.21962
290	1.3010	121.921	.23189	1.0792	121.736	.22872	.81386	123.075	.22592	.64518	122.707	.22191
300	1.3197	123.618	.23414	1.0949	123.438	.23098	.82586	124.795	.22817	.65497	124.439	.22417
310	1.3383	125.321	.23637	1.1106	125.147	.23321	.83781	126.521	.23039	.66472	126.176	.22641
320	1.3569	127.032	.23857	1.1262	126.863	.23543	.84973	128.253	.23260	.67444	127.917	.22863
330	1.3754	128.749	.24076	1.1418	128.585	.23762	.86161	129.990	.23479	.68411	129.665	.23083
340	.....	.....	.....	1.1574	130.313	.23980	.....	.....	.....	.....	.....	.....

TABLE A.11 (cont'd.)

Temp. °F t	Abs. Pressure 120 lb/in.² Gage Pressure 105.3 lb/in.² (Sat. Temp. 93.3°F)			Abs. Pressure 140 lb/in.² Gage Pressure 125.3 lb/in.² (Sat. Temp. 104.4°F)			Abs. Pressure 180 lb/in.² Gage Pressure 165.3 lb/in.² (Sat. Temp. 123.4°F)			Abs. Pressure 220 lb/in.² Gage Pressure 205.3 lb/in.² (Sat. Temp. 139.5°F)		
	v	h	s	v	h	s	v	h	s	v	h	s
Sat.	(0.33886)	(86.459)	(0.16340)	(0.28964)	(87.389)	(0.16299)	(0.22276)	(88.857)	(0.16228)	(0.17917)	(89.937)	(0.16161)
100	0.34655	87.675	0.16559	. . . . .	. . . . .	. . . . .	. . . . .	. . . . .	. . . . .	. . . . .	. . . . .	. . . . .
110	.35766	89.466	.16876	0.29548	88.448	0.16486	. . . . .	. . . . .	. . . . .	. . . . .	. . . . .	. . . . .
120	.36841	91.237	.17184	.30549	90.297	.16808	. . . . .	. . . . .	. . . . .	. . . . .	. . . . .	. . . . .
130	.37884	92.992	.17484	.31513	92.120	.17120	0.22863	90.179	0.16454	. . . . .	. . . . .	. . . . .
140	.38901	94.736	.17778	.32445	93.923	.17423	.23710	92.136	.16783	0.17957	90.043	0.16179
150	0.39896	96.471	0.18065	0.33350	95.709	0.17718	0.24519	94.053	0.17100	0.18746	92.156	0.16528
160	.40870	98.199	.18346	.34232	97.483	.18007	.25297	95.940	.17407	.19487	94.203	.16861
170	.41826	99.922	.18622	.35095	99.247	.18289	.26047	97.803	.17705	.20190	96.199	.17181
180	.42766	101.642	.18892	.35939	101.003	.18566	.26775	99.647	.17995	.20861	98.157	.17489
190	.43692	103.359	.19159	.36769	102.754	.18838	.27484	101.475	.18279	.21506	100.084	.17788
200	0.44606	105.076	0.19421	0.37584	104.501	0.19104	0.28176	103.291	0.18556	0.22130	101.986	0.18079
210	.45508	106.792	.19679	.38387	106.245	.19367	.28852	105.098	.18828	.22735	103.869	.18362
220	.46401	108.509	.19934	.39179	107.987	.19625	.29516	106.896	.19095	.23324	105.735	.18638
230	.47284	110.227	.20185	.39961	109.728	.19879	.30168	108.689	.19357	.23900	107.589	.18909
240	.48158	111.948	.20432	.40734	111.470	.20130	.30810	110.478	.19614	.24463	109.432	.19175
250	0.49025	113.670	0.20677	0.41499	113.212	0.20377	0.31442	112.263	0.19868	0.25015	111.267	0.19435
260	.49885	115.396	.20918	.42257	114.956	.20621	.32066	114.046	.20117	.25557	113.095	.19691
270	.50739	117.125	.21157	.43008	116.701	.20862	.32682	115.828	.20363	.26091	114.919	.19942
280	.51587	118.857	.21393	.43753	118.449	.21100	.33292	117.610	.20605	.26617	116.738	.20190
290	.52429	120.593	.21626	.44492	120.199	.21335	.33895	119.392	.20845	.27136	118.555	.20434
300	0.53267	122.333	0.21856	0.45226	121.953	0.21567	0.34492	121.174	0.21081	0.27648	120.369	0.20674
310	.54100	124.077	.22084	.45955	123.709	.21797	.35084	122.958	.21314	.28155	122.183	.20912
320	.54929	125.825	.22310	.46680	125.470	.22024	.35672	124.744	.21545	.28657	123.996	.21146
330	.55754	127.578	.22533	.47400	127.233	.22249	.36255	126.531	.21772	.29153	125.809	.21377
340	.56575	129.335	.22754	.48117	129.001	.22471	.36834	128.321	.21998	.29645	127.623	.21605
350	0.57393	131.097	0.22973	0.48831	130.773	0.22692	0.37409	130.113	0.22220	0.30134	129.438	0.21830
360	.58208	132.863	.23190	.49541	132.548	.22910	.37980	131.909	.22441	.30618	131.255	.22053
370	.59019	134.634	.23405	.50248	134.328	.23125	.38549	133.707	.22659	.31099	133.073	.22274
380	.59829	136.410	.23618	.50953	136.112	.23339	.39114	135.509	.22875	.31576	134.893	.22492
390	.60635	138.191	.23829	.51654	137.901	.23551	.39677	137.314	.23088	.32051	136.715	.22708
400	. . . . .	. . . . .	. . . . .	. . . . .	. . . . .	. . . . .	0.40237	139.122	0.23300	0.32523	138.540	0.22921
410	. . . . .	. . . . .	. . . . .	. . . . .	. . . . .	. . . . .	.40794	140.934	.23509	.32992	140.368	.23133

TABLE A.13
Gas Constant Values

Gas	Chemical formula	Molecular weight	English units				SI units		
			R $\left(\dfrac{\text{ft lb}}{\text{lb}°\text{R}}\right)$	c_p $\left(\dfrac{\text{Btu}}{\text{lb}°\text{R}}\right)$ at 77°F	c_v $\left(\dfrac{\text{Btu}}{\text{lb}°\text{R}}\right)$ at 77°F	$k,$ $\dfrac{c_p}{c_v}$	R $\left(\dfrac{\text{kJ}}{\text{kg}°\text{K}}\right)$	c_p $\left(\dfrac{\text{kJ}}{\text{kg}°\text{K}}\right)$	c_v $\left(\dfrac{\text{kJ}}{\text{kg}°\text{K}}\right)$
Acetylene	C_2H_2	26.02	59.39	0.361	0.285	1.27	0.31955	1.5116	1.1933
Air		28.97	53.34	0.240	0.171	1.40	0.28700	1.0052	0.7180
Ammonia	NH_3	17.024	90.7	0.52	0.404	1.29	0.48802	2.1773	1.6916
Argon	A	39.94	38.68	0.124	0.075	1.667	0.20813	0.5207	0.3124
Butane	C_4H_{10}	58.12	26.58	0.415	0.381	1.09	0.14304	1.7164	1.5734
Carbon dioxide	CO_2	44.01	35.10	0.202	0.157	1.29	0.18892	0.8464	0.6573
Carbon monoxide	CO	28.01	55.16	0.249	0.178	1.40	0.29683	1.0411	0.7441
Ethane	C_2H_6	30.07	51.38	0.427	0.361	1.18	0.27650	1.7662	1.4897
Ethylene	C_2H_4	28.052	55.07	0.411	0.340	1.21	0.29637	1.5482	1.2518
Helium	He	4.003	386.0	1.25	0.753	1.667	2.07703	5.1926	3.1156
Hydrogen	H_2	2.016	766.4	3.420	2.434	1.405	4.12418	14.3193	10.1919
Methane	CH_4	16.04	96.35	0.532	0.403	1.32	0.51835	2.2537	1.7354
Nitrogen	N_2	28.016	55.15	0.248	0.177	1.40	0.29680	1.0404	0.7434
Octane	C_8H_{18}	114.14	13.53	0.409	0.392	1.044	0.07279	1.7113	1.6385
Oxygen	O_2	32.000	48.28	0.219	0.157	1.39	0.25983	0.9190	0.6590
Propane	C_3H_8	44.094	35.04	0.407	0.362	1.124	0.18855	1.6794	1.4909
Water vapor	H_2O	18.016	85.76	0.445	0.335	1.33	0.46152	1.8649	1.4031

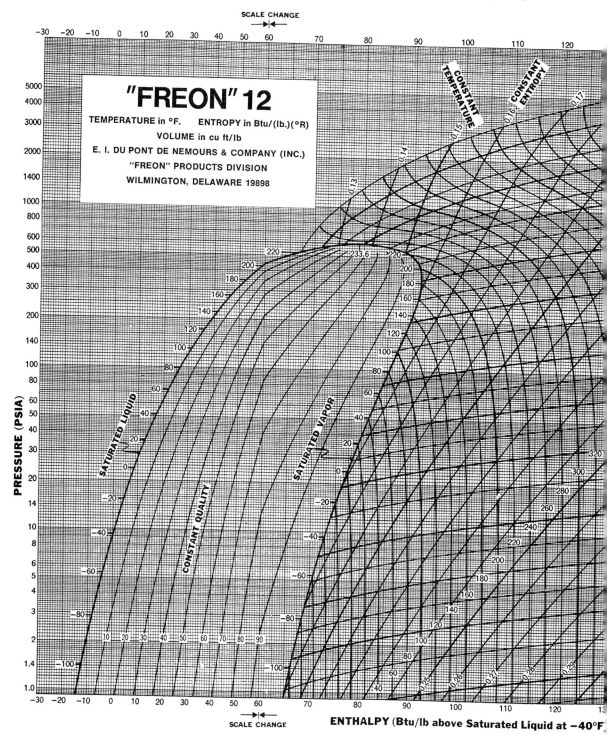

PRESSURE (PSIA)

SCALE CHANGE

ENTHALPY (Btu/lb above Saturated Liquid at −40°F

SCALE CHANGE

"FREON" 12

TEMPERATURE in °F.   ENTROPY in Btu/(lb.)(°R)

VOLUME in cu ft/lb

E. I. DU PONT DE NEMOURS & COMPANY (INC.)

"FREON" PRODUCTS DIVISION

WILMINGTON, DELAWARE 19898

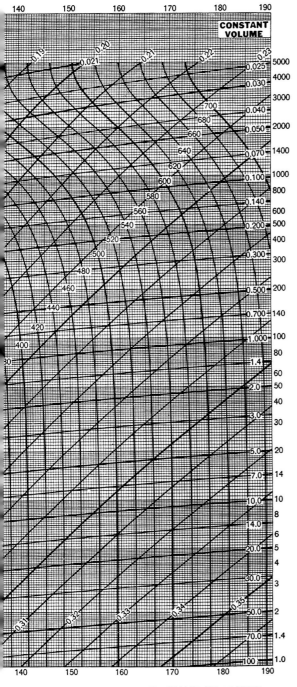

CONSTANT VOLUME

TABLE A.14
Critical Constants

Substance	Formula	Molecular Weight	Temperature °K	Temperature °R	Pressure atm	Pressure lb$_f$/in^2	Volume ft^3/lb mole
Ammonia	NH_3	17.03	405.5	729.8	111.3	1636	1.16
Argon	A	39.944	151	272	48.0	705	1.20
Bromine	Br_2	159.832	584	1052	102	1500	2.17
Carbon dioxide	CO_2	44.01	304.2	547.5	72.9	1071	1.51
Carbon monoxide	CO	28.01	133	240	34.5	507	1.49
Chlorine	Cl_2	70.914	417	751	76.1	1120	1.99
Deuterium (Normal)	D_2	4.00	38.4	69.1	16.4	241	....
Helium	He	4.003	5.3	9.5	2.26	33.2	0.926
Helium^3	He	3.00	3.34	6.01	1.15	16.9	....
Hydrogen (Normal)	H_2	2.016	33.3	59.9	12.8	188.1	1.04
Krypton	Kr	83.7	209.4	376.9	54.3	798	1.48
Neon	Ne	20.183	44.5	80.1	26.9	395	0.668
Nitrogen	N_2	28.016	126.2	227.1	33.5	492	1.44
Nitrous oxide	N_2O	44.02	309.7	557.4	71.7	1054	1.54
Oxygen	O_2	32.00	154.8	278.6	50.1	736	1.25
Sulfur dioxide	SO_2	64.06	430.7	775.2	77.8	1143	1.95
Water	H_2O	18.016	647.4	1165.3	218.3	3208	0.90
Xenon	Xe	131.3	289.75	521.55	58.0	852	1.90
Benzene	C_6H_6	78.11	562	1012	48.6	714	4.17
n-Butane	C_4H_{10}	58.120	425.2	765.2	37.5	551	4.08
Carbon tetrachloride	CCl_4	153.84	556.4	1001.5	45.0	661	4.42
Chloroform	$CHCl_3$	119.39	536.6	965.8	54.0	794	3.85
Dichlorodifluoromethane	CCl_2F_2	120.92	384.7	692.4	39.6	582	3.49
Dichlorofluoromethane	$CHCl_2F$	102.93	451.7	813.0	51.0	749	3.16
Ethane	C_2H_5	30.068	305.5	549.8	48.2	708	2.37
Ethylalcohol	C_2H_5OH	46.07	516.0	929.0	63.0	926	2.68
Ethylene	C_2H_4	28.052	282.4	508.3	50.5	742	1.99
n-Hexane	C_6H_{14}	86.172	507.9	914.2	29.9	439	5.89
Methane	CH_4	16.042	191.1	343.9	45.8	673	1.59
Methyl alcohol	CH_3OH	32.04	513.2	923.7	78.5	1154	1.89
Methyl chloride	CH_3Cl	50.49	416.3	749.3	65.9	968	2.29
Propane	C_3H_8	44.094	370.0	665.9	42.0	617	3.20
Propene	C_3H_6	42.078	365.0	656.9	45.6	670	2.90
Propyne	C_3H_4	40.062	401	722	52.8	776	....
Trichlorofluoromethane	CCl_3F	137.38	471.2	848.1	43.2	635	3.97

Source: G. J. Van Wylen and R. E. Sonntag, *Fundamentals of Classical Thermodynamics*, John Wiley & Sons, Inc., New York, 1965, p. 579, with permission.

TABLE A.15
HFC-134a Saturation Properties—Temperature Table*

TEMP. °F	PRESSURE psia	VOLUME ft³/lb LIQUID v_f	VOLUME ft³/lb VAPOR v_g	DENSITY lb/ft³ LIQUID $1/v_f$	DENSITY lb/ft³ VAPOR $1/v_g$	ENTHALPY Btu/lb LIQUID h_f	ENTHALPY Btu/lb LATENT h_{fg}	ENTHALPY Btu/lb VAPOR h_g	ENTROPY Btu/(lb)(°R) LIQUID s_f	ENTROPY Btu/(lb)(°R) VAPOR s_g	TEMP. °F
−40	7.417	0.0113	5.7904	88.31	0.1727	0.0	97.2	97.2	0.0000	0.2316	−40
−30	9.851	0.0115	4.4366	87.31	0.2254	3.0	95.7	98.7	0.0070	0.2297	−30
−20	12.885	0.0116	3.4471	86.30	0.2901	6.0	94.2	100.2	0.0139	0.2281	−20
−15	14.659	0.0117	3.0525	85.79	0.3276	7.5	93.4	100.9	0.0174	0.2274	−15
−10	16.620	0.0117	2.7115	85.28	0.3688	9.0	92.6	101.7	0.0208	0.2267	−10
−5	18.784	0.0118	2.4155	84.76	0.4140	10.6	91.8	102.4	0.0241	0.2261	−5
0	21.163	0.0119	2.1580	84.23	0.4634	12.1	91.0	103.1	0.0275	0.2255	0
5	23.772	0.0119	1.9327	83.70	0.5174	13.7	90.2	103.9	0.0308	0.2250	5
10	26.625	0.0120	1.7355	83.17	0.5762	15.2	89.4	104.6	0.0342	0.2244	10
15	29.739	0.0121	1.5620	82.63	0.6402	16.8	88.5	105.3	0.0375	0.2240	15
20	33.129	0.0122	1.4090	82.08	0.7097	18.4	87.7	106.0	0.0408	0.2235	20
25	36.810	0.0123	1.2737	81.52	0.7851	20.0	86.8	106.7	0.0440	0.2231	25
30	40.800	0.0124	1.1538	80.96	0.8667	21.6	85.9	107.4	0.0473	0.2227	30
35	45.115	0.0124	1.0472	80.40	0.9549	23.2	85.0	108.1	0.0505	0.2223	35
40	49.771	0.0125	0.9523	79.82	1.0501	24.8	84.1	108.8	0.0538	0.2220	40
45	54.787	0.0126	0.8675	79.24	1.1527	26.4	83.1	109.5	0.0570	0.2217	45
50	60.180	0.0127	0.7916	78.64	1.2633	28.0	82.1	110.2	0.0602	0.2214	50
55	65.963	0.0128	0.7234	78.04	1.3823	29.7	81.2	110.9	0.0634	0.2211	55
60	72.167	0.0129	0.6622	77.43	1.5102	31.4	80.2	111.5	0.0666	0.2208	60
65	78.803	0.0130	0.6069	76.81	1.6477	33.0	79.1	112.2	0.0698	0.2206	65
70	85.890	0.0131	0.5570	76.18	1.7952	34.7	78.1	112.8	0.0729	0.2203	70
75	93.447	0.0132	0.5119	75.54	1.9536	36.4	77.0	113.4	0.0761	0.2201	75
80	101.494	0.0134	0.4709	74.89	2.1234	38.1	75.9	114.0	0.0792	0.2199	80
85	110.050	0.0135	0.4337	74.22	2.3056	39.9	74.8	114.6	0.0824	0.2196	85
90	119.138	0.0136	0.3999	73.54	2.5009	41.6	73.6	115.2	0.0855	0.2194	90
95	128.782	0.0137	0.3690	72.84	2.7102	43.4	72.4	115.8	0.0886	0.2192	95
100	138.996	0.0139	0.3408	72.13	2.9347	45.1	71.2	116.3	0.0918	0.2190	100
105	149.804	0.0140	0.3149	71.40	3.1754	46.9	69.9	116.9	0.0949	0.2188	105
110	161.227	0.0142	0.2912	70.66	3.4337	48.7	68.6	117.4	0.0981	0.2185	110
115	173.298	0.0143	0.2695	69.89	3.7110	50.5	67.3	117.9	0.1012	0.2183	115
120	186.023	0.0145	0.2494	69.10	4.0089	52.4	65.9	118.3	0.1043	0.2181	120
130	213.572	0.0148	0.2139	67.46	4.6745	56.2	63.0	119.2	0.1106	0.2175	130
140	244.068	0.0152	0.1835	65.70	5.4491	60.0	59.9	119.9	0.1170	0.2168	140
150	277.721	0.0157	0.1573	63.81	6.3584	64.0	56.5	120.5	0.1234	0.2160	150
160	314.800	0.0162	0.1344	61.74	7.4390	68.1	52.7	120.8	0.1299	0.2150	160
170	355.547	0.0168	0.1143	59.45	8.7470	72.4	48.5	120.9	0.1366	0.2137	170
180	400.280	0.0176	0.0964	56.83	10.3750	76.9	43.7	120.7	0.1435	0.2119	180
190	449.384	0.0186	0.0800	53.73	12.4962	81.8	38.0	119.8	0.1508	0.2093	190
200	503.361	0.0201	0.0645	49.73	15.5155	87.3	30.7	118.0	0.1589	0.2054	200
210	563.037	0.0232	0.0474	43.19	21.1071	94.4	18.9	113.4	0.1693	0.1976	210
213	582.316	0.0259	0.0394	38.65	25.3583	98.4	11.1	109.5	0.1750	0.1915	213

*Tables A.15, A.16 and A.17, including SI, from "Thermodynamic Properties of HFC-134a," with permission of E. I. Du Pont de Nemours and Co.

TABLE A.16
HFC-134a Superheated Vapor—Constant Pressure Tables

V = Volume in ft³/lb   H = Enthalpy in Btu/lb   S = Entropy in Btu/(lb)(°R) v_s = Velocity of Sound in ft/sec
C_p = Heat Capacity at Constant Pressure in Btu/(lb)(°F) C_p/C_v = Heat Capacity Ratio (Dimensionless)

PRESSURE = 1.00 PSIA

TEMP °F		V	H	S	C_p	C_p/C_v	v_s
−97.6	SAT LIQ	0.01065	−16.6	−0.0426	0.2813	1.5295	3163.1
−97.6	SAT VAP	37.87879	88.5	0.2478	0.1570	1.1488	447.2
−90		38.61004	89.7	0.2511	0.1588	1.1462	451.6
−80		38.68254	91.3	0.2554	0.1612	1.1429	457.3
−70		40.81633	92.9	0.2596	0.1637	1.1398	463.0
−60		41.84100	94.6	0.2638	0.1662	1.1370	468.5
−50		42.91845	96.3	0.2679	0.1686	1.1342	474.0
−40		44.05286	98.0	0.2720	0.1711	1.1317	479.3
−30		45.04505	99.7	0.2761	0.1736	1.1292	484.7
−20		46.08295	101.4	0.2801	0.1760	1.1269	489.9
−10		47.16981	103.2	0.2841	0.1785	1.1247	495.1
0		48.30918	105.0	0.2880	0.1809	1.1226	500.2
10		49.26108	106.8	0.2919	0.1834	1.1206	505.3
20		50.25126	108.7	0.2958	0.1858	1.1187	510.3
30		51.28205	110.5	0.2997	0.1882	1.1169	515.2
40		52.35602	112.4	0.3035	0.1906	1.1151	520.1
50		53.47594	114.4	0.3073	0.1930	1.1134	525.0
60		54.64481	116.3	0.3111	0.1953	1.1118	529.8
70		55.55556	118.3	0.3148	0.1977	1.1103	534.5
80		56.81818	120.2	0.3186	0.2000	1.1088	539.3
90		57.80347	122.3	0.3223	0.2023	1.1073	543.9
100		58.82353	124.3	0.3259	0.2046	1.1059	548.5
110		59.88024	126.4	0.3296	0.2069	1.1046	553.1
120		60.97561	128.4	0.3332	0.2092	1.1033	557.7
130		62.11180	130.5	0.3368	0.2114	1.1020	562.2
140		62.89308	132.7	0.3404	0.2137	1.1008	566.6
150		64.10256	134.8	0.3439	0.2159	1.0997	571.1
160		64.93506	137.0	0.3474	0.2181	1.0985	575.5
170		66.22517	139.2	0.3509	0.2203	1.0974	579.8
180		67.11409	141.4	0.3544	0.2224	1.0964	584.2
190		68.49315	143.6	0.3579	0.2246	1.0953	588.5
200		69.44444	145.9	0.3613	0.2267	1.0943	592.7
210		70.42254	148.2	0.3648	0.2288	1.0934	597.0
200		—	—	—	—	—	—

PRESSURE = 5.00 PSIA

TEMP °F		V	H	S	C_p	C_p/C_v	v_s
−53	SAT LIQ	0.01116	−3.8	−0.0093	0.2929	1.5071	2768.7
−53	SAT VAP	8.37521	95.2	0.2343	0.1726	1.1471	466.8
−50		8.44595	95.7	0.2356	0.1732	1.1458	468.6
−40		8.66551	97.5	0.2398	0.1752	1.1418	474.4
−30		8.88889	99.2	0.2439	0.1772	1.1382	480.0
−20		9.10747	101.0	0.2480	0.1793	1.1349	485.6
−10		9.32836	102.8	0.2521	0.1814	1.1318	491.1
0		9.55110	104.7	0.2561	0.1835	1.1289	496.5
10		9.76563	106.5	0.2601	0.1857	1.1263	501.8
20		9.98004	108.4	0.2640	0.1879	1.1238	507.0
30		10.20408	110.3	0.2679	0.1901	1.1215	512.2
40		10.41667	112.2	0.2718	0.1923	1.1193	517.3
50		10.62699	114.1	0.2756	0.1945	1.1172	522.3
60		10.84599	116.1	0.2794	0.1968	1.1152	527.3
70		11.06195	118.0	0.2832	0.1990	1.1134	532.2
80		11.27396	120.0	0.2869	0.2012	1.1116	537.0
90		11.49425	122.1	0.2906	0.2034	1.1100	541.8
100		11.70960	124.1	0.2943	0.2056	1.1084	546.5
110		11.91895	126.2	0.2980	0.2078	1.1068	551.2
120		12.13592	128.3	0.3016	0.2100	1.1054	555.9
130		12.34568	130.4	0.3052	0.2122	1.1040	560.5
140		12.56281	132.5	0.3088	0.2144	1.1026	565.0
150		12.77139	134.7	0.3124	0.2166	1.1013	569.5
160		12.98701	136.8	0.3159	0.2187	1.1001	574.0
170		13.19261	139.0	0.3194	0.2209	1.0989	578.4
180		13.40483	141.3	0.3229	0.2230	1.0977	582.8
190		13.62398	143.5	0.3264	0.2251	1.0966	587.2
200		13.83126	145.8	0.3299	0.2272	1.0955	591.5
210		14.04494	148.0	0.3333	0.2293	1.0945	595.8
220		14.24501	150.3	0.3367	0.2313	1.0935	600.0
230		14.47178	152.7	0.3401	0.2334	1.0925	604.3
240		14.68429	155.0	0.3435	0.2354	1.0916	608.4
250		14.90313	157.4	0.3468	0.2374	1.0907	612.6
260		—	—	—	—	—	—

TABLE A.16 (cont'd.)

PRESSURE = 10.00 PSIA

TEMP °F	V	H	S	C_p	C_p/C_v	v_s
−29.5	0.01146	3.1	0.0074	0.2996	1.5038	2567.0
−29.5	4.37445	98.8	0.2297	0.1821	1.1503	474.4
−30	—	—	—	—	—	—
−20	4.48430	100.5	0.2336	0.1836	1.1457	480.1
−10	4.59770	102.3	0.2378	0.1852	1.1414	486.0
0	4.71254	104.2	0.2419	0.1869	1.1374	491.8
10	4.82393	106.1	0.2459	0.1888	1.1338	497.4
20	4.93583	108.0	0.2499	0.1906	1.1306	502.9
30	5.04796	109.9	0.2539	0.1926	1.1275	508.3
40	5.15996	111.8	0.2578	0.1945	1.1247	513.7
50	5.26870	113.8	0.2616	0.1966	1.1222	518.9
60	5.37924	115.8	0.2655	0.1986	1.1197	524.1
70	5.48847	117.8	0.2693	0.2007	1.1175	529.2
80	5.59910	119.8	0.2731	0.2027	1.1154	534.2
90	5.70776	121.8	0.2768	0.2048	1.1134	539.1
100	5.81734	123.9	0.2805	0.2069	1.1115	544.0
110	5.92417	125.9	0.2842	0.2090	1.1097	548.8
120	6.03500	128.0	0.2879	0.2111	1.1080	553.6
130	6.14251	130.2	0.2915	0.2132	1.1064	558.3
140	6.25000	132.3	0.2951	0.2153	1.1049	563.0
150	6.35728	134.5	0.2987	0.2174	1.1034	567.6
160	6.46412	136.7	0.3022	0.2195	1.1020	572.1
170	6.57030	138.9	0.3057	0.2216	1.1007	576.7
180	6.68003	141.1	0.3093	0.2237	1.0994	581.1
190	6.78887	143.3	0.3127	0.2257	1.0982	585.6
200	6.89180	145.6	0.3162	0.2278	1.0970	590.0
210	6.99790	147.9	0.3196	0.2298	1.0959	594.3
220	7.10732	150.2	0.3231	0.2319	1.0948	598.6
230	7.21501	152.5	0.3265	0.2339	1.0938	602.9
240	7.32064	154.9	0.3298	0.2359	1.0928	607.2
250	7.42942	157.2	0.3332	0.2379	1.0918	611.4
260	7.53580	159.6	0.3366	0.2398	1.0908	615.6
270	7.63942	162.0	0.3399	0.2418	1.0899	619.7
280	7.74593	164.5	0.3432	0.2437	1.0890	623.8

PRESSURE = 15.00 PSIA

	TEMP °F	V	H	S	C_p	C_p/C_v	v_s
SAT LIQ	−14.1	0.01167	7.8	0.0180	0.3043	1.5047	2437.1
SAT VAP	−14.1	2.98686	101.1	0.2273	0.1888	1.1543	478.1
	−10	3.01932	101.8	0.2290	0.1893	1.1520	480.7
	0	3.09885	103.7	0.2332	0.1906	1.1467	486.9
	10	3.17662	105.6	0.2373	0.1920	1.1420	492.8
	20	3.25415	107.6	0.2414	0.1935	1.1378	498.7
	30	3.33111	109.5	0.2454	0.1952	1.1340	504.4
	40	3.40716	111.5	0.2494	0.1969	1.1306	510.0
	50	3.48189	113.5	0.2533	0.1987	1.1274	515.5
	60	3.55745	115.4	0.2571	0.2005	1.1245	520.8
	70	3.63240	117.5	0.2610	0.2024	1.1218	526.1
	80	3.70645	119.5	0.2648	0.2043	1.1193	531.3
	90	3.78072	121.5	0.2686	0.2063	1.1169	536.4
	100	3.85356	123.6	0.2723	0.2082	1.1147	541.4
	110	3.92773	125.7	0.2760	0.2102	1.1127	546.4
	120	4.00160	127.8	0.2797	0.2122	1.1108	551.3
	130	4.07332	130.0	0.2833	0.2142	1.1090	556.1
	140	4.14594	132.1	0.2869	0.2163	1.1072	560.9
	150	4.21941	134.3	0.2905	0.2183	1.1056	565.6
	160	4.29185	136.5	0.2941	0.2203	1.1041	570.3
	170	4.36300	138.7	0.2977	0.2223	1.1026	574.9
	180	4.43656	140.9	0.3012	0.2244	1.1012	579.4
	190	4.50857	143.2	0.3047	0.2264	1.0999	584.0
	200	4.58085	145.5	0.3081	0.2284	1.0986	588.4
	210	4.65333	147.7	0.3116	0.2304	1.0974	592.8
	220	4.72367	150.1	0.3150	0.2324	1.0962	597.2
	230	4.79616	152.4	0.3184	0.2344	1.0951	601.6
	240	4.86618	154.7	0.3218	0.2363	1.0940	605.9
	250	4.93827	157.1	0.3252	0.2383	1.0929	610.2
	260	5.01002	159.5	0.3285	0.2403	1.0919	614.4
	270	5.08130	161.9	0.3319	0.2422	1.0910	618.6
	280	5.15198	164.4	0.3352	0.2441	1.0900	622.8
	290	5.22466	166.8	0.3385	0.2460	1.0891	626.9

TABLE A.16 (cont'd.)

PRESSURE = 30.00 PSIA

TEMP °F	V	H	S	C_p	C_p/C_v	v_s
SAT LIQ 15.4	0.01211	16.9	0.0377	0.3142	1.5132	2189.7
SAT VAP 15.4	1.54895	105.4	0.2239	0.2032	1.1670	482.1
20	1.56863	106.3	0.2259	0.2033	1.1635	485.3
30	1.61082	108.3	0.2301	0.2039	1.1566	491.9
40	1.65235	110.4	0.2342	0.2046	1.1505	498.4
50	1.69319	112.4	0.2383	0.2056	1.1451	504.6
60	1.73340	114.5	0.2423	0.2068	1.1403	510.7
70	1.77336	116.6	0.2462	0.2080	1.1360	516.6
80	1.81291	118.7	0.2501	0.2094	1.1321	522.4
90	1.85185	120.8	0.2540	0.2109	1.1285	528.0
100	1.89107	122.9	0.2578	0.2124	1.1253	533.5
110	1.92938	125.0	0.2616	0.2141	1.1223	538.9
120	1.96773	127.2	0.2653	0.2157	1.1196	544.2
130	2.00562	129.3	0.2690	0.2175	1.1171	549.4
140	2.04332	131.5	0.2727	0.2192	1.1147	554.6
150	2.08117	133.7	0.2764	0.2210	1.1125	559.6
160	2.11864	135.9	0.2800	0.2228	1.1105	564.6
170	2.15610	138.2	0.2835	0.2247	1.1085	569.5
180	2.19346	140.4	0.2871	0.2265	1.1067	574.3
190	2.23015	142.7	0.2906	0.2284	1.1050	579.0
200	2.26706	145.0	0.2941	0.2303	1.1034	583.7
210	2.30415	147.3	0.2976	0.2321	1.1019	588.4
220	2.34082	149.6	0.3011	0.2340	1.1004	593.0
230	2.37756	152.0	0.3045	0.2359	1.0990	597.5
240	2.41429	154.4	0.3079	0.2378	1.0977	602.0
250	2.45038	156.7	0.3113	0.2396	1.0965	606.5
260	2.48694	159.1	0.3147	0.2415	1.0953	610.9
270	2.52334	161.6	0.3180	0.2434	1.0941	615.2
280	2.55951	164.0	0.3213	0.2452	1.0930	619.5
290	2.59605	166.5	0.3246	0.2471	1.0920	623.8
300	2.63227	169.0	0.3279	0.2489	1.0910	628.0
310	2.66809	171.5	0.3312	0.2507	1.0900	632.2
320	2.70416	174.0	0.3344	0.2526	1.0890	636.4

PRESSURE = 20.00 PSIA

TEMP °F	V	H	S	C_p	C_p/C_v	v_s
SAT LIQ −2.4	0.01184	11.4	0.0259	0.3081	1.5071	2338.7
SAT VAP −2.4	2.27635	102.8	0.2258	0.1943	1.1585	480.2
0	2.29095	103.3	0.2268	0.1945	1.1569	481.8
10	2.35183	105.2	0.2310	0.1955	1.1510	488.2
20	2.41196	107.2	0.2351	0.1966	1.1457	494.3
30	2.47158	109.1	0.2392	0.1979	1.1410	500.3
40	2.52972	111.1	0.2432	0.1994	1.1368	506.2
50	2.58799	113.1	0.2472	0.2009	1.1329	511.9
60	2.64550	115.1	0.2511	0.2025	1.1295	517.5
70	2.70270	117.2	0.2550	0.2042	1.1263	523.0
80	2.76014	119.2	0.2588	0.2060	1.1233	528.4
90	2.81690	121.3	0.2626	0.2078	1.1206	533.6
100	2.87274	123.4	0.2664	0.2096	1.1181	538.8
110	2.92826	125.5	0.2701	0.2115	1.1158	543.9
120	2.98418	127.6	0.2738	0.2134	1.1136	549.0
130	3.03951	129.7	0.2775	0.2153	1.1116	553.9
140	3.09502	131.9	0.2811	0.2172	1.1097	558.8
150	3.14961	134.1	0.2847	0.2192	1.1078	563.6
160	3.20513	136.3	0.2883	0.2211	1.1061	568.4
170	3.26052	138.5	0.2918	0.2231	1.1045	573.1
180	3.31455	140.8	0.2954	0.2251	1.1030	577.7
190	3.36927	143.0	0.2989	0.2270	1.1015	582.3
200	3.42349	145.3	0.3024	0.2290	1.1002	586.9
210	3.47826	147.6	0.3058	0.2310	1.0988	591.4
220	3.53232	149.9	0.3093	0.2329	1.0976	595.8
230	3.58680	152.3	0.3127	0.2349	1.0964	600.2
240	3.64033	154.6	0.3161	0.2368	1.0952	604.6
250	3.69413	157.0	0.3195	0.2387	1.0941	608.9
260	3.74813	159.4	0.3228	0.2407	1.0930	613.2
270	3.80228	161.8	0.3261	0.2426	1.0920	617.5
280	2.85654	164.2	0.3295	0.2445	1.0910	621.7
290	3.91083	166.7	0.3328	0.2464	1.0901	625.9
300	3.96354	169.2	0.3360	0.2482	1.0891	630.0

TABLE A.16 (cont'd.)

PRESSURE = 40.00 PSIA

TEMP °F	V	H	S	C_p	C_p/C_v	v_s
SAT LIQ 29	0.01233	21.2	0.0467	0.3194	1.5200	2075.7
SAT VAP 29	1.17606	107.3	0.2228	0.2106	1.1756	482.3
30	1.17925	107.5	0.2232	0.2106	1.1748	483.0
40	1.21242	109.6	0.2275	0.2105	1.1663	490.2
50	1.24471	111.7	0.2316	0.2108	1.1589	497.0
60	1.27665	113.8	0.2357	0.2114	1.1525	503.6
70	1.30787	116.0	0.2398	0.2122	1.1468	510.0
80	1.33887	118.1	0.2437	0.2131	1.1417	516.2
90	1.36930	120.2	0.2477	0.2142	1.1372	522.2
100	1.39958	122.4	0.2515	0.2155	1.1331	528.1
110	1.42939	124.5	0.2554	0.2168	1.1294	533.8
120	1.45900	126.7	0.2592	0.2183	1.1260	539.4
130	1.48832	128.9	0.2629	0.2198	1.1229	544.9
140	1.51745	131.1	0.2666	0.2213	1.1201	550.3
150	1.54655	133.3	0.2703	0.2229	1.1174	555.5
160	1.57505	135.6	0.2739	0.2246	1.1150	560.7
170	1.60359	137.8	0.2775	0.2263	1.1127	565.8
180	1.63212	140.1	0.2811	0.2280	1.1106	570.8
190	1.66058	142.4	0.2847	0.2298	1.1086	575.7
200	1.68890	144.7	0.2882	0.2316	1.1068	580.6
210	1.71674	147.0	0.2917	0.2333	1.1050	585.4
220	1.74520	149.3	0.2952	0.2351	1.1034	590.1
230	1.77305	151.7	0.2986	0.2369	1.1018	594.8
240	1.80050	154.1	0.3020	0.2388	1.1003	599.4
250	1.82849	156.5	0.3054	0.2406	1.0989	604.0
260	1.85598	158.9	0.3088	0.2424	1.0976	608.5
270	1.88359	161.3	0.3122	0.2442	1.0963	612.9
280	1.91131	163.8	0.3155	0.2460	1.0951	617.4
290	1.93911	166.2	0.3188	0.2478	1.0939	621.7
300	1.96657	168.7	0.3221	0.2496	1.0928	626.1
310	1.99362	171.2	0.3254	0.2514	1.0917	630.3
320	2.02102	173.8	0.3286	0.2532	1.0907	634.6
330	2.04834	176.3	0.3319	0.2550	1.0897	638.8
340	—					—

PRESSURE = 50.00 PSIA

TEMP °F	V	H	S	C_p	C_p/C_v	v_s
SAT LIQ 40.2	0.01253	24.9	0.0539	0.3239	1.5271	1981.8
SAT VAP 40.2	0.94805	108.9	0.2220	0.2172	1.1844	481.7
50	0.97494	111.0	0.2262	0.2166	1.1748	489.0
60	1.00180	113.1	0.2304	0.2165	1.1663	496.2
70	1.02807	115.3	0.2345	0.2167	1.1589	503.1
80	1.05396	117.5	0.2386	0.2172	1.1524	509.8
90	1.07933	119.7	0.2426	0.2178	1.1467	516.3
100	1.10436	121.8	0.2465	0.2187	1.1416	522.5
110	1.12905	124.0	0.2504	0.2197	1.1370	528.6
120	1.15354	126.2	0.2542	0.2209	1.1329	534.5
130	1.17772	128.5	0.2580	0.2222	1.1292	540.2
140	1.20178	130.7	0.2617	0.2235	1.1258	545.9
150	1.22534	132.9	0.2655	0.2250	1.1226	551.4
160	1.24891	135.2	0.2691	0.2265	1.1198	556.8
170	1.27226	137.5	0.2728	0.2280	1.1171	562.1
180	1.29550	139.7	0.2764	0.2296	1.1147	567.3
190	1.31874	142.0	0.2799	0.2312	1.1124	572.4
200	1.34174	144.4	0.2835	0.2329	1.1103	577.4
210	1.36444	146.7	0.2870	0.2346	1.1083	582.3
220	1.38735	149.1	0.2905	0.2363	1.1064	587.2
230	1.41004	151.4	0.2940	0.2380	1.1046	592.0
240	1.43266	153.8	0.2974	0.2398	1.1030	596.8
250	1.45518	156.2	0.3008	0.2415	1.1014	601.5
260	1.47754	158.6	0.3042	0.2433	1.0999	606.1
270	1.50015	161.1	0.3076	0.2450	1.0985	610.7
280	1.52230	163.5	0.3109	0.2468	1.0972	615.2
290	1.54440	166.0	0.3142	0.2485	1.0959	619.6
300	1.56691	168.5	0.3176	0.2503	1.0947	624.1
310	1.58907	171.0	0.3208	0.2521	1.0935	628.4
320	1.61108	173.6	0.3241	0.2538	1.0924	632.8
330	1.63292	176.1	0.3273	0.2555	1.0913	637.0
340	1.65508	178.7	0.3306	0.2573	1.0903	641.3
350	1.67701	181.2	0.3338	0.2590	1.0893	645.5

TABLE A.16 (cont'd.)

PRESSURE = 60.00 PSIA

TEMP °F	V	H	S	C_p	C_p/C_v	v_s
49.8 SAT LIQ	0.01271	28.0	0.0601	0.3282	1.5343	1901.3
49.8 SAT VAP	0.79390	110.2	0.2214	0.2232	1.1934	480.5
50	0.79428	110.2	0.2215	0.2232	1.1932	480.6
60	0.81793	112.4	0.2258	0.2222	1.1820	488.5
70	0.84104	114.6	0.2300	0.2217	1.1725	496.0
80	0.86356	116.9	0.2342	0.2215	1.1643	503.2
90	0.88566	119.1	0.2382	0.2217	1.1572	510.1
100	0.90728	121.3	0.2422	0.2222	1.1509	516.7
110	0.92868	123.5	0.2462	0.2229	1.1453	523.2
120	0.94967	125.8	0.2500	0.2237	1.1403	529.4
130	0.97040	128.0	0.2539	0.2247	1.1359	535.5
140	0.99098	130.3	0.2577	0.2258	1.1318	541.4
150	1.01133	132.5	0.2614	0.2271	1.1282	547.1
160	1.03135	134.8	0.2651	0.2284	1.1248	552.8
170	1.05130	137.1	0.2688	0.2298	1.1218	558.3
180	1.07101	139.4	0.2724	0.2312	1.1189	563.7
190	1.09075	141.7	0.2760	0.2327	1.1163	569.0
200	1.11012	144.0	0.2796	0.2343	1.1139	574.2
210	1.12943	146.4	0.2831	0.2359	1.1117	579.3
220	1.14903	148.8	0.2866	0.2375	1.1096	584.3
230	1.16795	151.1	0.2901	0.2391	1.1076	589.2
240	1.18723	153.5	0.2936	0.2408	1.1057	594.1
250	1.20627	156.0	0.2970	0.2425	1.1040	598.9
260	1.22519	158.4	0.3004	0.2442	1.1024	603.7
270	1.24409	160.8	0.3038	0.2459	1.1008	608.4
280	1.26295	163.3	0.3071	0.2476	1.0993	613.0
290	1.28172	165.8	0.3105	0.2493	1.0979	617.5
300	1.30039	168.3	0.3138	0.2510	1.0966	622.0
310	1.31891	170.8	0.3171	0.2527	1.0953	626.5
320	1.33761	173.4	0.3204	0.2544	1.0941	630.9
330	1.35630	175.9	0.3236	0.2561	1.0930	635.3
340	1.37457	178.5	0.3268	0.2578	1.0919	639.6
350	1.39334	181.1	0.3301	0.2595	1.0908	643.9
360	—	—	—	—	—	—

PRESSURE = 70.00 PSIA

TEMP °F	V	H	S	C_p	C_p/C_v	v_s
58.3 SAT LIQ	0.01288	30.8	0.0655	0.3321	1.5417	1830.2
58.3 SAT VAP	0.68236	111.3	0.2209	0.2288	1.2026	478.9
60	0.68601	111.7	0.2217	0.2285	1.2002	480.3
70	0.70691	114.0	0.2260	0.2272	1.1880	488.5
80	0.72717	116.2	0.2302	0.2264	1.1777	496.3
90	0.74694	118.5	0.2344	0.2260	1.1688	503.7
100	0.76628	120.7	0.2385	0.2259	1.1611	510.8
110	0.78524	123.0	0.2425	0.2262	1.1543	517.6
120	0.80386	125.3	0.2464	0.2267	1.1484	524.2
130	0.82217	127.5	0.2503	0.2274	1.1431	530.6
140	0.84034	129.8	0.2541	0.2283	1.1383	536.8
150	0.85807	132.1	0.2579	0.2293	1.1341	542.8
160	0.87573	134.4	0.2617	0.2304	1.1302	548.7
170	0.89326	136.7	0.2654	0.2316	1.1266	554.4
180	0.91058	139.0	0.2690	0.2329	1.1234	560.0
190	0.92773	141.4	0.2726	0.2343	1.1205	565.5
200	0.94473	143.7	0.2762	0.2357	1.1177	570.9
210	0.96172	146.1	0.2798	0.2372	1.1152	576.2
220	0.97847	148.5	0.2833	0.2387	1.1128	581.4
230	0.99512	150.9	0.2868	0.2403	1.1106	586.4
240	1.01184	153.3	0.2903	0.2419	1.1086	591.5
250	1.02828	155.7	0.2937	0.2435	1.1067	596.4
260	1.04482	158.1	0.2971	0.2451	1.1049	601.2
270	1.06135	160.6	0.3005	0.2467	1.1032	606.0
280	1.07747	163.1	0.3039	0.2484	1.1015	610.8
290	1.09385	165.6	0.3072	0.2501	1.1000	615.4
300	1.11012	168.1	0.3106	0.2517	1.0986	620.0
310	1.12625	170.6	0.3139	0.2534	1.0972	624.6
320	1.14233	173.1	0.3172	0.2551	1.0959	629.1
330	1.15835	175.7	0.3204	0.2567	1.0947	633.5
340	1.17426	178.3	0.3237	0.2584	1.0935	637.9
350	1.19033	180.9	0.3269	0.2601	1.0923	642.3
360	1.20642	183.5	0.3301	0.2618	1.0913	646.6
370	—	—	—	—	—	—

TABLE A.16 (cont'd.)

PRESSURE = 80.00 PSIA

TEMP °F	V	H	S	C_p	C_p/C_v	v_s
SAT LIQ 65.9	0.01304	33.3	0.0703	0.3359	1.5492	1766.4
SAT VAP 65.9	0.59787	112.3	0.2205	0.2342	1.2120	477.0
70	0.60580	113.2	0.2224	0.2333	1.2058	480.7
80	0.62445	115.6	0.2267	0.2317	1.1928	489.1
90	0.64255	117.9	0.2309	0.2306	1.1818	497.1
100	0.66020	120.2	0.2351	0.2300	1.1723	504.7
110	0.67741	122.5	0.2392	0.2298	1.1642	511.9
120	0.69430	124.8	0.2432	0.2299	1.1571	518.9
130	0.71083	127.1	0.2471	0.2303	1.1508	525.6
140	0.72711	129.4	0.2510	0.2308	1.1453	532.1
150	0.74322	131.7	0.2548	0.2316	1.1403	538.4
160	0.75896	134.0	0.2586	0.2325	1.1358	544.6
170	0.77465	136.3	0.2623	0.2335	1.1318	550.5
180	0.79020	138.7	0.2660	0.2347	1.1281	556.4
190	0.80548	141.0	0.2696	0.2359	1.1248	562.0
200	0.82068	143.4	0.2733	0.2372	1.1217	567.6
210	0.83577	145.8	0.2768	0.2386	1.1188	573.0
220	0.85063	148.2	0.2804	0.2400	1.1162	578.4
230	0.86558	150.6	0.2839	0.2414	1.1138	583.6
240	0.88020	153.0	0.2874	0.2429	1.1115	588.8
250	0.89501	155.4	0.2908	0.2445	1.1094	593.8
260	0.90950	157.9	0.2943	0.2460	1.1074	598.8
270	0.92404	160.4	0.2977	0.2476	1.1056	603.7
280	0.93853	162.8	0.3011	0.2492	1.1038	608.5
290	0.95302	165.3	0.3044	0.2508	1.1021	613.3
300	0.96721	167.9	0.3078	0.2525	1.1006	618.0
310	0.98155	170.4	0.3111	0.2541	1.0991	622.6
320	0.99582	172.9	0.3144	0.2557	1.0977	627.2
330	1.01010	175.5	0.3176	0.2574	1.0964	631.8
340	1.02417	178.1	0.3209	0.2590	1.0951	636.2
350	1.03821	180.7	0.3241	0.2606	1.0939	640.7
360	1.05230	183.3	0.3273	0.2623	1.0927	645.0
370	1.06644	185.9	0.3305	0.2639	1.0916	649.4

PRESSURE = 90.00 PSIA

TEMP °F	V	H	S	C_p	C_p/C_v	v_s
SAT LIQ 72.8	0.01319	35.7	0.0747	0.3396	1.5568	1708.1
SAT VAP 72.8	0.53155	113.1	0.2202	0.2395	1.2217	475.0
80	0.54416	114.9	0.2234	0.2376	1.2100	481.6
90	0.56107	117.2	0.2278	0.2357	1.1964	490.2
100	0.57743	119.6	0.2320	0.2344	1.1849	498.3
110	0.59333	121.9	0.2361	0.2337	1.1751	506.0
120	0.60887	124.3	0.2402	0.2333	1.1666	513.4
130	0.62406	126.6	0.2442	0.2333	1.1592	520.5
140	0.63906	128.9	0.2481	0.2336	1.1527	527.4
150	0.65368	131.3	0.2520	0.2341	1.1470	534.0
160	0.66814	133.6	0.2558	0.2347	1.1418	540.4
170	0.68236	136.0	0.2596	0.2355	1.1372	546.6
180	0.69643	138.3	0.2633	0.2365	1.1331	552.6
190	0.71023	140.7	0.2670	0.2376	1.1293	558.5
200	0.72401	143.1	0.2706	0.2387	1.1258	564.2
210	0.73768	145.5	0.2742	0.2400	1.1227	569.9
220	0.75126	147.9	0.2778	0.2413	1.1197	575.4
230	0.76470	150.3	0.2813	0.2426	1.1170	580.8
240	0.77791	152.7	0.2848	0.2441	1.1145	586.0
250	0.79108	155.2	0.2883	0.2455	1.1122	591.2
260	0.80431	157.6	0.2917	0.2470	1.1101	596.3
270	0.81746	160.1	0.2951	0.2485	1.1080	601.4
280	0.83036	162.6	0.2985	0.2501	1.1061	606.3
290	0.84331	165.1	0.3019	0.2516	1.1043	611.2
300	0.85616	167.6	0.3052	0.2532	1.1026	616.0
310	0.86904	170.2	0.3086	0.2548	1.1010	620.7
320	0.88191	172.7	0.3119	0.2564	1.0995	625.4
330	0.89461	175.3	0.3151	0.2580	1.0981	630.0
340	0.90728	177.9	0.3184	0.2596	1.0967	634.5
350	0.91988	180.5	0.3216	0.2612	1.0955	639.0
360	0.93257	183.1	0.3249	0.2628	1.0942	643.5
370	0.94518	185.7	0.3280	0.2644	1.0930	647.9
380	0.95767	188.4	0.3312	0.2660	1.0919	652.3

TABLE A.16 (cont'd.)

PRESSURE = 100.00 PSIA

TEMP °F	V	H	S	C_p	C_p/C_v	v_s
SAT LIQ 79.1	0.01333	37.8	0.0787	0.3433	1.5646	1654.2
SAT VAP 79.1	0.47803	113.9	0.2199	0.2446	1.2317	472.8
80	0.47952	114.1	0.2203	0.2442	1.2300	473.6
90	0.49552	116.6	0.2248	0.2413	1.2129	482.9
100	0.51093	119.0	0.2291	0.2393	1.1989	491.7
110	0.52587	121.4	0.2333	0.2379	1.1871	499.9
120	0.54037	123.7	0.2375	0.2370	1.1770	507.8
130	0.55451	126.1	0.2415	0.2366	1.1683	515.3
140	0.56838	128.5	0.2455	0.2365	1.1608	522.5
150	0.58194	130.8	0.2494	0.2366	1.1541	529.4
160	0.59527	133.2	0.2533	0.2370	1.1482	536.1
170	0.60835	135.6	0.2570	0.2376	1.1430	542.5
180	0.62135	137.9	0.2608	0.2384	1.1383	548.8
190	0.63416	140.3	0.2645	0.2393	1.1340	554.9
200	0.64675	142.7	0.2682	0.2403	1.1301	560.9
210	0.65924	145.1	0.2718	0.2414	1.1266	566.7
220	0.67155	147.6	0.2754	0.2426	1.1234	572.3
230	0.68385	150.0	0.2789	0.2439	1.1204	577.9
240	0.69604	152.4	0.2825	0.2452	1.1177	583.3
250	0.70806	154.9	0.2859	0.2466	1.1151	588.6
260	0.72015	157.4	0.2894	0.2480	1.1128	593.9
270	0.73196	159.9	0.2928	0.2495	1.1106	599.0
280	0.74388	162.4	0.2962	0.2509	1.1085	604.1
290	0.75569	164.9	0.2996	0.2524	1.1066	609.0
300	0.76740	167.4	0.3030	0.2540	1.1047	613.9
310	0.77906	170.0	0.3063	0.2555	1.1030	618.7
320	0.79064	172.5	0.3096	0.2571	1.1014	623.5
330	0.80225	175.1	0.3129	0.2586	1.0999	628.2
340	0.81387	177.7	0.3162	0.2602	1.0984	632.8
350	0.82535	180.3	0.3194	0.2618	1.0970	637.4
360	0.83675	182.9	0.3226	0.2634	1.0957	641.9
370	0.84818	185.6	0.3258	0.2649	1.0945	646.4
380	0.85955	188.2	0.3290	0.2665	1.0933	650.8
390	—	—	—	—	—	—

PRESSURE = 120.00 PSIA

TEMP °F	V	H	S	C_p	C_p/C_v	v_s
SAT LIQ 90.5	0.01361	41.8	0.0858	0.3504	1.5808	1557.1
SAT VAP 90.5	0.39689	115.3	0.2194	0.2546	1.2528	467.9
100	0.41044	117.7	0.2238	0.2506	1.2326	477.6
110	0.42402	120.2	0.2282	0.2475	1.2153	487.1
120	0.43710	122.6	0.2324	0.2453	1.2010	496.0
130	0.44976	125.1	0.2366	0.2438	1.1890	504.4
140	0.46206	127.5	0.2407	0.2428	1.1788	512.4
150	0.47405	129.9	0.2447	0.2423	1.1700	520.0
160	0.48577	132.3	0.2487	0.2421	1.1623	527.3
170	0.49724	134.8	0.2525	0.2422	1.1555	534.3
180	0.50860	137.2	0.2564	0.2425	1.1495	541.1
190	0.51967	139.6	0.2601	0.2430	1.1442	547.6
200	0.53064	142.0	0.2639	0.2437	1.1394	554.0
210	0.54139	144.5	0.2675	0.2445	1.1351	560.1
220	0.55206	146.9	0.2712	0.2454	1.1311	566.2
230	0.56268	149.4	0.2747	0.2465	1.1275	572.0
240	0.57307	151.9	0.2783	0.2476	1.1243	577.8
250	0.58350	154.4	0.2818	0.2488	1.1212	583.4
260	0.59379	156.8	0.2853	0.2501	1.1184	588.9
270	0.60394	159.4	0.2888	0.2514	1.1158	594.2
280	0.61406	161.9	0.2922	0.2527	1.1134	599.5
290	0.62414	164.4	0.2956	0.2541	1.1112	604.7
300	0.63408	167.0	0.2990	0.2555	1.1091	609.8
310	0.64404	169.5	0.3023	0.2570	1.1071	614.8
320	0.65389	172.1	0.3057	0.2584	1.1052	619.8
330	0.66375	174.7	0.3090	0.2599	1.1035	624.6
340	0.67354	177.3	0.3122	0.2614	1.1019	629.4
350	0.68329	179.9	0.3155	0.2629	1.1003	634.1
360	0.69300	182.6	0.3187	0.2645	1.0988	638.8
370	0.70264	185.2	0.3220	0.2660	1.0974	643.4
380	0.71225	187.9	0.3252	0.2675	1.0961	648.0
390	0.72192	190.6	0.3283	0.2690	1.0948	652.5
400	0.73142	193.3	0.3315	0.2706	1.0936	656.9

TABLE A.16 (cont'd.)

PRESSURE = 140.00 PSIA

TEMP °F	V	H	S	C_p	C_p/C_v	v_s
100.5 SAT LIQ	0.01388	45.3	0.0921	0.3576	1.5980	1470.8
100.5 SAT VAP	0.33818	116.4	0.2190	0.2646	1.2757	462.7
110	0.35042	118.9	0.2234	0.2593	1.2512	473.2
120	0.36266	121.4	0.2279	0.2553	1.2307	483.3
130	0.37438	124.0	0.2322	0.2523	1.2140	492.8
140	0.38568	126.5	0.2364	0.2502	1.2001	501.7
150	0.39662	129.0	0.2406	0.2487	1.1884	510.1
160	0.40727	131.5	0.2446	0.2477	1.1784	518.1
170	0.41764	133.9	0.2486	0.2472	1.1697	525.7
180	0.42779	136.4	0.2524	0.2470	1.1622	533.0
190	0.43777	138.9	0.2563	0.2470	1.1555	540.1
200	0.44755	141.3	0.2601	0.2473	1.1496	546.9
210	0.45714	143.8	0.2638	0.2478	1.1443	553.5
220	0.46659	146.3	0.2675	0.2485	1.1395	559.9
230	0.47596	148.8	0.2711	0.2492	1.1352	566.1
240	0.48520	151.3	0.2747	0.2501	1.1313	572.1
250	0.49432	153.8	0.2782	0.2511	1.1277	578.0
260	0.50342	156.3	0.2818	0.2522	1.1244	583.8
270	0.51235	158.8	0.2853	0.2534	1.1214	589.4
280	0.52121	161.4	0.2887	0.2546	1.1186	594.9
290	0.53011	163.9	0.2921	0.2558	1.1160	600.3
300	0.53888	166.5	0.2955	0.2572	1.1136	605.6
310	0.54750	169.1	0.2989	0.2585	1.1113	610.9
320	0.55611	171.7	0.3022	0.2599	1.1092	616.0
330	0.56478	174.3	0.3056	0.2613	1.1073	621.0
340	0.57330	176.9	0.3089	0.2627	1.1054	626.0
350	0.58184	179.5	0.3121	0.2641	1.1036	630.9
360	0.59028	182.2	0.3154	0.2656	1.1020	635.7
370	0.59866	184.8	0.3186	0.2670	1.1004	640.4
380	0.60716	187.5	0.3218	0.2685	1.0989	645.1
390	0.61542	190.2	0.3250	0.2700	1.0975	649.7
400	0.62375	192.9	0.3282	0.2715	1.0962	654.3
410	0.63203	195.6	0.3313	0.2730	1.0949	658.8

PRESSURE = 160.00 PSIA

TEMP °F	V	H	S	C_p	C_p/C_v	v_s
109.5 SAT LIQ	0.01414	48.5	0.0977	0.3648	1.6163	1392.4
109.5 SAT VAP	0.29362	117.3	0.2186	0.2748	1.3006	457.2
110	0.29426	117.5	0.2188	0.2744	1.2988	457.9
120	0.30608	120.2	0.2235	0.2675	1.2685	469.6
130	0.31726	122.8	0.2281	0.2624	1.2448	480.4
140	0.32792	125.4	0.2324	0.2588	1.2258	490.4
150	0.33817	128.0	0.2367	0.2561	1.2101	499.7
160	0.34806	130.5	0.2409	0.2541	1.1970	508.5
170	0.35767	133.1	0.2449	0.2528	1.1859	516.8
180	0.36700	135.6	0.2489	0.2519	1.1764	524.7
190	0.37611	138.1	0.2528	0.2515	1.1681	532.3
200	0.38504	140.6	0.2566	0.2513	1.1608	539.6
210	0.39378	143.1	0.2604	0.2514	1.1544	546.7
220	0.40238	145.7	0.2641	0.2517	1.1486	553.4
230	0.41085	148.2	0.2678	0.2522	1.1435	560.0
240	0.41916	150.7	0.2715	0.2528	1.1388	566.4
250	0.42746	153.2	0.2751	0.2536	1.1346	572.6
260	0.43554	155.8	0.2786	0.2545	1.1308	578.6
270	0.44364	158.3	0.2821	0.2555	1.1273	584.5
280	0.45165	160.9	0.2856	0.2565	1.1241	590.3
290	0.45954	163.5	0.2891	0.2577	1.1211	595.9
300	0.46729	166.0	0.2925	0.2588	1.1183	601.4
310	0.47508	168.6	0.2959	0.2601	1.1158	606.9
320	0.48281	171.2	0.2992	0.2613	1.1134	612.2
330	0.49053	173.9	0.3026	0.2627	1.1112	617.4
340	0.49816	176.5	0.3059	0.2640	1.1091	622.5
350	0.50566	179.1	0.3092	0.2654	1.1071	627.6
360	0.51324	181.8	0.3124	0.2667	1.1053	632.5
370	0.52067	184.5	0.3157	0.2681	1.1035	637.4
380	0.52818	187.2	0.3189	0.2696	1.1019	642.2
390	0.53565	189.9	0.3221	0.2710	1.1003	647.0
400	0.54301	192.6	0.3253	0.2724	1.0988	651.7
410	0.55030	195.3	0.3284	0.2738	1.0974	656.3
420	—	—	—	—	—	—

TABLE A.16 (cont'd.)

PRESSURE = 180.00 PSIA

TEMP °F		V	H	S	C_p	C_p/C_v	v_s
117.7	SAT LIQ	0.01439	51.5	0.1029	0.3723	1.6360	1320.2
117.7	SAT VAP	0.25856	118.1	0.2182	0.2854	1.3280	451.5
120		0.26125	118.8	0.2193	0.2831	1.3184	454.6
130		0.27221	121.6	0.2241	0.2749	1.2838	467.0
140		0.28251	124.3	0.2287	0.2690	1.2572	478.3
150		0.29231	126.9	0.2331	0.2647	1.2361	488.7
160		0.30170	129.6	0.2374	0.2615	1.2189	498.4
170		0.31073	132.2	0.2415	0.2591	1.2046	507.5
180		0.31948	134.8	0.2456	0.2575	1.1925	516.1
190		0.32803	137.3	0.2496	0.2563	1.1822	524.3
200		0.33629	139.9	0.2535	0.2556	1.1732	532.2
210		0.34438	142.4	0.2573	0.2553	1.1654	539.7
220		0.35227	145.0	0.2611	0.2552	1.1586	546.9
230		0.36006	147.5	0.2648	0.2554	1.1524	553.8
240		0.36776	150.1	0.2685	0.2557	1.1470	560.6
250		0.37536	152.7	1.2722	0.2562	1.1420	567.1
260		0.38278	155.2	0.2757	0.2569	1.1376	573.4
270		0.39014	157.8	0.2793	0.2577	1.1335	579.6
280		0.39739	160.4	0.2828	0.2586	1.1298	585.6
290		0.40461	163.0	0.2863	0.2596	1.1264	591.5
300		0.41168	165.6	0.2897	0.2606	1.1233	597.2
310		0.41873	168.2	0.2931	0.2617	1.1204	602.8
320		0.42577	170.8	0.2965	0.2629	1.1177	608.3
330		0.43269	173.4	0.2999	0.2641	1.1152	613.7
340		0.43964	176.1	0.3032	0.2653	1.1129	619.0
350		0.44645	178.7	0.3065	0.2666	1.1107	624.2
360		0.45325	181.4	0.3098	0.2679	1.1086	629.3
370		0.46007	184.1	0.3131	0.2693	1.1067	634.4
380		0.46677	186.8	0.3163	0.2706	1.1049	639.3
390		0.47344	189.5	0.3195	0.2720	1.1032	644.2
400		0.48015	192.2	0.3227	0.2734	1.1016	649.0
410		0.48681	195.0	0.3259	0.2748	1.1000	653.8
420		0.49336	197.7	0.3290	0.2762	1.0986	658.5
430		—	—	—	—	—	—

PRESSURE = 200.00 PSIA

TEMP °F		V	H	S	C_p	C_p/C_v	v_s
125.2	SAT LIQ	0.01465	54.3	0.1076	0.3801	1.6574	1253.0
125.2	SAT VAP	0.23026	118.8	0.2178	0.2966	1.3582	445.6
130		0.23544	120.2	0.2202	0.2908	1.3352	452.3
140		0.24565	123.0	0.2250	0.2816	1.2969	465.3
150		0.25521	125.8	0.2296	0.2749	1.2678	477.0
160		0.26428	128.5	0.2340	0.2700	1.2449	487.8
170		0.27294	131.2	0.2383	0.2664	1.2263	497.8
180		0.28129	133.9	0.2425	0.2637	1.2110	507.2
190		0.28930	136.5	0.2466	0.2618	1.1981	516.1
200		0.29712	139.1	0.2505	0.2604	1.1872	524.5
210		0.30477	141.7	0.2545	0.2595	1.1777	532.5
220		0.31217	144.3	0.2583	0.2590	1.1694	540.2
230		0.31946	146.9	0.2621	0.2587	1.1622	547.5
240		0.32658	149.5	0.2658	0.2588	1.1557	554.6
250		0.33360	152.1	0.2695	0.2590	1.1500	561.5
260		0.34046	154.7	0.2731	0.2594	1.1448	568.1
270		0.34727	157.3	0.2767	0.2600	1.1401	574.6
280		0.35398	159.9	0.2802	0.2607	1.1359	580.9
290		0.36063	162.5	0.2837	0.2615	1.1320	587.0
300		0.36716	165.1	0.2872	0.2624	1.1285	593.0
310		0.37365	167.7	0.2906	0.2634	1.1252	598.8
320		0.38008	170.4	0.2941	0.2645	1.1222	604.5
330		0.38646	173.0	0.2974	0.2656	1.1194	610.1
340		0.39282	175.7	0.3008	0.2667	1.1168	615.5
350		0.39909	178.3	0.3041	0.2679	1.1144	620.9
360		0.40532	181.0	0.3074	0.2692	1.1121	626.2
370		0.41151	183.7	0.3107	0.2704	1.1100	631.3
380		0.41762	186.4	0.3139	0.2717	1.1080	636.4
390		0.42375	189.2	0.3171	0.2730	1.1061	641.5
400		0.42989	191.9	0.3203	0.2743	1.1044	646.4
410		0.43590	194.7	0.3235	0.2757	1.1027	651.3
420		0.44191	197.4	0.3267	0.2770	1.1011	656.1
430		0.44791	200.2	0.3298	0.2784	1.0996	660.8
440		—	—	—	—	—	—

TABLE A.16 (cont'd.)

PRESSURE = 240.00 PSIA

TEMP °F		V	H	S	C_p	C_p/C_v	v_s
138.7	SAT LIQ	0.01517	59.5	0.1162	0.3972	1.7064	1130.0
138.7	SAT VAP	0.18715	119.8	0.2169	0.3214	1.4297	433.3
140		0.18846	120.2	0.2176	0.3189	1.4201	435.4
150		0.19819	123.3	0.2227	0.3032	1.3589	450.9
160		0.20712	126.3	0.2276	0.2924	1.3154	464.6
170		0.21545	129.2	0.2322	0.2846	1.2829	476.9
180		0.22330	132.0	0.2366	0.2789	1.2575	488.1
190		0.23083	134.8	0.2409	0.2747	1.2371	498.6
200		0.23799	137.5	0.2451	0.2716	1.2204	508.3
210		0.24495	140.2	0.2492	0.2692	1.2064	517.5
220		0.25165	142.9	0.2531	0.2675	1.1945	526.2
230		0.25818	145.6	0.2570	0.2664	1.1843	534.5
240		0.26457	148.2	0.2609	0.2656	1.1754	542.5
250		0.27079	150.9	0.2646	0.2652	1.1677	550.1
260		0.27685	153.5	0.2683	0.2650	1.1608	557.4
270		0.28289	156.2	0.2720	0.2651	1.1546	564.4
280		0.28878	158.8	0.2756	0.2653	1.1491	571.3
290		0.29459	161.5	0.2792	0.2658	1.1441	577.9
300		0.30028	164.1	0.2827	0.2663	1.1396	584.3
310		0.30593	166.8	0.2862	0.2670	1.1355	590.6
320		0.31156	169.5	0.2896	0.2678	1.1318	596.7
330		0.31709	172.2	0.2931	0.2687	1.1283	602.7
340		0.32253	174.8	0.2965	0.2696	1.1251	608.5
350		0.32794	177.5	0.2998	0.2706	1.1222	614.2
360		0.33332	180.3	0.3031	0.2717	1.1194	619.8
370		0.33866	183.0	0.3064	0.2728	1.1169	625.3
380		0.34395	185.7	0.3097	0.2740	1.1145	630.6
390		0.34921	188.5	0.3130	0.2751	1.1122	635.9
400		0.35443	191.2	0.3162	0.2764	1.1101	641.1
410		0.35962	194.0	0.3194	0.2776	1.1081	646.2
420		0.36478	196.8	0.3226	0.2788	1.1063	651.2
430		0.36990	199.6	0.3257	0.2801	1.1045	656.2
440		0.37501	202.4	0.3289	0.2814	1.1028	661.0
450		—	—	—	—	—	—

PRESSURE = 280.00 PSIA

TEMP °F		V	H	S	C_p	C_p/C_v	v_s
150.6	SAT LIQ	0.01570	64.3	0.1238	0.4172	1.7668	1018.2
150.6	SAT VAP	0.15571	120.5	0.2160	0.3512	1.5212	420.5
160		0.16463	123.7	0.2211	0.3275	1.4317	437.5
170		0.17318	126.8	0.2262	0.3111	1.3684	453.2
180		0.18101	129.9	0.2310	0.2998	1.3235	467.1
190		0.18832	132.9	0.2356	0.2916	1.2899	479.6
200		0.19519	135.7	0.2400	0.2857	1.2637	491.1
210		0.20178	138.6	0.2443	0.2812	1.2427	501.7
220		0.20809	141.4	0.2484	0.2778	1.2254	511.7
230		0.21413	144.1	0.2525	0.2753	1.2110	521.0
240		0.22001	146.9	0.2564	0.2735	1.1988	529.8
250		0.22573	149.6	0.2603	0.2722	1.1882	538.3
260		0.23127	152.3	0.2641	0.2713	1.1791	546.3
270		0.23672	155.0	0.2678	0.2707	1.1711	554.0
280		0.24208	157.7	0.2715	0.2704	1.1639	561.5
290		0.24728	160.4	0.2751	0.2704	1.1576	568.6
300		0.25246	163.1	0.2787	0.2706	1.1519	575.6
310		0.25753	165.9	0.2823	0.2709	1.1468	582.3
320		0.26250	168.6	0.2858	0.2714	1.1422	588.9
330		0.26744	171.3	0.2892	0.2720	1.1379	595.2
340		0.27230	174.0	0.2927	0.2727	1.1340	601.4
350		0.27712	176.7	0.2960	0.2735	1.1305	607.5
360		0.28188	179.5	0.2994	0.2744	1.1272	613.4
370		0.28661	182.2	0.3027	0.2753	1.1241	619.2
380		0.29128	185.0	0.3060	0.2763	1.1213	624.8
390		0.29592	187.7	0.3093	0.2774	1.1186	630.4
400		0.30053	190.5	0.3126	0.2785	1.1162	635.8
410		0.30509	193.3	0.3158	0.2796	1.1138	641.2
420		0.30964	196.1	0.3190	0.2807	1.1117	646.4
430		0.31414	198.9	0.3222	0.2819	1.1096	651.6
440		0.31862	201.8	0.3253	0.2831	1.1077	656.7
450		0.32308	204.6	0.3285	0.2843	1.1059	661.7
460		0.32751	207.4	0.3316	0.2855	1.1042	666.6

TABLE A.16 (cont'd.)

PRESSURE = 320.00 PSIA

TEMP °F	V	H	S	C_p	C_p/C_v	v_s
161.3 SAT LIQ	0.01627	68.7	0.1308	0.4415	1.8437	914.5
161.3 SAT VAP	0.13160	120.9	0.2148	0.3889	1.6421	407.2
170	0.13984	124.1	0.2200	0.3548	1.5153	425.6
180	0.14815	127.5	0.2254	0.3308	1.4254	443.4
190	0.15562	130.7	0.2304	0.3152	1.3655	458.8
200	0.16251	133.8	0.2351	0.3043	1.3225	472.5
210	0.16893	136.8	0.2396	0.2964	1.2899	484.9
220	0.17500	139.7	0.2439	0.2905	1.2644	496.3
230	0.18080	142.6	0.2482	0.2861	1.2438	506.8
240	0.18638	145.5	0.2523	0.2828	1.2267	516.8
250	0.19175	148.3	0.2563	0.2803	1.2125	526.1
260	0.19694	151.1	0.2602	0.2784	1.2003	535.0
270	0.20198	153.8	0.2640	0.2771	1.1898	543.4
280	0.20692	156.6	0.2678	0.2762	1.1807	551.5
290	0.21174	159.4	0.2715	0.2756	1.1727	559.3
300	0.21648	162.1	0.2751	0.2753	1.1655	566.8
310	0.22111	164.9	0.2787	0.2752	1.1592	574.0
320	0.22567	167.6	0.2823	0.2753	1.1535	581.0
330	0.23016	170.4	0.2858	0.2756	1.1483	587.7
340	0.23459	173.1	0.2892	0.2760	1.1436	594.3
350	0.23895	175.9	0.2927	0.2766	1.1393	600.7
360	0.24327	178.7	0.2961	0.2772	1.1354	607.0
370	0.24754	181.5	0.2994	0.2780	1.1318	613.1
380	0.25176	184.2	0.3028	0.2788	1.1285	619.0
390	0.25594	187.0	0.3061	0.2797	1.1254	624.8
400	0.26008	189.8	0.3094	0.2806	1.1225	630.5
410	0.26419	192.6	0.3126	0.2816	1.1198	636.1
420	0.26827	195.5	0.3158	0.2827	1.1173	641.6
430	0.27232	198.3	0.3190	0.2837	1.1150	647.0
440	0.27633	201.1	0.3222	0.2848	1.1127	652.3
450	0.28032	204.0	0.3254	0.2860	1.1107	657.5
460	0.28428	206.9	0.3285	0.2871	1.1087	662.6
470	0.28823	209.7	0.3316	0.2883	1.1069	667.6

PRESSURE = 360.00 PSIA

TEMP °F	V	H	S	C_p	C_p/C_v	v_s
171 SAT LIQ	0.01689	72.9	0.1373	0.4727	1.9458	816.6
171 SAT VAP	0.11237	120.9	0.2135	0.4391	1.8089	393.7
180	0.12101	124.6	0.2193	0.3841	1.6069	415.6
190	0.12915	128.2	0.2249	0.3508	1.4840	435.3
200	0.13633	131.6	0.2301	0.3303	1.4071	452.1
210	0.14285	134.9	0.2350	0.3164	1.3540	466.8
220	0.14888	138.0	0.2396	0.3066	1.3149	480.0
230	0.15458	141.0	0.2440	0.2993	1.2847	492.1
240	0.15998	144.0	0.2483	0.2939	1.2608	503.2
250	0.16513	146.9	0.2524	0.2898	1.2413	513.6
260	0.17007	149.8	0.2565	0.2867	1.2251	523.4
270	0.17486	152.6	0.2604	0.2843	1.2114	532.6
280	0.17950	155.5	0.2643	0.2825	1.1997	541.4
290	0.18402	158.3	0.2680	0.2813	1.1895	549.8
300	0.18843	161.1	0.2718	0.2804	1.1806	557.9
310	0.19275	163.9	0.2754	0.2798	1.1728	565.6
320	0.19698	166.7	0.2790	0.2795	1.1658	573.1
330	0.20113	169.5	0.2826	0.2794	1.1596	580.3
340	0.20522	172.3	0.2861	0.2796	1.1539	587.2
350	0.20925	175.1	0.2896	0.2798	1.1488	594.0
360	0.21322	177.9	0.2930	0.2803	1.1442	600.6
370	0.21713	180.7	0.2964	0.2808	1.1399	607.0
380	0.22100	183.5	0.2998	0.2814	1.1360	613.2
390	0.22483	186.3	0.3031	0.2821	1.1324	619.3
400	0.22862	189.1	0.3064	0.2829	1.1291	625.3
410	0.23237	192.0	0.3097	0.2838	1.1260	631.1
420	0.23609	194.8	0.3130	0.2847	1.1231	636.9
430	0.23978	197.7	0.3162	0.2856	1.1205	642.5
440	0.24343	200.5	0.3194	0.2866	1.1180	648.0
450	0.24707	203.4	0.3226	0.2877	1.1156	653.4
460	0.25066	206.3	0.3257	0.2887	1.1134	658.7
470	0.25425	209.2	0.3288	0.2898	1.1113	663.9
480	0.25781	212.1	0.3319	0.2909	1.1093	669.0

TABLE A.16 (cont'd.)

PRESSURE = 500.00 PSIA

TEMP °F	V	H	S	C_p	C_p/C_v	v_s
199.4 SAT LIQ	0.02000	86.9	0.1584	0.7816	3.0254	496.6
199.4 SAT VAP	0.06538	118.1	0.2057	0.9822	3.6942	345.3
200	0.06646	118.7	0.2065	0.9106	3.4381	348.6
210	0.07860	125.3	0.2164	0.5274	2.0697	387.4
220	0.08669	130.0	0.2234	0.4325	1.7319	413.4
230	0.09326	134.1	0.2294	0.3868	1.5687	434.1
240	0.09896	137.8	0.2348	0.3596	1.4704	451.7
250	0.10410	141.3	0.2397	0.3416	1.4041	467.2
260	0.10883	144.6	0.2444	0.3289	1.3561	481.1
270	0.11326	147.9	0.2489	0.3195	1.3196	493.8
280	0.11744	151.0	0.2532	0.3125	1.2909	505.5
290	0.12142	154.1	0.2573	0.3072	1.2677	516.4
300	0.12523	157.2	0.2614	0.3030	1.2486	526.7
310	0.12891	160.2	0.2653	0.2998	1.2325	536.4
320	0.13248	163.2	0.2692	0.2974	1.2188	545.6
330	0.13594	166.1	0.2730	0.2955	1.2070	554.4
340	0.13930	169.1	0.2767	0.2940	1.1966	562.9
350	0.14260	172.0	0.2803	0.2930	1.1876	571.0
360	0.14582	174.9	0.2839	0.2923	1.1795	578.8
370	0.14898	177.9	0.2874	0.2918	1.1723	586.3
380	0.15208	180.8	0.2909	0.2916	1.1659	593.6
390	0.15513	183.7	0.2944	0.2916	1.1600	600.7
400	0.15813	186.6	0.2978	0.2917	1.1547	607.6
410	0.16110	189.5	0.3012	0.2920	1.1499	614.3
420	0.16402	192.5	0.3045	0.2923	1.1454	620.8
430	0.16691	195.4	0.3078	0.2928	1.1413	627.1
440	0.16976	198.3	0.3111	0.2934	1.1376	633.4
450	0.17259	201.2	0.3143	0.2940	1.1341	639.4
460	0.17538	204.2	0.3176	0.2947	1.1308	645.4
470	0.17815	207.1	0.3208	0.2955	1.1278	651.2
480	0.18090	210.1	0.3239	0.2963	1.1249	656.9
490	0.18362	213.1	0.3271	0.2971	1.1223	662.5
500	0.18632	216.0	0.3302	0.2981	1.1198	668.0

PRESSURE = 400.00 PSIA

TEMP °F	V	H	S	C_p	C_p/C_v	v_s
179.9 SAT LIQ	0.01759	76.9	0.1435	0.5155	2.0897	722.7
179.9 SAT VAP	0.09649	120.7	0.2119	0.5106	2.0521	380.0
180	0.09655	120.7	0.2119	0.5097	2.0487	380.2
190	0.10650	125.2	0.2190	0.4139	1.6998	408.0
200	0.11446	129.1	0.2249	0.3701	1.5401	429.3
210	0.12136	132.7	0.2303	0.3445	1.4457	447.2
220	0.12755	136.0	0.2352	0.3278	1.3828	462.7
230	0.13325	139.3	0.2399	0.3160	1.3375	476.6
240	0.13858	142.4	0.2444	0.3075	1.3031	489.2
250	0.14362	145.4	0.2487	0.3011	1.2762	500.8
260	0.14842	148.4	0.2529	0.2963	1.2544	511.6
270	0.15302	151.3	0.2570	0.2926	1.2364	521.7
280	0.15746	154.3	0.2609	0.2898	1.2214	531.2
290	0.16176	157.1	0.2648	0.2876	1.2085	540.3
300	0.16594	160.0	0.2686	0.2860	1.1974	548.9
310	0.17001	162.9	0.2724	0.2849	1.1878	557.2
320	0.17399	165.7	0.2760	0.2841	1.1793	565.1
330	0.17788	168.5	0.2796	0.2836	1.1717	572.8
340	0.18171	171.4	0.2832	0.2833	1.1650	580.2
350	0.18546	174.2	0.2867	0.2833	1.1590	587.3
360	0.18916	177.0	0.2902	0.2834	1.1535	594.3
370	0.19280	179.9	0.2937	0.2837	1.1485	601.0
380	0.19639	182.7	0.2971	0.2842	1.1440	607.5
390	0.19994	185.6	0.3004	0.2847	1.1399	613.9
400	0.20345	188.4	0.3038	0.2853	1.1360	620.1
410	0.20692	191.3	0.3071	0.2860	1.1325	626.2
420	0.21034	194.1	0.3103	0.2868	1.1292	632.2
430	0.21375	197.0	0.3136	0.2876	1.1262	638.0
440	0.21712	199.9	0.3168	0.2885	1.1233	643.7
450	0.22046	202.8	0.3200	0.2894	1.1207	649.3
460	0.22377	205.7	0.3232	0.2904	1.1182	654.8
470	0.22706	208.6	0.3263	0.2914	1.1159	660.2
480	0.23033	211.5	0.3294	0.2924	1.1137	665.5
490	—	—	—	—	—	—

TABLE A.15 (SI)

HFC-134a Saturation Properties—Temperature Table

TEMP. °C	PRESSURE kPa (abs)	VOLUME m³/kg LIQUID v_f	VOLUME m³/kg VAPOR v_g	DENSITY kg/m³ LIQUID $1/v_f$	DENSITY kg/m³ VAPOR $1/v_g$	ENTHALPY kJ/kg LIQUID h_f	ENTHALPY kJ/kg LATENT h_{fg}	ENTHALPY kJ/kg VAPOR h_g	ENTROPY kJ/(kg)(°K) LIQUID s_f	ENTROPY kJ/(kg)(°K) VAPOR s_g	TEMP. °C
−40	51.14	0.0007	0.3614	1414.6	2.767	148.4	225.9	374.3	0.7967	1.7655	−40
−35	66.07	0.0007	0.2843	1400.2	3.518	154.6	222.8	377.4	0.8231	1.7586	−35
−30	84.29	0.0007	0.2260	1385.7	4.424	160.9	219.6	380.6	0.8492	1.7525	−30
−25	106.32	0.0007	0.1817	1371.0	5.504	167.3	216.4	383.7	0.8750	1.747	−25
−20	132.67	0.0007	0.1474	1356.0	6.784	173.7	213.1	386.8	0.9005	1.7422	−15
−15	163.90	0.0007	0.1207	1340.8	8.288	180.2	209.7	389.8	0.9257	1.7379	−15
−10	200.60	0.0008	0.0996	1325.3	10.044	186.7	206.2	392.9	0.9507	1.7341	−10
−5	243.39	0.0008	0.0828	1309.4	12.082	193.3	202.5	395.9	0.9755	1.7308	−5
0	292.93	0.0008	0.0693	1293.3	14.435	200.0	198.8	398.8	1.0000	1.7278	0
5	349.87	0.0008	0.0583	1276.7	17.140	206.8	194.9	401.7	1.0244	1.7252	5
10	414.92	0.0008	0.0494	1259.8	20.236	213.6	190.9	404.5	1.0485	1.7229	10
15	488.78	0.0008	0.0421	1242.3	23.770	220.5	186.8	407.3	1.0726	1.7208	15
20	572.25	0.0008	0.0360	1224.4	27.791	227.5	182.5	410.0	1.0964	1.7189	20
25	666.06	0.0008	0.0309	1205.9	32.359	234.6	178.0	412.6	1.1202	1.7171	25
30	771.02	0.0008	0.0266	1186.7	37.540	241.8	173.3	415.1	1.1439	1.7155	30
35	887.91	0.0009	0.0230	1166.8	43.413	249.2	168.3	417.5	1.1676	1.7138	35
40	1017.61	0.0009	0.0200	1146.1	50.072	256.6	163.2	419.8	1.1912	1.7122	40
45	1161.01	0.0009	0.0174	1124.5	57.630	264.2	157.7	421.9	1.2148	1.7105	45
50	1319.00	0.0009	0.0151	1101.8	66.225	271.9	151.9	423.8	1.2384	1.7086	50
55	1492.59	0.0009	0.0132	1077.9	76.035	279.8	145.8	425.6	1.2622	1.7064	55
60	1682.76	0.0010	0.0115	1052.5	87.287	287.9	139.2	427.1	1.2861	1.7039	60
65	1890.54	0.0010	0.0100	1025.3	100.283	296.2	132.1	428.3	1.3102	1.7009	65
70	2117.34	0.0010	0.0087	995.9	115.442	304.8	124.4	429.1	1.3347	1.6971	70
75	2364.31	0.0010	0.0075	963.7	133.373	313.7	115.8	429.5	1.3597	1.6924	75
80	2632.97	0.0011	0.0065	927.8	155.010	322.9	106.3	429.2	1.3854	1.6863	80
85	2925.11	0.0011	0.0055	886.7	181.929	332.8	95.3	428.1	1.4121	1.6782	85
90	3242.87	0.0012	0.0046	837.3	217.162	343.4	82.1	425.5	1.4406	1.6668	90
95	3589.44	0.0013	0.0037	772.3	268.255	355.6	64.9	420.5	1.4727	1.6489	95
100	3969.94	0.0015	0.0027	651.4	375.503	373.2	33.8	407.0	1.5187	1.6092	100
101	4051.35	0.0018	0.0022	566.4	457.594	383.0	13.0	396.0	1.5447	1.5794	101

TABLE A.16 (SI)

HFC-134a Superheated Vapor—Constant Pressure Tables

V = Volume in m^3/kg H = Enthalpy in kJ/kg S = Entropy in kJ/(kg) (°K) v_s = Velocity of Sound in m/sec
C_p = Heat Capacity at Constant Pressure in kJ/(kg) (°C) C_p/C_v = Heat Capacity Ratio (Dimensionless)

	TEMP °C	PRESSURE = 40.00 kPa (abs)					
		V	H	S	C_p	C_p/C_v	v_s
SAT LIQ	−44.57	0.00070	142.8	0.7722	1.2311	1.5059	831.3
SAT VAP	−44.57	0.45496	371.4	1.7725	0.7301	1.1475	142.8
	−45	—	—	—	—	—	—
	−40	0.46490	374.8	1.7871	0.7364	1.1440	144.3
	−35	0.47574	378.5	1.8028	0.7437	1.1404	145.9
	−30	0.48662	382.2	1.8183	0.7511	1.1372	147.4
	−25	0.49727	386.0	1.8336	0.7587	1.1342	148.9
	−20	0.50787	389.8	1.8489	0.7665	1.1314	150.4
	−15	0.51867	393.6	1.8639	0.7745	1.1288	151.9
	−10	0.52910	397.5	1.8789	0.7825	1.1264	153.4
	−5	0.53967	401.5	1.8937	0.7906	1.1241	154.8
	0	0.55036	405.4	1.9083	0.7988	1.1219	156.2
	5	0.56085	409.4	1.9229	0.8070	1.1199	157.7
	10	0.57110	413.5	1.9374	0.8152	1.1180	159.0
	15	0.58173	417.6	1.9517	0.8235	1.1161	160.4
	20	0.59207	421.7	1.9659	0.8318	1.1144	161.8
	25	0.60241	425.9	1.9801	0.8401	1.1128	163.1
	30	0.61312	430.1	1.9941	0.8484	1.1112	164.4
	35	0.62344	434.4	2.0081	0.8566	1.1097	165.7
	40	0.63371	438.7	2.0219	0.8649	1.1082	167.0
	45	0.64392	443.0	2.0357	0.8731	1.1068	168.3
	50	0.65445	447.4	2.0494	0.8813	1.1055	169.6
	55	0.66489	451.9	2.0630	0.8895	1.1042	170.9
	60	0.67522	456.3	2.0765	0.8977	1.1030	172.1
	65	0.68540	460,8	2.0899	0.9058	1.1018	173.4
	70	0.69589	465.4	2.1033	0.9138	1.1006	174.6
	75	0.70621	470.0	2.1165	0.9219	1.0995	175.8
	80	0.71633	474.6	2.1297	0.9299	1.0985	177.0
	85	0.72674	479.3	2.1429	0.9378	1.0974	178.2
	90	0.73692	484.0	2.1559	0.9457	1.0964	179.4
	95	0.74738	488.7	2.1689	0.9535	1.0954	180.6
	100	0.75758	493.5	2.1818	0.9613	1.0945	181.8
	105	0.76805	498.3	2.1947	0.9691	1.0936	183.0
	110	0.77821	503.2	2.2074	0.9768	1.0927	184.1

	TEMP °C	PRESSURE = 80.00 kPa (abs)					
		V	H	S	C_p	C_p/C_v	v_s
SAT LIQ	−31.09	0.00072	159.5	0.8435	1.2606	1.5038	768.3
SAT VAP	−31.09	0.23747	379.9	1.7538	0.7718	1.1515	145.0
	−30	0.23872	380.7	1.7572	0.7729	1.1505	145.4
	−25	0.24438	384.6	1.7730	0.7783	1.1460	147.1
	−20	0.24994	388.5	1.7886	0.7842	1.1420	148.7
	−15	0.25549	392.4	1.8040	0.7904	1.1383	150.3
	−10	0.26103	396.4	1.8192	0.7969	1.1349	151.9
	−5	0.26645	400.4	1.8343	0.8037	1.1318	153.4
	0	0.27196	404.5	1.8492	0.8107	1.1290	154.9
	5	0.27732	408.5	1.8640	0.8179	1.1263	156.4
	10	0.28273	412.6	1.8786	0.8252	1.1238	157.8
	15	0.28810	416.8	1.8931	0.8326	1.1215	159.3
	20	0.29343	421.0	1.9075	0.8402	1.1193	160.7
	25	0.29878	425.2	1.9218	0.8478	1.1172	162.1
	30	0.30404	429.4	1.9359	0.8555	1.1153	163.5
	35	0.30941	433.7	1.9500	0.8632	1.1135	164.8
	40	0.31466	438.1	1.9640	0.8709	1.1118	166.2
	45	0.32000	442.4	1.9778	0.8787	1.1101	167.5
	50	0.32520	446.9	1.9916	0.8865	1.1085	168.8
	55	0.33047	451.3	2.0053	0.8944	1.1071	170.1
	60	0.33568	455.8	2.0188	0.9022	1.1056	171.4
	65	0.34095	460.3	2.0323	0.9100	1.1043	172.7
	70	0.34614	464.9	2.0457	0.9178	1.1030	173.9
	75	0.35137	469.5	2.0591	0.9255	1.1017	175.2
	80	0.35663	474.2	2.0723	0.9333	1.1005	176.4
	85	0.36179	478.8	2.0855	0.9410	1.0994	177.6
	90	0.36697	483.6	2.0986	0.9487	1.0982	178.9
	95	0.37216	488.3	2.1116	0.9564	1.0972	180.1
	100	0.37736	493.1	2.1246	0.9640	1.0961	181.3
	105	0.38256	498.0	2.1375	0.9716	1.0951	182.5
	110	0.38775	502.8	2.1503	0.9792	1.0942	183.6
	115	0.39293	507.8	2.1630	0.9867	1.0932	184.8
	120	0.39809	512.7	2.1757	0.9941	1.0923	186.0

TABLE A.16 (SI) (cont'd.)

PRESSURE = 100.00 kPa (abs)

TEMP °C	V	H	S	C_p	C_p/C_v	v_s
−26.34 (SAT LIQ)	0.00073	165.6	0.8681	1.2715	1.5046	746.2
−26.34 (SAT VAP)	0.19246	382.8	1.7484	0.7876	1.1539	145.7
−25	0.19372	383.9	1.7527	0.7888	1.1525	146.1
−20	0.19829	387.9	1.7685	0.7935	1.1477	147.8
−15	0.20284	391.8	1.7840	0.7988	1.1434	149.5
−10	0.20734	395.8	1.7994	0.8045	1.1395	151.1
−5	0.21182	399.9	1.8146	0.8106	1.1359	152.7
0	0.21626	404.0	1.8297	0.8169	1.1327	154.2
5	0.22065	408.1	1.8445	0.8235	1.1296	155.7
10	0.22502	412.2	1.8593	0.8304	1.1269	157.2
15	0.22941	416.4	1.8739	0.8373	1.1243	158.7
20	0.23370	420.6	1.8883	0.8445	1.1218	160.1
25	0.23804	424.8	1.9027	0.8517	1.1196	161.6
30	0.24231	429.1	1.9169	0.8591	1.1175	163.0
35	0.24661	433.4	1.9310	0.8665	1.1155	164.3
40	0.25088	437.7	1.9450	0.8741	1.1136	165.7
45	0.25517	442.1	1.9589	0.8816	1.1118	167.1
50	0.25940	446.6	1.9727	0.8892	1.1101	168.4
55	0.26364	451.0	1.9864	0.8968	1.1085	169.7
60	0.26788	455.5	2.0001	0.9045	1.1070	171.0
65	0.27203	460.1	2.0136	0.9121	1.1055	172.3
70	0.27624	464.7	2.0270	0.9198	1.1042	173.6
75	0.28043	469.3	2.0404	0.9274	1.1028	174.9
80	0.28466	473.9	2.0537	0.9350	1.1016	176.1
85	0.28885	478.6	2.0669	0.9427	1.1004	177.4
90	0.29300	483.4	2.0800	0.9503	1.0992	178.6
95	0.29718	488.1	2.0930	0.9578	1.0981	179.8
100	0.30139	492.9	2.1060	0.9654	1.0970	181.0
105	0.30553	497.8	2.1189	0.9729	1.0959	182.2
110	0.30969	502.7	2.1318	0.9804	1.0949	183.4
115	0.31387	507.6	2.1445	0.9878	1.0940	184.6
120	0.31797	512.5	2.1572	0.9952	1.0930	185.8
125	0.32216	517.5	2.1698	1.0026	1.0921	186.9

PRESSURE = 150.00 kPa (abs)

TEMP °C	V	H	S	C_p	C_p/C_v	v_s
−17.12 (SAT LIQ)	0.00074	177.4	0.9150	1.2939	1.5081	703.7
−17.12 (SAT VAP)	0.13123	388.5	1.7397	0.8201	1.1599	146.5
−15	0.13259	390.3	1.7464	0.8213	1.1574	147.3
−10	0.13576	394.4	1.7622	0.8247	1.1519	149.1
−5	0.13889	398.5	1.7778	0.8287	1.1470	150.8
0	0.14196	402.7	1.7931	0.8333	1.1426	152.4
5	0.14503	406.9	1.8083	0.8384	1.1386	154.1
10	0.14806	411.1	1.8233	0.8438	1.1350	155.7
15	0.15106	415.3	1.8381	0.8496	1.1316	157.2
20	0.15404	419.6	1.8528	0.8557	1.1285	158.7
25	0.15701	423.9	1.8673	0.8620	1.1257	160.2
30	0.15995	428.2	1.8817	0.8685	1.1231	161.7
35	0.16289	432.5	1.8959	0.8752	1.1206	163.2
40	0.16581	436.9	1.9101	0.8820	1.1183	164.6
45	0.16869	441.4	1.9241	0.8890	1.1162	166.0
50	0.17156	445.8	1.9380	0.8960	1.1142	167.4
55	0.17449	450.3	1.9518	0.9032	1.1123	168.7
60	0.17734	454.9	1.9656	0.9103	1.1105	170.1
65	0.18018	459.4	1.9792	0.9176	1.1088	171.4
70	0.18305	464.0	1.9927	0.9249	1.1072	172.8
75	0.18587	468.7	2.0061	0.9322	1.1057	174.1
80	0.18875	473.4	2.0195	0.9395	1.1043	175.3
85	0.19157	478.1	2.0327	0.9468	1.1029	176.6
90	0.19440	482.8	2.0459	0.9541	1.1016	177.9
95	0.19720	487.6	2.0590	0.9615	1.1003	179.1
100	0.20000	492.4	2.0720	0.9688	1.0991	180.4
105	0.20284	497.3	2.0850	0.9761	1.0979	181.6
110	0.20563	502.2	2.0978	0.9834	1.0968	182.8
115	0.20846	507.1	2.1106	0.9907	1.0958	184.0
120	0.21124	512.1	2.1234	0.9979	1.0947	185.2
125	0.21404	517.1	2.1360	1.0052	1.0937	186.4
130	0.21683	522.2	2.1486	1.0123	1.0928	187.6
135	0.21964	527.2	2.1611	1.0195	1.0919	188.7

TABLE A.16 (SI) (cont'd.)

PRESSURE = 200.00 kPa (abs)

TEMP °C	V	H	S	C_p	C_p/C_v	v_s	
−10.08	0.00075	186.6	0.9503	1.3124	1.5125	671.3	SAT LIQ
−10.08	0.09985	392.8	1.7342	0.8468	1.1661	146.9	SAT VAP
−10	0.09989	392.9	1.7344	0.8468	1.1660	146.9	
−5	0.10235	397.1	1.7504	0.8485	1.1595	148.8	
0	0.10478	401.4	1.7661	0.8510	1.1537	150.6	
5	0.10717	405.6	1.7815	0.8543	1.1485	152.3	
10	0.10953	409.9	1.7968	0.8583	1.1438	154.0	
15	0.11186	414.2	1.8118	0.8627	1.1396	155.7	
20	0.11417	418.5	1.8267	0.8676	1.1358	157.3	
25	0.11647	422.9	1.8414	0.8729	1.1323	158.9	
30	0.11874	427.3	1.8560	0.8784	1.1291	160.4	
35	0.12099	431.7	1.8704	0.8843	1.1261	162.0	
40	0.12324	436.1	1.8847	0.8904	1.1234	163.4	
45	0.12547	440.6	1.8989	0.8967	1.1208	164.9	
50	0.12767	445.1	1.9129	0.9031	1.1185	166.3	
55	0.12989	449.6	1.9268	0.9097	1.1163	167.8	
60	0.13207	454.2	1.9406	0.9164	1.1142	169.2	
65	0.13425	458.8	1.9543	0.9232	1.1122	170.5	
70	0.13643	463.4	1.9679	0.9301	1.1104	171.9	
75	0.13860	468.1	1.9814	0.9370	1.1087	173.2	
80	0.14075	472.8	1.9948	0.9440	1.1070	174.6	
85	0.14290	477.5	2.0082	0.9511	1.1055	175.9	
90	0.14505	482.3	2.0214	0.9581	1.1040	177.2	
95	0.14719	487.1	2.0345	0.9652	1.1026	178.4	
100	0.14932	491.9	2.0476	0.9723	1.1013	179.7	
105	0.15145	496.8	2.0606	0.9794	1.1000	181.0	
110	0.15359	501.7	2.0735	0.9865	1.0988	182.2	
115	0.15571	506.7	2.0864	0.9936	1.0976	183.4	
120	0.15783	511.7	2.0991	1.0007	1.0965	184.7	
125	0.15995	516.7	2.1118	1.0078	1.0954	185.9	
130	0.16207	521.8	2.1244	1.0148	1.0944	187.1	
135	0.16418	526.8	2.1370	1.0218	1.0934	188.2	
140	0.16628	532.0	2.1495	1.0288	1.0924	189.4	
145	—	—	—	—	—	—	

PRESSURE = 240.00 kPa (abs)

TEMP °C	V	H	S	C_p	C_p/C_v	v_s
−5.37	0.00076	198.2	0.9736	1.3254	1.5164	649.7
−5.37	0.08389	395.6	1.7310	0.8656	1.1711	147.0
−5	0.08405	396.0	1.7322	0.8657	1.1706	147.2
0	0.08615	400.3	1.7482	0.8663	1.1634	149.1
5	0.08821	404.6	1.7639	0.8680	1.1571	150.9
10	0.09024	409.0	1.7794	0.8705	1.1515	152.7
15	0.09224	413.3	1.7947	0.8738	1.1465	154.4
20	0.09422	417.7	1.8097	0.8776	1.1420	156.1
25	0.09618	422.1	1.8246	0.8820	1.1379	157.8
30	0.09812	426.5	1.8393	0.8867	1.1341	159.4
35	0.10004	431.0	1.8539	0.8919	1.1307	161.0
40	0.10194	435.4	1.8683	0.8973	1.1276	162.5
45	0.10383	439.9	1.8825	0.9030	1.1247	164.0
50	0.10571	444.5	1.8967	0.9090	1.1220	165.5
55	0.10757	449.0	1.9107	0.9151	1.1196	167.0
60	0.10942	453.6	1.9245	0.9214	1.1172	168.4
65	0.11127	458.3	1.9383	0.9278	1.1151	169.8
70	0.11311	462.9	1.9520	0.9344	1.1130	171.2
75	0.11494	467.6	1.9656	0.9410	1.1111	172.6
80	0.11675	472.3	1.9790	0.9478	1.1093	173.9
85	0.11857	477.1	1.9924	0.9546	1.1076	175.3
90	0.12038	481.9	2.0057	0.9614	1.1060	176.6
95	0.12219	486.7	2.0189	0.9683	1.1045	177.9
100	0.12398	491.5	2.0320	0.9752	1.1030	179.2
105	0.12577	496.4	2.0450	0.9821	1.1017	180.5
110	0.12755	501.4	2.0579	0.9891	1.1004	181.7
115	0.12935	506.3	2.0708	0.9960	1.0991	183.0
120	0.13111	511.3	2.0836	1.0030	1.0979	184.2
125	0.13291	516.4	2.0963	1.0099	1.0967	185.4
130	0.13466	521.4	2.1090	1.0168	1.0956	186.7
135	0.13646	526.5	2.1216	1.0237	1.0946	187.9
140	0.13824	531.7	2.1341	1.0306	1.0936	189.0
145	0.13998	536.8	2.1465	1.0375	1.0926	190.2
150	—	—	—	—	—	—

TABLE A.16 (SI) (cont'd.)

PRESSURE = 300.00 kPa (abs)

TEMP °C	V	H	S	C_p	C_p/C_v	v_s
SAT LIQ 0.66	0.00077	200.9	1.0032	1.3431	1.5225	622.1
SAT VAP 0.66	0.06770	399.2	1.7275	0.8912	1.1787	147.0
5	0.06921	403.1	1.7415	0.8902	1.1715	148.7
10	0.07092	407.5	1.7573	0.8904	1.1642	150.7
15	0.07260	412.0	1.7729	0.8915	1.1578	152.5
20	0.07424	416.4	1.7883	0.8936	1.1521	154.3
25	0.07587	420.9	1.8034	0.8964	1.1469	156.1
30	0.07748	425.4	1.8183	0.8998	1.1423	157.8
35	0.07906	429.9	1.8331	0.9038	1.1381	159.5
40	0.08063	434.4	1.8477	0.9082	1.1343	161.1
45	0.08219	439.0	1.8621	0.9130	1.1309	162.7
50	0.08372	443.6	1.8764	0.9181	1.1277	164.2
55	0.08525	448.2	1.8905	0.9236	1.1247	165.8
60	0.08678	452.8	1.9045	0.9292	1.1220	167.3
65	0.08829	457.5	1.9184	0.9350	1.1195	168.7
70	0.08978	462.1	1.9322	0.9411	1.1171	170.2
75	0.09128	466.9	1.9458	0.9472	1.1149	171.6
80	0.09276	471.6	1.9594	0.9535	1.1129	173.0
85	0.09424	476.4	1.9728	0.9599	1.1110	174.4
90	0.09571	481.2	1.9862	0.9664	1.1091	175.7
95	0.09718	486.1	1.9994	0.9730	1.1074	177.1
100	0.09863	490.9	2.0126	0.9796	1.1058	178.4
105	0.10009	495.9	2.0257	0.9862	1.1042	179.7
110	0.10153	500.8	2.0387	0.9929	1.1028	181.0
115	0.10299	505.8	2.0516	0.9996	1.1014	182.3
120	0.10444	510.8	2.0645	1.0064	1.1001	183.5
125	0.10585	515.9	2.0772	1.0131	1.0988	184.8
130	0.10730	520.9	2.0899	1.0199	1.0976	186.0
135	0.10873	526.1	2.1025	1.0266	1.0964	187.3
140	0.11017	531.2	2.1151	1.0334	1.0953	188.5
145	0.11158	536.4	2.1275	1.0401	1.0943	189.7
150	0.11301	541.6	2.1399	1.0468	1.0932	190.9
155	0.11442	546.9	2.1523	1.0535	1.0922	192.1

PRESSURE = 280.00 kPa (abs)

TEMP °C	V	H	S	C_p	C_p/C_v	v_s
−1.24	0.00077	198.3	0.9939	1.3374	1.5204	630.8
−1.24	0.07235	398.1	1.7285	0.8830	1.1762	147.0
0	0.07281	399.2	1.7325	0.8827	1.1741	147.5
5	0.07464	403.6	1.7485	0.8826	1.1665	149.5
10	0.07645	408.0	1.7643	0.8835	1.1598	151.3
15	0.07822	412.4	1.7797	0.8854	1.1539	153.2
20	0.07996	416.9	1.7950	0.8881	1.1486	154.9
25	0.08168	421.3	1.8100	0.8915	1.1438	156.7
30	0.08338	425.8	1.8249	0.8954	1.1395	158.3
35	0.08506	430.3	1.8396	0.8998	1.1356	160.0
40	0.08672	434.8	1.8541	0.9045	1.1320	161.6
45	0.08837	439.3	1.8685	0.9096	1.1288	163.1
50	0.09001	443.9	1.8827	0.9150	1.1258	164.7
55	0.09163	448.5	1.8968	0.9207	1.1230	166.2
60	0.09325	453.1	1.9108	0.9266	1.1204	167.6
65	0.09485	457.7	1.9246	0.9326	1.1180	169.1
70	0.09645	462.4	1.9383	0.9388	1.1158	170.5
75	0.09804	467.1	1.9520	0.9451	1.1137	171.9
80	0.09962	471.8	1.9655	0.9516	1.1117	173.3
85	0.10118	476.6	1.9789	0.9581	1.1098	174.7
90	0.10275	481.4	1.9922	0.9647	1.1081	176.0
95	0.10432	486.3	2.0055	0.9714	1.1064	177.3
100	0.10588	491.1	2.0186	0.9781	1.1049	178.7
105	0.10743	496.1	2.0317	0.9848	1.1034	180.0
110	0.10897	501.0	2.0447	0.9916	1.1020	181.2
115	0.11052	506.0	2.0576	0.9984	1.1006	182.5
120	0.11206	511.0	2.0704	1.0052	1.0993	183.8
125	0.11358	516.0	2.0831	1.0120	1.0981	185.0
130	0.11511	521.1	2.0958	1.0189	1.0969	186.2
135	0.11665	526.2	2.1084	1.0257	1.0958	187.5
140	0.11818	531.4	2.1209	1.0325	1.0947	188.7
145	0.11969	536.5	2.1334	1.0392	1.0937	189.9
150	0.12121	541.7	2.1458	1.0460	1.0927	191.0

SAT LIQ
SAT VAP

TABLE A.16 (SI) (cont'd.)

PRESSURE = 340.00 kPa (abs)

TEMP °C	V	H	S	C_p	C_p/C_v	v_s
4.18 (SAT LIQ)	0.00078	205.6	1.0204	1.3541	1.5266	605.9
4.18 (SAT VAP)	0.05998	401.2	1.7256	0.9069	1.1838	146.8
5	0.06024	402.0	1.7283	0.9064	1.1823	147.2
10	0.06180	406.5	1.7444	0.9047	1.1736	149.2
15	0.06333	411.0	1.7602	0.9043	1.1660	151.2
20	0.06483	415.5	1.7758	0.9050	1.1594	153.1
25	0.06630	420.1	1.7911	0.9066	1.1535	154.9
30	0.06776	424.6	1.8062	0.9091	1.1482	156.7
35	0.06919	429.2	1.8211	0.9122	1.1434	158.4
40	0.07060	433.7	1.8358	0.9158	1.1391	160.1
45	0.07200	438.3	1.8504	0.9200	1.1352	161.8
50	0.07339	442.9	1.8648	0.9245	1.1316	163.4
55	0.07476	447.6	1.8790	0.9294	1.1284	164.9
60	0.07612	452.2	1.8931	0.9346	1.1254	166.5
65	0.07747	456.9	1.9070	0.9400	1.1226	168.0
70	0.07880	461.6	1.9209	0.9456	1.1200	169.5
75	0.08013	466.4	1.9346	0.9515	1.1176	170.9
80	0.08146	471.1	1.9482	0.9575	1.1153	172.3
85	0.08280	475.9	1.9617	0.9636	1.1132	173.8
90	0.08410	480.8	1.9751	0.9698	1.1113	175.1
95	0.08540	485.6	1.9884	0.9761	1.1094	176.5
100	0.08669	490.5	2.0016	0.9826	1.1077	177.9
105	0.08800	495.5	2.0147	0.9890	1.1060	179.2
110	0.08929	500.4	2.0278	0.9955	1.1044	180.5
115	0.09058	505.4	2.0407	1.0021	1.1030	181.8
120	0.09184	510.4	2.0536	1.0087	1.1015	183.1
125	0.09314	515.5	2.0664	1.0153	1.1002	184.4
130	0.09440	520.5	2.0791	1.0220	1.0989	185.6
135	0.09568	525.7	2.0917	1.0286	1.0977	186.9
140	0.09696	530.9	2.1043	1.0352	1.0965	188.1
145	0.09822	536.1	2.1168	1.0419	1.0954	189.3
150	0.09948	541.3	2.1292	1.0485	1.0943	190.5
155	0.10074	546.6	2.1416	1.0551	1.0933	191.7
160	—	—	—	—	—	—

PRESSURE = 380.00 kPa (abs)

TEMP °C	V	H	S	C_p	C_p/C_v	v_s
7.4 (SAT LIQ)	0.00079	210.0	1.0360	1.3645	1.5308	591.1
7.4 (SAT VAP)	0.05384	403.1	1.7240	0.9218	1.1890	146.6
10	0.05459	405.5	1.7325	0.9200	1.1838	147.8
15	0.05600	410.1	1.7486	0.9178	1.1750	149.8
20	0.05738	414.6	1.7644	0.9170	1.1672	151.8
25	0.05874	419.2	1.7799	0.9174	1.1604	153.7
30	0.06007	423.8	1.7952	0.9188	1.1544	155.6
35	0.06138	428.4	1.8102	0.9209	1.1490	157.4
40	0.06267	433.0	1.8251	0.9238	1.1441	159.1
45	0.06395	437.7	1.8397	0.9272	1.1398	160.8
50	0.06521	442.3	1.8542	0.9311	1.1358	162.5
55	0.06645	447.0	1.8686	0.9354	1.1321	164.1
60	0.06769	451.7	1.8827	0.9401	1.1288	165.7
65	0.06892	456.4	1.8968	0.9451	1.1258	167.2
70	0.07014	461.1	1.9107	0.9503	1.1229	168.8
75	0.07135	465.9	1.9245	0.9558	1.1203	170.2
80	0.07254	470.7	1.9381	0.9615	1.1179	171.7
85	0.07374	475.5	1.9517	0.9673	1.1156	173.1
90	0.07492	480.3	1.9652	0.9733	1.1135	174.6
95	0.07611	485.2	1.9785	0.9794	1.1115	176.0
100	0.07729	490.1	1.9918	0.9856	1.1096	177.3
105	0.07846	495.1	2.0049	0.9919	1.1078	178.7
110	0.07962	500.1	2.0180	0.9982	1.1061	180.0
115	0.08078	505.1	2.0310	1.0046	1.1046	181.3
120	0.08193	510.1	2.0439	1.0111	1.1031	182.6
125	0.08308	515.2	2.0567	1.0175	1.1016	183.9
130	0.08423	520.3	2.0694	1.0241	1.1003	185.2
135	0.08538	525.4	2.0821	1.0306	1.0990	186.5
140	0.08653	530.6	2.0947	1.0371	1.0977	187.7
145	0.08767	535.8	2.1072	1.0437	1.0966	189.0
150	0.08881	541.0	2.1196	1.0502	1.0954	190.2
155	0.08994	546.3	2.1320	1.0567	1.0943	191.4
160	0.09107	551.6	2.1443	1.0632	1.0933	192.6

TABLE A.16 (SI) (cont'd.)

PRESSURE = 400.00 kPa (abs)

TEMP °C	V	H	S	C_p	C_p/C_v	v_s
8.91	0.00079	212.1	1.0433	1.3695	1.5329	584.1
8.91	0.05122	403.9	1.7234	0.9290	1.1916	146.5
10	0.05152	404.9	1.7269	0.9281	1.1893	147.0
15	0.05288	409.6	1.7432	0.9249	1.1797	149.1
20	0.05421	414.2	1.7590	0.9233	1.1714	151.2
25	0.05552	418.8	1.7747	0.9230	1.1641	153.1
30	0.05680	423.4	1.7900	0.9238	1.1576	155.0
35	0.05806	428.0	1.8051	0.9254	1.1519	156.9
40	0.05930	432.7	1.8201	0.9278	1.1467	158.6
45	0.06052	437.3	1.8348	0.9309	1.1421	160.4
50	0.06172	442.0	1.8493	0.9345	1.1379	162.0
55	0.06293	446.7	1.8637	0.9385	1.1341	163.7
60	0.06411	451.4	1.8779	0.9429	1.1306	165.3
65	0.06529	456.1	1.8920	0.9477	1.1274	166.9
70	0.06645	460.8	1.9059	0.9527	1.1244	168.4
75	0.06761	465.6	1.9198	0.9580	1.1217	169.9
80	0.06876	470.4	1.9335	0.9635	1.1192	171.4
85	0.06989	475.3	1.9471	0.9692	1.1168	172.8
90	0.07102	480.1	1.9605	0.9751	1.1146	174.3
95	0.07215	485.0	1.9739	0.9810	1.1125	175.7
100	0.07328	489.9	1.9872	0.9871	1.1106	177.1
105	0.07440	494.9	2.0004	0.9933	1.1087	178.4
110	0.07551	499.9	2.0134	0.9996	1.1070	179.8
115	0.07661	504.9	2.0264	1.0059	1.1054	181.1
120	0.07772	509.9	2.0394	1.0123	1.1038	182.4
125	0.07881	515.0	2.0522	1.0187	1.1023	183.7
130	0.07991	520.1	2.0649	1.0251	1.1009	185.0
135	0.08101	525.2	2.0776	1.0316	1.0996	186.3
140	0.08210	530.4	2.0902	1.0381	1.0983	187.5
145	0.08318	535.6	2.1028	1.0445	1.0971	188.8
150	0.08426	540.9	2.1152	1.0510	1.0960	190.0
155	0.08535	546.1	2.1276	1.0575	1.0949	191.2
160	0.08642	551.4	2.1399	1.0640	1.0938	192.4
165	—	—	—	—	—	—

PRESSURE = 500.00 kPa (abs)

TEMP °C	V	H	S	C_p	C_p/C_v	v_s
15.71 (SAT LIQ)	0.00081	221.5	1.0759	1.3937	1.5436	552.8
15.71 (SAT VAP)	0.04114	407.7	1.7205	0.9633	1.2049	145.8
20	0.04213	411.8	1.7347	0.9578	1.1947	147.8
25	0.04325	416.6	1.7508	0.9535	1.1844	150.0
30	0.04434	421.3	1.7667	0.9509	1.1755	152.1
35	0.04541	426.1	1.7822	0.9496	1.1677	154.1
40	0.04646	430.8	1.7975	0.9496	1.1609	156.1
45	0.04749	435.6	1.8125	0.9505	1.1548	158.0
50	0.04851	440.4	1.8274	0.9523	1.1493	159.8
55	0.04951	445.1	1.8420	0.9547	1.1444	161.5
60	0.05050	449.9	1.8565	0.9577	1.1400	163.3
65	0.05147	454.7	1.8708	0.9612	1.1360	164.9
70	0.05244	459.5	1.8849	0.9652	1.1323	166.6
75	0.05339	464.3	1.8989	0.9695	1.1290	168.2
80	0.05434	469.2	1.9128	0.9741	1.1259	169.7
85	0.05528	474.1	1.9265	0.9791	1.1230	171.3
90	0.05621	479.0	1.9401	0.9842	1.1204	172.8
95	0.05713	483.9	1.9536	0.9895	1.1179	174.2
100	0.05806	488.9	1.9670	0.9950	1.1156	175.7
105	0.05898	493.9	1.9803	1.0007	1.1134	177.1
110	0.05989	498.9	1.9934	1.0065	1.1114	178.5
115	0.06078	503.9	2.0065	1.0124	1.1095	179.9
120	0.06169	509.0	2.0195	1.0183	1.1077	181.3
125	0.06258	514.1	2.0324	1.0244	1.1060	182.6
130	0.06348	519.3	2.0452	1.0305	1.1044	184.0
135	0.06437	524.4	2.0580	1.0367	1.1029	185.3
140	0.06526	529.6	2.0707	1.0429	1.1015	186.6
145	0.06613	534.9	2.0832	1.0491	1.1001	187.8
150	0.06702	540.1	2.0957	1.0554	1.0988	189.1
155	0.06790	545.4	2.1082	1.0616	1.0975	190.4
160	0.06877	550.7	2.1205	1.0679	1.0964	191.6
165	0.06964	556.1	2.1328	1.0742	1.0952	192.8
170	0.07052	561.5	2.1450	1.0804	1.0941	194.0

TABLE A.16 (SI) (cont'd.)

PRESSURE = 700.00 kPa (abs)

TEMP °C		V	H	S	C_p	C_p/C_v	v_s
26.68	SAT LIQ	0.00083	237.0	1.1282	1.4385	1.5658	501.8
26.68	SAT VAP	0.02939	413.5	1.7166	1.0265	1.2332	144.0
30		0.02999	416.8	1.7278	1.0182	1.2223	145.7
35		0.03086	421.9	1.7444	1.0085	1.2081	148.2
40		0.03171	426.9	1.7606	1.0015	1.1960	150.6
45		0.03253	431.9	1.7764	0.9967	1.1857	152.8
50		0.03333	436.9	1.7919	0.9935	1.1768	155.0
55		0.03412	441.9	1.8071	0.9918	1.1689	157.0
60		0.03488	446.8	1.8221	0.9913	1.1620	159.0
65		0.03564	451.8	1.8369	0.9917	1.1559	160.9
70		0.03638	456.7	1.8515	0.9929	1.1504	162.8
75		0.03711	461.7	1.8658	0.9949	1.1454	164.6
80		0.03784	466.7	1.8801	0.9974	1.1409	166.3
85		0.03855	471.7	1.8941	1.0005	1.1369	168.0
90		0.03925	476.7	1.9080	1.0040	1.1331	169.7
95		0.03995	481.7	1.9217	1.0078	1.1297	171.3
100		0.04064	486.8	1.9354	1.0120	1.1266	172.9
105		0.04133	491.9	1.9489	1.0165	1.1236	174.5
110		0.04201	496.9	1.9622	1.0212	1.1209	176.0
115		0.04269	502.1	1.9755	1.0261	1.1184	177.5
120		0.04336	507.2	1.9887	1.0312	1.1161	179.0
125		0.04402	512.4	2.0018	1.0365	1.1139	180.4
130		0.04468	517.6	2.0147	1.0419	1.1118	181.8
135		0.04535	522.8	2.0276	1.0473	1.1099	183.2
140		0.04600	528.0	2.0404	1.0529	1.1081	184.6
145		0.04665	533.3	2.0531	1.0586	1.1063	186.0
150		0.04730	538.6	2.0657	1.0643	1.1047	187.3
155		0.04795	544.0	2.0782	1.0701	1.1032	188.7
160		0.04859	549.3	2.0907	1.0760	1.1017	190.0
165		0.04923	554.7	2.1031	1.0818	1.1003	191.3
170		0.04987	560.2	2.1154	1.0877	1.0990	192.5
175		0.05051	565.6	2.1276	1.0936	1.0977	193.8
180		0.05114	571.1	2.1398	1.0996	1.0965	195.0

PRESSURE = 600.00 kPa (abs)

TEMP °C		V	H	S	C_p	C_p/C_v	v_s
21.54	SAT LIQ	0.00082	229.7	1.1038	1.4165	1.5545	525.8
21.54	SAT VAP	0.03432	410.8	1.7183	0.9954	1.2187	145.0
25		0.03502	414.2	1.7299	0.9890	1.2090	146.7
30		0.03600	419.2	1.7463	0.9819	1.1967	149.0
35		0.03695	424.1	1.7623	0.9771	1.1862	151.3
40		0.03787	428.9	1.7780	0.9740	1.1771	153.4
45		0.03878	433.8	1.7934	0.9723	1.1692	155.5
50		0.03967	438.7	1.8086	0.9718	1.1622	157.4
55		0.04054	443.5	1.8235	0.9724	1.1560	159.3
60		0.04140	448.4	1.8382	0.9738	1.1505	161.2
65		0.04224	453.3	1.8527	0.9759	1.1455	163.0
70		0.04308	458.1	1.8671	0.9786	1.1410	164.7
75		0.04390	463.0	1.8813	0.9818	1.1369	166.4
80		0.04472	468.0	1.8953	0.9854	1.1331	168.1
85		0.04552	472.9	1.9092	0.9895	1.1297	169.7
90		0.04633	477.9	1.9229	0.9938	1.1265	171.3
95		0.04712	482.8	1.9365	0.9985	1.1236	172.8
100		0.04790	487.8	1.9500	1.0033	1.1209	174.3
105		0.04868	492.9	1.9634	1.0084	1.1184	175.8
110		0.04945	497.9	1.9767	1.0137	1.1161	177.3
115		0.05023	503.0	1.9899	1.0191	1.1139	178.7
120		0.05100	508.1	2.0030	1.0247	1.1118	180.1
125		0.05175	513.3	2.0159	1.0303	1.1099	181.5
130		0.05252	518.4	2.0288	1.0361	1.1081	182.9
135		0.05327	523.6	2.0417	1.0419	1.1063	184.3
140		0.05402	528.8	2.0544	1.0478	1.1047	185.6
145		0.05477	534.1	2.0670	1.0538	1.1032	186.9
150		0.05552	539.4	2.0796	1.0598	1.1017	188.2
155		0.05626	544.7	2.0921	1.0658	1.1003	189.5
160		0.05700	550.0	2.1045	1.0719	1.0990	190.8
165		0.05774	555.4	2.1168	1.0780	1.0977	192.0
170		0.05848	560.8	2.1291	1.0840	1.0965	193.3
175		0.05921	566.3	2.1413	1.0901	1.0954	194.5

TABLE A.16 (SI) (cont'd.)

PRESSURE = 800.00 kPa (abs)

TEMP °C		V	H	S	C_p	C_p/C_v	v_s
31.29	SAT LIQ	0.00085	243.7	1.1500	1.4602	1.5775	480.2
31.29	SAT VAP	0.02565	415.7	1.7150	1.0569	1.2485	142.9
30		—	—	—	—	—	—
35		0.02626	419.6	1.7278	1.0453	1.2344	145.0
40		0.02705	424.8	1.7445	1.0332	1.2184	147.6
45		0.02782	430.0	1.7608	1.0243	1.2049	150.1
50		0.02856	435.1	1.7767	1.0178	1.1935	152.4
55		0.02928	440.2	1.7923	1.0133	1.1836	154.7
60		0.02998	445.2	1.8076	1.0105	1.1750	156.8
65		0.03067	450.3	1.8226	1.0089	1.1674	158.8
70		0.03135	455.3	1.8374	1.0085	1.1607	160.8
75		0.03202	460.4	1.8520	1.0090	1.1548	162.7
80		0.03267	465.4	1.8664	1.0103	1.1494	164.6
85		0.03331	470.5	1.8806	1.0122	1.1446	166.4
90		0.03395	475.5	1.8947	1.0147	1.1402	168.1
95		0.03457	480.6	1.9086	1.0177	1.1362	169.8
100		0.03520	485.7	1.9223	1.0212	1.1325	171.5
105		0.03581	490.8	1.9360	1.0250	1.1292	173.1
110		0.03642	495.9	1.9494	1.0291	1.1261	174.7
115		0.03703	501.1	1.9628	1.0334	1.1232	176.3
120		0.03763	506.3	1.9761	1.0380	1.1205	177.8
125		0.03822	511.5	1.9892	1.0428	1.1181	179.3
130		0.03881	516.7	2.0023	1.0478	1.1157	180.8
135		0.03940	522.0	2.0152	1.0529	1.1136	182.2
140		0.03998	527.2	2.0281	1.0582	1.1115	183.6
145		0.04056	532.5	2.0408	1.0636	1.1096	185.0
150		0.04114	537.9	2.0535	1.0690	1.1078	186.4
155		0.04171	543.2	2.0661	1.0745	1.1061	187.8
160		0.04229	548.6	2.0786	1.0801	1.1045	189.1
165		0.04285	554.0	2.0910	1.0858	1.1029	190.5
170		0.04342	559.5	2.1034	1.0915	1.1015	191.8
175		0.04398	565.0	2.1157	1.0972	1.1001	193.1
180		0.04455	570.5	2.1279	1.1030	1.0988	194.3
185		0.04511	576.0	2.1400	1.1087	1.0975	195.6

PRESSURE = 1000.00 kPa (abs)

TEMP °C		V	H	S	C_p	C_p/C_v	v_s
39.35	SAT LIQ	0.00087	255.6	1.1881	1.5035	1.6025	442.1
39.35	SAT VAP	0.02034	419.5	1.7124	1.1177	1.2817	140.6
40		0.02044	420.2	1.7147	1.1144	1.2782	141.0
45		0.02114	425.7	1.7322	1.0928	1.2546	144.1
50		0.02181	431.2	1.7491	1.0766	1.2354	146.9
55		0.02246	436.5	1.7655	1.0644	1.2195	149.6
60		0.02308	441.8	1.7816	1.0552	1.2061	152.1
65		0.02368	447.1	1.7972	1.0485	1.1947	154.4
70		0.02427	452.3	1.8126	1.0438	1.1848	156.7
75		0.02485	457.5	1.8277	1.0406	1.1762	158.9
80		0.02541	462.7	1.8425	1.0388	1.1687	160.9
85		0.02596	467.9	1.8571	1.0381	1.1619	162.9
90		0.02650	473.1	1.8715	1.0383	1.1559	164.9
95		0.02703	478.3	1.8857	1.0393	1.1506	166.8
100		0.02756	483.5	1.8997	1.0409	1.1457	168.6
105		0.02807	488.7	1.9136	1.0432	1.1413	170.4
110		0.02859	493.9	1.9273	1.0459	1.1372	172.1
115		0.02909	499.1	1.9409	1.0490	1.1335	173.8
120		0.02959	504.4	1.9543	1.0525	1.1301	175.4
125		0.03008	509.7	1.9676	1.0563	1.1270	177.0
130		0.03058	515.0	1.9809	1.0604	1.1241	178.6
135		0.03106	520.3	1.9940	1.0647	1.1214	180.1
140		0.03155	525.6	2.0070	1.0692	1.1189	181.7
145		0.03203	531.0	2.0198	1.0739	1.1165	183.1
150		0.03250	536.3	2.0326	1.0787	1.1143	184.6
155		0.03298	541.8	2.0453	1.0837	1.1122	186.0
160		0.03345	547.2	2.0579	1.0888	1.1103	187.5
165		0.03392	552.6	2.0705	1.0940	1.1084	188.9
170		0.03438	558.1	2.0829	1.0993	1.1067	190.2
175		0.03485	563.6	2.0953	1.1046	1.1050	191.6
180		0.03531	569.2	2.1076	1.1100	1.1035	192.9
185		0.03577	574.7	2.1198	1.1154	1.1020	194.2
190		0.03623	580.3	2.1319	1.1209	1.1006	195.6

TABLE A.16 (SI) (cont'd.)

PRESSURE = 1500.00 kPa (abs)

TEMP °C	V	H	S	C_p	C_p/C_v	v_s
55.2 SAT LIQ	0.00093	280.1	1.2632	1.6205	1.6778	365.0
55.2 SAT VAP	0.01308	425.7	1.7063	1.2844	1.3876	134.2
60	0.01363	431.7	1.7246	1.2358	1.3416	138.0
65	0.01417	437.8	1.7427	1.1989	1.3056	141.7
70	0.01468	443.7	1.7601	1.1715	1.2776	145.0
75	0.01516	449.5	1.7769	1.1507	1.2552	148.0
80	0.01562	455.2	1.7932	1.1349	1.2368	150.9
85	0.01606	460.9	1.8090	1.1228	1.2214	153.6
90	0.01649	466.4	1.8245	1.1136	1.2084	156.1
95	0.01690	472.0	1.8397	1.1067	1.1972	158.5
100	0.01731	477.5	1.8546	1.1016	1.1875	160.8
105	0.01770	483.0	1.8692	1.0982	1.1790	163.0
110	0.01809	488.5	1.8837	1.0959	1.1714	165.2
115	0.01847	494.0	1.8978	1.0948	1.1647	167.2
120	0.01885	499.4	1.9119	1.0945	1.1587	169.2
125	0.01921	504.9	1.9257	1.0950	1.1532	171.1
130	0.01958	510.4	1.9394	1.0962	1.1483	173.0
135	0.01993	515.9	1.9529	1.0979	1.1438	174.8
140	0.02028	521.4	1.9663	1.1001	1.1397	176.6
145	0.02063	526.9	1.9795	1.1027	1.1360	178.3
150	0.02098	532.4	1.9926	1.1057	1.1325	180.0
155	0.02132	537.9	2.0056	1.1090	1.1293	181.6
160	0.02166	543.5	2.0185	1.1125	1.1263	183.2
165	0.02199	549.1	2.0313	1.1163	1.1235	184.8
170	0.02233	554.7	2.0440	1.1203	1.1209	186.3
175	0.02265	560.3	2.0566	1.1244	1.1185	187.9
180	0.02298	565.9	2.0691	1.1288	1.1162	189.3
185	0.02331	571.6	2.0815	1.1332	1.1141	190.8
190	0.02363	577.2	2.0939	1.1378	1.1121	192.3
195	0.02395	582.9	2.1061	1.1425	1.1102	193.7
200	0.02427	588.7	2.1183	1.1472	1.1084	195.1
205	0.02459	594.4	2.1303	1.1520	1.1067	196.5
210	0.02491	600.2	2.1424	1.1569	1.1050	197.8

PRESSURE = 2000.00 kPa (abs)

TEMP °C	V	H	S	C_p	C_p/C_v	v_s
67.47 SAT LIQ	0.00099	300.4	1.3223	1.7690	1.7844	302.2
67.47 SAT VAP	0.00931	428.8	1.6991	1.5055	1.5484	127.2
70	0.00959	432.5	1.7101	1.4452	1.4943	129.9
75	0.01009	439.5	1.7303	1.3600	1.4173	134.7
80	0.01055	446.1	1.7493	1.3019	1.3638	138.9
85	0.01097	452.5	1.7673	1.2601	1.3242	142.7
90	0.01137	458.8	1.7845	1.2291	1.2936	146.2
95	0.01175	464.8	1.8011	1.2055	1.2692	149.3
100	0.01211	470.8	1.8173	1.1874	1.2493	152.3
105	0.01246	476.7	1.8330	1.1733	1.2327	155.1
110	0.01279	482.6	1.8483	1.1625	1.2186	157.7
115	0.01312	488.4	1.8633	1.1542	1.2066	160.2
120	0.01344	494.1	1.8781	1.1479	1.1961	162.6
125	0.01374	499.8	1.8925	1.1433	1.1869	164.9
130	0.01405	505.5	1.9068	1.1401	1.1789	167.1
135	0.01434	511.2	1.9208	1.1380	1.1717	169.2
140	0.01463	516.9	1.9347	1.1369	1.1652	171.3
145	0.01492	522.6	1.9483	1.1366	1.1594	173.3
150	0.01520	528.3	1.9619	1.1370	1.1541	175.2
155	0.01547	534.0	1.9752	1.1380	1.1493	177.1
160	0.01575	539.7	1.9884	1.1395	1.1449	178.9
165	0.01602	545.4	2.0015	1.1415	1.1409	180.7
170	0.01629	551.1	2.0145	1.1439	1.1372	182.4
175	0.01655	556.8	2.0274	1.1466	1.1337	184.1
180	0.01681	562.6	2.0401	1.1496	1.1305	185.7
185	0.01707	568.3	2.0527	1.1528	1.1276	187.4
190	0.01733	574.1	2.0653	1.1563	1.1248	189.0
195	0.01758	579.9	2.0777	1.1599	1.1222	190.5
200	0.01784	585.7	2.0900	1.1638	1.1198	192.1
205	0.01809	591.5	2.1023	1.1678	1.1175	193.6
210	0.01834	597.4	2.1145	1.1719	1.1153	195.0
215	0.01858	603.2	2.1265	1.1761	1.1133	196.5
220	0.01883	609.1	2.1386	1.1804	1.1114	197.9

TABLE A.16 (SI) (cont'd.)

PRESSURE = 3000.00 kPa (abs)

TEMP °C	V	H	S	C_p	C_p/C_v	v_s
SAT LIQ 86.22	0.00114	335.3	1.4188	2.3843	2.2770	195.8
SAT VAP 86.22	0.00528	427.6	1.6758	2.5300	2.3769	112.1
90	0.00575	436.1	1.6992	2.0156	1.9317	119.1
95	0.00624	445.3	1.7245	1.7214	1.6778	126.1
100	0.00665	453.5	1.7466	1.5654	1.5427	131.7
105	0.00701	461.1	1.7667	1.4676	1.4573	136.6
110	0.00733	468.2	1.7855	1.4007	1.3977	140.9
115	0.00764	475.1	1.8034	1.3523	1.3537	144.8
120	0.00792	481.8	1.8204	1.3161	1.3197	148.4
125	0.00819	488.3	1.8369	1.2884	1.2926	151.7
130	0.00845	494.7	1.8528	1.2669	1.2704	154.7
135	0.00869	501.0	1.8683	1.2500	1.2520	157.6
140	0.00893	507.2	1.8835	1.2366	1.2363	160.4
145	0.00916	513.3	1.8983	1.2262	1.2229	163.0
150	0.00939	519.4	1.9128	1.2179	1.2113	165.5
155	0.00960	525.5	1.9271	1.2116	1.2011	167.9
160	0.00982	531.6	1.9411	1.2068	1.1921	170.2
165	0.01003	537.6	1.9549	1.2032	1.1841	172.4
170	0.01023	543.6	1.9686	1.2007	1.1770	174.5
175	0.01043	549.6	1.9820	1.1992	1.1705	176.6
180	0.01063	555.6	1.9953	1.1984	1.1646	178.6
185	0.01083	561.6	2.0085	1.1983	1.1593	180.5
190	0.01102	567.6	2.0215	1.1988	1.1544	182.4
195	0.01121	573.6	2.0344	1.1998	1.1499	184.3
200	0.01139	579.6	2.0471	1.2012	1.1458	186.1
205	0.01158	585.6	2.0598	1.2030	1.1420	187.8
210	0.01176	591.6	2.0723	1.2051	1.1384	189.6
215	0.01194	597.6	2.0847	1.2075	1.1351	191.3
220	0.01212	603.7	2.0970	1.2102	1.1320	192.9
225	0.01230	609.7	2.1092	1.2130	1.1291	194.5
230	0.01248	615.8	2.1214	1.2161	1.1264	196.1
235	0.01265	621.9	2.1334	1.2194	1.1239	197.7
240	0.01283	628.0	2.1454	1.2227	1.1215	199.2

PRESSURE = 4000.00 kPa (abs)

TEMP °C	V	H	S	C_p	C_p/C_v	v_s
SAT LIQ 100.37	0.00158	375.6	1.5250	28.1470	24.2211	95.7
SAT VAP 100.37	0.00254	404.4	1.6022	42.1018	35.2394	95.0
100	—	—	—	—	—	—
105	0.00376	433.7	1.6804	3.1309	2.8151	111.6
110	0.00429	446.7	1.7143	2.2216	2.0540	120.4
115	0.00468	456.8	1.7406	1.8780	1.7663	127.1
120	0.00501	465.7	1.7634	1.6927	1.6106	132.7
125	0.00530	473.8	1.7840	1.5762	1.5117	137.5
130	0.00556	481.5	1.8031	1.4963	1.4429	141.8
135	0.00580	488.8	1.8212	1.4384	1.3921	145.8
140	0.00603	495.9	1.8385	1.3950	1.3530	149.4
145	0.00624	502.8	1.8550	1.3615	1.3219	152.8
150	0.00645	509.5	1.8710	1.3354	1.2965	155.9
155	0.00664	516.2	1.8866	1.3146	1.2755	158.9
160	0.00683	522.7	1.9018	1.2981	1.2577	161.7
165	0.00702	529.2	1.9166	1.2849	1.2424	164.4
170	0.00719	535.6	1.9311	1.2743	1.2293	167.0
175	0.00737	541.9	1.9454	1.2658	1.2177	169.4
180	0.00753	548.2	1.9594	1.2591	1.2076	171.8
185	0.00770	554.5	1.9732	1.2539	1.1985	174.1
190	0.00786	560.8	1.9867	1.2499	1.1904	176.3
195	0.00802	567.0	2.0001	1.2470	1.1831	178.4
200	0.00817	573.2	2.0134	1.2450	1.1765	180.5
205	0.00833	579.4	2.0265	1.2438	1.1705	182.5
210	0.00848	585.7	2.0394	1.2432	1.1650	184.5
215	0.00863	591.9	2.0522	1.2431	1.1600	186.4
220	0.00877	598.1	2.0649	1.2436	1.1554	188.3
225	0.00892	604.3	2.0774	1.2445	1.1511	190.1
230	0.00906	610.5	2.0898	1.2458	1.1471	191.9
235	0.00920	616.8	2.1022	1.2475	1.1434	193.6
240	0.00934	623.0	2.1144	1.2494	1.1400	195.3
245	0.00948	629.3	2.1265	1.2515	1.1368	197.0
250	0.00962	635.5	2.1386	1.2539	1.1337	198.6

TABLE A.18
Thermodynamic Properties of Air at Low Pressure

T, °R	h, Btu/lb	p_r	u, Btu/lb	v_r	ϕ Btu/lb. °R
200	47.67	0.04320	33.96	1714.9	0.36303
220	52.46	0.06026	37.38	1352.5	0.38584
240	57.25	0.08165	40.80	1088.8	0.40666
260	62.03	0.10797	44.21	892.0	0.42582
280	66.82	0.13986	47.63	741.6	0.44356
300	71.61	0.17795	51.04	624.5	0.46007
320	76.40	0.22290	54.46	531.8	0.47550
340	81.18	0.27545	57.87	457.2	0.49002
360	85.97	0.3363	61.29	396.6	0.50369
380	90.75	0.4061	64.70	346.6	0.51663
400	95.53	0.4858	68.11	305.0	0.52890
420	100.32	0.5760	71.52	270.1	0.54058
440	105.11	0.6776	74.93	240.6	0.55172
460	109.90	0.7913	78.36	215.33	0.56235
480	114.69	0.9182	81.77	193.65	0.57255
500	119.48	1.0590	85.20	174.90	0.58233
520	124.27	1.2147	88.62	158.58	0.59173
540	129.06	1.3860	92.04	144.32	0.60078
560	133.86	1.5742	95.47	131.78	0.60950
580	138.66	1.7800	98.90	120.70	0.61793
600	143.47	2.005	102.34	110.88	0.62607
620	148.28	2.249	105.78	102.12	0.63395
640	153.09	2.514	109.21	94.30	0.64159
660	157.92	2.801	112.67	87.27	0.64902
680	162.73	3.111	116.12	80.96	0.65621
700	167.56	3.446	119.58	75.25	0.66321
720	172.39	3.806	123.04	70.07	0.67002
740	177.23	4.193	126.51	65.38	0.67665
760	182.08	4.607	129.99	61.10	0.68312
780	186.94	5.051	133.47	57.20	0.68942
800	191.81	5.526	136.97	53.63	0.69558
820	196.69	6.033	140.47	50.35	0.70160
840	201.56	6.573	143.98	47.34	0.70747
860	206.46	7.149	147.50	44.57	0.71323

Source: Abridged from Table 1 in Joseph H. Keenan and Joseph Kaye, *Gas Tables*, John Wiley & Sons, Inc., New York, 1948.

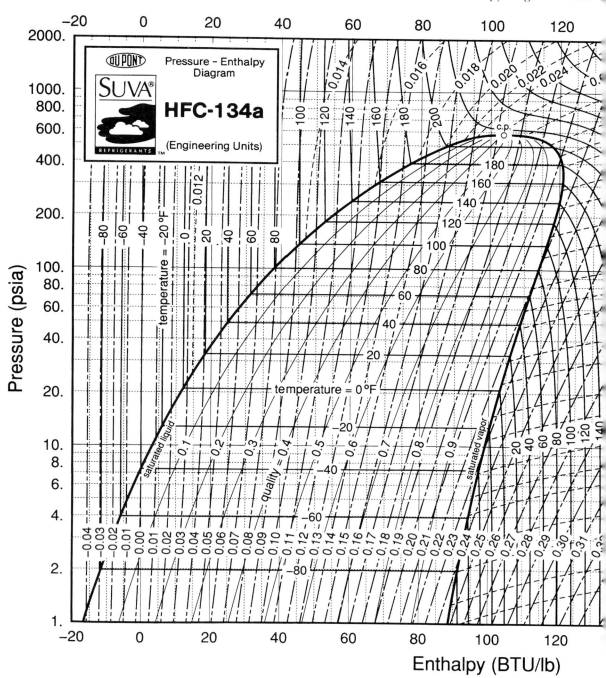

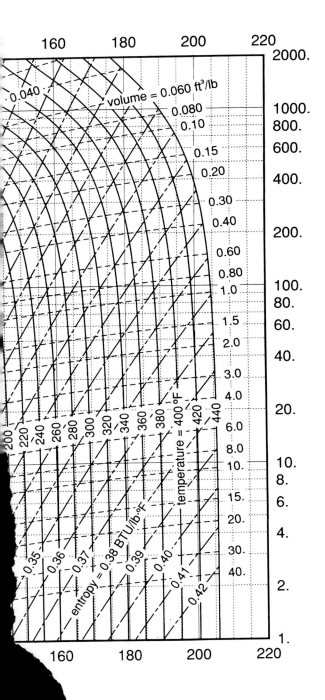

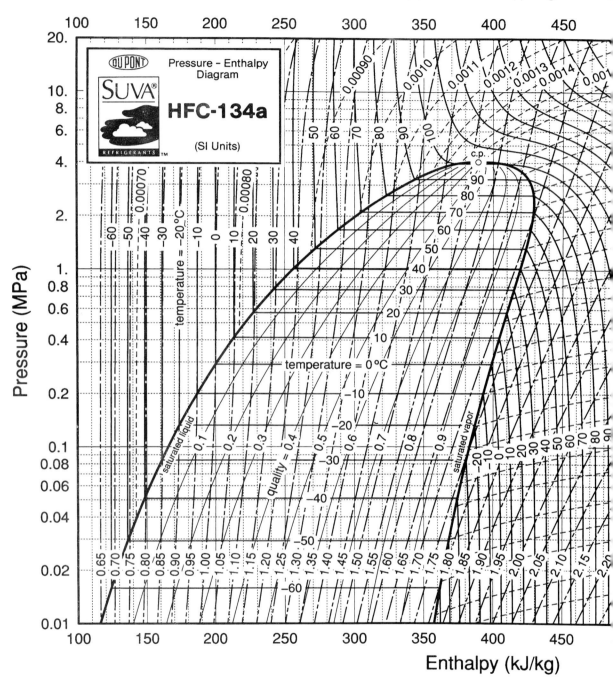

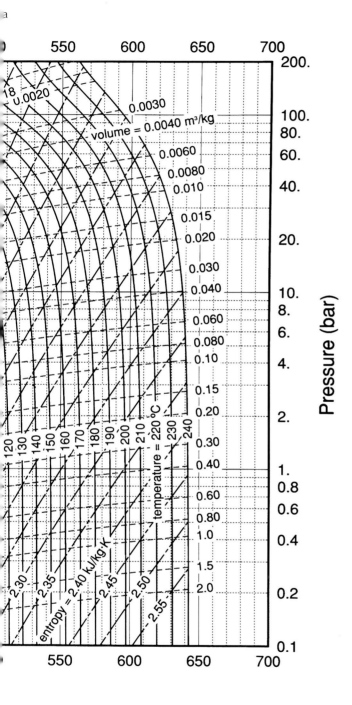

T, °R	h, Btu/lb	p_r	u, Btu/lb	v_r	ϕ Btu/lb·°R
880	211.35	7.761	151.02	42.01	0.71886
900	216.26	8.411	154.57	39.64	0.72438
920	221.18	9.102	158.12	37.44	0.72979
940	226.11	9.834	161.68	35.41	0.73509
960	231.06	10.610	165.26	33.52	0.74030
980	236.02	11.430	168.83	31.76	0.74540
1000	240.98	12.298	172.43	30.12	0.75042
1020	245.97	13.215	176.04	28.59	0.75536
1040	250.95	14.182	179.66	27.17	0.76019
1060	255.96	15.203	183.29	25.82	0.76496
1080	260.97	16.278	186.93	24.58	0.76964
1100	265.99	17.413	190.58	23.40	0.77426
1120	271.03	18.604	194.25	22.30	0.77880
1140	276.08	19.858	197.94	21.27	0.78326
1160	281.14	21.18	201.63	20.293	0.78767
1180	286.21	22.56	205.33	19.377	0.79201
1200	291.30	24.01	209.05	18.514	0.79628
1220	296.41	25.53	212.78	17.700	0.80050
1240	301.52	27.13	216.53	16.932	0.80466
1260	306.65	28.80	220.28	16.205	0.80876
1280	311.79	30.55	224.05	15.518	0.81280
1300	316.94	32.39	227.83	14.868	0.81680
1320	322.11	34.31	231.63	14.253	0.82075
1340	327.29	36.31	235.43	13.670	0.82464
1360	332.48	38.41	239.25	13.118	0.82848
1380	337.68	40.59	243.08	12.593	0.83229
1400	342.90	42.88	246.93	12.095	0.83604
1420	348.14	45.26	250.79	11.622	0.83975
1440	353.37	47.75	254.66	11.172	0.84341
1460	358.63	50.34	258.54	10.743	0.84704
1480	363.89	53.04	262.44	10.336	0.85062
1500	369.17	55.86	266.34	9.948	0.85416
1520	374.47	58.78	270.26	9.578	0.85767
1540	379.77	61.83	274.20	9.226	0.86113
1560	385.08	65.00	278.13	8.890	0.86456
1580	390.40	68.30	282.09	8.569	0.86794
1600	395.74	71.73	286.06	8.263	0.87130
1620	401.09	75.29	290.04	7.971	0.87462

$T,$ °R	$h,$ Btu/lb	p_r	$u,$ Btu/lb	v_r	ϕ Btu/lb.°R
1640	406.45	78.99	294.03	7.691	0.87791
1660	411.82	82.83	298.02	7.424	0.88116
1680	417.20	86.82	302.04	7.168	0.88439
1700	422.59	90.95	306.06	6.924	0.88758
1720	428.00	95.24	310.09	6.690	0.89074
1740	433.41	99.69	314.13	6.465	0.89387
1760	438.83	104.30	318.18	6.251	0.89697
1780	444.26	109.08	322.24	6.045	0.90003
1800	449.71	114.03	326.32	5.847	0.90308
1820	455.17	119.16	330.40	5.658	0.90609
1840	460.63	124.47	334.50	5.476	0.90908
1860	466.12	129.95	338.61	5.302	0.91203
1880	471.60	135.64	342.73	5.134	0.91497
1900	477.09	141.51	346.85	4.974	0.91788
1920	482.60	147.59	350.98	4.819	0.92076
1940	488.12	153.87	355.12	4.670	0.92362
1960	493.64	160.37	359.28	4.527	0.92645
1980	499.17	167.07	363.43	4.390	0.92926
2000	504.71	174.00	367.61	4.258	0.93205
2020	510.26	181.16	371.79	4.130	0.93481
2040	515.82	188.54	375.98	4.008	0.93756
2060	521.39	196.16	380.18	3.890	0.94026
2080	526.97	204.02	384.39	3.777	0.94296
2100	532.55	212.1	388.60	3.667	0.94564
2120	538.15	220.5	392.83	3.561	0.94829
2140	543.74	229.1	397.05	3.460	0.95092
2160	549.35	238.0	401.29	3.362	0.95352
2180	554.97	247.2	405.53	3.267	0.95611
2200	560.59	256.6	409.78	3.176	0.95868
2220	566.23	266.3	414.05	3.088	0.96123
2240	571.86	276.3	418.31	3.003	0.96376
2260	577.51	286.6	422.59	2.921	0.96626
2280	583.16	297.2	426.87	2.841	0.96876
2300	588.82	308.1	431.16	2.765	0.97123
2320	594.49	319.4	435.46	2.691	0.97369
2340	600.16	330.9	439.76	2.619	0.97611
2360	605.84	342.8	444.07	2.550	0.97853
2380	611.53	355.0	448.38	2.483	0.98092
2400	617.22	367.6	452.70	2.419	0.98331

TABLE A.19
Properties of Some Gases at Low Pressure[a]

Temp °R	Products of Combustion, 400% Theoretical Air		Products of Combustion, 200% Theoretical Air		Nitrogen		Oxygen	
	$\bar{h}$	$\bar{\phi}$	$\bar{h}$	$\bar{\phi}$	$\bar{h}$	$\bar{\phi}$	$\bar{h}$	$\bar{\phi}$
537	3746.8	46.318	3774.9	46.300	3729.5	45.755	3725.1	48.986
600	4191.9	47.101	4226.3	47.094	4167.9	46.514	4168.3	49.762
700	4901.7	48.195	4947.7	48.207	4864.9	47.588	4879.3	50.858
800	5617.5	49.150	5676.3	49.179	5564.4	48.522	5602.0	51.821
900	6340.3	50.002	6413.0	50.047	6268.1	49.352	6337.9	52.688
1000	7072.1	50.773	7159.8	50.833	6977.9	50.099	7087.5	53.477
1100	7812.9	51.479	7916.4	51.555	7695.0	50.783	7850.4	54.204
1200	8563.4	52.132	8683.6	52.222	8420.0	51.413	8625.8	54.879
1300	9324.1	52.741	9461.7	52.845	9153.9	52.001	9412.9	55.508
1400	10095.0	53.312	10250.7	53.430	9896.9	52.551	10210.4	56.099
1500	10875.6	53.851	11050.2	53.981	10648.9	53.071	11017.1	56.656
1600	11665.6	54.360	11859.6	54.504	11409.7	53.561	11832.5	57.182
1700	12464.3	54.844	12678.6	55.000	12178.9	54.028	12655.6	57.680
1800	13271.7	55.306	13507.0	55.473	12956.3	54.472	13485.8	58.155
1900	14087.2	55.747	14344.1	55.926	13741.6	54.896	14322.1	58.607
2000	14910.3	56.169	15189.3	56.360	14534.4	55.303	15164.0	59.039
2100	15740.5	56.574	16042.4	56.777	15334.0	55.694	16010.9	59.451
2200	16577.1	56.964	16902.5	57.177	16139.8	56.068	16862.6	59.848
2300	17419.8	57.338	17769.3	57.562	16951.2	56.429	17718.8	60.228
2400	18268.0	57.699	18642.1	57.933	17767.9	56.777	18579.2	60.594
2500	19121.4	58.048	19520.7	58.292	18589.5	57.112	19443.4	60.946
2600	19979.7	58.384	20404.6	58.639	19415.8	57.436	20311.4	61.287
2700	20842.8	58.710	21293.8	58.974	20246.4	57.750	21182.9	61.616
2800	21709.8	59.026	22187.5	59.300	21081.1	58.053	22057.8	61.934
2900	22581.4	59.331	23086.0	59.615	21919.5	58.348	22936.1	62.242
3000	23456.6	59.628	23988.5	59.921	22761.5	58.632	23817.7	62.540
3100	24335.5	59.916	24895.3	60.218	23606.8	58.910	24702.5	62.831
3200	25217.8	60.196	25805.6	60.507	24455.0	59.179	25590.5	63.113
3300	26102.9	60.469	26719.2	60.789	25306.0	59.442	26481.6	63.386
3400	26991.4	60.734	27636.4	61.063	26159.7	59.697	27375.9	63.654
3500			28556.8	61.329	27015.9	59.944	28273.3	63.914
3600			29479.9	61.590	27874.4	60.186	29173.9	64.168
3700			30406.0	61.843	28735.1	60.422	30077.5	64.415
3800			31334.8	62.091	29597.9	60.652	30984.1	64.657
3900			32266.2	62.333	30462.8	60.877	31893.6	64.893
4000					31329.4	61.097	32806.1	65.123
4100					32198.0	61.310	33721.6	65.350
4200					33068.1	61.520	34639.9	65.571
4300					33939.9	61.726	35561.1	65.788
4400					34813.1	61.927	36485.0	66.000
4500					35687.8	62.123	37411.8	66.208
4600					36563.8	62.316	38341.4	66.413
4700					37441.1	62.504	39273.6	66.613
4800					38319.5	62.689	40208.6	66.809
4900					39199.1	62.870	41146.1	67.003
5000					40079.8	63.049	42086.3	67.193
5100					40961.6	63.223	43029.1	67.380
5200					41844.4	63.395	43974.3	67.562
5300					42728.3	63.563	44922.2	67.743

Source: Abridged from Tables 4, 7, 11, 13, 15, 17, 19, and 21 in Joseph H. Keenan and Joseph Kaye, *Gas Tables*, John Wiley & Sons, Inc., New York, 1948.

[a] $\bar{h}$ = enthalapy, Btu/lb mole; $\phi = \int_{T=0}^{T} c_{p_0} \dfrac{dT}{T}$, Btu/lb mole °R; ϕ is essentially equal to the absolute entropy at 1 atm pressure, Btu/lb mole °R.

Temp °R	Water Vapor		Carbon Dioxide		Hydrogen		Carbon Monoxide	
	$\bar{h}$	$\bar{\phi}$	$\bar{h}$	$\bar{\phi}$	$\bar{h}$	$\bar{\phi}$	$\bar{h}$	$\bar{\phi}$
537	4258.3	45.079	4030.2	51.032	3640.3	31.194	3729.5	47.272
600	4764.7	45.970	4600.9	52.038	4075.6	31.959	3168.0	48.044
700	5575.4	47.219	5552.0	53.503	4770.2	33.031	4866.0	49.120
800	6396.9	48.316	6552.9	54.839	5467.1	33.961	5568.2	50.058
900	7230.9	49.298	7597.6	56.070	6165.3	34.784	6276.4	50.892
1000	8078.9	50.191	8682.1	57.212	6864.5	35.520	6992.2	51.646
1100	8942.0	51.013	9802.6	58.281	7564.6	36.188	7716.8	52.337
1200	9820.4	51.777	10955.3	59.283	8265.8	36.798	8450.8	52.976
1300	10714.5	52.494	12136.9	60.229	8968.7	37.360	9194.6	53.571
1400	11624.8	53.168	13344.7	61.124	9673.8	37.883	9948.1	54.129
1500	12551.4	53.808	14576.0	61.974	10381.5	38.372	10711.1	54.655
1600	13494.9	54.418	15829.0	62.783	11092.5	38.830	11483.4	55.154
1700	14455.4	54.999	17101.4	63.555	11807.4	39.264	12264.3	55.628
1800	15433.0	55.559	18391.5	64.292	12526.8	39.675	13053.2	56.078
1900	16427.5	56.097	19697.8	64.999	13250.9	40.067	13849.8	56.509
2000	17439.0	56.617	21018.7	65.676	13980.1	40.441	14653.2	56.922
2100	18466.9	57.119	22352.7	66.327	14714.5	40.799	15463.3	57.317
2200	19510.8	57.605	23699.0	66.953	15454.4	41.143	16279.4	57.696
2300	20570.6	58.077	25056.3	67.557	16199.8	41.475	17101.0	58.062
2400	21645.7	58.535	26424.0	68.139	16950.6	41.794	17927.4	58.414
2500	22735.4	58.980	27801.2	68.702	17707.3	42.104	18758.8	58.754
2600	23839.5	59.414	29187.1	69.245	18469.7	42.403	19594.3	59.081
2700	24957.2	59.837	30581.2	69.771	19237.8	42.692	20434.0	59.398
2800	26088.0	60.248	31982.8	70.282	20011.8	42.973	21277.2	59.705
2900	27231.2	60.650	33391.5	70.776	20791.5	43.247	22123.8	60.002
3000	28386.3	61.043	34806.6	71.255	21576.9	43.514	22973.4	60.290
3100	29552.8	61.426	36227.9	71.722	22367.7	43.773	23826.0	60.569
3200	30730.2	61.801	37654.7	72.175	23164.1	44.026	24681.2	60.841
3300	31918.2	62.167	39086.7	72.616	23965.5	44.273	25539.0	61.105
3400	33116.0	62.526	40523.6	73.045	24771.9	44.513	26399.3	61.362
3500	34323.5	62.876	41965.2	73.462	25582.9	44.748	27261.8	61.612
3600	35540.1	63.221	43411.0	73.870	26398.5	44.978	28126.6	61.855
3700	36765.4	63.557	44860.6	74.267	27218.5	45.203	28993.5	62.093
3800	37998.9	63.887	46314.0	74.655	28042.8	45.423	29862.3	62.325
3900	39240.2	64.210	47771.0	75.033	28871.1	45.638	30732.9	62.551
4000	40489.1	64.528	49231.4	75.404	29703.5	45.849	31605.2	62.772
4100	41745.4	64.839	50695.1	75.765	30539.8	46.056	32479.1	62.988
4200	43008.4	65.144	52162.0	76.119	31379.8	46.257	33354.4	63.198
4300	44278.0	65.444	53632.1	76.464	32223.5	46.456	34231.2	63.405
4400	45553.9	65.738	55105.1	76.803	33070.9	46.651	35109.2	63.607
4500	46835.9	66.028	56581.0	77.135	33921.6	46.842	35988.6	63.805
4600	48123.6	66.312	58059.7	77.460	34775.7	47.030	36869.3	63.998
4700	49416.9	66.591	59541.1	77.779	35633.0	47.215	37751.0	64.188
4800	50715.5	66.866	61024.9	78.091	36493.4	47.396	38633.9	64.374
4900	52019.0	67.135	62511.3	78.398	37356.9	47.574	39517.8	64.556
5000	53327.4	67.401	64000.0	78.698	38223.3	47.749	40402.7	64.735
5100	54640.3	67.662	65490.9	78.994	39092.8	47.921	41288.6	64.910
5200	55957.4	67.918	66984.0	79.284	39965.1	48.090	42175.5	65.082
5300	57278.7	68.172	68479.1	79.569	40840.2	48.257	43063.2	65.252

TABLE A.20
One-Dimensional Isentropic Compressible Flow Functions for an Ideal Gas with Constant Specific Heat and Molecular Weight and $k = 1.4$

M	M^*	$\dfrac{A}{A^*}$	$\dfrac{P}{P_o}$	$\dfrac{\rho}{\rho_o}$	$\dfrac{T}{T_o}$
0	0	∞	1.00000	1.00000	1.00000
0.10	0.10943	5.8218	0.99303	0.99502	0.99800
0.20	0.21822	2.9635	0.97250	0.98027	0.99206
0.30	0.32572	2.0351	0.93947	0.95638	0.98232
0.40	0.43133	1.5901	0.89562	0.92428	0.96899
0.50	0.53452	1.3398	0.84302	0.88517	0.95238
0.60	0.63480	1.1882	0.78400	0.84045	0.93284
0.70	0.73179	1.09437	0.72092	0.79158	0.91075
0.80	0.82514	1.03823	0.65602	0.74000	0.88652
0.90	0.91460	1.00886	0.59126	0.68704	0.86058
1.00	1.00000	1.00000	0.52828	0.63394	0.83333
1.10	1.08124	1.00793	0.46835	0.58169	0.80515
1.20	1.1583	1.03044	0.41238	0.53114	0.77640
1.30	1.2311	1.06631	0.36092	0.48291	0.74738
1.40	1.2999	1.1149	0.31424	0.43742	0.71839
1.50	1.3646	1.1762	0.27240	0.39498	0.68965
1.60	1.4254	1.2502	0.23527	0.35573	0.66138
1.70	1.4825	1.3376	0.20259	0.31969	0.63372
1.80	1.5360	1.4390	0.17404	0.28682	0.60680
1.90	1.5861	1.5552	0.14924	0.25699	0.58072
2.00	1.6330	1.6875	0.12780	0.23005	0.55556
2.10	1.6769	1.8369	0.10935	0.20580	0.53135
2.20	1.7179	2.0050	0.09352	0.18405	0.50813
2.30	1.7563	2.1931	0.07997	0.16458	0.48591
2.40	1.7922	2.4031	0.06840	0.14720	0.46468
2.50	1.8258	2.6367	0.05853	0.13169	0.44444
2.60	1.8572	2.8960	0.05012	0.11787	0.42517
2.70	1.8865	3.1830	0.04295	0.10557	0.40684
2.80	1.9140	3.5001	0.03685	0.09462	0.38941
2.90	1.9398	3.8498	0.03165	0.08489	0.37286
3.00	1.9640	4.2346	0.02722	0.07623	0.35714
3.50	2.0642	6.7896	0.01311	0.04523	0.28986
4.00	2.1381	10.719	0.00658	0.02766	0.23810
4.50	2.1936	16.562	0.00346	0.01745	0.19802
5.00	2.2361	25.000	$189(10)^{-5}$	0.01134	0.16667
6.00	2.2953	53.180	$633(10)^{-6}$	0.00519	0.12195
7.00	2.3333	104.143	$242(10)^{-6}$	0.00261	0.09259
8.00	2.3591	190.109	$102(10)^{-6}$	0.00141	0.07246
9.00	2.3772	327.189	$474(10)^{-7}$	0.000815	0.05814
10.00	2.3904	535.938	$236(10)^{-7}$	0.000495	0.04762
∞	2.4495	∞	0	0	0

Source: Abridged from Table 30 in Joseph H. Keenan and Joseph Kaye, *Gas Tables*, John Wiley & Sons, Inc., New York, 1948.

TABLE A.21
Normal Total Emissivity of Various Surfaces

Surfaces	°F	ϵ
A. Metals and Their Oxides		
Aluminum:		
Highly polished plate, 98.3% pure	440, 1070	0.039, 0.057
Polished plate	73	0.040
Rough plate	78	0.055
Oxidized at 1110°F	390, 1110	0.11, 0.19
Al-surfaced roofing	110	0.216
Al-treated surfaces, heated at 1110°F:		
Copper	390, 1110	0.18, 0.19
Steel	390, 1110	0.52, 0.57
Brass:		
Highly polished:		
73.2% Cu, 26.7% Zn, by weight	476, 674	0.028, 0.031
62.4% Cu, 36.8% Zn, 0.4% Pb, 0.3% Al, by weight	494, 710	0.0388, 0.037
82.9% Cu, 17.0% Zn, by weight	530	0.030
Hard-rolled, polished, but direction of polishing visible	70	0.038
But somewhat attacked	73	0.043
But traces of stearin from polish left on	75	0.053
Polished	100, 600	0.096, 0.096
Rolled plate:		
Natural surface	72	0.06
Rubbed with coarse emery	72	0.20
Dull plate	120, 660	0.22
Oxidized by heating at 1110°F	390, 1110	0.61, 0.59
Chromium:		
See Nickel Alloys for Ni-Cr steels		
Copper:		
Carefully polished electrolytic Cu	176	0.018
Commercial, emeried, polished, but pits remaining	66	0.030
Scraped shiny, but not mirrorlike	72	0.072
Polished	242	0.023
Plate heated at 1110°F	390, 1110	0.57, 0.57
Cuprous oxide	1470, 2010	0.66, 0.54
Plate, heated for a long time, covered with thick oxide layer	77	0.78
Molten copper	1970, 2300	0.16, 0.13
Gold:		
Pure, highly polished	440, 1160	0.018, 0.035
Iron and steel:		
Metallic surfaces (or very thin oxide layer):		
Electrolytic iron, highly polished	350, 440	0.052, 0.074
Polished iron	800, 1880	0.144, 0.377
Iron freshly emeried	68	0.242
Cast iron, polished	392	0.21
Wrought iron, highly polished	100, 480	0.28
Cast iron, newly turned	72	0.435
Polished steel casting	1420, 1900	0.52, 0.56
Ground sheet steel	1720, 2010	0.55, 0.61
Smooth sheet iron	1650, 1900	0.55, 0.60
Cast iron, turned on lathe	1620, 1810	0.60, 0.70
Oxidized surfaces:		
Iron plate, pickled, then rusted red	68	0.612
Then completely rusted	67	0.685
Rolled sheet steel	70	0.657
Oxidized iron	212	0.736
Cast iron, oxidized at 1100°F	390, 1110	0.64, 0.78
Steel oxidized at 1100°F	390, 1110	0.79, 0.79
Smooth, oxidized electrolytic iron	260, 980	0.78, 0.82
Iron oxide	930, 2190	0.85, 0.89
Rough ingot iron	1700, 2040	0.87, 0.95

Source: From A. I. Brown and S. M. Marco, *Introduction to Heat Transfer*, 3rd ed., McGraw-Hill Book Company, New York, 1958, pp. 54–58, with permission.

Surfaces	°F	ϵ
Sheet steel, strong rough oxide layer	75	0.80
Dense shiny oxide layer	75	0.82
Cast plate:		
Smooth	73	0.80
Rough	73	0.82
Cast iron, rough, strongly oxidized	100, 480	0.95
Wrought iron, dull-oxidized	70, 680	0.94
Steel plate, rough	100, 700	0.94, 0.97
High-temperature alloy steels; *see* Nickel alloys		
Molten metals:		
Molten cast iron	2370, 2550	0.29, 0.29
Molten mild steel	2910, 3270	0.28, 0.28
Lead:		
Pure (99.96%) unoxidized	260, 440	0.057, 0.075
Gray oxidized	75	0.281
Oxidized at 390°F	390	0.63
Mercury, pure clean	32, 212	0.09, 0.12
Molybdenum filament	1340, 4700	0.096, 0.292
Ni-Cu alloy, oxidized at 1110°F	390, 1110	0.41, 0.46
Nickel:		
Electroplated on polished iron, then polished	74	0.045
Technically pure (98.9% Ni by weight, +Mn), polished	440, 710	0.07, 0.087
Electroplated on pickled iron, not polished	68	0.11
Wire	368, 1844	0.096, 0.186
Plate, oxidized by heating at 1110°F	390, 1110	0.37, 0.48
Nickel oxide	1200, 2290	0.59, 0.86
Nickel alloys:		
Cr-Ni alloy	125, 1894	0.64, 0.76
(18–32% Ni, 55–68% Cu, 20% Zn by weight), gray oxidized	70	0.262
Alloy steel (8% Ni, 18% Cr); light silvery, rough, brown after heating	420, 914	0.44, 0.36
Same, after 24 hr heating at 980°F	420, 980	0.62, 0.73
Alloy (20% Ni, 25% Cr), brown, splotched, oxidized from service	420, 980	0.90, 0.97
Alloy (60% Ni, 12% Cr), smooth, black, firm adhesive oxide coat from service	520, 1045	0.89, 0.82
Platinum:		
Pure, polished plate	440, 1160	0.054, 0.104
Strip	1700, 2960	0.12, 0.17
Filament	80, 2240	0.036, 0.192
Wire	440, 2510	0.073, 0.182
Silver:		
Polished, pure	440, 1160	0.0198, 0.0324
Polished	100, 700	0.0221, 0.0312
Steel, *see* Iron		
Tantalum filament	2420, 4580	0.193, 0.31
Tin, bright, tinned iron sheet	76	0.043, 0.064
Tungsten:		
Filament, aged	80, 6000	0.032, 0.35
Filament	6000	0.39
Zinc:		
Commercial, 99.1% pure, polished	440, 620	0.045, 0.053
Oxidized by heating at 750°F	750	0.11
Galvanized sheet iron:		
Fairly bright	82	0.228
Gray, oxidized	75	0.276

B. Refractories, Building Materials, Paints, and Miscellaneous		
Asbestos board	74	0.96
Asbestos paper	100, 700	0.93, 0.945
Brick:		
Red, rough, but no gross irregularities	70	0.93
Silica unglazed, rough	1832	0.80

Surfaces	°F	ϵ
Silica glazed, rough	2012	0.85
Grog brick, glazed	2012	0.75
See Refractory materials, below		
Carbon:		
T-carbon, 0.9% ash...	260, 1160	0.81, 0.79
Carbon filament	1900, 2560	0.526
Candle soot..	206, 520	0.952
Lampblack:		
Water-glass coating	209, 362	0.959, 0.947
Water-glass coating	260, 440	0.957, 0.952
Thin layer on iron plate	69	0.927
Thick coat..	68	0.967
0.003 in. or thicker	100, 700	0.945
Enamel, white, fused on iron	66	0.897
Glass, smooth	72	0.937
Gypsum, 0.02 in. thick or smooth on blackened plate....................	70	0.903
Marble, light gray, polished......................................	72	0.931
Oak, planed..	70	0.895
Oil layers on polished nickel (lubricating oil):		
Polished surface alone	68	0.045
+0.001 in. oil..	68	0.27
+0.002 in. oil..	68	0.46
+0.005 in. oil..	68	0.72
+∞..	68	0.82
Oil layers on aluminum foil (linseed oil):		
Aluminum foil..	212	0.087
+1 coat oil...	212	0.561
+2 coats oil	212	0.574
Paints, lacquers, varnishes:		
Snow-white enamel varnish on rough iron plate	73	0.906
Black shiny lacquer, sprayed on iron	76	0.875
Black shiny shellac on tinned iron sheet..........................	70	0.821
Black-matte shellac..	170, 295	0.91
Black lacquer	100, 200	0.80, 0.95
Flat black lacquer	100, 200	0.96, 0.98
White lacquer...	100, 200	0.80, 0.95
Oil paints, 16 different, all colors	212	0.92, 0.96
Aluminum paints and lacquers:		
10% Al, 22% lacquer body, on rough or smooth surface............	212	0.52
26% Al, 27% lacquer body, on rough or smooth surface............	212	0.30
Other aluminum paints, varying age and Al content................	212	0.27, 0.67
Aluminum lacquer, varnish binder, on rough plate	70	0.39
Aluminum paint, after heating to 620°F	300, 600	0.35
Paper, thin:		
Pasted on tinned iron plate	66	0.924
Pasted on rough iron plate....................................	66	0.929
Pasted on black lacquered plate	66	0.944
Plaster, rough, lime	50, 190	0.91
Porcelain, glazed	72	0.924
Quartz, rough, fused	70	0.932
Refractory materials, 40 different..................................	1110, 1830	
Poor radiators..	....	0.65, 0.75
		0.70
Good radiators..	....	0.80, 0.85
		0.85, 0.90
Roofing paper...	69	0.91
Rubber:		
Hard, glossy plate	74	0.945
Soft, gray, rough (reclaimed)	76	0.859
Serpentine, polished	74	0.900
Water..	32, 212	0.95, 0.963

Index